Isoflavones
Chemistry, Analysis, Function and Effects

Food and Nutritional Components in Focus

Series Editor:
Professor Victor R Preedy, *School of Medicine, King's College London, UK*

Titles in the Series:
1: Vitamin A and Carotenoids: Chemistry, Analysis, Function and Effects
2: Caffeine: Chemistry, Analysis, Function and Effects
3: Dietary Sugars: Chemistry, Analysis, Function and Effects
4: B Vitamins and Folate: Chemistry, Analysis, Function and Effects
5: Isoflavones: Chemistry, Analysis, Function and Effects

How to obtain future titles on publication:
A standing order plan is available for this series. A standing order will bring delivery of each new volume immediately on publication.

For further information please contact:
Book Sales Department, Royal Society of Chemistry, Thomas Graham House, Science Park, Milton Road, Cambridge, CB4 0WF, UK
Telephone: +44 (0)1223 420066, Fax: +44 (0)1223 420247
Email: booksales@rsc.org
Visit our website at http://www.rsc.org/Shop/Books/

Isoflavones
Chemistry, Analysis, Function and Effects

Edited by

Victor R Preedy
School of Medicine, King's College London, UK

RSCPublishing

Food and Nutritional Components in Focus No. 5

ISBN: 978-1-84973-419-6
ISSN: 2045-1695

A catalogue record for this book is available from the British Library

© The Royal Society of Chemistry 2013

All rights reserved

Apart from fair dealing for the purposes of research for non-commercial purposes or for private study, criticism or review, as permitted under the Copyright, Designs and Patents Act 1988 and the Copyright and Related Rights Regulations 2003, this publication may not be reproduced, stored or transmitted, in any form or by any means, without the prior permission in writing of The Royal Society of Chemistry or the copyright owner, or in the case of reproduction in accordance with the terms of licences issued by the Copyright Licensing Agency in the UK, or in accordance with the terms of the licences issued by the appropriate Reproduction Rights Organization outside the UK. Enquiries concerning reproduction outside the terms stated here should be sent to The Royal Society of Chemistry at the address printed on this page.

The RSC is not responsible for individual opinions expressed in this work.

Published by The Royal Society of Chemistry,
Thomas Graham House, Science Park, Milton Road,
Cambridge CB4 0WF, UK

Registered Charity Number 207890

For further information see our web site at www.rsc.org

Printed in the United Kingdom by CPI Group (UK) Ltd, Croydon, CR0 4YY, UK

Preface

In the past three decades there have been major advances in our understanding of the chemistry and function of nutritional components. This has been enhanced by rapid developments in analytical techniques and instrumentation. Chemists, food scientists and nutritionists are, however, separated by divergent skills, and professional disciplines. Hitherto this transdisciplinary divide has been difficult to bridge.

The series **Food and Nutritional Components in Focus** aims to cover in a single volume the chemistry, analysis, function and effects of single components in the diet or its food matrix. Its aim is to embrace scientific disciplines so that information becomes more meaningful and applicable to health in general.

The series **Food and Nutritional Components in Focus** covers the latest knowledge base and has a structured format.

Isoflavones has four major sections, namely:

Isoflavones in Context
Chemistry and Biochemistry
Analysis
Function and Effects

The first section covers phytoestrogens in health and plants, then material on isoflavones in foods and the diet. The chemistry and biochemistry section covers structures, bioconversion and biotransformation, the human estrogen receptor, homonuclear NMR spectroscopy, genistein, daidzein, xenoestrogens, methylated derivatives and non-natural isoflavonoids. The section on analysis includes foods, beverages, nuts, traditional medicines, herbs, pharmacologically active isoflavones, plasma and urine. Methodology encompasses microwave-extraction, HPLC, LC-MS/MS and LC-UV/PDA and many other techniques. Finally, the section on function and effects covers ingestion of isoflavones by different populations, isoflavones in beverages, inherited metabolic diseases, mucopolysaccharidoses, clinical trials, obesity, inflammation in adipose tissue, menopausal vasomotor syndrome, estrogenic activity, testicular function, thyroid function, gastric cancer, learning and memory, prenatal exposure, cell proliferation, bone, NMDA and GABA receptors, and insulin secretion. Individual isoflavones are also described, such as daidzein, genistein and glycitein, as well as their derivatives, such as equol and tetrahydroxyisoflavone. As isoflavones occur within a complex plant matrix, other estrogenic and bioactive compounds in isoflavones-rich foods, *e.g.* coumestrol, are described for comparative reference.

Each chapter transcends the intellectual divide with a novel cohort of features namely by containing:

- *Abstract*
- *Summary Points*
- *Key Facts* (areas of focus explained for the lay person)
- *Definitions of Words and Terms*

It is designed for chemists, food scientist and nutritionists, as well as healthcare workers and research scientists. Contributions are from leading national and international experts, including contributions from world-renowned institutions.

Professor Victor R Preedy
King's College London

Contents

Isoflavones in Context

Chapter 1 Phytoestrogens in Health: The Role of Isoflavones   3
Rodney J. Baber

     1.1 Introduction   3
     1.2 Absorption, Metabolism and Excretion of Isoflavones   4
     1.3 Mechanisms of Action of Isoflavones   5
     1.4 Clinical Effects of Isoflavones   6
         1.4.1 Cardiovascular Health   6
         1.4.2 The Brain   6
         1.4.3 Bone Health   7
         1.4.4 Breast Cancer   7
         1.4.5 Prostate Cancer   8
         1.4.6 Menopausal Symptoms   8
     1.5 Areas in Dispute   8
     Summary Points   9
     Key Facts   9
     List of Abbreviations   10
     References   10

Chapter 2 Phytoestrogens in Plants: With Special Reference to Isoflavones   14
Franz Bucar

     2.1 Characterization of Phytoestrogens   14
     2.2 Distribution of Phytoestrogens in the Plant Kingdom   15
         2.2.1 Phytoestrogens other than Isoflavones   15
         2.2.2 Isoflavones   15

Food and Nutritional Components in Focus No. 5
Isoflavones: Chemistry, Analysis, Function and Effects
Edited by Victor R Preedy
© The Royal Society of Chemistry 2013
Published by the Royal Society of Chemistry, www.rsc.org

2.3	Chemical Features of Plant Isoflavones	17
2.4	Biosynthesis of Isoflavones	18
2.5	Localisation of Isoflavones in Plants	19
2.6	Biological Function of Isoflavones in Plants	22
	2.6.1 Interaction of Isoflavones with Micro-organisms of the Rhizosphere	22
	2.6.2 Isoflavones as Phytoalexins and Phytoanticipins	22
Summary Points		23
Key Facts		24
Definitions of Words and Terms		24
List of Abbreviations		25
References		26

Chapter 3 Isoflavones in Foods and Ingestion in the Diet — 28
Baskaran Stephen Inbaraj and Bing Huei Chen

3.1	Introduction	28
3.2	An Overview of Structure and Analysis of Dietary Isoflavones	29
	3.2.1 Structure	29
	3.2.2 Analysis	29
3.3	Isoflavone Content in Selected Foods	31
3.4	Growth, Variety, Environmental and Post-harvest Storage Conditions Affecting Isoflavone Composition in Soybeans	31
3.5	Isoflavone Composition as Affected by Processing	36
Summary Points		41
Key Facts		41
Definitions of Words and Terms		41
List of Abbreviations		42
References		43

Chemistry and Biochemistry

Chapter 4 The Chemistry/Biochemistry of the Bioconversion of Isoflavones in Food Preparation — 49
Kashif Ghafoor, Fahad Y. Al-Juhaimi and Jiyong Park

4.1	Introduction	49
4.2	Bioconversion of Isoflavones from Glycosides to Aglycones	53
4.3	Enzymatic Transformation of Isoflavone Isomers in Fermented Soymilk	54
4.4	β-Glycosidase Reaction during Bioconversion of Isoflavones	56

	Summary Points	57
	Key Facts	57
	Definitions of Word and Terms	57
	List of Abbreviations	58
	References	59
Chapter 5	**Chemistry and Synthesis of Daidzein and its Methylated Derivatives: Formononetin, Isoformononetin, and Dimethyldaidzein**	**61**

Vincent M. Carroll, Jeffrey D. St. Denis, Kyle F. Biegasiewicz and Ronny Priefer

5.1	Introduction	61
5.2	Synthesis of Daidzein (1)	63
	5.2.1 Synthesis of Daidzein: $BF_3 \cdot OEt_2$-catalyzed Friedel–Crafts Acylation and Ring Closure	63
	5.2.2 Synthesis of Daidzein: RCM with Grubbs' Catalyst	64
	5.2.3 Synthesis of Daidzein: I_2-Mediated Cyclization	64
	5.2.4 Synthesis of Daidzein: Oxidative Rearrangement with Thallium(III) Nitrate	67
5.3	Synthesis of Formononetin (2)	67
	5.3.1 Carbonyls as One-carbon Electrophiles	69
	5.3.2 Activated Aldehyde Surrogates as One-carbon Surrogates	70
5.4	Synthesis of Isoformononetin (3)	73
	5.4.1 Selective Methylation of Daidzein (1)	73
	5.4.2 Suzuki-mediated Synthesis	74
5.5	Synthesis of Dimethyldaidzein (4)	74
	5.5.1 Synthesis of Dimethyldaidzein: Polymer-supported Iodobenzene Diacetate (PSIBD)-promoted Oxidative Rearrangement	76
	5.5.2 Synthesis of Dimethyldaidzein: Organolead-mediated Arylation	76
	Summary Points	78
	Key Facts	79
	Definitions of Words and Terms	79
	List of Abbreviations	79
	References	80

Chapter 6	**Non-natural Isoflavonoids**	**83**

Namita Bhan and Mattheos Koffas

6.1	Introduction	83
6.2	Phenylpropanoid and Isoflavonoid Pathways	83
6.3	Isoflavanoids	85

		6.4	Non-natural Isoflavonoids	85

      6.4   Non-natural Isoflavonoids   85
           6.4.1   Semi-synthesis of Non-natural Isoflavonoids   85
           6.4.2   Chemical Synthesis of Non-natural Flavanones   86
           6.4.3   ER Assay   86
           6.4.4   Protein Engineering of Uridine Diphosphate Glycoslytransferase (UGT)   89
      6.5   Structural and Functional Studies of IFS   89
      Summary Points   90
      Key Facts   90
      Definitions of Words and Terms   91
      List of Abbreviations   91
      References   92

Chapter 7   The Structure of Isoflavones by 1D and 2D Homonuclear and Heteronuclear NMR Spectroscopy   94
Kristiina Wähälä, Somdatta Deb and Tapio Hase

      7.1   Introduction   94
      7.2   Structure Elucidation: Establishing Spectral Correlations   96
      7.3   Chromatography-NMR   101
      7.4   Conformation and NMR   102
      7.5   Substituent Effects and Prediction of Structures   103
      7.6   Isoflavone Metabolites   103
      7.7   Solvent Effects   107
      Summary Points   110
      Key Facts   110
      Definitions of Words and Terms   111
      List of Abbreviations   112
      Acknowledgement   112
      References   112

Chapter 8   Biotransformation and Transfer of Genistein: a Comparison with Xenoestrogens and a Focus on the Human Placenta   115
Hsiu-Wen Chan, Greg E. Rice and Murray D. Mitchell

      8.1   Biotransformation of Genistein Compounds in the Gastrointestinal Tract   115
           8.1.1   Deglycosylation of Genistin in the Small Intestine   116
           8.1.2   Glucuronidation and Sulfation of Genistein   117
           8.1.3   Hydroxylation of Genistein in the Liver   119
           8.1.4   Breakdown of Genistein and Luminal Excretion   119
           8.1.5   Transfer of Genistein in Placenta   121
      8.2   A Comparison with Other Xenoestrogens   122
           8.2.1   Chemical Structure of BPA and NP   123

Contents　　　　　　　　　　　　　　　　　　　　　　　　　　　　　　　　　xi

	8.2.2　Biotransformation of Xenoestrogens	123
	8.2.3　Placenta Transfer	124
8.3	Conclusion	125
Summary Points		125
Key Facts		126
Definitions of Words and Terms		126
List of Abbreviations		126
References		127

Chapter 9　The Biological Effects of Genistein and its Intracellular Metabolite, 5,7,3′,4′-Tetrahydroxyisoflavone　131
Giulia Corona and Jeremy P. E. Spencer

9.1	Genistein and its Biological Activities	131
	9.1.1　Oestrogenic Activity	132
	9.1.2　Anticancer Effects	133
	9.1.3　Antioxidant Action	134
	9.1.4　Anti-inflammatory Activity	134
9.2	Intracellular Formation of the Genistein Metabolite 5,7,3′,4′-Tetrahydroxyisoflavone (THIF)	135
9.3	THIF and Cancer	137
9.4	THIF and Endothelial Function	139
9.5	Summary	140
Summary Points		141
Key Facts		141
List of Abbreviations		142
References		142

Chapter 10　Genistein Chemistry and Biochemistry　148
Francesco Squadrito and Alessandra Bitto

10.1	Introduction	148
10.2	Chemical Structure and Interaction with Estrogen Receptors	149
10.3	Metabolism	152
10.4	Conclusion	153
Summary Points		153
Key Facts		154
Definitions of Words and Terms		154
List of Abbreviations		155
References		155

Chapter 11　Isoflavones and Human Estrogen Receptor: When Plants Synthesize Mammalian Hormone Mimetics　157
Patricia de Cremoux and Yves Jacquot

11.1	Introduction	157
11.2	From Plant Biosynthesis to Mammalian Biosynthesis	158

		11.2.1	Biosynthesis of Isoflavonoids in Leguminous Cells	159
		11.2.2	Biosynthesis of Isoflavonoids in ERs	163
	11.3	Interaction of Isoflavonoids with ERs		163
	11.4	Mammalian Metabolism and Bioavailability		165
	11.5	Biochemical and Physiological Functions of Isoflavonoids		167
		11.5.1	Estrogenic-related Effects	168
		11.5.2	Non-estrogenic-related Effects	169
	11.6	Conclusion		170
	Summary Points			170
	Key Facts			171
	Definitions of Words and Terms			172
	List of Abbreviations			173
	References			174

Analysis

Chapter 12 Continuous Microwave-assisted Isoflavone Extraction   **179**
Dorin Boldor and Cristina Mirela Sabliov

	12.1	Introduction		179
	12.2	Microwave Extraction System Designs		181
	12.3	Microwave-assisted Extraction Process		183
	12.4	Oil Separation and Isoflavone Purification		183
	12.5	Significant Results and Discussion		184
		12.5.1	Isoflavones (and Oil) Yields	184
		12.5.2	Influence of Time and Temperature on the Oil and Isoflavones Extraction Yield – Reaction Kinetics	186
	12.6	Conclusion		189
	Summary Points			190
	Key Facts			190
	Definitions of Words and Terms			191
	List of Abbreviations			192
	Acknowledgements			192
	References			192

Chapter 13 Isoflavones: High-performance Liquid Chromatographic Analysis of Glucuronic Acid- and Sulfuric Acid-conjugated Metabolites of Daidzein and Genistein in Human Plasma and Urine   **196**
Kazuo Ishii, Kaori Hosoda and Takashi Furuta

	13.1	Introduction	196
	13.2	Conventional Methods	198

		13.2.1	Selective Enzymatic Hydrolysis	198
		13.2.2	Fractionation by Ion-exchange Chromatography	203
		13.2.3	Direct Assay	204
	13.3	A Simple Direct Assay of Conjugated Metabolites Using HPLC-UV		205
		13.3.1	Reference Compounds	206
		13.3.2	SPE	207
		13.3.3	Chromatographic Conditions	209
		13.3.4	Application	210
	Summary Points			212
	Key Facts			213
	Definitions of Words and Terms			213
	List of Abbreviations			215
	References			216

Chapter 14 High-throughput Quantification of Pharmacologically Active Isoflavones using LC-UV/PDA and LC-MS/MS — 218
Wahajuddin and Sumit Arora

	14.1	Introduction		218
	14.2	High-throughput HPLC-UV/MS Analytical/Bioanalytical Methods		222
		14.2.1	High-throughput HPLC Methods using UV Detection	222
		14.2.2	High-throughput HPLC Methods using MS	224
	14.3	Conclusions		238
	Summary Points			238
	Key Facts			239
	Definitions of Words and Terms			239
	List of Abbreviations			240
	References			240

Chapter 15 Methods for Isoflavones: A Focus on Beverage Analysis — 244
Rita C. Alves and M. Beatriz P. P. Oliveira

	15.1	Introduction		244
	15.2	Methods of Extraction		245
		15.2.1	Direct Extraction	245
		15.2.2	Hydrolysis	255
		15.2.3	Internal Standards	257
	15.3	Chromatographic Methods		258
	15.4	Conclusion		258
	Summary Points			259
	Key Facts			259

Definitions of Words and Terms		260
List of Abbreviations		260
Acknowledgements		260
References		260

Chapter 16 The Determination of Isoflavones in Supplemented Foods: An Overview   263
Alberto Zafra-Gómez, Sonia Capel-Cuevas and Noemí I. Dorival-García

16.1	Introduction		263
16.2	Isoflavone Isolation		266
16.3	Techniques for the Analysis of Isoflavones in Supplemented Food		268
	16.3.1	Liquid Chromatography (LC)	269
	16.3.2	Gas Ghromatography (GC)	271
	16.3.3	Capillary Electrophoresis (CE)	271
	16.3.4	Mass Spectrometry (MS)	272
	16.3.5	Immunoanalysis	273
16.4	Conclusions		273
Summary Points			274
Key Facts			274
Definition of Words and Terms			275
List of Abbreviations			276
References			276

Chapter 17 Isoflavones: LC-MS/MS Profiling of Isoflavone Glycosides and Other Conjugates   280
Piotr Kachlicki and Maciej Stobiecki

17.1	Introduction	280
17.2	Chromatographic Separation of Flavonoids	281
17.3	Mass Spectrometry (MS) as a Tool for Identification of Isoflavone Glycoconjugates	281
17.4	Sources of Structural Variability of Flavonoid Compounds	283
17.5	Differentiation of Isoflavone and Flavone Glycoconjugates with Instrumental Methods	285
17.6	Differentiation of C-Glycosides and O-Glycosides of Isoflavonoids	287
17.7	Differentiation of Acylated Glycoconjugates of Isoflavonoids	288
Summary Points		291
Key Facts		291
Definitions of Word and Terms		291

	List of Abbreviations	292
	References	292

Chapter 18 *Puerariae radix* Isoflavones   **294**
Lei Wan and Chia-Hung Lin

	18.1	Introduction	294
	18.2	Use of Resources	296
	18.3	Chemical Composition	296
	18.4	Pharmacology and Potential Applications	298
		18.4.1 Anti-tumor Activity	298
		18.4.2 Anti-inflammation	302
		18.4.3 Anti-hypertension	302
		18.4.4 Anti-atherosclerosis	303
		18.4.5 Neuroprotection	304
		18.4.6 Anti-allergy	305
		18.4.7 Anti-hyperglycemic Effect	306
	Conclusions		307
	Summary Points		308
	Key Facts		308
	Definitions of Words and Terms		309
	List of Abbreviations		310
	References		310

Chapter 19 Pattern Profiling and Quantitative Determination of Isoflavones in Herbal Chemotypes using Liquid Chromatography Tandem Mass Spectrometry   **316**
Lakshmi Manickavasagam, Smriti Mishra and Girish Kumar Jain

	19.1	Introduction	316
	19.2	Herbal Preparations for Osteoporosis	317
		19.2.1 Pattern Profiling and Quantitative Analysis of Herbal Fractions	317
		19.2.2 Osteogenic Herbal Fractions and Compounds of *Butea monosperma*	317
	19.3	LC-MS/MS Method for Qualitative and Quantitative Analysis of Herbal Fractions	319
		19.3.1 Chromatographic Conditions	319
		19.3.2 Mass Spectrometric Conditions	321
		19.3.3 Validation of the LC-MS/MS Method	322
		19.3.4 Pattern Profiling and Quantitative Analysis of Isoflavones	325
	19.4	Conclusion	325
	Summary Points		326

Key Facts		327
Definition of Words and Terms		327
List of Abbreviations		329
Acknowledgements		329
References		329

Chapter 20 Analysis of Novel Isoflavone Digycoside in Nuts — 333
Kazuhiro Nara

20.1	Introduction	333
20.2	Isoflavones	334
20.3	Isoflavones in Groundnuts	334
	20.3.1 Plant Material and Preparation of Groundnut Extract	334
	20.3.2 HPLC Analysis of Isoflavones	334
	20.3.3 Isolation and Purification of a Novel Isoflavone from Groundnut Extract	335
	20.3.4 Content of a Novel Isoflavone in Groundnuts	335
20.4	Analytical Profiles of Groundnuts	336
	20.4.1 Isoflavone Profile of Groundnuts	336
	20.4.2 Influence of the Extraction Temperature on the Isoflavone Composition of Groundnut Extract	337
	20.4.3 Purification of a Novel Isoflavone (Peak-1) from the Groundnut Extract	339
	20.4.4 Structural Analysis of a Novel Isoflavone from the Groundnut Extract	340
	20.4.5 Content of a Novel Isoflavone from the Groundnut Extract	341
20.5	Conclusion	342
Summary Points		343
Key Facts		343
Definitions of Words and Terms		344
List of Abbreviations		345
References		345

Function and Effects

Chapter 21 Isoflavone Ingestion by Multiethnic Populations: Implications for Health — 349
Baskaran Stephen Inbaraj and Bing Huei Chen

21.1	Introduction	349
21.2	Isoflavone Ingestion by Multiethnic Populations	350
	21.2.1 Australia and the UK	350
	21.2.2 Brazil	351
	21.2.3 Canada	354

Contents   xvii

        21.2.4   China and Hong Kong   354
        21.2.5   Italy   355
        21.2.6   Japan   356
        21.2.7   Korea   356
        21.2.8   Singapore and Indonesia   357
        21.2.9   Taiwan   358
        21.2.10 USA   358
    Summary Points   359
    Key Facts   360
    Definitions of Words and Terms   360
    List of Abbreviations   361
    References   361

Chapter 22 Isoflavones in Beverages   **365**
Rita C. Alves and M. Beatriz P. P. Oliveira

    22.1   Introduction   365
    22.2   Isoflavone Sources   366
        22.2.1   Beverages as Isoflavone Sources   367
    22.3   Conclusion   376
    Summary Points   376
    Key Facts   377
    Definitions of Words and Terms   377
    Acknowledgements   378
    References   378

Chapter 23 Use of Isoflavones in Inherited Metabolic Diseases: A Focus on Mucopolysaccharidoses   **381**
Daniel Scherman, Audrey Arfi and Corinne Marie

    23.1   General Introduction   381
    23.2   Introduction on Mucopolysaccharidoses (MPS)   382
        23.2.1   Clinical Information, Genetics and Biochemistry   382
        23.2.2   Pathophysiology   383
    23.3   Introduction on Glycosaminoglycan (GAGs) and Isoflavone Families   384
        23.3.1   GAGs: Structure, Function and Life Cycle   384
        23.3.2   Isoflavone Family: Structure, Function and Life Cycle   387
    23.4   Use of Isoflavones for Substrate Reduction Therapy (SRT)   387
        23.4.1   Substrate Reduction Therapy (SRT): Definition   388
        23.4.2   Isoflavones and Analogs are Potent Agents for SRT   388

		23.4.3	Isoflavone Action on GAG Synthesis and GAG Accumulation	391
	23.5		Conclusions and Perspectives for the Clinic	392
	Summary Points			394
	Key Facts			394
	Definitions of Words and Terms			394
	List of Abbreviations			395
	References			395

Chapter 24 Optimizing Isoflavone-rich Food Delivery Systems for Human Clinical Trials   399
Jennifer Ahn-Jarvis, Steven Schwartz and Yael Vodovotz

	24.1	Introduction		399
		24.1.1	Delivery of Isoflavone Dosage without Compromising Palatability	400
		24.1.2	Processing Effects on Soy Isoflavones	401
		24.1.3	Relevance of Isoflavone Chemical Composition	402
	24.2	Materials and Methods		403
		24.2.1	Preparation of Fermented and Thermally Treated Soy Ingredients	403
		24.2.2	Soy-almond Bread (SAB) Preparation	403
		24.2.3	Isoflavone Analysis	404
		24.2.4	Physicochemical Experiments	405
		24.2.5	Characterization of Organoleptic Properties	406
	24.3	Results and Discussion		407
		24.3.1	Process of Selecting the Optimal Phytochemical Delivery Vehicle	407
		24.3.2	Isoflavone Analysis	409
		24.3.3	Physicochemical Properties	410
		24.3.4	Characterization of Organoleptic Properties	413
	24.4	Conclusions		416
	Summary Points			417
	Key Facts			417
	Definitions of Words and Terms			418
	List of Abbreviations			419
	References			420

Chapter 25 Isoflavones and Thyroid Function: An Overview   423
Francesco Squadrito and Alessandra Bitto

	25.1	Introduction		423
		25.1.1	Goitrogenic Effects of Flavonoids	425
	25.2	The Case of Genistein		425
		25.2.1	Genistein and Thyroid Function	426

25.3	Conclusions	430
	Summary Points	431
	Key Facts	432
	Definitions of Words and Terms	433
	List of Abbreviations	433
	References	434

Chapter 26 Isoflavones against Gastric Cancer: Function and Effects 438
Sue K. Park and Kwang-Pil Ko

26.1	Introduction	438
26.2	Absorption, Metabolism, and Excretion	439
26.3	Function and Effect of Isoflavones against Gastric Cancer	441
	26.3.1 Anti-bacterial Effects	441
	26.3.2 Antioxidant Activity	442
	26.3.3 Anti-inflammation Effect	442
	26.3.4 Inhibition of Enzymes	444
	26.3.5 Inhibition of Angiogenesis	444
	26.3.6 Enhancement of Apoptosis	445
	26.3.7 Evidence from Population Studies	445
	Summary Points	447
	Key Facts	447
	Definitions of Words and Terms	447
	List of Abbreviations	448
	References	448

Chapter 27 Dietary Isoflavones and Learning and Memory 451
Craige C. Wrenn

27.1	Introduction	451
27.2	Pre-clinical Studies of the Deleterious, Endocrine-disrupting Effects of Dietary Isoflavones on Learning and Memory	452
27.3	Pre-clinical Studies of the Beneficial Effects of Dietary Isoflavones in Rodent Models of Age-associated Cognitive Deficits	454
27.4	Pre-clinical Studies of Beneficial Effects of Dietary Isoflavones in Ovariectomized Rats	455
27.5	Human Clinical Studies of Beneficial Effects of Dietary Isoflavones on Learning and Memory	457
27.6	Summary	459
	Summary Points	459
	Key Facts	460
	Definition of Words and Terms	460

List of Abbreviations	461
References	461

Chapter 28 Glycitein in Health   465
Brian R. Stephens and Joshua A. Bomser

28.1	Introduction	465
28.2	Absorption	466
28.3	Glycitein and Bone Health	467
28.4	Glycitein and Cancer	467
28.5	Glycitein and Cardiovascular Health	470
28.6	Glycitein as an Estrogen Receptor (ER) Agonist	471
28.7	Glycitein and Memory and Mood	471
28.8	Glycitein and Oxidative Stress	472
	Summary Points	473
	Key Facts	474
	Definitions of Words and Terms	474
	List of Abbreviations	475
	References	476

Chapter 29 Isoflavones and Prenatal Exposure to Equol   480
Edwin D. Lephart

29.1	Background: Polyphenols and Isoflavonoids		480
29.2	Isoflavonoid Metabolism in Mammals: Postnatal Development		481
29.3	Equol: Biosynthesis (Plants and Mammals), Antioxidant Activity, and Safety Data in Postnatal Development		482
29.4	Comparison between Human and Rat: Newborn and Adult Parameters		485
29.5	Perinatal Studies and Transplacental Transport of Isoflavonoids		487
	29.5.1	Isoflavonoids: Perinatal Rodent Studies	487
	29.5.2	Equol: Perinatal Rodent Studies	488
	29.5.3	Isoflavonoids: Perinatal Human Studies	490
29.6	Summary and Considerations		492
	Summary Points		493
	Key Facts		494
	Definition of Words and Terms		495
	List of Abbreviations		496
	References		496

Contents   xxi

Chapter 30 Genistein: $GABA_A$ and NMDA Receptors   500
Renqi Huang and Glenn H. Dillon

    30.1 Introduction   500
    30.2 $GABA_A$ Receptor Function and Structure   501
        30.2.1 $GABA_A$ Receptor Function   501
        30.2.2 $GABA_A$ Receptor Structure   502
        30.2.3 $GABA_A$ Receptor Modulation   503
    30.3 Genistein Modulation of GABA Receptors   503
        30.3.1 Genistein as an Inhibitor of Protein Tyrosine Kinases (PTKs)   503
        30.3.2 Direct Inhibition of $GABA_A$ Receptors   504
    30.4 NMDA Receptor Function, Structure and Modulation   505
        30.4.1 NMDA Receptor Function   505
        30.4.2 NMDA Receptor Structure   506
        30.4.3 NMDA Receptor Modulation   508
    30.5 Genistein Modulates NMDA Receptor Modulation   508
        30.5.1 Genistein as an Inhibitor of Protein Tyrosine Kinases (PTKs)   508
        30.5.2 Direct Inhibition of NMDA Receptors   509
    30.6 Genistein Modulation of $GABA_A$ and NMDA Receptors *via* the Estrogenic Pathway   509
    30.7 Functional Implications   511
    30.8 Conclusion   511
    Summary Points   512
    Key Facts   512
    Definitions of Words and Terms   512
    List of Abbreviations   513
    References   513

Chapter 31 Estrogenic Activity and Molecular Mechanisms of Coumestrol-induced Biological Effects   518
Kenneth Ndebele, Barbara Graham and Paul Tchounwou

    31.1 Introduction   518
    31.2 Animal Studies   519
    31.3 Mechanisms of the Toxic Action   520
    31.4 Human Studies   521
    31.5 Interpretation of Data in Relation to Human Exposure   524
    Summary Points   525
    Key Facts   525
    Definitions of Words and Terms   526
    List of Abbreviations   526
    Acknowledgements   526
    References   526

Chapter 32	**Genistein and Insulin Secretory Function** *Dongmin Liu*	**529**
	32.1 Introduction	529
	32.2 Genistein may have Anti-diabetic Effects	530
	32.3 Genistein at Physiological Concentrations Augments Glucose-stimulated Insulin Secretion (GSIS) in Beta-cells and Pancreatic Islets	532
	Summary Points	537
	Key Facts	537
	Definitions of Words and Terms	537
	List of Abbreviations	538
	References	538
Chapter 33	**Prevention and Management of Obesity by Isoflavones** *Barbara B. Doonan, Erxi Wu and Joseph M. Wu*	**541**
	33.1 Introduction	541
	33.2 Evidence that Dietary Factors and Soy Isoflavones Play a Role in Obesity	543
	33.2.1 Introduction to Soy Phytochemicals and Isoflavones	543
	33.2.2 Population-based Studies	544
	33.2.3 Animal Studies	545
	33.2.4 Laboratory Studies	546
	33.3 Identification and Mechanism of Soy Isoflavones with Anti-obesity Activities	547
	33.3.1 Modulation of Fat Deposition and Adipogenesis by Soy Isoflavones	547
	33.3.2 A Facile Strategy for the Identification of Soy Isoflavones with Potential for Regulating Visceral Fat	549
	33.3.3 Discovery of Proteins Targeted by Soy Isoflavones with a Regulatory Role in Adipogenesis	551
	33.3.4 Candidate Molecular Targets and Mechanisms of Anti-obesity Isoflavones: Interplay Between NQO1 and p53 in Regulation of Adipogenesis	552
	33.3.5 Role of PPARα, PPARγ, C/EBPα and Other Candidate Molecular Sensors/Effectors of Adipogenesis	553
	33.4 Conclusions	555
	Summary Points	555
	Key Facts	555

Definitions of Words and Terms		556
List of Abbreviations		556
References		557

Chapter 34 Soy Isoflavones and Testicular Function — 562
Benson T. Akingbemi

34.1	Introduction	562
34.2	Dietary Sources of Isoflavones	563
34.3	Hormonal Regulation of Reproductive Tract Development	563
34.4	Regulation of Gonadal Function	565
34.5	Isoflavone Action in Testicular Cells	566
	34.5.1 Effects on Steroidogenesis	567
	34.5.2 Effects on Sperm Production	568
	34.5.3 Effects on Developing *versus* Mature Stages of Development	569
34.6	Mechanisms of Isoflavone Action	570
34.7	Isoflavone Action in the Human Testis	572
34.8	Conclusion	573
Summary Points		574
Key Facts		574
Definitions of Words and Terms		574
List of Abbreviations		575
References		575

Chapter 35 Equol and Cell Proliferation — 580
Zhong Li, Caiyun Zhong and Chunyan Hu

35.1	Introduction	580
35.2	Equol and Hormone-positive Cell Proliferation	581
	35.2.1 Equol Binds to ERs with a Greater Affinity than Daidzein	581
	35.2.2 Estrogenic Activity and Breast Cancer	581
	35.2.3 Equol and Prostate Cancer	583
	35.2.4 Equol and the Reproductive System	586
	35.2.5 Equol and Neuroprotection	586
35.3	Equol and Other Cancer Cell Proliferation	586
	35.3.1 Equol and Pancreatic Cancer	586
	35.3.2 Equol and Colon Cancer	587
35.4	Equol and CVDs	587
35.5	Equol and Immune Functions	588
35.6	Equol and Genotoxicity	588
35.7	Equol and Cell Invasion	589
35.8	Equol and Cancer Chemoprevention	589

35.9	Equol and Drug Metabolism	590
35.10	General Conclusions	591
Summary Points		591
Key Facts		592
Definitions of Words and Terms		592
List of Abbreviations		593
References		593

Chapter 36 Bone, Genistein, Daidzein and Equol — 597
Marina Komrakova, Ewa Klara Stuermer, Klaus Michael Stuermer and Stephan Sehmisch

36.1	Introduction	597
36.2	Bone and Estrogen	598
36.3	Bone and Soy Isoflavones	599
36.4	Bone and Purified Genistein, Daidzein and Equol	600
36.5	Conclusion	605
Summary Points		606
Key Facts		606
Definition of Words and Terms		606
List of Abbreviations		607
References		607

Chapter 37 Isoflavones and Inflammation in Adipose Tissue and Implications for Health — 611
Maria Teresa Blay, Montserrat Pinent and Anna Ardévol

37.1	An Introduction to Inflammation in Adipose Tissue	611
37.2	An Overview of Isoflavones as Anti-inflammatory Agents	615
	37.2.1 Isoflavones can Modulate Proinflamatory Factors in Macrophages and Adipocytes and Down-regulate Gene Expression of Proinflammatory Mediators	616
	37.2.2 Isoflavones can Modulate Proinflamatory Factors in Adipose Tissue by a Decrease in Adipose Tissue Mass and/or Number	617
37.3	What We Know from Animal Models and Isoflavone Effects on Adipose Tissue Inflammation and Health	619
	37.3.1 The Effects of Isoflavones in Animal Models of Obesity: Effect of Isoflavones in Rats with High-fat Diet (HFD)-induced Obesity/Insulin Resistance and the Analysis of Inflammatory Factors	619

	37.4	What We Know from Human Studies and the Effects of Isoflavone on Adipose Tissue Inflammation	621
		37.4.1 Studies in Obese Postmenopausal Women	621
		37.4.2 Studies with Genistein (Gen) in Human Cell Inflammation	621
	Summary Points		622
	Key Facts		622
	Definitions of Words and Terms		623
	List of Abbreviations		624
	References		624

Chapter 38 Isoflavones for Menopausal Vasomotor Syndrome   **627**
Rafael Bolaños Diaz and Juan Carlos Zavala Gonzales

38.1	Pharmacology Aspects	627
38.2	Clinical Evidence	630
38.3	Conclusions	640
Summary Points		640
Key Facts		640
Definition of Words and Terms		641
List of Abbreviation		641
References		641

Subject Index   **646**

Isoflavones in Context

CHAPTER 1
Phytoestrogens in Health: The Role of Isoflavones

RODNEY J. BABER

Medical Suites, Level 3, North Shore Private Hospital, Westbourne St., St Leonards, NSW 2065, Australia
Email: rbaber@med.usyd.edu.au

1.1 Introduction

Phytoestrogens are diphenolic, non-steroidal estrogen-like substances found in all plants, the highest concentrations being in legumes (Baber 2010; Nelson 2008), and are classified according to their chemical structure. The greatest estrogenic activity is found in flavones, flavonols, flavanones, lignans, chalcones and isoflavones. The most common of these are the isoflavones and lignans, which are found mainly in fruit, vegetables and whole grains (Lethaby *et al.* 2007). These compounds have a steric structure similar to that of steroidal estrogens, allowing them to bind to the human estrogen receptor (ER) and they are therefore capable of exerting various estrogenic or anti-estrogenic effects (Duncan *et al.* 2003). Isoflavones bind with a greater affinity to ERβ than to ERα (Kuiper *et al.* 1996). The binding affinity of isoflavones for ERs has been estimated to be between 10^{-2} and 10^{-4} of that of 17β-estradiol (Collins *et al.* 1997; Miksicek 1994), however, these substances can be present in the blood at levels up to 10 000 times that of steroidal estrogens (Adlercreutz *et al.* 1991).

Over 10 000 peer-reviewed papers have been published on the role of isoflavones in health. Amongst commonly consumed foods, isoflavones are found in physiologically relevant amounts in soybeans and foods derived from soy

Food and Nutritional Components in Focus No. 5
Isoflavones: Chemistry, Analysis, Function and Effects
Edited by Victor R Preedy
© The Royal Society of Chemistry 2013
Published by the Royal Society of Chemistry, www.rsc.org

(Adlercreutz and Mazur 1997). The greater reliance on vegetables and particularly legumes such as soy for dietary protein in Asian and Central American communities means that those communities typically have substantially higher dietary isoflavone intake than those found in Western countries (Goldin *et al.* 1986).

Soybeans contain three primary isoflavones in their glycoside form: genistin, daidzin and glycitin. Digestion leads to the cleavage of the sugar moiety and the formation of the respective aglycones: genistein, daidzein and glycitein. Red clover has also been used to manufacture supplements for human use. Red clover contains four isoflavones: formononetin, biochanin, daidzein and genistein. At physiological concentrations, formononetin does not bind to the ER but is metabolized to daidzein and then by intestinal bacteria to equol, both of which have been shown to have estrogenic properties (Baber 2010).

Interest in these compounds began over 20 years ago with research by the US National Cancer Institute exploring possible chemo-preventive properties for phytoestrogens, especially in regard to breast cancer. Ongoing research led to further investigation of these compounds in relation to cardiovascular and skeletal health, cognitive function and the alleviation of menopausal symptoms.

Despite continuing research, the role of isoflavones in health remains controversial, with concerns expressed that the estrogen-like effects of these compounds may pose a risk to certain individuals, especially women with breast cancer or at high risk of that disease. Clinical trials, although numerous, have typically been small, short, of variable quality, have tested different (and often impure) substances and have chosen different endpoints to measure usefulness, efficacy and safety.

There are substantial inconsistencies between results from Asian epidemiologic studies, which are generally viewed as supportive of health benefits, and Western clinical trials, which are much less so. The epidemiologic data are based on the intake of traditional soy foods, such as tofu, miso and soymilk, whereas the intervention trials have usually utilized isolated soy protein and soy or isoflavone supplements. The outcome differences may be due to differences between traditional isoflavone-rich foods and prepared isoflavone supplements, to individual differences in absorption of isoflavones, the timing and duration of exposure to isoflavones and ethnic differences or to a healthy user effect.

1.2 Absorption, Metabolism and Excretion of Isoflavones

Isoflavone aglycones are absorbed in the upper small intestine by passive diffusion, peaking in the blood within 1 h of being ingested (Sfakianos *et al.* 1997). In contrast, the β-glucosides are not passively absorbed. They are hydrolyzed by β-glucosidases from intestinal bacteria or an intestinal enzyme, lactase-phlorizin hydrolase. Isoflavone aglycones are converted into their β-glucuronides by UDP-glucuronyltransferases in gut mucosal cells (King *et al.* 1996) and to a lesser extent to sulfate esters catalyzed by 3′-phosphoadenosine

5′-phosphosulfate (PAPS)-sulfotransferases (Ronis et al. 2006). Glucuronidation and sulfation also occur in the liver. These phase II metabolites are excreted in the bile and are deconjugated in the lower bowel, allowing them to be reabsorbed again, creating an enterohepatic circulation (Sfakianos et al. 1997). Daidzein is metabolized to dihydrodaidzein, which is further metabolized to equol and O-desmethylangolensin (O-DMA). Genistein is transformed to dihydrogenistein and then metabolized to 6-hydroxy-O-DMA. Human urinary excretion of these metabolites is variable, and only approx. 30–40% of subjects excrete significant quantities of equol after isoflavone consumption (Kurzer et al. 1997). The same is not true of animals in which equol excretion appears quite consistent. Based on many studies, the consensus is that only 25–30% of the adult population of Western countries excrete equol when fed soy foods. This is significantly lower than the reported 50–60% frequency of equol-producers in adults from Japan, Korea or China, or in Western adult vegetarians. This regional or ethnic difference in equol production has led to the hypothesis that equol production is necessary for an individual to derive the predicted health benefits from isoflavone consumption (Setchell and Clerici 2010).

1.3 Mechanisms of Action of Isoflavones

The ability of Isoflavones to bind to to mammalian ERs has been known for over 40 years, although, compared with 17β-estradiol, isoflavones have approx. 100 times weaker affinity (Kuiper et al. 1997).

A greater understanding of estrogen action began in 1996 with the discovery of ERβ (Kuiper et al. 1996). Although related to ERα, which is located on chromosome 6, ERβ is located on chromosome 14. The ligand-binding sites are highly homologous between ERα and ERβ. However, the few amino acid differences result in isoflavones exerting preferential binding affinity to ERβ. ERα and ERβ are expressed at various concentrations in different organs and different cell types. This has led to the development of compounds, known as selective estrogen receptor modulators (SERMs), which selectively target receptors and which may be antagonistic in some and agonistic in others. Some have suggested that isoflavone binding to ERβ is an antagonistic process, not an agonistic one, and that hence isoflavones may be 'natural SERMs' (Matthews and Gustafsson 2003). The mechanism of action of isoflavones may thus be estrogenic or anti-estrogenic in different tissues depending upon the concentrations of ERα and ERβ, the concentration of the isoflavone and the concentration of endogenous sex hormones. Isoflavones also have non-genomic activity. Genestein, in vitro, will inhibit the activity of tyrosine protein kinase (Akiyama et al. 1987). Genistein also affects genes involved in the control of cell growth via effects on natural killer cell function, as well as enzyme inhibition and the peroxisome proliferator regulator (Sarkar and Li 2003). Overall, evidence suggests that many pathways, not just estrogen-dependent events, mediate the biological effects of isoflavones and their metabolites. Future studies may utilize data that have been obtained in DNA microarray experiments. Lastly, studies have suggested that some of the benefits of dietary isoflavones observed in other populations may depend on early life exposure (Korde et al. 2009), which may involve their impact on gene expression at an epigenetic level.

1.4 Clinical Effects of Isoflavones

1.4.1 Cardiovascular Health

Isoflavones may exert an effect on the cardiovascular system by three major mechanisms:

(1) directly through ER-mediated effects;
(2) through ER-independent effects directly on cardiovascular risk factors and putative atherogenic risk factors;
(3) indirectly through the displacement of animal protein intake.

Isoflavones exert estrogenic and anti-estrogenic effects (Kuiper et al. 1998). In animal models, isoflavones have been shown to require the presence of the ER to exert anti-atherogenic efffects. The vascular endothelium is a rich site of ERβ expression and the preferential binding of isoflavones for ERβ suggests they may exert anti-atherogenic effects in vascular tissue (Makela et al. 1999). However, binding affinities for the ER do not explain the functional complexity of isoflavones. For example, although genistein and equol have a 20-fold greater affinity for ERβ than ERα, the transcriptional expression is greatest for equol relative to all other isoflavones. The contribution of the protein portion of soy *versus* the isoflavone portion to the cardiovascular effects reported also remains unclear.

At present any cardiovascular effects of isoflavones are thought to be related to their effects on lipid metabolism. Results are awaited from trials examining the effects of isoflavones on progression of atherosclerosis and whether the capacity to convert daidzein into equol confers greater cardiovascular benefits. No randomized clinical trials examining clinical endpoints of cardiovascular health are available. Therefore, although there is some evidence of a beneficial effect of isoflavones on lipids and vascular function, and while a healthy diet should be encouraged as a general health measure, it is not appropriate to recommend phytoestrogen supplements or diets as a primary cardiovascular disease preventive intervention.

1.4.2 The Brain

Both ERα and ERβ are abundantly expressed in brain and exhibit a pattern of distribution consistent with their roles in reproductive and cognitive function (Spencer et al. 2008). As expected, ERα occurs in brain regions involved in regulation of reproduction but both occur, particularly ERβ, in brain regions involved in cognition. The expression and localization of ERs are dynamic and can vary depending upon brain region, cell type, hormonal status and neurological condition. The effects of isoflavones in the brain may be due to genomic or non-genomic actions. Some clinical trials also point to different effects on cognition in women depending on the age since the menopause, a situation analogous to the 'critical window' hypothesis for the effects of postmenopausal estrogens on the cardiovascular and cognitive health of women.

1.4.3 Bone Health

A number of genomic and non-genomic mechanisms exist by which phytoestrogens may affect bone metabolism. Isoflavones increase the synthesis of vitamin D in a number of non-renal cell types, have been shown to stimulate calcium uptake in bone and to increase bone cell proliferation and differentiation in animal studies. These effects may lead to improved bone health consistent with data on postmenopausal estrogen therapy and bone health (Writing Group for the Women's Health Initiative Investigators 2002). More than 25 studies have examined the effects of isoflavones on bone mineral density (BMD) in postmenopausal women most, as usual, being small and short. A 2-year trial in postmenopausal osteopenic women given 54 mg day^{-1} genistein found that spinal and hip BMD increased significantly compared with placebo (Marini *et al.* 2008). However, two long-term trials (Brink *et al.* 2008; Vupadhyayula *et al.* 2009) showed no significant effect. Possible explanations for this inconsistency include chronological differences in exposure (Asian adult soy intake assessed in the epidemiologic studies may reflect lifelong intake) and the ability to convert daidzein into equol. There may be a critical dose of isoflavones required to achieve an effect on BMD, and there may be differences in bioavailability between various supplements, foods and purified compounds. Consequently, it remains unclear whether isoflavones have a beneficial effect on BMD and, importantly, no randomized trials have demonstrated efficacy of phytoestrogens on fracture prevention (Baber 2010).

1.4.4 Breast Cancer

Isoflavones have been identified as putative chemopreventives (Messina 2010). Most research interest has focused on breast and prostate cancer because of the difference in incidence of these cancers in Western and Asian communities. Isoflavones may alter the metabolism of endogenous estrogens, potentially producing indirect effects on estrogenic pathways. Other mechanisms by which isoflavones may be cancer protective include antiproliferative effects, tyrosine kinase inhibition, modulation of steroid hormone-metabolizing enzyme activity, induction of apoptosis and inhibition of angiogenesis. Dietary isoflavones also reduce circulating and intra-breast estradiol concentrations in monkeys, with a corresponding decrease in uterine and breast tissue proliferation (NAMS 2011).

A recent meta-analysis (Wu *et al.* 2008) found that high soy intake was associated with an odds ratio (OR) of 0.71 [95% confidence interval

(CI) = 0.60–0.85] for breast cancer. However, there is an intriguing body of evidence suggesting that to derive protection against breast cancer, soy consumption must occur during childhood and/or adolescence (Messina and Hilakivi-Clarke 2009). The current operating hypothesis for this observation is that isoflavone exposure stimulates differentiation of breast tissue leading to a reduction in the anatomical structures that give rise to cancer cells.

1.4.5 Prostate Cancer

A meta-analysis (Yan and Spitznagel 2009) reported a combined relative risk/OR of 0.74 ($P = 0.01$) for prostate cancer when comparing high with low soy intake. Animal studies generally show isoflavone-containing diets retard the development of prostate cancer (Pollard and Suckow 2006). There is also evidence that isoflavones inhibit prostate tumour spread *via* the non-genomic mechanisms.

1.4.6 Menopausal Symptoms

Adlercreutz *et al.* (1992) suggested that the low prevalence of hot flushes in Japanese postmenopausal women might be partially due to their high consumption of soy foods. It was speculated that the estrogen-like effects of isoflavones might mitigate the drop in estrogen levels. This hypothesis encouraged the development of a vast complementary therapy industry marketing various mixtures of isoflavones, sometimes combined with other botanicals, as treatments for menopausal vasomotor symptoms. More than 50 clinical trials evaluating the efficacy of isoflavone-containing products have been conducted. Most trials have failed to show a significant benefit (Baber 2010; Lethaby *et al.* 2007) and there is increasing evidence that women who lack the capacity to convert daidzein into equol may derive little benefit. Placebo response rates in clinical trials of interventions to alleviate vasomotor symptoms typically show a 40–50% response (Baber 2010) and if only 30% of isoflavone users are able to convert them into the more estrogenic equol, it is not surprising that results have been disappointing. To test whether conversion into equol is critical, a trial was conducted on 96 healthy postmenopausal women (Jou *et al.* 2008). Volunteers were randomized to the isoflavone or placebo group. The isoflavone group was further divided into equol producers ($n = 34$) and non-producers ($n = 32$) based on urinary equol levels after consuming 135 mg of isoflavones day^{-1} for a week. Equol producers showed significantly greater reduction in some categories of Kupperman menopausal symptom scores than the placebo group.

1.5 Areas in Dispute

Most concern regarding high level isoflavone consumption has focused on the estrogenicity of isoflavones and potential harm when used by breast cancer patients and women at high risk of developing breast cancer. Although isoflavones *in vitro* may stimulate breast cancer cells, observational studies have found that isoflavones do not adversely affect markers of breast cancer risk, including breast tissue density, breast cell proliferation or circulating estrogen levels, and epidemiologic studies report the consumption of traditional soy foods after a diagnosis of breast cancer has no effect on or improves prognosis (Baber 2010; Guha et al. 2009; Shu et al. 2009). Isoflavones may affect thyroid function in patients receiving thyroid replacement therapy, but this effect is due to an impact on drug absorption rather than a direct effect on thyroid function (Villar et al. 2007).

A Cochrane Review found no adverse effects of phytoestrogens on endometrial pathology for up to 1 year (Lethaby et al. 2007). However, a 5-year placebo-controlled study of the effects of soy isoflavone (150 mg daily) on endometrial tissue in 376 women found isoflavone use increased the incidence of endometrial hyperplasia (Unfer et al. 2004). The increased incidence was small and there were no cases of endometrial cancer identified. These findings have not been confirmed by other studies and hence the long-term endometrial safety of high doses of phytoestrogen supplements is not fully established.

Summary Points

- Isoflavones are diphenolic plant chemicals structurally similar to 17β-estradiol, which bind selectively but weakly to mammalian ERs with a preference for ERβ.
- Isoflavones are capable of exerting receptor-mediated estrogenic and anti-estrogenic effects, as well as non-genomic effects.
- The effects on human health predicted from epidemiological studies of Asian communities consuming substantial dietary isoflavones have not been born out in clinical trials using dietary supplements containing equivalent amounts of isoflavones on Western populations.
- The discord seen between these results may be due to a number of factors:
- Lifelong, or at least pre-pubertal, exposure may be required to manifest some biological effects.
- It may be that biological effects achieved from whole food sources of isoflavones are different to those achieved with the use of extracts or supplements.
- The effects attributed to isoflavones may be due to some as yet unidentified substance found in popular dietary sources of isoflavones.
- Biological effects may only be seen in those people who are capable of converting isoflavones into equol.

Key Facts

Key Facts for Isoflavones

- Isoflavones are diphenolic compounds of plant origin, which bind weakly but selectively to mammalian ERS. Isoflavones exert biological effects *via* genomic and non-genomic mechanisms.

- Epidemiological studies in Asian communities, which consume substantial quantities of dietary isoflavones, suggest a beneficial effect of isoflavones on cardiovascular health, menopausal vasomotor symptoms and some hormone-dependent cancers.
- The promise of these benefits has not been born out in clinical trials on Western populations.
- This may be due to timing and duration of exposure, the type of isoflavones ingested, variable absorption, equol-producer status or other factors. Further research is required.

Key Facts for Estrogen Receptors

- ERs are hormone receptors, which are activated by the hormone 17β-estradiol and potentially by compounds of similar stearic structure. There are at least two estrogen receptors: ERα and ERβ.
- ERs are predominantly cytoplasmic receptors in the un-liganded state, although a portion resides in the nucleus.
- Although widely distributed, ERα is preferentially expressed in endometrium, breast, ovarian stroma and hypothalamus, and ERβ in kidney, brain, bone, heart, lungs and mucosa.
- 17β-Estradiol binds equally well to both receptors. Estrone and raloxifene bind preferentially to the α receptor, whereas estriol and isoflavones, such as genistein, bind preferentially to the β receptor.

List of Abbreviations

BMD	bone mineral density
ER	estrogen receptor
O-DMA	*O*-desmethylangolensin
OR	odds ratio
SERM	selective estrogen receptor modulator

References

Adlercreutz, H., and Mazur, W., 1997. Phytoestrogens and Western diseases. *Annals of Medicine*. 29: 95–120.

Adlercreutz, H., Honjo, H., Higashi, A., Fotsis, T., Hamalainen, E., Hasegawa, T., and Okada, H., 1991. Urinary excretion of lignans and isoflavonoid phytoestrogen and oestrogens in Japanese men and women consuming a traditional Japanese diet. *American Journal of Clinical Nutrition*. 54(6): 1093–1100.

Adlercreutz, H., Hamalainen, E., Gorbach, S., and Goldin, B., 1992. Dietary phytoestrogens and the menopause in Japan. *Lancet*. 339: 1233.

Akiyama, T., Ishida, J., Nakagawa, S., Ogawara, H., Watanabe, S., Itoh, N., Shibuya, M., and Fukami, Y., 1987. Genistein, a specific inhibitor of tyrosine specific protein kinases. *Journal of Biological Chemistry*. 262: 5592–5595.

Baber, R., 2010. Phytoestrogens in post reproductive health. *Maturitas*. 66: 344–394.

Baber, R., 2010. *Side effects of phytoestrogens*. Drug Safety. *Maturitas*. 66: 344–349.

Brink, E., Coxam, V., Robins, S., Wahala, K., Cassidy, A., and Branca, F., 2008. Long-term consumption of isoflavone-enriched foods does not affect bone mineral density, bone metabolism, or hormonal status in early post-menopausal women: a randomized, double-blind, placebo controlled study. *American Journal of Clinical Nutrition*. 87: 761–770.

Collins, B., McLaughlin, J., and Arnold, S., 1997. The anti-oestrogenic activities of phytochemicals with human oestrogen receptors expressed in yeast. *Steroids*. 62: 365–372.

Duncan, A.M., Phipps, W.R., and Coetze, M.S., 2003. Phytoestrogens: best practice. *Research in Clinical Endocrinology and Metabolism*. 17: 253–271.

Goldin, B., Adlercreutz, H., Gorbach, S., Woods, M., Dwyer, J., Conlon, T., Bohn, E., and Gershoff, S., 1986. The relationship between oestrogen levels and diets of Caucasian American and oriental immigrant women. *American Journal of Clinical Nutrition*. 44: 945–953.

Guha, N., Kwan, M.L., Quesenberry, Jr, C.P., Weltzien, E.K., Castillo, A.L., and Caan, B.J., 2009. Soy isoflavones and risk of cancer recurrence in a cohort of breast cancer survivors: the Life After Cancer Epidemiology study. *Breast Cancer Research and Treatment*. 118: 395–405.

Jou, H.J., Wu, S.C., Chang, F.W., Ling, P.Y., Chu, K.S., and Wu, W.H., 2008. Effect of intestinal production of equol on menopausal symptoms in women treated with soy isoflavones. *International Journal of Gynaecology and Obstetrics*. 102: 44–49.

King, R.A., Broadbent, J.L., and Head, R.J., 1996. Absorption and excretion of the soy isoflavone genistein in rats. *Journal of Nutrition*. 126: 176–182.

Korde, L., Wu, A., Fears, T., Nomura, A., West, D., Kolonel, L., Pike, M., Hoover, R., and Ziegler, R., 2009. Childhood soy intake and breast cancer risk in Asian American women. Cancer Epidemiology. *Biomarkers and Prevention*. 18: 1050–1059.

Kuiper, G.G., Enmark, E., Pelto-Huikko, M., Nilsson, S., and Gustafsson, J.A., 1996. Cloning of a novel receptor expressed in rat prostate and ovary. *Proceedings of the National Academy of Science U.S.A.* 93: 5925–5930.

Kuiper, G.G., Carlsson, B., Grandien, K., Enmark, E., Haggblad, J., Nilsson, S., and Gustafsson, J.A., 1997. Comparison of the ligand binding specificity and transcript tissue distribution of estrogen receptors α and β. *Endocrinology*. 138: 863–870.

Kuiper, G., Lemmen, J., Carlsson, B., Corton, J., Safe, S., Van der Saag, P., Van der Burg, B., and Gustafsson, J., 1998. Interaction of estrogenic chemicals and phytoestrogens with estrogen receptor β. *Endocrinology*. 139: 4252–4263.

Kurzer, M.S., and Xu, X., 1997. Dietary phytoestrogens. *Annual Review of Nutrition*. 17: 353–381.

Lethaby, A., Brown, J., Marjoribanks, J., Kronenberg, F., Roberts, H., and Eden, J., 2007. Phytoestrogens for vasomotor menopausal symptoms. *Cochrane Database of Systematic Reviews*. Oct 17;(4):CD001395.

Makela, S., Savolainen, H., Aavik, E., Myllarniemi, M., Strauss, L., Taskinen, E., Gustafsson, J., and Hayry, P., 1999. Differentiation between vasculoprotective and uterotrophic effects of ligands with different binding affinities to estrogen receptors α and β. *Proceedings of the National Academy of Science U.S.A*. 96: 7077–7082.

Marini, H., Levy, R., D'Anna, R., Frisini, N., Mazzaferro, S., Cancellieri, F., Cannata, M., Corrado, F., Frisina, A., Adamo, V., Lubrano, C., Sansotta, C., Marini, R., Adamo, E., and Squadrito, F., 2008. Breast safety and efficacy of genistein aglycone for postmenopausal bone loss: a follow-up study. *Journal of Clinical Endocrinology and Metabolism*. 93: 4787–4796.

Matthews, J., and Gustafsson, J.A., 2003. Estrogen signalling: a subtle balance between ERα and ERβ. *Molecular Interventions*. 3: 281–292.

Messina, M., 2010. Insights gained from 20 years of soy research. *Journal of Nutrition*. 140: 2289S–2295S.

Messina, M., and Hilakivi-Clarke, L., 2009. Early intake appears to be the key to the proposed protective effects of soy intake against breast cancer. *Nutrition and Cancer*. 61: 792–798.

Miksicek, R.J., 1994. Interaction of naturally occurring no steroidal oestrogens with expressed recombinant human oestrogen receptor. *Journal of Steroid Biochemistry and Molecular Biology*. 49: 153–160.

NAMS, 2011. Isoflavones report. *Menopause*. 18: 732–753.

Nelson, H.D., 2008. Menopause. *Lancet*. 371: 760–770.

Pollard, M., and Suckow, M.A., 2006. Dietary prevention of hormone refractory prostate cancer in Lobund–Wistar rats: a review of studies in a relevant animal model. *Comparative Medicine*. 56: 461–467.

Ronis, M.J., Little, J.M., Barone, G.W., Chen, G., Radominska-Pandya, A., and Badger, T.M., 2006. Sulfation of the isoflavones genistein and daidzein in human and rat liver and gastrointestinal tract. *Journal of Medicinal Food*. 9: 348–355.

Sarkar, F.H., and Li, Y., 2003. Soy isoflavones and cancer prevention. *Cancer Investigation*. 21: 744–757.

Setchell, K.D., and Clerici, C.J., 2010. Equol: history, chemistry and formation. *J. Nutr*. 140: 1355S–1362S.

Sfakianos, J., Coward, L., Kirk, M., and Barnes, S., 1997. Intestinal uptake and biliary excretion of the isoflavone genistein in the rat. *Journal of Nutrition*. 127: 1260–1268.

Shu, X.O., Zheng, Y., Cai, H., Gu, K., Chen, Z., Zheng, W., and Lu, W., 2009. Soy food intake and breast cancer survival. *Journal of the American Medical Association*. 302: 2437–2443.

Spencer, J.L., Waters, E.M., Romeo, R.D., Wood, G.E., Milner, T.A., and McEwen, B.S., 2008. Uncovering the mechanisms of estrogen effects on hippocampal function. *Frontiers in Neuroendocrinology*. 29: 219–237.

Unfer, V., Casini, M.L., Costabile, L., Mignosa, M., Gerli, S., and Di Renzo, G.C., 2004. Endometrial effects of long-term treatment with phytoestrogens: a randomized, double blind, placebo-controlled study. *Fertility and Sterility*. 82: 145–148, quiz 265.

Villar, H.C., Saconato, H., Valente, O. and Atallah, A.N., 2007. Thyroid hormone replacement for subclinical hypothyroidism. Cochrane Database Systematic Reviews CD003419.

Vupadhyayula, P.M., Gallagher, J.C., Templin, T., Logsdon, S.M., and Smith, L.M., 2009. Effects of soy protein isolate on bone mineral density and physical performance indices in postmenopausal women – a 2-year randomized, double-blind, placebo-controlled trial. *Menopause*. 16: 320–328.

Writing Group for the Women's Health Initiative Investigators, 2002. Risks and benefits of estrogen plus progestin in healthy postmenopausal women: principal results from the Women's Health Initiative randomized controlled trial. *Journal of the American Medical Association*. 288: 321–333.

Wu, A.H., Yu, M.C., Tseng, C.C., and Pike, M.C., 2008. Epidemiology of soy exposures and breast cancer risk. *British Journal of Cancer*. 98: 9–14.

Yan, L., and Spitznagel, E.L., 2009. Soy consumption and prostate cancer risk in men: a revisit of a meta-analysis. *American Journal of Clinical Nutrition*. 89: 1155–1163.

CHAPTER 2
Phytoestrogens in Plants: With Special Reference to Isoflavones

FRANZ BUCAR

Institute of Pharmaceutical Sciences, Department of Pharmacognosy, University of Graz, Universitätsplatz 4, A-8010 Graz, Austria
Email: franz.bucar@uni-graz.at

2.1 Characterization of Phytoestrogens

Natural non-steroidal plant compounds exerting, usually weak, estrogenic and/or anti-estrogenic activity to mammals are generally designated as "phytoestrogens".

According to their phytochemistry, the majority of these compounds can be classed with flavonoids. The latter include isoflavones, coumestanes, pterocarpanes and rotenoides, to mention a few isoflavonoid subclasses (Andersen and Markham 2006). Non-flavonoidal phytoestrogens are present within lignans, such as secoisolariciresinol and stilbenes [*e.g. trans*-resveratrol (Dinelli *et al.* 2006)]. Fungi have also been reported to produce compounds of similar biological activity, which are known as mycoestrogens, such as zealarenone, a macrocyclic β-resorcinolic acid lactone (Metzler *et al.* 2010).

The biological activities of phytoestrogens in mammals are closely related to their structural similarity to 17β-estradiol, the most important endogenous estrogen, and their affinity to the estrogen receptor (ER) subtypes ERα and ERβ (Collins *et al.* 1997; Leclercq *et al.* 2011). Due to the selective ER subtype affinities, phytoestrogens are also designated as selective estrogen receptor modulators (SERMs). Certain structural features have been identified as being essential for estradiol-mimicking activity, *i.e.* not all isoflavonoids can be assumed to have ER modulating effects and the *in vivo* potency of individual compounds has to be evaluated in each case, not to mention their interactions with a number of other biological systems (Leclercq *et al.* 2011). Figure 2.1 exemplifies structures of phytoestrogens and mycoestrogens.

2.2 Distribution of Phytoestrogens in the Plant Kingdom

2.2.1 Phytoestrogens other than Isoflavones

Whereas isoflavones are found mainly in the Leguminosae, subfamily Papilionideae (see Section 2.2.2), a more diverse distribution of other phytoestrogens among non-related plant species can be observed (Dinelli *et al.* 2006). Examples of flavonoidal phytoestrogens are prenylated flavanones in hops (*Humulus lupulus* L., Cannabaceae) or phloretin (dihydronaringenin chalcone) and its 2′-*O*-glucoside phlorizin in leaves of apple trees (*Malus* sp., Rosaceae). High concentrations of secoisolariciresinol and matairesinol glycosides, two representatives of lignans which exert estrogenic activity *via* their metabolites enterodiol and enterolactone (Adlercreutz 2007), have been reported for flaxseed (*Linum usitatissimum* L., Linaceae) (Figure 2.2); however, lignan phytoestrogens are distributed in various non-related plant genera (Cassidy *et al.* 2000; Dixon 2004).

Stilbene monomers, dimers and polymers occur widely among liverworts and higher plants (Cassidy *et al.* 2000). *trans*-Resveratrol, the most representative monomeric stilbene, and related compounds are well known as phytoalexins in grapes (*Vitis vinifera* L., Vitaceae). Significant amounts are also found in peanuts (*Arachis hypogaea* L., Leguminosae) and above all in rhizomes and roots of different species of the Polygonaceae (Cassidy *et al.* 2000) (Figure 2.3). Extracts of *Fallopia japonica* (Houtt.) Ronse Decr. (syn. *Polygonum cuspidatum*) are used in nutraceuticals as a resveratrol source, however, it has to be taken into account that they may also contain significant amounts of anthraquinones such as emodin (Bucar and Knauder 2005).

The distribution of isoflavonoid subtypes in non-leguminous taxa has been evaluated in detail by Lapcik (2007) and Reynaud *et al.* (2005). As outlined by Lapcik (2007), isoflavonoids besides isoflavones have been found in 49 non-leguminous plant families.

2.2.2 Isoflavones

Isoflavones are mainly found in the subfamily Papilionideae of the Leguminosae plant family.

Excellent reviews by Veitch (2007; 2009) summarize recent reports of isoflavonoids, insights into biosynthesis of isoflavonoids, and hyphenated instrumental techniques for isoflavonoid analysis, as well as structure elucidation. Isoflavones exceed by far any other subtype of isoflavonoids concerning the number of reported compounds. Isoflavones have also been found in at least

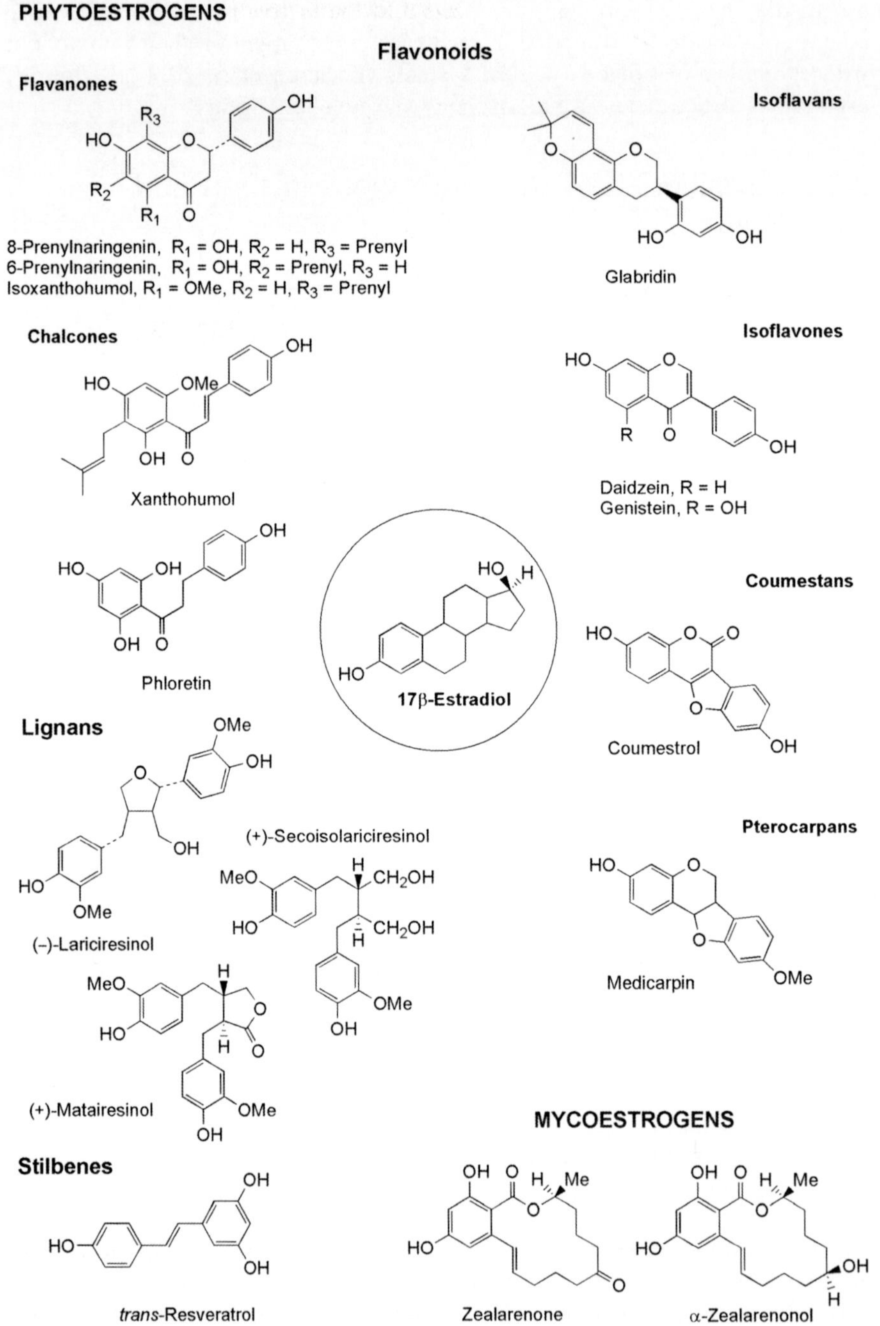

PHYTOESTROGENS

Flavonoids

Flavanones

8-Prenylnaringenin, R_1 = OH, R_2 = H, R_3 = Prenyl
6-Prenylnaringenin, R_1 = OH, R_2 = Prenyl, R_3 = H
Isoxanthohumol, R_1 = OMe, R_2 = H, R_3 = Prenyl

Chalcones

Xanthohumol

Phloretin

Lignans

(+)-Secoisolariciresinol

(−)-Lariciresinol

(+)-Matairesinol

Stilbenes

trans-Resveratrol

Isoflavans

Glabridin

Isoflavones

Daidzein, R = H
Genistein, R = OH

17β-Estradiol

Coumestans

Coumestrol

Pterocarpans

Medicarpin

MYCOESTROGENS

Zealarenone

α-Zealarenonol

Figure 2.1 Selected phytoestrogens and mycoestrogens. Chemical structures of phytoestrogens (from plants) and mycoestrogens (from fungi) are shown in this Figure. For comparison, in the centre the structure of the steroid hormone 17β-estradiol is presented.
(Unpublished Figure.)

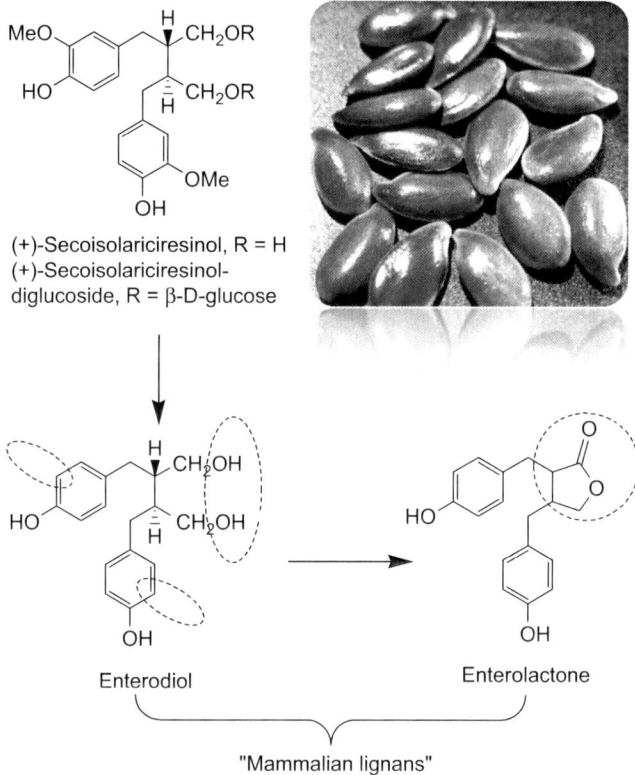

Figure 2.2 Degradation of secoisolariciresinol and its diglucoside in the human intestine. Secoisolariresinol and its diglucoside, the major lignans in flaxseed (*L. usitatissimum*), are degraded by the micro-organisms in the gut to enterodiol (deglucosidation, demethoxylation) and further to enterolactone (dehydration). In contrast to plant lignans, these metabolites are also referred to as "mammalian lignans" (Adlercreutz 2007; Dixon 2004). Sites of modification are indicated by the dotted lines.
(Photo: F. Bucar, unpublished.)

59 families outside the Leguminosae, however, the relationships between isoflavonoid-producing families still await clarification (Lapcik 2007).

2.3 Chemical Features of Plant Isoflavones

Isoflavones differ from the corresponding flavones by an aromatic ring migration from position C-2 to C-3. Major structural diversity arises from hydroxylation, methoxylation or methylenedioxy substitution, prenylation and glycosidation (Veitch 2007). Glucose is the predominant sugar moiety, although in some cases rhamnose, galactose, xylose, apiose, arabinose and cymarose could be identified as glycosidic moieties (Andersen and Markham 2006; Veitch 2009). A remarkable feature of isoflavone glucosides is their

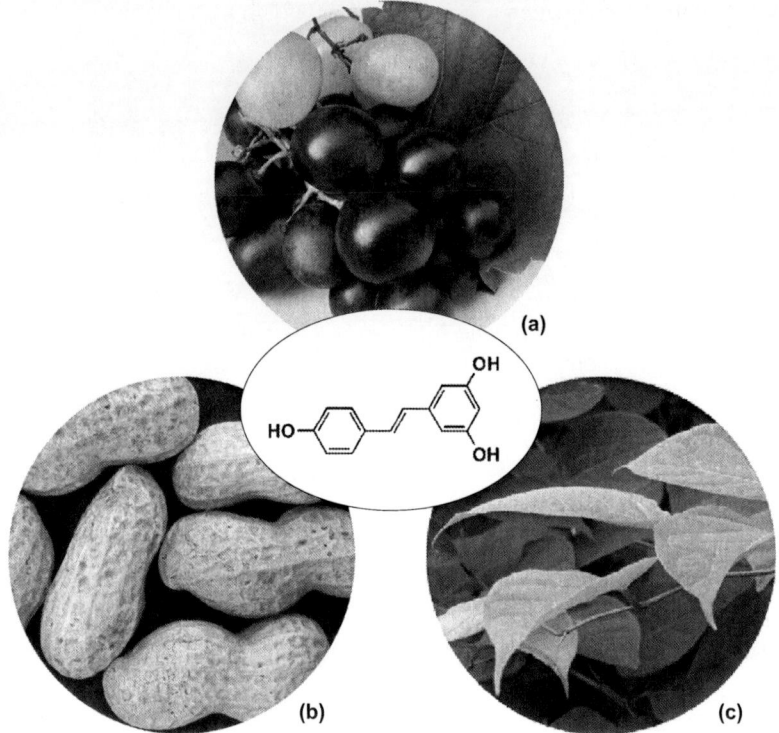

Figure 2.3 Major plant sources of *trans*-resveratrol. Sources of *trans*-resveratrol, a stilbene with phytoestrogenic activity, are presented: (a) grapes and wine leaves (*V. vinifera*), (b) peanuts (*A. hypogaea*) and (c) Japanese knotweed (*F. japonica*).
(Photos: F. Bucar, unpublished.)

conjugation with malonyl residues. In plant extracts they can be found together with the corresponding glucosides and acetylglucosides, respectively (Figure 2.4). However, the latter are suspected to be decarboxylation products of the naturally occurring 6″-malonylglucosides (Veitch 2009). Inappropriate treatment of plant material can lead to significant losses of the conjugated acids (Toebes *et al.* 2005; Tsao *et al.* 2006), hence a higher number of naturally occurring malonyl conjugates are expected if their labile nature is considered during the workup procedure and by application of on-line coupling techniques such as liquid chromatography–electrospray–mass spectrometry (multiple stage) LC-ESI-MS^n analysis.

2.4 Biosynthesis of Isoflavones

Isoflavones are biosynthetically derived from the phenylpropanoid pathway. Isoflavonoid biosynthesis has been thoroughly reviewed (Du *et al.* 2010; Lapcik

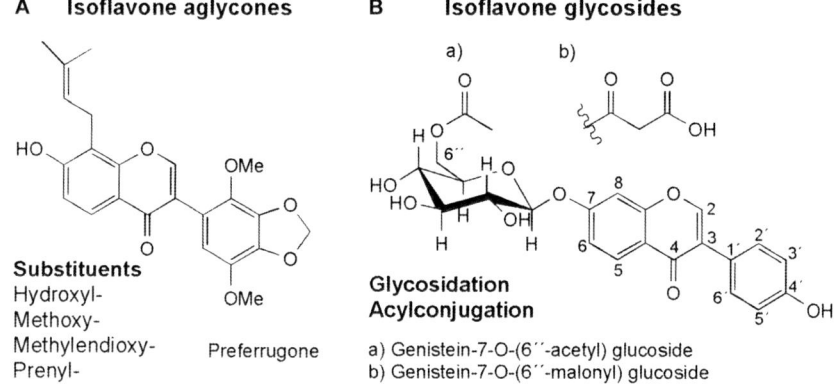

Figure 2.4 Main structural modifications of isoflavones. The structural features of isoflavones in plants are presented: (A) frequently occurring modifications of the aglycones (according to Veitch 2007) and (B) examples of isoflavone glycosides with acetyl or malonyl conjugates.
(Unpublished Figure.)

2007; Veitch 2007; 2009; Wang 2011). The crucial step in isoflavone biosynthesis is the conversion of flavanones into the respective isoflavanones by 2-hydroxyisoflavanone synthase (2HIS) and the subsequent dehydration to isoflavones, for which 2-hydroxyisoflavanone dehydrase (HID) has been characterized (He et al. 2011). However, this step still needs further clarification since the reaction might also occur spontaneously. The term "isoflavone synthase" (IFS) is used either to cover both the conversion and dehydration reactions, or synonymously for 2HIS. 5-Deoxyisoflavones are markers of the Leguminosae and arise from the activity of chalcone reductase, which catalyses the synthesis of 6′-deoxychalcone precursors of 5-deoxyisoflavones (Veitch 2009) (Figure 2.5).

Modification of basic structures resulting in alterations in solubility, stability and biological activity primarily concern O-methylation, glycosylation and acyl conjugation. The detailed knowledge of the phenylpropanoid pathway, including the isoflavone specific branch, opens a plethora of possibilities for the metabolic engineering of plants and plant tissues in order to improve yields and chemical diversity of isoflavones (Du et al. 2010; Wang 2011).

2.5 Localisation of Isoflavones in Plants

Certainly, isoflavones are not equally distributed in all tissues of a specific plant and their content depends on the type of tissue, the ontogenic stage of the respective plant and modulation by biotic or abiotic stress factors. For selected plant sources, such as soybean (*Glycine max* L.), red clover (*Trifolium pratense* L.) or white lupine (*Lupinus albus* L.), investigations on the distribution of isoflavones and their fluctuation during plant development have been performed (for structures see Figure 2.6).

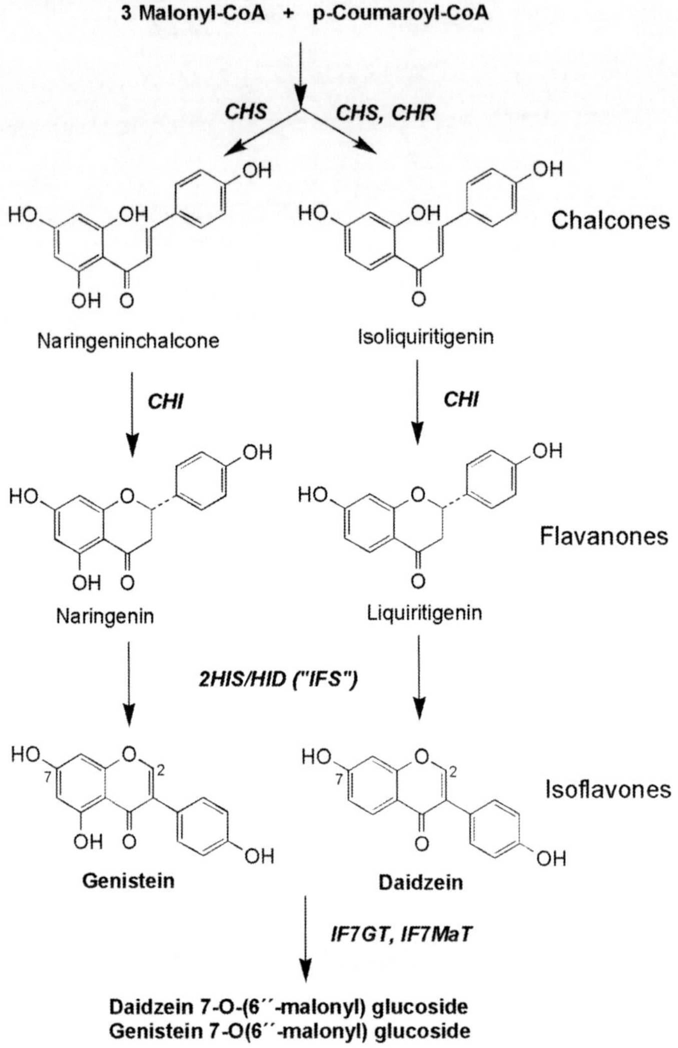

Figure 2.5 Biosynthesis of daidzein and genistein and their malonylconjugates, The biosynthesis of isoflavones is illustrated by the key steps of the generation of daidzein (a 5-deoxyisoflavone), genistein (a 5-hydroxyisoflavone) and their respective malonylglucosides. CHI: chalcone isomerase; CHS: chalcone synthase; CHR: chalcone reductase; HID: 2-hydroxyisoflavanone dehydrase; 2HIS: 2-hydroxyflavanone isomerase; IF7GT: UDP-glucose:isoflavone 7-O-glucosyltransferase; IF7MaT: isoflavone 7-O-malonyltransferase; IFS: isoflavone synthase.
Data are summarized from Veitch 2007, 2009; with permission.

Reviews of the distribution of isoflavones with respect to localization in plant organs, plant tissues and plant cells covering more than a few species have not been undertaken until now. Graham (1991) performed a detailed study on

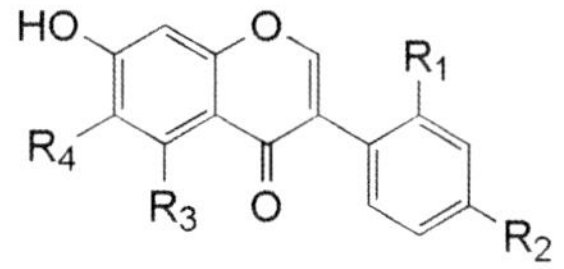

Phytoestrogens in Plants: With Special Reference to Isoflavones

Compound	R₁	R₂	R₃	R₄
Biochanin A	H	OMe	OH	H
Daidzein	H	OH	H	H
Formononetin	H	OMe	H	H
Genistein	H	OH	OH	H
Glycitein	H	OH	H	OMe
Luteone	OH	OH	OH	Prenyl
Wighteone	H	OH	OH	Prenyl

Figure 2.6 Selected isoflavone aglycones. The chemical structures of selected isoflavone aglycones which occur in one or more of the plants discussed in Chapter 2.5 (soybean, red clover and white lupine) are presented. OMe, methoxyl.
(Unpublished Figure.)

flavonoid and isoflavonoid distribution in seedling organs and root exudates of soybean seedlings. In 7-day-old light-grown seedlings, daidzein, its 7-*O*-glucoside daidzin and the malonylated daidzin (DMG) were predominantly found in the roots, especially the root tips, regions of root growth and the cotyledons. On the other hand, genistein malonylglucoside (GMG), although also present in significant amounts in the roots, was found in higher concentrations in the cotyledons. Growing the seedling in the dark led to reduced amounts of daidzein and genistein, including their glycosides, in the roots, but to significantly higher levels in the cotyledons. The isoflavone profiles of seed and root exudates and that of the seedlings did not match (Graham 1991).

Studies of Kim and Chung (2007) revealed that in all maturity stages of soy seeds, malonyl conjugates were predominant, whereas acetyl glucosides were absent or only found in traces. This supports the assumption that decarboxylation of malonylglucosides is the main source of acetylglucosides if found in soy products in significant amounts.

Investigations by Tsao *et al.* (2006) of red clover plant organs of a number of cultivars at early bud stage and late flowering stage showed highest total isoflavone contents in the leaves, but only low amounts in the flowers (late flowering stage). Formononetin and biochanin A were predominant in all plant organs, but at different ratios. Levels of these compounds were followed during different maturity stages and a clear decline of levels of both isoflavones in inflorescences was recognised just before flowering.

D'Agostina *et al.* (2008) investigated the isoflavone pattern fluctuation during a growth cycle of 1 year in white lupine (*L. albus*) in different cultivars and two different sowing times (autumn and early spring, respectively). Significantly higher amounts of genistein 7-*O*-glucoside were detected in leaves and stems of spring-sown lupine compared to autumn-sown plants, in which genistein was predominantly found, independent of the type of cultivar. Total isoflavones in leaves and stems were much higher than those in roots. No malonyl or acetyl conjugates were found in roots, in contrast to other studies (Bednarek *et al.* 2001), and it was concluded that malonyl conjugates might be only present in high amounts in lupine roots at early developmental stages of the plants (D'Agostina *et al.* 2008). Eliciting leaves and stems of white lupine with a yeast cell wall preparation did not have a remarkable influence on levels of malonyl conjugates of genistein glucosides, but there was a significant increase in prenylated isoflavones in elicited leaves and stems as compared with the control (Bednarek *et al.* 2001), supporting their role as phytoanticipins (see Section 2.6.2).

2.6 Biological Function of Isoflavones in Plants

2.6.1 Interaction of Isoflavones with Micro-organisms of the Rhizosphere

Sound evidence exists that isoflavones play an important role at the plant–microbe interface of leguminous plants. They have been recognised as signaling compounds in the early phase of the host-specific nitrogen-fixing symbiont attraction and as nod gene inducers, *i.e.* part of the nodulation process, which is essential for the maintenance of the rhizobium–plant symbiosis. A stimulating review by Shaw *et al.* (2006) discusses a number of questions regarding the role of flavonoids, including isoflavones, as chemical tools to form the rhizospherical microflora. Studies at the proteomic level of root–microbe interactions have been extensively reviewed by Mathesius (2009). Specific legume isoflavones excreted by the plant roots, which seems to involve active transport by efflux pumps (Sugiyama *et al.* 2007), interact with NodD, a protein which induces the biosynthesis of a series of bacterial nodulation signals (lipochitin oligosaccharides). In response to these signals, morphological changes such as root-hair curling or cortical cell division are initiated that are the basis of nitrogen-fixing nodule formation in legumes (Aoki *et al.* 2000; Mathesius 2009) (Figure 2.7). A detailed review of rhizobium–plant symbiosis is given by Van Rhijn and Vanderleyden (1995), who propose also other possible infection routes.

Flavonoids, including isoflavones, are also involved in the modulation of auxin transport, which is essential for nodulation (Grunewald *et al.* 2009). It seems that isoflavones are also part of the signaling between mycorrhizal fungi and their respective host plants (Shaw *et al.* 2006).

Due to the presence of microbes or their products, isoflavonoid biosynthetic steps could be influenced and alterations of isoflavonoid signals could also arise from degradation reactions catalysed by micro-organisms in the rhizosphere (Cooper 2004; Shaw *et al.* 2006).

2.6.2 Isoflavones as Phytoalexins and Phytoanticipins

Defence against microbial plant pathogens is another area where isoflavonoids exert an important impact. Although the target compounds are more

Figure 2.7 Root nodules of *T. repens* L. (white clover). Nodules at the roots of various legumes (here: white clover, *T. repens*) are formed through colonization by bacteria such as *Rhizobium* spp. capable of nitrogen fixation. (Photos: F. Bucar, unpublished.)

frequently found within the pterocarpan or isoflavan subclass of isoflavonoids and less within the isoflavones, the latter have a key function as precursors of pterocarpans and isoflavans, frequently lacking a hydroxyl at position C-5. In contrast to phytoalexins, which are biosynthesized on demand in the case of a pathogen attack, phytoanticipins are constitutively present in the plant cells to prevent infection by a microbial plant pathogen (Shaw *et al.* 2006). This differentiation, however, has not been strictly followed in the literature and it might be difficult to distinguish clearly between phytoalexins and phytoanticipins. The function as phytoalexin or phytoanticipin has been deduced from the fact that isoflavonoids are present in infected plant cells at concentrations which make an antimicrobial activity *in vivo* plausible, as well as from a number of elicitation studies (Aoki *et al.* 2000).

Summary Points

- Phytoestrogens are mainly found within compounds derived from the phenylpropanoid pathway: flavonoids (isoflavonoids), lignans and stilbenes.
- The natural occurrence of isoflavones is predominantly in the Leguminosae subfamily Papilionoideae, but they have also been found in other plant families outside the Leguminosae.

- Isoflavone biosynthesis is catalysed by "isoflavone synthase" (2-hydroxyflavanone isomerase and 2-hydroxyisoflavanone dehydrase) from flavanone precursors.
- The structural diversity of plant isoflavones is mainly derived from the introduction of hydroxy, methoxy, methylenedioxy or prenyl groups, as well as glycosidation with possible further acyl conjugation.
- In legumes, high levels of isoflavones occur in roots, seeds and leaves, but depend on the ontogenical stage of the individual plant as well as the influences of biotic and abiotic stress factors.
- Isoflavones play an important role in rhizobium–legume symbiosis, as well as in plant pathogen defence mechanisms.

Key Facts

The key facts of the plant family of Leguminosae (Fabaceae) and why legumes have a significant impact on other plants, animals and humans.

- Legumes (in botanical terms Leguminosae or Fabaceae) are the third largest plant family of flowering plants, including about 19 000 plant species.
- Many plants of economic importance are legumes including food plants (*e.g.* soy beans, peas, pea nuts, chick peas and beans), but also plants used as forage (*e.g.* clover and alfalfa). Their seeds are a valuable source of proteins and fixed oils.
- Leguminous plants are able to use nitrogen in the air to build up organic nitrogen compounds (*e.g.* amino acids) by symbiosis with specific soil bacteria. Nodules are formed at the roots of many leguminous plants which host the symbiotic bacteria.
- Through nitrogen fixation, the content of nitrogenous compounds, which serve as nutrients for other plants, in the soil is restored.
- Legumes are a pivotal source of isoflavone phytoestrogens.

Definitions of Words and Terms

Elicitors: Cellular or acellular structures (*e.g.* cell wall fragments of plant pathogens) which induce the biosynthesis of phytoalexins or phytoanticipins (see ***Phytoalexins***). In the context of plant cell cultures elicitation is also used in a wider sense for other compounds or factors (UV light, heavy metal salts *etc.*) which increase or induce biosynthesis of secondary metabolites.

Estrogens: Female sex hormones with a steroid structure. Their activity is mediated by binding to the estrogen receptor. One of the most important estrogens is 17β-estradiol.

Flavonoids: Natural phenolic compounds having a C_{15} structural core consisting of two aromatic rings linked *via* a C_3 bridge. Through ring closure, a heterocyclic ring system is created (flavones, flavanones, anthocyanidins *etc.*); open ring structures are chalcones. Isoflavonoids represent a specific subclass of flavonoids (see **Isoflavonoids**). Flavonoids are ubiquitous in the plant kingdom.

Glycosides: Natural products which consist of a non-sugar moiety (= aglycone) that is attached to one or more sugars *via* an oxygen bridge ("O-glycosides") or by a C–C bonding ("C-glycosides"). The biological activity is mainly derived from the type of aglycone which serves as criterion for classification (flavonoid glycosides, coumarin glycosides, saponins, *etc.*).

Isoflavonoids: A subclass of flavonoids (see **Flavonoids**) having the second aromatic ring attached to C-3 instead of C-2. Structural modifications with additional ring closures lead to further subgroups such as pterocarpans, coumestans, rotenones *etc.*

Lignans: A group of natural compounds that are derived from the binding of two phenylpropane (C_6–C_3) units *via* two specific carbons (C-2–C-2'). Further reactions (cyclisation, isomerisation, *etc.*) lead to a wide variety of structures. Lignans can occur as aglycones, but also as glycosides in plants.

Nitrogen fixation: Conversion of elemental inorganic nitrogen from the air or soil to organic nitrogen compounds. Above all, bacteria such as *Rhizobium*, *Brazorhizobium* or *Azorhizobium* spp. are able to convert elemental nitrogen into ammonia, which is further used by plants to form organic nitrogen compounds.

Nodulation: Formation of specific spherical structures ("nodules") at the roots or stems of legumes which enable symbiotic bacteria (*Rhizobium* spp. *etc.*) to exist within the plant tissue.

Ontogenesis: Development of an individuum from origin to the adult stage.

Phytoalexins: Plant compounds which are formed as response to biotic (*e.g.* fungal infection) or abiotic (*e.g.* wounding, UV light) stress. The term "phytoalexin" occasionally is restricted to those compounds which are newly biosynthesized, whereas compounds for which concentration is increased in response to stress are designated as "phytoanticipins".

Stilbenes: Plant phenolics with a diaryl-1,2-ethene structure. Stilbenes such as *trans*-resveratrol are associated with resistance to fungal infections, either inducible or constitutively.

List of Abbreviations

CHI	chalcone isomerase
CHR	chalcone reductase
CHS	chalcone synthase
DMG	daidzein malonylglucoside
ER	estrogen receptor
ERα, ERβ	estrogen receptor subtypes
GMG	genistein malonylglucoside
HID	2-hydroxyisoflavanone dehydrase
2HIS	2-hydroxyisoflavanone synthase
IF7GT	UDP-glucose:isoflavone 7-O-glucosyltransferase

IF7MaT	isoflavone 7-O-malonyltransferase
IFS	isoflavone synthase
LC-ESI-MS^n	liquid chromatography-electrospray mass spectrometry (multiple stage)

References

Adlercreutz, H., 2007. Lignans and human health. *Critical Reviews in Clinical Laboratory Sciences*. 44: 483–525.

Andersen, O.M., and Markham, K.R., 2006. (ed.), Flavonoids. *Chemistry, Biochemistry and Applications*. Taylor & Francis, Boca Raton, FL, USA.

Aoki, T., Akashi, T., and Ayabe, S., 2000. Flavonoids of leguminous plants: structure, biological activity, and biosynthesis. *Journal of Plant Research*. 113: 475–488.

Bednarek, P., Franski, R., Kerhoas, L., Einhorn, J., Wojtaszek, P., and Stobiecki, M., 2001. Profiling changes in metabolism in *Lupinus albus* treated with biotic elicitor. *Phytochemistry*. 56: 77–85.

Bucar, F. and Knauder, E., 2005. Analysis of resveratrol containing extracts and botanical dietary supplements. In: *Book of Abstracts, 53rd Annual Congress of the Society for Medicinal Plant Research*. August 21–25, 2005. Florence, Italy, p. 387.

Cassidy, A., Hanley, B., and Lamuela-Raventos, R.M., 2000. Isoflavones, lignans and stilbenes – origins, metabolism and potential importance to human health. *Journal of the Science of Food and Agriculture*. 80: 1044–1062.

Collins, B.M., McLachlan, J.A., and Arnold, S.F., 1997. The estrogenic and antiestrogenic activities of phytochemicals with the human estrogen receptor expressed in yeast. *Steroids*. 62: 365–372.

Cooper, J.E., 2004. Multiple responses of rhizobia to flavonoids during legume root infection. *Advances in Botanical Research*. 41: 1–62.

D'Agostina, A., Boschin, G., Resta, D., Annicchiarico, P., and Arnoldi, A., 2008. Changes of isoflavones during the growth cycle of Lupinus albus. *Journal of Agricultural and Food Chemistry*. 56: 4450–4456.

Dinelli, G., Bonetti, A., Elementi, S., D'Antuono, L.F. and Catizone, P., 2006. In: Yldiz, F. (ed.) *Phytoestrogens in Functional Foods*. Taylor & Francis, Boca Raton, FL, USA, pp. 19–80.

Dixon, R.A., 2004. Phytoestrogens. *Annual Reviews of Plant Biology*. 55: 225–261.

Du, H., Huang, Y., and Tang, Y., 2010. Genetic and metabolic engineering of isoflavonoid biosynthesis. *Applied Microbiology and Biotechnology*. 86: 1293–1312.

Graham, T.L., 1991. Flavonoid and isoflavonoid distribution in developing soybean seedling tissues and in seed and root exudates. *Plant Physiology*. 95: 594–603.

Grunewald, W., van Norden, G., Van Isterdael, G., Beeckman, T., Gheysen, G., and Mathesius, U., 2009. Manipulation of auxin transport in plant roots during Rhizobium symbiosis and nematode parasitism. *The Plant Cell.* 21: 2553–2562.

He, X., Blount, J.W., Ge, S., Tang, Y., and Dixon, R.A., 2011. A genomic approach to isoflavone biosynthesis in kudzu (Pueraria lobata). *Planta.* 233: 843–855.

Kim, J.-A., and Chung, I.-M., 2007. Change in isoflavone concentration of soybean (Glycine max L.) seeds at different growth stages. *Journal of the Science of Food and Agriculture.* 87: 496–503.

Lapcik, O., 2007. Isoflavonoids in non-leguminous taxa: a rarity or a rule? *Phytochemistry.* 68: 2909–2916.

Leclercq, G., de Cremoux, P., This, P., and Jacquot, Y., 2011. Lack of sufficient information on the specificity and selectivity of commercial phytoestrogens preparations for therapeutic purposes. *Maturitas.* 68: 56–64.

Mathesius, U., 2009. Comparative proteomic studies of root–microbe interactions. *Journal of Proteomics.* 72: 353–366.

Metzler, M., Pfeiffer, E., and Hildebrand, A.A., 2010. Zearalenone and its metabolites as endocrine disrupting chemicals. *World Mycotoxin Journal.* 3: 385–401.

Reynaud, J., Guilet, D., Terreux, R., Lussignol, M., and Walchshofer, N., 2005. Isoflavonoids in non-leguminous families: an update. *Natural Product Reports.* 22: 504–515.

Shaw, L.J., Morris, P., and Hooker, J.E., 2006. Perception and modification of plant flavonoid signals by rhizosphere microorganisms. *Environmental Microbiology.* 8: 1867–1880.

Sugiyama, A., Shitan, N., and Yazaki, K., 2007. Involvement of a soybean ATP-binding cassette-type transporter in the secretion of genistein, a signal flavonoid in legume–Rhizobium symbiosis. *Plant Physiology.* 144: 2000–2008.

Toebes, A.H.W., De Boer, V., Verkleij, J.A.C., Lingeman, H., and Ernst, W.H.O., 2005. Extraction of isoflavone malonylglucosides from Trifolium pratense L. *Journal of Agricultural and Food Chemistry.* 53: 4660–4666.

Tsao, R., Papadopoulos, Y., Yang, R., Young, J.C., and McRae, K., 2006. Isoflavone profiles of red clovers and their distribution in different parts harvested at different growing stages. *Journal of Agricultural and Food Chemistry.* 54: 5797–5805.

Van Rhijn, P., and Vanderleyden, J., 1995. The rhizobium–plant symbiosis. *Microbiological Reviews.* 59: 124–142.

Veitch, N.C., 2007. Isoflavonoids of the Leguminosae. *Natural Product Reports.* 24: 417–464.

Veitch, N.C., 2009. Isoflavonoids of the Leguminosae. *Natural Product Reports.* 26: 776–802.

Wang, X., 2011. Structure, function, and engineering of enzymes in isoflavonoid biosynthesis. *Functional and Integrative Genomics.* 11: 13–22.

CHAPTER 3

Isoflavones in Foods and Ingestion in the Diet

BASKARAN STEPHEN INBARAJ AND
BING HUEI CHEN*

Department of Food Science, Fu Jen University, Taipei 242, Taiwan
*Email: 002622@mail.fju.edu.tw

3.1 Introduction

Several epidemiological and intervention studies have shown an inverse relationship between consumption of soy foods and incidence of chronic diseases such as cancer, which may be due to the presence of high amounts of isoflavones in soybean (*Glycine max* L.) (Ikeda *et al.* 2006; Wu *et al.* 2008). Isoflavones are a class of isoflavonoids possessing estrogenic activity similar to estradiol hormones (Albulescu and Popovici 2007). They are present in abundant quantities in legumes and exist predominantly (97–98%) as glucosides (Kao and Chen 2006). Among various legumes, soybean is the major dietary source of isoflavones. The key facts regarding soybean, including its origin, prevalence, annual production, physical and chemical composition, common forms of consumption and biological significance, are summarized at the end of this Chapter. Several possible mechanisms for the health-promoting effect of soy isoflavones include estrogenic or anti-estrogenic activity, antiproliferation of cancer cells, antioxidant activity, anti-inflammatory effect, regulation of the immune system, alteration in cellular signaling, reduction in cardiovascular disease, prevention of osteoporosis and alleviation of postmenopausal syndrome (Albulescu and Popovici 2007; Barnes 2010; Kao and Chen 2006;

Kao *et al.* 2007; Pan *et al.* 2010; Wang *et al.* 2009). The isoflavone content in soybean ranges from 0.4–9.5 mg g^{-1} (Wang and Murphy 1994a), however, it can vary depending on genetics, crop year, site of cultivation, growth and environmental conditions (Hoeck *et al.* 2000; Lee *et al.* 2003a). In addition, various processing conditions involved in the conversion of soybeans into consumable soy or soy-based foods can also affect the composition and distribution of isoflavone in foods as well as their ingestion in the diet (Chien *et al.* 2005). This Chapter reviews the isoflavone content in different varieties of foods and the stability of isoflavone as affected by growth, environmental and processing conditions.

3.2 An Overview of Structure and Analysis of Dietary Isoflavones

3.2.1 Structure

Isoflavones are actually the signals released by the soybean to attract the rhizobial bacteria for formation by the same biosynthetic pathway as flavonoids (Albulescu and Popovici 2007). The basic structure of isoflavones is characterized by a flavone nucleus composed of two benzene rings (A and B) linked to a heterocyclic ring C. Dietary isoflavones are mainly composed of 12 different chemical forms classified into four groups, namely aglycone (genistein, daidzein and glycitein), β-glucosides (genistin, daidzin and glycitin), acetylglucosides (6″-acetylgenistin, 6″-acetyldaidzin and 6″-acetylglycitin), malonylglucosides (6″-malonylgenistin, 6″-malonyldaidzin and 6″-malonylglycitin) (Kao and Chen 2006) (Figure 3.1). The presence of hydroxyl group at carbon-5 of genistein makes it more hydrophilic than daidzein, conferring genistein with some unique therapeutic properties, such as enhancement in antioxidant activity (Kao and Chen 2006). Thus, the number of phenolic hydroxyl groups, the nature of substitutions and the specific position in the ring structures can all influence the biological functions of isoflavones (Albulescu and Popovici 2007).

3.2.2 Analysis

Five steps are often involved in the analysis of dietary soy isoflavones, including sampling, sample pretreatment, extraction, a post-extraction step and analysis. A critical comparison of various extraction methods was performed by Rostagno *et al.* (2009). In terms of extraction efficiency, ultrasound-assisted extraction (UAE), pressurized liquid extraction (PLE) and microwave-assisted extraction (MAE) have proved to be efficient, but supercritical fluid extraction (SFE) can result in low recovery (Rostagno *et al.* 2009). Nevertheless, SFE is deemed to be an appropriate technique as it is environmentally friendly and the low recovery can be enhanced by incorporating polar modifier such as ethanol into supercritical fluid (CO_2) (Kao *et al.* 2008). Comparatively, solvent extraction is more applicable to the extraction of malonylglycoside and

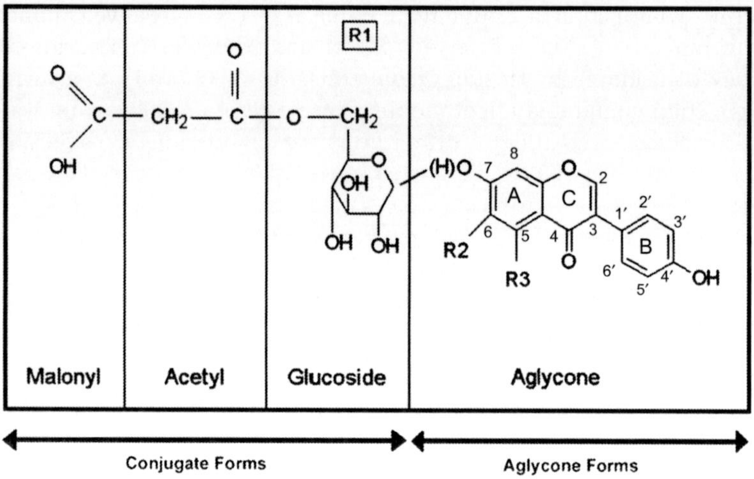

Isoflavone	Symbol	R1	R2	R3
Aglycones				
Genistein	Ge	H	H	OH
Daidzein	De	H	H	H
Glycitein	Gle	H	OCH_3	H
Glucosides				
Genistin	Gi	$C_6O_5H_{11}$	H	OH
Daidzin	Di	$C_6O_5H_{11}$	H	H
Glycitin	Gli	$C_6O_5H_{11}$	OCH_3	H
Acetylglucosides				
Acetylgenistin	AGi	$C_6O_5H_{11}+COCH_3$	H	OH
Acetyldaidzin	ADi	$C_6O_5H_{11}+COCH_3$	H	H
Acetylglycitin	AGli	$C_6O_5H_{11}+COCH_3$	OCH_3	H
Malonylglucosides				
Malonylgenistin	MGi	$C_6O_5H_{11}+COCH_2+COOH$	H	OH
Malonyldaidzin	MDi	$C_6O_5H_{11}+COCH_2+COOH$	H	H
Malonylglycitin	MG	$C_6O_5H_{11}+COCH_2+COOH$	OCH_3	H

Figure 3.1 Structures of 12 commonly occurring dietary soy isoflavones. This Figure shows the chemical structures of common dietary isoflavones categorized into four groups, namely aglycones, glucosides, acetylglucosides and malonylglucosides.
These structures are from Kao and Chen (2006), with permission from the publishers.

glycoside, whereas SFE is used for the extraction of acetylglycoside and aglycone (Kao et al. 2008). Various analytical methods [capillary electrophoresis (CE), gas chromatography–mass spectrometry (GC-MS), high-performance liquid chromatography (HPLC) and immunoassay] employed for the separation and quantification of isoflavones have been recently reviewed, of which HPLC is now the most frequently used method for analyzing 12 isoflavones in soybeans and soy foods because of short separation time, high sensitivity and high-resolution power (Hsu et al. 2010). More specifically, by employing a Vydac 201TP54 C18 column and a gradient mobile phase of acetonitrile and water with a flow rate of $2\,mL\,min^{-1}$, Hsieh et al. (2004) resolved all the 12 soy isoflavones within 15 min. In a later study, a relatively faster separation within 10 min was achieved by Rostagno et al. (2007) using a monolithic Chromolith RP-18e column and a gradient mobile phase of 0.1% acetic acid in water and 0.1% acetic acid in methanol with a flow rate of $4\,mL\,min^{-1}$.

3.3 Isoflavone Content in Selected Foods

The wide variation in isoflavone content in foods can be attributed to many factors, including raw material variety, growth and environmental conditions, methods of processing and storage, as well as analysis (Caldwell et al. 2005; Hsu et al. 2010; Lee et al. 2003a; Rostagno et al. 2009). Owing to the difficulty in separating malonyl and acetyl isoflavones from isoflavone glucosides and aglycones, not all of the 12 dietary isoflavones are quantified. Instead, most studies focused on analysis of biologically active aglycone forms by hydrolysis of conjugated isoflavones using acids, bases or enzymes. A comprehensive database with 99 references has been established by Nutrient Data Laboratory (NDL) of Agricultural Research Service/United States Department of Agriculture (ARS/USDA) in collaboration with Iowa State University (USDA 2008). Table 3.1 summarizes the mean values of the total isoflavones on a wet weight basis in soybean and some selected foods. Nevertheless, some studies report the levels of the 12 individual isoflavones in some commercial soybean, soy foods and soy-based infant formulas (Murphy et al. 1997; Wang and Murphy 1994b) (Table 3.2). An internet source reports a daily amount of consumption of soy products for cholesterol reduction (25–50 g of soy protein), cancer (breast, prostate and colon) prevention (20–40 g of soy protein), alleviation of hot flushes (45 g of soy flour day^{-1}) and osteoporosis, as well as postmenopausal women's syndrome (40 g of soy protein day^{-1}) (MedCom Resource 2000). Furthermore, it is recommended that consumers should intake four servings of at least 6.25 g of soy protein each serving for a total of 25 g each day along with a diet low in saturated fat and cholesterol to reduce the risk of heart disease (USFDA 1999).

3.4 Growth, Variety, Environmental and Post-harvest Storage Conditions Affecting Isoflavone Composition in Soybeans

Early in 1983, Eldridge and Kwolek (1983) showed a variation in isoflavone levels as affected by four different cultivars grown in different locations of Illinois for 4 years (1975–1979). On a weight basis, the isoflavones were highest in hypocotyls (1400–1750 mg 100 g^{-1}), followed by in cotyledon (319–808 mg 100 g^{-1}) and hull (10–20 mg 100 g^{-1}). Identical distribution patterns for isoflavones were observed among the American cultivar soybeans grown within three crop years (1989–1991) in Iowa, with $6''$-O-malonylgenistin,

Table 3.1 Mean values of total isoflavones in soybean and some selected foods. The total isoflavone contents on the wet weight basis in soybean and some selected foods are shown. It is a part of a comprehensive database prepared from 99 reported research articles by converting the values of glycoside forms into aglycone forms for solid foods and adjusting by specific gravity for beverages. Data are from USDA database obtained from the website: http://www.ars.usda.gov/SP2UserFiles/Place/12354500/Data/isoflav/Isoflav_R2.pdf.

Food item	Total isoflavones (mg 100 g of edible^{-1})
Egg, whole, raw, fresh	0.05
Non-dairy creamer with added soy flour or soy protein	0.21
Different infant formula	2.21–28.01
Mayonnaise, made of tofu	16.80
Chicken breast tenders, uncooked	0.55
Chicken patty, frozen, uncooked	0.55
Chicken nuggets, meatless, canned	14.60
Black bean, sauce	10.26
Miso	41.45
Miso soup	1.52
Miso soup mix, dry	69.84
Different sauce mix	0.20–9.90
Soup, cream of chicken, canned, condensed	0.10
Soup, ramen noodle, dry, beef and chicken flavor	1.23 and 0.40
Different breakfast cereals	0.01–93.90
Different fruits and fruit juices	0.01–0.08
Beans, raw and differently processed	0.02–0.59
Broadbeans, raw and fried	0.63 and 1.29
Raw potatoes/frozen green peas/raw garlic	0.01–0.02
Chickpeas, mature seeds, raw and boiled	0.38 and 0.02
Alfafa seeds, raw broccoli, mung bean and raw clover (sprouted, raw)	0.04, 0.04, 0.10 and 0.25
Clover, red	21.00
Soybeans, green, raw and boiled	48.95 and 17.92
Soybeans, mature seeds, sprouted, raw and steamed	34.39 and 12.50
Pistachio nuts, raw	3.63
Other nuts	0.01–0.03
Different beef products	1.14–1.86
Beverages (Coffee, tea)	0.02–0.05
Tuna fish, without salt, drained solids, canned in water and oil	0.09 and 0.28
Different baked products	0.02–14.67
Cereal grains and pasta	0.01–0.07
Pizza, fast food	0.01–0.47
Dessert and pudding made with soymilk	14.00 and 9.13
Bacon, meatless, whole and bits	9.36 and 118.50
Instant beverage, soy powder, non-reconstituted	109.51
Natto	82.29
Oncom	9.70
Peanut butter, smooth, reduced fat	2.09
Different sausage varieties	3.75–14.34

Table 3.1 (*Continued*)

Food item	Total isoflavones (mg 100 g of edible^{-1})
Different soy cheese varieties	6.02–25.72
Soy fiber	44.43
Textured, defatter, full-fat (raw) and full-fat (roasted) Soy flour	172.55, 150.94, 178.10 and 165.04
Soy hot dog, frozen, unprepared	1.00
Soy lecithin	15.70
Soy meal, defatted, raw	209.58
Soy noodles, flat	8.50
Soy paste	38.24
Water-washed and alcohol-extracted soy protein concentrate	94.65 and 11.49
Soy protein drink	81.65
Soy yogurt	33.17
Soybean chips	54.16
Soybean, curd, fermented	34.68
Soybeans, flakes, full-fat and defatted	62.31 and 131.53
Raw, canned, boiled and dry roasted mature soybeans seeds	154.53, 52.82, 65.11 and 148.50
Mature soybeans seeds from Australia, Brazil, China, Europe, Japan, Korea, Taiwan and United States	120.84, 99.82, 118.28, 103.56, 130.65, 178.81, 85.68 and 159.98
Soy protein isolate and soy protein drink	91.05 and 81.65
Raw and cooked soymilk skin or film	196.05 and 44.67
Soymilk curd, dried	83.30
Other soymilk varieties	2.56–10.73
Sufu, traditional Chinese soybean curd	13.75
Tempeh, a traditional Indonesian staple food	60.61
Fresh, braised, cooked, fried, pressed, dried-frozen, smoked, silken and salted tofu	22.73, 16.79, 22.05, 34.78, 33.91, 83.20, 13.10, 18.04 and 48.51
Tofu yogurt	16.30

6″-*O*-malonyldaidzin, genistin and daidzin accounting for approx. 83–93% of total isoflavone (Wang and Murphy 1994a). However, the isoflavone content varied from cultivar-to-cultivar (205.3–421.6 mg 100 g^{-1}) and year-after-year (117.6–330.9 mg 100 g^{-1}). Likewise, among the three Japanese varieties grown in 1991 and 1992, the isoflavone levels ranged from 204.1–234.3 and 126.1–141.7 mg 100 g^{-1}, respectively, with crop year showing a more pronounced effect than site of cultivation. Compared to the American varieties, the Japanese varieties contained higher levels of 6″-*O*-malonylglycitin and the ratio of malonylglucosides to glucosides was 2–4 times greater than the former, implying genetics plays a significant role in soy isoflavone distribution (Wang and Murphy 1994a).

The aforementioned outcomes were further shown in 15 Korean soybean cultivars cultivated in three different locations and crop years (Lee *et al.* 2003a). Total isoflavones ranged from 188.4–685.6 mg 100 g^{-1} in 1998, 218.8–948.9 mg 100 g^{-1}

Table 3.2 Content (mg 100 g^{-1}) of individual isoflavones for selected soybean, soy foods and infant formulas. Isoflavone levels of commonly occurring 12 dietary isoflavones are shown for some selected varieties of soybean, soy foods and infant formulas. Data are the means of 3 replicates. For the expansion of abbreviations, refer to Figure 3.1. nd, not detected; tr, traces. Data are from Wang and Murphy (1994b) and Murphy et al. (1997), with permission from the publishers.

Sample	Di	Gi	Gli	MDi	MGi	MGli	ADi	AGi	AGli	De	Ge	Gle	Total
Soybean	69.0	85.2	5.6	30.0	74.3	5.0	0.1	0.9	nd	2.6	2.9	2.0	277.6
Green soybean	45.1	43.0	4.8	51.5	85.1	5.7	tr	0.2	nd	1.0	1.6	1.8	239.8
Roasted soybeans	46.0	55.1	6.8	4.5	6.3	7.2	39.7	74.3	102	3.9	6.9	5.2	266.1
Soy flour	14.7	40.7	4.1	26.1	102.3	5.7	tr	0.1	32	0.4	2.2	1.9	201.4
Soy granule	727	870	132	106	193	60	72	135	48	12	27	22	240.4
Textured vegetable protein	507	634	146	93	192	60	187	320	90	12	29	25	229.5
Soy isolate	13.3	38.2	5.5	1.9	9.5	3.7	3.6	12.2	40	1.2	3.6	2.2	98.9
Soymilk	8.2	10.5	1.5	1.0	1.4	0.3	0	0.6	0	0.2	0.2	0	23.9
Instant beverage	525	745	75	98	259	44	12	26	33	30	50	21	191.8
Tofu	2.5	8.4	0.8	15.9	10.8	nd	0.8	0.1	29	4.6	5.2	1.2	53.2
Tempeh	0.2	6.5	1.4	25.5	16.4	nd	1.1	nd	nd	13.7	19.3	2.4	86.5
Miso	7.2	12.3	1.8	nd	nd	2.2	0.1	1.1	nd	3.4	9.3	1.5	38.9
Bean paste	nd	9.6	2.1	nd	nd	1.9	0.1	0.2	nd	27.1	18.3	5.4	64.7
Fermented bean curd	nd	tr	nd	nd	nd	nd	nd	nd	nd	14.3	22.3	2.3	38.9
Soy hot dog	3.5	6.7	1.5	1.2	4.2	1.5	tr	0.4	14	0.8	1.6	0.8	23.6
Soy bacon	tr	2.7	1.4	tr	0.5	1.2	tr	0.3	nd	2.6	4.8	0.9	14.4
Tempeh burger	3.6	15.8	1.8	2.5	Nd	nd	nd	0.1	nd	3.4	9.6	1.8	38.6
Tofu yogurt	4.2	8.0	1.2	6.1	7.9	nd	nd	tr	nd	tr	3.0	0.5	28.2
Alsoy[a]	3.7	8.3	1.7	4.8	9.2	1.9	4.0	8.4	14	1.0	1.1	0	45.5
Gerber[a]	3.8	6.5	1.8	3.0	6.6	1.8	3.2	7.1	12	1.1	1.3	3.0	37.7

[a]Soy-based infant formulas.

in 1999 and 293.1–483.0 mg $100\,g^{-1}$ in 2000. However, by comparing different locations, it ranged from 240.9–948.9 mg $100\,g^{-1}$ in Seoul, 483–724.5 mg $100\,g^{-1}$ in Suwon and 188.4–811.1 mg $100\,g^{-1}$ in Kyongsan. This result revealed the significant difference in isoflavone levels could be affected by different crop years, sites of cultivation and cultivars. Several studies have shown that the interactions between the genotypes, genotype–crop year, genotype–site of cultivation and genotype–crop year–site of cultivation could greatly influence the soy isoflavone contents (Hoeck et al. 2000; Lee et al. 2003b). In another study, the wild soybean species in Korea was shown to contain the highest level of isoflavones (97.6–222.3 mg $100\,g^{-1}$), followed by cultivars (56.0–197.9 mg $100\,g^{-1}$) and landraces (62.4–233.0 mg $100\,g^{-1}$) (Choi et al. 2000). The growth period in plants can cause a variation in isoflavone content, with a level being higher in middle/late maturing cultivars than in early maturing ones (Choi et al. 2000). Additionally, the early maturing soybean variety could produce a large amount of daidzein.

Also, crop season can significantly affect the isoflavone level in soybeans. For instance, the isoflavone contents in three cultivar varieties of Taiwan grown during fall and spring seasons from 2001–2005 showed a marked difference, with the fall crop (204.9–723.9 mg $100\,g^{-1}$) showing a higher level than the spring crop (90.1–241.5 mg $100\,g^{-1}$) (Tsai et al. 2007). A 10 °C difference in temperature between soybeans harvested in June and December can cause a variation in isoflavone levels between spring and fall crops. Tsukamoto et al. (1995) observed a 14–16-fold higher isoflavone yield from soybeans grown in growth chambers at a low temperature (25 °C during the day and 10 °C at night) than at a high temperature (38 °C during the day and 28 °C at night). The isoflavone content in cotyledon was largely affected by high temperature, but it remained intact in hypocotyls. Caldwell et al. (2005) reported that raising the temperature from 18–23 °C and 18–28 °C during seed development in dwarf soybean in environmentally controlled chambers mitigated the total isoflavone content by 65 and 30% for CO_2 level at 400 ppm and 87 and 74% for CO_2 level at 700 ppm, respectively. All 11 isoflavones, except 6″-O-malonylglycitin, contributed to the decline in isoflavone level, with the decline being inconsistent owing to different metabolic responses to hyperthermia. A combination of drought (water stress), elevated CO_2 level (700 ppm) and a temperature of 23 °C during seed development could reverse the temperature-dependent decrease to produce the isoflavone content equivalent at 18 °C (control). Furthermore, the variation in CO_2 levels showed a greater impact on relative proportions of isoflavones than drought (Caldwell et al. 2005). Post-harvest storage of soybean seeds at room temperature (25 °C) for 1 year increased the total isoflavone content (477.2 mg $100\,g^{-1}$) than those stored for 2 years (429.3 mg $100\,g^{-1}$) or 3 years (425.4 mg $100\,g^{-1}$) (Lee et al. 2003a). More specifically, the levels of glycosides and aglycones soared 3–4-fold after storage for 2 years, whereas malonylglycosides declined by approx. 2-fold. Thus, the magnitude of soybean plant responses to the aforementioned factors suggests that the subtle changes may affect the isoflavone contents in commercially grown soybeans.

3.5 Isoflavone Composition as Affected by Processing

In unprocessed foods, isoflavones exist mainly as glucosides. However, the isoflavone glucosides breakdown during various processing steps involving heat, acid and enzymes and thus many processed foods possess higher amounts of biologically active aglycone forms. Isoflavones are quite stable at temperatures up to 260 °C and are not destroyed by heat, but are rather subjected to interconversion between different forms (Chien et al. 2005). Figure 3.2 summarizes the chemical reactions of isoflavones during food processing as affected by soaking, acid, base, heat, enzyme and fermentation.

Acid hydrolysis of isoflavone glucosides at three HCl concentrations (1, 2 and 3 M) showed 1 M HCl to be the most effective in the conversion of glucosides into aglycones (Wang et al. 1990); but for treatment at <1 M for 2 h, the aglycone concentration declined and boiling in acid for 2 h degrades both genistein and daidzein. However, acid or alkaline treatment at high temperature can hydrolyze both malonyl and acetyl ester linkages. Mathias et al. (2006) reported malonyl and acetyl forms to be more stable at pH 2 and 7, respectively, with their interconversion being the most efficient at pH 10 and 100 °C for the former and pH 10 and 80 °C for the latter. Regardless of pH, the stability of both malonyl and acetyl forms decreased with a rise in temperature. Malonylglucosides were more stable than acetylglucosides, especially under acidic conditions and loss in isoflavone derivatives were significantly higher for daidzin than for genistin at all pH and temperature. Upon heating at 185 °C for 3 min, the percentage loss of daidzin, glycitin and genistin from soy flour were 26, 27 and 27%, respectively, which rose to 65, 98 and 74% at 215 °C for 3 min and 91, 99 and 94% at 215 °C for 15 min, with the thermal stability following the order: daidzin > genistin > glycitin (Xu et al. 2002). Heating at temperatures above 135 °C produced acetyldaidzin, acetylgenistin, daidzein, glycitein and genistein, and the rate of formation of acetyl forms from daidzin and genistin was higher than the rate of loss of a glucoside moiety. Nevertheless, the levels of acetyldaidzin and acetylgenistin declined sharply at temperatures above 200 °C, whereas the aglycones were relatively stable with the stability of daidzein being higher than glycitein or genistein (Xu et al. 2002). Chien et al. (2005) determined the kinetic rates for conversion and degradation of isoflavones during dry or moist heating at 100, 150 and 200 °C for varied lengths of time (Table 3.3). For dry heating, the conversion of malonylgenistin into genistin showed the highest rate constant, followed by malonylgenistin into acetylgenistin, acetylgenistin into genistin, acetylgenistin into genistein, genistin into genistein and malonylgenistin into genistein. Moist heating also followed a similar trend with the exception that the last three conversions did not occur. Comparatively, the isoflavone glucosides were degraded faster by moist heating than by dry heating under the same temperature (Chien et al. 2005). To account for the 20% loss unexplained by mass balance study, Yerramsetty et al. (2011) monitored the formation and interconversion of malonylglucoside isomers in heated buffer and soymilk systems at 100 °C and found two stereoisomers of malonylgenistin and malonyldaidzin, with the stereoisomer of the former

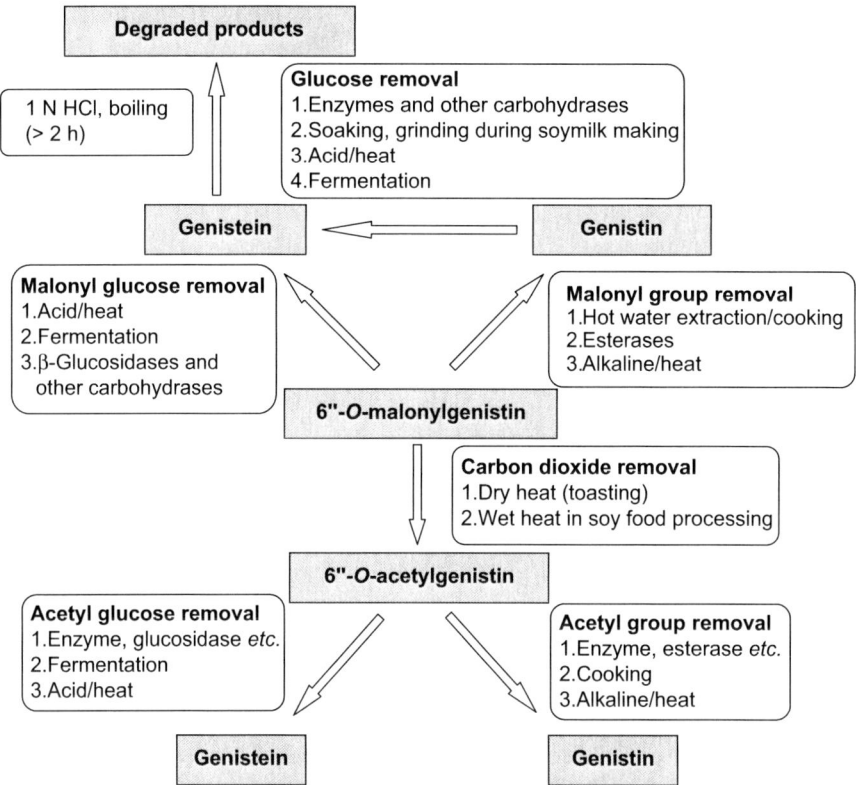

Figure 3.2 Different chemical reactions of isoflavones during food processing as affected by acid, heat, enzymes and fermentation. The interconversion between different conjugated and unconjugated forms of isoflavones as affected by heat, acid, base, enzyme and fermentation is shown. (Unpublished Figure.)

($4''$-O-malonylgenistin) accounting for 6–9% of total genistein in soymilk. Thus, more studies are necessary to monitor isoflavone loss in heated systems and the content of unknown isomers should also be taken into account to reflect actual isoflavone amounts.

Wang and Murphy (1996) studied the various processing steps affecting the isoflavone level and found a 12% loss occurring during soaking, a 49% loss during heating of tempeh production, a 44% loss in coagulation during tofu processing and a 53% loss in alkaline extraction during protein isolate production. Malonylglucosides followed a declined trend after soaking and cooking during the processing of tempeh, soymilk and tofu, with a concomitant rise in the content of acetyl forms. Both daidzein and genistein levels soared during fermentation of tempeh and alkaline extraction of the protein isolate, which may be attributed to enzymatic and alkaline hydrolysis, respectively. Kao *et al.* (2004a) evaluated the stability of isoflavone glucosides during

Table 3.3 Conversion and degradation rate constants of four isoflavones during dry and moist heating. Isoflavone standards are used in a simulated system and the rate constants are based on a complex consecutive reaction models. D, unknown degradation products; for the expansion of other abbreviations, refer to Figure 3.1. Data are from Chien et al. (2005), with permission from the publishers.

Conversion/ degradation	Rate constants (h^{-1}) at different temperatures (°C)[a]			Ea^b (kcal mol^{-1})
	100	150	200	
Dry heating				
MGi → AGi	0.045 ± 0.003	1.430 ± 0.006	7.860 ± 0.015	18.30
MGi → Gi	–	2.310 ± 0.014	9.710 ± 0.059	11.95
MGi → Ge	–	–	0.000 ± 0.099	–
MGi → D	–	0.206 ± 0.000	5.808 ± 0.000	27.76
AGi → Gi	–	0.221 ± 0.001	0.585 ± 0.018	8.10
AGi → Ge	–	–	0.277 ± 0.000	–
AGi → D	–	0.219 ± 0.001	–	–
Gi → Ge	–	–	0.016 ± 0.003	–
Gi → D	–	0.037 ± 0.003	4.610 ± 1.000	40.15
Ge → D	–	0.034 ± 0.002	0.853 ± 0.065	26.84
Moist heating				
MGi → AGi	0.773 ± 0.001	46.70 ± 0.84	76.00 ± 0.96	16.6
MGi → Gi	1.800 ± 0.003	265.0 ± 5.4	318.0 ± 4.0	18.9
MGi → D	0.422 ± 0.007	0.012 ± 0.000	–	–
AGi → Gi	0.211 ± 0.011	2.540 ± 0.006	6.600 ± 0.000	12.2
AGi → D	–	4.400 ± 0.000	19.00 ± 0.00	12.1
Gi → D	–	1.230 ± 0.074	11.60 ± 0.53	18.6
Ge → D	0.074 ± 0.005	5.530 ± 0.290	12.60 ± 0.44	18.5

[a] Mean of triplicate measurements ± S.D. [b] Activation energy.

soaking of soybean and processing of soymilk and tofu. After soaking of soybeans at 45 °C for 12 h, the levels of daidzein, glycitein and genistein increased by 403.1, 190.9 and 779.9 µg g^{-1}, respectively, with genistein being more susceptible to formation during soaking than daidzein and glycitein (Figure 3.3). The increase in aglycone during soaking may be attributed to high β-glucosidase activity at elevated temperature. In contrast, the amounts of daidzin, glycitin, genistin, malonyldaidzin, malonylglycitin and malonylgenistin declined during soaking, which should be due to their conversion into aglycone/glucoside or leaching into water. Likewise, both acetyldaidzin and acetylglycitin were converted into glucoside or aglycone after soaking at 25, 35 or 45 °C for 4 h, implying soaking soybeans at high temperature for prolonged period of time can reduce the total isoflavones. For processing of soymilk, cooking at 100 °C for 30 min was shown to be the optimum condition for preserving the nutritional value and flavor (Kao et al. 2004a). During heating at 100 °C for 30 min, the amount of isoflavone aglycone did not change, but the levels of daidzin, glycitin, genistin, malonyldaidzin, malonylglycitin, malonylgenistin and acetylglycitin decreased by 94.9, 15.9, 104.9, 254.2, 79.9, 546.7 and 5.4 µg g^{-1}, respectively (Kao et al. 2004a). The formation of

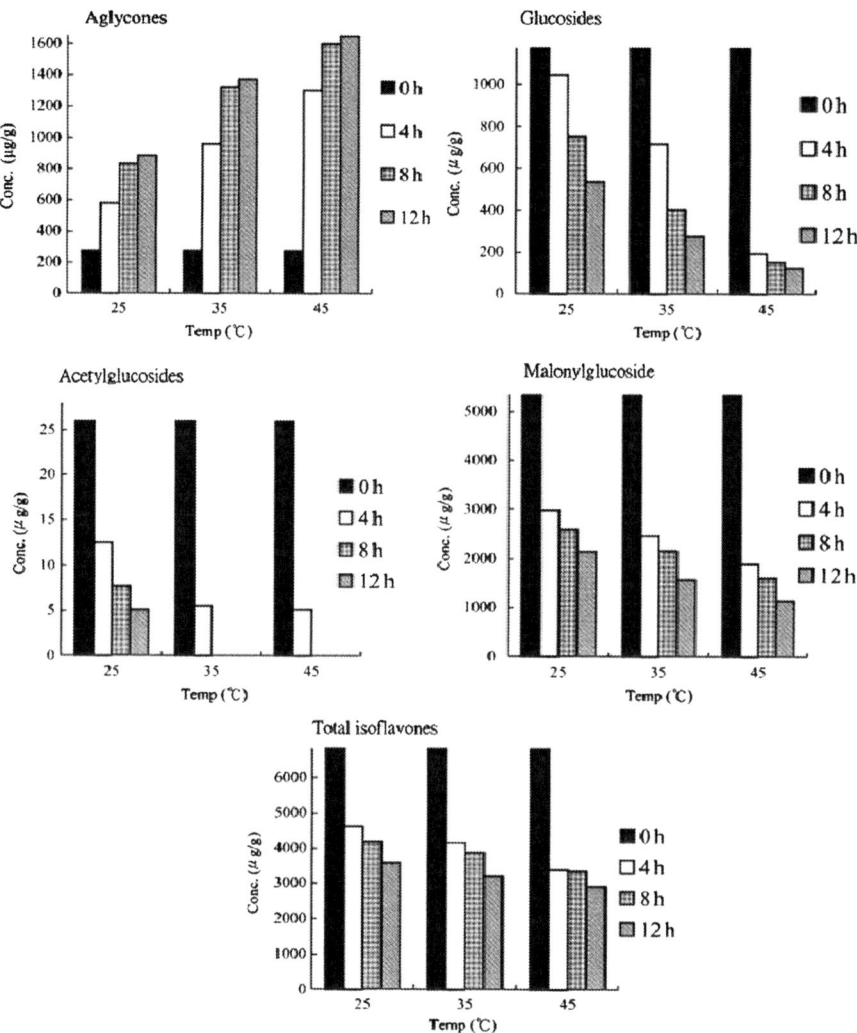

Figure 3.3 Influence of temperature and soaking time on content of different isoflavone forms and total isoflavones in soybean on a dry weight basis. Raw soybeans were washed with tap water and soaked in 300 mL of deionized water preheated to 25, 35 or 45 °C in a water bath with soaking times at each temperature of 0, 4, 8 or 12 h. Data are from Kao *et al.* (2004), with permission from the publishers.

glucoside derivatives can be attributed to the hot aqueous environment (100 °C) during processing of soymilk. On the contrary, soymilk processed at a similar heating condition (96 °C for 20 min) yielded a 2–4-fold more free isoflavones than soybean, which may be due to enzymatic hydrolysis of glucoside and malonylglucoside (Kao *et al.* 2004b). Apparently, moist heating can induce the hydrolysis of malonyl glucoside to glucosides, whereas dry heating promotes the formation of acetylglucoside from malonyglucoside (Barnes et al. 1998). Following a rise in water-to-bean ratio from 59, amounts of both daidzin and genistin in soymilk increased and reached a peak at 9–11, but the aglycone remained unaffected at 5–11 (Kao et al. 2004b).

For tofu making, the optimum cooking condition of soybean was suggested to be 100 °C for 10–20 min with calcium sulfate showing a higher yield of isoflavones than calcium chloride. Specifically, tofu processed with 0.3% calcium sulfate could generate a high amount of isoflavones (2272.3 µg g^{-1}), but declined to 1956.6 µg g^{-1} with 0.7% calcium sulfate (Kao et al. 2004a). Obviously, the incorporation of coagulants at a high level can shorten the coagulation time, yet the syneresis occurring during coagulation can result in considerable loss of isoflavones. Tofu made with water-to-bean ratios of 9 and 10 resulted in a maximum amount of daidzin and genistin, which may be due to a homogeneous network resulting from hydrophilic interaction with proteins (Kao et al. 2004b). Similarly, with water-to-bean ratios of 5 and 7, both daidzein and genistein were retained at a higher amount in tofu, which should be due to strong hydrophobic interaction between proteins and aglycones. However, their levels dropped significantly at water-to-bean ratios of 7–11 owing to formation of many bigger holes in the microstructure of tofu, and, in turn, facilitate the draining of daidzein and genistein (Kao et al. 2004b). Based on the yield of daidzin, genistin, daidzein and genistein, the water-to-bean ratios of 9, 10 and 11 for soymilk and 9 and 10 for tofu making were considered to be optimum (Kao et al. 2004b). Despite yielding higher levels of aglycone, the water-to-bean ratios of 5 and 7 should be avoided as both genistein and daidzein can impart strong objectionable flavor to soymilk and tofu.

In summary, the extent of isoflavone loss during the processing of soy foods depends mainly on the difference in polarity between different chemical forms. Nonetheless, with enzymatic conversion of isoflavones glucosides into their corresponding aglycones, the bioavailability can be greatly enhanced. The effect of enzyme pretreatment of soymilk for subsequent preparation of tofu by adding koji enzyme extract into soymilk and incubated at 35 °C for various length of time was studied by Wu et al. (2004), and the amounts of daidzein and genistein in soymilk and tofu increased following a rise in enzyme pretreatment time, accompanied by a decline in both daidzin and genistin. A similar phenomenon was illustrated by Ismail and Hayes (2005) for β-glycosidase, however, it was ineffective towards conjugated β-glucoside, even under high concentration and a prolonged treatment time. Recently, Tipkanon et al. (2010) evaluated the effects of enzyme concentration, substrate concentration, pH, incubation temperature and incubation time on the production of soy germ flour containing high amounts of daidzein, glycitein and genistein by employing factorial design and response surface methodology, and the optimum conditions were: soy germ/deionized water at 1:5 (w/v), β-glucosidase at 1 unit g^{-1} of soy germ flour, pH at 5, incubation temperature at 45 °C and the time was 5 h.

Summary Points

- Isoflavones are phytoestrogens present in abundant quantities in soybean and soybean-based products.
- Fast analysis of 12 common dietary isoflavones in soybean and soy foods can be achieved within 10 or 15 min with adequate resolution and sensitivity.
- The content and daily ingestion of isoflavones are largely affected by soybean variety, growth and environmental factors and processing methods.
- In soybean products, the total genistein contributes mainly to total isoflavones, followed by total daidzein and glycitein.
- Isoflavone conjugates are the major forms in non-fermented soy foods and unprocessed foods, whereas aglycones predominate in fermented and processed foods.
- Instead of only three aglycones, the development of a database to investigate the levels of all the 12 dietary isoflavones is necessary. Also, more studies are essential to elucidate unexplained isoflavone loss during heating.

Key Facts

- Soybean (*Glycine max* L.) is a legume with a long history as a domestic plant since 11th century BC in China, 3rd and 4th centuries AD in Korea and Japan, 1739 in Europe and 1765 in USA.
- Its annual world production in 2007 was 206.4 million tons, with both Brazil and Argentina contributing almost 90 million tons. The major producers of soybean include USA (35%), Brazil (27%), Argentina (19%), China (6%) and India (4%).
- In soybeans, proteins contributes to the largest proportion (41%) by dry weight, followed by carbohydrates (35%), oils (19%) and ash (5%).
- Soybeans contain phytic acid, α-linolenic acid and isoflavones. Among legumes, the soybean is the only significant dietary source of isoflavones containing four groups of isoflavones, namely aglycones, glucosides, acetylglucosides and malonylglucosides.
- Soybeans are processed into non-fermented foods such as soymilk and tofu, as well as fermented foods including fermented bean paste, natto, soy sauce, tempeh and miso. They are also incorporated as food additives to enrich nutritional value.
- Epidemiological and clinical studies have shown the isoflavones in soybean or soy-based foods/products to be responsible for the alleviation of postmenopausal syndrome and the prevention of osteoporosis, cancer and cardiovascular disease.

Definitions of Words and Terms

Bioavailability: The rate at which an administered drug or any functional compound is available at the site of physiological activity. Alternatively, it is the percentage of drug available in the systemic circulation in active form through a given route.

C18 column: A thin film of stationary phase which is chemically anchored to an inert material such as silica gel particles and made up of hydrophobic alkyl chains of 18 carbon atoms that can interact with the analyte in liquid chromatographic separation.

Cotyledon: A cotyledon is a vital part of the embryo within the seed of a plant which becomes leaves of the seedling on germination.

Fermentation: A process of conversion of sugars into ethanol. In food processing, it is typically the conversion of carbohydrates into alcohols, carbon dioxide or organic acids by using micro-organisms such as yeasts and/or bacteria.

Genotype: The genotype is the genetic design of a cell or organism.

β-Glucosidase: A glucosidase enzyme that can hydrolyze $\beta 1 \rightarrow 4$ bonds between two glucose or glucose-substituted molecules. It has specificity for a variety of β-D-glycoside compounds for catalyzing the hydrolysis of the terminal non-reducing residues to release glucose.

Hypocotyl: A hypocotyl is the stem of a germinating seedling found below the seed leaves (cotyledons) and above the root.

Mobile phase: A part of a chromatographic system that carries the analyte compounds through the stationary phase and is modified to effect chromatographic separation with variable retention.

Osteoporosis: Osteoporosis is the medical condition of bones becoming brittle and fragile due to the loss of tissue resulting from hormonal changes and/or deficiency in calcium or vitamin D.

Phenotype: The phenotype is the morphological or observable characteristics.

Postmenopausal syndrome: Symptoms that occur in women after menopause including bleeding, hot flushes and osteoporosis.

Stereoisomers: These are isomers with the same molecular formula and connection of bonds between atoms, but differ in their spatial arrangement.

Supercritical fluid: A material which can either be liquid or gas used in a state above its critical temperature and critical pressure where gas and liquid coexist. They are suitable substitutes for organic solvents and can be fine-tuned by changing temperature and pressure.

List of Abbreviations

CE	capillary electrophoresis
GC-MS	gas chromatography-mass spectrometry
HPLC	high-performance liquid chromatography
MAE	microwave-assisted extraction
NDL	nutrient data laboratory
PLE	pressurized liquid extraction
SFE	supercritical fluid extraction
UAE	ultrasound-assisted extraction
USDA	United States Department of Agriculture

References

Albulescu, M., and Popovici, M., 2007. Isoflavones – biochemistry, pharmacology and therapeutic use. *Revue Roumaine de Chimie*. 52: 537–550.

Barnes, S., 2010. The biochemistry, chemistry and physiology of the isoflavones in soybeans and their food products. *Lymphatic Research and Biology*. 8: 89–98.

Barnes, S., Coward, L., Kirk, M., and Sfakianos, J., 1998. HPLC-mass spectrometry analysis of isoflavones. *Proceedings of the Society for Experimental Biology and Medicine*. 217: 254–262.

Caldwell, C.R., Britz, S.J. and Mirecki, R.M., 2005. Effect of temperature, elevated carbon dioxide, and drought during seed development on the isoflavone content of dwarf soybean (*Glycine max* L. Merrill) grown in controlled environments. *Journal of Agricultural and Food Chemistry*. 53: 1125–1129.

Chien, J.T., Hsieh, H.C., Kao, T.H., and Chen, B.H., 2005. Kinetic model for studying the conversion and degradation of isoflavones during heating. *Food Chemistry*. 91: 425–434.

Choi, Y.S., Lee, B.H., Kim, J.H., and Kim, N.S., 2000. Concentration of phytoestrogens in soybeans and soybean products in Korea. *Journal of the Science of Food and Agriculture*. 80: 1709–1712.

Eldridge, A.C., and Kwolek, W.F., 1983. Soybean isoflavones: effect of environment and variety on composition. *Journal of Agricultural and Food Chemistry*. 31: 394–396.

Hoeck, J.A., Fehr, W.R., Murphy, P.A., Cook, L., and Hendrich, S., 2000. Influence of genotype and environment on isoflavone contents of soybean. *Crop Science*. 40: 48–51.

Hsieh, H.C., Kao, T.H., and Chen, B.H., 2004. A fast HPLC method for analysis of isoflavones in soybean. *Journal of Liquid Chromatography and Related Techniques*. 27: 315–324.

Hsu, B.Y., Inbaraj, B.S., and Chen, B.H., 2010. Analysis of soy isoflavones in foods and biological fluids: an overview. *Journal of Food and Drug Analysis*. 18: 141–154.

Ikeda, Y., Iki, M., Morita, A., Kajita, E., Kagamimori, S., Kagawa, Y., and Yoneshima, H., 2006. Intake of fermented soybeans, natto, is associated with reduced bone loss in postmeanopausal women: Japanese population-based osteroporosis (JPOS) study. *The Journal of Nutrition*. 136: 1323–1328.

Ismail, B., and Hayes, K., 2005. β-glycosidase activity toward different glycosidic forms of isoflavones. *Journal of Agricultural and Food Chemistry*. 53: 4918–4924.

Kao, T.H., and Chen, B.H., 2006. Functional components in soybean cake and their effects on antioxidant activity. *Journal of Agricultural and Food Chemistry*. 54: 7544–7555.

Kao, T.H., Lu, Y.F., Hsieh, H.C., and Chen, B.H., 2004a. Stability of isoflavone glucosides during processing of soymilk and tofu. *Food Research International*. 37: 891–900.

Kao, F.J., Su, N.W., and Lee, M.H., 2004b. Effect of water-to-bean ratio on the contents and compositions of isoflavones in tofu. *Journal of Agricultural and Food Chemistry.* 52: 2277–2281.

Kao, T.H., Wu, W.M., Hung, C.F., Wu, W.B., and Chen, B.H., 2007. Anti-inflammatory effects of isoflavone powder produced from soybean cake. *Journal of Agricultural and Food Chemistry.* 55: 11068–11079.

Kao, T.H., Chien, J.T., and Chen, B.H., 2008. Extraction yield of isoflavones from soybean cake as affected by solvent and supercritical carbon dioxide. *Food Chemistry.* 107: 1728–1736.

Lee, S.J., Ahn, J.K., Kim, S.H., Kim, J.T., Han, S.J., Jung, M.Y., and Chung, I.M., 2003a. Variation in isoflavone of soybean cultivars with location and storage duration. *Journal of Agricultural and Food Chemistry.* 51: 3382–3389.

Lee, S.J., Yan, W., Ahn, J.K., and Chung, I.M., 2003b. Effects of year, site, genotype and their interactions on various soybean isoflavones. *Fields Crops Research.* 81: 181–192.

Mathias, K., Ismail, B., Corvalan, C.M., and Hayes, K.D., 2006. Heat and pH effects on the conjugated forms of genistin and daidzin isoflavones. *Journal of Agricultural and Food Chemistry.* 54: 7495–7502.

MedCom Resource, 2000. Concentrations of isoflavones in Food. Available at: http://medcomres.com/articles/soy_isoflavones.htm. Accessed 24 August 2011.

Murphy, P.A., Song, T., Buseman, G., and Barua, K., 1997. Isoflavones in soy-based infant formulas. *Journal of Agricultural and Food Chemistry.* 45: 4635–4638.

Pan, M.H., Lai, C.S., and Ho, C.T., 2010. Anti-inflammatory activity of natural dietary flavonoids. *Food and Function.* 1: 15–31.

Rostagno, M.A., Palma, M., and Barroso, C.G., 2007. Fast analysis of soy isoflavones by high-performance liquid chromatography with monolithic columns. *Analytica Chimica Acta.* 582: 243–249.

Rostagno, M.A., Villares, A., Guillamón, E., García-Lafuente, A., and Martínez, J.A., 2009. Sample preparation for the analysis of isoflavones from soybean and soy foods. *Journal of Chromatography A.* 1216: 2–29.

Tipkanon, S., Chompreeda, P., Haruthaithanasan, V., Suwonsichon, T., Prinyawiwatkul, W., and Xu, Z., 2010. Optimizing time and temperature of enzymatic conversion of isoflavone glucosides to aglycones in soy germ flour. *Journal of Agricultural and Food Chemistry.* 58: 11340–11345.

Tsai, H.S., Huang, L.J., Lai, Y.H., Chang, J.C., Lee, R.S., and Chiou, R.Y.Y., 2007. Solvent effects on extraction and HPLC analysis of soybean isoflavones and variations of isoflavone compositions as affected by crop season. *Journal of Agricultural and Food Chemistry.* 55: 7712–7715.

Tsukamoto, C., Shimada, S., Igita, K., Kudou, S., Kokubun, M., Okubo, K., and Kitamura, K., 1995. Factors affection isoflavone content in soybean seeds: changes in isoflavones, saponins, and composition of fatty acids at different temperatures during seed development. *Journal of Agricultural and Food Chemistry.* 43: 1184–1192.

United States Food and Drug Administration (USDA), 1999. The FDA soy health claim. Available at: http://www.mnsoybean.org/all-about-soy/soyfoods-and-health/health-fact-sheets/the-fda-soy-health-claim/. Accessed 24 August 2011.

United States Department of Agriculture (USDA), 2008. USDA database for the isoflavone content of selected foods. Release 2.0. Available at: http://www.ars.usda.gov/SP2UserFiles/Place/12354500/Data/isoflav/Isoflav_R2.pdf. Accessed 24 August 2011.

Wang, H.J., and Murphy, P.A., 1994a. Isoflavone composition of American and Japanese soybeans in Iowa: effects of variety, crop year and location. *Journal of Agricultural and Food Chemistry*. 42: 1674–1677.

Wang, H.J., and Murphy, P.A., 1994b. Isoflavone content in commercial soybean foods. *Journal of Agricultural and Food Chemistry*. 42: 1666–1673.

Wang, H.J., and Murphy, P.A., 1996. Mass balance study of isoflavones during soybean processing. *Journal of Agricultural and Food Chemistry*. 44: 2377–2383.

Wang, G., Kuan, S.S., Francis, O.J., Ware, G.M., and Carman, A.S., 1990. A simplified HPLC method for the determination of phytoestrogens in soybean and its processed products. *Journal of Agricultural and Food Chemistry*. 38: 185–190.

Wang, B.F., Wang, J.S., Lu, J.F., Kao, T.H., and Chen, B.H., 2009. Antiproliferation effect and mechanism of prostate cancer cell lines as affected by isoflavones from soybean cake. *Journal of Agricultural and Food Chemistry*. 57: 2221–2232.

Wu, M.L., Chang, J.C., Lai, Y.H., Cheng, S.L., and Chiou, R.Y.Y., 2004. Enhancement of tofu isoflavone recovery by pretreatment of soy milk with koji enzyme extract. *Journal of Agricultural and Food Chemistry*. 52: 4785–4790.

Wu, A.H., Yu, M.C., Tseng, C.C., and Pike, M.C., 2008. Epidemiology of soy exposures and breast cancer risk. *British Journal of Cancer*. 98: 9–14.

Xu, Z., Wu, Q., and Godber, J.S., 2002. Stabilities of daidzein, glycitin, genistin, and generation of derivatives during heating. *Journal of Agricultural and Food Chemistry*. 50: 7402–7406.

Yerramsetty, V., Mathias, K., Bunzel, M., and Ismail, B., 2011. Detection and structural characterization of thermally generated isoflavone malonylglucoside derivatives. *Journal of Agricultural and Food Chemistry*. 59: 174–183.

Chemistry and Biochemistry

CHAPTER 4

The Chemistry/Biochemistry of the Bioconversion of Isoflavones in Food Preparation

KASHIF GHAFOOR,[a] FAHAD Y. AL-JUHAIMI[a] AND JIYONG PARK*[b]

[a] Department of Food and Nutrition Sciences, College of Food and Agricultural Sciences, King Saud University, P.O. Box 2460, Riyadh 11451, Saudi Arabia; [b] Biomaterials Process Engineering Laboratory, Department of Biotechnology, Yonsei University, 134 Sinchon Dong, Seodaemun-gu, Seoul 120-749, Republic of Korea
*Email: foodpro@yonsei.ac.kr

4.1 Introduction

The soybean, *Glycine max*, has had a long history as a domesticated plant, with records of its use as far back as the eleventh century BC in China. Missionaries took it into Korea and Japan in the third and fourth centuries AD (Hymowitz 1990). Soy is the principal plant that produces isoflavones (1–2 mg g^{-1}) (Esaki *et al.* 1999). Generally, only small amounts are found in most other plants (Dixon 2004). The chickpea has 1/100th of the amount in soy. These are a subclass of the much more common flavonoids. These in turn are included in the large family of polyphenols that are widely available in plants. Isoflavonoids are formed by the same biosynthetic pathway for flavonoids

(Deavours and Dixon 2005). Daidzein 8-C-glucoside is known as puerarin (Figure 4.1) and it is often the major constituent of isoflavone dietary supplements. Isoflavones in the soybean are converted into 7-*O*-β-glucosides by a glucosyltransferase and then to their 6″-*O*-malonate forms by a malonyl transferase. This chemical form is stored in cell vacuoles until used by the plant and it is also the major form in harvested soybeans.

Fermented soyfoods, such as miso and tempeh, contain the unconjugated isoflavone aglucones, whereas non-fermented soyfoods, such as soymilk, tofu, soy flour, soy-protein concentrate and isolated soy protein, contain their β-glucoside conjugates. Isoflavones in the past were recovered for analysis by extraction into hot, aqueous organic solvents such as acetonitrile, ethanol or methanol. However, in 1991, it was found that extraction of soybeans with these solvents without the use of heat led to a different set of isoflavone glucosides, which were identified as the 6″-*O*-malonyl-β-glucoside (6MalGlc) conjugates (Kudou *et al*. 1991). Careful investigation of soyfoods with reversed phase high-performance liquid chromatography–mass spectrometry (HPLC-MS) and revised extraction protocols showed that most soyfoods contained mixtures of the β-glucoside, 6″-*O*-malonyl-β-glucoside (6OMalGlc) and 6″-*O*-acetyl-β-glucoside (6OAcGlc) conjugates (Barnes *et al*. 1994; Wang and Murphy 1994). The 6OAcGlc conjugates had been identified in toasted soy flour (Farmakalidis and Murphy 1985). The principal chemical forms of isoflavones in soybean are their 6OMalGlc conjugates. Processing soybean into different food products results in changes in the chemistry and biochemistry of its constituents (Figure 4.2). In general, fermentation causes the removal of the glucosidic group, releasing the isoflavone aglucone (Kuo *et al*. 2006). Bioconversion, also known as biotransformation, refers to the use of live organisms, often micro-organisms, to carry out a chemical reaction. If the fermentation is a lengthy process (for miso or some forms of soy sauce this can be up to 9 months), additional oxidative metabolism can occur, introducing

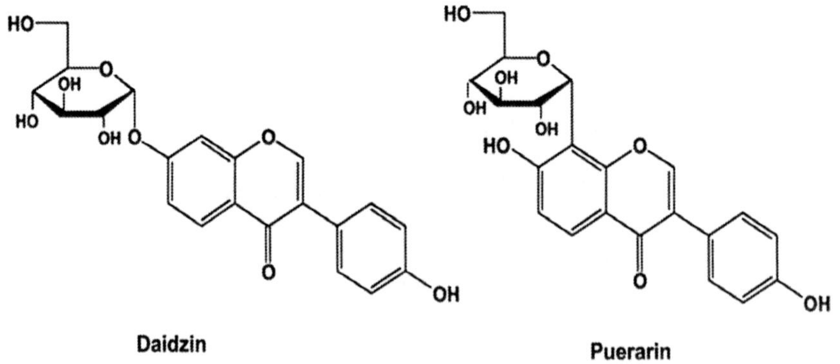

Figure 4.1 The structure of C- and O-glucosides of daidzein. Daidzein undergoes conjugation with glucose either to form the 7-*O*-glucoside daidzin (as in soybeans) or the 8-C-glucoside puerarin.

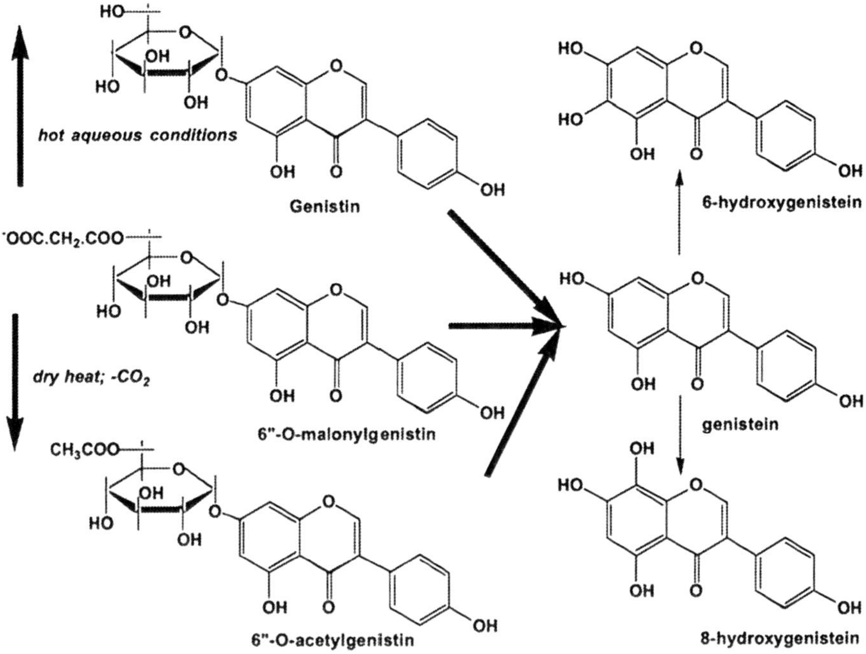

Figure 4.2 Effect of processing on soy chemistry. The 6″-O-malonate ester of genistin is either hydrolyzed (hot water or aqueous solvent) to genistein or decarboxylated by dry heat to 6″-O-acetylgenistin. Fermentation to release genistein, the aglycone, can also be accompanied by 6- or 8-hydroxylation.

hydroxyl groups into the 6- and 8-positions on the A-ring (Figure 4.2) (Esaki et al. 1999). Hexane extraction to recover oil components does not alter the composition. However, boiling water extraction of soybeans to produce soymilk causes the hydrolysis of the malonyl group, yielding simple β-glucosides. When soy is heated in its dry form (extrusion of soy protein concentrate or toasting of soybeans, soy flour or the hypocotyls), the malonyl group is decarboxylated to form the 6″-O-acetyl-7-O-β-glucoside (Barnes et al. 1994). These purified protein products can also be treated at the last stage of manufacturing with micro-organisms to generate a soy protein material containing unconjugated isoflavones.

Table 4.1 shows the content and composition of isoflavones in whole soybean flour (WSF) and autoclaved whole soybean flour (AWSF). The total isoflavone content was obtained from the sum of the glycoside (daidzin + glycitin + genistin) and aglycone (daidzein + glycitein + genistein) (da Silva et al. 2011). The concentration of total isoflavones was observed to range from 459.19 µg g^{-1} in the WSF to 506.41 µg g^{-1} in the AWSF. A significant decrease in the concentration of isoflavones glycosides was observed, from 97.33% of the total isoflavones in the WSF to 93.06%. In the AWSF, a rise in the isoflavones aglycones, from 2.67% of the total isoflavones to 6.94%, was observed.

Table 4.1 Concentrations (μg g^{-1}) of the isoflavones in WSF, AWSF and fermented AWSF with *Aspergillus oryzae*. Arithmetic means of three replicates ± S.D. followed by the same letter in the same line do not differ according to the Tukey's test ($P<0.05$). nd, not detected.

Isoflavones	WSF μg g^{-1}	%	AWSF μg g^{-1}	%	FAWSF (24 h) μg g^{-1}	%	FAWSF (48 h) μg g^{-1}	%
Daidzin	143.65 ± 2.39^b	31.28	153.21 ± 3.10^a	30.25	150.49 ± 0.40^a	22.24	26.30 ± 1.88^c	4.44
Glycitin	23.95 ± 0.71^b	5.22	nd	–	33.38 ± 1.62^a	4.93	24.49 ± 0.12^b	4.14
Genistin	279.32 ± 3.76^b	60.83	318.07 ± 3.12^a	62.81	247.44 ± 1.19^c	36.57	94.15 ± 1.00^d	15.91
Total glycosides	446.91 ± 2.28^b	97.33	471.28 ± 3.11^a	93.06	431.31 ± 1.07^c	63.75	144.94 ± 1.00^d	24.49
Daidzein	nd	–	nd	–	74.45 ± 0.22^b	11.00	133.07 ± 3.09^a	22.48
Glycitein	nd	–	16.16 ± 0.39^c	3.19	27.80 ± 0.71^b	4.11	35.56 ± 1.65^a	6.01
Genistein	12.27 ± 0.35^d	2.67	18.97 ± 0.40^c	3.75	142.98 ± 0.54^b	21.13	278.27 ± 3.86^a	47.02
Total aglycones	12.27 ± 0.35^d	2.67	35.13 ± 0.40^c	6.94	245.23 ± 0.49^b	36.25	446.90 ± 2.89^a	75.51
Total isoflavones	459.19 ± 1.80^d	100	506.41 ± 1.56^c	100	676.54 ± 0.78^a	100	591.84 ± 2.00^b	100

The heat treatment resulted in an increase in the levels of isoflavone aglycones of approximately 1.9 times as compared with the WSF. This type of behavior was also observed by Park *et al.* (2002). Genistin was the isoflavone isomer found in greatest amounts in WSF (279.32 µg/g), with a significant increase in AWSF (318.07 µg/g). Studies show that the glycoside forms are found in higher concentrations in soybeans and these are heat sensitive, being converted into malonyl-β-glycosylated isoflavone when heated (Coward *et al.* 1998). On the other hand, according to studies carried out by Setchell *et al.* (2002), the aglycone forms are quite stable at high temperatures.

4.2 Bioconversion of Isoflavones from Glycosides to Aglycones

Soy isoflavones are found predominantly in the glycoside form and in low concentrations as aglycones (Figure 4.3) (Izumi *et al.* 2000). The content and composition of isoflavones in soy-based products depend on the techniques used in processing, such as heat treatment, cooking, enzymatic hydrolysis and fermentation (Wang and Murphy 1994).

Studies have shown that the fermentation process of soybean promotes changes in the phytochemical compounds, causing changes in the isoflavone forms, hydrolysing the proteins and reducing the anti-nutritional factors, by

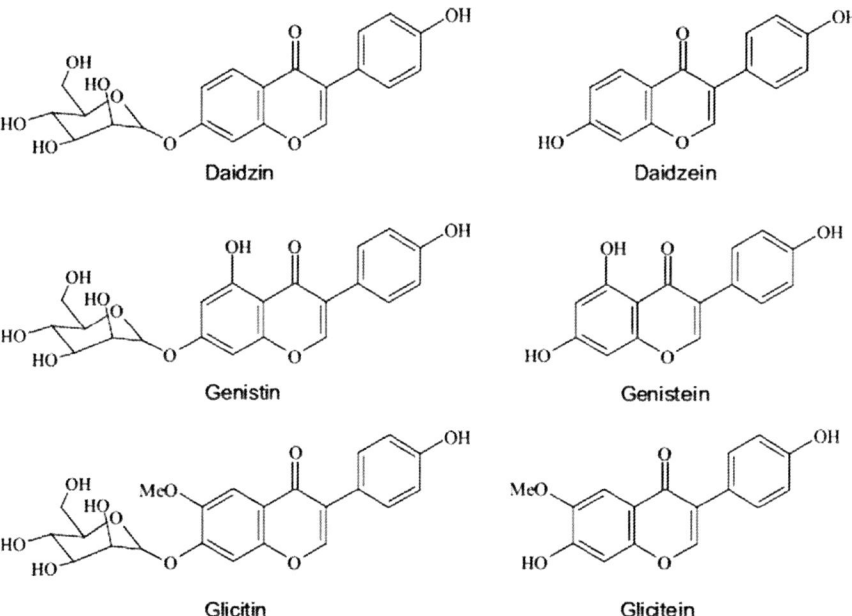

Figure 4.3 Chemical structures of some isoflavones, namely daidzin, genistin, glicitin, daidzein, genistein and glicitein.

reducing trypsin inhibitor content (Molteni *et al.* 1995; Zhu *et al.* 2005). The development of products with fermented soy may increase the functional components, such as increased isoflavone aglycones and active peptides that have greater potential health benefits (Hong *et al.* 2004; Mejia and Lumen 2006).

4.3 Enzymatic Transformation of Isoflavone Isomers in Fermented Soymilk

Rekha and Vijayalakshmi (2010) utilized five isolates of probiotic lactic acid bacteria (LAB), *Lactobacillus acidophilus* B4496 (La), *Lactobacillus bulgaricus* CFR2028 (Lb), *Lactobacillus casei* B1922 (Lc), *Lactobacillus plantarum* B4495 (Lp) and *Lactobacillus fermentum* B4655 (Lf) with the yeast *Saccharomyces boulardii* (Sb) to ferment soymilk to obtain the bioactive isoflavones, genistein and daidzein. Bioactive aglycones genistein and daidzein after 24 and 48 h of fermentation ranged from 97.49 to 98.49% and 62.71 to 92.31%, respectively with different combinations of LAB with yeast. The elution profile of glucosidic isoflavones genistin and daidzin and aglyconic isoflavones, namely daidzein and genistein, in soymilk fermented with the Sb + La combination is shown in Figure 4.4. Changes occurred in the concentration of β-glucoside and aglycone isoflavone isomers in soymilk fermented by La, Lb, Lc, Lp and Lf in combination with Sb for 24 and 48 h at 37 °C. Soymilk fermented with five different strains of LAB individually with yeast Sb contained a total amount of isoflavones in the range of 26.82–34.23 mg 100 ml^{-1}. After 24 h of incubation, the concentration of aglycones in soymilk fermented with the Sb + Lb, Sb + Lc, Sb + Lp and Sb + Lf combinatios showed marginal increase of 6.81, 1.04, 2.64 and 3.63 mg 100 ml^{-1} than Sb + La (28.53 mg 100 ml^{-1}). However, after 48 h of fermentation, the content of daidzein and genistein in soymilk fermented with Sb + La increased from 28.53 to 31.21 mg 100 ml^{-1}. The significant bioconversion of the glucoside isoflavones into their corresponding aglycones is represented in Figure 4.5. The reduction in the content of b-glucosides (daidzin and genistin) and the increase in the content of their respective aglycones are based on the hydrolytic reaction catalysed by β-glucosidase produced by each bacterial strain. All the combinations in fermented soymilk showed highest bioconversion at 24 h rather than 48 h of fermentation, except Sb + Lf. Of the bioactive aglycone isomers, the concentration of genistein (12.50–25.01 mg 100 ml^{-1}) was significantly higher than daidzein (4.32–6.23 mg 100 ml^{-1}) (Rekha and Vijayalakshmi 2010). β-Glucoside forms of isoflavones comprised greater than 80% of the total isoflavone concentration of soymilk (King and Bignell 2000). Tsangalis *et al.* (2003) reported that the concentration of genistein was higher than the aglycone forms of daidzein in soymilk. Thus, the viability, β-glucosidase activity of LAB and yeast used by Rekha and Vijayalakshmi (2010) were associated with the extent of isoflavones bioconversion in fermented soymilk. Yeast possess stability-enhancing effects on LAB, and the specific effects of yeast on LAB stability vary with the type of yeast as well as with the LAB (Shao-Quan and Marlene 2009).

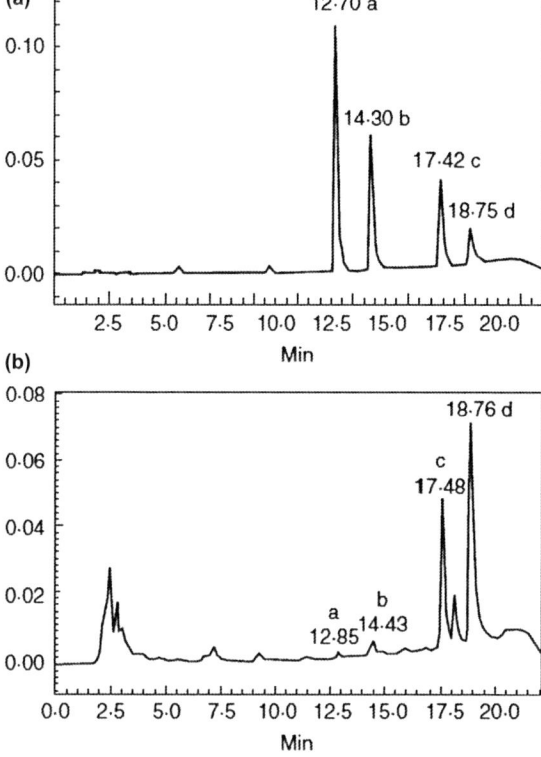

Figure 4.4 HPLC chromatogram showing the elution profile of standard isoflavones. (a) Daidzin, a; genistin, b; daidzein, c; genistein, d. (b) Soymilk fermented with the Sb + La combination: daidzin, a; genistin, b; daidzein, c; genistein, d.

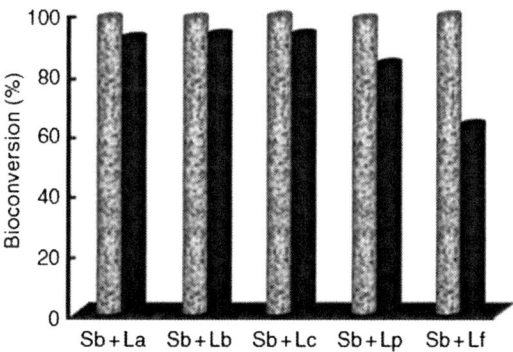

Figure 4.5 Bioconversion of glucosides into aglycones in soymilk fermented with LAB with Sb for 24 and 48 h at 37 °C. Grey columns, 24 h; black columns, 48 h.

It has been found that the glucosidic isoflavones are the predominant isomeric forms in non-fermented soymilk, and glucosides present in soymilk require bacterial induced hydrolytic deconjugation for transformation into a bioavailable aglycone form (King and Bignell 2000). Pham and Shah (2008) reported that supplementation of 0.5% lactulose to soymilk enhances the biotransformation of glucoside to aglycone by all four *Lactobacillus* spp. studied. This demonstrated that the β-glycosyl bond of isoflavones glycosides is cleaved to produce their aglycones by microbes during fermentation. After 24 and 48 h of fermentation, genistein and daidzein ranged from 97.49 to 98.49% to 62.71 to 92.31% with different combinations of LAB with yeast. However, after 48 h of fermentation, the content of daidzein and genistein in soymilk fermented with Sb + La increased from 28.53 to 31.21 mg 100 ml^{-1}. A similar observation was reported by Tsangalis *et al.* (2003). They found that there was very little bioconversion of isoflavone glucosides into aglycones after 24 h of soymilk fermentation with *Bifidobacteria animalis* Bb-12, hence limiting the fermentation period to 24 h would have little effect on the extent of enrichment of aglycone forms.

4.4 β-Glycosidase Reaction during Bioconversion of Isoflavones

Ryu *et al.* (2010) found by using a β-glycosidase (0.02%, w/v) reaction at 55 °C for 30 min that a maximum conversion of isoflavone-glycoside into isoflavone-aglycone of 84.5% was obtained in soft tofu (Table 4.2). There were no

Table 4.2 Effects of the β-glycosidase reaction time on the conversion of glycoside isoflavone into aglycone isoflavone in soft tofu. Data are the means ± S.D. ($n=3$). Reaction temperature: 55 °C. β-Glycosidase concentration: 0.06%.

Enzyme reaction time (min)	Type	Isoflavone contents (mg 100 g^{-1}, dry weight)			
		Daidzein	Glycitein	Genistein	Total
0	Isoflavone-aglycone	8.42 ± 0.67	1.58 ± 0.13	9.78 ± 0.78	19.78 ± 1.57
	Total isoflavone	73.97 ± 0.79	31.47 ± 0.34	77.19 ± 0.82	182.63 ± 1.95
	Conversion rate (%)				10.8
10	Isoflavone-aglycone	30.96 ± 1.25	16.14 ± 0.65	50.70 ± 2.05	97.79 ± 3.96
	Total isoflavone	71.99 ± 2.48	30.61 ± 1.05	74.83 ± 2.58	177.43 ± 6.11
	Conversion rate (%)				55.1
30	Isoflavone-aglycone	55.50 ± 1.62	23.34 ± 0.68	78.62 ± 2.30	157.46 ± 4.61
	Total isoflavone	75.04 ± 1.29	31.58 ± 0.54	76.88 ± 1.32	183.50 ± 3.15
	Conversion rate (%)				85.8
60	Isoflavone-aglycone	51.77 ± 1.79	22.77 ± 0.75	76.66 ± 2.51	154.19 ± 5.05
	Total isoflavone	75.41 ± 2.19	31.78 ± 0.92	76.07 ± 2.20	183.27 ± 5.31
	Conversion rate (%)				84.1

significant changes in the conversion rates by raising the reaction time beyond 30 min.

This study also reported that the bioconversion reaction in part depended on enzyme concentration. The β-glycosidase had a higher affinity toward daidzein and genistein than glycitein, and the results are in accord with those reported by Ismail and Hayes (2005). Soybean isoflavones are generally present in the glycoside form, which has a sugar component bound to a hydroxyl group of the aglycone (Walter 1941; Ito *et al.* 2008). The isoflavone glycoside is poorly absorbed in the small intestine and is less bioactive than the aglycone form due to its higher molecular weight and hydrophilicity (Piskula *et al.* 2008). The conversion of the isoflavone glycosides to isoflavone aglycones in soy products can increase the bioavailability and hence improve the isoflavone-associated health benefits.

Summary Points

- Isoflavones in the soybean are converted into 7-*O*-β-glucosides by a glucosyltransferase and then into their 6″-*O*-malonate forms by a malonyltransferase.
- Fermented soy foods contain the unconjugated isoflavone aglucones, whereas non-fermented soy foods contain their β-glucoside conjugates.
- Soy isoflavones are found predominantly in the glycoside form and in low concentrations as aglycones.
- Fermentation causes the removal of the glucosidic group, releasing the isoflavone aglucone.
- The fermentation process carries out bioconversion of isoflavones to increase their bioavailability, such as in the form of genistein and daidzein, which are bioactive isoflavones.
- The reaction for the bioconversion of isoflavone-glycoside into isoflavone-aglycone can be enhanced by the use of enzymes such as β-glycosidase.

Key Facts

- Isoflavones are the most common group of phytoestrogens which are present in significantly large amounts in soybean and soy products.
- Isoflavones occur naturally in glycoside forms having lower bioavailability than their aglycone forms.
- The bioconversion of isoflavone glycosides to isoflavone aglycones raises the bioavailability of isoflavones and this can be done by bacterial and enzyme action.

Definitions of Word and Terms

Bioavailability (BA): In pharmacology, bioavailability is a subcategory of absorption and is used to describe the fraction of an administered dose of unchanged drug that reaches the systemic circulation, one of the principal pharmacokinetic properties of drugs.

Bioconversion: It is also known as biotransformation and refers to the use of live organisms, often micro-organisms, to carry out a chemical reaction.

High-performance liquid chromatography–mass spectrometry (HPLC-MS): This is an analytical chemistry technique that combines the physical separation capabilities of liquid chromatography (or HPLC) with the mass analysis capabilities of mass spectrometry. It is a powerful technique, which has very high sensitivity and selectivity, used for many applications.

Hydrophilicity: It is the tendency of a molecule to be solvated by water.

Lactic acid bacteria (LAB): They comprise of Gram-positive, low-GC, acid-tolerant, generally non-sporulating, non-respiring rod or cocci that are associated by their common metabolic and physiological characteristics. These bacteria, usually found in decomposing plants and lactic products, produce lactic acid as the major metabolic end-product of carbohydrate fermentation.

Malonyltransferase: An enzyme that catalyzes the following chemical reaction:

$$\text{malonyl-CoA} + [\text{acyl-carrier protein}] \rightleftharpoons \text{CoA} + \text{malonyl-[acyl-carrier protein]}$$

Thus, the two substrates of this enzyme are malonyl-CoA and acyl-carrier protein, whereas its two products are CoA and malonyl-acyl-carrier protein.

Soft tofu: It is a type of tofu which is a food made by coagulating soy milk and then pressing the resulting curds into soft white blocks. Soft tofu has a fine texture with 88–90% water content and it is packed in containers together with water.

Soymilk: Sometimes referred to as soy drink/beverage, it is a beverage made from soybeans. A stable emulsion of oil, water and protein, it is produced by soaking dry soybeans and grinding them with water.

Whole soybean flour (WSF): Whole soybean flour is obtained from the milling of the whole soybean grain. It usually has a protein content between 35 and 40% and lipid content from 15 to 20%. It is classified as a raw flour, enzymatically active and used for bread whitening or as a toasted flour for general use.

List of Abbreviations

AWSF	autoclaved whole soybean flour
La	*Lactobacillus acidophilus*
LAB	lactic acid bacteria
Lb	*Lactobacillus bulgaricus*
Lc	*Lactobacillus casei*
Lf	*Lactobacillus fermentum*
Lp	*Lactobacillus plantarum*
6OAcGlc	$6''$-O-acetyl-β-glucoside
6OMalGlc	$6''$-O-malonyl-β-glucoside
Sb	*Saccharomyces boulardii*
WSF	whole soybean flour

References

Barnes, S., Kirk, M., and Coward, L., 1994. Isoflavones and their conjugates in soy foods: extraction conditions and analysis by HPLC mass spectrometry. *Journal of Agricultural and Food Chemistry.* 42: 2466–2474.

Coward, L., Smith, M., Kirk, M., and Barnes, S., 1998. Chemical modification of isoflavones in soy foods during cooking and processing. *American Journal of Clinical Nutrition.* 68: 1486S–1491S.

da Silva, L.H., Celeghini, R.M.S., and Chang, Y.K., 2011. Effect of the fermentation of whole soybean flour on the conversion of isoflavones from glycosides to aglycones. *Food Chemistry.* 128: 640–644.

Deavours, B.E., and Dixon, R.A., 2005. Metabolic engineering of isoflavonoid biosynthesis in alfalfa. *Plant Physiology.* 138: 2245–2259.

Dixon, R.A., 2004. Phytoestrogens. *Annual Reviews of Plant Biology.* 55: 225–261.

Esaki, H., Kawakishi, S., Morimitsu, Y., and Osawa, T., 1999. New potent antioxidative *o*-dihydroxyisoflavones in fermented Japanese soybean products. *Bioscience Biotechnology and Biochemistry.* 63: 1637–1639.

Farmakalidis, E., and Murphy, P.A., 1985. Isolation of 6″-O-acetyidaidzein and 6″-O-acetylgenistein from toasted defatted soy flakes. *Journal of Agricultural and Food Chemistry.* 33: 385–389.

Hong, K.J., Lee, C.H., and Kim, S.W., 2004. *Aspergillus oryzae* GB-107 fermentation improves nutritional quality of food soybeans and feed soybean meal. *Journal of Medicinal Food.* 7: 430–436.

Hymowitz, T., 1990. Soybeans: the success story. In: J. Janick and J.E. Simon (ed.) *Advances in New Crops*. Timber Press, Portland, OR, USA, pp. 159–163.

Ismail, B., and Hayes, K., 2005. β-Glycosidase activity toward different glycosidic forms of isoflavones. *Journal of Agricultural and Food Chemistry.* 53: 4918–4924.

Ito, J., Sahara, H., Kaya, M., Hata, Y., Shibasaki, S., and Kawata, K., 2008. Characterization of yeast cell surface displayed *Aspergillus oryzae* β-glucosidase 1 high hydrolytic activity for soybean isoflavone. *Journal of Molecular Catalysis B: Enzymatic.* 55: 69–75.

Izumi, T., Piskula, M.K., Osawa, S., Obata, A., Tobe, K., and Saito, M., 2000. Soy isoflavone aglycones are absorbed faster and in higher amounts than their glucosides in humans. *Journal of Nutrition.* 130: 1695–1699.

King, R.A., and Bignell, C.M., 2000. Concentrations of isoflavones phytoestrogens and their glucosides in Australian soya beans and soya foods. *Australian Journal of Nutrition and Dietetics.* 57: 70–78.

Kudou, S., Fleury, Y. and Welti, D., 1991. Malonyl isoflavone glycosides in soybean seeds (Glycine max Merrill). *Agricultural and Biological Chemistry.* 55: 2227–2233.

Kuo, L.C., Cheng, W.Y., Wu, R.Y., Huang, C.J., and Lee, K.T., 2006. Hydrolysis of black soybean isoflavone glycosides by *Bacillus subtilis* natto. *Applied Microbiology and Biotechnology.* 73: 314–320.

Mejia, E., and Lumen, B.O., 2006. Soybean bioactive peptides: a new horizon in preventing chronic diseases. *Sexuality, Reproduction and Menopause.* 4: 91–95.

Molteni, A., Brizio-Molteni, L., and Persky, V., 1995. *In-vitro* hormonal effects of soybean isoflavones. *Journal of Nutrition.* 125: 751S–756S.

Park, Y.K., Aguiar, C.L., Alencar, S.M., Mascarenhas, H.A.A., and Scamparini, A.R.P., 2002. Conversão de malonil-b-glicosil isoflavonas em isoflavonas glicosiladas presentes em algumas cultivares de soja brasileira. *Ciência e Tecnologia de Alimentos.* 22: 130–135.

Pham, T.T., and Shah, N.P., 2008. Effect of lactulose on biotransformation of isoflavone glycosides to aglycones in soymilk by *Lactobacilli*. *Journal of Food Science.* 73: M158–M165.

Piskula, M.K., Yamakoshi, J., and Iwai, Y., 1999. Daidzein and genistein but not their glucosides are absorbed from the rat stomach. *FEBS Letters.* 447: 287–291.

Rekha, C.R., and Vijayalakshmi, G., 2010. Bioconversion of isoflavone glycosides to aglycones, mineral bioavailability and vitamin B complex in fermented soymilk by probiotic bacteria and yeast. *Journal of Applied Microbiology.* 109: 1198–1208.

Ryu, Y.G., Won, B., Park, H., Ghafoor, K., and Park, J., 2010. Effects of β-glycosidase reaction on bio-conversion of isoflavones and quality during tofu processing. *Journal of the Science of Food and Agriculture.* 90: 843–849.

Setchell, K.D.R., Brown, N.M., Zimmer-Nechemias, L., Brashear, W.T., Wolfe, B.E., and Kirschner, A.S., 2002. Evidence for lack of absorption of soy isoflavones glycosides in humans, supporting the crucial role of intestinal metabolism for bioavailability. *American Journal of Clinical Nutrition.* 76: 447–453.

Shao-Quan, L., and Marlene, T., 2009. Enhancement of survival of probiotic and non-probiotic lactic acid bacteria by yeasts in fermented milk under non-refrigerated conditions. *International Journal of Food Microbiology.* 135: 34–38.

Tsangalis, D., Ashton, J.F., McGill, A.E.J., and Shah, N.P., 2003. Biotransformation of isoflavones by bifidobacteria in fermented soymilk supplemented with D-glucose and L-cysteine. *Journal of Food Science.* 68: 623–631.

Walter, E.D., 1941. Genistin (an isoflavone glucoside) and its aglycone, genistein, from soybeans. *Journal of American Chemical Society.* 63: 3273–3276.

Wang, H., and Murphy, P.A., 1994. Isoflavone content in commercial soybean foods. *Journal of Agricultural and Food Chemistry.* 42: 1666–1673.

Zhu, D., Hettiarachchy, N.S., Horax, R., and Chen, P., 2005. Isoflavone contents in germinated soybean seeds. *Plant Foods for Human Nutrition.* 60: 147–151.

CHAPTER 5

Chemistry and Synthesis of Daidzein and its Methylated Derivatives: Formononetin, Isoformononetin, and Dimethyldaidzein

VINCENT M. CARROLL,[b] JEFFREY D. ST. DENIS,[b] KYLE F. BIEGASIEWICZ[b] AND RONNY PRIEFER*[a,b]

[a]Department of Pharmaceutical & Administrative Science, College of Pharmacy, Western New England University, CSP326-1215 Wilbraham Rd., Springfield, MA, USA; [b]Department of Chemistry, Niagara University, 206 DePaul Hall, Niagara University, NY, USA
*Email: ronny.priefer@wne.edu

5.1 Introduction

The isoflavones, daidzein (4′,7-hydroxyisoflavone, **1**) and its methylated derivatives formononetin (4′-methoxy-7-hydroxyisoflavone, **2**), isoformononetin (4′-hydroxy-7-methoxyisoflavone, **3**), and dimethyldaidzein (4′-methoxy-7-methoxyisoflavone, **4**) (Figure 5.1), are found in a variety of geographical locations and in a myriad of plant sources.

Among the variety of these locations, each has either been extracted or detected in an array of specific sources. Daidzein (**1**), one of the most abundant

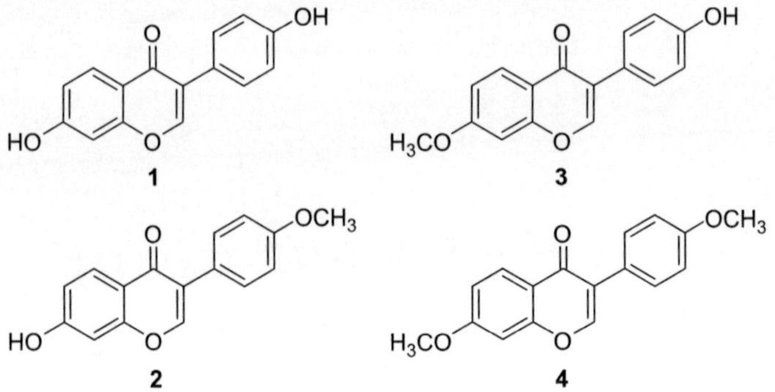

Figure 5.1 Isoflavones: daidzein (**1**), formononetin (**2**), isoformononetin (**3**), and dimethyldaidzein (**4**).

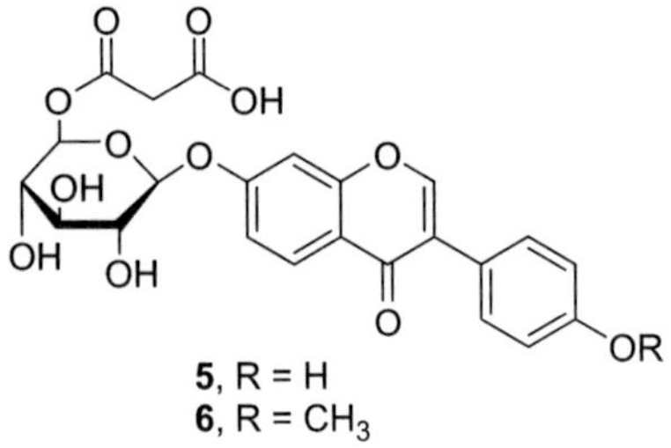

5, R = H
6, R = CH$_3$

Figure 5.2 Molecular structure of daidzein 7-*O*-glucoside-6″-*O*-malonate (**5**) and formononetin 7-*O*-glucoside-6″-*O*-malonate (**6**).

isoflavones in legumes, can be extracted by various methods, such as the utilization of the non-ionic surfactant oligo(ethylene glycol) monoalkyl ether (Genapol X-080) with the Chinese herb *Puerariae Radix* (Luthria and Natarajan 2009). In addition, solid-phase microextraction in combination with liquid chromatography–mass spectrometry (LC-MS) analysis has been used for its detection in human urine (Stalikas 2007). Most notably, daidzein (**1**) can be extracted from a plethora of soy products including soymilk, tofu, and soy molasses. It is also commonly isolated, along with its methylated derivative, formononetin (**2**), in a variety of other legume sources including kidney beans, navy beans, and Japanese arrowroot (Tolleson *et al.* 2002). Typical extraction involves the employment of a hot aqueous polar solvent, such as acetonitrile, or polar protic solvent, such as methanol, through mixing or Soxhlet extraction (Barnes *et al.* 1994). It should be noted that both daidzein and formononetin are stored in the vacuoles of legumes as malonylglucoside conjugates, daidzein 7-*O*-glucoside-6″-*O*-malonate (**5**) and formononetin 7-*O*-glucoside-6″-*O*-malonate (**6**), respectively (Figure 5.2). These are typically difficult to extract in this form due to their degradation by the enzymes isoflavone malonylglucoside malonyleterase and isoflavone glucoside glucosidase (Toebes *et al.* 2005).

5.2 Synthesis of Daidzein (1)

Similar to other flavonoids, daidzein (Figure 5.3, **1**) is characterized by the presence of a three-ring system, but differs from the various classes in the degree of oxidation and pattern of B-ring substitution to form the phenylchromenone as its basic chemical scaffold (Leonarduzzi *et al.* 2010). The key to the synthesis of daidzein (**1**) is the construction of the challenging benzopyran ring and thus serves as the focal point of many synthetic approaches. To date, four main strategies have been developed towards addressing this structural challenge: (1) $BF_3 \cdot OEt_2$-catalyzed Friedel–Crafts acylation and ring closure, (2) ring-closing metathesis (RCM) with Grubbs' catalyst, (3), I_2-mediated cyclization, and (4) oxidative rearrangement with thallium (III) nitrate.

5.2.1 Synthesis of Daidzein: $BF_3 \cdot OEt_2$-catalyzed Friedel–Crafts Acylation and Ring Closure

The most widely utilized method for preparation of the desired chromenone framework has been through the use of $BF_3 \cdot OEt_2$ as first reported by Bass (1976). Starting from commercially available resorcinol (**7**) and 4-hydroxyphenyl acetic acid (**8**), the corresponding deoxybenzoin **9** can be prepared in good yields by a Friedel–Crafts reaction using $BF_3 \cdot OEt_2$ as both the catalyst and the solvent (Goto *et al.* 2009), as shown in Scheme 5.1. Under refluxing conditions, significant rate enhancements (<10 min) can be achieved that allows for rapid access to a variety of potential daidzein derivatives, since the

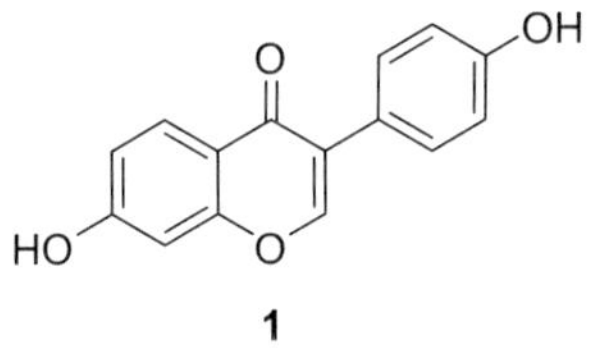

Figure 5.3 Structure of daidzein (**1**).

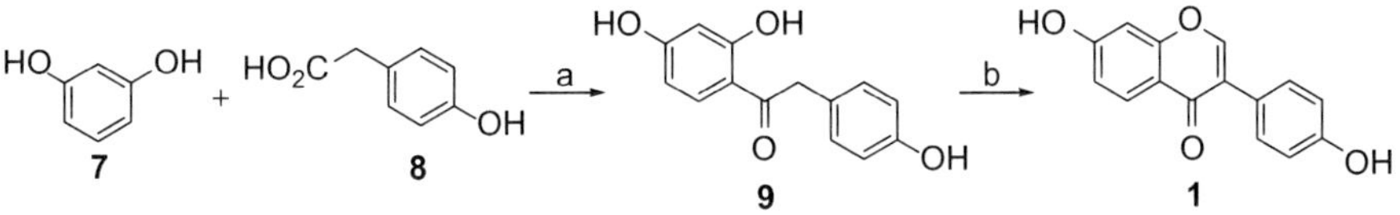

Scheme 5.1 Synthesis of **1** with $BF_3 \cdot OEt_2$ and one-carbon electrophiles. Reagents and conditions: (a) $BF_3 \cdot OEt_2$, 75 °C, 4 h, 64%; (b) $BF_3 \cdot OEt_2$, MsCl, DMF, 75 °C, 1.5 h, 35%.

initial diphenol **7** and acid **8** can be readily substituted to form a variety of isoflavones, lending value to the method. Subsequent additional equivalences of $BF_3 \cdot OEt_2$ in the presence of one-carbon sources [*i.e.* *N*,*N*-dimethylformamide (DMF)], and elimination of the key intermediate with methanesulfonyl chloride (MsCl) installs the requisite isoflavone **1** scaffold in a rapid and efficient manner without the need for protecting groups. Further investigation into the reaction sequence has revealed microwave-assisted rate and yield enhancements (Chang *et al.* 1994), as well as synthetically useful one-pot procedures (Goto *et al.* 2009) in the conversion of **8** into **1**.

5.2.2 Synthesis of Daidzein: RCM with Grubbs' Catalyst

In an attempt to address the deficiencies of previous routes, Wang and co-workers relied on RCM chemistry to rapidly access daidzein (**1**), in addition to other interesting flavonoids *via* a common intermediate (Li *et al.* 2009). As shown in Scheme 5.2, a Wittig reaction with 4-benzyloxysalicyclic aldehyde (**10**) proceeds by the initial *in situ* formation of methylene(triphenyl)phosphorane (MTPP) from methyltriphenylphosphine bromide (MTPPB) and potassium *tert*-butoxide and subsequent reaction to form the intermediate phenolate **11**, which undergoes *O*-alkylation in refluxing tetrahydrofuran (THF) with 2-bromo-4'-methoxyacetophenone to produce the second intermediate **12**. Without isolation, the intermediate underwent a second Wittig reaction with MTPP to furnish the desired diene **13** in a respectable one-pot yield. The corresponding diene **13** was further transformed into isoflavene **14** using Grubbs' second-generation catalyst and served as a key intermediate towards the construction of several interesting isoflavonoids including daidzein, equol, and formononetin.

Next, **14** was converted into the di-protected isoflavone derivative **16** through a two-step sequence involving classic hydroboration–oxidation chemistry to afford the corresponding chroman-4-ol **15** which further underwent a second oxidation–dehydrogenation process to yield isoflavone **16** in a 69% yield over two steps. Finally, deprotection of the benzyl and methyl ether protecting groups using $EtSH/AlCl_3$ gave daidzein (**1**) a 92% yield.

5.2.3 Synthesis of Daidzein: I$_2$-Mediated Cyclization

Inspired by the power of palladium-mediated processes for the construction of carbon–carbon bond formation, two groups have recently incorporated such strategies within their novel approaches (Biegasiewicz *et al.* 2010; Matin *et al.* 2009), with only one of the two described below (Scheme 5.3) (Biegasiewicz *et al.* 2010). The synthesis commenced with a base-catalyzed Claisen–Schmidt condensation of *N*,*N*-dimethylformamide dimethylacetal (DMF-DMA) and 2,4-dihydroxyacetophenone (**17**). Initial attempts to append only the enamine moiety to the starting acetophenone **17** proved futile as undesired *O*-methylation on the 4-OH group could not be controlled. Nevertheless, exposure of

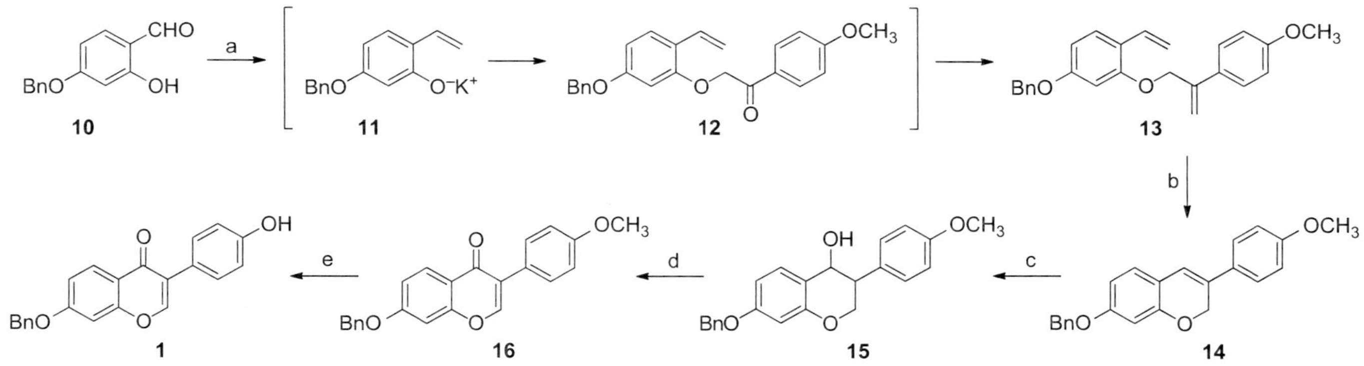

Scheme 5.2 Synthesis of **1** with Grubbs' catalyst. Reagents and conditions: (a) MTPPB, *t*-BuOK, THF, 0 °C, 2 h; then 2-bromo-4′-acetophenone, reflux, 1 h; then MTPPB, *t*-BuOK, THF, 0 °C, 1 h, 74%; (b) Grubbs' second catalyst, dichloromethane (DCM), 40 °C, 8 h, 82%; (c) BH$_3$-SMe$_2$, THF, 0 °C, 4 h; then H$_2$O, 10% NaOH, 37% H$_2$O$_2$, 30 min, 60%; (d) 2,3-dichloro-5,6-dicyano-1,4-benzoquinone (DDQ), 1,4-dioxane, reflux, 8 h, 78%; (e) AlCl$_3$, EtSH, DCM, 0 °C, 30 min, 92%. Bn, benzyl.

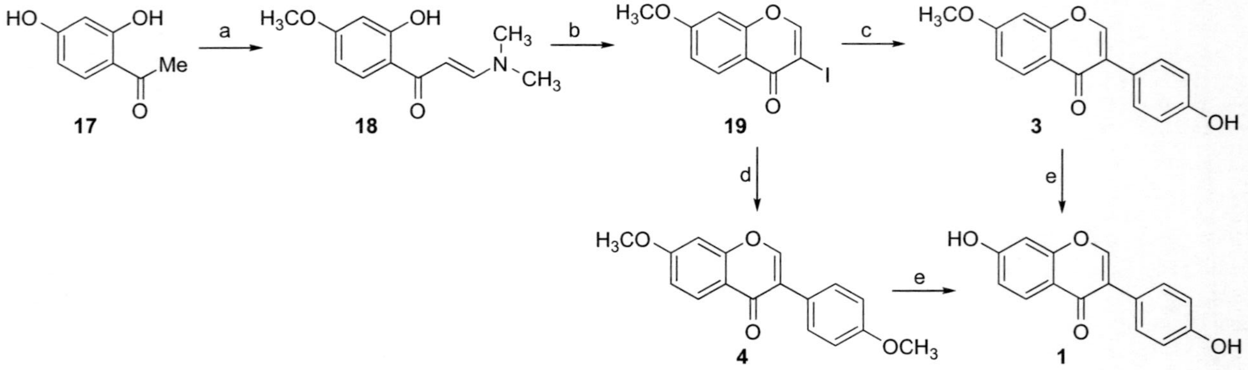

Scheme 5.3 Synthesis of **1** *via* I₂-mediated cyclization and Suzuki coupling. Reagents and conditions: (a) DMF-DMA, DMF, 80 °C, 24 h, 89%; (b) I₂, MeOH, room temperature (rt), 24 h, 81%; (c) 4-hydroxyphenylboronic acid, Na₂CO₃, Pd(OAc)₂, PEG 10000, 50 °C, CH₃OH, 3 h, 98%; (d) 4-methoxyphenylboronic acid, Na₂CO₃, Pd(OAc)₂, PEG 10000, 50 °C, MeOH, 3 h, 90%; HI, reflux, 4 h, 89%.

the monoprotected species (**18**) to iodine affected a smooth cyclization with concomitant loss of dimethyl amine to provide exclusively the 3-iodo-4*H*-chromen-4-one (**19**).

With the closed pyran ring in hand, the iodobenzopyranone (**19**) served as an ideal substrate for the Suzuki coupling reaction, as the oxidative addition step in the sp^2-hybridized system is enhanced by the presence of an ortho ketone group to facilitate the addition of the key third ring. Typical Suzuki coupling often necessitates the use of expensive and, at times, toxic phosphine-based ligands that enhance the oxidative addition step, but can also complicate purification methods (Ren and Meng 2008). Priefer and co-workers utilized a recently described green approach (Liu *et al.* 2005) to this reaction through the utilization of poly(ethylene glycol) 10000 (PEG 10000) as the ligand instead of phosphines to provide exceptional yields of the desired products (**3, 4**) under mild conditions upon using the appropriate phenylboronic acid derivatives. Moreover, this coupling continued to proceed effectively upon recycling of the palladium without further addition of the expensive palladium catalyst to subsequent reactions. Finally, global deprotection proceeded cleanly to provide daidzein (**1**) in high yields.

5.2.4 Synthesis of Daidzein: Oxidative Rearrangement with Thallium(III) Nitrate

The final strategy for the construction the isoflavone scaffold has been through the use of thallium trinitrate (TTN) to promote an oxidative rearrangement of chalcones to isoflavones involving a 1,2-aryl migration (Jung *et al.* 2003). As shown in Scheme 5.4, base condensation of an acetophenone (**20**) and a benzaldehyde (**21**) affords the key chalcone intermediate **22** that, in the presence of TTN, form a phenol (**23**) upon hydrolysis. Finally, a palladium-mediated hydrogenolysis cleaved the benzyl-protecting group yielding daidzein (**1**).

Unfortunately, this method is complicated by two significant drawbacks: sensitivity to the nature of substitution present on the aryl ring and undesirable rearranged products. In this case, the daidzein scaffold is sufficiently electron-rich to affect the 1,2-aryl migration, but the oxythalliation does not yield desired product when the hydroxyl group *para* to the ketone is unprotected. Suppression of this unproductive pathway was achieved through protection of the phenolic moiety with a benzyl group (Jung *et al.* 2003).

5.3 Synthesis of Formononetin (2)

Isoflavones are a class of polycyclic natural products mainly occurring in the Leguminosae family of plants and, due to their intriguing biological activity, have received a large amount of attention from the pharmaceutical/agricultural industry (Balasubramanian and Nair 2000). Formononetin (Figure 5.4, **2**) and its naturally occurring glycoside, ononin, is a commonly occurring isoflavone from this family. Due to its agricultural significance, formononetin has been

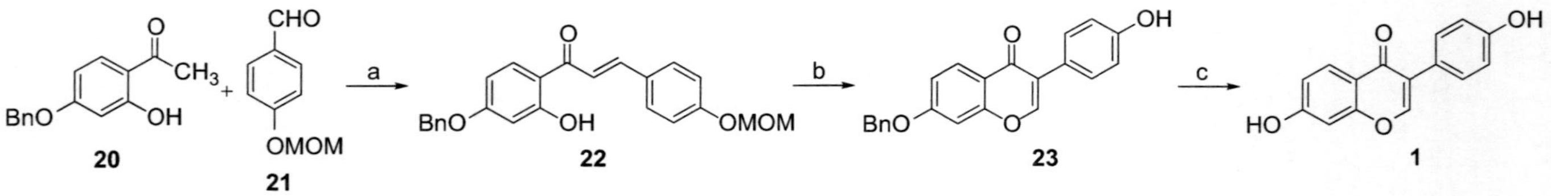

Scheme 5.4 Synthesis of **1** with TTN-promoted oxidative rearrangement. Reagents and conditions: (a) 9% NaOH in 90% EtOH, 50 °C, 2 h, 78%; (b) TTN, CH$_3$OH, room temperature (rt), 1 day; then 2 M HCl, 50 °C, 5 h, 51%; (c) H$_2$, 10% Pd/C, rt, 2 h, 64%. MOM, methoxymethyl ether.

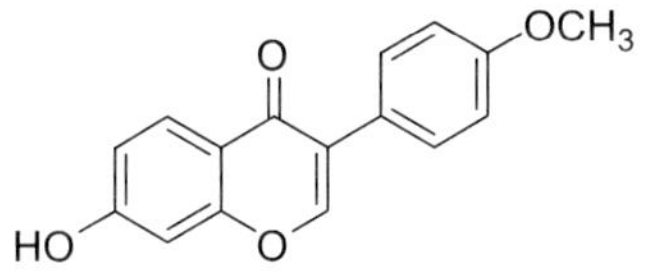

Figure 5.4 Structure of formononetin (**2**).

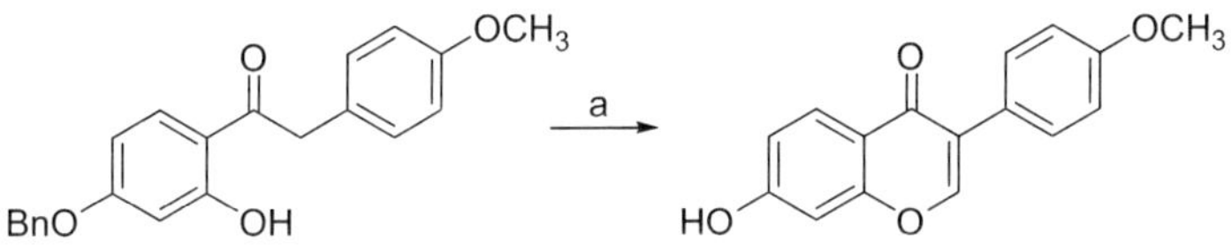

Scheme 5.5 Venkataraman synthesis of **2**. Reagents and conditions: (a) Na dust, ethyl formate; then debenzylation. Bn, benzyl.

developed commercially under the trade name Mycoform® as a compound to stimulate the growth of both vesicular arbuscular mycorrihzal fungi and the plants that play host to the fungi (Balasubramanian and Nair 2000).

5.3.1 Carbonyls as One-carbon Electrophiles

The first reported synthesis of formononetin (**2**) was by Wessely *et al.* (1933). This was performed by heating a sealed tube with 2,4-dihydroxy-4'-methoxy deoxybenzoin (**24**), Na°, and ethyl formate. Formononetin (**2**) was only obtained in a 19% yield. In the case of this synthesis, as with all related syntheses utilizing this methodology, cyclization is inefficient or fails when there are two or more free hydroxyl groups present (Baker *et al.* 1953).

In 1934, Venkataraman and co-workers demonstrated a higher yielding method to form **2** using a 2-hydroxy-4-benzyloxyphenyl benzyl ketone **24** scaffold in the presence of sodium dust and ethyl formate (Scheme 5.5) (Mahal *et al.* 1934). It was necessary to protect the 7-hydroxy group as the benzyl ether in order for the cyclization to proceed more efficiently, producing *O*-benzyl formononetin in a 33% yield (as compared to 19% for the Wessely synthesis). The protected isoflavone was then subjected to debenzylation to produce **2** in an undisclosed yield.

A later synthesis of formononetin (**2**), using amides (formamide or formanilide) instead of ethyl formate, proved fruitful in affording the isoflavone in a 50% yield. In this case it was also found that prior protection of all hydroxyl groups, except for the 2-position was necessary, with the optimal protecting groups being electron-withdrawing (Gowan *et al.* 1958).

5.3.2 Activated Aldehyde Surrogates as One-carbon Surrogates

The deoxybenzoin scaffold has afforded the library synthesis of numerous varieties of intriguing isoflavones. The methods previously described in Section 5.3.1 are generally limited by low yields, long reaction times, and tedious protection/deprotection sequences. The use of activated one-carbon electrophiles generally increases the recovered yields and decreases the reaction times. As highlighted above (Sections 5.2.1 and 5.2.3), DMF or DMF-DMA are commonly employed one-carbon surrogates. Other reagents have been employed for similar types of reactions in the formation of formononetin. All of these reactions rely on the formation of iminium cations, or other activated nitrogen intermediates.

5.3.2.1 1,3,5-Triazine as C_1-Synthon

A similar method to the synthesis of isoflavones utilizing a different C_1-synthon was described by Zilliken and co-workers in 1981 (Jha et al. 1981). This method used 1,3,5-triazine and a one-carbon electrophile to produce **2** from **25** (Scheme 5.6). The initial step involves enol attack upon 1,3,5-triazine to produce α-1,3,5-triazinylbenzoin (**26**) followed by cyclization, eliminating 1-amino-2,4-diaza-1,3-butadiene and producing **2** in a 91% yield (Jha et al. 1981). Although this approach does suffer from poor atom economy and the high cost of 1,3,5-triazine, all of the previously described syntheses were all developed as separate reactions; first to produce the deoxybenzoin, followed by the addition of a C_1-source. One-pot reactions are both valuable and more cost efficient than other methods, due to reduced solvent and purification costs (Balasubramanian and Nair 2000).

5.3.2.2 N,N'-Dimethyl(chloromethylene)ammonium chloride as Active One-carbon Donor

Another one-pot procedure, put forth by Balasubramanian and Nair (2000) to synthesize **2**, used an active one-carbon donor derived from DMF (**27**).

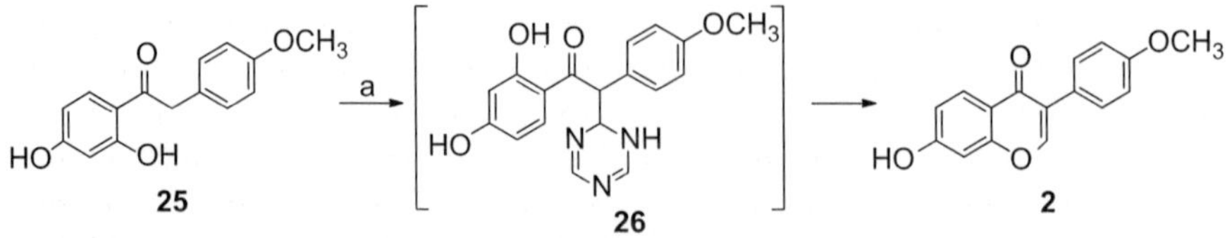

Scheme 5.6 The use of 1,3,5-triazine to produce **2**. Reagents and conditions: (a) $BF_3 \cdot OEt_2$, Ac_2O, AcOH, reflux, 2 h.

The synthesis of **2** was accomplished through a Friedel–Crafts acylation on the phenylacetic acid, **28**, followed by the addition of the C_1-synthon. The electrophile was formed by the action of PCl_5 on **27** to form N,N'-dimethyl(chloromethylene)ammonium chloride (**29**) and was added to **25**, which, after 1 h at room temperature, was quenched with methanolic HCl to produce a yellow precipitate. The solid cyclizes to produce **2** in a 90% yield (Scheme 5.7). Additionally, this methodology is also useful in producing the isomer isoformononetin **3** in a 82% yield (Balasubramanian and Nair 2000).

5.3.2.3 Palladium Catalysis

Palladium catalysis is ubiquitous in every aspect of organic chemistry and isoflavone synthesis is no exception (Hoshino *et al.* 1988). Every example described above uses a deoxybenzoin precursor then through addition of a one-carbon unit, followed by cyclization, yields the desired isoflavone. To increase the modular aspect of isoflavone synthesis, availability of derivatives, and decrease the necessity of rigorously anhydrous conditions, coupling reactions were the optimal tool to use. As published by Hoshino *et al.* (1988), an isoflavone scaffold can be synthesized in good to high yields under the standard conditions.

Felpin (2005) put forth an inexpensive and high-yielding reaction pathway for the synthesis of a variety of isoflavones, including **2**. The synthesis begins with the protection of the 4-position of 2,4-dihydroxyacetophenone (**17**) as the tetrahydropyran (THP)-ether, followed by the addition of DMF-DMA to produce the vinylogous amide (**30**), as seen in Scheme 5.8. Subsequent exposure of **30** to a mixture of I_2 in DMF promotes the cyclization to compound **31**. The cyclization proceeds in a two-step sequence, whereby cyclization is followed by iodination (St. Denis *et al.* 2010). A two-step coupling and deprotection ultimately secures **2**. An extremely useful and applicable aspect of this method is that it utilizes inexpensive Pd/C, which on a solid support is more amenable to industrial scale synthesis and can be recycled to further decrease the cost of the final product (Felpin 2005). This method has also been used to synthesize a library of potential peroxisome-proliferator-activated receptor agonists by Matin *et al.* (2009).

A directed investigation by Granados-Covarrubias and Maldonado (2009) to use a protected cyanohydrin (**32**) for the synthesis of **2** is illustrated in Scheme 5.9. The penultimate step involves a Wacker–Cook tandem cyclization which entails deprotection–cyclization–oxidation. It begins first with the conjugate addition on to a β-nitro styrene in the presence of KH to yield an α-methylene deoxybenzoin (**33, 34**) in a 65% yield. The mixture of protected **33** and **34** could then be subjected to Wacker–Cook tandem reaction conditions to produce **2** in a 58% yield (overall 37%). Although this is an extended reaction scheme, it does provide mechanistic value.

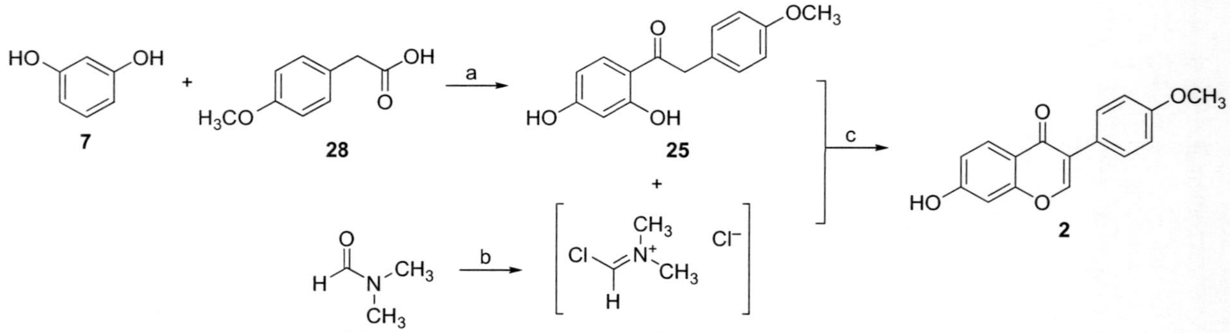

Scheme 5.7 One-pot synthesis of **2** using *N*,*N'*-dimethyl(chloromethylene)ammonium chloride (**29**). Reagents and conditions: (a) BF$_3 \cdot$ OEt$_2$, 85 °C, 90 min; (b) PCl$_5$; (c) BF$_3 \cdot$ OEt$_2$, room temperature (rt), 1 h.

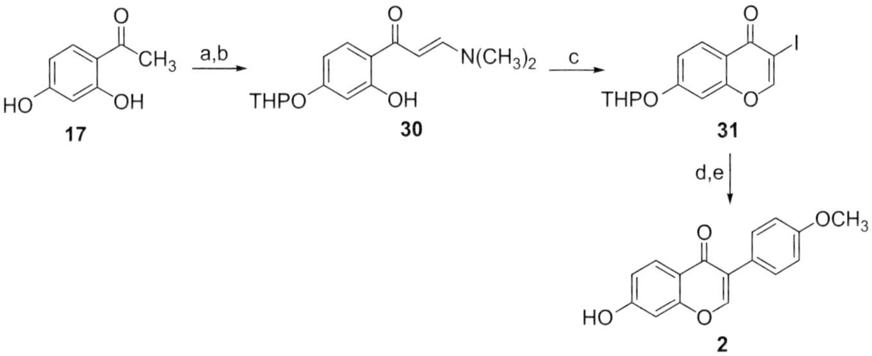

Scheme 5.8 Suzuki coupling to produce **2**. Reagents and conditions: (a) 3,4-dihydro-2*H*-pyran (DHP), pyridinium *p*-toluenesulfonate (PPTS), dichloromethane (DCM), room temperature (rt); (b) DMF-DMA, 95 °C; (c) I$_2$ pyridine, DCM, rt; (d) 4-methoxyphenyl boronic acid, 10%, Pd/C, Na$_2$CO$_3$, dimethoxyethane (DME)/H$_2$O, 45 °C; (e) H$^+$; THP, tetrahydropyran.

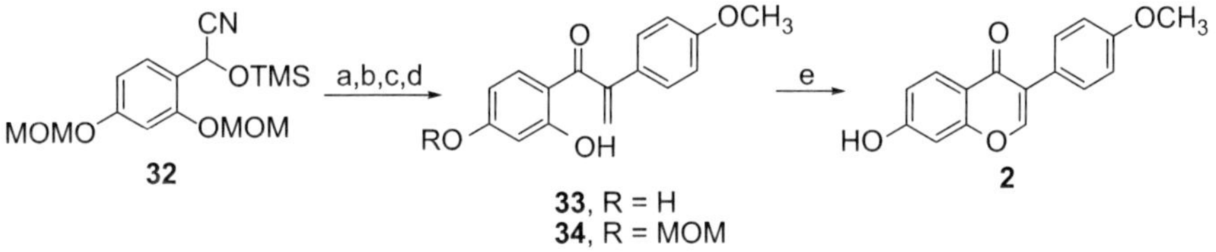

Scheme 5.9 Wacker–Cook tandem reaction to produce **2**. Reagents and conditions: (a) KH, dimethoxyethane (DME), then (*E*)-4-methoxy-nitrostyrene; (b) 5% H$_2$SO$_4$, 50 °C; (c) Et$_3$N, acetone; (d) BCl$_3$ or 20% H$_2$SO$_4$, 55 °C; (e) Na$_2$PdCl$_4$, *t*-BuOOH, NaOAc, *t*-BuOH, AcOH, H$_2$O, 80 °C, then 10% HCl. MOM, methoxymethyl ether; TMS, trimethylsilyl.

5.4 Synthesis of Isoformononetin (3)

Isoformononetin (Figure 5.5, **3**) is a rare isoflavone isolated from few natural sources (Ingham *et al.* 1981). Owing to its similarity with formononetin (**2**) and other more commonly encountered isoflavones, little research has been devoted to isoformononetin (**3**).

5.4.1 Selective Methylation of Daidzein (1)

Typical syntheses of **3** involve starting with daidzein and selective methylation of the 7-position or starting with the appropriately substituted deoxybenzoin (Scheme 5.10). Selective methylation of daidzein (**1**) stems from the difference in pK_a of the two free phenolic groups. The pK_a of the hydroxyl groups at both the 4′- and 7-position of daidzein was measured to be 9.65 ± 0.07 and 7.47 ± 0.02 respectively. The difference of two pK_a units allows for the selective

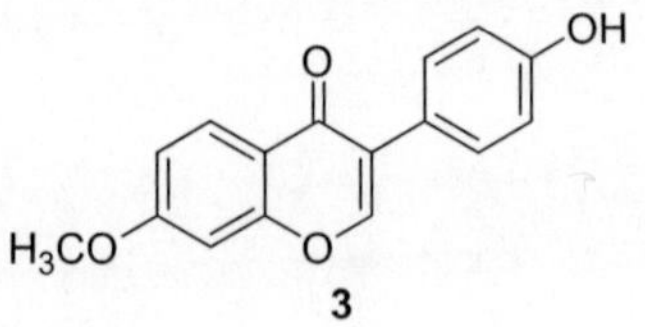

Figure 5.5 Structure of isoflavone isoformononetin (**3**).

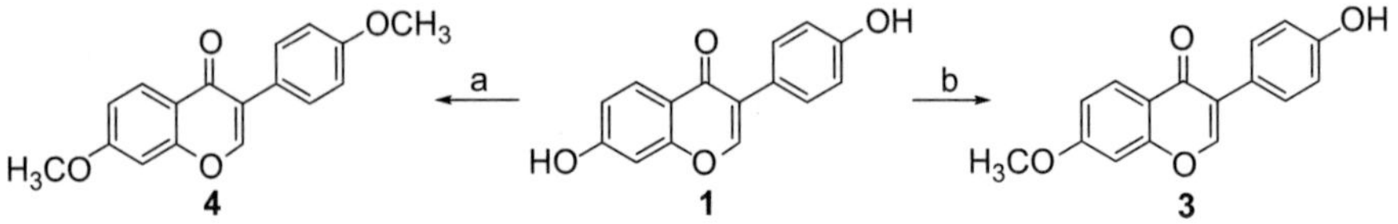

Scheme 5.10 Selective methylation of **1** for the formation of either **3** or **4**. Reagents and conditions: (a) 2.5 equiv. NaOH, 2.0 equiv. MeI, 18%; (b) 1.0 equiv. NaOH, 2.0 equiv. MeI, 68%.

deprotonation and methylation in moderate yield (68%) (Liang *et al.* 2008). The synthesis of **3** can also be affected through the use of diazomethane on daidzein (Lapcík *et al.* 2005).

5.4.2 Suzuki-mediated Synthesis

A variety of isoflavones, including **3** was synthesized by Priefer and co-workers (Biegasiewicz *et al.* 2010). One highlighting aspect is that no separate protection step of the 4-hydroxyl group was necessary as it is methylated with DMF-DMA to compound **19** (Scheme 5.11) (Sinkevich *et al.* 2007). Capitalizing on the venerable Suzuki reaction and having a practical concern with recycling the palladium catalyst, they immobilized the catalyst on a PEG 10000 support (Vasselin *et al.* 2006). This allowed for recovery of the palladium catalyst, which could be reused in further transformations. After coupling of the aryl boronic acid, **3** was obtained in an overall three-step sequence in a 71% overall yield. This is the most efficient and mild procedure published to date for the synthesis of isoformononetin **3** and allows for more detailed investigations into its biological properties as well as other 7-alkyl isoflavones.

5.5 Synthesis of Dimethyldaidzein (4)

Dimethyldaidzein (Figure 5.6, **4**) is the rarest of the methylated daidzein series. A majority of the reported syntheses have relied on utilizing commercially available daidzein (**1**) and performing a basic methylation step to obtain the desired product as shown in Scheme 5.12. Although this strategy gains rapid access to **4**, the starting material can be quite expensive especially when large

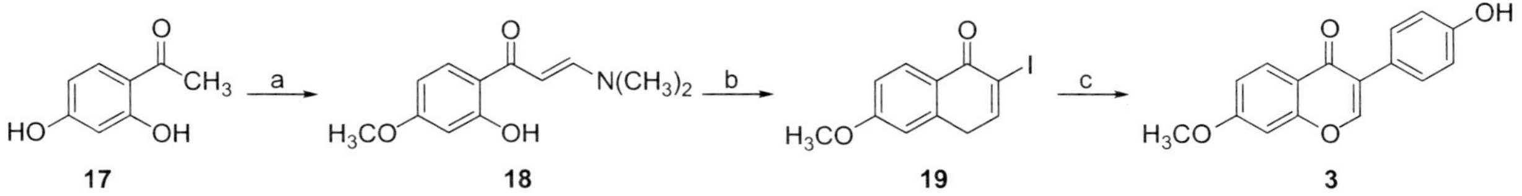

Scheme 5.11 Three-step synthesis of **3**. Reagents and conditions: (a) DMF-DMA, DMF, 80 °C, 24 h; (b) I$_2$, CH$_3$OH, 24 h; (c) 4-hydroxybenzene boronic acid, Na$_2$CO$_3$, Pd(OAc)$_2$, PEG 10000, CH$_3$OH, 50 °C, 3 h.

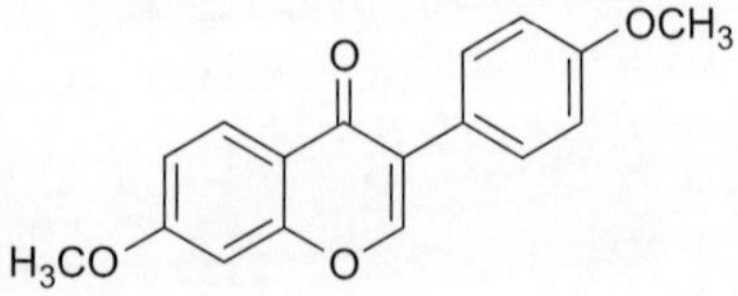

Figure 5.6 Structure of dimethyldaidzein (**4**).

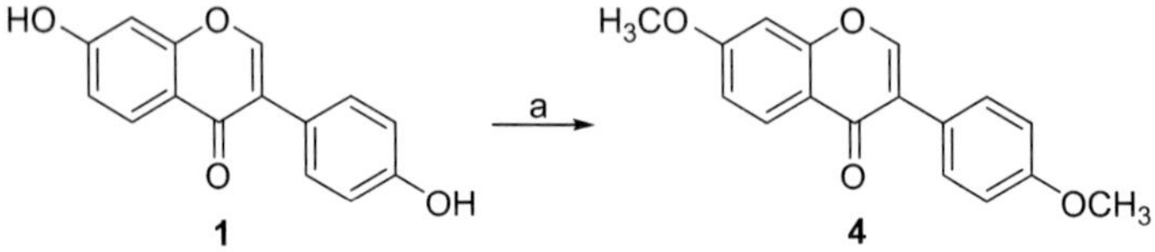

Scheme 5.12 Synthesis of **4** from daidzein (**1**). Reagents and conditions: (a) MeI, NaH, DMF, room temperature (rt), 12 h.

quantities are needed, and thus have stimulated efforts toward more cost-effective approaches. As described earlier (Section 5.2.3), the use of I_2-mediated cyclization followed by Suzuki coupling has produced **4** in an excellent yield (Biegasiewicz et al. 2010).

5.5.1 Synthesis of Dimethyldaidzein: Polymer-supported Iodobenzene Diacetate (PSIBD)-promoted Oxidative Rearrangement

To address the toxicity and cost factors associated with TTN, Kawamura et al. (2003) reported an alternative approach utilizing an inexpensive and safer hypervalent iodine reagent that promotes a similar oxidative rearrangement process as depicted in Scheme 5.13. Although monomeric species such as Koser's reagent and iodobenzene diacetate were discovered to promote the rearrangement process efficiently, a polymeric reagent (PSIBD) was also developed for practical value within combinatorial synthesis. Merits of the process include ease of product separation from reaction mixture, no release of iodobenzene, and recycling of the reagent to afford **36** in high yields, which was subsequently cyclized in basic conditions to yield dimethyldaidzein (**4**).

5.5.2 Synthesis of Dimethyldaidzein: Organolead-mediated Arylation

Since the difficulty associated with the formation of the closed pyran ring is well known, an ideal substrate would contain the preformed ring system which

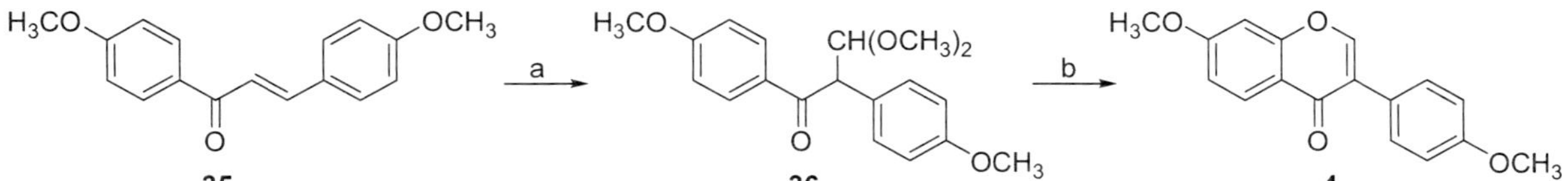

35 36 4

Scheme 5.13 Synthesis of **2** *via* PSIBD-promoted oxidative rearrangement. Reagents and conditions: PSIBD, *p*-toluenesulfonic acid (TsOH), DCM/CH_3OH, room temperature (rt), 24 h, 72%; (b) NaOH, H_2O/CH_3OH, rt, 24 h, 61%.

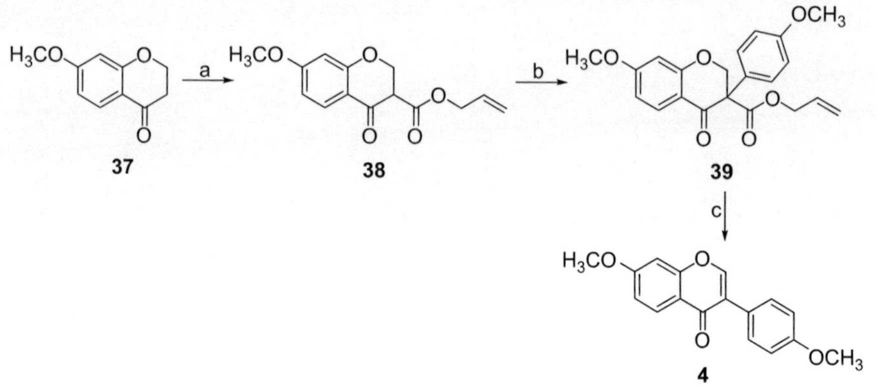

Scheme 5.14 Synthesis of **4** via organolead-mediated arylation. Reagents and conditions: LiHMDS, THF, −78 °C, 30 min; then NCCO$_2$CH$_2$CHCH$_2$, THF, 78 °C, 15 min, 72%; (b) ArPb(OAc)$_3$, CHCl$_3$, pyridine, room temperature (rt), 2 h, 87%; (c) Pd(OAc)$_2$, ethylenebis(diphenylphosphine) (DPPE), acetonitrile (ACN), reflux, 73%.

would be further elaborated to **4**. Rattigan and co-workers (Donnelly *et al.* 1993) have described an α-arylation approach of ketonic substrates, exhibiting similar scaffolds to those present isoflavanones and isoflavones, with lead(IV) in high yields, with the application towards the synthesis of **4** shown in Scheme 5.14. Preparation of **4** began with enolization of ketone **37** with lithium bis(trimethylsilyl)amide (LiHMDS), followed by addition of allyl cyanoformate to yield the 3-allyloxycarbonylchroman-4-one **38** in a 72% yield. Installation of this moiety activates the ketone towards the key lead(IV)-mediated α-arylation of **38** with a preformed ArPb(OAc)$_3$ species to afford high yields of the coupled product **39**. Cleavage of the allyl ester with catalytic Pd(OAc)$_2$ under Tsuji conditions provided dimethyldaidzein (**4**) in a rapid and concise manner.

Summary Points

- The isoflavone scaffold has been synthesized in numerous ways.
- The use of deoxybenzoin in BF$_3$·OEt$_2$-mediated Baker–Venkataraman rearrangement and activated aldehyde surrogate chemistry has furnished a number of the isoflavones.
- Thallium(III) nitrate has been exploited in the synthesis of isoflavones.
- Wacker–Cook and Grubbs' chemistry has successfully been utilized in isoflavone synthesis.
- Suzuki-mediated cross-coupling provides access to most of aforementioned isoflavone derivatives as well as a convenient means for further diversification.

Key Facts

- Daidzein (**1**) possesses an estrogen-like molecular structure which allows for alteration of target tissues that control growth, development, and function of the respective tissues.
- Formononetin (**2**) has been shown to increase allergic responses through enrichment of interleukin-4 (IL-4) production in T-cells.
- Isoformononetin (**3**) is not typically found in legume sources, but has been detected as a plant secondary metabolite in a variety of plant sources including pea hairy roots, red clover, and chickpeas.
- Dimethyldaidzein (**4**) was recently isolated from the Thai traditional herb *Butea superba*.

Definitions of Words and Terms

Benzopyran: A fused benzene and pyran ring, also known as a chromene.

Electrophile: A neutral or positively charged compound that is prone to attack by electrons (*i.e.* nucleophile).

Friedel–Crafts acylation: An electrophilic aromatic substitution between an aromatic and an acid chloride to yield a ketone.

Iminium cation: A charged imine (C=N), whether by protonation or substitution.

Isomers: Molecules that have different structural arrangements but possess identical molecular formulas.

Polyphenols: An organic molecule that possesses multiple hydroxyl groups (-OH) directly attached to an aromatic group.

Pyran: A six-membered ring containing two double bonds and one oxygen.

Ring-closing metathesis (RCM): Developed by the Nobel Laureate Robert H. Grubbs, this reaction couples two alkenes to produce a cycloalkene structure.

Soxhlet extraction: A purification technique whereby a mixture of solids are heated so that the soluble part is concentrated out in a hot liquid.

Suzuki coupling: Developed by the Nobel Laureate Akira Suzuki, this palladium(0) catalyzed reaction between an aryl or vinyl boronic acid and an aryl or vinyl halide forms a new C–C bond.

List of Abbreviations

DMF	*N,N*-dimethylformamide
DMF-DMA	*N,N*-dimethylformamide dimethylacetal
LiHMDS	lithium bis(trimethylsilyl)amide
MsCl	methanesulfonyl chloride
MTPP	methylene(triphenyl)phosphorane
MTPPB	methyltriphenylphosphine bromide

PEG	poly(ethylene glycol)
PSIBD	polymer-supported iodobenzene diacetate
RCM	ring closing metathesis
THF	tetrahydrofuran
TTN	thallium trinitrate

References

Baker, W., Chadderton, J., Harborne, J.B., and Ollis, W.D., 1953. A new synthesis of isoflavones. Part I. *Journal of the Chemical Society*. 1852–1860.

Balasubramanian, S., and Nair, M.G., 2000. An efficient "one pot" synthesis of isoflavones. *Synthetic Communications: An International Journal for Rapid Communication of Synthetic Organic Chemistry*. 30: 469–484.

Barnes, S., Kirk, M., and Coward, L., 1994. Isoflavones and their conjugates in soy foods: extraction conditions and analysis by HPLC-mass spectrometry. *Journal of Agricultural and Food Chemistry*. 42: 2466–2474.

Bass, R.J., 1976. Synthesis of chromones by cyclization of 2-hydroxyphenyl ketones with boron trifluoride-diethyl ether and methanesulphonyl chloride. *Journal of the Chemical Society, Chemical Communications*. 78–79.

Biegasiewicz, K.F., St. Denis, J.D., Carroll, V.M., and Priefer, R., 2010. An efficient synthesis of daidzein, dimethyldaidzein, and isoformononetin. *Tetrahedron Letters*. 51: 4408–4410.

Chang, Y., Muraleedharan, G.N., Santell, R.C., and Helferich, W.G., 1994. Microwave-mediated synthesis of anticarcinogenic isoflavones from soybeans. *Journal of Agricultural and Food Chemistry*. 42: 1869–1871.

Donnelly, D.M.X., Finet, J.P., and Rattigan, B.A., 1993. Organolead-mediated arylation of allyl β-ketoesters: a selective synthesis of isoflavanones and isoflavones. *Journal of the Chemical Society, Perkin Transactions 1: Organic and Bio-Organic Chemistry*. 1729–1735.

Felpin, F.-X., 2005. Practical and efficient Suzuki – Miyaura cross-coupling of 2-iodocycloenones with arylboronic acids catalyzed by recyclable Pd(0)/C. *The Journal of Organic Chemistry*. 70: 8575–8578.

Goto, H., Yoshiyasu, T., and Akai, S., 2009. Synthesis of various kinds of isoflavones, isoflavanes, and biphenyl-ketones and their 1,1-diphenyl-2-picrylhydrazyl radical-scavenging activities. *Chemical and Pharmaceutical Bulletin*. 57: 346–360.

Gowan, J.E., Lynch, M.F., O'Connor, N.S., Philbin, E.M., and Wheeler, T.S., 1958. The synthesis of isoflavones. *Journal of the Chemical Society*. 2495–2499.

Granados-Covarrubias, E.H., and Maldonado, L.A., 2009. A Wacker–Cook synthesis of isoflavones: formononetine. *Tetrahedron Letters*. 50: 1542–1545.

Hoshino, Y., Miyaura, N., and Suzuki, A., 1988. Novel synthesis of isoflavones by the palladium-catalyzed cross-coupling reaction of 3-bromochromones with arylboronic acids or their esters. *Bulletin of the Chemical Society of Japan*. 61: 3008–3010.

Ingham, J.L., Keen, N.T., Mulheirn, L.J., and Lyne, R.L., 1981. Inducibly-formed isoflavonoids from leaves of soybean. *Phytochemistry*. 20: 795–798.

Jha, H.C., Zilliken, F., and Breitmaier, E., 1981. Isoflavone synthesis with 1,3,5-triazine. *Angewandte Chemie (International Edition in English)*. 20: 102–103.

Jung, S.-H., Cho, S.-H., The, H.D., Lee, J.-H., Ju, J.-H., Kim, M.-K., Lee, S.-H., Ryu, J.-C., and Kim, Y., 2003. Structural requirement of isoflavonones for the inhibitory activity of interleukin-5. *European Journal of Medicinal Chemistry*. 38: 537–545.

Kawamura, Y., Maruyama, M., Yamashita, K., and Tsukayama, M., 2003. Environmentally benign synthesis of isoflavone derivatives using polymer-supported hypervalent iodine(III) reagent. *International Journal of Modern Physics B: Condensed Matter Physics, Statistical Physics, Applied Physics*. 17: 1482–1486.

Lapcík, O., Klejdus, B., Kokoska, L., Davidová, M., Afandi, K., Kubán, V., and Hampl, R., 2005. Identification of isoflavones in *Acca sellowiana* and two *Psidium* species (Myrtaceae). *Biochemical Systematics and Ecology*. 33: 983–992.

Leonarduzzi, G., Testa, G., Sottero, B., Gamba, P., and Poli, G., 2010. Design and development of nanovehicle-based delivery systems for preventive or therapeutic supplementation with flavonoids. *Current Medicinal Chemistry*. 17: 74–95.

Li, S.-R., Chen, P.-Y., Chen, L.-Y., Lo, Y.-F., Tsai, I.-L., and Wang, E.-C., 2009. Synthesis of haginin E, equol, daidzein, and formononetin from resorcinol via an isoflavene intermediate. *Tetrahedron Letters*. 50: 2121–2123.

Liang, J., Tian, Y.-X., Fu, L.-M., Wang, T.-H., Li, H.-J., Wang, P., Rui-Min, H., Zhang, J.-P., and Skibsted, L.H., 2008. Daidzein as an antioxidant of lipid: effects of the microenvironment in relation to chemical structure. *Journal of Agricultural and Food Chemistry*. 56: 10376–10383.

Liu, L., Zhang, Y., and Wang, Y.J., 2005. Phosphine-free palladium acetate catalyzed Suzuki reaction in water. *Journal of Organic Chemistry*. 70: 6122–6125.

Luthria, D.L., and Natarajan, S.S., 2009. Influence of sample preparation on the assay of isoflavones. *Planta Medica*. 75: 704–710.

Mahal, H.S., Rai, H.S., and Venkataraman, K., 1934. Synthetical experiments in the chromone group. Part XV. A synthesis of formononetin, daidzein, and φ-baptigenin. *Journal of the Chemical Society*. 1769–1771.

Matin, A., Gavande, N., Kim, M.S., Yang, N.X., Salam, N.K., Hanrahan, J. R., Roubin, R.H., and Hibbs, D.E., 2009. 7-Hydroxy-benzopyran-4-one derivatives: a novel pharmacophore of peroxisome proliferator-activated receptor α and γ (PPARα and γ) dual agonists. *Journal of Medicinal Chemistry*. 52: 6835–6850.

Ren, L.Z., and Meng, L.J., 2008. Suzuki coupling reactions catalyzed by poly(*N*-ethyl-4-vinylpyridinium) bromide stabilized palladium nanoparticles in aqueous solution. *Express Polymer Letters*. 2: 251–255.

Sinkevich, Y., Shchekotikhin, A., Luzikov, Y., Buyanov, V., and Kovalenko, L., 2007. Synthesis of thiopheno-quinizarine derivatives. *Chemistry of Heterocyclic Compounds*. 43: 1252–1259.

Stalikas, C.D., 2007. Extraction, separation, and detection methods for phenolic acids and flavanoids. *Journal of Separation Science*. 30: 3268–3295.

St. Denis, J.D., Gordon, IV, J.S., Carroll, V.M., and Priefer, R., 2010. Novel synthesis of the isoflavone genistein. *Synthesis*. 1590: 1592.

Toebes, A.H.W., Boer, V., Verkleu, J.A.C., Lingeman, H., and Ernst, W.H., 2005. Extraction of isoflavone malonylglucosides from *Trifolium pratense* L. *Journal of Agricultural and Food Chemistry*. 53: 4660–4666.

Tolleson, W.H., Doerge, D.R., Churchwell, M.L., Marques, M.M., and Roberts, D.W., 2002. Metabolism of biochanin A and formononetin by human liver microsomes *in vitro*. *Journal of Agricultural and Food Chemistry*. 50: 4783–4790.

Vasselin, D.A., Westwell, A.D., Matthews, C.S., Bradshaw, T.D., and Stevens, M.F.G., 2006. Structural studies on bioactive compounds. 40.1 Synthesis and biological properties of fluoro-, methoxyl-, and amino-substituted 3-phenyl-4h-1-benzopyran-4-ones and a comparison of their antitumor activities with the activities of related 2-phenylbenzothiazoles. *Journal of Medicinal Chemistry*. 49: 3973–3981.

Wessely, F., Kornfield, L., and Lechner, F., 1933. Synthesis of daidzein and of 7-hydroxy-4'-methoxyisoflavone. *Berichte der Deutschen Chemischen Gesellschaft [Abteilung] B: Abhandlungen*. 66B: 685–687.

CHAPTER 6
Non-natural Isoflavonoids

NAMITA BHAN[a] AND MATTHEOS KOFFAS*[b]

[a] Department of Chemical and Biological Engineering, Rensselaer Polytechnic Institute, 110 8th Street, Troy, NY 12180, USA; [b] Center for Biotechnology and Interdisciplinary Studies, Room 4005D, Rensselaer Polytechnic Institute, 110 8th Street, Troy, NY 12180, USA
*Email: koffam@rpi.edu

6.1 Introduction

Flavonoids are plant secondary metabolites produced in response to biotic or abiotic stresses, such as pathogenic attack, UV radiation and injury. Being secondary metabolites, they are produced in very small quantities and play important roles in plant physiology and ecology, and are also of intrinsic value as food ingredients. Approx. 6000 different varieties of these compounds have been identified, which can be divided into six categories: isoflavones, flavanones, flavones, catechins and anthocyanins based on bond saturation, hydroxylation and ring position.

6.2 Phenylpropanoid and Isoflavonoid Pathways

Flavonoids are synthesized *via* the phenylpropanoid pathway from the common precursor phenylalanine or tyrosine. Phenylalanine is converted into cinnamic acid by phenylalanine ammonia-lyase (PAL), which is converted into *p*-coumaric acid by cinnamate-4-hydroxylase (C4H). Alternatively, tyrosine is converted into *p*-coumaric acid by tyrosine ammonia-lyase (TAL) or PAL. 4-Coumaroly-CoA ligase accepts cinnamic acid, *p*-coumaric acid or caffeic acid as the substrate to form the acid–CoA complex. Furthermore, three acetyl groups from malonly-CoA are added to the acid–CoA complex by chalcone synthase (CHS) to from chalcones, which act as the direct precursors for biosynthesis of flavonoids. Flavonoids are the substrates for isoflavone synthase (IFS), flavanol-3-hydroxylase, and flavone synthase I and flavone synthase II. The compounds we are going to discuss in this Chapter are the isoflavonoids, which are synthesized by IFS. IFS performs the novel aryl ring transfer of the

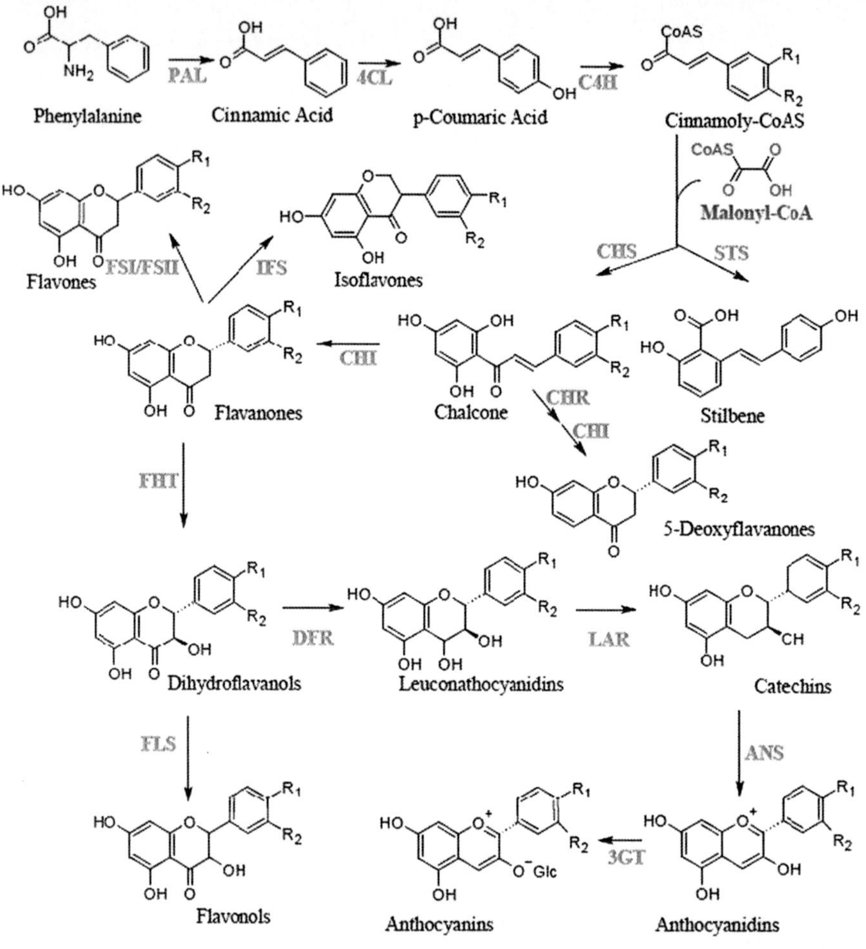

Figure 6.1 The phenylpropanoid pathway and the isoflavonoid pathway. Phenylpropanoid biosynthetic pathway. All the enzymes are shown in green. Malonly-CoA (purple) is the limiting precursor for the over-expression of the pathway in a heterologous host. The abbreviations are as follows: phenylalanine ammonia-lyase (PAL), 4-coumaroyl-CoA (4CL), coumarate-4-hydroxylase (C4H), stilbene synthase (STS), chalcone synthase (CHS), chalcone reductase (CHR), chalcone isomerase (CHI), isoflavone synthase (IFS), flavone synthase I or flavone synthase II (FSI/FSII), flavonol 3b-hydroxylase (FHT), dihydroflavonol reductase (DFR), leucocyanidin reductase (LAR), anthocyanin synthase (ANS), 3-glucosyltransferase (3GT).

aromatic B-ring from position C-2 in flavanones to C-3 and hydroxylation in position C-2 to form 2-hydroxyisoflavanone. Furthermore, 2-hydroxyisoflavanone dehydratase (HID) catalyzes the dehydration of 2-hydroxyisoflavanones; this reaction might also occur spontaneously to generate isoflavones, *e.g.*, daidzein and genistein. Isoflavonoid derivatives, such as medicarpin, pisatin, and maackiain, are generated through the legume-specific branch pathways and isoflavonoid-specific enzymes (Figure 6.1).

6.3 Isoflavanoids

Plant secondary metabolites, isoflavonoids, possess valuable health-promoting activities, in relation to heart diseases and cancer, and some display striking affinity for steroid receptors. Leguminous plants are the primary sources of isoflavonoids, where they are present usually in variable but small quantities. They play significant roles as antimicrobial phytoalexins (Dixon 1999). Induction of *Rhizobium* nodulation genes is also attributed to their activities (Dixon 1999).

Similar to other phenylpropanoids, isoflavonoids have been associated with an array of biological activities against human diseases, specifically estrogenic, anti-angiogenic, antioxidant and anticancer effects (Dixon 1999; Dixon and Sumner 2003). A high soy diet leads to a lower risk of cancer; as found out by epidemiological studies indicating the anti-cancer effects of dietary isoflavones (Banerjee 2008; Sarkar and Li 2004). Genistein inhibits growth of various cancer cells through modulation of genes that are involved in the control of cell cycle, apoptosis and cell signaling pathways. It has also been reported to reduce angiogenesis, metastasis and oxidative stress (Sarkar and Li 2004) and have a high affinity for human estrogen receptors (ERs).

6.4 Non-natural Isoflavonoids

Non-natural isoflavonoids have been created recently *via* various processes, a combination of chemical and biological synthesis, and also by the application of protein engineering to the uridine diphosphate glycosyltransferase (UGT). The high affinity of genistein for human estrogen receptors α and β (hERα and hERβ) (Zhao and Brinton 2005) has been used to identify non-natural isoflavonoids that can potentially be used to promote the benefits of ER modification therapy (Hollmer 2006; Osborne *et al.* 2000).

6.4.1 Semi-synthesis of Non-natural Isoflavonoids

Non-natural isoflavonoids have been created *via* a combined metabolic engineering and chemical synthesis approach (Chemler *et al.* 2010). IFS is the key enzyme for the biosynthesis of isoflavanones. The entire isoflavonoid pathway was cloned into the yeast *Saccharomyces cerevisiae*. IFS from *Glycine max* (soy bean) and *Tirfolium pratense* (red clover) was expressed along with the cytochrome P450 reductase (CPR) and HID genes. Different combinations of IFS with only CPR, only HID or with both CPR and HID were attempted:

T. pratense IFS, *G. max* HID, *G. max* CPR; *T. pratense* IFS, *G. max* HID, *Catharanthus roseus* CPR; *G. max* IFS, *G. max* HID, *G. max* CPR; *G. max* IFS, *G. max* HID, *C. roseus* CPR all gave approx. 35 mg L^{-1} final production, although they had different rates of production. The three enzyme combinations always showed higher production as compared with two enzymes (IFAs and CPR or IFS and HID) or only IFS expression (Figure 6.2).

Five different IFS enzymes [*G. max*, *T. pratense*, *Glycyrrhiza echinata* (licorice), *Pisum sativum* (pea) and *Medicago truncatula* (alfalfa)] were fed with dihydroxyl-, dimethoxy-, chlorine-, fluorine- and bromine-substituted flavanones. Some of these substrates were chemically synthesized. All the different enzymes showed similar promiscuity to novel substrates. This indicates that the enzymes are evolutionarily very similar, and altering substrate specificity by modifying enzyme structure could be a difficult task. However, certain criteria were important for accepting the flavanones as substrates by all IFSs, for example, a C7 position hydroxyl group is necessary to bind to the enzyme and a C5 hydroxyl group is expendable. Substitutions at C2' or C6' were not tolerated, whereas C3' and/or C5 substitutions by small groups were accepted. Finally, an hydroxyl group at C4' is essential for ring migration, forming 2-hydroxyisoflavanones.

Novel isoflavonoids obtained showed that a hydroxyl group at the C5 position enhances the binding affinity for both ERα and ERβ. Four-fold lower IC_{50} values were observed for ERβ in the case of isoflavanones with a hydroxyl group at C5, whereas the IC_{50} values were reduced a few fold for ERα; in comparison with isoflavonoids without the hydroxyl group at C5 position. The IC_{50} values of some of these compounds are shown in Table 6.1. These values indicate that the new novel isoflavonoids should have small substituents at the C3' position giving improved interaction with ERs. A variety of analogues can be created also by modifying the hydroxyl groups, many of the natural isoflavonoids are glycosylated or methylated. Altering the C7 or C4' hydroxyl reduces the affinity to both ERα and ERβ and increased IC_{50} values by a factor of 10 with respect to genistein (Chemler *et al.* 2010).

6.4.2 Chemical Synthesis of Non-natural Flavanones

Flavanones were formed by carrying out Claisen–Shmidt condensation of the appropriate acetophenone and benzaldehyde. The first step involves utilization of methylchloromethyl ether to selectively bis-protect the acetophenone. Then the methylchloromethyl ether is used for methoxymethyl ether (MOM) protection of the benzaldehyde. The intermediates are then reacted *via* the Claisen–Shmidt condensation reaction. The final step involves chalcone isomerization and MOM group removal (Urgaonkar *et al.* 2005).

6.4.3 ER Assay

Competitive binding of the isoflavonoid to ERα or ERβ is used as the basis of calculating the affinity of the compound to the ER. Terbium labeled anti-glutathione S-transferase (GST) antibody binds to the ER at the GST tag,

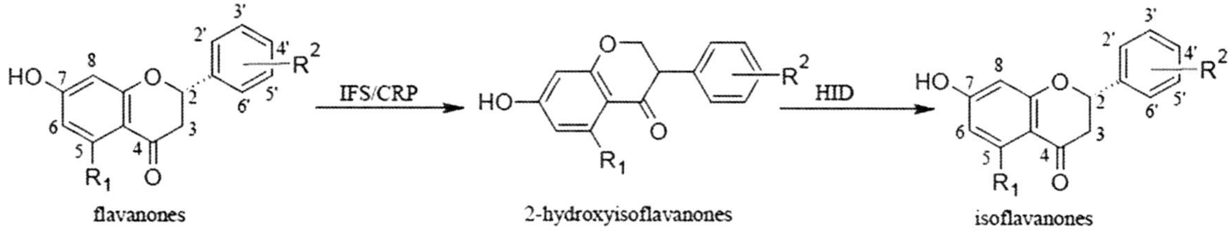

Figure 6.2 Reactions catalyzed by IFS, CRP and HID enzymes. Isoflavone synthase, IFS; 2-hydroxyisoflavanone dehydratase, HID; cytochrome P450 reductase, CPR.

Table 6.1 Non-natural isoflavonoids and their estrogenic activities. Summary of some of the non-natural isoflavonoids and their IC_{50} values against ERα and ERβ. NA, not available.

Isoflavonoid	R_1 5	7	R_2 2'	3'	4'	5'	ERα IC_{50} (nM)	ERβ IC_{50} (nM)	ERα/ERβ
Estradiol							0.1 ± 0.0	0.2 ± 0.1	0.4
Genistein	OH	OH	H	H	OH	H	34.8 ± 12.0	5.4 ± 0.6	6.4
Daidzein	H	OH	H	H	OH	H	485.4 ± 145.4	26.3 ± 6.7	18.5
Orobol	OH	OH	H	OH	OH	H	70.8 ± 10.5	16.5 ± 5.6	4.3
3',4',7-Trihydroxyisoflavone	H	OH	H	OH	OH	H	>1000	187 ± 96.4	NA
3'-Methoxy-4',5,7-rihydroxyisoflavone	OH	OH	H	OCH$_3$	OH	H	880.3 ± 287.0	449.8 ± 74.0	2.0
4',7-Dihydroxy-3'-methoxyisoflavone	H	OH	H	OCH$_3$	OH	H	>1000	706.4 ± 153	NA
3'-Ethoxy-4',5,7-trihydroxyflavanone	OH	OH	H	OCH$_2$CH$_3$	OH	H	>1000	>1000	NA
4',7-Dihydroxy-3'-ethoxyflavanone	H	OH	H	OCH$_2$CH$_3$	OH	H	>1000	>1000	NA
3'-Methyl-4',5,7-trihydroxyisoflavone	OH	OH	H	CH$_3$	OH	H	69.3 ± 12.4	30.8 ± 9.5	2.2
4',7-Dihydroxy-3'-methylisoflavone	H	OH	H	CH$_3$	OH	H	540.9 ± 159.5	70.8 ± 20.9	7.6
3',5'-Dimethyl-4',5,7-trihydroxyisoflavone	OH	OH	H	CH$_3$	OH	H	>1000	155.4 ± 55.5	NA
4',7-Dihydroxy-3',5'-dimethylisoflavone	H	OH	H	CH$_3$	OH	CH$_3$	>1000	>1000	NA
3'-Chloro-4',5,7-trihydroxyisoflavone	OH	OH	H	Cl	OH	H	101.4 ± 32.3	39.5 ± 6.0	2.6
3'-Chloro-4',7-dihydroxyisoflavone	H	OH	H	Cl	OH	H	104.4 ± 41.7	208.7 ± 49.8	0.5
3'-Bromo-4',5,7-trihydroxyisoflavone	OH	OH	H	Br	OH	H	28.4 ± 10.5	17.1 ± 4.2	1.7
3'-Bromo-4',7-dihydroxyisoflavone	OH	OH	H	Br	OH	H	>1000	841.5 ± 183.7	NA
Genistin (genistein-7-O-glucoside)	–	–	–	–	–	–	>1000	>1000	NA
Prunetin (4',5-dihydroxy-7-methoxyisoflavone)	–	–	–	–	–	–	413.3 ± 91.5	299.4 ± 51.7	1.4
Biochantin A (5,7-dihydroxy-4'-methoxyisoflavone)	–	–	–	–	–	–	465.4 ± 115.2	157.1 ± 24.1	3.0
Naringenin	–	–	–	–	–	–	328.8 ± 127.2	186.4 ± 32.6	1.8

thus labeling it. In the absence of affinity of the compound to the ER, tracer, a small fluorescent molecule is bound to the ER, bringing it into close proximity of the terbium, leading to a fluorescent signal as a result of Föster resonance energy transfer (FRET). Replacement of the tracer by the isoflavonoid leads to the loss of fluorescence signal between the labeled, bound antibody and the tracer. FRET is based on the principal of transfer of energy from one fluorophore (donor) upon excitation to another fluorophore (acceptor) in close proximity to it, leading to an increase in fluorescence of the acceptor, and decrease in fluorescence of the donor.

6.4.4 Protein Engineering of Uridine Diphosphate Glycoslytransferase (UGT)

Protein engineering of glycosyltransferases can be a very interesting way of altering the activity of the isoflavonoids. Glycosylation of isoflavonoids adds various sugars to the compounds enhancing their stability, solubility, and facilitating their storage in and accumulation in plant cells. Uracil diphosphate glycosyltransferases (UGTs), members of the family 1 glycosyltransferases are responsible for the glycosylation of isoflavonoids (Wang 2009). Several isoflavonoid-specific UGTs have been identified. Detailed structural studies of these have led to specific amino acid changes resulting in novel glycosylated isoflavonoid compounds (Wang 2011). For example, manipulation around the acceptor binding pocket regions affects the UGT enzyme activity and alters the acceptor substrate specificity and product regio-selectivity; UGT71G1 is able to glycosylate the A-ring 7-hydroxyl of the isoflavone genistein. Its mutant Y202A converted genistein to two products, the 7-O-glucoside and 5-O-glucoside (He et al. 2006).

6.5 Structural and Functional Studies of IFS

Previously attempts have been made to alter the specificity, and catalytic activity of the enzyme IFS by site-directed mutagenesis. Since the crystal structure of IFS has not been elucidated yet, homology modeling was used to identify catalytically important residues for the *G. enchinata* IFS based on the structure of P450BM3 (Sawada et al. 2002). Serine 310 (Ser 310) and Lysine 375 (Lys 375) were identified to play an important role in the aryl migration from C2 to C3 position. 2-hydroxyflavonol forming activity was lost upon altering Lys 375. Further, this residue was deduced to interact with the C7 hydroxyl group on flavanones substrate, and that it is indispensable for the aryl ring migration. Leucine 371 (Leu 371) has been identified as important for the catalytic activity and the protein stability (Sawada and Ayabe 2005). With regard to the studies conducted by Chemler et al. (2010) the C4' hydroxyl group is necessary for formation of isoflavones, also absence of B ring substituents results in just dihydroflavonols. On the basis of these results it was suggested that the Lys 375 binds to the C4' hydroxyl group and that is how it is responsible for the aryl ring migration of the B ring. Also they suggested that the C7 hydroxyl group binds to the Ser 310 position, since in absence of it no product is formed, and the Ser 310 CYP98C2 mutant lowered the amount of 2-hydroxyisoflavanone formed (Sawada et al. 2002).

Summary Points

- Isoflavonoids are produced as secondary metabolites *via* the phenylpropanoid and the isoflavonoid pathways in mostly leguminous plants.
- Isoflavonoids show several pharmaceutically important properties such as cardioprotective and anti-cancer effects and a high affinity for estrogen receptors.
- Non-natural isoflavonoids have been produced *via* semi-synthesis.
- A screening assay based on estrogen affinity led to the identification of non-natural isoflavonoids with a higher efficiency as compared with natural isoflavonoids
- Protein engineering of certain glycosyltranferases has also been employed to form non-natural isoflavonoids.

Key Facts

Key facts for uridine diposphate glucosyltranferases (UGTs)

- UGTs are responsible for the addition of a glycosyl group to a small hydrophobic molecule.
- They are a superfamily of enzymes found in both eukaryotes and prokaryotes.
- Glycosylation of a compound can alter its activity and solubility, and so lead to diversification of the secondary metabolites.
- Protein engineering of several UGTs has been attempted to alter their functions.

Key Facts for Estrogen Receptor (ER) Modification Therapy

- This is mainly employed for treatment of breast cancer.
- Tamoxifen has been used as an anti-estrogen compound to treat breast cancer for over 30 years.
- Tamoxifen has been associated with several side effects, such as the increased risk of uterine and other related cancers.
- Selective estrogen receptor modulators (SERMs) have found new success as a replacement for tamoxifen as they do not show these side effects.
- SERMs can be developed with either estrogenic activity or anti-estrogenic activities.
- SERMs can be used in the treatment of various illnesses, such as osteoporosis, and as hormone replacement therapy.

Definitions of Words and Terms

Angiogenic effects: Angiogenesis involves the growth of new blood vessels from already existing ones. It is a common phenomenon that is important for wound healing, but is also the step at which a tumor can turn cancerous.

Antioxidants: Any species that can quench a radical ion is known as an antioxidant. These are pharmaceutically important compounds as radical ions can lead to oxidative stress in the cells causing damage.

Estrogen receptors (ERs): Receptors are compounds that bind to the cell surface leading to a signal. Estrogen receptors are activated by the hormone estrogen.

Föster resonance energy transfer (FRET): When a fluorophore is excited by light within its excitation range, it absorbs energy and releases it in the form of fluorescence at a different wavelength from the excitation wavelength. If another flurophore (acceptor) which has the excitation wavelength within the range of the emission wavelength of the first one (donor) is brought into close proximity with the donor, it gets excited and emits light at its emission wavelength. This transfer of energy, leading to a shift in the emission wavelength due to proximity of the two fluorophores, is known as FRET.

Methoxymethyl ether (MOM) protection of hydroxyl groups: MOM is basically reacted with the hydroxyl (OH) containing R group (R-OH), so that it replaces the hydroxyl group (R-MOM). Then the modifications to be carried out on the R group are carried out (R'-MOM), and finally the MOM group is removed and replaced by the OH (R'-OH).

Phenylpropanoic compounds: These are compounds composed of a phenolic ring and a propene chain, and are all formed from the amino acid phenylalanine or sometimes tyrosine.

Phytoalexins: These are antimicrobials produced by plants in response to pathogenic attacks. They act as toxins and are usually small molecule secondary metabolites.

Protein Engineering: It involves the identification of catalytically or structurally important amino acid residues in a protein, and then carrying out site-directed mutagenesis to alter the chemistry or solubility of the enzyme.

Secondary metabolites: These are compounds that are produced by secondary metabolic pathways in very small quantities for specific functions in an organism.

Semi-synthesis: This involves the use of both chemical synthesis and biocatalysis for the production of a compound. For example, part of the process utilizes chemical synthesis, such as providing the substrate for the biosynthetic pathway.

List of Abbreviations

ANS	anthocyanin synthase
C4H	coumarate-4-hydroxylase
CHI	chalcone Isomerase

CHR	chalcone reductase
CHS	chalcone synthase
4CL	4-coumaroyl-CoA
CPR	cytochrome P450 reductase
DFR	dihydroflavonol reductase
ER	estrogen receptor
FHT	flavonol 3b-hydroxylase
FRET	Föster resonance energy transfer
FSI/FSII	flavone synthase I or flavone synthase II
3GT	3-glucosyltransferase
hERα	human estrogen receptor α
hERβ	human estrogen receptor β
HID	2-hydroxyisoflavanone dehydratase
IC_{50}	inhibitory concentration at which 50% inhibition is observed
IFS	isoflavone synthase
LAR	leucocyanidin reductase
PAL	phenylalanine ammonia-lyase
SERM	selective estrogen replacement modulator
STS	stilbene synthase
UGT	uridine diphosphate glycosyltransferase
UV rays	ultraviolet rays

References

Banerjee, S., Li, Y., Wang, Z., and Sarkar, F.H., 2008. Multi-targeted therapy of cancer by genistein. *Cancer Letters*. 269: 226–242.

Chemler, J.A., Lim, C.G., Daiss, J.L., and Koffas, M.A., 2010. A versatile microbial system for biosynthesis of novel polyphenols with altered estrogen receptor binding activity. *Chemical Biology*. 17: 392–401.

Dixon, R., 1999. Isoflavonoids: biochemistry, molecular biology and biological functions. In Sankawa, U. (ed.) *Comprehensive Natural Products Chemistry* Pergamon: Oxford, UK.

Dixon, R., and Sumner, L.W., 2003. Legume natural products: understanding and manipulating complex pathways for human and animal health. *Plant Physiology*. 131: 878–885.

He, X.Z., Wang, X., and Dixon, R.A., 2006. Mutational analysis of the Medicago glycosyltransferase UGT71G1 reveals residues that control regioselectivity for (iso)flavonoid glycosylation. *Journal of Biological Chemistry*. 281: 34441–34447.

Hollmer, M., 2006. Biotech takes aim at large anti-estrogen markets. *Nature Biotechnology*. 24: 1455–1456.

Osborne, C.K., Zhao, H., and Fuqua, S.A., 2000. Selective estrogen receptor modulators: structure, function, and clinical use. *Journal of Clinical Oncology*. 18: 3172–3186.

Sarkar, F.H., and Li, Y., 2004. The role of isoflavones in cancer chemoprevention. *Frontiers in Bioscience*. 9: 2714–2724.

Sawada, Y., and Ayabe, S., 2005. Multiple mutagenesis of P450 isoflavonoid synthase reveals a key active-site residue. *Biochemical Biophysical Research Communications*. 330: 907–913.

Sawada, Y., Kinoshita, K., Akashi, T., Aoki, T., and Ayabe, S., 2002. Key amino acid residues required for aryl migration catalysed by the cytochrome P450 2-hydroxyisoflavanone synthase. *Plant Journal*. 31: 555–564.

Urgaonkar, S., La Pierre, H.S., Meir, I., Lund, H., Chaudhuri, D.R., and Shaw, J.T., 2005. Synthesis of antimicrobial natural products targeting FtsZ: (+/−)-dichamanetin and (+/−)-2″hydroxy-5″benzylisouvarinol-B. *Organic Letters*. 7: 5609–5612.

Wang, X., 2009. Structure, mechanism and engineering of plant natural product glycosyltransferases. *FEBS Letters*. 538: 3303–3309.

Wang, X., 2011. Structure, function, and engineering of enzymes in isoflavonoid biosynthesis. *Funct Integr Genomics*. 11: 13–22.

Zhao, L., and Brinton, R.D., 2005. Structure-based virtual screening for plantbased ER beta-selective ligands as potential preventative therapy against age-related neurodegenerative diseases. *Journal of Medicinal Chemistry*. 48: 3463–3466.

CHAPTER 7

The Structure of Isoflavones by 1D and 2D Homonuclear and Heteronuclear NMR Spectroscopy

KRISTIINA WÄHÄLÄ,* SOMDATTA DEB AND TAPIO HASE

Department of Chemistry, Laboratory of Organic Chemistry, University of Helsinki, P. O. Box 55, FIN-00014, Finland
*Email: kristiina.wahala@helsinki.fi

7.1 Introduction

In previous times, detailed structural work on newly isolated natural products often relied on extensive degradation procedures and painstaking chemical correlation with compounds whose structures were already known. However, towards the end of the previous century, spectroscopy, particularly nuclear magnetic resonance (NMR), has completely changed the situation as regards the expediency and reliability of such structural work. In addition to one-dimensional (1D) ^{1}H and 1D ^{13}C, a number of two-dimensional (2D) NMR techniques are now routinely applied for the structure determination of natural products. Whereas 1D NMR provides plausible identities of specific atoms, 2D NMR spectra yield valuable information on atom-to-atom connectivity (through bond or through space). Recent examples of the application of various NMR techniques [correlation spectroscopy (COSY), distortionless enhancement by polarization transfer (DEPT), double quantum filtered COSY (DQFCOSY), heteronuclear multiple bond correlation (HMBC), heteronuclear single quantum correlation (HSQC), heteronuclear multiple quantum correlation (HMQC), nuclear Overhauser enhancement spectroscopy (NOESY) and rotating frame Overhauser effect spectroscopy (ROESY)] in the structure elucidation and study of equilibria of isoflavonoids are discussed below.

Most of the compounds of interest in this connection are derivatives of either isoflavone (**1**), such as daidzein (**3**) and genistein (**4**), or of the ring C reduced isoflavan (**2**) such as equol (**5**) (Figure 7.1). Daidzein and genistein are present in soy and other edible beans of the Leguminosae family, whereas equol is the principal isoflavone metabolite in humans (Heinonen *et al.* 2002; 2003).

The 2D NMR spectra are exemplified in Figures 7.2 and 7.3. The former shows the HMBC NMR spectrum of genistein (**4**) (horizontal) and the corresponding ^{13}C spectrum (vertical). Figure 7.3 presents similarly ^1H and ^{13}C NMR HSQC spectra of equol. In simple terms, the HSQC technique gives the ability to identify, in a molecule, the *one-bond* correlations between a carbon atom and an attached H atom, whereas HMBC correlates H nuclei with C nuclei that are separated by *more than one bond* from the H (also across non-protonated carbon or oxygen bridges). Thus, as seen in Figure 7.2, in the

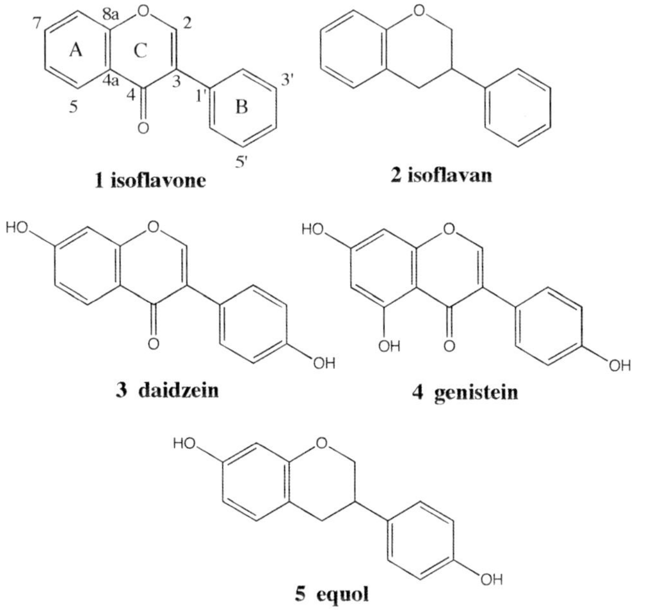

Figure 7.1 Basic isoflavonoid structures (**1**, **2**) and examples of the most common isoflavonoids daidzein (**3**), genistein (**4**), and equol (**5**). Numbering of the ring system, used for all isoflavonoids, is shown for **1**. Actual naturally occurring compounds are derivatives of **1** or **2**, typically carrying hydroxy (OH) or alkoxy (OR) groups at C-5, C-7 and/or C-4′.

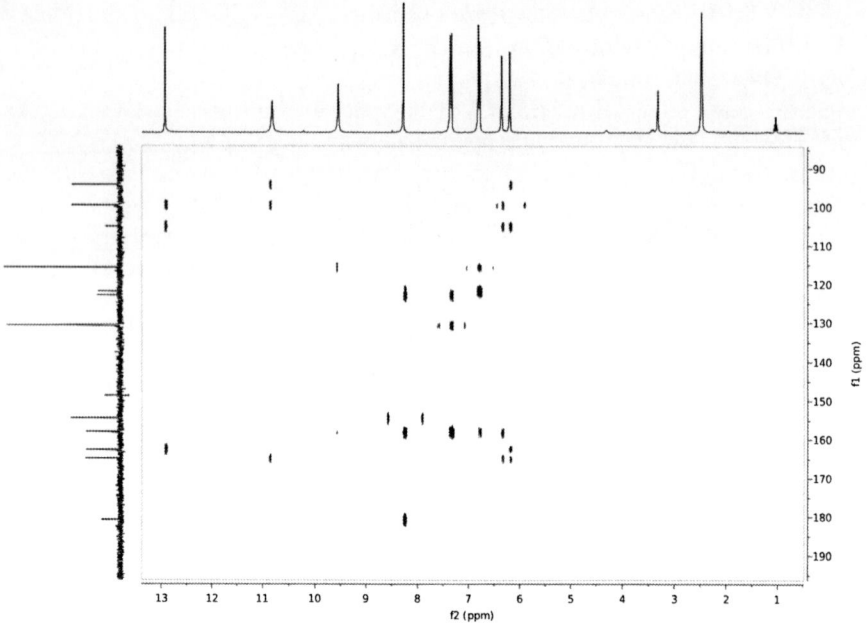

Figure 7.2 HMBC spectra of genistein (**4**) (horizontal, ^{1}H NMR spectrum, showing signals from eight unlike H atoms; vertical, ^{13}C spectrum showing signals from 13 unlike C atoms). For the ^{1}H NMR spectrum, the δ scale is given at the bottom; the ^{13}C spectrum is shown vertically at left with the δ scale at right. The H–H couplings between nearby hydrogen atoms cause the splitting of signals in the ^{1}H spectrum such as the doublets seen at ca. δ 6.8 and 7.3. To identify which proton(s) are attached to which carbon atom(s) over two or more bonds, one goes, from each "smudge" in the rectangular central area, directly upwards to locate a H signal, and to the left to locate the corresponding carbon atom(s).

HMBC spectrum each H coincides with two or more C atoms, whereas in the HSQC spectrum (Figure 7.3) each H is shown coinciding with just its "own" C atom.

7.2 Structure Elucidation: Establishing Spectral Correlations

The structure of daidzein has been resolved by NMR (Rasku *et al*. 1999). We found that previously published work (Chang *et al*. 1994; Jha *et al*. 1980) was not in agreement in assigning the 3- and 1′-C atoms of daidzein (**3**). Using HMBC we ascertained that C-1′ appears at δ 122.5 and C-3 at 123.4, based on the observed couplings between C-1′ and H-3′(5′), and between C-3 and H-2′(6′).

Seventeen isoflavones were obtained and identified from the leaves of *Ateleia herbert-smithii*, six of which were reported for the first time (Veitch *et al*. 2003) (Figure 7.4). For unambiguous assignments of NMR signals and structure

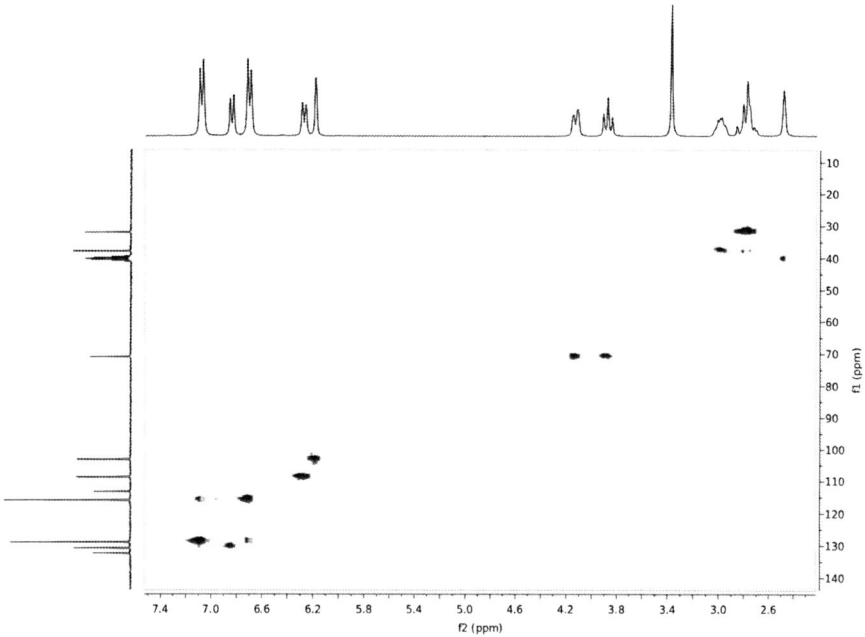

Figure 7.3 HSQC spectra of equol (**5**) (horizontal, ^{1}H NMR spectrum; vertical, ^{13}C spectrum). Connecting a "smudge" with a H signal (up) and a C signal (to the left) identifies a C–H moiety. The peaks at δ 2.5, 3.4 and 6.2 are due to solvent (d$_6$-DMSO), residual water and equol OH groups, respectively.

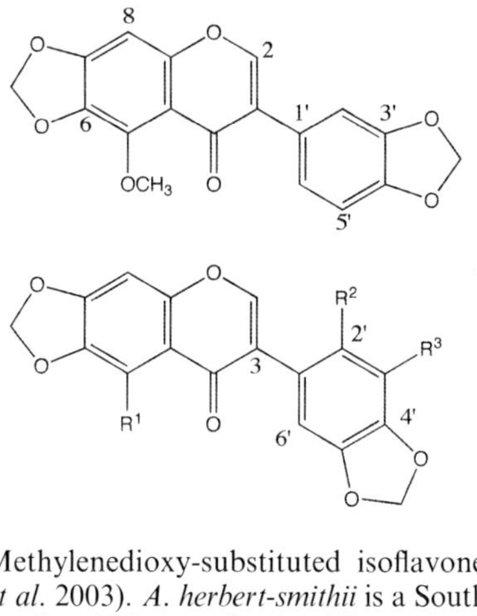

7 $R^1 = R^2 = H; R^3 = OCH_3$
8 $R^1 = R^2 = OCH_3; R^3 = H$
9 $R^1 = R^3 = OCH_3; R^2 = H$

Figure 7.4 Methylenedioxy-substituted isoflavones from *A. herbert-smithii* (Veitch *et al.* 2003). *A. herbert-smithii* is a South American tree of the Leguminosae family which contains 17 different isoflavones, **6–9** among them, obtained by ether extraction of the leaves. Further components are shown in Figure 7.5.

Table 7.1 HSQC and HMBC correlations of protons in compound **6**.

	δ	HSQC (ppm)	HMBC
H-2	7.76	150.4	C-4, C-9, C-1'
H-8	6.63	93.3	C-9, C-6, C7, C-10
OCH$_2$-6,7	6.06	102.2	C-6, C-7
OCH$_2$-3'4'	5.97	101.1	C-3', C-4'
OCH$_3$-5'	4.08	61.3	C-5

elucidation, ^{1}H, ^{13}C, DEPT, DQFCOSY, HSQC, HMBC, NOESY and ROESY NMR measurements were carried out. In compound **6** three multiplets in the aromatic region are assigned for the aromatic protons in ring B. The two singlets at δ 7.76 and 6.63 are due to H-2 and H-8, respectively. By means of HSQC (see Table 7.1), the proton at δ 7.76, attached to the non-aromatic C at 150.4, is assigned as H-2. The other singlet is therefore H-8, as was further confirmed by HMBC. Table 7.1 shows the HSQC and HMBC correlations. In HMBC, the proton at δ 7.76 shows correlation with C-1'. Therefore this proton must be H-2, H-8 being too far from C-1' to show any correlation. Likewise, H-2 is too distant to show any correlation with C-6 and C-7, so δ 6.63 corresponds to H-8. In a similar manner the methylenedioxy protons were assigned from HMBC.

Compound **7** has the same substituents as **6** but at different locations. The sites of the methylenedioxy groups were determined by 1D ^{1}H, 1D ^{13}C and HSQC, whereas those of the methoxy groups were assigned with the assistance of site-selective ROESY. Site-selective excitation of OCH$_3$ at δ 3.94 gave correlations only with H-2', therefore OCH$_3$ must be located *ortho* to H-2' and there is no other proton *ortho* to OCH$_3$.

Compound **8** is similar to **6**. The aromatic protons in ring B are *para* related and the OCH$_3$ group in ring B shows 1D nuclear Overhauser enhancement (NOE) with one of them. Correlations of H-6' with C-2' and C-3 in HMBC further confirm the structure. Similarly in **9**, in which the ring A is as in **8**, and in ring B where the aromatic protons are in a *meta* relationship, NOE correlation of H-2' with OCH$_3$ and HMBC with C-3, C-3' and C-4 establishes the assignment of H-2'. H-6' was assigned by HMBC correlation with C-3, C-4' and C-5'.

In compound **10**, the aromatic protons and H-2 were assigned by their coupling pattern and HSQC (Figure 7.5). The OCH$_3$ groups were assigned by 1D ROESY measurements. The rotating frame Overhauser effect (ROE) connectivities of each OCH$_3$ group with neighbouring protons (Table 7.2) are useful in working out the position of OCH$_3$ groups.

In compound **11**, OCH$_3$-7, OCH$_3$-3' and OCH$_3$-4' were assigned by NOE correlations between the OCH$_3$ group and neighbouring proton. OCH$_3$-5 and OCH$_3$-6 were assigned by HMBC. Structure elucidation of **6–10** was performed similarly (Veitch *et al.* 2003).

The 3-arylcoumarin derivatives **12a** and **12b** were isolated from *Glycyrrhiza glabra* root. The ^{1}H spectrum of the former shows a chromene ring [δ 1.47 (6H, *s*),

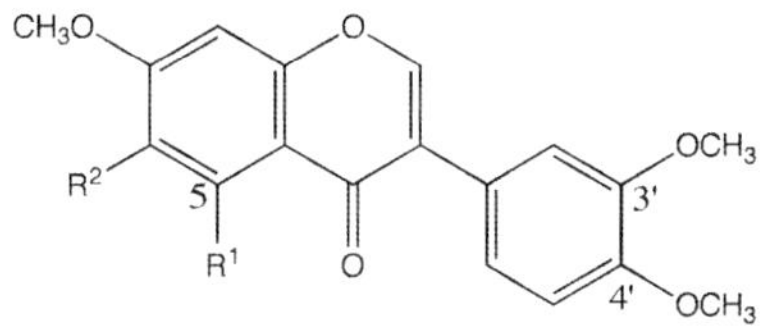

10 $R^1 = OH; R^2 = H$
11 $R^1 = R^2 = OCH_3$

Figure 7.5 Polymethoxylated isoflavones from *A. herbert-smithii* (Veitch *et al.* 2003). The positions of the methoxy (OCH_3) groups were worked out by first assigning the aromatic ring protons followed by employing ROESY to determine which aromatic protons carried an adjacent methoxy group.

Table 7.2 ROE connectivities of protons with OCH_3 groups in compound **10**.

	OCH_3-7, δ 3.87	OCH_3-4′, δ 3.91	OCH_3-3′, δ 3.92
ROE connectivity	H-6, H-8	H-5	H-2′

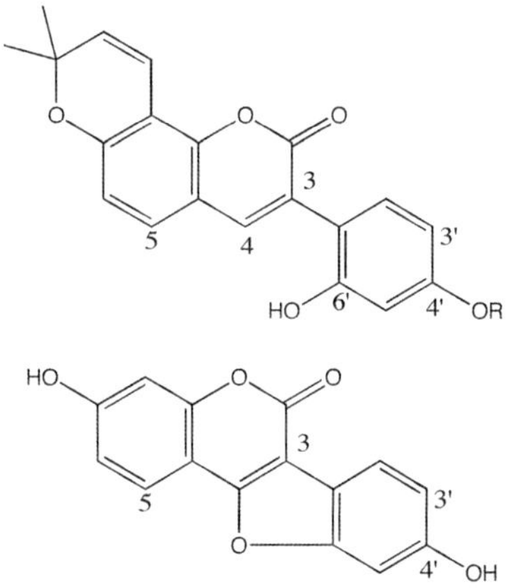

12a R = H
12b $R = CH_3$

12c

Figure 7.6 3-Arylcoumarins (**12a**, **12b**) from *G. glabra* (Kinoshita *et al.* 1997) and coumestrol (**12c**) (Rasku *et al.* 1999). The licorice-producing *G. glabra* roots are a rich source of aromatic compounds including isoflavans, flavanones, chalcones, *etc.*, and the 3-arylcoumarins **12a–12c**. These compounds possess a range of pharmaceutical and antioxidant properties. Examples of isoflavones from the same source are shown in Figure 7.7.

5.89 (1H, *d*, $J = 10.3$ Hz), 6.83 (1H, *d*, $J = 10.3$ Hz). An AB-quartet (δ 6.76 and 7.44, $J = 8.4$ Hz) and an ABX system (δ 6.49, *d*, $J = 2.2$ Hz; δ6.44, *dd*, $J = 2.2$ Hz and 8.4 Hz; δ 7.20, *d*, $J = 8.4$ Hz)] indicate that two aromatic system are present. (Figure 7.6).

The location of the dimethylpyrane ring was established by NOESY. Additionally, NOE between H-4 (δ 7.86, s) and H-5 (δ 7.44, d, J = 8.4 Hz) suggests that the AB system is present in ring A and the ABX system in ring B, and that the dimethylpyrane ring is fused to ring A. In compound **12b**, a monomethyl ether of **12a**, the location of the methoxy group was determined by long range ^{13}C-^{1}H COSY spectral analysis. In confirmation, the structures **12a** and **12b** were prepared by chemical synthesis (Kinoshita *et al.* 1997).

However, if the dimethylpyrane ring were fused to ring B at 3'-4' and the hydroxyls were at C-2' and C-7 (see structure **13**), an AB quartet and ABX system would also be present. Furthermore, yet another alternative having the AB quartet and ABX system aromatic substitution pattern (structure **14**) exists. NOESY alone is inadequate for unequivocal structure elucidation in these cases, as NOESY only shows interaction through space (not through bonds), and such structures are better elucidated using HMBC spectra (Figure 7.7). In fact, **13** and **14** are also naturally occurring *Glycyrrhiza* isoflavones The structure of isoglabrone (**14**) was unequivocally elucidated by HMBC.

^{1}H and ^{13}C signals of **13** were assigned by ^{1}H-^{1}H and ^{1}H-^{13}C COSY. Spectral data [δ 1.46 (6H, s), 6.81 (1H, d, J = 10.2 Hz), 5.94 (1H, d, J = 10.2 Hz)], similar to that of glabrone (**14**) and parvisoflavone (**15**), indicated the presence of a dimethylpyrane ring. In HMBC correlations of H-5″ with C-8, and H-4″ with C-7 and C-9, confirmed that the dimethylpyrane ring is fused with ring A. ^{1}H NMR showed a pair of *ortho*-coupled aromatic protons, and an ABX aromatic proton system is present. Since the dimethylpyrane ring is present in ring A, the *ortho*-coupled aromatic protons must be in ring A. HMBC correlations of H-5 with C-7 and C-4 confirm the structure of **13** (Baba *et al.* 2008). These results served to correct certain previously assigned structures of *G. glabra* origin (Kinoshita *et al.* 1997). The related coumarin derivative coumestrol (**12c** in Figure 7.6) has also been analyzed by HMBC (Rasku *et al.* 1999).

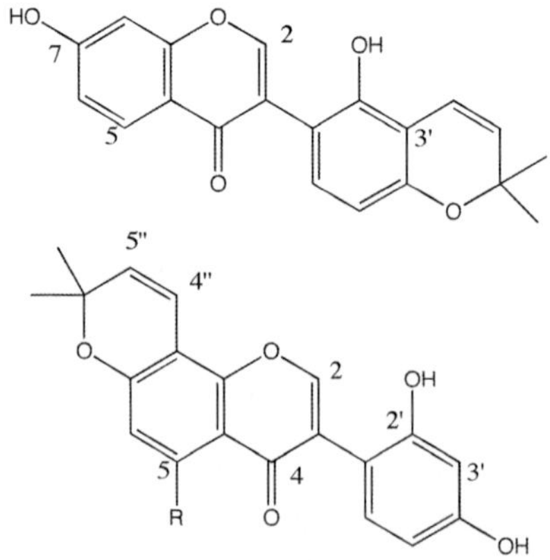

Figure 7.7 Dimethylpyrano-fused isoflavones from *G. glabra* (Kinoshita *et al.* 1997).

7.3 Chromatography-NMR

Isoflavonoids are often obtained from plants and foodstuffs as a mixture of closely related compounds. Accordingly, prior separation and purification are necessary, making the characterization procedure laborious and time consuming. A hyphenated technique, *i.e.* the combination of chromatography with on-line NMR, is useful for the analysis of components in a complex mixture. Using liquid chromatography (LC)-NMR together with LC-mass spectrometry (MS) five isoflavonoids were identified directly in a 70 min run from ethanol extracts of *Belamcanda chinensis* root. 5,6,7,3'-Tetrahydroxy-4'-methoxyisoflavone (**16**), tectorigenin (**17**), iristectorigenin A (**18**), irigenin (**19**) and irisflorentine (**20**) were separated and identified by ^{1}H NMR (Figure 7.8).

Reverse-phase chromatography was carried out with an acetonitrile/D_2O eluent. An isocratic solvent system (30% acetonitrile/70% D_2O) was used to keep a constant solvent composition in an NMR flow cell. Varian WET solvent suppression was used to eliminate the acetonitrile and residual water peaks (Kang *et al.* 2008). The solvent suppression technique usually reduces the intensities of neighboring peaks also and distorts the baseline. Direct NMR of the compounds eluted from LC columns has several drawbacks. High performance liquid chromatography (HPLC) should then be performed with costly deuterated solvents. If on the other hand the NMR is run in the HPLC solvent, it is very likely not the best solvent choice. These limitations can be overcome by introducing a solid phase extraction (SPE) step between HPLC and NMR. In HPLC-SPE-NMR the bands eluted from HPLC are trapped on SPE cartridges. The cartridge is then dried to remove the solvent and eluted with a small amount of an appropriate solvent to the NMR flow cell. The process may be fully automated. Apart from flexibility in choice of solvent for HPLC and NMR, the advantage of this procedure is that multiple trapping

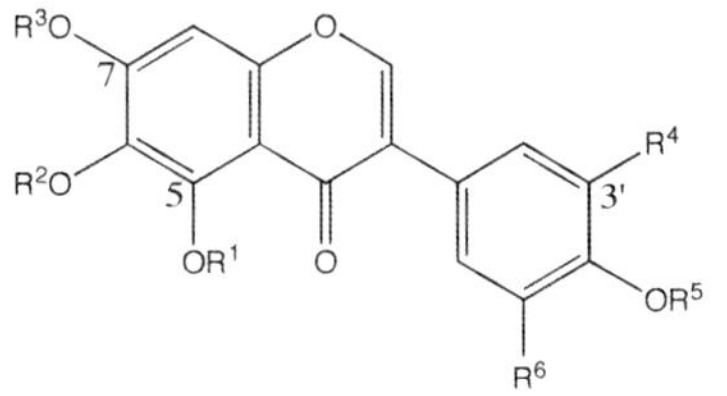

	R^1	R^2	R^3	R^4	R^5	R^6
16	H	H	H	OH	CH_3	H
17	H	CH_3	H	H	H	H
18	H	CH_3	H	OCH_3	H	H
19	H	CH_3	H	OCH_3	CH_3	OCH_3
20	CH_3	$-CH_2-$		OCH_3	CH_3	OCH_3

Figure 7.8 Isoflavones from *B. chinensis*, identified by NMR directly from LC fractions (Kang *et al.* 2008). The "hyphenated" technique that uses NMR (or MS) for the detection and identification of LC fractions is extremely useful for the rapid analysis of complex mixtures, without the need for extensive prior purification.

on SPE is possible, making a more concentrated NMR sample accessible (Lambert et al. 2005). Using this technique, isoflavonoids were identified from the root of *Smirnowia iranica*. After drying, the SPE cartridge was eluted with acetonitrile-d_3 and 1D and 2D NMR spectra were run. 1D NMR alone was not sufficient for the determination of positions of substituents, therefore HSQC, HMBC, COSY and NOESY measurements were performed. For minor chromatographic peaks, larger amounts of material required for 2D studies were obtained by repeated injections and repeated trapping. Finally, 17 isoflavonoids were identified from 10 chromatographic bands. Many HPLC peaks were comprised of two or more co-eluting compounds. Sometimes these fractions were again subjected to HPLC-NMR for isolation and unambiguous structure determination. NMR data were sometimes sufficient for partial structure elucidation only. HPLC-MS data were acquired separately and were used to assist structure elucidation (Lambert et al. 2005).

7.4 Conformation and NMR

Trans-Isoflavan-4-ol (**21**) and *trans*-4′,7-dihydroxyisoflavan-4-ol (**22**) exist as mixture of diaxial and diequitorial half-chair conformations (Figure 7.9). The conformational equilibria in solution were assessed by NMR in combination with molecular modeling calculations. ^{1}H and ^{13}C NMR spectra were acquired over the temperature range 178–323 K. The two conformers are interconverting rapidly on the NMR time scale. From coupling constant thermodynamic data, derived by molecular modeling calculations, the ratio of diequatorial to diaxial conformer at 298 K was found equal to 73:27 and 66:34 for **21** and **22**, respectively (Pihlaja et al. 2003; Wähälä et al. 1997).

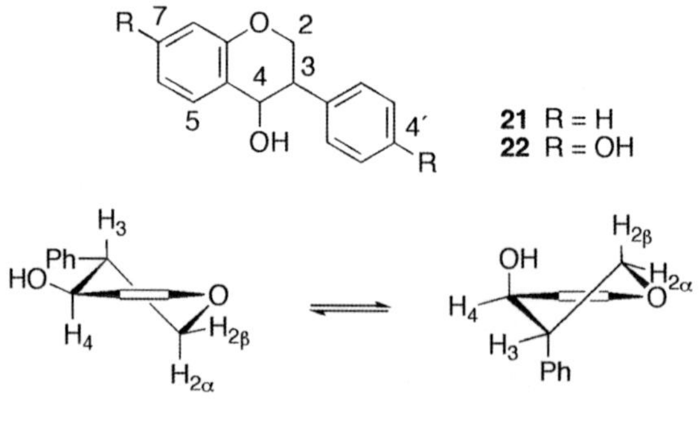

Figure 7.9 Equilibria of the diequatorial and diaxial half-chair conformations of *trans*-isoflavan-4-ol (**21**) and *trans*-4′,7-dihydroxyisoflavan-4-ol (**22**). The conformational equilibria of the heterocyclic C ring in the isoflavan-4-ols were assessed by NMR spectroscopy and molecular modeling, showing that the diaxial/diequatorial ratio is *ca*. 3:1 for **21** and 2:1 for **22**.

7.5 Substituent Effects and Prediction of Structures

Substitution in one position of an isoflavone often causes substantial changes in the chemical shifts of other protons in a pattern. Thus it is possible to predict the location of a substituent by observing the change in δ value of other protons.

When a methoxy group in aromatic ring of isoflavone is replaced with an acetoxy group, the signal of the *para*-H shows a downfield shift. If the *para* position is substituted, the *ortho*-H shows a similar trend. A methoxy group in an aromatic ring increases electron density at the *ortho* and *para* positions. When the methoxy group is replaced with an acetoxy group, electron density decreases, therefore the *ortho* and *para* proton signals shift downfield. A similar effect is seen when a glycosyl group is replaced with an acetoxy group. Thus the change in proton δ value is an indicator of the position of substituent (Kalidhar 1990).

Lambert *et al.* (2005) observed that changes in chemical shifts of aromatic protons due to an *O*-methyl group are dependent on the *ortho* substitution at OCH_3. When both *ortho* positions of OCH_3 are substituted, the shift in the *para*-H signal is larger than in the case of a hydroxyl group with one (or no) substitution. This type of behavior is helpful in structure elucidation.

In isoflavones an OH-2′ substituent has an influence on the chemical shift of OH-5. Tahara *et al.* (1991) have shown that the presence of OH-2′ shifts the OH-5 signal to a higher field consistently in a number of isoflavones. Thus the chemical shift of OH-5 is an indicator of whether there is a hydroxyl group at C-2′ present in a compound. Presumably, the OH-2′ interacts with C=O-4 and thus reduces the hydrogen-bonding energy between C=O-4 and OH-5, resulting in increased shielding of OH-5 (Figure 7.10). On the other hand, 6-methoxylation or prenylation has a deshielding effect on OH-5 and will shift the signal to a lower field. As OH-5 gives an easily distinguishable sharp singlet at δ 12.5–13.5, the chemical shift of OH-5 can be used as an indicator of the presence or absence of substitution at C-2′ and C-6. These arguments were employed to correct the structure of 6′-prenylpiscerythrone, isolated from *Piscidia erythrina*, from the original **23** to **24**.

In the 3-isoflavene field, glabrene had originally been assigned the 2′-OH structure (**25**), but NOESY work on the dimethyl ether caused the replacement of structure **25** by **26**. On methoxy group irradiation NOE was observed for H-6 (12%), H-8 (19%), H-5′ (16%) and H-4″ (27%). NOE between H-5′ and OCH_3 unambiguously proved that the methoxy group is located at C-4′. Thus glabrene is 4′,7-dihydroxy-[(6″,6″-dimethylpyrano- (2″,3″:2′,3′)]-isoflav-3-ene **26** (Kinoshita *et al.* 1997) (Figure 7.11).

7.6 Isoflavone Metabolites

Soidinsalo and Wähälä (2004) synthesized a number of isoflavone (**27**) and isoflavan di-*O*-sulfates, which are metabolites of human dietary phytoestrogens. The presence of sulfate groups was confirmed by positive electrospray ionization time-of-flight mass spectrometry (ESI-TOF). Proton and carbon

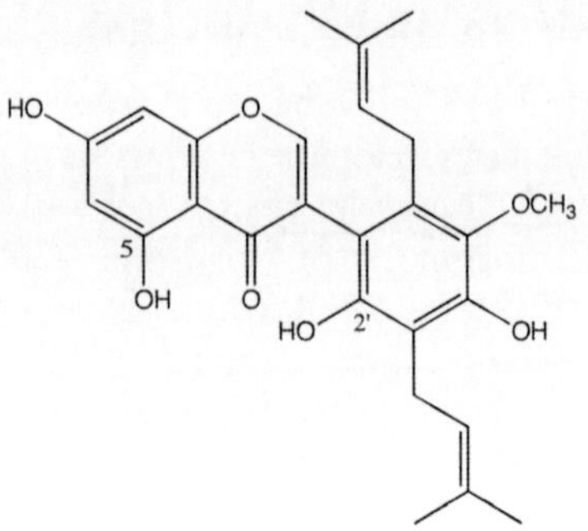

Figure 7.10 Erroneous structure (**23**) of a natural isoflavone corrected to **24** (Tahara et al., 1991). The authors reported a systematic dependence, due to hydrogen bonding, of the δ value of the 5-OH proton on the presence or absence of another OH at C-2′: in the latter case, the 5-OH will appear at ca. δ 13.0, whereas, if there is an OH group at C-2′, the 5-OH is seen at ca. 12.5. Obviously, this is a very useful tool for the structural elucidation of highly ring B-substituted isoflavones.

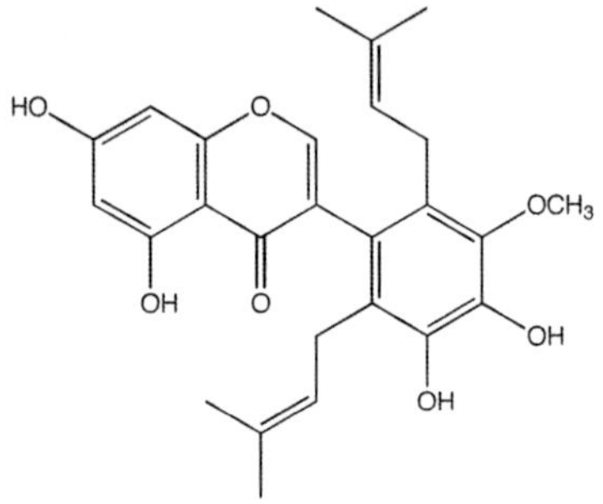

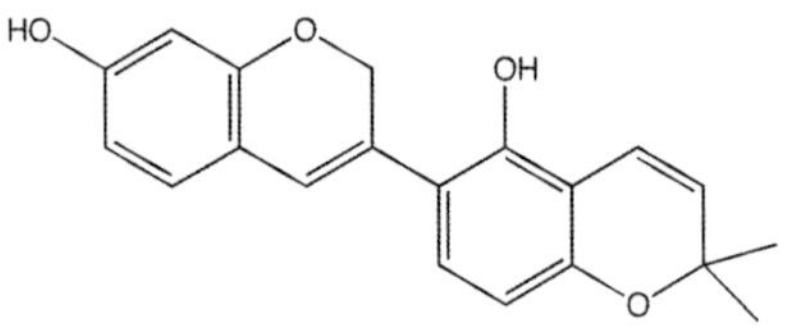

Figure 7.11 Erroneous structure (**25**) of an isoflav-3-ene corrected to (**26**) (Kinoshita et al. 1997). The nuclear Overhauser effect measurement of glabrene dimethyl ether (OCH_3 groups instead of OH groups) showed that the 4′-methoxy group is much closer to H-5′ than to H-4″.

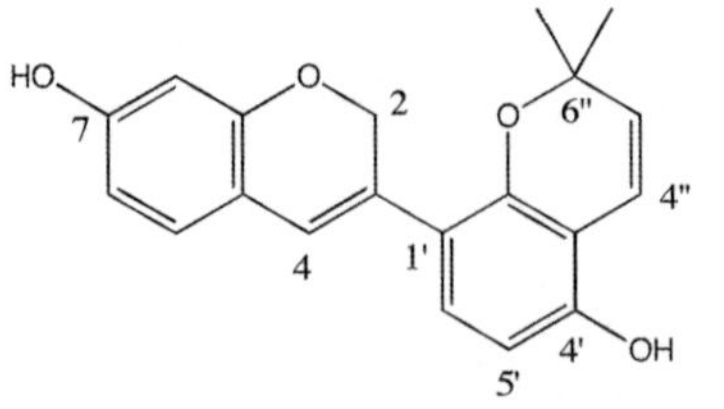

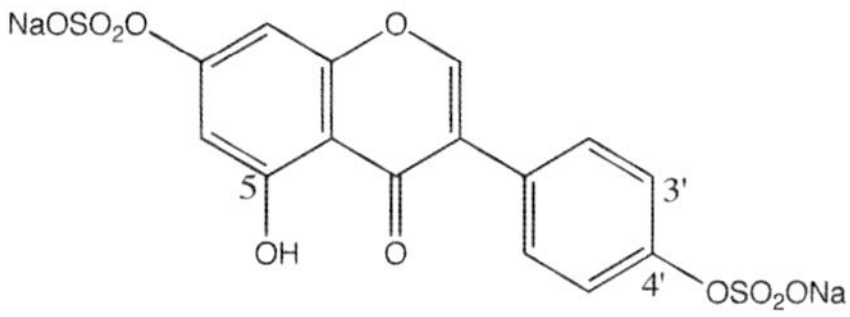

Figure 7.12 Genistein 7,4′-di-O-sulfate (**27**) (Soidinsalo and Wähälä 2004). Disulfation of genistein (**4**) leaves the 5-OH free, as was proven by the characteristic shifts observed in the ^{1}H and ^{13}C NMR spectra of the disulfate (**27**).

signals of NMR spectra were assigned by COSY, HSQC and HMBC spectra. Sulfation causes considerable downfield shifts, up to 0.35–0.59 ppm, of the protons *ortho* to sulfate group. ^{13}C signals of the carbon atoms *ortho* to sulfate are also shifted 2.8–7.1 ppm downfield, whereas the carbon atoms carrying a sulfate group showed an upfield shift of 1.9–4.6 ppm (Figure 7.12).

Thirteen metabolites of kakkalide (a major isoflavonoid from *Pueraria lobata* flowers) were found in rat urine. Among these, the new phase-II metabolites (**28**–**31**) were identified by NMR and with other spectroscopic methods [ultraviolet (UV), infrared (IR) and MS] (Figure 7.13). **28** was identified as the glucuronide conjugate from MS. In ^{1}H and ^{13}C NMR spectra the presence of an anomeric proton and carboxylic group further confirmed the presence of the glucuronic acid moiety. Enzymatic hydrolysis of **28** gave irisolidone, identified by HPLC-UV. NMR of the aglycone moiety also closely resembled that of irisolidone. Position of the conjugation in **28** was determined by NMR. An upfield shift of C-7 (−0.8 ppm) and downfield shift of the neighboring C-6 (+1.1 ppm) and C-8 (+0.2 ppm), compared with the unconjugated compound (irisolidone), indicate that the glucuronic acid moiety is present at C-7-O.

IR spectra of **29**–**31** (Figure 7.13) showed that they are sulfate conjugates. **29** and **30** had an IR absorption band at 1038 cm^{-1}, and **31** absorbed at 1045 cm^{-1}. [M-SO$_3$H]$^-$ ion peak in negative ESI-MS further confirmed the presence of the sulfate group. All three metabolites were enzymatically hydrolyzed and the isoflavones identified by co-chromatography with tectorigenin and 6-hydroxybiochanin A as standards using HPLC-UV. The aglycone of **29** and **30** was identified as tectorigenin and that of **31** as 6-hydroxybiochanin A. ^{1}H and ^{13}C NMR of **29**–**31** were assigned based on the literature values of the corresponding aglycones. Sites of the sulfate group were determined by relative chemical shifts in NMR. Upfield and downfield shifts of the signals relative to the unconjugated compounds (Table 7.3) indicate the site of conjugation. The aglycones (irisolidone, tectorigenin, tectoridin, 5,7-dihydroxy-8,4′-dimethoxyisoflavone, isotectorigenin, biochanin-A, genistein, daidzein and equol) were identified by comparison of the spectral data with the literature values (Bai *et al.* 2010).

Three glycosides **32**–**34** were isolated and identified from a soybean broth fermented by *Paecilomyces militaris* (Figure 7.14). The structures of the compounds were elucidated by ^{1}H, ^{13}C, H-H COSY, HMBC and HMQC spectra. The NMR spectra were similar to those of daidzein, glycitein and genistein.

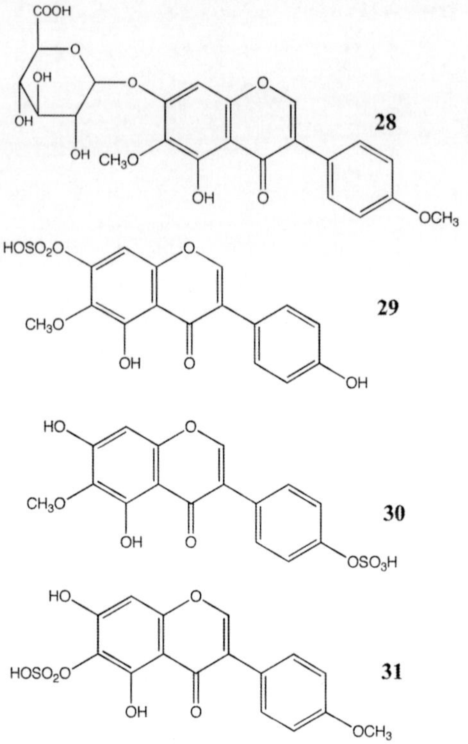

Figure 7.13 Kakkalide metabolites **28–31** (Bai et al. 2010). The site of attachment of the glucuronic acid moiety on the parent 5,7-dihydroxy compound was established by comparing the respective ^{13}C NMR shifts in the two compounds. Similar analysis of the trihydroxy compounds and the derived monosulfates (**29–31**) led to the latter structures.

Table 7.3 Upfield and downfield shifts of the signals relative to the unconjugated compounds.

Compound	Upfield shift (ppm)	Downfield shift (ppm)
30	C-7 (−4.5)	C-6 (+2.2), C-8 (+4), C-10 (+2.1), H-8 (+0.76),
31	C-4′ (−4.2)	C-3′ and C-5′ (+4.9)
32	C-6 (−4.1)	C-7 (+4.1), C-5 (+6.8)

The position of the methoxy group on the sugar moiety was determined by HMBC, correlations between OCH$_3$ and C-4″ showing that the methoxy group is on C-4″ (Hu et al. 2009).

The reported literature NMR data of equol (**5**) are summarized in Table 7.4, mainly to indicate the large variations in values measured by different workers, and the general unreliability of literature spectral data which is *not* simply a result of different NMR solvents having been used. In the spectra reported by

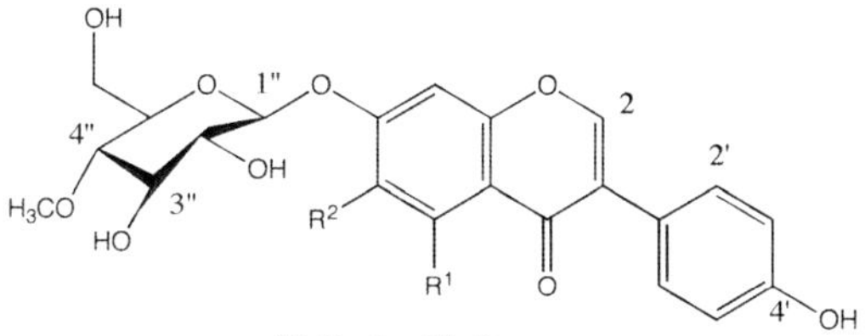

Figure 7.14 Isoflavone metabolites from fermented soybean broth (Hu et al. 2009). The non-sugar part of the NMR spectra showed that the compounds are derived from daidzein (**3**), glycitein (=6-OCH$_3$-daidzein) and genistein (**4**). The CH$_3$ group on the sugar moiety was placed on the 4″-OH based on HMBC evidence.

Kim et al. (2010) and by Wang et al. (2005), respectively, the data for H-8 and H-6 are missing. Surprisingly, the δ value 6.16 for H-7 (there is no such H in **5**) was reported by Kim et al. (2010) [purportedly for enantiopure (3S)-equol]. We are omitting an additional report by Alda et al. (1996), in which the ring numbering is quite inappropriate, and certainly not helpful in facilitating the comparison with spectra from other sources. In any case, the δ values reported here do not conform with those seen in the actual spectrum. Thus it not clear which values the authors consider correct, and why there are two conflicting sets of δ values in the first place. The 2D NMR spectra of equol run by us is shown in Figure 7.3 for comparison.

7.7 Solvent Effects

A change of NMR solvent will not only alter the chemical shifts of signals but multiplicities and coupling constants may vary from solvent to solvent as well. In the ^{1}H NMR spectrum of isoflavone **35** in DMSO-d$_6$ solvent the coupling pattern of the ring B protons suggests that three *meta* coupled protons are present in ring B [6.96 (*d*, 1.5 Hz, H-2′), 7.06 (*t*, 1.5 Hz, H-4′), 6.96 (*d*, 1.5 Hz, H-6′)] and thus indicative of structure **36** (Figure 7.15). However, subsequent analysis of 2D NMR spectra revealed that the compound is most likely **35** and not **36** (Figure 7.15). NMR spectra recorded in pyridine-d$_5$ or acetone-d$_6$ clearly showed that two *ortho* coupled protons are present and the coupling pattern is consistent with compound **35** (Du et al. 2006).

As is well known, keto-enol tautomerism is highly solvent dependent. Pyridine-d$_5$ as NMR solvent may cause enolisation. When 12 mg of **37** was dissolved in 0.75 ml of pyridine-d$_5$ and 0.1 ml of HOAc-d$_4$ was added, compound **37** instantly enolized to **39** which the NMR spectra obtained is consistent with. Surprisingly the isoflavone **38** does (Figure 7.16) not enolise under similar conditions. It is possible that the 5-OH in **37** is complexing with pyridine and inducing enolisation, or alternatively, being hydrogen bonded to the carbonyl, the 5-OH facilitates the enolization of the carbonyl (Chang et al. 1995) (Figure 7.16).

Table 7.4 Literature ^{1}H NMR data (δ values, in **bold**) of equol; J values in Hz, in *italics*; d = doublet, t = triplet, s = singlet, m = multiplet, ? = questionable (see text).

Reference	Rafii et al. (2003) (500 MHz)	Kim et al. (2010) (400 MHz)	Bai et al. (2010)	Wang et al. (2005) (400 MHz)	Li et al. (2009) (400 MHz)
H-2	**3.88** (2H *d*)	**3.88** (1H *dd 10.5 & 10.5*); **4.12** (1H *dddd 10.5 & 3.7 & 2.3*)	**3.89** (1H *t 10.5*); **4.17** (1H *10.5 & 1.8*)	**3.92** (1H m); **4.17** (1H m)	**3.89** (1H *t 10.4*); **4.14** (1H *d 10.4*)
H-3	**3.03** (1H m)	**2.97** (1H *dddd 10.5 & 10.5 & 5.5 & 3.7*)	**2.96–3.04** (1H m)	**3.06** (1H m)	**2.80** (1H m);
H-4	**2.88** (2H *d*)	**2.75** (1H *dd 15.6 & 5.5 & 2.3*); **2.81** (1H *dd 15.6 & 10.5*)	**2.78–2.88** (1H m)	**2.91** (2H m)	**2.80** (1H m); **3.01** (1H m)
H-5	**6.83** (1H *d 8.4*)	**6.84** (1H *d 8.4*)	**6.86** (1H *d 8.4*)		**6.86** (1H *d 8.4*)
H-6	**6.28** (1H *d 8.2*)	**6.27** [1H *dd(?) 2.1*]	**6.29** (1H *dd 8.4 & 2.1*)	**6.35** (1H *dd(?) 8.3*	**6.29** (1H *dd 8.4 & 2.4*)
H-7(?)		**6.16** (1H *d 2.4*)		?	
H-8	**6.19** (1H *s*)	?	**6.19** (1H *d 2.1*)	**6.27** (1H *d 2.4*)	**6.20** (1H *d 2.4*)
H-2',6'	**7.06** (2H *d 8.5*)	**7.09** (2H *d 8.4*)	**7.10** (2H *d 8.4*)	**7.14** (2H *d 8.5*)	**7.10** (2H *d 8.4*)
H-3',5'	**6.71** (2H *d 8.6*)	**6.71** (2H *d 8.4*)	**6.72** (2H *d 8.4*)	**6.82** (2H *d 8.5*)	**6.72** (2H *d 8.4*)
OH		**9.15** (1H *s*); **9.25** (1H *s*)		**8.08** (1H *s*); **8.22** (1H *s*)	**9.18** (1H *s*); **9.29** (1H *s*)
Solvent	CH$_3$OD	DMSO-d$_6$	DMSO-d$_6$	acetone-d$_6$	DMSO-d$_6$

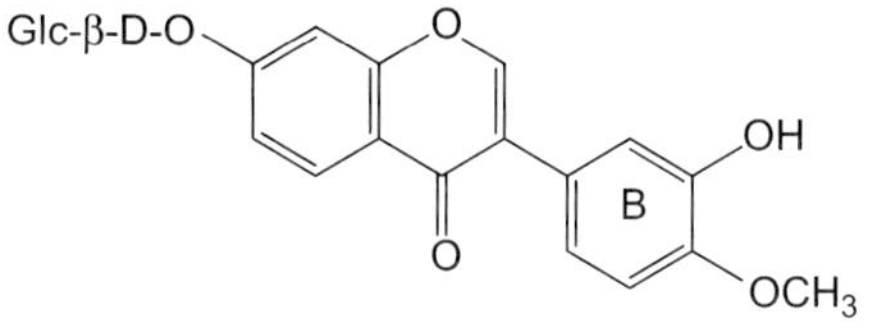

Figure 7.15 A misleading solvent effect (Du *et al.* 2006). Recognition of aromatic substituent patterns relies heavily on being able to discern *ortho*, *meta* and *para* couplings between two aromatic H atoms (7-10, 2-3 and 0-1, respectively). In DMSO-d_6 solvent, a ring B 1,3,5-trisubstitution pattern (as in **36**) was indicated (because of certain overlapping peaks) but in acetone-d_6 all peaks were resolved showing that ring B is actually 1,2,4-substituted (**35**).

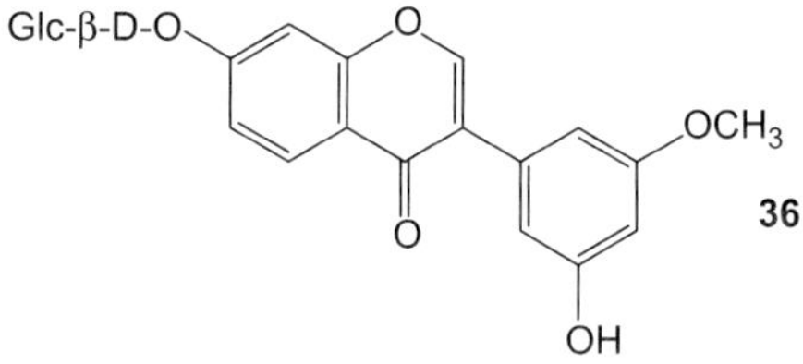

Table 7.5 NMR data of compound **35** in different NMR solvents (Du *et al.* 2006). br, broad.

proton	DMSO-d_6	methanol-d_4	pyridine-d_5	acetone-d_6
H-2′	7.06 (br s)	7.04 (d, J = 1.2 Hz)	7.77 (d, J = 1.8 Hz)	7.19 (d, J = 1.8 Hz)
H-5′	6.96 (br s)	6.97 (s)	7.03 (d, J = 8.1 Hz)	6.98 (d, J = 8.1 Hz)
H-6′	6.96 (br s)	6.97 (d, J = 1.2 Hz)	7.29 (dd, J = 1.8 Hz, 8.1 Hz)	7.07 (dd, J = 2.1 Hz, 8.1 Hz)

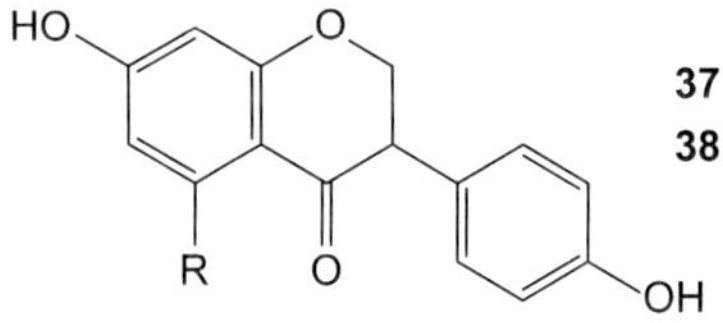

Figure 7.16 Keto–enol tautomerism in isoflavanones (Chang and Nair 1995). **37** in pyridine exists as the enol **39**, whereas **38** stays in the keto form. The enolization may be due to a hydrogen-bond formation from 5-OH to C=O.

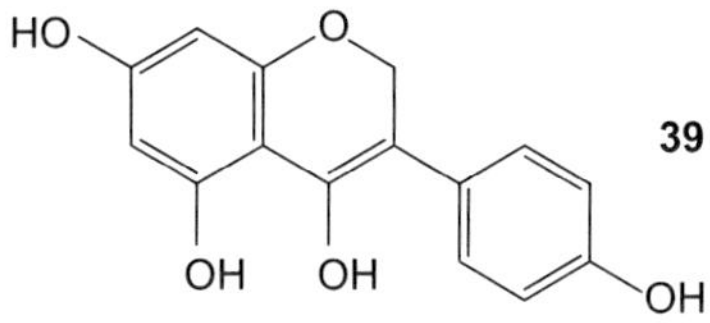

Summary Points

- This Chapter focuses on NMR of isoflavones.
- Isoflavones are produced in plants and are common food constituents.
- Isoflavones are metabolized in humans by gut microflora into physiologically active compounds.
- Most isoflavones contain two aromatic and one oxygen heterocyclic ring, substituted by OH, alkoxy and other groups.
- The structure determination of isoflavones and isoflavonoids relies mainly on 1D and 2D NMR spectroscopy.
- Some earlier studied isoflavone structures have been corrected by the modern 2D techniques.
- The available NMR techniques applied for isoflavone and isoflavonoid structural elucidation are COSY, HSQC, HMBC, DEPT, ROESY and NOESY.

Key Facts

- Nuclear magnetic resonance (NMR) spectroscopy is currently the most important analytical tool the organic chemist has available for elucidating the structure of an organic compound (without the use of a reference compound of known structure).
- Nuclei such as ^{1}H and ^{13}C, common in organic compounds, respond when experiencing a strong magnetic field, by aligning themselves in relation to the field. Irradiation at radio frequencies causes energy absorption (resonance) at wavelengths characteristic of the relative electronegativity of the nuclei in question. Thus the more electronegative a proton's (or carbon atom's) surroundings are, the larger the shift from an internal standard [tetramethylsilane (TMS)] becomes. This is the so-called *chemical shift* or δ value in each case (for TMS, $\delta = 0.00$).
- By a rule of thumb, protons bonded to saturated carbon atoms appear at δ 0-3, or up to δ 5 if an electronegative atom is in the vicinity (*e.g.*, H-C-O-); protons at unsaturated carbon atoms typically have δ values 5–10.
- Similar trends exist for ^{13}C but the δ scale is then much wider, from 0 to 250. Note that in ^{1}H NMR, natural abundance (*ca.* 100%) hydrogen atoms are observed, whereas that of ^{13}C is only 1.1% which makes the detection of their signals more difficult (^{12}C is not NMR active).
- The resonance signals (peaks) of protons have fine structure, caused by two different states for each H atom on a neighboring C atom. Thus the signal for H^B in $>CH^ACH_3^B$ will be a 1:1 doublet, whereas the H^A will be a 1:3:3:1 quartet. Although such 'multiplets' are not always readily discerned, they often give valuable information in establishing the structure of an unknown.
- In ^{13}C NMR, similar interactions will cause the formation of corresponding multiplets for the ^{13}C signals as well. Because the spectrum will then become very complex, such couplings are usually removed from the ^{13}C spectrum by electronic means.

- Ordinary [one-dimensional (1D)] NMR spectra are plots of frequency *vs.* intensity; in two-dimensional (2D) NMR two frequencies are plotted against each other. For definitions and explanations of the use of the common 2D NMR techniques, see the Definitions of Words and Terms Section.

Definitions of Words and Terms

AB quartet: Two coupled protons having closely similar δ values, giving rise to a four peak pattern with weaker outer peaks and stronger inner peaks.

ABX system: An AB-quartet where both A and B are coupled to a remote X proton; the result is an AB quartet with every peak split into two, plus four peaks for the X proton.

Chemical shift: The δ value (the resonance frequency relative to a standard) of a nucleus (commonly proton or a carbon) in the NMR spectrum.

Conformations: Spatial forms of organic compounds that can change into each other by rotation about a single C-C bond.

Correlation spectroscopy (COSY): This identifies protons that are spin–spin coupled with each other. In a 1D spectrum with overlapping peaks COSY can reveal the hidden couplings also. Thus COSY gives information about the connectivity of protons.

d (in italics, following a δ value): Doublet; (upright) abbreviation for deuterium.

δ value: Chemical shift (of a proton or a carbon atom in an NMR spectrum; in ppm units).

Distortionless enhancement by polarization transfer (DEPT): This measurement distinguishes between primary, secondary and tertiary carbon atoms.

Double quantum filtered COSY (DQFCOSY): This simplifies a complex spectrum by filtering out protons that are not coupled.

Heteronuclear multiple bond correlation (HMBC): This determines long range carbon–proton connectivity. The spectra show correlations between protons and carbons that are not directly attached but 2 to 4 bonds apart. Thus HMBC is effective in mapping partial skeletons.

Heteronuclear single quantum correlation (HSQC) and heteronuclear multiple quantum correlation (HMQC): These show the correlation between protons and carbons that are directly chemically bonded to each other (*i.e.*, separated by one bond). Thus HSQC allows the assignment of a carbon if the assignment of a proton is known and *vice versa*.

Hyphenated techniques: Organic chemistry analytical methods whereby one instrument (usually a gas chromatograph or LC for component separation) is directly combined with a spectroscopic instrument such as MS or NMR spectrometer.

J: Couplings constant (given in Hz).

Nuclear magnetic resonance (NMR): This is used for the identification and characterization of H or C atoms in an organic compound.

Nuclear Overhauser enhancement spectroscopy (NOESY) and rotating frame Overhauser effect spectroscopy (ROESY): These provide information of the spatial distance between two atoms. Peaks appear between atoms which are in close proximity through space.

q (following a δ value): Quartet.

Reverse-phase: Chromatography employing a hydrophobic non-polar stationary phase.

s (following a δ value): Singlet.

Structure elucidation: This involves finding the correct structure of an organic compound, usually by spectroscopic means

t (following a δ value): Triplet.

List of Abbreviations

COSY	correlation spectroscopy
1D	one-dimensional
2D	two-dimensional
DEPT	distortionless enhancement by polarization transfer
DMSO	dimethyl sulfoxide
DQFCOSY	double quantum filtered COSY
HMBC	heteronuclear multiple bond correlation
HMQC	heteronuclear multiple quantum correlation
HPLC	high performance (or pressure) liquid chromatography
HSQC	heteronuclear single quantum correlation
IR	infrared
LC	liquid chromatograph, liquid chromatography
MS	mass spectroscopy, mass spectrometer
NMR	nuclear magnetic resonance
NOE	nuclear Overhauser enhancement (or effect)
NOESY	nuclear Overhauser enhancement spectroscopy
ROESY	rotating frame Overhauser effect spectroscopy
SPE	solid phase extraction
UV	ultraviolet

Acknowledgement

We thank Gudrun Silvennoinen for running the NMR spectra.

References

Alda, J.O., Mayoral, J.A., Lou, M., Gimenez, I., Martinez, R.M., and Garay, R.P., 1996. Purification and chemical characterization of a potent inhibitor of the Na-K-Cl cotransport system in rat urine. *Biochemical and Biophysical Research Communications*. 221: 279–285.

Baba, M., Sumi, S., Iwasaki, N., Kai, H., Sakano, M., Okada, Y., and Okuyama, T., 2008. Studies of the Egyptian traditional folk medicines. IV. New isoflavonoid isolated from Egyptian licorice. *Heterocycles*. 75: 3085–3089.

Bai, X., Xie, Y., Liu, J., Kano, Y., and Yuan, D., 2010. Isolation and identification of urinary metabolites of kakkalide in rats. *Drug Metabolism and Disposition*. 38: 281–286.

Chang, Y.-C., and Nair, M.G., 1995. Metabolites of daidzein and genistein and their biological activities. *Journal of Natural Products*. 58: 1901–1905.

Chang, Y.-C., Nair, M.G., Santell, R.C., and Helferich, W.G., 1994. Microwave-mediated synthesis of anticarcinogenic isoflavones from soybeans. *Journal of Agricultural and Food Chemistry*. 42: 1869–1871.

Du, X., Bai, Y., Liang, H., Wang, Z., Zhao, Y., Zhang, Q., and Huang, L., 2006. Solvent effect in ^{1}H NMR spectra of 3'-hydroxy-4'-methoxy isoflavonoids from *Astragalus membranaceus* var. *mongholicus*. *Magnetic Resonance in Chemistry*. 44: 708–712.

Heinonen, S.-M., Wähälä, K., and Adlercreutz, H., 2002. Metabolism of isoflavones in human subjects. *Phytochemistry Reviews*. 1: 175–182.

Heinonen, S.-M., Hoikkala, A., Wähälä, K., and Adlercreutz, H., 2003. Metabolism of the soy isoflavones daidzein, genistein and glycitein in human subjects. Identification of new metabolites having an intact isoflavonoid skeleton. *Journal of Steroid Biochemistry and Molecular Biology*. 87: 285–299.

Hu, F.-L., He, Y.-Q., Huang, B., Li, C.-R., Fan, M.-Z., and Li, Z.-Z., 2009. Secondary metabolites in a soybean fermentation broth of *Paecilomyces militaris*. *Food Chemistry*. 116: 198–201.

Jha, H.C., Zilliken, F., and Breitmaier, E., 1980. Carbon-13 chemical shift assignments of chromones and isoflavones. *Canadian Journal of Chemistry*. 58: 1211–1219.

Kalidhar, S.B.1990. Structural studies in isoflavones using ^{1}H NMR alkoxylation and glycosyloxylation shifts. *Proceedings of the Indian National Science Academy - Part A: Physical Sciences*. 56: 217–223.

Kang, S.W., Kim, M.C., Kim, C.Y., Jung, S.H., and Um, B.H., 2008. The rapid identification of isoflavonoids from *Belamcanda chinensis* by LC-NMR and LC-MS. *Chemical and Pharmaceutical Bulletin*. 56: 1452–1454.

Kim, M., Marsh, E.N.G., Kim, S.-U., and Han, J., 2010. Conversion of (3S,4R)-Tetrahydrodaidzein to (3S)-equol by THD reductase: proposed mechanism involving a radical intermediate. *Biochemistry*. 49: 5582–5587.

Kinoshita, T., Tamura, Y., and Mizutano, K., 1997. Isolation and synthesis of two new 3-arylcoumarin derivatives from the root of *Glycyrrhiza glabra* (licorice), and structure revision of an antioxidant isoflavonoid glabrene. *Natural Product Letters*. 9: 289–296.

Lambert, M., Staerk, D., Hansen, S.H., Sairafianpour, M., and Jaroszewski, J.W., 2005. Rapid extract dereplication using HPLC-SPE-NMR: analysis of isoflavonoids from *Smirnowia iranica*. *Journal of Natural Products*. 68: 1500–1509.

Li, S.-R., Chen, P.-Y., Chen, L.-Y., Lo, Y.-F., Tsai, I.L., and Wang, E.-C., 2009. Synthesis of haginin E, equol, daidzein and formononetin from resorcinol *via* an isoflavene intermediate. *Tetrahedron Letters*. 50: 2121–2123.

Pihlaja, K., Tähtinen, P., Klika, K.D., Jokela, T., Salakka, A., and Wähälä, K., 2003. Experimental and DFT ^{1}H NMR study of conformational equilibria in *trans*-4′,7-dihydroxyisoflavan-4-ol and *trans*-isoflavan-4-ol. *Journal of Organic Chemistry*. 68: 6864–6869.

Rafii, F., Davis, C., Park, M., Heinze, T.M., and Beger, R.D., 2003. Variations in metabolism of the soy isoflavonoid daidzein by human intestinal microfloras from different individuals. *Archives of Microbiology*. 180: 11–16.

Rasku, S., Wähälä, K., Koskimies, J., and Hase, T., 1999. Synthesis of isoflavonoid deuterium labeled polyphenolic phytoestrogens. *Tetrahedron*. 55: 3445–3454.

Soidinsalo, O., and Wähälä, K., 2004. Synthesis of phytoestrogenic isoflavonoid disulfates. *Steroids*. 69: 613–616.

Tahara, S., Ingham, J.L., Hanawa, F., and Mizutani, J., 1991. ^{1}H NMR chemical shift value of the isoflavone 5-hydroxyl protons as a convenient indicator of 6-substitution or 2′-hydroxylation. *Phytochemistry*. 30: 1683–1689.

Veitch, N.C., Sutton, P.S.E., Kite, G.C., and Ireland, H.E., 2003. Six new isoflavones and a 5-deoxyflavonol glycoside from the leaves of *Ateleia herbert-smithii*. *Journal of Natural Products*. 66: 210–216.

Wähälä, K., Koskimies, J.K., Mesilaakso, M., Salakka, A.K., Leino, T.K., and Adlercreutz, H., 1997. The synthesis, structure, and anticancer activity of *cis*- and *trans*-4′,7-dihydroxyisoflavan-4-ols. *Journal of Organic Chemistry*. 62: 7690–7693.

Wang, X.-L., Hur, H.-G., Lee, J.-H., Kim, K.-T., and Kim, S-I., 2005. Enantioselective synthesis of *S*-equol from dihydrodaidzein by a newly isolated anaerobic human intestinal bacterium. *Applied and Environmental Microbiology*. 71: 214–219.

CHAPTER 8

Biotransformation and Transfer of Genistein: a Comparison with Xenoestrogens and a Focus on the Human Placenta

HSIU-WEN CHAN, GREG E. RICE AND
MURRAY D. MITCHELL*

University of Queensland Centre for Clinical Research, Building 71/918, Royal Brisbane and Women's Hospital Campus, Herston, QLD 4029, Australia
*Email: murray.mitchell@uq.edu.au

8.1 Biotransformation of Genistein Compounds in the Gastrointestinal Tract

Genistein (4′,5,7-trihydroxyisoflavone, **1**) comprises two phenolic rings (A and B) linked to a three-carbon bridge, which in this case is an oxygenated heterocyclic ring (C), with three hydroxyl groups at carbons 4′, 5, and 7 (Figure 8.1). In legumes, genistein exists in four chemical forms, as shown in Table 8.1. Of the four versions of genistein, glucose-conjugated genistin (**2**) is the most abundant form found in legumes; its concentration is up to 4 times higher than genistein (Hu *et al.* 2010). A large proportion of genistein and related compounds are transported to the colon (see the Key Facts at the end of the Chapter); during their passage through the small and large intestine, they are

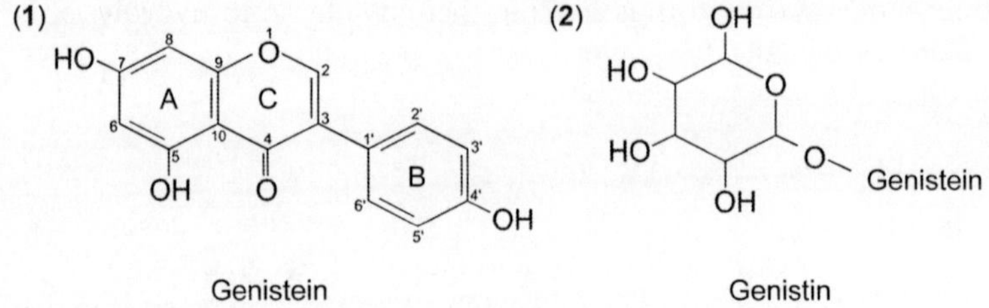

Figure 8.1 Chemical structures of naturally-occurring genistein compounds. The chemical structures are shown for genistein (**1**) and genistin (**2**). (Adapted from Mathias et al. 2006.)

Table 8.1 Chemical forms of genistein. The naturally occurring genistein and genistein-conjugated compounds found in plants are listed in this Table. The type of chemical form, their common names, and chemical names are detailed for each one.

Category	Name	Chemical name
Aglycone	Genistein	4′,5,7-Trihydroxyisoflavone
Glucoside	Genistin	Genistein-7-O-glucoside
Malonylglucoside	Malonyl genistin	6″-O-Malonylgenistin
Acetylglucoside	Acetyl genistin	6″-O-Acetylgenistin

readily hydrolysed by the intestinal microflora to simpler chemical compounds and eliminated in the faeces. A large proportion of genistein and genistin (40.1–58.3% of total genistein or genistin administered) is eliminated *via* the luminal effluent following their administration (Andlauer et al. 2000). It seems, however, that a fraction of the total isoflavone is absorbed, given that genistein, genistin and their metabolites are detectable in the small intestine, vascular tissue, and plasma. This presumably occurs *via* the enterohepatic system (Shelnutt et al. 2002).

8.1.1 Deglycosylation of Genistin in the Small Intestine

In addition to its physicochemical properties, the poor absorption demonstrated by genistin is likely to be attributed to its deglycosylation by intestinal β-glycosidases and gut microflora. Two broad specificity enzymes responsible for deglycosylating flavonoids in the small intestine have been identified as lactase-phlorizin hydrolase (LPH) and the cytosolic β-glucosidase (CBG) (Day et al. 1998; Nemeth et al. 2003). LPH is located on the apical membrane of human small intestine epithelial cells and promotes absorption by localising β-glucosides such as genistin to the membrane, and increasing their lipophilicity. In contrast, CBG hydrolyses β-glucosides that have already been transported into the epithelial cells. Sodium-dependent glucose transporter 1 present at the intestinal brush border is also capable of transporting glucosides

(Walgren *et al.* 2000) and thus making them available to hydrolysis by CBG. There is strong evidence, however, that this may be a secondary mechanism in the event that LPH is inhibited (Day *et al.* 2003).

Genistin reportedly also undergoes rapid deglycosylation mediated by the intestinal microflora. When genistin was incubated with cultures of rat caecal (beginning of the large intestine) and human faecal bacteria, genistin was 100% hydrolysed to genistein within 15 min (Coldham *et al.* 2002). Hur *et al.* (2000) extended this observations by identifying two strains of bacteria isolated from the human intestinal wall capable of converting genistin into genistein: *Escherichia coli* HGH21 and HGH6.

8.1.2 Glucuronidation and Sulfation of Genistein

Within the small and large intestine and liver, genistein is conjugated with glucuronide and/or sulfate groups at the 4′- and 7-hydroxyl groups (Figure 8.2, 3–6) by phase II metabolic enzymes, and as many as seven different genistein conjugates may be produced (Hosada *et al.* 2011). These enzymes are sulfotransferases (SULTs) and uridine-5′-disphosphate glucuronyltransferases (UGTs). UGTs can conjugate genistein with the glucuronide moiety from a nucleotide sugar. The result of this reaction is usually a more polar molecule that is more readily excreted. In humans, 15 isoforms of UGT enzymes have been identified and are derived from either the UGT1 or UGT2 gene *via* alternative splicing or differential promoter usage. SULTs, on the other hand, catalyse the transfer of sulfate groups from the donor substrate 3′-phosphadenosine-5′-phosphosulfate to genistein. Thirteen isoforms of SULTs have been identified in human tissues to date and can be cytosolic or membrane-bound (Lindsay *et al.* 2008).

After the ingestion of soy milk or a soy-based drink, genistein glucuronides are the major genistein constituent in human plasma and urine (up to 78% of total isoflavones), whereas both genistein aglycone and genistein sulfate are not as abundant (less than 5% and 25% of total isoflavone, respectively; Shelnutt *et al.* 2002). A possible reason for this is that genistein may be rapidly biotransformed in the intestine and that the catalysis of genistein glucuronides occurs more quickly than genistein sulfates. Zhu *et al.* (2010) measured the rate of genistein metabolite production in isolated rodent intestine (small and large) perfused with genistein. They discovered that intestinal mucosal cells generated genistein glucuronide at a rate that was approximately 25-fold faster than genistein sulfate in mice. Remarkably, it was measured at up to 1000-fold faster in rats. Although all these studies have been informative, the experimental technique chosen [β-glucuronidase or sulfatase hydrolysis coupled with high-performance liquid chromatography (HPLC)] is not able to distinguish between the different genistein metabolites shown in Figure 8.2. The positioning of the glucuronide or sulfate group(s) on one genistein molecule cannot be determined. Recently, Hosada *et al.* (2010) showed that, with the use of validated HPLC standards for all genistein metabolites, genistein-7-glucuronide-4′-sulfate and genistein-4′,7-glucuronide were, by far, in greatest

(3) Genistein-7-glucuronide

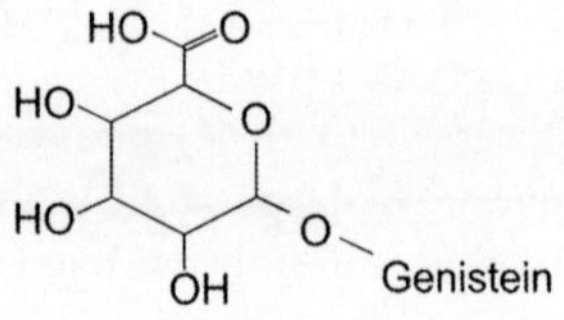

(4) Genistein-4'-glucuronide

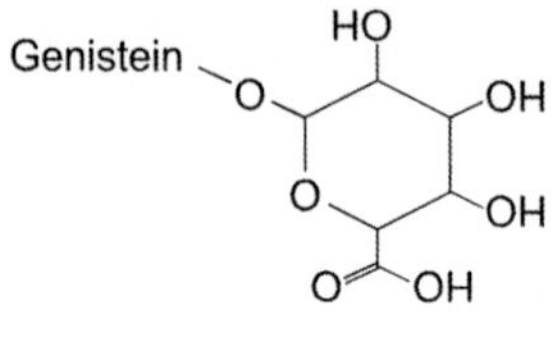

(5) Genistein-4'-7-glucuronide

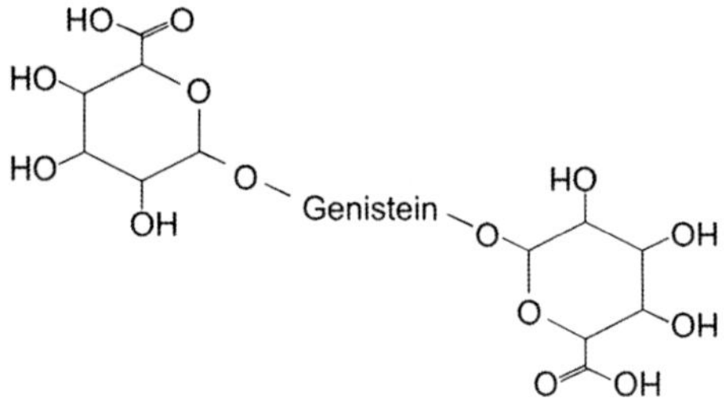

(6) Genistein-7-glucuronide-4'-sulfate

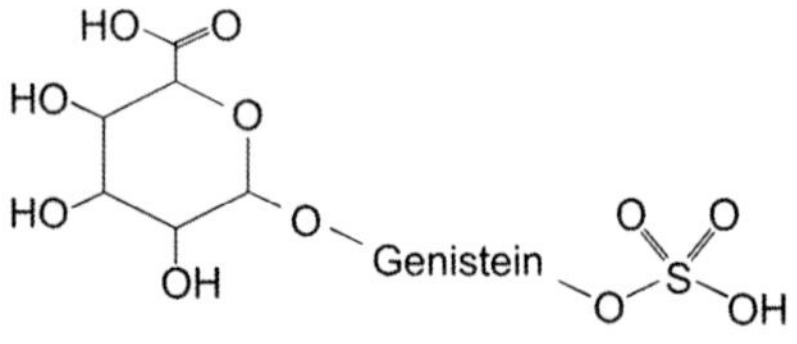

Figure 8.2 Conjugation of genistin during phase II metabolism. In the small and large intestine and liver, genistein undergoes extensive phase II metabolism. Seven distinct metabolites are produced as the result of glucuronide and/or sulfate conjugation in the 4'- and 7-positions of genistein. Glucuronide conjugation is shown here (**3–5**), as well as the glucuronide sulfate (**6**).

abundance (1–2.5 µg ml^{-1}) in human plasma following kinako (baked soybean powder) ingestion. Genistein-7-glucuronide (**3**), genistein-4'-glucuronide (**4**), and genistein-7-sulfate were all detected at lower concentrations (<500 ng ml^{-1}), whereas genistein-4'-sulfate, genistein-4',7-disulfate, and genistein aglycone (**1**) were below the level of detection. Overall, these data are consistent with observations from other studies that glucuronides are produced in higher concentrations than sulfates.

In humans, genistein sulfation occurs at similar rates in the liver and small intestine (approximately 12 pmol mg^{-1} min^{-1}; Ronis et al. 2006). When the sulfation was assayed according to intestinal segment, it was found to be highest in the mid-jejunum section. Another study identified sulfotransferases, SULT1A1 and SULT1E1, as the key enzymes with high affinity for genistein (K_m = 0.3 and 0.7 μM, respectively) (Nishiyama et al. 2002). When the formation of genistein-7-sulfate and genistein-4′-sulfate by these two enzymes was analysed, SULT1E1 did not demonstrate a preference for attaching a sulfate group to either the 7- or 4′-position, whereas, for SULT1A1, its 7-/4′-sulfation ratio was 8.8 (Nakano et al. 2004). Interestingly, SULT1E1 most efficiently catalyses disulfation (K_{cat} 2.1–3.6 compared with SULT1A1 K_{cat} 0.07–0.4).

The UGT isoforms that have been identified as the major catalysts of genistein glucuronidation in the liver and intestine are UGT1A1, UGT1A8, UGT1A9, and UGT1A10 (Doerge et al. 2000; Tang et al. 2009). When the rate of glucuronidation was determined in human liver and intestinal microsomes, it was found that genistein was conjugated more quickly in the intestine (approximately 5 nmol min^{-1} mg^{-1} compared to 1 nmol min^{-1} mg^{-1} in the liver) regardless of genistein concentration.

8.1.3 Hydroxylation of Genistein in the Liver

Similar to the intestines, biotransformation of genistein in the liver leads to the production of glucuronide and sulfate metabolites. The conversion here is reduced as most aglycones have already been conjugated in the small intestines. Analysis of rat and human liver microsomes, cryopreserved hepatocytes, and human urine has revealed that genistein can be hydroxylated at any of the positions on the A and B rings, often in multiple locations (Bursztyka et al. 2008; Kulling et al. 2001). The oxidative metabolism of genistein is mediated by cytochrome P450 enzymes, which are known to metabolise xenobiotics, endogenous hormones, carcinogens, and other isoflavones. It is thought that cytochrome P450 activity leads to the formation of metabolites that are less biologically active and more easily excreted than the parent compound.

8.1.4 Breakdown of Genistein and Luminal Excretion

While the majority of genistein compounds pass through the intestinal lumen without being absorbed, there is some evidence that these compounds are significantly biotransformed into less complex molecules for excretion (Figure 8.3). Coldham and colleagues (1999) administered radiolabelled genistein to rats via oral gavage and used mass spectrometry to analyse genistein metabolites in luminal contents. They found that the major metabolites dihydrogenistein (**7**) and 6′-hydroxy-*O*-desmethylangolensin (**8**), and were able to deduce from the other identified molecules that genistein was metabolised.

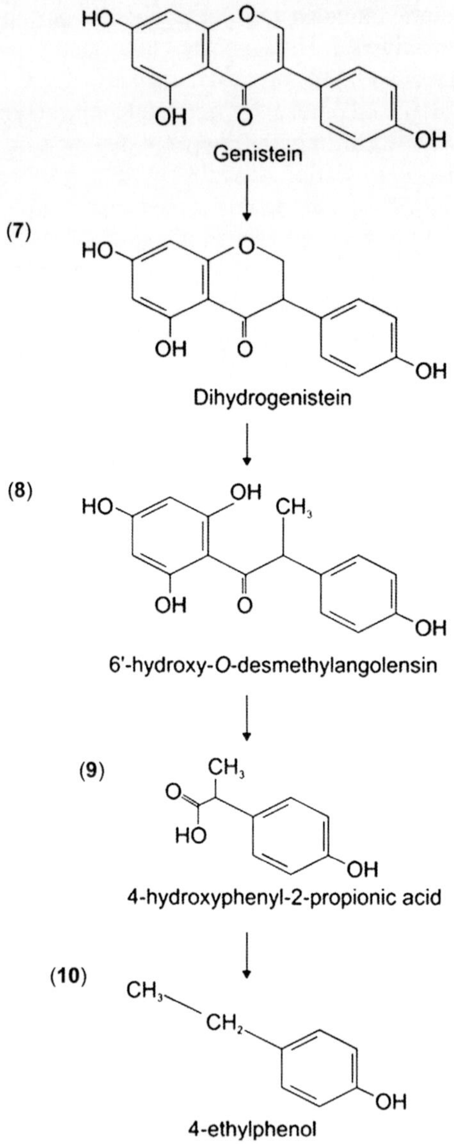

Figure 8.3 Hydroxylation of genistin in the liver. Genistein is biotransformed in the intestinal lumen and urine into dihydrogenistein, 6-hydroxy-O-desmethyl-anagolensin, and 4-hydroxyphenyl-2-propionic acid. This presumably occurs to aid excretion of genistein.
This figure was adapted from Coldham and colleagues (Coldham et al. 1999; 2002).

Genistein firstly undergoes a reduction of the C-ring double bond by intestinal microflora to form dihydrogenistein (7), which is then subjected to two heterocyclic ring fission events: of the C ring to open the ether bridge and form

6′-hydroxy-O-desmethylangolensin (**8**), then of the A ring to form 4-hydroxyphenyl-2-propionic acid (**9**). The same research group also found that when radiolabelled genistein was incubated with rat caecal and human faecal microflora, the same metabolites were produced (Coldham et al. 2002). Others have shown in rat urine that 4-hydroxyphenyl-2-propionic acid may be further decarboxylated to 4-ethylphenol (**10**) and that this molecule is the major end-product (King 1998). Genistein, dihydrogenistein, and 6′-hydroxy-O-desmethylangolensin has also been detected in healthy human volunteers on a soy diet, suggesting that genistein may undergo similar biotransformation events in rats and humans (Joannou et al. 1995).

The catalyst for these metabolic reactions is not well characterised, however, one study has shown that *Eubacterium ramulus*, which accounts for approximately 0.16% of the total intestinal bacteria, can degrade genistein to 6′-hydroxy-O-desmethylangolensin and, subsequently, 4-hydroxyphenyl-2-propionic acid (Schoefer et al. 2002). A *Clostridium*-like bacterium, TM-40, isolated from human faeces has also been shown to convert genistein into dihydrogenistein (Tamura et al. 2007).

8.1.5 Transfer of Genistein in Placenta

The placenta is exposed to endogenous and exogenous molecules *via* the maternal blood circulation. For genistein transfer in the placenta, it must first be absorbed *via* the gastrointestinal tract, transported into the liver, and then released *via* the hepatic vein where it enters the peripheral circulation. While a proportion of genistein is excreted by the kidneys (Dalais et al. 1998a; 1998b), evidence suggests that a small fraction (both conjugated and unconjugated) reaches the placenta and crosses the placental barrier. In pregnant Japanese women whose normal daily consumption of bean products (mainly soy beans, tofu, and miso) was averaged to be 77 g, 84% of genistein measured in amniotic fluid was glucuronidated (Adlercreutz et al. 1999). Conversely, when pregnant Sprague–Dawley rats were administered genistein by oral gavage, only a small fraction of the genistein detected in the placenta was glucuronidated, whereas sulfate concentrations were not detectable (Soucy et al. 2006). In the amniotic fluid, genistein aglycone was the major analyte at early time points, however, the concentrations of both glucuronides and sulfates increased over time. This suggests that some conjugation of genistein occurs after it is transferred from the mother.

Our laboratory used an *ex vivo* placental perfusion model to provide direct evidence that genistein readily transfers across the human placenta at environmentally relevant concentrations (Balakrishnan et al. 2010a). Genistein was perfused through the maternal compartment over a 3 h time period. It was observed that the concentration of genistein decreased in the maternal compartment over time, as the appearance and increase of this compound was detected in the fetal compartment. The majority of genistein in both the maternal and fetal compartments of the placentae was unconjugated. However, a greater proportion of genistein in the fetal compartment was conjugated compared with the maternal side. This supports the theory that the placenta is capable of conjugating genistein and that conjugated genistein can transfer across the placenta.

The human placenta expresses both the UGT and SULT enzymes and the presence of specific isoforms are differentially regulated throughout pregnancy. The human first trimester placenta expresses UGT1A, 2B, 2B4, and 2B7, whereas term placenta expresses primarily UGT2B isoforms but also UGT1A4 (Collier *et al.* 2002a, 2002b; Reimers *et al.* 2011). SULT expression, on the other hand, is not as well characterised in placenta. It seems that cytotrophoblasts, considered the barrier for placental transfer, express SULT1A1 and SULT1A3 (Mitra and Audus 2009). SULT2B1 has also been detected in human term placenta (He *et al.* 2004). The activity of these biotransforming enzymes would have consequences on the actions of genistein in the placenta and the overall effects on fetal growth.

8.2 A Comparison with Other Xenoestrogens

Xenoestrogens are exogenous compounds that, like genistein, mimic the effects of endogenous estrogen in the body. These chemicals are cause for concern as they potentially interfere with the normal endocrine processes, which may lead to adverse health effects in the exposed individual and/or their offspring. As such, they are classified as endocrine disrupting chemicals and have been the source of great interest in the last three decades. Synthetic and natural xenoestrogens are ubiquitous in the environment. The best-studied xenoestrogens are shown in Table 8.2. In most cases, xenoestrogens enter the body *via* ingestion, inhalation or cutaneously. Currently, two synthetic xenoestrogens, bisphenol A (BPA) and *p*-nonylphenol (NP), are generating global interest due to their potential adverse effects *in utero*. This Section will briefly describe the structural characteristics of these xenoestrogens, their biotransformation in the gastrointestinal and enterohepatic systems, and evidence of their transfer into the placenta in comparison with genistein.

Table 8.2 Natural and synthetic xenoestrogens and their sources. This table lists the best-studied xenoestrogens, categorised by type (natural or synthetic) and source.

Source	Xenoestrogen
Natural	
Soybeans	Genistein, daidzein, biochanin A
Many fruit, vegetable, leaves, and grains	Quercetin
Clover, alfalfa, spinach	Coumestrol
Synthetic	
Plastic, dental sealant	BPA
Industrial detergent, pesticides	NP
Anti-abortion drug	Diethylbestrol

Figure 8.4 Biotransformation of genistein for excretion. BPA (**11**) and NP (**12**) are two common environmental xenoestrogens. They both share structural similarities: phenolic compound with at least one hydroxyl group. These allow the compounds to bind estrogen receptors, albeit at different affinities, and are the basis for their estrogenic actions.

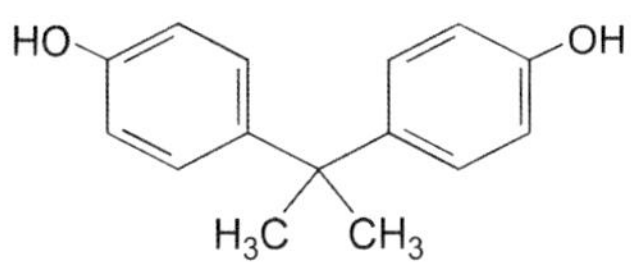

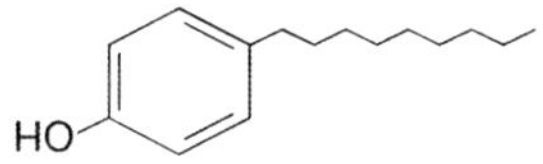

8.2.1 Chemical Structure of BPA and NP

BPA (4,4′-dihydroxy-2,2-diphenylpropane) is produced by combining acetone and phenol. The resultant product is a diphenol compound connected with a methyl bridge and two opposing hydroxyl groups (Figure 8.4, **11**). As such, it is a polar and lipophilic molecule. BPA is used in the production of polycarbonate plastics, and epoxy resins (reviewed in Kang *et al.* 2006). NP comprises one phenol ring and a long alkyl chain. It is classified as an alkylphenol and is widely used in industrial surfactants (Figure 8.4, **12**). BPA and NP are more hydrophobic and lipophilic compared with genistein. Genistein, because it more closely mimics 17β-estradiol in terms of structure, interacts with estrogen receptors with greater binding strength. Finally, BPA and NP are synthetic compounds and, therefore, do not exist in glycosylated or other conjugated forms prior to absorption in the gastrointestinal tract, similar to genistein.

8.2.2 Biotransformation of Xenoestrogens

8.2.2.1 Glucuronidation and Sulfation of BPA and NP

Conjugation of BPA and NP is less extensive than of genistein. Oral administration of BPA or NP in rats led to the appearance of conjugated products in their plasma. BPA was found to be primarily mono-glucuronidated and, to a lesser extent, mono-sulfated (Pottenger *et al.* 2000). BPA glucuronide-sulfate was also detected in F344 rat hepatocytes treated with BPA, but has not been observed elsewhere (Pritchett *et al.* 2002). In contrast, glucuronidation of NP was apparent in liver and serum following administration in rats *via* oral gavage or diet (Doerge *et al.* 2002), although its sulfation has only been suggested *via* indirect evidence. SULT1A1 appears to be the key sulfotransferase for many xenoestrogens and genistein as it was found to attach sulfate groups to BPA, genistein and NP more quickly than other SULTs in HepG2 human hepatoma cells (Suiko *et al.* 2000). SULT1C also demonstrated activity towards both BPA and NP, albeit at one-ninth the rate of SULT1A1.

8.2.2.2 Hydroxylation of BPA and NP

Where genistein is polyhydroxylated in the liver, it has been found that some metabolism of BPA and NP may involve the addition of one hydroxyl group. The introduction of a hydroxyl group at the 3-position of BPA reduces its hydrophobicity (3-OH-BPA $\log p = 3.16$ *vs.* BPA $\log p = 3.64$) (Nakagawa and Suzuki 2001). NP can undergo simultaneous hydroxylation and glucuronidation (Doerge *et al.* 2002). Further analysis showed that NP was glucuronidated on its existing hydroxyl group and an additional hydroxyl group was attached to an adjacent carbon. Very little is known about these minor metabolites, however, based on the metabolism of other xenoestrogens, it may aid excretion due to the increased hydrophilicity. BPA is readily excreted once it undergoes biotransformation (glucuronidation, sulfation, and hydroxylation). Unlike genistein, it is not metabolised into simpler molecules. The excretion process of NP is less defined.

8.2.3 Placenta Transfer

Efforts have been made to determine the extent of xenoestrogen transfer through the placenta, and initial reports of the characterisation of this process for BPA and genistein have been reported. The effects of NP in the placenta have not yet been elucidated. Using a placental perfusion model, our laboratory demonstrated trans-placental transfer of BPA (Balakrishnan *et al.* 2010b). Unlike genistein, most of the perfused compound remained in the unconjugated form. Furthermore, Schönfelder *et al.* (2002) found that BPA was detectable in both human maternal plasma, umbilical cord blood (umbilical vein), and placenta tissue (0.3–18.9 ng ml^{-1}, 0.2–9.2 ng ml^{-1}, and 1–104.9 ng g^{-1}, respectively). They did not delineate between conjugated and unconjugated BPA, however, the majority of xenoestrogens in plasma are conjugated. Indeed, there is evidence that the glucuronidated BPA easily passes through the placental barrier. Nishikawa *et al.* (2010) found that uterine perfusion of 2 µM BPA-glucuronide in the rat led to its appearance [at a small fraction (25 nmol)] in the fetus, but not in the placenta or amniotic fluid. They did detect a small amount (<32 pmol) of BPA in the amniotic fluid and fetus, which suggests that BPA-glucuronide is deconjugated in the placenta.

The only direct evidence for NP placental transfer is a study conducted in Taiwan, in which NP content was analysed in maternal and umbilical cord plasma (Chen *et al.* 2008). NP was not detected in approximately 74.2 and 24.0% of the cord blood samples obtained from Central and Northern Taiwan, respectively. Northern Taiwan, which includes the capital city, Taipei, has a higher population density, more infrastructure, and industry and therefore its inhabitants are more likely to be exposed to xenoestrogens. In the rest of the cohort, however, up to $268\,\text{ng}\,\text{g}^{-1}$ and $100\,\text{ng}\,\text{g}^{-1}$ of maternal and umbilical cord plasma, respectively, was discovered. It is unknown whether NP was biotransformed and the consequences of this modification.

8.3 Conclusion

Genistein is produced in abundance in soybeans and may be ingested as an aglycone or glucoside. It undergoes extensive biotransformation (deglycosylation, glucuronidation, sulfation, and hydroxylation) in the gastrointestinal tract and enters the circulation mainly as a glucuronide. Other xenoestrogens, such as BPA and NP, are also subjected to similar biotransformation processes, however, the range of resulting metabolites is not as numerous. The conjugating moieties described in this Chapter increase the hydrophilicity of these compounds, most likely to allow their excretion. It appears, however, that organs are exposed to both the circulating conjugated and unconjugated xenoestrogens. There is some evidence that both forms may cross the placenta. The consequences of fetal exposure to xenoestrogens have been of great interest in the last two decades. Animal studies have shown that maternal exposure to environmentally relevant doses of genistein and BPA can promote pathology development in the offspring, such as proliferative lesions in mammary glands, impaired neuronal development, and uterine adenocarcinomas (reviewed in Soto and Sonnenschein 2010). The effects of conjugated xenoestrogens on placental function and fetal development is not known. With the advancement of technology, future studies will need to characterise the biotransformation of estrogenic compounds, determine the amount of exposure to tissues, and how this affects their function.

Summary Points

- This Chapter focuses on the biotransformation of genistein.
- Genistein is naturally abundant in legumes, particularly soybeans and clovers.
- Genistein has estrogenic properties and can be classified as a phytoestrogen and xenoestrogen.
- In the body, genistein undergoes extensive biotransformation, enters the enterohepatic system, and subsequently circulates the body in the plasma.
- Other xenoestrogens, such as BPA and p-nonylphenol NP, are synthetic and are present in industrial chemicals, plastics, and many other products.
- BPA and NP are structurally distinct from genistein, however, their estrogenicity is derived from the possession of a phenol ring and hydroxyl group similar to the A ring of 17β-estradiol.
- The biotransformation of BPA and NP is similar to genistein, albeit less characterised.
- An important commonality between the different xenoestrogens is their ability to transfer across the placenta, which may have deleterious effects on fetal growth and well-being.

Key Facts

- Digestion occurs in the gastrointestinal tract, which comprises the oesophagus, stomach, small intestine (duodenum, jejunum, and ileum), large intestine (cecum, colon, and rectum), and anus.
- The complete digestion and excretion process takes up to 30 h.
- Once the ingested food has been digested, nutrients and other molecules are absorbed into the small intestinal wall, where they are biotransformed.
- The biotransformed molecules may be transported back into the intestinal lumen for excretion, however, they may also be absorbed into the blood circulation.
- Once in the bloodstream, they may be transported to different organs around the body. This is the mechanism by which the placenta becomes exposed to nutrients and environmental chemicals.

Definitions of Words and Terms

Bioavailability: The amount of drug or substance that enters the circulation and is exposed to tissue and organs.
Biotransformation: The chemical alteration of a molecule within the body.
Chemopreventive agent: A drug, substance, or food supplement that is used to prevent the development of cancer.
Deglycosylation: The removal of a sugar ($C_6H_{12}O_6$) group from a molecule.
Glucuronidation: The addition of glucuronic acid ($C_6H_{10}O_7$) to a molecule.
Hydrophilic: Water-loving property; readily dissolves in water.
Hydroxylation: The addition of a hydroxyl (-OH) group to a molecule.
Lipophilic: A molecule that readily dissolves or combines with lipids.
Phytoestrogen: A molecule that is naturally produced in plants and binds to estrogen receptors and activates similar molecular processes as the estrogen hormone.
Sulfation: The addition of a sulfate group to a molecule.
Xenoestrogen: A synthetic molecule that binds to estrogen receptors and activates similar molecular processes as the estrogen hormone.

List of Abbreviations

BPA	bisphenol A
CBG	cytosolic β-glycosidase
HPLC	high-performance liquid chromatography
LPH	lactose-phlorizin hydrolase
NP	*p*-nonylphenol
SULT	sulfotransferase
UGT	uridine-5′-disphosphate glucuronyltransferase

References

Adlercreutz, H., Yamada, T., Wahala, K., and Watanabe, S., 1999. Maternal and neonatal phytoestrogens in Japanese women during birth. *American Journal of Obstetrics and Gynecology.* 180: 737–743.

Andlauer, W., Kolb, J., Stehle, P., and Furst, P., 2000. Absorption and metabolism of genistein in isolated rat small intestine. *Journal of Nutrition.* 130: 843–846.

Balakrishnan, B., Thorstensen, E.B., Ponnampalam, A.P., and Mitchell, M.D., 2010a. Transplacental transfer and biotransformation of genistein in human placenta. *Placenta.* 31: 506–511.

Balakrishnan, B., Henare, K., Thorstensen, E.B., Ponnampalam, A.P., and Mitchell, M.D., 2010b. Transfer of bisphenol A across the human placenta. *American Journal of Obstetrics and Gynecology.* 202(393), e391–e397.

Bursztyka, J., Perdu, E., Tulliez, J., Debrauwer, L., Delous, G., Canlet, C., De Sousa, G., Rahmani, R., Benfenati, E., and Cravedi, J.P., 2008. Comparison of genistein metabolism in rats and humans using liver microsomes and hepatocytes. *Food and Chemical Toxicology.* 46: 939–948.

Chen, M.L., Chang, C.C., Shen, Y.J., Hung, J.H., Guo, B.R., Chuang, H.Y., and Mao, I.F., 2008. Quantification of prenatal exposure and maternal–fetal transfer of nonylphenol. *Chemosphere.* 73: S239–S245.

Coldham, N. G., Howells, L.C., Santi, A., Montesissa, C., Langlais, C., King, L.J., Macpherson, D.D., and Sauer, M.J., 1999. Biotransformation of genistein in the rat: elucidation of metabolite structure by product ion mass fragmentology. *Journal of Steroid Biochemistry and Molecular Biology.* 70: 169–184.

Coldham, N.G., Darby, C., Hows, M., King, L.J., Zhang, A.Q., and Sauer, M.J., 2002. Comparative metabolism of genistin by human and rat gut microflora: detection and identification of the end-products of metabolism. *Xenobiotica.* 32: 45–62.

Collier, A.C., Ganley, N.A., Tingle, M.D., Blumenstein, M., Marvin, K.W., Paxton, J.W., Mitchell, M.D., and Keelan, J.A., 2002a. UDP-glucuronosyltransferase activity, expression and cellular localization in human placenta at term. *Biochemical Pharmacology.* 63: 409–419.

Collier, A.C., Tingle, M.D., Paxton, J.W., Mitchell, M.D., and Keelan, J.A., 2002b. Metabolizing enzyme localization and activities in the first trimester human placenta: the effect of maternal and gestational age, smoking and alcohol consumption. *Human Reproduction.* 17: 2564–2572.

Dalais, F.S., Rice, G.E., Wahlqvist, M.L., Grehan, M., Murkies, A.L., Medley, G., Ayton, R., and Strauss, B.J., 1998a. Effects of dietary phytoestrogens in postmenopausal women. *Climacteric.* 1: 124–129.

Dalais, F.S., Rice, G.E., Wahlqvist, M.L., Hsu-Hage, B.H.H., and Wattanapenpaiboon, N., 1998b. Urinary excretion of isoflavonoid phytoestrogens in Chinese and Anglo-Celtic populations in Australia. *Nutrition Research.* 18: 1703–1709.

Day, A.J., DuPont, M.S., Ridley, S., Rhodes, M., Rhodes, M.J., Morgan, M.R., and Williamson, G., 1998. Deglycosylation of flavonoid and isoflavonoid glycosides by human small intestine and liver beta-glucosidase activity. *FEBS Letters*. 436: 71–75.

Day, A.J., Gee, J.M., DuPont, M.S., Johnson, I.T., and Williamson, G., 2003. Absorption of quercetin-3-glucoside and quercetin-4'-glucoside in the rat small intestine: the role of lactase phlorizin hydrolase and the sodium-dependent glucose transporter. *Biochemical Pharmacology*. 65: 1199–1206.

Doerge, D.R., Chang, H.C., Churchwell, M.I., and Holder, C.L., 2000. Analysis of soy isoflavone conjugation *in vitro* and in human blood using liquid chromatography–mass spectrometry. *Drug Metabolism and Disposition*. 28: 298–307.

Doerge, D.R., Twaddle, N.C., Churchwell, M.I., Chang, H.C., Newbold, R.R., and Delclos, K.B., 2002. Mass spectrometric determination of *p*-nonylphenol metabolism and disposition following oral administration to Sprague–Dawley rats. *Reproduction and Toxicology*. 16: 45–56.

He, D., Meloche, C.A., Dumas, N.A., Frost, A.R., and Falany, C.N., 2004. Different subcellular localization of sulphotransferase 2B1b in human placenta and prostate. *Biochemistry Journal*. 379: 533–540.

Hosada, K., Furuta, T., and Ishii, K., 2010. Simultaneous determination of glucuronic acid and sulfuric acid conjugated metabolites of daidzein and genistein in human plasma by high-performance liquid chromatography. *Journal of Chromatography B*. 878: 628–636.

Hosada, K., Furuta, T. and Ishii, K., 2011. Metabolism and disposition of isoflavone conjugated metabolites in humans after ingestion of kinako. *Drug Metabolism and Disposition*. 39: 1762–1767.

Hu, Y., Ge, C., Yuan, W., Zhu, R., Zhang, W., Du, L., and Xue, J., 2010. Characterization of fermented black soybean natto inoculated with *Bacillus* natto during fermentation. *Journal of the Science of Food and Agriculture*. 90: 1194–1202.

Hur, H.G., Lay, Jr, J.O., Beger, R.D., Freeman, J.P., and Rafii, F., 2000. Isolation of human intestinal bacteria metabolizing the natural isoflavone glycosides daidzin and genistin. *Archives of Microbiology*. 174: 422–428.

Joannou, G.E., Kelly, G.E., Reeder, A.Y., Waring, M., and Nelson, C., 1995. A urinary profile study of dietary phytoestrogens. The identification and mode of metabolism of new isoflavonoids. *Journal of Steroid Biochemistry and Molecular Biology*. 54: 167–184.

Kang, J.H., Kondo, F., and Katayama, Y., 2006. Human exposure to bisphenol A. *Toxicology*. 226: 79–89.

King, R.A., 1998. Daidzein conjugates are more bioavailable than genistein conjugates in rats. *American Journal of Clinical Nutrition*. 68: 1496S–1499S.

Kulling, S.E., Honig, D.M., and Metzler, M., 2001. Oxidative metabolism of the soy isoflavones daidzein and genistein in humans *in vitro* and *in vivo*. *Journal of Agricultural and Food Chemistry*. 49: 3024–3033.

Lindsay, J., Wang, L.L., Li, Y., and Zhou, S.F., 2008. Structure, function and polymorphism of human cytosolic sulfotransferases. *Current Drug Metabolism*. 9: 99–105.

Mathias, K., Ismail, B., Corvalan, C.M., and Hayes, K.D., 2006. Heat and pH effects on the conjugated forms of genistin and daidzin isoflavones. *Journal of Agriculture and Food Chemistry*. 54: 7495–7502.

Mitra, P., and Audus, K.L., 2009. Expression and functional activities of selected sulfotransferase isoforms in BeWo cells and primary cytotrophoblast cells. *Biochemical Pharmacology*. 78: 1475–1482.

Nakagawa, Y., and Suzuki, T., 2001. Metabolism of bisphenol A in isolated rat hepatocytes and oestrogenic activity of a hydroxylated metabolite in MCF-7 human breast cancer cells. *Xenobiotica*. 31: 113–123.

Nakano, H., Ogura, K., Takahashi, E., Harada, T., Nishiyama, T., Muro, K., Hiratsuka, A., Kadota, S., and Watabe, T., 2004. Regioselective monosulfation and disulfation of the phytoestrogens daidzein and genistein by human liver sulfotransferases. *Drug Metabolism and Pharmacokinetics*. 19: 216–226.

Nemeth, K., Plumb, G.W., Berrin, J.G., Juge, N., Jacob, R., Naim, H.Y., Williamson, G., Swallow, D.M., and Kroon, P.A., 2003. Deglycosylation by small intestinal epithelial cell β-glucosidases is a critical step in the absorption and metabolism of dietary flavonoid glycosides in humans. *European Journal of Nutrition*. 42: 29–42.

Nishikawa, M., Iwano, H., Yanagisawa, R., Koike, N., Inoue, H., and Yokota, H., 2010. Placental transfer of conjugated bisphenol A and subsequent reactivation in the rat fetus. *Environmental Health Perspectives*. 118: 1196–1203.

Nishiyama, T., Ogura, K., Nakano, H., Kaku, T., Takahashi, E., Ohkubo, Y., Sekine, K., Hiratsuka, A., Kadota, S., and Watabe, T., 2002. Sulfation of environmental estrogens by cytosolic human sulfotransferases. *Drug Metabolism and Pharmacokinetics*. 17: 221–228.

Pottenger, L.H., Domoradzki, J.Y., Markham, D.A., Hansen, S.C., Cagen, S.Z., and Waechter, Jr, J.M., 2000. The relative bioavailability and metabolism of bisphenol A in rats is dependent upon the route of administration. *Toxicological Sciences*. 54: 3–18.

Pritchett, J.J., Kuester, R.K., and Sipes, I.G., 2002. Metabolism of bisphenol A in primary cultured hepatocytes from mice, rats, and humans. *Drug Metabolism and Disposition*. 30: 1180–1185.

Reimers, A., Ostby, L., Stuen, I., and Sundby, E., 2011. Expression of UDP-glucuronosyltransferase 1A4 in human placenta at term. *European Journal of Drug Metabolism and Pharmacokinetics*. 35: 79–82.

Ronis, M.J., Little, J.M., Barone, G.W., Chen, G., Radominska-Pandya, A., and Badger, T.M., 2006. Sulfation of the isoflavones genistein and daidzein in human and rat liver and gastrointestinal tract. *Journal of Medicinal Food*. 9: 348–355.

Schoefer, L., Mohan, R., Braune, A., Birringer, M., and Blaut, M., 2002. Anaerobic C-ring cleavage of genistein and daidzein by *Eubacterium ramulus*. *FEMS Microbiology Letters*. 208: 197–202.

Schönfelder, G., Wittfoht, W., Hopp, H., Talsness, C.E., Paul, M., and Chahoud, I., 2002. Parent bisphenol A accumulation in the human maternal–fetal–placental unit. *Environmental Health Perspectives*. 110: A703–A707.

Shelnutt, S.R., Cimino, C.O., Wiggins, P.A., Ronis, M.J., and Badger, T.M., 2002. Pharmacokinetics of the glucuronide and sulfate conjugates of genistein and daidzein in men and women after consumption of a soy beverage. *American Journal of Clinical Nutrition.* 76: 588–594.

Soto, A.M., and Sonnenschein, C., 2010. Environmental causes of cancer: endocrine disruptors as carcinogens. *Nature Reviews Endocrinology.* 6: 363–370.

Soucy, N.V., Parkinson, H.D., Sochaski, M.A., and Borghoff, S.J., 2006. Kinetics of genistein and its conjugated metabolites in pregnant Sprague–Dawley rats following single and repeated genistein administration. *Toxicological Sciences.* 90: 230–240.

Suiko, M., Sakakibara, Y., and Liu, M.C., 2000. Sulfation of environmental estrogen-like chemicals by human cytosolic sulfotransferases. *Biochemical Biophysical Research Communications.* 267: 80–84.

Tamura, M., Tsushida, T., and Shinohara, K., 2007. Isolation of an isoflavone-metabolizing, *Clostridium*-like bacterium, strain TM-40, from human faeces. *Anaerobe.* 13: 32–35.

Tang, L., Singh, R., Liu, Z., and Hu, M., 2009. Structure and concentration changes affect characterization of UGT isoform-specific metabolism of isoflavones. *Molecular Pharmacology.* 6: 1466–1482.

Walgren, R.A., Lin, J.T., Kinne, R.K., and Walle, T., 2000. Cellular uptake of dietary flavonoid quercetin 4′-β-glucoside by sodium-dependent glucose transporter SGLT1. *Journal of Pharmacology and Experimental Therapeutics.* 294: 837–843.

Zhu, W., Xu, H., Wang, S.W., and Hu, M., 2010. Breast cancer resistance protein (BCRP) and sulfotransferases contribute significantly to the disposition of genistein in mouse intestine. *American Association of Pharmaceutical Scientists Journal.* 12: 525–536.

CHAPTER 9
The Biological Effects of Genistein and its Intracellular Metabolite, 5,7,3',4'-Tetrahydroxyisoflavone

GIULIA CORONA* AND JEREMY P. E. SPENCER

Molecular Nutrition Group, Food and Nutritional Sciences Department, School of Chemistry, Food and Pharmacy, University of Reading, Reading RG6 6AP, UK
*Email: g.corona@reading.ac.uk

9.1 Genistein and its Biological Activities

The isoflavone genistein (5,7,4'-trihydroxyisoflavone) (Figure 9.1) is present primarily in soybeans and a variety of legumes (Reinli and Block 1996), where its content can differ depending on the variety of soybean, the year harvested, geographic location and plant part (Delmonte and Rader 2006). Non-soy legumes such as lentils and other types of beans do not contain appreciable amounts of genistein (Liggins *et al.* 2000). However, genistein can also occur in soy products in the form of its glycoside (Figure 9.1) genistin (Liggins *et al.* 2000; Reinli and Block 1996). Genistein exhibits a wide range of biological effects that contribute to its potential health benefits (Head 1998), such as oestrogenic activity (Cassidy *et al.* 1994), anticancer effects (Messina and Bennink 1998), antioxidant actions (Cai and Wei 1996) and anti-inflammatory activity (Fanti *et al.* 2006).

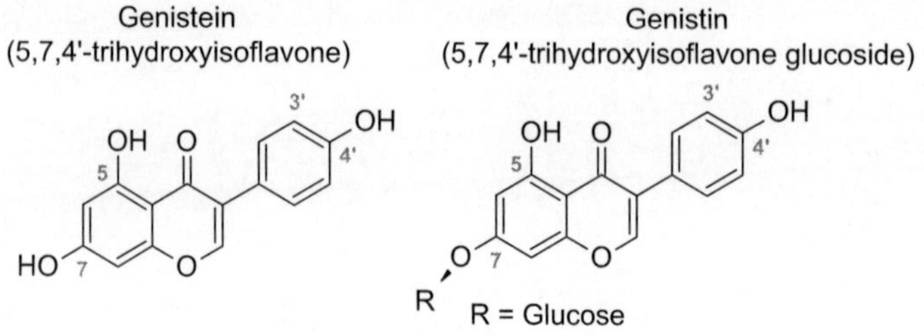

Figure 9.1 Chemical structures of genistein and genistin.

9.1.1 Oestrogenic Activity

Isoflavones are structurally similar to oestrogens (Miksicek 1995) and can function both as oestrogen agonists and antagonists depending on the hormonal milieu and the target tissue and species under investigation (Molteni *et al.* 1995). The anti-oestrogenic effects of genistein are due to its ability to bind to the oestrogen receptor (Wang *et al.* 1996; Yearley *et al.* 2007), therefore suppressing the more harmful effects of oestrogens, such as their effects on oestrogen-sensitive cancers (Anthony *et al.* 1998; Goodman *et al.* 1997; Horn-Ross *et al.* 2003). For example, in MCF-7 breast cancer cells, which are known to be oestrogen receptor-positive, genistein is able to stimulate oestrogen-responsive pS2 mRNA expression and compete with [^{3}H]estradiol binding to the oestrogen receptor with 50% inhibition (Wang *et al.* 1996). Furthermore, isoflavones have been found to exert an oestrogenic effect in both the absence or presence of endogenous oestrogen (Adlercreutz and Mazur 1997). Alternatively, genistein may reduce oestrogenic activity *via* its ability to induce the production of sex hormone-binding globulin (SHBG), thus leading to a faster clearance of sex hormones (including oestrogens) (Mousavi and Adlercreutz 1993) and a reduced risk of hormone-sensitive breast and prostate cancer (Messina and Hilakivi-Clarke 2009; Smith *et al.* 2008). Similarly, genistein is known to inhibit many of the enzymes involved in the biosynthesis and metabolism of steroid hormones (Head 1998). For example, genistein has direct effects on cellular progesterone synthesis which involve the inhibition of hydroxysteroid dehydrogenase/isomerase (3β-HSD) enzyme activity across the post-cAMP pathway. It directly affects cellular progesterone synthesis through its inhibition of 3β-HSD gene expression and down-regulation of its transcription (Tiemann *et al.* 2007).

There is convincing evidence that a diet rich in soy protein, which contains genistein, has an impact on the hormonal status and the regulation of the menstrual cycle of premenopausal women (Cassidy *et al.* 1994). A daily intake of 60 g of soy protein (containing 45 mg of isoflavones) for 1 month significantly increased follicular phase length and/or delayed menstruation, significantly suppressed the mid-cycle surges of luteinizing hormone and follicle-stimulating hormone and increased plasma oestradiol concentrations in the follicular phase (Cassidy et al. 1994). These effects were potentially associated with lower risk of breast cancer (Cassidy et al. 1994; Henderson et al. 1985).

9.1.2 Anticancer Effects

Interest in genistein as a anticancer agent arose due to population-based data indicating a link between genistein consumption and a decreased risk of mortality from several types of cancer, in particular prostate and breast cancer (Pavese et al. 2010). Epidemiological studies indicate a protective effect of isoflavone ingestion against breast cancer in premenopausal women and women who had high soy intakes during adolescence (Peeters et al. 2003; Piller et al. 2006), whereas in postmenopausal women the data vary depending on the study (Jones et al. 2002; Ju et al. 2006; 2002). Genistein is a potent inhibitor of the growth of breast cancer cells, whereas its β-glucoside genistin has little effect (Peterson and Barnes 1991; 1993). The effects of genistein on breast cancer cell growth and proliferation were studied in oestrogen-receptor negative (MDA-468) and positive (MCF-7 and MCF-7-D-40) cell lines and it was ascertained that the presence of the oestrogen receptor is not required for genistein to inhibit the cancer cell growth (Peterson and Barnes 1991). The oestrogen-independent effects of genistein can be due to their ability, or the ability of one of their metabolites, to block cell cycle progression through direct effects on intracellular signalling (Nguyen et al. 2006). For example, genestein has been shown to be capable of causing a cell cycle block in the G_2–M phase *in vitro* (Cappelletti et al. 2000; Santell et al. 2000) and an overexpression of the cyclin-dependent kinase inhibitor p21^{WAF1} in breast cancer cells, leading to cell cycle arrest (Chinni et al. 2003). Other mechanisms whereby genistein can exert its anticancer properties include the induction of apoptosis (Sergeev 2004; Xu and Loo 2001), inhibition of tyrosine kinases (Mitropoulou et al. 2002; Morton et al. 1999), modulation of mitogen-activated protein kinase (MAPK) signalling (Li et al. 2006), alterations in the phosphatidylinositol 3-kinase cascade (Lee et al. 2001) and inhibition of DNA topoisomerases (McCabe and Orrenius 1993).

The relationship between isoflavone supplementation and prostate cancer has also been extensively investigated. Male rats fed with genistein (0, 25 and 250 mg kg^{-1}) over their lifetime led to an inhibition of the development of *N*-methylnitrosourea-induced prostate invasive adenocarcinomas, in a dose-dependent manner (Wang et al. 2002). It has been proposed that the regulation of the androgen receptor (AR)–Akt–phosphatase and tensin homolog deleted on chromosome 10 (PTEN) pathway by genistein may be one of the molecular mechanisms by which it inhibits proliferation and induces apoptosis in prostate cancer cells (Wang et al. 2009). The Akt–glycogen synthase kinase-3 (GSK-3) pathway and its downstream effectors have also been identified as targets for the chemopreventive action of genistein in a transgenic adenocarcinoma mouse prostate model (TRAMP/FVB) mice, in which incorporation of genistein in the diet significantly inhibited the activation of Akt, restored the activation of

GSK-3β, reduced cyclin D1 levels post-transcriptionally and maintained the expression of the cadherin-1 complex *via* down-regulation of snail-1, decreasing the proliferative potential, retarding cancer progression and maintaining the integrity of the prostatic epithelial cells *in vivo* (El Touny and Banerjee 2007). Another mechanism of increased prostate cancer cell death by genistein is proposed to occur *via* its ability to inhibit nuclear factor-κB (NF-κB) signalling, leading to altered expression of regulatory cell cycle proteins such as cyclin B and/or $p21^{WAF1}$/Cip1, thus promoting G_2–M arrest. These findings support the important and novel strategy of combining genistein with radiation for the treatment of prostate cancer (Raffoul *et al.* 2006).

9.1.3 Antioxidant Action

Genistein, similar to other isoflavones, possesses a relatively strong antioxidant potential *in vitro* (Han *et al.* 2009), and exerts its antioxidant actions by scavenging free radicals (Kim *et al.* 2010), chelating metals (Dowling *et al.* 2010), inhibiting the production of oxidizing species, such as H_2O_2 (Sethy-Coraci *et al.* 2005), and by enhancing the activity of endogenous antioxidant enzymatic systems, such as catalase (Cai and Wei 1996). Although direct antioxidant actions *in vivo* are highly unlikely (Williams *et al.* 2004), physiologically achievable concentrations of 5 nM genistein increased intracellular reduced glutathione (GSH) levels by approximately 10%, whereas cellular α-tocopherol and uric acid remained unchanged (Guo *et al.* 2002). The mechanisms behind increases in GSH may include the effect of isoflavones on enzymes involved in the synthesis of GSH, such as γ-glutamyl cysteine synthetase (γ-GCS) (Guo *et al.* 2002). Dietary administration of genistein for 30 days in mice increases the activity of antioxidant enzymes such as catalase, superoxide dismutase (SOD), glutathione peroxidase (GPx) and glutathione in various organs, including skin, small intestine, liver, kidney and lung (Cai and Wei 1996). Furthermore, treatment of macrophages with genistein reduces lipopolysaccharide (LPS)-induced GSH depletion and induces SOD and catalase (Choi *et al.* 2003). Similar effects have been reported in rats, where feeding the animals with an isoflavone-rich diet increased GPx and glutathione reductase activities, blood glutathione levels and glutathione *S*-transferase levels in kidney (Appelt and Reicks 1999).

9.1.4 Anti-inflammatory Activity

Genistein has been shown to have an effect on the production of inflammatory mediators in human peripheral blood mononuclear and polymorphonuclear leukocytes stimulated with LPS and interferon-γ (IFNγ) (Richard *et al.* 2005). The anti-inflammatory activity of genistein has also been noted in human brain microvascular endothelial cells, where it dose-dependently inhibited cytokine-induced up-regulation of pro-inflammatory mediators such as tumour necrosis factor-α (TNF-α), interleukin-1β (IL-1β), monocyte chemoattractant protein-1 (MCP-1), IL-8 and intercellular adhesion molecule-1 (ICAM-1), and cytokine-induced transmigration of blood leukocytes (Lee and Lee 2008). Exposure to high amounts of genistein, as occurs in traditional East Asian diets, where the mean total intake of isoflavones is estimated between 17 and 47 mg day^{-1} (Vergne *et al.* 2009), may be associated with a lower incidence of inflammatory bowel disease (IBD) (Loftus 2004). Indeed, the risk of IBD appears to increase in Asian immigrants adopting Western lifestyles, suggesting a protective role for isoflavones (Loftus 2004). To confirm this hypothesis, the effect of orally administered genistein on the inflammatory response to 2,4,6-trinitrobenzenesulfonic acid-induced chronic colitis in rats has been investigated and indicates that genistein exerts beneficial anti-inflammatory effects by inhibiting molecular and biochemical inflammatory markers in the colon, specifically cyclo-oxygenase-2 (COX-2) and myeloperoxidase (MPO) (Seibel *et al.* 2009). Diabetic retinopathy (DR) is also associated with microglial activation and increased levels of inflammatory cytokines, and the efficacy of genistein for alleviation of diabetes-induced retinal inflammation has been studied in an animal model of diabetes. Genistein was found to be effective in dampening diabetes-induced retinal inflammation by reducing TNF-α release and inhibiting extracellular-signal-regulated kinase (ERK) and p38 phosphorylation in activated microglial cells (Ibrahim *et al.* 2010).

9.2 Intracellular Formation of the Genistein Metabolite 5,7,3′,4′-Tetrahydroxyisoflavone (THIF)

The biological effects of genistein are undoubtedly affected be its metabolism by phase II metabolic conjugation to glucuronic acid or sulphuric acid, catalysed by UDP-glucuronyl transferase or sulfotransferase enzymes in the intestinal epithelium and in the liver. In addition, intracellular metabolism of genistein may also act to affect it cellular actions. Genistein undergoes further transformation intracellularly to yield novel bioactive metabolites. For example, in tumorigenic breast epithelial cells, genistein is selectively taken up into the cell and is subjected to significant intracellular metabolism by cytochrome P450 (CYP450) enzymes leading to the formation of both THIF (orobol) and two glutathionyl conjugates of THIF (Nguyen *et al.* 2006). A scheme of the described metabolic pathways is presented in Figure 9.2. It is also shown that the co-treatment with cimetidine prevents the conversion of genistein into THIF, and because cimetidine is known to inhibit the CYP450 isoforms 1A2, 2C9, 2C19, 2D6 and 3A4, it is likely that one of these isoenzymes is responsible for the conversion of genistein into THIF in T47D cells (Nguyen *et al.* 2006). Previously, it has been suggested that isoflavone metabolism in transformed but not non-transformed breast epithelial cells may modulate the growth inhibitory effects of genistein (Peterson *et al.* 1996). Glutathionyl conjugates of THIF in cells may be formed either enzymatically *via* the action of glutathione *S*-transferase or non-enzymatically *via* oxidative metabolism of THIF and subsequent reaction of THIF *o*-quinone with the cellular thiol GSH. However, the inhibition of glutathione *S*-transferase does not block glutathione conjugate

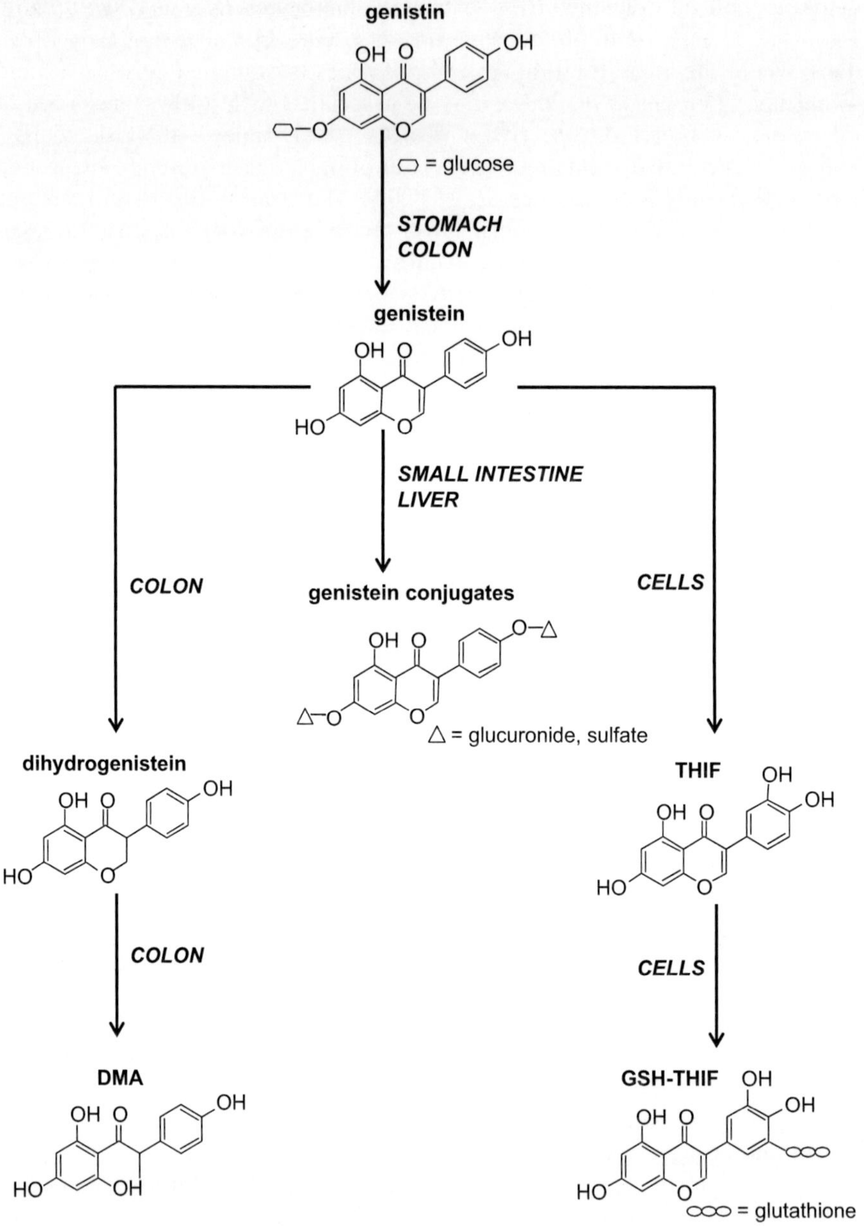

Figure 9.2 Genistein metabolic transformations. The glycoside genistin is not absorbed intact but hydrolyzed in the stomach and in the colon. Genistein can be absorbed by passive diffusion in the small intestine or undergo further biotransformation to a range of metabolites, such as dihydrogenistein and 6'-hydroxy-O-demethylangolensin (DMA) formed in the colon. Genistein can also undergo metabolic conjugation to glucuronic acid or sulphuric acid in the intestinal epithelium and in the liver. Intracellularly, genistein is hydroxylated to THIF (orobol) and two glutathionyl conjugates of THIF can also be formed.

formation, indicating that the latter is more likely to occur (Nguyen et al. 2006). Conjugations with thiols, such as glutathione, represent a major target for quinones, and the detoxification of quinones by GSH conjugation is generally considered to be cytoprotective (Monks and Lau 1997). Glutathionyl conjugates from a variety of polyphenol quinones have been observed in cellular systems and display a wide array of biological activities (Corona et al. 2006; Monks and Lau 1997; Spencer et al. 2003). Indeed, the redox activity of polyphenols is frequently enhanced following conjugation with GSH (Monks and Lau 1997) and, as GSH conjugation is often coupled to the subsequent export of the adduct from cells, the conjugation of THIF appears to represent a detoxification pathway (Monks and Lau 1997; Spencer et al. 2003).

9.3 THIF and Cancer

The exposure of breast epithelial cells to physiological concentrations of genistein selectively induces growth arrest and G_2–M phase cell cycle block in tumorigenic (T47D) but not non-tumorigenic (MCF10A) breast epithelial cells (Nguyen et al. 2006). These effects of cancer cell proliferation were paralleled by significant differences in the association of genistein with cells and in particular its intracellular metabolism (Nguyen et al. 2006). Previously, it had been suggested that isoflavone metabolism in transformed but not non-transformed breast epithelial cells may modulate the growth inhibitory effects of genistein (Peterson et al. 1996). In agreement with this hypothesis, genistein appeared to be selectively taken up into tumorigenic breast epithelial cells and subjected to significant intracellular metabolism leading to the formation of THIF and two glutathionyl conjugates of THIF (Nguyen et al. 2006). In contrast, there was minimal cell association of genistein with MCF10A cells and no subsequent formation of free THIF (Nguyen et al. 2006). In breast cancer cells THIF formation triggers the activation of the MAPK p38 (Figure 9.3). Active p38 prevents the phosphorylation of cyclin B1 and hence its transport to the nucleus, an event essential for correct functioning of the cdc2–cyclin B1 complex. In addition, active p38 may undergo translocation to the nucleus where it directly inhibits the phosphorylation/activation of cdc2, thereby blocking entry of cells into mitosis (G_2–M block) (Nguyen et al. 2006). These data suggest that the formation of THIF is crucial in driving the anticancer effects of genistein *in vivo*, in particular by inhibiting the proliferation of cancer cells but not affecting normal cell function. Further evidence relating to the cellular actions of the genistein metabolite THIF were elucidated in human breast carcinoma cells and summarized in Figure 9.3 (Vauzour et al. 2007). Here, THIF induced a G_2–M cell cycle arrest in T47D tumorigenic breast epithelial cells, which was mediated by the activation of ataxia telangiectasia and Rad3-related kinase (ATR) *via* its phosphorylation at Ser^{428} (Vauzour et al. 2007). This activation of ATR appeared to result from THIF-induced increases in intracellular oxidative stress, a depletion of cellular GSH and an increase in DNA strand breakage (Vauzour et al. 2007). These events led to the downstream inhibition of cdc2, which was accompanied by the phosphorylation of both p53 (Ser^{15})

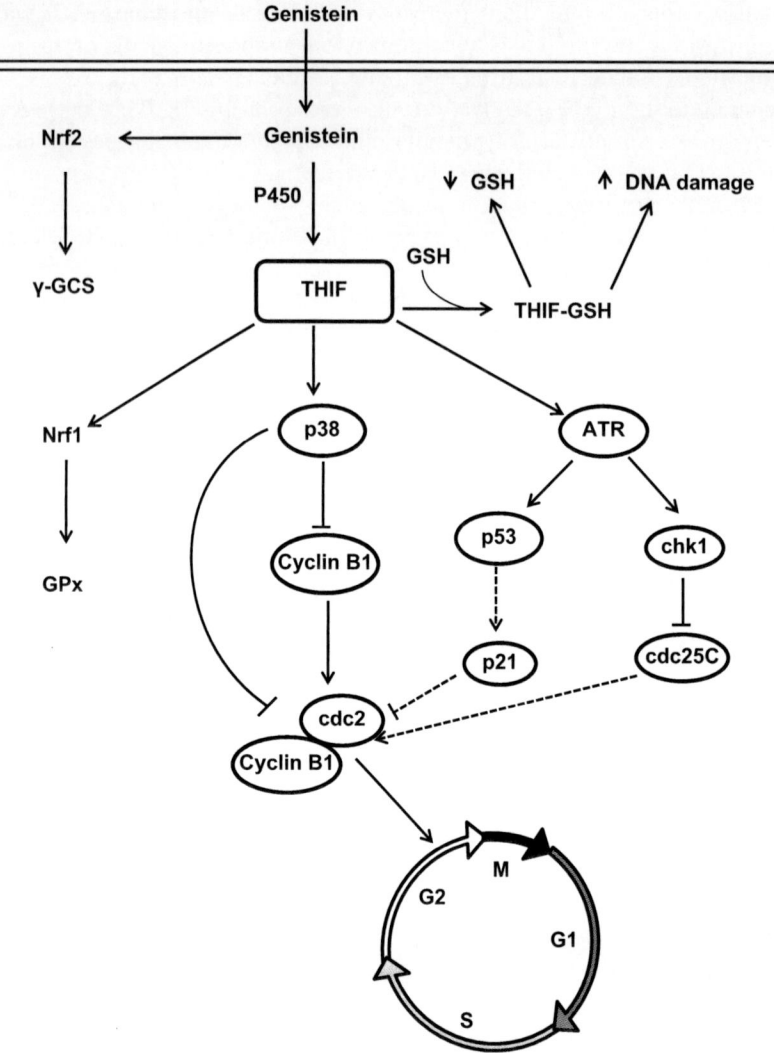

Figure 9.3 Genistein and THIF: intracellular mechanism of action. Genistein enters cells where it is subject the CYP450-induced intracellular metabolism, yielding THIF. THIF may also react with intracellular GSH to form the adduct THIF–GSH, inducing a reduction of intracellular GSH and an increased DNA damage. Genistein and THIF might induce the transcription of antioxidant enzymes: more specifically, genistein directly activates the translocation of Nrf2, which will increase the transcription of γ-GCS, leading to increases in GSH synthesis, whereas THIF stimulates the release of Nrf1 with the subsequent translocation into the nucleus where it may interact with the electrophile response element and induce the transcription of GPx. THIF triggers the activation of the MAPK p38. Active p38 prevents the phosphorylation of cyclin B1 and hence its transport to the nucleus, an event essential for correct functioning of the cdc2–cyclin B1 complex. In addition, active p38 may

and Chk1 (Ser[296]) and the de-activation of cdc25C phosphatase. It was suggested by the authors that the anti-proliferative actions of THIF may be mediated by initial oxidative DNA damage, activation of ATR and downstream regulation of the p53 and Chk1 pathways, leading to cell cycle arrest in G_2–M (Vauzour *et al.* 2007). These data sets suggest that the formation of THIF may mediate the effects of genistein on cancer cells. This hypothesis is in agreement with other investigations, which reported the effects of THIF (orobol) and other isoflavones (genistein, daidzein and 7,8,4'-trihydroxyisoflavone) on angiogenesis and endothelial cell proliferation (Kiriakidis *et al.* 2005). In a chicken chorioallantoic membrane assay, all compounds had the capacity to inhibit angiogenesis, albeit with different potencies [genistein > THIF > daidzein (48.98%) and 7,8,4'-trihydroxyisoflavone (24.42%)], and also inhibited endothelial cell proliferation, with THIF causing the greatest inhibition at lower concentrations (Kiriakidis *et al.* 2005).

9.4 THIF and Endothelial Function

Genistein has been shown to be able to protect endothelial cells against damage induced by oxidative stress and it decreases intracellular glutathione levels (Hernandez-Montes *et al.* 2006). These effects appear to also be mediated by genistein intracellular metabolism. When genistein enters cells, it is subject the CYP450-induced intracellular metabolism, yielding THIF. THIF may also react with intracellular GSH to form the adduct THIF–GSH (Figure 9.3), inducing a reduction of intracellular GSH and increased DNA damage (Hernandez-Montes *et al.* 2006). Genistein and THIF are also able to induce the transcription of antioxidant enzymes: more specifically genistein directly activates the translocation of Nrf2, which will increase the transcription of γ-GCS, leading to increases in GSH synthesis, whereas THIF stimulates the release of Nrf1 with the subsequent translocation into the nucleus where it may interact with the electrophile response element and induce the transcription of GPx (Hernandez-Montes *et al.* 2006). The intracellular formation of glutathionyl conjugates of THIF may result either from the action of glutathione *S*-transferase or from the autoxidation of THIF and subsequent reaction of THIF *o*-quinine with GSH (Hernandez-Montes *et al.* 2006). These finding are

---

undergo translocation to the nucleus where it directly inhibits the phosphorylation/activation of cdc2, thereby blocking entry of cells into mitosis (G_2–M block). THIF formation also induces the activation of ATR, accompanied by the phosphorylation of both p53 and Chk1, downstream signals playing an important role in DNA damage checkpoint control. One of p53 downstream targets is the tumour suppressor protein p21$^{Waf1/Cip1}$, which can act as an inhibitor of cell cycle progression *via* its ability to inhibit cdc2. THIF-induced activation of Chk1 is paralleled by an inactivation of cdc25C phosphatase, causing it to be sequestered it in the cytoplasm and preventing it from de-phosphorylating cdc2.

in agreement with other studies that indicate that glutathionyl conjugates from a variety of polyphenol quinones have been observed in cellular systems and display a wide array of biological activities (Corona *et al.* 2006; Monks and Lau 1997; Spencer *et al.* 2003). It appeared that the protective effects observed in endothelial cells in response to genistein exposure were dependent on the induction of the enzyme GPx, due to increases in both GPx mRNA and enzyme activity. It was suggested that the protective effects of genistein and its intracellular metabolites on endothelial cells were depend primarily on the activation of GPx mediated by Nrf1 activation, and not on Nrf2 activation or increases in glutathione synthesis (Hernandez-Montes *et al.* 2006).

9.5 Summary

The cellular mechanism of action of genistein is dependent on the cellular metabolism and uptake of the compound. Genistein enters the cells where it is hydroxylated to form THIF. It is suggested that THIF formation is crucial for its cellular activity and can be mediated by interactions of the newly formed cellular metabolite with signalling pathways. THIF is generated by the action of CYP450 enzymes in cells (Nguyen *et al.* 2006), and the catechol group-containing metabolite is observed to persist within the cells for up to 24 h without undergoing O-methylation (Vauzour *et al.* 2007), as has been observed for other catechol-containing polyphenols (Corona *et al.* 2006; Spencer *et al.* 2003). Indeed, cellular actions of isoflavones are cell-type specific and also depend on its CYP450-related metabolism, oxidative metabolism and GSH conjugation (Hernandez-Montes *et al.* 2006). Experiments conducted by testing both genistein and daizein in endothelial cells showed that only genistein significantly protected against the oxidative injury induced by H_2O_2, whereas daizein had no protective effect. In the same cell line, both genistein and daidzein significantly increased γ-GCS levels and were able to induce increases in cytosolic accumulation and nuclear translocation of Nrf2. In contrast, GSH levels increased only in the cells treated with daidzein; whereas, with genistein, they were significantly lower, due to its sequestration by metabolism. In addition, cytosolic levels of Nrf1, and the degree to which Nrf1 underwent nuclear translocation, were significantly higher in cells exposed to genistein than in those exposed to daidzein. Therefore it appeared that protection by genistein depended on its induction of GPx and changes in Nrf1 activation, suggesting that they may underlie its protective effects in endothelial cells, and be mediated by intracellular metabolism to THIF that is only relevant to genistein but not daidzein, for which hydroxylation cannot occur. Future work would be needed to fully investigate these metabolic differences in view of a more complete understanding on their molecular mechanism of action, and ascertain the presence of intracellular metabolites in plasma and urine following genistein intake.

The Biological Effects of Genistein

Summary Points

- The isoflavone genistein is present in soy products, mainly in the form of glycosides such as genistin.
- Genistein consumption is associated to numerous potential health benefits, such as estrogenic activity, anticancer effects, antioxidant actions, anti-inflammatory activity and cardiovascular effects.
- The biological effects of genistein in cells and tissues will be dependent on its absorption, metabolism and distribution.
- The most prevalent forms in the diet do not necessarily give rise to the highest concentrations *in vivo* and metabolites may be present in relevant amounts.
- Glycosilated forms of genistein are not absorbed intact in humans and their bioavailability requires hydrolysis to occur in the stomach and/or large intestine.
- Genistein can be absorbed by passive diffusion in the small intestine, or can be converted to dihydrogenistein and 6′-hydroxy-*O*-demethylangolensin (DMA) in the colon.
- Genistein can undergo phase II metabolic conjugation to glucuronic acid or sulphuric acid, catalysed by UDP-glucuronyl transferase or sulfotransferase enzymes in the intestinal epithelium and in the liver.
- Intracellularly, genistein is subjected to metabolic transformation by CYP450 enzymes, leading to the formation of both THIF (orobol) and two glutathionyl conjugates of THIF.
- THIF formation is crucial for cellular activity, acts *via* a mechanism involving oxidative DNA oxidation and the activation of downstream signalling pathways, and may mediate the effects of genistein intracellularly.

Key Facts

- The biological properties of flavonoids in the diet and their activity *in vivo* are dependent on the extent of their biotransformation and conjugation during absorption.
- The most prevalent flavonoids in the diet do not necessarily correspond to the most bioactive forms *in vivo*.
- Glycosylation has a great influence on flavonoid absorption and glycosylated forms may not be absorbed intact in humans.
- Dietary flavonoids are subjected to extensive phase I (oxidation, reduction and hydrolysis) and phase II (conjugation to glucuronic acid, sulphate and glutathione) metabolic reactions catalysed by enzymes, such as CYPP450, found both in the small intestine and the liver.
- Further transformations of flavonoids may occur in the colon, where bacterial enzymes may catalyse many reactions including hydrolysis, dehydroxylation, demethylation, ring cleavage and decarboxylation, as well as rapid de-conjugation, and are also able to catalyse the breakdown of the flavonoid backbone itself into simpler molecules such as phenolic acids.

List of Abbreviations

ATR	ataxia telangiectasia and Rad3-related kinase
CYP450	cytochrome P450
γ-GCS	γ-glutamyl cysteine synthetase
GPx	glutathione peroxidase
GSH	reduced glutathione
GSK-3	glycogen synthase kinase-3
3β-HSD	hydroxysteroid dehydrogenase/isomerase
IBD	inflammatory bowel disease
IL	interleukin
LPS	lipopolysaccharide
MAPK	mitogen-activated protein kinase
SOD	superoxide dismutase
THIF	5,7,3′,4′-tetrahydroxyisoflavone
TNF-α	tumour necrosis factor-α

References

Adlercreutz, H., and Mazur, W., 1997. Phyto-oestrogens and Western diseases. *Annals of Medicine.* 29(2), 95–120.

Anthony, M.S., Clarkson, T.B., and Williams, J.K., 1998. Effects of soy isoflavones on atherosclerosis: potential mechanisms. *American Journal of Clinical Nutrition.* 68(Suppl. 6), 1390S–1393S.

Appelt, L.C., and Reicks, M.M., 1999. Soy induces phase II enzymes but does not inhibit dimethylbenz[a]anthracene-induced carcinogenesis in female rats. *Journal of Nutrition.* 129(10), 1820–1826.

Cai, Q., and Wei, H., 1996. Effect of dietary genistein on antioxidant enzyme activities in SENCAR mice. *Nutrition and Cancer.* 25(1), 1–7.

Cappelletti, V., Fioravanti, L., Miodini, P., and Di Fronzo, G., 2000. Genistein blocks breast cancer cells in the G2-M phase of the cell cycle. *Journal of Cellular Biochemistry.* 79(4), 594–600.

Cassidy, A., Bingham, S., and Setchell, K.D., 1994. Biological effects of a diet of soy protein rich in isoflavones on the menstrual cycle of premenopausal women. *American Journal of Clinical Nutrition.* 60(3), 333–340.

Chinni, S.R., Alhasan, S.A., Multani, A.S., Pathak, S., and Sarkar, F.H., 2003. Pleotropic effects of genistein on MCF-7 breast cancer cells. *International Journal of Molecular Medicine.* 12(1), 29–34.

Choi, C., Cho, H., Park, J., Cho, C., and Song, Y., 2003. Suppressive effects of genistein on oxidative stress and NF-κB activation in RAW 264.7 macrophages. *Bioscience, Biotechnology and Biochemistry.* 67(9), 1916–1922.

Corona, G., Tzounis, X., Assunta Dessi, M., Deiana, M., Debnam, E.S., Visioli, F., and Spencer, J.P., 2006. The fate of olive oil polyphenols in the gastrointestinal tract: implications of gastric and colonic microflora-dependent biotransformation. *Free Radical Research.* 40(6), 647–658.

Delmonte, P., and Rader, J.I., 2006. Analysis of isoflavones in foods and dietary supplements. *Journal of AOAC International.* 89(4), 1138–1146.

Dowling, S., Regan, F., and Hughes, H., 2010. The characterisation of structural and antioxidant properties of isoflavone metal chelates. *Journal of Inorganic Biochemistry.* 104(10), 1091–1098.

El Touny, L.H., and Banerjee, P.P., 2007. Akt GSK-3 pathway as a target in genistein-induced inhibition of TRAMP prostate cancer progression toward a poorly differentiated phenotype. *Carcinogenesis.* 28(8), 1710–1717.

Fanti, P., Asmis, R., Stephenson, T.J., Sawaya, B.P., and Franke, A.A., 2006. Positive effect of dietary soy in ESRD patients with systemic inflammation -- correlation between blood levels of the soy isoflavones and the acute-phase reactants. *Nephrology, Dialysis, Transplantation.* 21(8), 2239–2246.

Goodman, M.T., Wilkens, L.R., Hankin, J.H., Lyu, L.C., Wu, A.H., and Kolonel, L.N., 1997. Association of soy and fiber consumption with the risk of endometrial cancer. *Amercican Journal of Epidemiology.* 146(4), 294–306.

Guo, Q., Rimbach, G., Moini, H., Weber, S., and Packer, L., 2002. ESR and cell culture studies on free radical-scavenging and antioxidant activities of isoflavonoids. *Toxicology.* 179(1–2), 171–180.

Han, R.M., Tian, Y.X., Liu, Y., Chen, C.H., Ai, X.C., Zhang, J.P., and Skibsted, L.H., 2009. Comparison of flavonoids and isoflavonoids as antioxidants. *Journal of Agriculture and Food Chemistry.* 57(9), 3780–3785.

Head, K.A., 1998. Isoflavones and other soy constituents in human health and disease. *Alternative Medicine Review.* 3(1), 433–450.

Henderson, B.E., Ross, R.K., Judd, H.L., Krailo, M.D., and Pike, M.C., 1985. Do regular ovulatory cycles increase breast cancer risk?. *Cancer.* 56(5), 1206–1208.

Hernandez-Montes, E., Pollard, S.E., Vauzour, D., Jofre-Montseny, L., Rota, C., Rimbach, G., Weinberg, P.D., and Spencer, J.P., 2006. Activation of glutathione peroxidase via Nrf1 mediates genistein's protection against oxidative endothelial cell injury. *Biochemical and Biophysical Research Communications.* 346(3), 851–859.

Horn-Ross, P.L., John, E.M., Canchola, A.J., Stewart, S.L., and Lee, M.M., 2003. Phytoestrogen intake and endometrial cancer risk. *Journal of the National Cancer Institute.* 95(15), 1158–1164.

Ibrahim, A.S., El-Shishtawy, M.M., Pena, Jr, A., and Liou, G.I., 2010. Genistein attenuates retinal inflammation associated with diabetes by targeting of microglial activation. *Molecular Vision.* 16: 2033–2042.

Jones, J.L., Daley, B.J., Enderson, B.L., Zhou, J.R. and Karlstad, M.D., 2002. Genistein inhibits tamoxifen effects on cell proliferation and cell cycle arrest in T47D breast cancer cells. *American Surgeon.* 68(6): 575–577; discussion 577–578.

Ju, Y.H., Doerge, D.R., Allred, K.F., Allred, C.D., and Helferich, W.G., 2002. Dietary genistein negates the inhibitory effect of tamoxifen on growth of estrogen-dependent human breast cancer (MCF-7) cells implanted in athymic mice. *Cancer Research.* 62(9), 2474–2477.

Ju, Y.H., Allred, K.F., Allred, C.D., and Helferich, W.G., 2006. Genistein stimulates growth of human breast cancer cells in a novel, postmenopausal animal model, with low plasma estradiol concentrations. *Carcinogenesis*. 27(6), 1292–1299.

Kim, S.J., Kwon do, Y., Kim, Y.S., and Kim, Y.C., 2010. Peroxyl radical scavenging capacity of extracts and isolated components from selected medicinal plants. *Archives of Pharmacal Research*. 33(6), 867–873.

Kiriakidis, S., Hogemeier, O., Starcke, S., Dombrowski, F., Hahne, J.C., Pepper, M., Jha, H.C., and Wernert, N., 2005. Novel tempeh (fermented soyabean) isoflavones inhibit *in vivo* angiogenesis in the chicken chorioallantoic membrane assay. *British Journal of Nutrition*. 93(3), 317–323.

Lee, J.M., Hanson, J.M., Chu, W.A., and Johnson, J.A., 2001. Phosphatidylinositol 3-kinase, not extracellular signal-regulated kinase, regulates activation of the antioxidant-responsive element in IMR-32 human neuroblastoma cells. *Journal of Biological Chemistry*. 276(23), 20011–20016.

Lee, Y.W., and Lee, W.H., 2008. Protective effects of genistein on proinflammatory pathways in human brain microvascular endothelial cells. *Journal of Nutritional Biochemistry*. 19(12), 819–825.

Li, J., Li, Z., and Mo, B.Q., 2006. Effects of ERK5 MAPK signaling transduction pathway on the inhibition of genistein to breast cancer cells. *Wei Sheng Yan Jiu*. 35(2), 184–186.

Liggins, J., Bluck, L.J., Runswick, S., Atkinson, C., Coward, W.A., and Bingham, S.A., 2000. Daidzein and genistein contents of vegetables. *British Journal of Nutrition*. 84(5), 717–725.

Loftus, Jr, E.V., 2004. Clinical epidemiology of inflammatory bowel disease: Incidence, prevalence, and environmental influences. *Gastroenterology*. 126(6), 1504–1517.

McCabe, Jr, M.J., and Orrenius, S., 1993. Genistein induces apoptosis in immature human thymocytes by inhibiting topoisomerase-II. *Biochemical and Biophysical Research Communications*. 194(2), 944–950.

Messina, M., and Bennink, M., 1998. Soyfoods, isoflavones and risk of colonic cancer: a review of the *in vitro* and *in vivo* data. *Baillière's Clinical Endocrinology and Metabolism*. 12(4), 707–728.

Messina, M., and Hilakivi-Clarke, L., 2009. Early intake appears to be the key to the proposed protective effects of soy intake against breast cancer. *Nutrition and Cancer*. 61(6), 792–798.

Miksicek, R.J., 1995. Estrogenic flavonoids: structural requirements for biological activity. *Proceedings of the Society of Experimental Biology and Medicine*. 208(1): 44–50.

Mitropoulou, T.N., Tzanakakis, G.N., Nikitovic, D., Tsatsakis, A., and Karamanos, N.K., 2002. *In vitro* effects of genistein on the synthesis and distribution of glycosaminoglycans/proteoglycans by estrogen receptor-positive and -negative human breast cancer epithelial cells. *Anticancer Research*. 22(5), 2841–2846.

Molteni, A., Brizio-Molteni, L., and Persky, V., 1995. *In vitro* hormonal effects of soybean isoflavones. *Journal of Nutrition*. 125(3 Suppl), 751S–756S.

Monks, T.J., and Lau, S.S., 1997. Biological reactivity of polyphenolic–glutathione conjugates. *Chemical Research in Toxicology*. 10(12), 1296–1313.

Morton, M.S., Turkes, A., Denis, L., and Griffiths, K., 1999. Can dietary factors influence prostatic disease?. *BJU International*. 84(5), 549–554.

Mousavi, Y., and Adlercreutz, H., 1993. Genistein is an effective stimulator of sex hormone-binding globulin production in hepatocarcinoma human liver cancer cells and suppresses proliferation of these cells in culture. *Steroids*. 58(7), 301–304.

Nguyen, D.T., Hernandez-Montes, E., Vauzour, D., Schonthal, A.H., Rice-Evans, C., Cadenas, E., and Spencer, J.P., 2006. The intracellular genistein metabolite 5,7,3′,4′-tetrahydroxyisoflavone mediates G2–M cell cycle arrest in cancer cells via modulation of the p38 signaling pathway. *Free Radical Biology and Medicine*. 41(8), 1225–1239.

Pavese, J.M., Farmer, R.L., and Bergan, R.C., 2010. Inhibition of cancer cell invasion and metastasis by genistein. *Cancer Metastasis Reviews*. 29(3), 465–482.

Peeters, P.H., Keinan-Boker, L., van der Schouw, Y.T., and Grobbee, D.E., 2003. Phytoestrogens and breast cancer risk. Review of the epidemiological evidence. *Breast Cancer Research and Treatment*. 77(2), 171–183.

Peterson, G., and Barnes, S., 1991. Genistein inhibition of the growth of human breast cancer cells: independence from estrogen receptors and the multi-drug resistance gene. *Biochemical and Biophysical Research Communications*. 179(1), 661–667.

Peterson, G., and Barnes, S., 1993. Genistein and biochanin A inhibit the growth of human prostate cancer cells but not epidermal growth factor receptor tyrosine autophosphorylation. *Prostate*. 22(4), 335–345.

Peterson, T.G., Coward, L., Kirk, M., Falany, C.N., and Barnes, S., 1996. The role of metabolism in mammary epithelial cell growth inhibition by the isoflavones genistein and biochanin A. *Carcinogenesis*. 17(9), 1861–1869.

Piller, R., Chang-Claude, J., and Linseisen, J., 2006. Plasma enterolactone and genistein and the risk of premenopausal breast cancer. *European Journal of Cancer Prevention*. 15(3), 225–232.

Raffoul, J.J., Wang, Y., Kucuk, O., Forman, J.D., Sarkar, F.H., and Hillman, G.G., 2006. Genistein inhibits radiation-induced activation of NF-κB in prostate cancer cells promoting apoptosis and G2/M cell cycle arrest. *BMC Cancer*. 6: 107.

Reinli, K., and Block, G., 1996. Phytoestrogen content of foods – a compendium of literature values. *Nutrition and Cancer*. 26(2), 123–148.

Richard, N., Porath, D., Radspieler, A., and Schwager, J., 2005. Effects of resveratrol, piceatannol, tri-acetoxystilbene, and genistein on the inflammatory response of human peripheral blood leukocytes. *Molecular Nutrition and Food Research*. 49(5), 431–442.

Santell, R.C., Kieu, N., and Helferich, W.G., 2000. Genistein inhibits growth of estrogen-independent human breast cancer cells in culture but not in athymic mice. *Journal of Nutrition*. 130(7), 1665–1669.

Seibel, J., Molzberger, A.F., Hertrampf, T., Laudenbach-Leschowski, U., and Diel, P., 2009. Oral treatment with genistein reduces the expression of molecular and biochemical markers of inflammation in a rat model of chronic TNBS-induced colitis. *European Journal of Nutrition*. 48(4), 213–220.

Sergeev, I.N., 2004. Genistein induces Ca^{2+}-mediated, calpain/caspase-12-dependent apoptosis in breast cancer cells. *Biochemical and Biophysical Research Communications*. 321(2), 462–467.

Sethy-Coraci, I., Crock, L.W., and Silverstein, S.C., 2005. PAF-receptor antagonists, lovastatin, and the PTK inhibitor genistein inhibit H_2O_2 secretion by macrophages cultured on oxidized-LDL matrices. *Journal of Leukocyte Biology*. 78(5), 1166–1174.

Smith, S., Sepkovic, D., Bradlow, H.L., and Auborn, K.J., 2008. 3,3'-Diindolylmethane and genistein decrease the adverse effects of estrogen in LNCaP and PC-3 prostate cancer cells. *Journal of Nutrition*. 138(12), 2379–2385.

Spencer, J.P., Kuhnle, G.G., Williams, R.J., and Rice-Evans, C., 2003. Intracellular metabolism and bioactivity of quercetin and its *in vivo* metabolites. *Biochemical Journal*. 372(Pt 1), 173–181.

Tiemann, U., Schneider, F., Vanselow, J., and Tomek, W., 2007. *In vitro* exposure of porcine granulosa cells to the phytoestrogens genistein and daidzein: effects on the biosynthesis of reproductive steroid hormones. *Reproductive Toxicology*. 24(3–4), 317–325.

Vauzour, D., Vafeiadou, K., Rice-Evans, C., Cadenas, E., and Spencer, J.P., 2007. Inhibition of cellular proliferation by the genistein metabolite 5,7,3',4'-tetrahydroxyisoflavone is mediated by DNA damage and activation of the ATR signalling pathway. *Archives of Biochemistry and Biophysics*. 468(2), 159–166.

Vergne, S., Sauvant, P., Lamothe, V., Chantre, P., Asselineau, J., Perez, P., Durand, M., Moore, N., and Bennetau-Pelissero, C., 2009. Influence of ethnic origin (Asian v. Caucasian) and background diet on the bioavailability of dietary isoflavones. *British Journal of Nutrition*. 102(11), 1642–1653.

Wang, T.T., Sathyamoorthy, N., and Phang, J.M., 1996. Molecular effects of genistein on estrogen receptor mediated pathways. *Carcinogenesis*. 17(2), 271–275.

Wang, J., Eltoum, I.E., and Lamartiniere, C.A., 2002. Dietary genistein suppresses chemically induced prostate cancer in Lobund–Wistar rats. *Cancer Letters*. 186(1), 11–18.

Wang, J., Eltoum, I.E., Carpenter, M., and Lamartiniere, C.A., 2009. Genistein mechanisms and timing of prostate cancer chemoprevention in Lobund–Wistar rats. *Asian Pacific Journal of Cancer Prevention*. 10(1), 143–150.

Williams, R.J., Spencer, J.P., and Rice-Evans, C., 2004. Flavonoids: antioxidants or signalling molecules?. *Free Radical Biology and Medicine*. 36(7), 838–849.

Xu, J., and Loo, G., 2001. Different effects of genistein on molecular markers related to apoptosis in two phenotypically dissimilar breast cancer cell lines. *Journal of Cellular Biochemistry*. 82(1), 78–88.

Yearley, E.J., Zhurova, E.A., Zhurov, V.V., and Pinkerton, A.A., 2007. Binding of genistein to the estrogen receptor based on an experimental electron density study. *Journal of the American Chemical Society*. 129(48), 15013–15021.

CHAPTER 10
Genistein Chemistry and Biochemistry

FRANCESCO SQUADRITO* AND ALESSANDRA BITTO

Department of Clinical and Experimental Medicine and Pharmacology, University of Messina, Messina, Italy
*Email: Francesco.Squadrito@unime.it

10.1 Introduction

In most populations, the major dietary source of isoflavones and phytoestrogens is soy. Soybeans and their products are consumed by humans in many forms, including whole soybeans, soy milk, tofu and tempeh. Soy proteins, primarily in the form of non-toasted defatted soybean flakes, can be isolated from the whole bean for consumption through different processes, and ingested through a variety of products, such as texturized foods, baked goods, confections, meat products and nutritional supplements; remarkably, up to 50% of the phytoestrogens found in unprocessed soybeans can be contained in this kind of products.

Other plant-derived foods contain irrelevant amounts of phytoestrogens compared with soy. The concentrations of isoflavones in soy products vary considerably, ranging in most soy foods between 0.1–3.0 mg g^{-1} (Setchell 1998). Soy oil or sauce contains only negligible amounts of isoflavones due to the water solubility of the isoflavone glycosides.

The isoflavones have a restricted distribution in the plant kingdom, and are mostly limited to the subfamily Papilionoideae of the Leguminosae. Their surprisingly large structural variation involves not only the number and complexity of substituents on the 3-phenylchroman framework, but also the presence of additional heterocyclic rings and different oxidation levels of the heterocycle. In processed soy products, such as soy hot dogs and tofu yogurt, isoflavones are only present at $1/10^{th}$ the isoflavone content of whole soy beans [0.2–0.3 vs. 2–4 (mg of isoflavone) g^{-1}] (Wang and Murphy 1994). However, even these processed soy products contain larger amounts of isoflavones than do most other legumes or non-leguminous plant foods, such as the raw green beans. In addition, the contents of isoflavones in different soy products (*e.g.* tofu and soy protein concentrates) vary substantially based on processing steps.

Genistein (4′,5,7-trihydroxyisoflavone) is the most widely studied isoflavone, ubiquitously found in soybeans with daidzein (4′,7-dihydroxyisoflavone) and smaller quantities of glycitein (4′,7- dihydroxy-6-methoxyisoflavone). More precisely, the concentration of genistein in most soy food ranges from 1–2 mg (g of protein)$^{-1}$. Soy milk and infant formulas have lower concentrations of isoflavones but are consumed in higher quantities, leading to high intake.

Genistein is a major subject of discussion in the context of nutraceuticals and functional foods, and may soon provide a case study for evaluating the delivery of health-promoting compounds through genetically modified plants. Genistein has been the focus of scientific research since 1966, as a consequence of the previously observed reproductive problems and infertility in Australian sheep after grazing in pastures with the flvonoids containing clover *Trifolium subterraneum* (commonly known as red clover) reported by Bennetts and Underwood (1951).

The many biological activities of genistein made it a subject in over 8770 published studies in the last 60 years, mostly focusing on the pharmacological features of genistein as a tyrosine kinase inhibitor, its chemoprotectant characteristics against cancer and cardiovascular diseases, and its estrogen-like activity.

10.2 Chemical Structure and Interaction with Estrogen Receptors

Genistein is considered the simplest isoflavonoid compound found in Leguminosae. It is a central intermediate in the biosynthesis of more complex isoflavones with roles in the establishment or inhibition of interactions between plants and microbes. As found for other isoflavones, it exerts a broad-spectrum antimicrobial activity and is therefore believed to help the plant fight microbial disease. Antimicrobial isoflavones can be classified as pre-formed "phytoanticipins" or inducible "phytoalexins" and genistein may be considered both as a phytoalexin and as a phytoanticipin. The majority have the A-ring substitution pattern of daidzein (4,7-dihydroxisoflavone), and genistein is therefore not an intermediate in their biosynthesis. The prenylated genisteins of lupin are derivatives synthesized during seedling development and may also function as phytoanticipins. However, only a low number of 5-hydroxyisoflavonoid phytoalexins derived from genistein has been reported in the Leguminosae.

The isoflavone genistein was first isolated in 1899 from the dyer's broom (*Genista tinctoria*); hence, the chemical name derived from the generic name. The compound nucleus was established in 1926, when it was found to be identical with prunetol, and it was chemically synthesized in 1928.

Genistein and its parent compound daidzein are metabolized from their plant precursors, biochanin A and formononetin, respectively. Like the other isoflavones, they have a common diphenolic structure that resembles the structure of the potent synthetic estrogens diethylstilbesterol and hexestrol and, due to these similarities, they are commonly referred to as phytoestrogens. The similar distance (11.5 Å) between the OH groups on the opposite sides of genistein and estradiol molecules makes genistein capable of binding the estrogen receptor subtypes α and β (ER-α and ER-β) and sex hormone-binding proteins, exerting both estrogenic and anti-estrogenic activity, the latter by competing with estradiol for receptor binding. Genistein, however, shows higher affinity to ER-β than to ER-α (Kuiper *et al.* 1997) and increases the binding of ER-β to a genomic estrogen response element (ERE) more strongly than with ER-α (Kostelac *et al.* 2003).

Since its discovery in 1996, the estrogen receptor isoform known as ER-β has become the focus of several studies investigating its potential drug target role. The clear differences in ER-β and ER-α expression suggests that tissues could be differentially targeted with ligands selective for either isoform (Couse *et al.* 1997). More precisely, the fact that ER-β is widely expressed but not the primary estrogen receptor in the reproductive tract (where estrogenic effects are mediated *via* ER-α) (Harris *et al.* 2002) offers the possibility of targeting other tissues while avoiding certain classical estrogenic effects.

As reported by Pike and co-workers (1999), the interactions between genistein and estrogen receptors are mediated by the phenol group, which mimics the estradiol "A-ring", with the phenolic hydroxyl (4'-OH) involved in a hydrogen-bonding network between the ER-α residues Glu-353 and Arg-394 (ER-β residues Glu-305 and Arg-346) and a highly ordered water molecule.

Another hydrogen bond is formed between the isoflavone 7-OH and Nδ1 of ER-α His-524 (ER-β His-475). The core scaffold fills the remainder of the primarily hydrophobic pocket. Interestingly, despite the nearly 40-fold ER-β selectivity of genistein, the ligand-binding modes are almost identical in both complexes.

The genistein γ-pyrone "B-ring" is very close to the ER-α Leu-384 → ER-β Met-336 residue substitution, with the B-ring centroid approximately 4.0–4.5 Å (measuring distances in both monomer units) from the ER-β Met-336 C atom *versus* 6.2 Å from the ER-α Leu-384 Cδ1 atom. The ER-β selectivity, however, is not dependent on this differential distance alone.

In addition, the carbon in position 3 of the isoflavone ring is approximately 4.2–4.6 Å from the ER-β Met-336 Sδ atom *versus* 4.8 Å from the ER-α Leu-384 Cδ2 atom, although it is not immediately obvious how the methyl → sulfur difference also affects the interaction.

Methionine–aromatic interactions are thought to stabilize proteins (Pal and Chakrabarti 2001), however, the magnitude of an interaction relative to the leucine–aromatic interaction, particularly, given the specific protein–ligand contacts observed in ER-α/β is unclear. ER-β Met-336 clearly has a greater potential to achieve an attractive interaction with an aryl ring than ER-α Leu-384. Furthermore, it is possible that the "B-ring" interaction with the ER-β Met-336 relative to the ER-α Leu-384 can lead to a differential interaction of approximately 1.2–2.2 kcal mol^{-1} which can be responsible for an 8- to 41-fold contribution to selectivity.

The 5-OH group of genistein is involved in an intramolecular hydrogen bond with the 4-keto moiety (Fang *et al.* 2001). This group is in close proximity to the ER-α Met-421→ER-β Ile-373 residue substitution. The methionine sulfur atom could be involved in a repulsive interaction, since oxygen is relatively non-polarizable and possesses a partial negative charge. However, from the genistein–ER-α structure, it seems clear that the ER-α Met-421 adopts a different rotamer state, where the Sδ atom of the ER-α Met421 is approximately 5.2 Å from the 5-OH oxygen of genistein, with the sulphur lone pairs facing away from the oxygen atom. It can be hypothesized that a repulsive interaction between the 5-OH oxygen and ER-α Met-421 Sδ contributes to the change in rotamer state. Anyway, as mentioned above, this residue seems able to use an alternate rotamer to eliminate any repulsive interaction. In contrast, the 5-OH group is approximately 4.0–4.2 Å from the ER-β Ile-373 Cγ1 atom (closest distance), and therefore any repulsive interaction with this residue has to be expected.

The repulsion between the 5-OH group and the ER-α Met-421 can be a further explanation for ER-β selectivity by preventing an interaction from occurring that would otherwise improve binding to ER-α relative to ER-β. At the same time, it can contribute to both ER-α and ER-β potency by forming a tight intramolecular hydrogen bond.

In addition to the different tissue expression of the estrogen receptor subtypes, ER-β and ER-α often mediate opposite functions (Gustafsson 2006). This may explain why the physiological net effect of genistein seems to be partly agonist and partly antagonist, initiating some estrogen-like actions and inhibiting others.

An explanation for this partial agonist action of genistein relies on the three-dimensional structure of the ligand-binding domain (LBD) of the complex formed by the interaction with ER-β, where the transactivation Helix 12 does not adopt the agonist position, but, instead, lies in a similar (not identical) orientation induced by raloxifene, a common selective estrogen receptor modulator (SERM) (Pike *et al.* 1999). This peculiar alignment of Helix 12 suggests that genistein acts as a partial agonist when bound to ER-β.

The conformation of Helix 12 is the same as that observed for other bound agonists (Brzozowski *et al.* 1997; Pike *et al.* 2000), in contrast to the previously reported antagonist-like conformation of Helix 12 for ER-β–genistein (Pike *et al.* 1999). These different findings can be explained by taking in consideration that the relative free energies of the two Helix 12 conformations are similar when genistein is bound and the complex is crystallized, with the "antagonist-like" state being slightly more stable. It is likely that the binding of the coactivator fragment to the ER-β–genistein complex provides the additional stabilization required to observe an "agonist-like" state.

Nevertheless, in human estrogen receptor subtype-specific reporter cells, genistein is a stronger agonist of ER-α (Barkhem *et al.* 1998). In addition to the different tissue expression of the estrogen receptor subtypes, ER-β and ER-α often mediate opposite functions (Gustafsson 2006). This may explain why the physiological net effect of genistein seems to be partly agonist and partly antagonist, initiating some estrogen-like actions and inhibiting others, alhough genistein has been classified to be a pure estrogen agonist in human cell lines. Determining the actual biological net effect of phytoestrogens is complicated by different factors: the route of administration, bioavailability and metabolism, timing and level of exposure, and endogenous estrogen state, as well as the various non-hormonal effects (Cassidy 1999).

10.3 Metabolism

In unfermented soy foods, genistein can be found conjugated to sugars forming β-glycosides, which cannot be absorbed as such by the human gastrointestinal tract (Setchell *et al.* 2002).

O-Glycosides, such as genistin (Figure 10.1) (genistein 7-*O*-β-D-glucopyranoside), appear to be the majority but a relevant number of C-glycosides has also been documented. The malonyl glucosides of daidzein and genistein (Figure 10.2) found in soybean are labile and rapidly degraded to the non-acylated glycosides after reaching high temperatures (*i.e.* cooking). Once ingested, the β-glycoside of genistein (genistin) is hydrolyzed by intestinal β-glucosidases to the respective aglycone form genistein, which is probably absorbed by non-ionic passive diffusion from the jejunum (Setchell *et al.* 2002). The free aglycones, but not the glycosides, are absorbed from stomach, at least in rats. In humans, indeed, isoflavones appear in blood plasma at a more rapid rate and higher levels after oral administration of aglycones as compared with glycosides, and genistein and daidzein, but not their glycosides, are promptly carried through human intestinal epithelial cell monolayers. However, once in the small intestine, brush border lactase-phlorizin hydrolase can effectively hydrolyze these glycosides forms of isoflavones.

Fermented soy products (miso, tempeh and natto) contain larger amounts of isoflavone aglycones due to bacterial hydrolysis, which may influence the

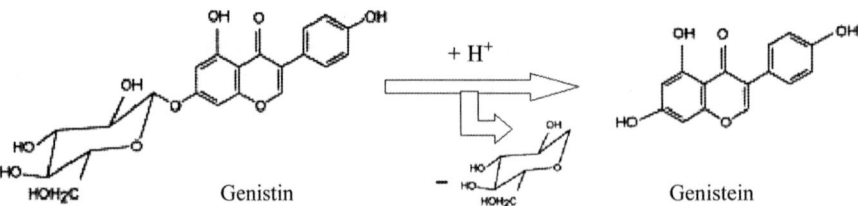

Figure 10.1 Chemical structures of genistin and genistein. The glucose bond to the oxygen atom in position 7 (genistin) is removed by hydrolysis and substituted with a hydrogen atom (genistein).

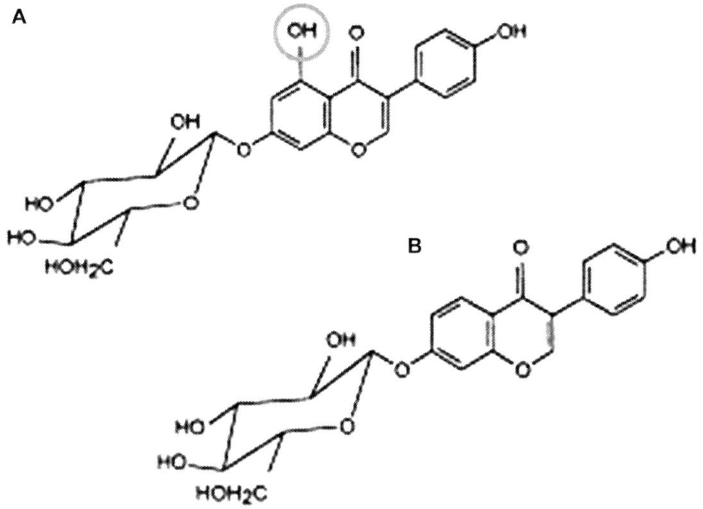

Figure 10.2 Chemical structures of genistin and daidzin. Genistin (A) and daidzin (B) are the glucoside forms in which the glucose bond to the oxygen atom in position 7 is removed before absorption of the aglycone forms (genistein and daidzein). The circle indicates the difference in the two structures.

bioavailability (Setchell 1998). In the intestinal wall and liver, isoflavone aglycones are mainly glucuronidated, while a smaller amount is sulfated. Like estrogens (endogenous or esogenous), a small fraction of isoflavones undergoes enterohepatic circulation, thus increasing the circulating levels hours after the first absorption. Due to their chemical structure and liver metabolism, in blood, the majority of isoflavones occurs as glucuronide and sulfate conjugates, and a smaller quantity of aglycones is present as free as well as bound to plasma proteins. The main metabolites of genistein in humans are 7-OH glucuronic acid and 4′-OH sulfate which are eventually excreted in the urine.

10.4 Conclusion

Genistein can be considered to be a peculiar isoflavone that shares many of the chemical and biological features of other similar compounds, as well as the ability to bind intracellular receptors and to modulate multiple kinase-dependent pathways. These multifaceted effects can be only partly explained by its chemical structure and are mostly due to the biomolecular interactions with different cellular systems.

Summary Points

- Genistein was identified in the early 1900s.
- Genistein shares chemical similarities with 17-β-estradiol.

- Genistin is converted into genistein by gut microflora.
- Genistein binds to estrogen receptors, especially ER-β due to its chemical interactions with the LBD.
- The biological activities of genistein are related to its bind to intracellular receptors and to the modulation of multiple kinase-dependent pathways.

Key Facts

- Genistein is an isoflavone from soy with estrogen-like properties.
- The glucoside genistin is converted into the aglycone genistein in the gut before absorption, however, many other facts (*i.e.* malabsorption syndromes) may affect the uptake and in turn the plasmatic concentration.
- Genistein binds mainly to ER-β, promoting positive effects on responsive tissues. ER-β without its AF-1 domain is able to suppress ER-α-mediated growth signallig.
- Genistein tends to bind preferentially to ER-β, hence its frequent description as a natural SERM. The mode of action of genistein depends upon the estrogen status. Genistein might act as an agonist of estrogen receptors in a low estrogen environment, but as a partial agonist or antagonist in presence of endogenous estrogens.
- The biological effect effects of genistein may also be mediated by other receptors such as peroxisome-proliferator-activated receptors (PPARs), thus affecting several other intracellular pathways.

Definitions of Words and Terms

Aglycone: This is the glucose-free form of an isoflavone, following hydrolyzation of the glycosidyc bound where glucose is replaced by a hydrogen atom.

Estrogen receptors: This a subgroup of the intracellular receptors that belong to the nuclear hormone receptor superfamily and commonly bind to estradiol. They can also bind phytoestrogens and other similar compounds.

Genistein: It is a common isoflavone derived from the hydrolysis of the glycoside genistin. Because of its similar structure to that of human estrogen, it is also called a phytoestrogen.

Glucoside: A glucoside is the common chemical form of plant-derived isoflavones, in which the molecule is bound to glucose.

Growth factors: These are small molecules released by different cell types that act by modulating cell division, cell differentiation and tissue growth.

Isoflavones: These are dietary-occurring polyphenolic compounds that are abundant in beans as soy and its derivatives.

Nutraceutical: This is the term indicating a naturally occurring compound when administered in pharmacologically active doses.

Partial agonist: This is the case of when genistein binds estrogen receptors, and is the peculiar action of certain compounds that are unable to elicit the maximal response of the receptor system.
Peroxisome-proliferator-activated receptors (PPARs): These are a class of ligand-inducible transcription factors that belong to the nuclear hormone receptor superfamily.
Selective estrogen receptor modulators (SERMs): Aclass of compounds that may bind to the estrogen receptors, and, due to their chemical structure, modify the conformational status of the receptor, thus exerting inhibitory or stimulating effects in various tissues.

List of Abbreviations

ERα estrogen receptor α
ERβ estrogen receptor β
LBD ligand-binding domain
SERM selective estrogen receptor modulator

References

Barkhem, T., Carlsson, B., Nilsson, Y., Enmark, E., Gustafsson, J., and Nilsson, S., 1998. Differential response of estrogen receptor α and estrogen receptor β to partial estrogen agonists/antagonists. *Molecular Pharmacology*. 54: 105–112.
Bennetts, H.W., and Underwood., E.J., 1951. The oestrogenic effects of subterranean clover (*Trifolium subterraneum*); uterine maintenance in the ovariectomized ewe on clover grazing. *The Australian Journal of Experimental Biology and Medical Science*. 29: 249–253.
Brzozowski, A.M., Pike, A.C.W., Dauter, Z., Hubbard, R.E., Bonn, T., Engstrom, O., Ohman, L., Greene, G.L., Gustafsson, J.A., and Carlquist, M., 1997. Molecular basis of agonism and antagonism in the oestrogen receptor. *Nature*. 389: 753–758.
Cassidy, A., 1999. Potential tissue selectivity of dietary phytoestrogens and estrogens. *Current Opinion in Lipidology*. 10: 47–52.
Couse, J.F., Lindzey, J., Grandien, K., Gustafsson, J.A., and Korach, K.S., 1997. Tissue distribution and quantitative analysis of estrogen receptor-α (ERα) and estrogen receptor-β (ERβ) messenger ribonucleic acid in the wild-type and ERα-knockout mouse. *Endocrinology*. 138: 4613–4621.
Fang, H., Tong, W., Shi, L.M., Blair, R., Perkins, R., Branham, W., Hass, B.S., Xie, Q., Dial, S.L., Moland, C.L., and Sheehan, D.M., 2001. Structure–activity relationships for a large diverse set of natural, synthetic, and environmental estrogens. *Chemical Research in Toxicology*. 14: 280–294.
Gustafsson, J.A., 2006. ERβ – scientific visions translate to clinical uses. *Climacteric*. 9: 156–160.

Harris, H.A., Katzenellenbogen, J.A., and Katzenellenbogen, B.S., 2002. Characterization of the biological roles of the estrogen receptors, ERα and ERβ, in estrogen target tissues *in vivo* through the use of an ERα-selective ligand. *Endocrinology*. 143: 4172–4177.

Kostelac, D., Rechkemmer, G., and Briviba, K., 2003. Phytoestrogens modulate binding response of estrogen receptors α and β to the estrogen response element. *Journal of Agriculture and Food Chemistry*. 51: 7632–7635.

Kuiper, G.G., Carlsson, B., Grandien, K., Enmark, E., Häggblad, J., Nilsson, S., and Gustafsson, J.A., 1997. Comparison of the ligand binding specificity and transcript tissue distribution of estrogen receptors α and β. *Endocrinology*. 138: 863–870.

Pal, D., and Chakrabarti, P., 2001. Non-hydrogen bond interactions involving the methionine sulfur atom. *Journal of Biomolecular Structure and Dynamics*. 19: 115–128.

Pike, A.C.W., Brzozowski, A.M., Hubbard, R.E., Bonn, T., Thorsell, A.G., Engstrom, O., Ljunggren, J., Gustafsson, J.K., and Carlquist, M., 1999. Structure of the ligand-binding domain of oestrogen receptor β in the presence of a partial agonist and a full antagonist. *EMBO Journal*. 18: 4608–4618.

Pike, A.C.W., Brzozowski, A.M., and Hubbard, R.E., 2000. A structural biologist's view of the oestrogen receptor. *Journal of Steroid Biochemistry and Molecular Biology*. 74: 261–268.

Setchell, K.D.R., 1998. Phytoestrogens: the biochemistry, physiology, and implications for human health of soy isoflavones. *American Journal of Clinical Nutrition*. 68: 1333S–1346S.

Setchell, K.D., Brown, N.M., Zimmer-Nechemias, L., Brashear, W.T., Wolfe, B.E., Kirschner, A.S., and Heubi, J.E., 2002. Evidence for lack of absorption of soy isoflavone glycosides in humans, supporting the crucial role of intestinal metabolism for bioavailability. *American Journal of Clinical Nutrition*. 76: 447–453.

Wang, H., and Murphy, P.A., 1994. Isoflavone content in commercial soybean foods. *Journal of Agriculture and Food Chemistry*. 42: 1666–1673.

CHAPTER 11

Isoflavones and Human Estrogen Receptor: When Plants Synthesize Mammalian Hormone Mimetics

PATRICIA DE CREMOUX[a] AND YVES JACQUOT*[b]

[a] Molecular Oncology Unit, Department of Biochemistry, Saint-Louis Hospital, University Paris-Diderot, 1 avenue Claude Vellefaux, 75010 Paris, France; [b] Laboratoire des BioMolécules – CNRS / UMR 7203, Department of Chemistry, Ecole Normale Supérieure (ENS), University Pierre & Marie Curie, 24 rue Lhomond, 75231 Paris Cedex 05, France
*Email: yves.jacquot@upmc.fr

11.1 Introduction

The nuclear estrogen receptor (ER) accepts a variety of non-steroidal chemicals belonging to the estrogen disruptor family, including environmental estrogens. Environmental estrogens encompass two classes of compounds: (i) xenoestrogens, which comprise non-natural compounds (*e.g.*, organophosphorus derivatives or pesticides) and (ii) phytoestrogens, which are produced in plants (*i.e.*, flavanones, flavonoids, isoflavonoids, coumestans, chalcones, stilbenes or lignans) (Figure 11.1).

According to structure–activity relationships (SARs) and in connection with ER targeting, the presence of at least one phenolic motif is absolutely required

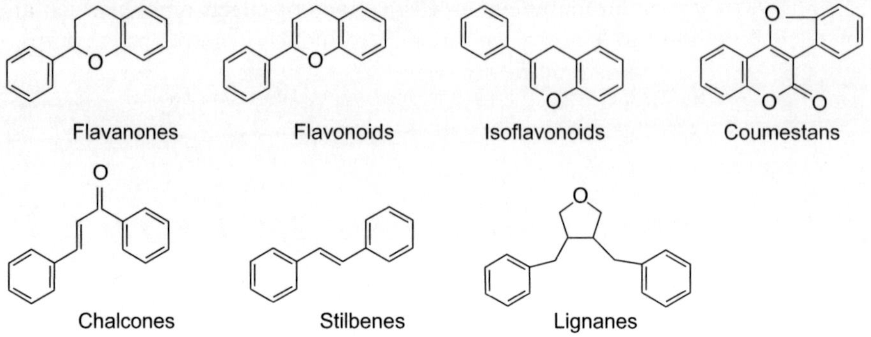

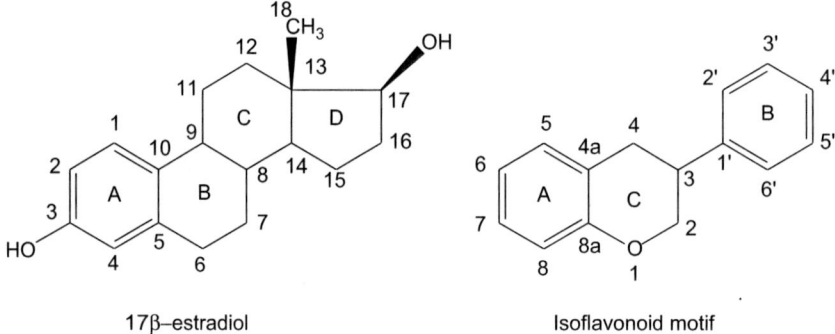

Figure 11.1 General structures of the different classes of phytoestrogens.

Figure 11.2 General structures and carbon atom numbering of 17β-estradiol and isoflavonoid motif.

for estrogenicity. This is the case of isoflavonoids (Figure 11.2), which constitute the most important group of phytoestrogens (Jacquot *et al.* 2003; Miksicek 1995).

The present Chapter deals with the biosynthesis of isoflavonoids and their structural link with estradiol, as well as their interaction patterns with ER. Their human metabolism and bioavailability are also tackled. Lastly, we will discuss the effects of isoflavonoids on human health.

11.2 From Plant Biosynthesis to Mammalian Biosynthesis

Isoflavonoids, which are present in several food sources, are principally isolated from leguminous plants, where they are at low concentrations. Fabaceae (*e.g.*, *Glycine max*, *Trifolium pratense*) contain the highest amount of isoflavonoids. These molecules contribute to symbiotic relationships between plants and bacteria, as well as plant stress/disease resistance responses (Pueppke 1996).

Strikingly, they exert in mammalian cells a range of effects among which are included the ability to bind to the ER and to induce ligand-dependent transcription. Furthermore, isoflavonoids induce expression of *Rhizobium* soil bacteria NodD transcriptional proteins, which share structural analogies with the ER and which assist plants in fixing atmospheric nitrogen (Fox 2004; Gyorgypal and Kondorsi 1991). Hence, NodD might constitute, from a phylogenic point of view, the missing link between the production of isoflavonoids in plants and their hormonal action in mammalians.

11.2.1 Biosynthesis of Isoflavonoids in Leguminous Cells

Derived from the shikimate cascade (Figure 11.3), chorismate is transformed into prephenate, phenylpyruvate and, in turn, into L-phenylalanine through the action of a chorismate mutase (CHM), a prephenate dehydratase and a transaminase, respectively. A subsequent deamination of L-phenylalanine, catalyzed by phenylalanine ammonia lyase (PAL), produces *trans*-cinnamate, which is, in turn, transformed into *p*-coumaric acid by a cinnamate 4-hydroxylase (C4H). Finally, *p*-coumaroyl-S–coenzyme A (CoA) is obtained through the action of a 4-coumarate:CoA ligase (4CL) (Figure 11.4).

Strikingly, *p*-coumaroyl-S–CoA is the starter for isoflavonoid biosynthesis and more specifically for two distinct branches that are in competition, leading to either daidzein or genistein (Figure 11.4). Responsible for the formation of the B ring and part of the C ring of isoflavones, *p*-coumaroyl S–CoA reacts with three malonyl S–CoA units in the presence of chalcone synthase (CHS) and chalcone reductase (CHR) or in the presence of CHS alone, to produce 4,2′,4′-trihydroxychalcone (isoliquiritigenin) and 4,2′,4′,6′-tetrahydroxychalcone (naringenin chalcone), respectively. Then, the cytochrome P450 mono-oxygenase chalcone isomerase (CHI) catalyzes stereospecifically the cyclization of isoliquiritigenin into 7,4′-dihydroxyflavanone (liquiritigenin) and naringenin chalcone into 5,7,4′-trihydroxyflavanone (naringenin). Daidzein and genistein are obtained after the action of isoflavone synthase (IFS), a microsomal cytochrome P450 enzyme (CYP93C) that is responsible for the formation of a 2–3 double bond and the regioselective migration of the B phenolic ring from position 2 to position 3. In fact, an intermediate, a 2-hydroxyisoflavanone, is dehydrated (spontaneously or through a dehydratase), allowing thereafter the formation of a 2–3 double bond. According to the "spirodienone" model, a Fe(III)-mediated process involving the formation of a nucleophilic reactive quinone occurs. Briefly, a hydroxyisoflavanone-4′-*O*-methyltransferase (HI4′OMT) catalyzes the reaction of this reactive quinone with a methyl carbocation radical resulting from SAM (*S*-adenosylmethionine) to produce, after a dehydratation step, a 3-phenyl-4′-methoxy derivative, as observed with liquiritigenin (Figure 11.5a) (Akashi *et al.* 2000; Crombie and Whiting 1992; Deavours and Dixon 2005). Thus, liquiritigenin leads to formononetin. Following the same process, naringenin leads to biochanin A (Figure 11.5b). It is of note that methylation can occur at other positions. Finally, a glycosylation

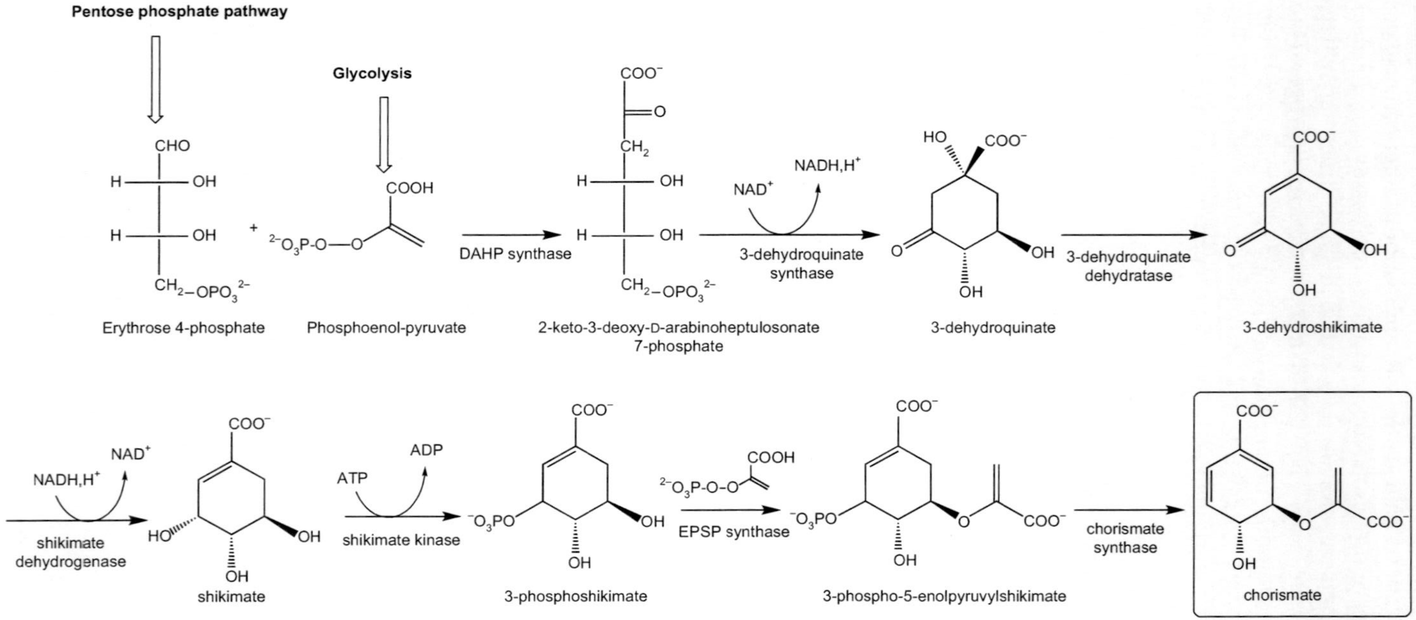

Figure 11.3 Biosynthesis of chorismate from the pentose phosphate and glycolysis pathways. DAHP synthase, 3-deoxy-D-arabino-7-heptulosonate-7-phosphate synthase; EPSP synthase, 5-enolpyruvylshikimate-3-phosphate synthase; NAD, nicotinamide adenine dinucleotide.

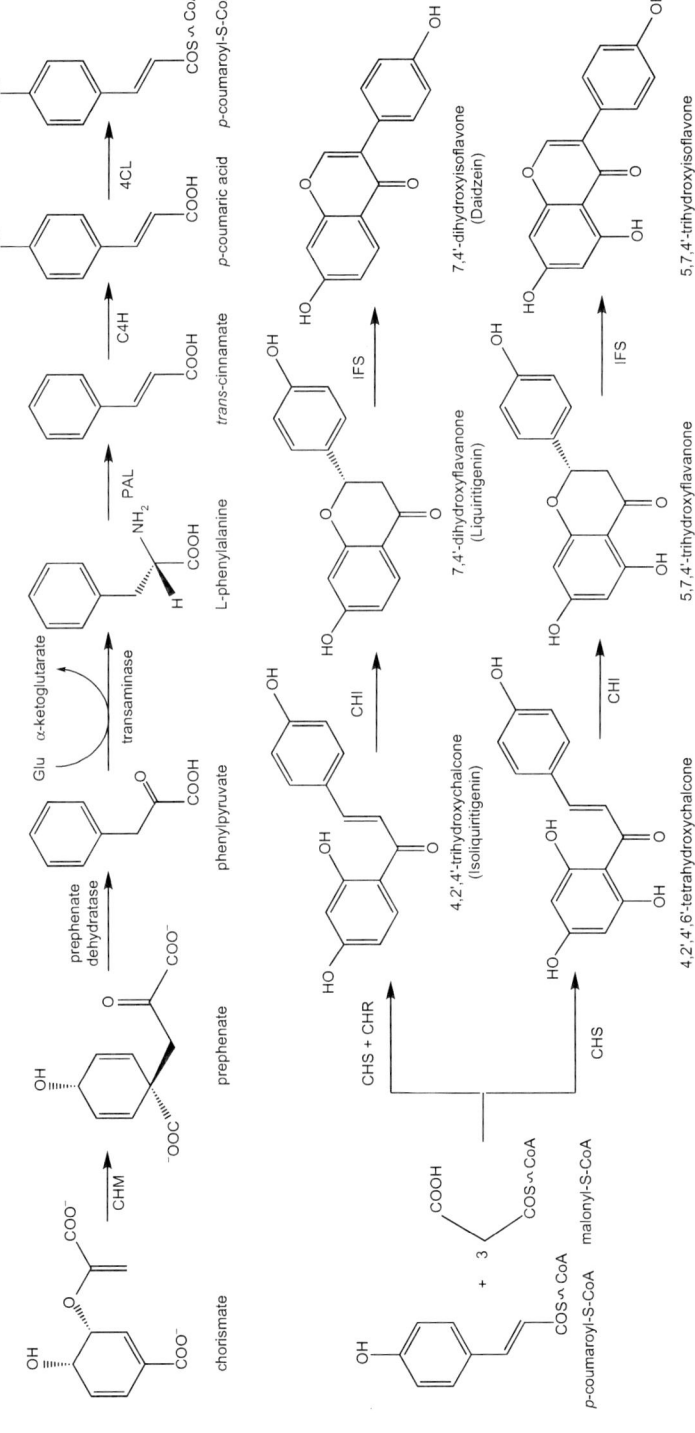

Figure 11.4 Biosynthesis of isoflavonoids from chorismate. C4H, cinnamate 4-hydroxylase; CHI, chalcone isomerase; CHR, chalcone reductase; CHS, chalcone synthase; 4CL, 4 coumarate:CoA lyase; CMH, chorismate mutase; IFS, isoflavone synthase; PAL, phenylalanine ammonia lyase.

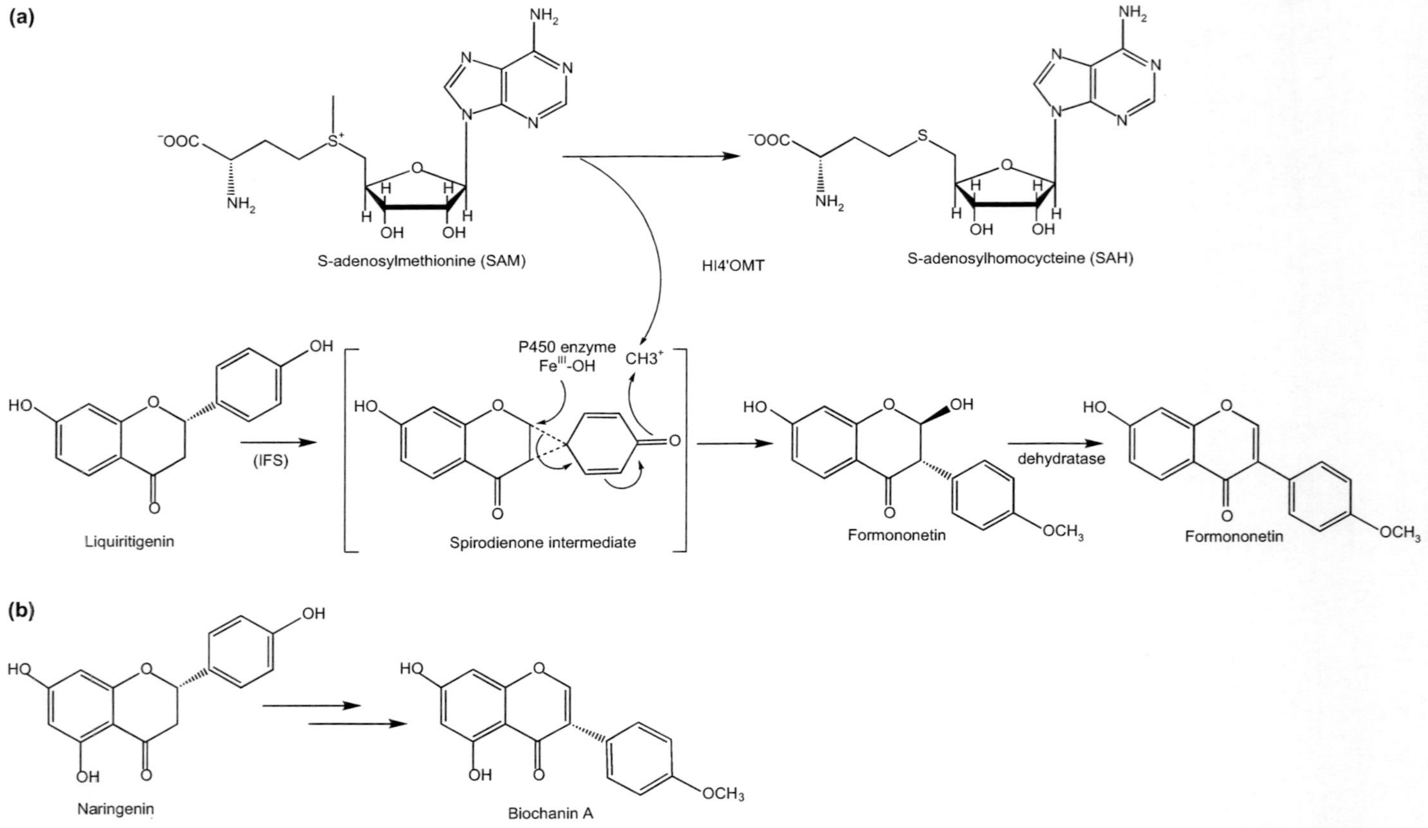

Figure 11.5 Mechanism of the 4′ O-methylation and 2→3 migration of the B ring of liquiritigenin, a flavanone, to form the isoflavone formononetin. (a) Catalyzed by an isoflavone synthase (IFS) and with the assistance of Fe(III)-OH (cytochrome P450 enzyme), a spirodienone intermediate is synthesized. Thus, a nucleophilic attack by this spirodienone on a CH_3^+ obtained from S-adenosylmethionine SAM [which is transformed into S-adenosylhomocysteine (SAH)] occurs. Finally, a dehydratase allows the formation of a 2–3 double bond. (b) A similar mechanism occurs from naringenin, allowing the formation of biochanin A.

step at position 7 through the action of 7-*O*-glycosyltransferase is observed, leading to glycone form.

11.2.2 Biosynthesis of Isoflavonoids in ERs

In humans, isoflavonoids can be produced from flavanones through cytochrome P450 enzymes, and more specifically through CYP1A1 (also known as aryl hydrocarbon hydrolase) and CYP2B6 (a phenobarbital-inducible enzyme that is involved in the biosynthesis of steroids and the metabolism of a number of xenobiotics).

11.3 Interaction of Isoflavonoids with ERs

Biochemical and structural studies have shown the direct interaction of isoflavonoids with ERs. Two distinct genes (ERα, chromosome 6; ERβ, chromosome 14) code for two ER isoforms, *i.e.*, ERα [595 residues, molecular weight (MW) ∼ 66 kDa] and ERβ (530 residues, MW ∼ 60 kDa). Both ER subtypes share 53% sequence homology in their ligand-binding domain and 94% sequence homology in their DNA-binding domain. Their respective ligand-binding domains have similar secondary and tertiary structures, *i.e.*, twelve α-helices, which are arranged in an antiparallel three-layer sandwich topology. ERα and ERβ are characterized by four key domains: (i) an activation function 1 (AF1), which is ligand-independent, (ii) a DNA-binding domain, (iii) a hinge region [that supports, at least partially, the autonomous activation function (AF2a)] and (iv) an activation function 2 (AF2) (that encompasses the ligand-binding domain), and that operates in a ligand-dependent manner. Despite these strong similarities, they differ in terms of tissue distribution, transcription and ligand selectivity, suggesting that they would be responsible for distinct physiological actions.

The principal ER endogenous ligands are estradiol and its two metabolites estrone and estriol. The ER-binding motifs of estradiol consist of a hydrophobic tetracyclic cyclopentanoperhydrophenanthrenic motif flanked by two hydroxyls in position 3 and 17β, ∼ 10.8 Å apart (Pike *et al.* 1999). Also, estradiol encompasses an aromatic A ring contributing to stabilizing π-stacking contacts with a ER phenylalanine residue (Phe-356 in ERβ and Phe-404 in ERα) (Figures 11.6a and 11.6b). Fortuitously, isoflavonoids share similar structural features with estradiol, *i.e.*, two phenolic hydroxyls that are separated by ∼ 12.1 Å through aromatic cycles constituting, partially, a benzopyran (γ-pyrone) scaffold (Jacquot *et al.* 2003; Turner *et al.* 2007). Hence, isoflavonoids can be considered structurally as estradiol mimetics.

Accordingly, genistein associates through its 4′- and 7-phenolic hydroxyls across the pocket of the receptor usually occupied by estradiol and delimited by the helices H3 and H6, as well as a β-hairpin (Manas *et al.* 2004; Seo *et al.* 2006). The 4′ hydroxyl (which corresponds to the A ring phenolic hydroxyl of estradiol) is involved in hydrogen bonds with Glu-305/Arg-346 of ERβ (Glu-353/Arg-394 in ERα), as well as with a buried molecule of water (Manas *et al.* 2004; Pike *et al.* 1999). The phenolic hydroxyl in position 7 associates through a

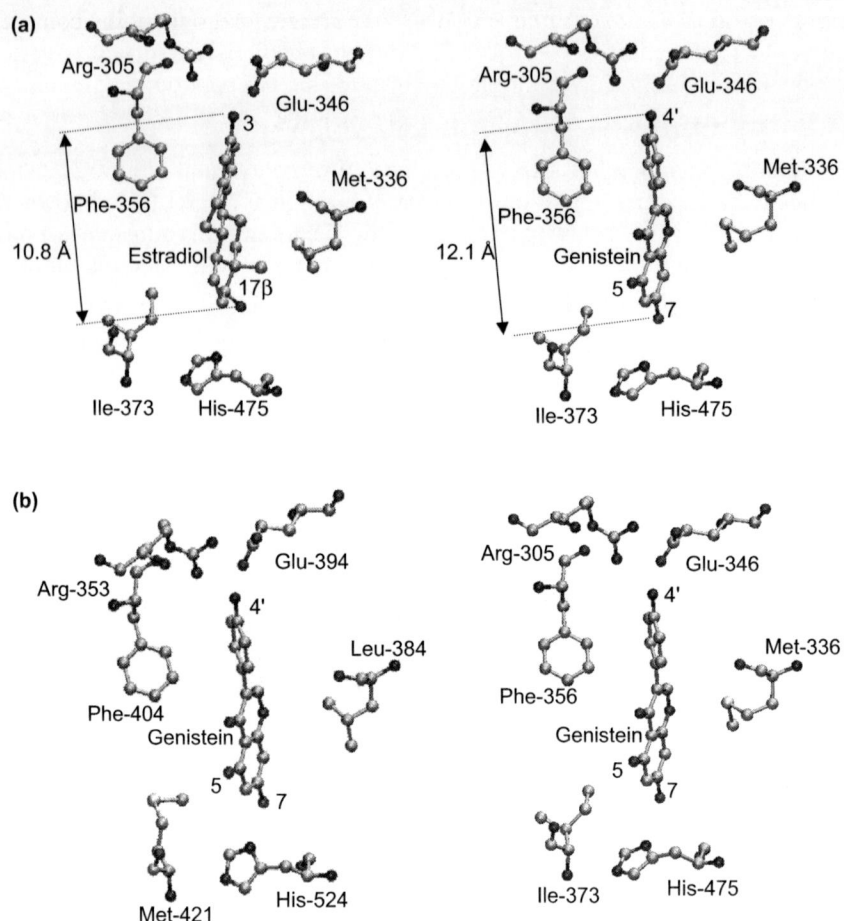

Figure 11.6 Mode of interaction of Genistein within the estradiol binding-pocket of ERα and ERβ. (a) Genistein interacts within the same binding pocket as estradiol. X-ray structure of the complex estradiol–ERβ [PDB code 3OLS (Mocklinghoff et al. 2010)] (left-hand panel) and genistein–ERβ [PDB code 1X7J (Manas et al. 2004)] (right-hand panel). (b) Genistein interacts within the pocket of ERα [PDB code 1X7R (Manas et al. 2004)] (left-hand panel) and ERβ [PDB code 1X7J (Manas et al. 2004)] (right-hand panel).

hydrogen interaction with His-475 of ERβ (His-524 in ERα) and occupies thereby the same region as the D ring 17β-hydroxyl of estradiol (Figure 11.6a).

Isoflavonoids exhibit an intriguing preference for ERβ in terms of binding affinity (Kuiper et al. 1998). For example, genistein displays 8- to 40-fold selectivity for ERβ when compared with ERα, corresponding to an increase of the binding energy difference between 1.2 kcal mol^{-1} and 2.2 kcal mol^{-1} (Manas et al. 2004). Met-336 of ERβ (Leu-384 in ERα) and Met-421 of ERα (Ile-373 in ERβ) may be responsible, through attractive potential, for such a preference (Manas et al. 2004; Pike et al. 1999) (Figure 11.6b). In addition, the B ring of genistein is in close proximity from Met-336. Furthermore, one should consider the volume differences between the binding-pocket of ERβ and that of ERα. Indeed, the Met-336 side chain may be responsible for the volume differences between ERβ and ERα. The volume of the binding cavity is 390 Å^3 for ERβ (complex with genistein) and 490 Å^3 in the case of ERα (complex with estradiol). Thus, not only electronic effects but also steric effects may explain the preference of isoflavonoids for ERβ (Manas *et al.* 2004; Pike *et al.* 1999).

Likewise, Met-421 of ERα, which is in the proximity of the 5 hydroxyl of genistein, may discriminate between the two ER subtypes. By interacting intramolecularly through a hydrogen bond with its 4-keto group, the hydroxyl in position 5, which is present in a number of isoflavonoids, is weakly polarizable. Thus, repulsion forces may occur between this keto group and Met-421. Strikingly, this residue corresponds in ERβ to an isoleucine residue (Ile-373), which is more adapted in this context (Figure 11.6b). Finally, appropriate spatial orientations of ERβ residue side chains covering the wall of the pocket and corresponding to low-energy rotamers are not excluded (Manas *et al.* 2004).

When associated with estrogens, ERs adopt an agonist conformation allowing their dimerization and phosphorylation as well as their ability to recruit coactivators, which participate in the interaction of the ER with DNA. Coactivators are engulfed, through an LxxLL motif (where L corresponds to leucine and x to any amino acid), in a shallow hydrophobic binding site located at the surface of ER, between the helices H3 and H5. The recruitment of LxxLL-containing proteins is closely related to the orientation of helix H12 of the ER. In the presence of antagonists, helix H12 occupies this LxxLL-binding site, thereby preventing the recruitment of coactivators. In contrast, in the presence of agonists, helix H12 is projected away from the LxxLL-binding site, allowing the recruitment of coactivators. Hence the ligand-dependent conformational status of ER is closely related to its ability to recruit coactivators and to induce transcription (Leclercq *et al.* 2011a; Pike *et al.* 2000).

Very surprisingly, when genistein, which is by nature an estrogen agonist, associates with ERβ, it adopts an antagonist-like conformation (Manas *et al.* 2004; Pike *et al.* 1999). In fact, the free energy ΔG ($\Delta G = -RT$ Ln $[K_1/K_2]$) associated with the antagonist conformation of ERβ would be, in this context, slightly more favorable than that calculated for the agonist conformation. Hence, the conformation adopted by ERβ when complexed with isoflavonoids is confusing and must be carefully considered, since it is not relevant, for thermodynamic reasons, to pharmacological and biochemical data. However, when associated with isoflavonoids, unconventional conformations seem likely, possibly modifying the formation of ER–estrogen-response element (ERE) complexes, as shown with genistein and daidzein (Nikov *et al.* 2000).

11.4 Mammalian Metabolism and Bioavailability

Isoflavonoids are claimed to display beneficial effects against climacteric syndromes. In this context, they are subject of self-medication and, therefore, escape medical control. Thus, they are consumed in postmenopausal women. The most active derivatives are metabolites. Two types of endogenous metabolism occur in mammals: (i) type I metabolism, which is responsible for oxidative/reductive reactions, dealkylation and deglycosylation, and (ii) type II metabolism, which is responsible for conjugation reactions. Metabolism can be endogenous (mammalian-mediated catalysis) or exogenous (gut microflora-mediated catalysis) (Atkinson *et al.* 2005; de Cremoux *et al.* 2010).

Endogenous lactate-phlorizin hydrolases, present in the small intestine brush border, as well as jejunum, liver and kidney, and gut microflora β-glycosidases hydrolyze the sugar moiety of isoflavonoids (type I metabolism), particularly in position 7, to produce the aglycone (free) forms (Day *et al.* 2000). These are absorbed through stomach, duodenum and proximal jejunum biological barriers and are detected in plasma after a short period, ranging between 30 min and 2 h (plasmatic concentrations decrease after 12 to 24 h to be finally undetectable after 48 h).

In our context, type II metabolism occurs principally in the intestine and requires specific enzymes such as catechol-*O*-methyltransferase (COMT), sulfotransferase (SULT) and β-glucuronidases. Even if COMT-mediated *O*-methylation is a minor route for type II metabolism, it constitutes an alternative to sulfation, which is catalyzed by cytosolic SULTs. Briefly, SULT enzymes catalyze the transfer of sulfate ions from 3′-phosphoadenosine-5′-phosphosulfate to a phenolic hydroxyl present on isoflavonoids (Pasqualini 2009). Remarkably, sulfation sites are decisive for estrogenicity: sulfation at position 4′ leads to a modest estrogenicity, whereas sulfation in position 7 induces a stronger estrogenicity. Also, mono- and di-sulfation reactions could have significant repercussions on transcription. Likewise, activated SULT isoforms would depend on isoflavonoid plasmatic concentrations. Another type II metabolism concerns endogenous endoplasmic reticulum uridine diphosphoglucuronyltransferases, which are present in intestinal mucosa and liver and that facilitate the urinary and biliary elimination of compounds after monoglucuronidation in the position 7. Finally, isoflavonoids are easily metabolized by intestine bacteria through *O*-demethylation, dehydroxylation, hydrogenation, lactonisation or ring cleavage reactions.

The most active metabolites are equol and *O*-desmethylangolensin (O-Dma). O-Dma is produced by 80–90% of the population, whereas equol is produced by 30–40% of the population, suggesting inter-individual variations with respect to responsiveness. Thus, isoflavone consumers could be classified as "equol producers" (plasma concentration $> 20\,\mu g\,L^{-1}$), "equol non-producers" (plasma concentration $< 10\,\mu g\,L^{-1}$), "O-Dma producers" and "O-Dma non-producers" (Atkinson *et al.* 2005). The exact reasons for which equol and O-Dma are variably produced in the body are still obscure. However, gram-positive and -negative intestinal bacteria are strongly suspected. In this regard, gut microflora population can be modified by host genetic or non-genetic factors (including lifestyle and diet habits), a feature that may have deep consequences, not only for the production of equol and O-Dma, but also on ERE-dependent transcription (de Cremoux *et al.* 2010).

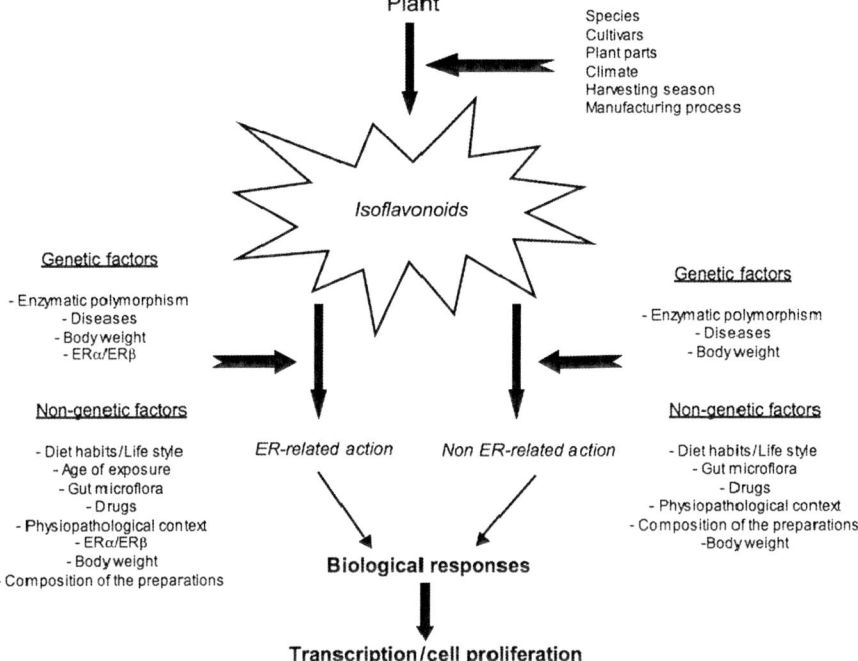

Figure 11.7 Factors influencing the biosynthesis of isoflavones in plants, as well as their action in mammals.

The bioavailability of isoflavonoids is closely related to their metabolism and to the bacteria species composing the gut microflora (de Cremoux et al. 2010). Accordingly, deglycosylation, which is crucial for observing a cross through the intestinal barrier, is a limiting step. In this context, isoflavonoid bioavailability depends on age, sex, the physiopathological context (gastric pH, stress, bowel diseases...), diet habits (e.g., dietary fibers modify the reabsorption and excretion of isoflavonoids as they influence on the activity of β-glucosidases and β-glucuronidases) and drug intake (e.g., antibiotics modify the gut microflora). Finally, glucuronid (>80% of the eliminated form) or sulfate-conjugated aglycones are easily eliminated in urines and bile. However, bacteria can catalyze deconjugation, leading to reabsorption through entherohepatic recycling. Accordingly, a second plasma peak that corresponds to a reabsorption in bile is usually observed. Finally, it should be noted that isoflavonoid aglycones distribute easily in fat tissues, suggesting that retention can occur in women with high body mass index (BMI) (Figure 11.7).

11.5 Biochemical and Physiological Functions of Isoflavonoids

Based on epidemiological evidence that Japanese women are approximately 4-fold less likely to have breast cancer than their Caucasian and European congeners, the isoflavonoid story started at the beginning of the 1980s. Diet habits were rapidly suspected to be responsible, exclusively, for such a difference, with Asian populations consuming a traditional phytoestrogen-rich diet (typically 20 to 150 mg day^{-1} during their whole life including their childhood). Accordingly, when mothers migrate in the United States, the prevalence of breast tumors in Asian daughter generations is as high as in American women.

Even if these observations seem multifactorial (de Cremoux *et al.* 2010; Leclercq *et al.* 2011b), isoflavones can display opposing effects on cell growth, depending on the composition of the preparations (Setchell *et al.* 2001) and inter-individual variations in plasma concentrations, as well as individual genetic and non-genetic factors (Figure 11.7).

11.5.1 Estrogenic-related Effects

11.5.1.1 Antiproliferative Effects

Among the clinical trials devoted to isoflavonoids, only a few show significant protective effects against breast cancer. The situation appears strongly paradoxical and, therefore, extremely complex with respect to the agonist activities of phytoestrogens.

The reasons for which isoflavonoids exert such protective effects may not be limited to their content in diet or food supplements and their sole interaction with ERs. It could extend to other food and genetic factors, as well as interactions with other targets and biochemical cascades (Leclercq *et al.* 2011a; Rice and Whitehead 2006). Possibly due to a low fat and fiber-rich diet rather than the presence of isoflavonoids, a decrease of circulating estradiol has been found in Asian women when compared with European women (Goldin *et al.* 1986; Key *et al.* 1990). Likewise, isoflavones increase the level of sex hormone-binding globulin (SHBG) (Pino *et al.* 2000). Also, metabolic differences related to gut microflora modifications, diet habits (Xu *et al.* 1995) and the age of isoflavonoid exposure, as well as intra- and inter-ethnic enzymatic polymorphism (Sangrajrang *et al.* 2009), must be considered (de Cremoux *et al.* 2010; Leclercq *et al.* 2011b and references herein). The ability of isoflavonoids to inhibit the enzymes involved in steroid biogenesis and metabolism, including those that are responsible for estradiol production, is likely (Rice and Whitehead 2006). More recently, a correlation between phytoestrogen-induced proliferation and the ERα/ERβ ratio (Sotoca *et al.* 2008), which would vary in function of the physiopathological and more specifically the tumoral context, has been reported (Figure 11.7).

It should be stressed that the antiproliferative effects of isoflavonoids have been observed in experimental models and are irrelevant to female physiological conditions, the latter occurring at concentrations $> 10\,\mu M$, which cannot be easily achieved in women consuming phytoestrogen-rich diet or with isoflavonoid supplementation. Hence doubts are raised concerning such protective effects *in vivo*.

11.5.1.2 Proliferative Effects

Strikingly, the action exerted by isoflavonoids is biphasic. At concentrations consistent with human plasma ($<10\,\mu M$), they stimulate cell growth (Allred et al. 2001) in an ER-dependent or -independent manner and through "direct" (nuclear) or "indirect" (membrane) events, leading possibly to the activation of the Ras cascade, phosphoinositide 3-kinase (PI3K)/Akt, and the oncoproteins c-fos and c-jun.

Even if isoflavonoids are not *per se* hormone-dependent tumor initiators, they induce, in an appropriate context, the proliferation of pre-existing tumor cells, raising serious doubts, again, concerning their preventive effects against breast cancer, their specificity of action, as well as their safety. In this regard, genistein antagonizes the effects induced by tamoxifen, justifying scepticism concerning a beneficial contribution from isoflavonoids in women with breast cancer (Limer et al. 2006; Seo et al. 2007). Also, several phytoestrogens are clastogenic and induce DNA damage (Kulling et al. 1999; Stopper et al. 2005), necessitating utmost care in their use.

According to their estrogenic properties and despite an evident lack of specificity and selectivity (Leclercq et al. 2011b), which contrast with their protective effects, isoflavone extracts are commonly used for their possible benefits against climacteric complaints (hot flushes and osteoporosis). Escaping medical review, they are freely commercialized as pill forms and doses ranging typically from 40 to $80\,\text{mg day}^{-1}$ and are used for long periods (typically from 1 to 5 years).

11.5.2 Non-estrogenic-related Effects

In comparison with the estrogen-like effects displayed by isoflavonoids, few literature data are available concerning non-estrogenic-related effects. Isoflavonoids interact with a number of proteins, altering therefore several cellular functions including signalling pathways (Leclercq et al. 2011b; Rice and Whitehead 2006). Independently from their direct interaction with ERs, additional mechanisms related to anti-proliferative (inhibition of metalloproteinases, tyrosine kinases and DNA topoisomerase), anti-angiogenic and anti-oxidant effects, cell cycle arrest, apoptosis (through p53-dependent and -independent mechanisms) and cell differentiation have been proposed.

Isoflavonoids act in vagina, uterus, thyroid, bone as well as the cardiovascular system.

Their estrogenic properties seem to have significant effects on the vagina and uterus, with anatomopathological studies revealing desquamation and cornification of both (Unfer et al. 2004).

Due to enzymatic inhibitory effects, isoflavonoids modify the functions of the thyroid (Haselkorn et al. 2003). Indeed, thyroid peroxidase, which is in charge of the iodination of the tyrosine moiety of thyroglobulin, is inhibited by genistein. Likewise, an increase in the levels of the hormones tri-iodothyronine (T3) and thyroxine (T4), as well as of thyroid-stimulating hormone (TSH), occurs.

The potential benefits of isoflavonoids on the cardiovascular system seem to be due to their action on the plasma concentration of high-density (HDLs), low-density (LDLs) and very-low-density (VLDLs) lipoproteins. While they increase HDL ("good" cholesterol), they decrease LDL ("bad" cholesterol) and VLDL (Wangen *et al.* 2001; Yousef *et al.* 2004). Lastly, they stimulate the production of nitric oxide (NO), participating therefore in vasodilation. These actions could explain the preventive effects of isoflavonoids against atherosclerosis. Here again, very few clinical data have been published, and no accurate conclusions may be drawn.

11.6 Conclusion

To summarize, isoflavones (and more generally phytoestrogens), which are biosynthesized in plants, behave in the mammalian tissues and cellular levels as estradiol mimetics. According to *in vitro* assays, they participate in cell proliferation by associating preferentially with the ERβ. However, the action of isoflavonoids is not limited to estrogenic mechanisms, raising doubts concerning their selectivity and specificity. It has been claimed that they display various benefits, even if these are poorly significant and disparate *in vivo*. The reasons for which isoflavonoids protect Asian women against breast cancer seem multifactorial and irreconcilable with non-Asian populations. According to literature data, they may exert antagonist effects on estrogenic pathways, preventing therefore against symptoms associated with menopause. Hence, the situation appears highly confused and far from conclusive enough for safe medical applications. In the light of recent literature data, the agonist and antagonist activities of isoflavones remain uncertain as the results of preclinical and clinical studies are heterogeneous and, therefore, not comparable. Accordingly, their efficacy and safety are strongly subject to debate, particularly when they are used in long-term treatments without medical control. Hence, the best recommendation would be prudence. Thus, women who want to ingest phyto-hormones must take medical advice, particularly those with a hormone-dependent cancer risk.

Summary Points

- This Chapter focuses on the links existing between the synthesis of isoflavonoids in plants and their estrogenic activity in mammalian.
- Isoflavonoids are principally produced in leguminous plants from phenylalanine.
- They have key functions in plants (plant–bacteria symbiotic relationships, stress and disease resistance).
- In mammals, they are bioavailable after deglycosylation and become active (estrogenic).
- In mammalians, they directly activate the estrogen receptor (ER).
- The ER–isoflavonoid complex activates transcription and estrogenic activity.

- Automedication with isoflavonoids is observed, with the expectation of benefits against menopause syndromes (osteoporosis and hot flushes).
- The biological action of isoflavonoids in women appears extremely complex and their benefits uncertain. Due to their poor specificity and selectivity, their use is not safe.

Key Facts

Key facts of Isoflavonid Biosynthesis

- Plants synthesize a number of molecules that display physiological functions by participating, for example, in stress resistance and in symbiosis with bacteria. Isoflavonoids are polyphenolic heterocyclic compounds sharing one heteroatom of oxygen.
- Isoflavonoids, which are principally produced in leguminous plants, are synthesized from two biochemical pathways, *i.e.*, the pentose phosphate pathway and glycolysis. These pathways produce two key precursors, erythrose-4-phosphate and phosphoenol-pyruvate, respectively, that are responsible for the production of chorismate. Chorismate is transformed into L-phenylalanine and, after, into isoflavonoids.
- Finally, isoflavonoids are produced from plants under a conjugated (glycosylated) form.

Key Facts of Interaction of Isoflavonoids with Estrogen Receptors (ERs)

- Isoflavonoids interact with NodD, a protein isolated from bacteria found in plant nodules. NodD has structural similarities with the human ER.
- Also, isoflavonoids have strong structural similarities with estradiol, the natural ligand of the ER, *i.e.*, two hydroxyls that are approximately 10 Å apart through a hydrophobic block.
- Accordingly, isoflavonoids interact with weak affinity within the ER pocket that is occupied by estradiol. Thus, the ER adopts an agonist conformation, interacts with specific DNA regions and activates specific biological responses. Thus, isoflavonoids are weak estrogens.

Key Facts of Metabolism of Isoflavonoids in Mammals

- Most compounds are metabolized in the body by specific endogenous or bacterial enzymes, producing active or inactive derivatives.
- Isoflavonoids are primarily metabolized into active aglycone (isoflavones devoid of their sugar moiety). This structural transformation is necessary to make isoflavonoids bioavailable.
- Among the structural modifications catalyzed by metabolic enzymes, several lead to active derivatives. Accordingly, isoflavonoids lead to equol

and *O*-desmethylangolensin (O-Dma), which are highly estrogenic when compared with the initial aglycone.
- Metabolic pathways depend on a number of uncontrolled parameters such as age, enzyme polymorphism, the physiopathological context, gut microflora, drug intake or diet habits. Moreover, the quantitative and qualitative composition of isoflavonoid preparations, which is ignored by suppliers, would be crucial in this context.

Key Facts of Biochemical and Physiological Functions of Isoflavonoids

- Isoflavonoids exert two types of action in women. At low concentrations, which are compatible with plasma concentrations, they induce cell proliferation. At higher concentrations, they exert an antiproliferative effect.
- These effects are not specific, are dependent or independent from estradiol, and are mediated through several targets.
- The situation is highly confused and complex. The effects of isoflavonoids on human estrogen-dependent tissues would depend on the qualitative and quantitative composition of the preparations as well as a number of genetic and non-genetic parameters, raising doubts about the legitimacy of isoflavonoids in the context of human health. Because of their origin from plants, they are wrongly considered as food supplements, and are freely available, escaping medical review.

Definitions of Words and Terms

Climacteric syndromes: When approximately 51.4 years old, women are subjected to a drastic decrease of the production of estradiol by the ovaries. This event is the cause of the end of their reproductive period and is associated with underlying metabolic upheaval. Osteoporosis, hot flushes and atherosclerosis are the principal disorders observed. They are called climacteric syndromes.

Entherohepatic recycling: Molecules that are metabolized into conjugated forms are not bioavailable and are, therefore, directed into urinary and biliary pathways for elimination. However, several bacterial processes that are relevant to deconjugation mechanisms can occur. Thus, the molecules are reabsorbed from the intestine to the blood circulation and therefore become bioavailable.

Enzyme: Similarly to the classical receptor, an enzyme is a protein accepting ligands that are called in this case substrates. In the case of enzymes, the substrates interact in a pocket that is called the catalytic site and that is in charge of specific chemical transformations to afford a molecule devoted to some biological functions.

Estrogen receptor (ER): Among the nuclear hormone receptors, several accept steroids as ligands. It is the case for the androgen, glucocorticoid, progesterone or estrogen receptors. The ER, which exists in two principal isoforms (α and β) and for which the main ligand is estradiol, is a transcription factor that is responsible for a number of actions contributing to the development and the maintenance of the primary and secondary sexual characteristics in women. Also, it participates in neuroprotection and has beneficial effects against atherosclerosis and osteoporosis.

Lipoproteins: Lipids, being insoluble in water, are carried in the blood by specific proteins that are called lipoproteins. Three different types of lipoproteins are usually considered: low-density lipoproteins (LDLs) transport cholesterol from the liver to the cells, whereas high-density lipoproteins (HDLs) do the contrary. Very-low-density lipoproteins (VLDLs) are in charge of the transport of triglycerides from the liver to fat tissues.

Nitric oxide (NO): Nitric oxide is crucial for many physiological processes and is a key intracellular messenger. It is particularly important in vascular tone and in some pathologies including cancer, diabetes or inflammation.

Nuclear hormone receptors: Signal transduction is mediated in cells through specific receptors. Such proteins are composed at their surface or in their core of specific regions that interact with partners (proteins, peptides, small molecules, hormones...) responsible for their physiological action. These proteins are located in different compartments of the cell, depending of a number of factors. Nuclear hormone receptors are located in the nucleus of the cell and accept hormones as ligands.

Secondary and tertiary structures: A protein is a flexible biopolymer constituted of amino acids. Depending on the nature of the latter, they adopt different spatial orientations (folding). A first degree of complexity is relevant to secondary structures (regular conformation), which are stabilized through $-C=O\ H-N-$ hydrogen bonds. α-Helices, β-sheets or turns are regular secondary structures. The tertiary structure is a spatial folding of the secondary structures. They are stabilized through covalent bonds, hydrogen bonds, electrostatic or van der Waals interactions.

Structure–activity relationships (SARs): The pharmacological profile displayed by an active compound is strongly related to its chemical structure. The approaches that consist of exploring the structural requirements necessary for an activity of interest are called structure–activity relationship studies.

Thyroglobulin: This is a protein that is in charge of the storage of iodine in the thyroid. It plays a role of prime importance in the biosynthesis of the two main thyroid hormones, *i.e.*, tri-iodothyronine (T3) and thyroxine (T4). They are crucial in metabolism as well as in several cardiovascular and respiratory functions.

List of Abbreviations

CHS	chalcone synthase
CoA	coenzyme A
COMT	catechol-*O*-methyltransferase
ER	estrogen receptor
ERE	estrogen-response element

HDL	high-density lipoprotein
LDL	low-density lipoprotein
MW	molecular weight
O-Dma	*O*-desmethylangolensin
SULT	sulfotransferase
VLDL	very-low-density lipoprotein

References

Akashi, T., Sawada, Y., Aoki, T., and Ayabe, S.I., 2000. New scheme of the biosynthesis of formononetin involving 2,7,4′-trihydroxyisoflavone but not daidzein as the methyl acceptor. *Bioscience, Biotechnology, and Biochemistry.* 64: 2276–2279.

Allred, C.D., Allred, K.F., Ju, Y.H., Virant, S.M., and Helferich, W.G., 2001. Soy diets containing varying amounts of genistein stimulate growth of estrogen-dependent (MCF-7) tumors in a dose-dependent manner. *Cancer Research.* 61: 5045–5050.

Atkinson, C., Frankenfeld, C.L., and Lampe, J.W., 2005. Gut bacteria metabolism of the soy isoflavone daidzein: Exploring the relevance to human health. *Experimental Biology and Medicine.* 230: 155–170.

Crombie, L., and Whiting, D.A., 1992. The mechanism of the enzymic induced isoflavone – isoflavone change. *Tetrahedron Letters.* 33: 3663–3666.

Day, A.J., Canada, F.J., Diaz, J.C., Kroon, P.A., McLauchlan, R., Faulds, C.B., Plumb, G.W., Morgan, M.R., and Williamson, G., 2000. Dietary flavonoid and isoflavone glycosides are hydrolysed by the lactate site of lactate phlorizin hydrolase. *FEBS Letters.* 468: 166–170.

Deavours, B.E., and Dixon, R.A., 2005. Metabolic engineering of isoflavonoid biosynthesis in alfalfa. *Plant Physiology.* 138: 2245–2259.

de Cremoux, P., This, P., Leclercq, G., and Jacquot, Y., 2010. Controversies concerning the use of phytoestrogens in menopause management: Bioavailability and metabolism. *Maturitas.* 65: 334–339.

Fox, J.E., 2004. Chemical communication threatened by endocrine-disrupting chemicals. *Environmental Health Perspectives.* 112: 648–653.

Goldin, B.R., Adlercreutz, H., Gorbach, S.L., Woods, M.N., Dwyer, J.T., Conlon, T., Bohn, E., and Gershoff, S.N., 1986. The relationship between estrogen levels and diet habits of Caucasian American and Oriental immigrant women. *The American Journal of Clinical Nutrition.* 44: 945–953.

Gyorgypal, Z., and Kondorosi, A., 1991. Homology of the ligand-binding regions of Rhizobium symbiotic regulatory protein NodD and vertebrate nuclear receptors. *Molecular and General Genetics.* 226: 337–340.

Haselkorn, T., Stewart, S.L., and Horn-Ross, P.L., 2003. Why are thyroid cancer rates so high in southeast Asian women living in United States? The Bay Area Thyroid Cancer Study. *Cancer Epidemiology, Biomarkers and Prevention.* 12: 144–150.

Jacquot, Y., Rojas, C., Refouvelet, B., Robert, J.F., Leclercq, G., and Xicluna, A., 2003. Recent advances in the development of phytoestrogens and derivatives: An update of the promising perspectives in the prevention of postmenopausal diseases. *Mini-Review in Medicinal Chemistry*. 3: 333–346.

Key, T.J.A., Chen, J., Wang, D.Y., Pike, M.C., and Boreham, J., 1990. Sex hormones in women in rural China and in Britain. *British Journal of Cancer*. 62: 631–636.

Kuiper, G.G.J.M., Lemmen, J.G., Carlsson, B., Corton, J.C., Safe, S.H., van der Saag, P.T., van der Burg, B., and Gustafsson, J. Å., 1998. Interaction of estrogenic chemicals and phytoestrogens with estrogen receptor β. *Endocrinology*. 139: 4252–4263.

Kulling, S.E., Rosenberg, B., Jacobs, E., and Metzler, M., 1999. The phytoestrogens coumoestrol and genistein induce structural chromosomal aberrations in cultured human peripheral blood lymphocytes. *Archives of Toxicology*. 73: 50–54.

Leclercq, G., Gallo, D., Cossy, J., Laïos, I., Larismont, D., Laurent, G. and Jacquot, Y., 2011a. Peptides targeting estrogen receptor α – potential applications for breast cancer treatment. *Current Pharmaceutical Design*. 17: 2632–2653.

Leclercq, G., de Cremoux, P., This, P., and Jacquot, Y., 2011b. Lack of sufficient information on the specificity and selectivity of commercial phytoestrogens preparations for therapeutic purposes. *Maturitas*. 68: 56–64.

Limer, J.L., Parkes, A.T., and Speirs, V., 2006. Differential response to phytoestrogens in endocrine sensitive resistant breast cancer cells *in vitro*. *International Journal of Cancer*. 119: 515–521.

Manas, E.S., Xu, Z.B., Unwalla, R.J., and Somers, W.S., 2004. Understanding the selectivity of genistein for human estrogen receptor-β using X-ray crystallography and computational methods. *Structure*. 12: 2197–2207.

Miksicek, R.J., 1995. Estrogenic flavonoids: structural requirements for biological activity. *Proceedings of the Society for Experimental Biology and Medicine*. 208: 44–50.

Mocklinghoff, S., Rose, R., Carraz, M., Visser, A., Ottman, C., and Brunsveld, L., 2010. Synthesis and crystal structure of a phosphorylated estrogen receptor ligand binding domain. *ChemBiochem*. 11: 2251–2254.

Nikov, G.N., Hopkins, N.E., Boue, S., and Alworth, W.L., 2000. Interactions of dietary estrogens with human estrogen receptors and the effects on estrogen receptor–estrogen response element complex formation. *Environmental Health Perspectives*. 108: 867–872.

Pasqualini, J.R., 2009. Estrogen sulfotransferases in breast and endometrial cancers. *Annals of the New York Academy of Sciences*. 1155: 88–98.

Pike, A.C.W., Brzozowski, A.M., Hubbard, R.E., Bonn, T., Thorsell, A.G., Engström, O., Ljunggren, J., Gustafsson, J.Å., and Carlquist, M., 1999. Structure of the ligand-binding domain of oestrogen receptor β in the presence of a partial agonist and a full antagonist. *EMBO Journal*. 18: 4608–4618.

Pike, A.C.W., Brzozowski, A.M., and Hubbard, R.E., 2000. A structural biologist's view of the oestrogen receptor. *Journal of Steroid and Molecular Biology*. 74: 261–268.

Pino, A.M., Valladares, L.E., Palma, M.A., Mancilla, A.M., Yáñez, M., and Albala, C., 2000. Dietary isoflavones affect sex hormone-binding globulin levels in postmenopausal women. *Journal of Clinical Endocrinology and Metabolism*. 85: 2797–2800.

Pueppke, J.L., 1996. The genetics and biochemical basis for nodulation of legumes by rhizobia. *Critical Reviews in Biotechnology*. 16: 1–51.

Rice, S., and Whitehead, S.A., 2006. Phytoestrogens and breast cancer – promoters or protectors?. *Endocrine-Related Cancer*. 13: 995–1015.

Sangrajrang, S., Sato, Y., Sakamoto, H., Ohnami, S., Laird, N.M., Khuhaprema, T., Brennan, P., Boffetta, P., and Yoshida, T., 2009. Genetic polymorphisms of estrogen metabolizing enzyme and breast cancer risk in Thai women. *International Journal of Cancer*. 125: 837–843.

Seo, H.S., DeNardo, D.G., Jacquot, Y., Laïos, I., Vidal, D.S., Zambrana, C.R., Leclercq, G., and Brown, P.H., 2006. Stimulatory effect of genistein and apigenin on the growth of breast cancer cells correlate with their ability to activate ERα. *Breast Cancer Research and Treatment*. 99: 121–134.

Setchell, K.D.R., Brown, N.M., Desai, P., Zimmer-Nechemias, L., Wolfe, B.E., Brashear, W.T., Kirschner, A.S., Cassidy, A., and Heubi, J.E., 2001. Bioavailability of pure isoflavones in healthy humans and analysis of commercial soy isoflavone supplements. *Journal of Nutrition*. 131: 1362S–1375S.

Sotoca, A.M., Ratman, D., van der Saag, P., Stroem, A., Gustafsson, J.A., Vervoort, J., Rietjens, I.M.C.M., and Murk, A.J., 2008. Phytoestrogen-mediated inhibition of proliferation of the human T47D breast cancer cells depends on the ERα/ERβ ratio. *Journal of Steroid Biochemistry and Molecular Biology*. 112: 171–178.

Stopper, H., Schmitt, E., and Kobras, K., 2005. Genotoxicity of phytoestrogens. *Mutation Research*. 574: 139–155.

Turner, J.V., Agatonovic-Kustrin, S., and Glass, B.D., 2007. Molecular aspects of phytoestrogen selective binding at estrogen receptor. *Journal of Pharmaceutical Sciences*. 96: 1879–1885.

Unfer, V., Casini, M.L., Costabile, L., Mignosa, M., Gerli, S., and Di Renzo, G.C., 2004. Endometrial effects of long-term treatment with phytoestrogens: a randomized, double-blind, placebo-controlled study. *Fertility and Sterility*. 82: 145–148.

Wangen, K.E., Duncan, A.M., Xu, X., and Kurzer, M.S., 2001. Soy isoflavones improve plasma lipids in normocholesterolemic and mildly hypercholesterolemic postmenopausal women. *American Journal of Clinical Nutrition*. 73: 225–231.

Yousef, M.I., Kamel, K.I., Esmail, A.M., and Baghdadi, H.H., 2004. Antioxidant activities and lipid lowering effects of isoflavone in male rabbits. *Food and Chemical Toxicology*. 42: 1497–1503.

Xu, X., Harris, K.S., Wang, H.J., Murphy, P.A., and Hendrich, S., 1995. Bioavailability of soybean isoflavones depends upon gut microflora in women. *Journal of Nutrition*. 125: 2307–2315.

Analysis

CHAPTER 12
Continuous Microwave-assisted Isoflavone Extraction

DORIN BOLDOR* AND CRISTINA MIRELA SABLIOV

Department of Biological and Agricultural Engineering, Louisiana State University Agricultural Center and College of Agriculture, Louisiana State University A&M College, 149 E. B. Doran Bldg, LSU AgCenter, Baton Rouge, LA 70803, USA
*Email: dboldor@agcenter.lsu.edu

12.1 Introduction

Isoflavone, proteins, and lecithin are all found in significant quantities in many plants and plant products. Soybeans are a major source for all these substances and, due to the many nutritional and health benefits of these compounds, the annual worldwide consumption of soybeans has increased by approximately 50% (from 114 million to 170 million tons) in a span of 10 years (Klejdus *et al.* 2004). Isoflavones are a class of antioxidants compounds which can be found in many plants and have been demonstrated to assist in the prevention of several ailments and diseases including cancer of prostate, breast, thyroid, and colon; these compounds also assist in increasing bone density and therefore preventing osteoporosis, decrease cholesterol levels in blood, and reduce many of the symptoms associated with menopause in women (Ali *et al.* 2004; 2005; Lo *et al.* 2007; Mazur *et al.* 1998; Nahas *et al.* 2007; Phrakonkham *et al.* 2007; Wuttke *et al.* 2007). The preponderant isoflavones in soybeans are genistein, daidzein, glycitein, genistin, daidzin, and glycitin (Nahas *et al.* 2007), with their specific proportions depending on climatic conditions, agronomic practices, maturity at harvest, and post-harvest management and storage conditions (Mazur et al. 1998). Although isoflavones can be extracted from other plants, the information presented in this Chapter refers mostly to those isoflavones extracted from soybeans. However, regardless of their source, the methodology and results described herein can be applied to virtually any number of potential agricultural and biological feedstock that contains isoflavones.

Whereas different methods have been used for isoflavones extraction (pressurized solvent extraction, ultrasonication, stirring and shaking, and Soxhlet (Lee and Lin, 2007; Luthria et al. 2007; Rostagno et al. 2004), we describe here an alternative, recently developed method based on microwave-assisted solvent extraction in a continuous system. Microwave-assisted extraction (MAE) can be a more environmentally and economically friendly process (Pan et al. 2002), and its efficiency depends on processing time, temperature, solid/liquid ratio, and solvent type and particular composition (mixture vs. pure) (Grigonis et al. 2005; Hemwimon et al. 2007; Kwon et al. 2003; Pan et al. 2002). Electromagnetic energy, including in the microwave region of the spectrum (approximately 300 to 6000 MHz), interacts with all materials (in this case agricultural and biological materials of interest) directly at the molecular and atomic level. In order to be effective, a microwave extraction system has to be designed such that the processing enclosure (processing tube) has minimal to no dielectric interaction with the microwaves, which allows for the energy to be efficiently transferred to the processed material.

In microwave-based extraction processes, the feedstock material (solvent–oil–food matrix) behaves as a dielectric which interacts with the energy via two main mechanisms (Metaxas and Meredith 1983): vibration/rotation of polar molecules and oscillations of ionic components. The rotations of polar molecules occur as these are trying to align themselves with the changing polarity of the oscillating electric field component of the energy. Simultaneously, the charged ions are trying to move back and forth according to the same polarity of the electric field. These oscillations, rotations, and vibrations occur at frequencies in the order of MHz and GHz (depending on the specific microwave operating frequency), and result in intermolecular friction and near-instantaneous volumetric heating (Metaxas and Meredith 1983). Targeted polar molecules heat extremely rapid above their boiling point (and above that of the liquid existent in the cells), resulting in high local pressures at microscopic levels, rupturing cell walls (Choi et al. 2006). High, localized electric fields (depending on a specific system configuration) may also generate a transmembrane potential high enough to open permanent pores in the membranes of cells (electroporation), compounding with the thermal effects to increase the efficiency of the microwave extraction process (Choi et al. 2006).

The specific rate of heating and maximum temperatures achieved in a continuous system depend on several factors including frequency and power of the electromagnetic field, dielectric properties of the material, and specific geometry of the microwave system (Boldor et al. 2004; Nelson 1980). Microwave heating systems can be designed to be very versatile, and have been used to inactivate bacteria and enzymes in foods (Canumir et al. 2002; Huang et al.

2007; Kozempel *et al.* 1997; 1998; Yaghmaee and Durance, 2005) and to extract anthraquinones from *Morinda citrifolia* (Hemwimon *et al.* 2007).

From soybeans, isoflavones are usually extracted with water, ethanol, methanol or other more polar solvents (Luthria *et al.* 2007), as most isoflavones are polar phenolic compounds. Methods using non-polar compounds, including those using supercritical fluids (such as CO_2), are not as effective for isoflavones extraction unless modifiers are used to increase the polarity of the medium to improve extraction yields (Grigonis *et al.* 2005). Considerabe debate still exists regarding the optimum solvent or solvent mixtures that maximize isoflavone yields (Luthria *et al.* 2007; Rostagno *et al.* 2004). When using microwaves to enhance extraction, the polarity of the solvent plays a particularly important role as it affects its interaction not only with the isoflavones, but also with the microwave energy.

12.2 Microwave Extraction System Designs

Conceptually, a microwave system is composed of a microwave generator (which includes a power supply, a magnetron, and the ancillary control and cooling systems), waveguides that transport the energy to the applicator cavity, and the applicator cavity itself in which the material to be processed is located. Depending on the complexity and power capacity of the system, additional components may include a circulator plus a water load that isolate and protect the magnetron from reflected energy, a directional coupler to measure power, and a tuning section that is used to match the impedance of the load (the material to be processed) to the impedance of the waveguides in order to increase system's performance. There are two different microwave cavity designs that can be used for microwave extraction: a multi-mode cavity (similar with the home microwave unit), and a single mode cavity, which allow for more control over the distribution of the microwave energy. The multi-mode cavities are usually of low power (up to 1.6–2 kW), have short waveguides, and usually lack circulators, tuning, and directional couplers, as the magnetrons are robust enough to sustain a certain degree of exposure to reflected microwaves. Both multi-mode and single-mode cavities can be further customized to allow for continuous processing, and permit the control of various operating parameters such as temperature and pressure. Multi-mode cavities are characterized by non-uniform distribution of microwave energy, as the waves reflect off the cavity walls and intersect at different physical locations (similar with light beams in a mirror box). These distributions can be accurately predicted and calculated based on the solutions to Maxwell's electromagnetic wave equations. In single-mode cavity, the microwave energy can be better controlled, but these systems are much more expensive.

A higher-power low-cost system can be designed to operate in continuous mode by stacking three household microwave units in series (labeled M_i). The units are modified by cutting holes in their top and bottom sections to allow insertion of $3/8''$ Teflon® tubing in the vertical position to serve as an

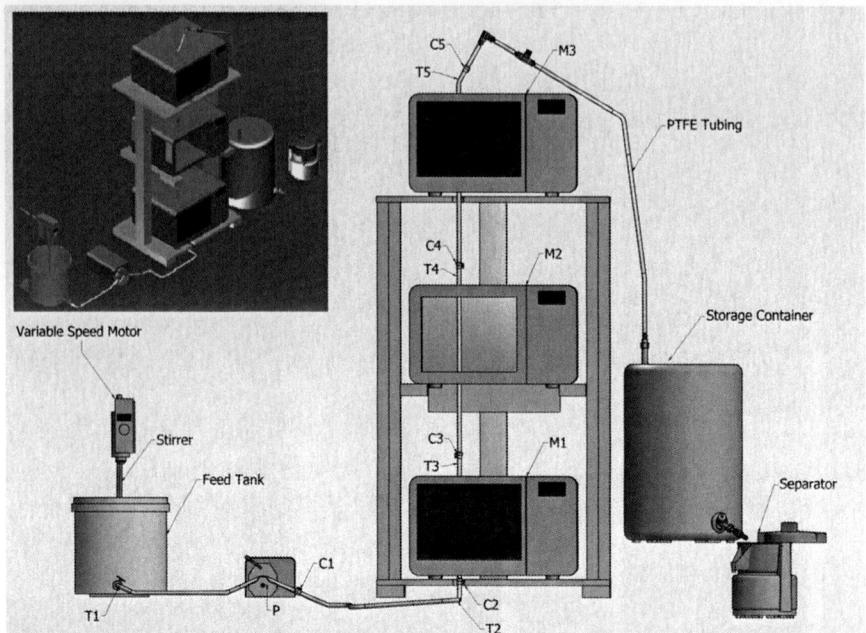

Figure 12.1 Schematic of the low-cost multiple-stage microwave extraction system (three-dimensional inset in the top left-hand corner). $M_{1,2,3}$ – microwave units, C_{1-5} – connectors, P – peristaltic pump, T_{1-5} – thermocouples.

applicator tube (Figure 12.1). Between the units, the tubes are connected using quick disconnect fittings (labeled C_i) to facilitate easy dismantling and cleaning, and the mixture of soybean flower and solvent is pumped using peristaltic pumps (P). The amount of energy delivered into the applicator cavity can be selected in one of two ways, depending on the type of microwave generator used. In traditional system (which currently cover the vast majority of the market, >90–95%), the energy output of the magnetron is constant, and the amount of energy delivered into the applicator is modulated by periodically turning the microwave on and off. A novel technology, which is slowly gaining market share, uses a new inverter technology which keeps the magnetrons operating continuously at lower power levels.

For continuous-flow operation at constant temperatures, both the power level and flow rate need to be controlled and monitored. However, in the case of traditional systems (those not using inverter technology), it is difficult to control the power level. As such, the temperature control is dictated by the flow rate of the soybean/solvent mixture and the number of microwaves in operation (*i.e.* for maximum exposure all three microwaves are operated simultaneously at full power). The material is fed into the tube from a container retrofitted with a stirrer that maintains the soybean particles in suspension to prevent clogging. The temperature is monitored at various points (inlet, outlet, and in between microwave units) with thermocouples (labeled T_i) *via* a data logger that may be connected to a computer. It is also critical that the microwave system be monitored for energy leaks using an appropriate detector, especially around the tube insertion points. The maximum permissible value is $5\,\text{mW}\,\text{cm}^{-2}$. Alternatively, if the microwave system is more complex and allows for real-time control of the generated power, the same temperature monitoring setup can be used to control the power levels in order to maintain a desired temperature.

12.3 Microwave-assisted Extraction Process

Soybeans (preferably freshly harvested) with a moisture content of between 5 and 10% is milled into a flower using a kitchen mill such as that provided by K-Tec® or a commercial mill for larger quantities. Various mixture ratios of solvent (in this case 200 proof ethanol) to soybean flour can be used, but for practical purposes (maintaining the particle in suspension and prevent clogging during pumping) a mixture of 3 parts solvent to 1 part soybean flour (wet weight basis) is pumped at a flow rate of at least $300\,\text{mL}\,\text{min}^{-1}$. At this flow rate, operating one microwave unit results in a final temperature of $55\,°\text{C}$ in 7 s, and operating two units simultaneously yields an average temperature of $73\,°\text{C}$ after approximately 14–15 s. Operating a third microwave at the same flow rate would result in temperatures above the boiling point of ethanol ($76\,°\text{C}$) in approximately 20–21 s, which would impair the overall process by increasing the viscosity of the mixture, the settling of solids, and loss of isoflavones in the ethanol vapors. The third microwave can be operated either by applying backpressure which increases the ethanol boiling point, by increasing the flow rate, or when using different solvents and feedstock that may require higher temperatures and/or longer heating time. The extraction process is finalized by maintaining the final temperature for certain periods of time *via* a container placed in a controlled temperature water or oil bath (times of 0 to 16 min in 4 min increments). The microwave extraction method is compared with the conventional heating method in which the same mixtures are heated in a round-bottomed flask equipped with a water condenser placed on a plate heater. In this case, the mixture is continuously mixed with magnetic stirrers driven by a rotating magnetic field underneath the heating plate.

12.4 Oil Separation and Isoflavone Purification

After microwave extraction, the mixture is vacuum filtered through a $\Phi = 47$ paper filter to remove the solids, then the oil/solvent/isoflavone liquid mixture is placed in an vacuum evaporator to remove the solvent. The separation of isoflavones from the oil is made *via* chromatographic methods.

In order to evaluate the performance of the microwave system in extracting isoflavones, high-performance liquid chromatography (HPLC) analysis is used to quantify isoflavone components by comparison with authentic standards. For chromatographic analysis, a volume of 2 mL of the oil/solvent mixture is transferred into HPLC vials, and the analysis is performed in the ultraviolet (UV) spectrum at 254 nm. The system used can be an Agilent 1200 Series, consisting of an autosampler (G1329B ALS SL), pump (SL Bin Pump G1312B) and a photodiode array detector (G1315C DAD SL), but similar systems can also be employed. The analysis is performed in reverse phase on a Zorbax Eclipse XDB-C18 column (3.5 μm × 4.6 mm × 100 mm; Agilent, USA) fitted with a guard column, with a linear gradient at 1.3 ml min^{-1} flow rate. The injection volume is 25 μL, and elution is achieved using a linear gradient of mixture of methanol and water as a mobile phase, with the percentage of methanol increasing from 10 to 50% in 45 min.

Isoflavones are identified by comparison of retention times and UV spectra of separated compounds with authentic standards of genistin, genistein, daidzin, and daidzein, and their quantification [amount of isoflavones extracted (g of oil)$^{-1}$] is performed by integration of the peak areas at 254 nm. The response should be linear between 10 and 100 μg L^{-1} (four points) for all isoflavones and the regression coefficients (r^2) should be relatively close to unity.

12.5 Significant Results and Discussion

12.5.1 Isoflavones (and Oil) Yields

Compared with conventional heating, microwave extraction obtains significantly higher yields at all temperatures. In the case of oil, the significantly higher yields are obtained not only for soybeans (Terigar *et al.* 2010), but also for rice bran (Kanitkar *et al.* 2011; Terigar *et al.* 2011). Most interesting, the efficiency of oil extraction increases as the scale of the process increases (Terigar *et al.* 2011). In the case of a continuous, scaled-up version of the microwave-based oil extraction technology, virtually all of the extractable oil is removed from the matrix in as little as 5 min.

Similarly for isoflavones, at a given temperature (73 °C), the amount of the individual components present in the microwave-extracted oil was significantly higher than what can be obtained *via* conventional heating (Figure 12.2). In addition to the inherent higher value of separated compounds, the quality of oil is also increased as the oxidative stability is increased due to higher antioxidant capacity.

The higher yields for both isoflavones and oil are mainly due to the specific interaction of microwaves with the biological matrix holding them. When the feedstock is heated with microwave, water and other polar molecules (such as the ethanol solvent) are initially targeted. The targeted molecules, which are usually localized in the matrix, heat up rapidly and evaporate in relative short time (ms). The highly (micro)localized rapid evaporation generates an extremely large pressure gradient across the cell walls, which is disrupted as the pressure is released towards the exterior of the cell. The cell wall alteration changes the permeability, resulting in its intracellular contents spilling out and increasing the rate of mass transfer (Chemat *et al.* 2005). The rate of extraction

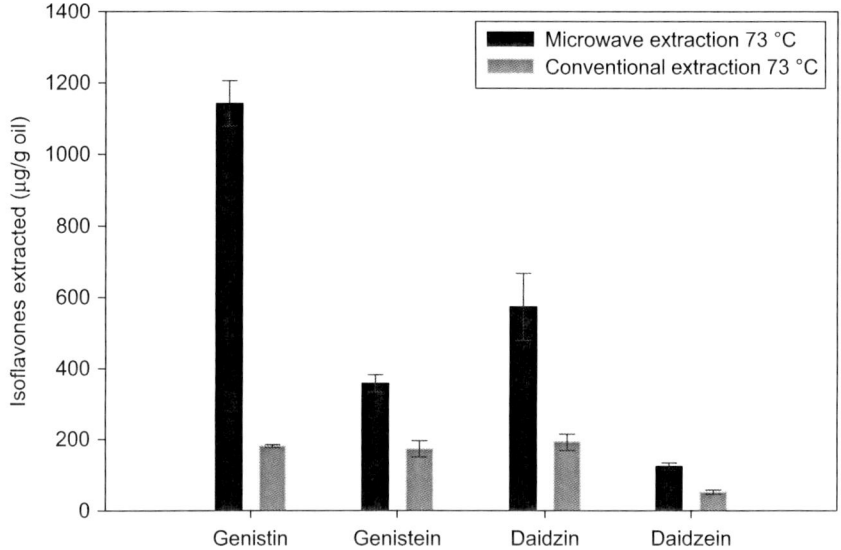

Figure 12.2 Isoflavone compound yields obtained using microwave and conventional heating at 73 °C. Microwave heating results in significantly higher yields. Reproduced from Terigar et al. (2010), with permission from Elsevier B.V.

for oil and other intracellular compounds from soybean plant material is therefore significantly increased in a microwave heating system compared with a conventional heating method (Choi et al. 2006). When comparing a microwave-assisted oil extraction in the batch vs. the continuous system, the continuous system performs better (Boldor et al. 2010). The improvement in yields in this case can be attributed to the continuous exposure of the plant matrix to the microwave energy (in the batch system the microwaves turn on and off in order to maintain the process temperature) (Boldor et al. 2010).

The improvement in extraction yields when microwaves are used is significant not only in the total amounts extracted, but in the rate at which this extraction happens (for more details see the discussion related to determination of extraction rate kinetics in the next Sections). Overall, at 73 °C, substantial increases for all individual isoflavones can be obtained with the microwave-assisted method compared with the conventional heating, as follow: genistin, 5-fold increase; daidzin, 3-fold increase; genistein, 60% increase; and daidzein, 100% increase.

The increased performance of the microwave process for genistin and daidzin is due to their particular molecular structures. Belonging to the glycosides class, based on their polarity and molecular weight, these two compounds are more soluble in ethanol compared with genistein and daidzein, which belong to the aglycones class. The higher solubility in ethanol (the chosen extraction medium) leads to better extraction performance due to better interdiffusion of molecules. In addition, the higher polarity of the glycosides makes much better

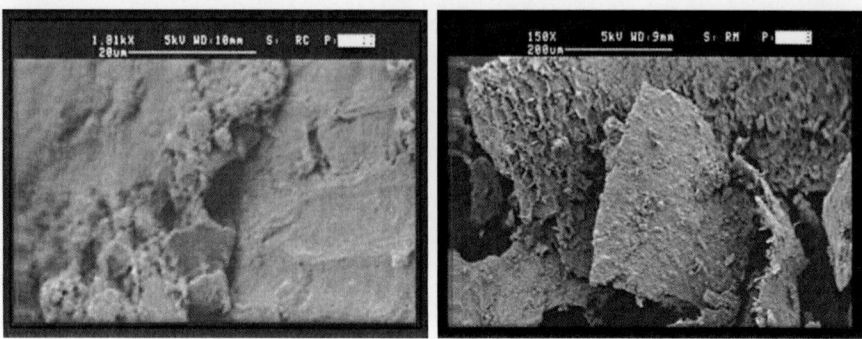

Figure 12.3 Scanning electron microscopy (SEM) image of soybeans subjected to conventional (left) and microwave (right) extraction at 120 °C for 16 min. The cellular damage is clearly visible in the microwave method *vs.* the conventional heating method.
Reproduced from Terigar *et al.* (2010), with permission from Elsevier B.V.

targets for the oscillating electric field component of the microwave field, as discussed above, thus enhancing their extraction rate. In contrast, conventional mass transfer processes (driven by conventional heating methods) may in general favor the transport of the less polar aglycones through cellular membranes, as these compounds diffuse faster through the lipid portion of the membranes. In the case of microwave heating, the normal diffusion processes are rendered obsolete by the rapid and severe damage to the cell walls (Figure 12.3), which allow for a solubility-driven extraction based on the polarity of the solvent, which benefits the more polar glycosides. This phenomenon is not limited only to these isoflavones extracted from soybeans. Increased yields of active compounds extracted using microwave heating are also reported for terpenes from caraway seeds (Chemat *et al.* 2005), oil from soybean (Cravotto *et al.* 2008; Terigar *et al.* 2011), vitamin E from rice bran (Duvernay *et al.* 2005), and antioxidants from rice bran oil (Zigoneanu *et al.* 2008).

12.5.2 Influence of Time and Temperature on the Oil and Isoflavones Extraction Yield – Reaction Kinetics

As the temperature increases, the oil and isoflavones yields in both microwave and conventional heating extraction increases significantly. For oil, the increase is more pronounced for the microwave system [from 10.0 to 11.7% (g of soy flour)$^{-1}$] compared with conventional heating (8.0 to 9.4%). For isoflavones, genistin and daidzin concentrations tripled at the higher temperature, whereas genistein and daidzein showed increases of 75 and 110%, respectively. The increase in yield with increased temperature is a direct consequence of an increase in cell damage and higher diffusion and solubilization rates, which increase mass transfer. This is observed not only for isoflavones in soybeans, but also in the extraction of oil and vitamin E from rice bran (Duvernay *et al.* 2005; Zigoneanu *et al.* 2008).

As the extraction time is increased, the oil yields increase for both microwave and conventional heating methods. In the case of microwave extraction, oil yields stabilized in a relatively short time (approximately 8 min). However, in conventional heating the yields continue to increase throughout the extraction process (up to the total time of 24 min investigated), and it can be presumed that these yields will continue to increase until they reach the same values as for the microwave extraction process. In microwave oil extraction, in general, the extraction tine can be as little as 1/8th of the time required for the Soxhlet method (Cravotto *et al.* 2008).

In addition to higher overall oil yields as time is increased, the specific isoflavone yields also increased (Figure 12.4). Using data from Terigar *et al.* (2010), the maximum theoretical yields at 73 °C, based on the assumption that microwave extraction follows a first-order rate, are estimated to be 1703 µg g^{-1} oil for genistin ($R^2 = 0.959$), 413 µg g^{-1} oil for genistein ($R^2 = 0.980$), 703 µg g^{-1} oil for daidzin ($R^2 = 0.962$), and 153 µg g^{-1} oil for daidzein ($R^2 = 0.985$), respectively. Using these maximum obtainable yields, the extraction rate constants at individual temperatures and specific extraction method can be calculated using eqn (1) for each isoflavone:

$$C(t) = C_f \cdot \left(1 - e^{-k \cdot t}\right) \tag{1}$$

where: $C(t)$ = concentration of respective isoflavone at time t (%); C_f = final isoflavone concentration assumed to be 100% (complete extraction); k = first-order extraction rate constant (s^{-1}); and t = time (s).

Analyzing yields at individual temperatures, it can be said that at both 55 °C and 73 °C genistin yields tripled after approximately 8 min in the microwave process. Genisteine yields also tended to equalize in the microwave process after increasing spectacularly in the first few minutes of the process. The increase in yields *vs.* time in the conventional process paled in comparison with what is achievable in the microwave-based extraction. Both daidzin and daidzein followed similar time-based patterns in term of yields. It is noted here that, according to Terigar *et al.* (2010), there is no statistical differences in the yields obtained after approximately 8 min in the microwave process, although the first-order curve fitting seems to indicate the likely existence of a maximum value which is higher than what is obtained after these 8 min.

Probably, the easiest way to compare the performance of the two extraction systems (microwave *vs.* conventional heating) is to compare their respective extraction rate constants (and corresponding R^2) as determined by curve-fitting of eqn (1) to the existent literature data (Table 12.1) (Terigar *et al.* 2010). At 55 °C, the extraction rate constants in the microwave-based process are orders of magnitude higher than in the conventional heating (~2.5× for daidzein and genistein, ~5× for daidzin, and ~25× for genistin). At 73 °C, the extraction rate constants in the microwave process were approximately between 5 and 10 times higher than in the conventional heating process. It is worth mentioning here that the limited data available in literature for this particular process (only two temperature values), does not allow reliable computations of

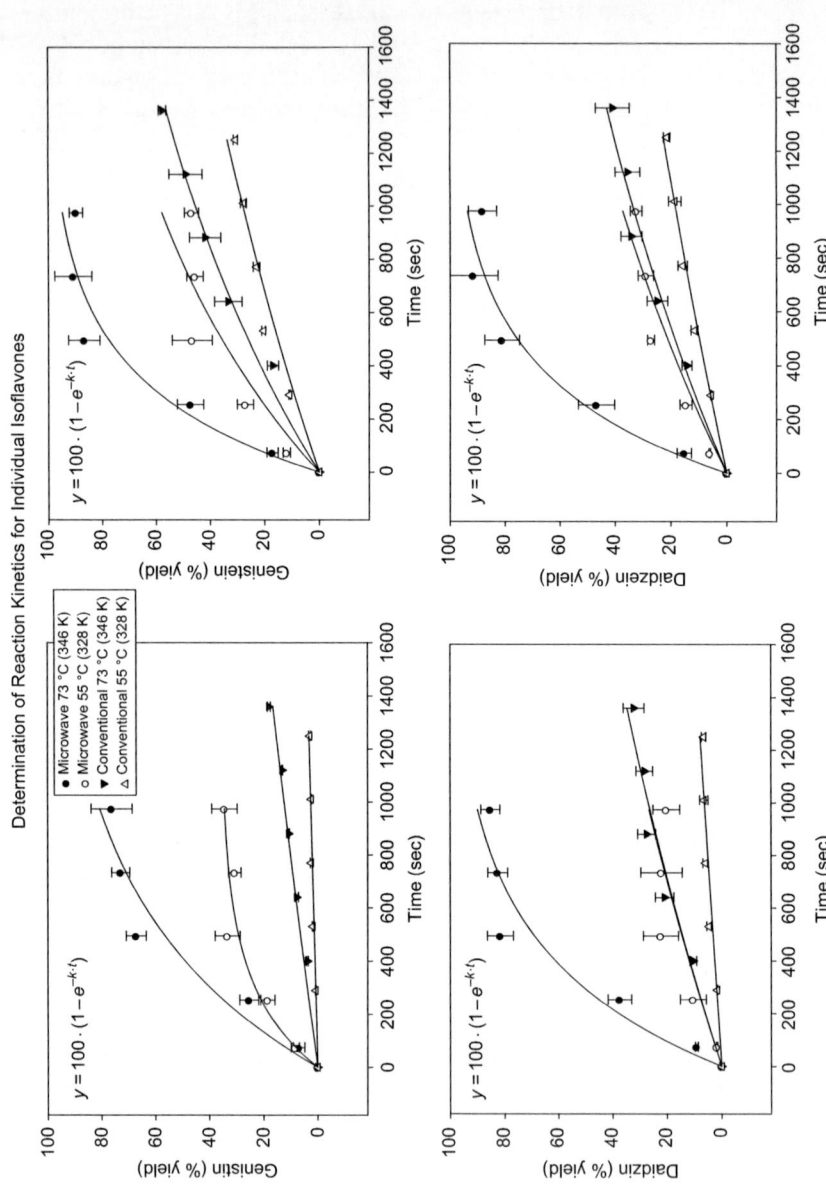

Figure 12.4 Determination of reaction kinetics for individual isoflavones. Individual isoflavones yields as a function of time, temperature, and method of extraction at two different temperatures.

Table 12.1 Extraction rate constants for individual compounds at each temperature and method.

Compound	Temperature (°C)	55		73	
		$k \times 10^{-3}$ (s^{-1})	R^2	$k \times 10^{-3}$ (s^{-1})	R^2
Genistin	Microwave	0.5453	0.977	1.6820	0.959
	Conventional	0.0255	0.910	0.1314	0.808
Genistein	Microwave	0.8870	0.989	3.0153	0.980
	Conventional	0.3280	0.950	0.6065	0.839
Daidzin	Microwave	0.3171	0.963	2.3625	0.962
	Conventional	0.0656	0.888	0.3122	0.789
Daidzein	Microwave	0.4778	0.980	2.8215	0.985
	Conventional	0.2058	0.987	0.4129	0.913

respective activation energies and/or pre-exponential factors according to Arrhenius's equation. In order to accurately determine the values of these parameters, more data need to be collected at several other individual temperatures.

The results of the continuous microwave extraction system validate those obtained in batch processes (Rostagno et al. 2007), and it has been demonstrated that the continuous microwave technology can be successfully scaled up for the extraction of oil from soybeans and other feedstock (Terigar et al. 2011). In fact, the continuous microwave system were found to perform better as the scale is increased, mostly due to the lower frequency (and in therefore higher penetration depth in the target material) and especially due to a certain specific configuration of the microwave cavity which allows for the energy to be directed only on the processing tube. The most significant advantages of the continuous microwave extraction system as described in this Chapter are as follows: 1. a dramatic decrease in the extraction time required to obtain significant amounts of isoflavones compared with conventional heating (from hours to less than 10 min), and even compared with a batch microwave system (which required approximately 30 min for similar yields); 2. a significantly reduced amount of solvent (1:3 vs 1:20, w/w, ratios) as compared with conventional heating methods, which simplifies downstream processing and reduces solvent recovery costs; 3. an inherent advantage of using a continuous operation, which allows for more material to be processed in a much shorter time; 4. the relatively low cost of building and operating the continuous microwave extraction system, which can be designed and built using low-cost individual components; and 5. the system is extremely scalable, either by utilizing a higher power, industrial-type microwave, or by increasing the number of smaller units operating simultaneously.

12.6 Conclusion

To summarize, continuous microwave extraction can be successfully deployed to extract isoflavones from soybeans with maximum efficiency and at a much faster rate than using conventional heating methods. As it is generally expected, higher temperatures and longer extraction times yield maximum values of isoflavones, but even these longer times are much shorter (on the order of minutes) than the hours required for similar results using conventional heating. The amount of solvent used can also be significantly reduced, with the limiting value on this parameter being imposed only by the flow requirements of the solvent/soybean mixture. The advantage of the continuous microwave methods stems from the specific interaction of the microwave energy with the soybean matrix, in which polar molecules and ionic compounds rapidly oscillate under the influence of the electric field component of the microwave. As a consequence, these molecules and ions heat extremely rapidly inside the matrix, disrupting cell walls and virtually eliminating the diffusion-based limitations existent in extraction processes that use conventional heating.

Summary Points

- This Chapter focuses on isoflavone extraction using continuous microwave technology.
- Microwave technology has emerged in recent years as a valuable tool for rapid extraction of different value-added compounds from many agricultural commodities.
- The low-cost system presented here is designed to operate on a modular concept (multiple microwave units stacked on top of each other) allowing operation at multiple flow rates and residence times.
- Isoflavones are extracted from soybean flower mixed with ethanol and passed through the system while being exposed to microwave energy.
- The isoflavone yield increases up to 2.5 fold for individual isoflavones compared with the conventional heating method at the same time and temperature parameters.
- For reaction kinetics, extraction rate constants in the microwave system are several times higher than those determined for conventional heating (up to 25 times higher in the case of genistin).

Key Facts

- The disruptive effect of the microwave on biological matrices is due to direct molecular-level energy transfer to polar compounds and charged ions.
- Microwave energy has two active fields, an oscillating electric field component and an oscillating magnetic field component.
- The magnetic field component in a microwave does not interact to any significant degree with biological matrices, as it is very small.
- The electric field component, on the other hand, is much larger and interacts strongly with the molecules and ions that make up the biological matrix.

- Polar molecules spin and create molecular friction, whereas ions oscillate rapidly and dissipate their kinetic energy as heat, leading to rapid microscopic heating and large temperature and pressure gradients that disrupt cell structure.
- Increased cellular damage significantly increases extraction rates and yields compared with conventional heating.

Definitions of Words and Terms

Circulator: A magnetic device placed in a waveguide that bends the microwaves toward a specific direction. It is used together with a water load to protect the magnetron from microwaves reflected from the applicator.

Dielectric material: A material that is neither a pure electrical conductor, nor a perfect electrical insulator (virtually all biological and agricultural materials are dielectric in nature). These materials are able to both store electrical energy and dissipate it as heat.

Directional coupler: A special section of a waveguide that directs a fraction of the incoming energy toward a sensor able to measure and quantify the amount of power in the incident microwave field.

Electroporation: A phenomenon in which an applied electric field is strong enough to overcome the normally occurring trans-membrane electrical potential established across a cellular membrane, and a dielectric breakdown occurs which ruptures the membrane.

Extraction: A processing operation in which a target compound is removed from the biological matrix without damaging its functionality.

Magnetron: A device that converts electrical energy into microwaves of specific frequency.

Maxwell's equations: A set of partial differential equations governing the transmission and dissipation of electromagnetic energy. These are named after the Scottish physicist and mathematician James Clerk Maxwell who developed them in the mid-19th century.

Microwave applicator: A metallic cavity of a certain geometric configuration where the material to be processed is placed prior to microwave exposure (or, in the case of continuous-flow systems, the cavity through which the processed material flows).

Microwave energy: Electromagnetic energy in the frequency range of 300 to 6000 MHz, composed of a sinusoidal electric field and a sinusoidal magnetic field. Most of the frequencies in this range are reserved for communication, military, and radar applications. There are only a few frequencies in this range reserved for industrial, scientific, and medical (ISM) applications.

Multimode cavity: A metallic cavity in which the incoming microwave field generates standing patterns (static regions of high, respectively low, microwave energy) due to multiple reflections of the walls.

Polar molecules: Molecular structures that have a permanent electric dipole (*i.e.* water molecules are very polar) and able to interact strongly with microwave and electric fields.

Post-harvest management: A set of transportation, logistic, processing, and storage operations undertaken for a crop after harvesting until undergoing processing in order to minimize quantitative and qualitative losses, or enhance certain qualities in the end product.

Single mode cavity: A metallic cavity in which a single region (or a very small number of regions) of high electric field is allowed to exists at a precise location.

Tuning section: A section of waveguide with metallic inserts located at specific distances, which allow the system to be tuned such that it minimizes microwave power losses at the transition between the waveguide and the applicator.

Water load: A waveguide end-section that has water continuously flowing through it to dissipate microwave energy. As water is a strong microwave absorber and has high heat capacity, it is used in combination with a circulator to harmlessly dump excess microwave energy from the system.

Waveguide: Hollow metallic rectangular pipes that are used to transport microwaves of specific frequencies without losses from the generator to the dielectric material being processed.

List of Abbreviations

HPLC   high-performance liquid chromatography
UV   ultraviolet

Acknowledgements

The authors would also like to thank the many people that contributed to the research in microwave-based extraction at Louisiana State University Agricultural Center, including, but not limited to, Beatrice Terigar, Akanksha Kanitkar, Laura Picou, Sundar Balasubramanian, Pranjali Muley, Marybeth Lima, Casey McMann, Mark Gabriel, Carlos Astete, Imola Zigoneanu, and many more. The authors are also acknowledging the multiple entities that contributed financial resources dedicated to the development of this field, including Louisiana Board of Regents – Industrial Ties Research Subprogram [award number LEQSF (2007-10)-RD-B-01], Louisiana Board of Regents Enhancement Grant [LEQSF(2006-2008)-ENH-TR-04] and Louisiana Soybean and Grain Research and Promotion Board. This Chapter is published with the approval of the Director of the Louisiana Agricultural Experiment Station as publication no. 2011-232-6389.

References

Ali, A.A., Velasquez, M.T., Hansen, C.T., Mohamed, A.I., and Bhathena, S.J., 2004. Effects of soybean isoflavones, probiotics, and their interactions on lipid metabolism and endocrine system in an animal model of obesity and diabetes. *Journal of Nutritional Biochemistry*. 15: 583–590.

Ali, A.A., Velasquez, M.T., Hansen, C.T., Mohamed, A.I., and Bhathena, S.J., 2005. Modulation of carbohydrate metabolism and peptide hormones by soybean isoflavones and probiotics in obesity and diabetes. *Journal of Nutritional Biochemistry*. 16: 693–699.

Boldor, D., Kanitkar, A., Terigar, B.G., Leonard, C., Lima, M., and Breitenbeck, G.A., 2010. Microwave assisted extraction of biodiesel feedstock from the seeds of invasive Chinese tallow tree. *Environmental Science and Technology*. 44: 4019–4025.

Boldor, D., Sanders, T.H., and Simunovic, J., 2004. Dielectric properties of in-shell and shelled peanuts at microwave frequencies. *Transactions of the ASAE*. 47: 1159–1169.

Canumir, J.A., Celis, J.E., de Bruijn, J., and Vidal, L.V., 2002. Pasteurisation of apple juice by using microwaves. *Lebensmittel-Wissenschaft Und-Technologie – Food Science and Technology*. 35: 389–392.

Chemat, S., Ait-Amar, H., Lagha, A., and Esveld, D.C., 2005. Microwave-assisted extraction kinetics of terpenes from caraway seeds. *Chemical Engineering and Processing*. 44: 1320–1326.

Choi, I., Choi, S.J., Chun, J.K., and Moon, T.W., 2006. Extraction yield of soluble protein and microstructure of soybean affected by microwave heating. *Journal of Food Processing and Preservation*. 30: 407–419.

Cravotto, G., Boffa, L., Mantegna, S., Perego, P., Avogadro, M., and Cintas, P., 2008. Improved extraction of vegetable oils under high-intensity ultrasound and/or microwaves. *Ultrasonics Sonochemistry*. 15: 898–902.

Duvernay, W.H., Assad, J.M., Sabliov, C.M., Lima, M., and Xu, Z., 2005. Microwave extraction of antioxidant components from rice bran. *Pharmaceutical Engineering*. 25: 5.

Grigonis, D., Venskutonis, P.R., Sivik, B., Sandahl, M., and Eskilsson, C.S., 2005. Comparison of different extraction techniques for isolation of antioxidants from sweet grass (*Hierochloe odorata*). *Journal of Supercritical Fluids*. 33: 223–233.

Hemwimon, S., Pavasant, P., and Shotipruk, A., 2007. Microwave-assisted extraction of antioxidative anthraquinones from roots of *Morinda citrifolia*. *Separation and Purification Technology*. 54: 44–50.

Huang, Y., Sheng, J., Yang, F., and Hu, Q., 2007. Effect of enzyme inactivation by microwave and oven heating on preservation quality of green tea. *Journal of Food Engineering*. 78: 687–692.

Kanitkar, A., Sabliov, C.M., Balasubramanian, S., Lima, M., and Boldor, D., 2011. Microwave-assisted extraction of soybeans and rice bran oil: yield and extraction kinetics. *Transactions of the ASABE*. 54: 8.

Klejdus, B., Mikelová, R., Adam, V., Zehnálek, J., Vacek, J., Kizek, R., and Kubán, V., 2004. Liquid chromatographic–mass spectrometric determination of genistin and daidzin in soybean food samples after accelerated solvent extraction with modified content of extraction cell. *Analytica Chimica Acta*. 517: 1–11.

Kozempel, M., Scullen, O.J., Cook, R., and Whiting, R., 1997. Preliminary investigation using a batch flow process to determine bacteria destruction by microwave energy at low temperature. *Food Science and Technology – Lebensmittel-Wissenschaft und Technologie.* 30: 691–696.

Kozempel, M.F., Annous, B.A., Cook, R.D., Scullen, O.J., and Whiting, R.C., 1998. Inactivation of microorganisms with microwaves at reduced temperatures. *Journal of Food Protection.* 61: 582–585.

Kwon, J.H., Belanger, J.M.R., and Pare, J.R.J., 2003. Optimization of microwave-assisted extraction (MAP) for ginseng components by response surface methodology. *Journal of Agricultural and Food Chemistry.* 51: 1807–1810.

Lee, M.-H., and Lin, C.-C., 2007. Comparison of techniques for extraction of isoflavones from the root of *Radix puerariae*: ultrasonic and pressurized solvent extractions. *Food Chemistry.* 105: 223–228.

Lo, F.H., Mak, N.K., and Leung, K.N., 2007. Studies on the anti-tumor activities of the soy isoflavone daidzein on murine neuroblastoma cells. *Biomedicine and Pharmacotherapy.* 61: 591–595.

Luthria, D.L., Biswas, R., and Natarajan, S., 2007. Comparison of extraction solvents and techniques used for the assay of isoflavones from soybean. *Food Chemistry.* 105: 325–333.

Mazur, W.M., Duke, J.A., Wahala, K., Rasku, S., and Adlercreutz, H., 1998. Isoflavonoids and lignans in legumes: nutritional and health aspects in humans. *Journal of Nutritional Biochemistry.* 9: 193–200.

Metaxas, A.C. and Meredith, R.J., 1983. Industrial microwave heating. *P. Peregrinus on behalf of the Institution of Electrical Engineers.* London, UK.

Nahas, E.A.P., Nahas-Neto, J., Orsatti, F.L., Carvalho, E.P., Oliveira, M.L.C.S., and Dias, R., 2007. Efficacy and safety of a soy isoflavone extract in postmenopausal women: a randomized, double-blind, and placebo-controlled study. *Maturitas.* 58: 249–258.

Nelson, S.O., 1980. Microwave dielectric properties of fresh fruits and vegetables. *Transactions of the ASAE.* 23: 1314–1317.

Pan, X.J., Niu, G.G., and Liu, H.Z., 2002. Comparison of microwave-assisted extraction and conventional extraction techniques for the extraction of tanshinones from *Salvia miltiorrhiza* bunge. *Biochemical Engineering Journal.* 12: 71–77.

Phrakonkham, P., Chevalier, J., Desmetz, C., Pinnert, M.-F., Berges, R., Jover, E., Davicco, M.-J., Bennetau-Pelissero, C., Coxam, V., Artur, Y., and Canivenc-Lavier, M.-C., 2007. Isoflavonoid-based bone-sparing treatments exert a low activity on reproductive organs and on hepatic metabolism of estradiol in ovariectomized rats. *Toxicology and Applied Pharmacology.* 224: 105–115.

Rostagno, M.A., Palma, M., and Barroso, C.G., 2004. Pressurized liquid extraction of isoflavones from soybeans. *Analytica Chimica Acta.* 522: 169–177.

Rostagno, M.A., Palma, M., and Barroso, C.G., 2007. Microwave assisted extraction of soy isoflavones. *Analytica Chimica Acta.* 588: 274–282.

Terigar, B.G., Balasubramanian, S., Boldor, D., Xu, Z., Lima, M., and Sabliov, C.M., 2010. Continuous microwave-assisted isoflavone extraction system: design and performance evaluation. *Bioresource Technology*. 101: 2466–2471.

Terigar, B.G., Balasubramanian, S., Sabliov, C.M., Lima, M., and Boldor, D., 2011. Soybean and rice bran oil extraction in a continuous microwave system: from laboratory- to pilot-scale. *Journal of Food Engineering*. 104: 208–217.

Wuttke, W., Jarry, H., and Seidlova-Wuttke, D., 2007. Isoflavones – Safe food additives or dangerous drugs? *Ageing Research Reviews*. 6: 150–188.

Yaghmaee, P., and Durance, T.D., 2005. Destruction and injury of *Escherichia coli* during microwave heating under vacuum. *Journal of Applied Microbiology*. 98: 498–506.

Zigoneanu, I.G., Wilhams, L., Xu, Z., and Sabliov, C.M., 2008. Determination of antioxidant components in rice bran oil extracted by microwave-assisted method. *Bioresource Technology*. 99: 4910–4918.

CHAPTER 13

Isoflavones: High-performance Liquid Chromatographic Analysis of Glucuronic Acid- and Sulfuric Acid-conjugated Metabolites of Daidzein and Genistein in Human Plasma and Urine

KAZUO ISHII,*[a] KAORI HOSODA[a] AND
TAKASHI FURUTA[b]

[a] Department of Clinical Pharmacology, School of Health Sciences, Kyorin University, 476 Miyashita, Hachioji, Tokyo 192-8508, Japan; [b] Department of Medicinal Chemistry and Clinical Pharmacy, School of Pharmacy, Tokyo University of Pharmacy and Life Sciences, 1432-1 Horinouchi, Hachioji, Tokyo 192-0392, Japan
*Email: ishiikaz@ks.kyorin-u.ac.jp

13.1 Introduction

The effect on human health of the isoflavones daidzein and genistein has been widely studied. These isoflavones are thought to play a role in the prevention of various hormone-dependent cancers such as breast and prostate cancers (Adlercreutz 2002). Similar to other flavonoids, isoflavones are present as glycosides in the plant kingdom. After ingestion, isoflavone glycosides are converted into their aglycones, such as daidzein and genistein, and sugar moieties by enteral micro-organisms. Most aglycones that are absorbed from the gastrointestinal tract circulate as glucuronides and sulfates of phase II metabolites (Figure 13.1). Only a few percent of the absorbed aglycones are present as free-form aglycones in the systemic circulation. These aglycones, in addition to equol produced from daidzein by gut bacteria, have been recognized as physiologically active compounds with estrogenic effects. The glucuronide and sulfate conjugates also possess weakly estrogenic activities. It has been reported that the extent of biological activity is dependent on the positions of the conjugates on the isoflavone skeleton. For example, sulfation at the 4'-position of daidzein results in a modest reduction in estrogen agonist activity, whereas sulfation at the 7-position of daidzein increases estrogen agonist

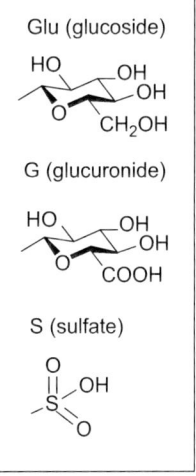

	Abbreviation	R_1	R_2
daidzin	-	Glu	H
daidzein	-	H	H
daidzein-4', 7-diglucuronide	D-4', 7-diG	G	G
daidzein-4', 7-disulfate	D-4', 7-diS	S	S
daidzein-7-glucuronide-4'-sulfate	D-7G-4'S	G	S
daidzein-7-glucuronide	D-7-G	G	H
daidzein-4'-glucuronide	D-4'-G	H	G
daidzein-7-sulfate	D-7-S	S	H
daidzein-4'-sulfate	D-4'-S	H	S

	Abbreviation	R_3	R_4
genistin	-	Glu	H
genistein	-	H	H
genistein-4', 7-diglucuronide	G-4', 7-diG	G	G
genistein-4', 7-disulfate	G-4', 7-diS	S	S
genistein-7-glucuronide-4'-sulfate	G-7G-4'S	G	S
genistein-7-glucuronide	G-7-G	G	H
genistein-4'-glucuronide	G-4'-G	H	G
genistein-7-sulfate	G-7-S	S	H
genistein-4'-sulfate	G-4'-S	H	S

Figure 13.1 Structures of isoflavones and their conjugated metabolites. Daidzein group and genistein group.

activity (Pugazhendhi *et al.* 2008). Therefore, the identification and quantification of conjugated metabolites and aglycones are essential for our understanding of the pharmacological and medicinal properties, as well as the metabolic pathways, of isoflavones in humans.

This Chapter describes: (1) conventional assay methods used for isoflavone conjugates, and (2) a simple direct assay of 16 isoflavone metabolites using high-performance liquid chromatography–ultraviolet (HPLC-UV). These methods are summarized in Table 13.1.

13.2 Conventional Methods

The choice of analytical method is the key to the successful study of bioavailability and pharmacokinetics. In this Section, representative methods for the analysis of conjugated metabolites are introduced. The major isoflavones daidzein and genistein exist as glycosides such as daidzin, genistin, their $6''$-O-malonylglycosides, and so on, in the plant kingdom. These isoflavone glycosides are absorbed as aglycones after deconjugation by β-glucosidase. The majority of the absorbed aglycones then circulate as conjugated metabolites in the body. For example, isoflavones are present as both glucuronide conjugates and sulfate conjugates in human plasma and urine.

HPLC methods are frequently used for the assay of isoflavones; however, because soybeans and soy foods have a large quantity of polar isoflavone glycosides, in general, the aglycones obtained after removal of the polar groups by acid or enzymatic hydrolysis are first extracted and then analyzed. Similarly, for the analysis of highly polar glucuronides and sulfates in plasma and urine, the aglycones arising from these metabolites after general enzymatic (β-glucuronidase/sulfatase; *Helix pomatia*) hydrolysis are analyzed. The quantities of free aglycones, total metabolites, glucuronides and sulfates are then calculated in terms of aglycone equivalents.

13.2.1 Selective Enzymatic Hydrolysis

Selective enzymatic hydrolysis (Cimino *et al.* 1999; Shelnutt *et al.* 2000; 2002) has most frequently been used as a means of typical sample pre-treatment. Using β-glucuronidase/sulfatase (from *H. pomatia*), sulfatase (type VIII) and β-glucuronidase (type B-1; bovine liver), the aglycones liberated from glucuronides and sulfates are analyzed. From the three types of enzymatic hydrolysis, the quantity of total aglycones, that of sulfates plus free aglycones and that of glucuronides plus free aglycones, respectively, are evaluated as aglycone equivalents. The quantity of free aglycones is evaluated without enzymatic hydrolysis.

For enzymatic hydrolysis, β-glucuronidase (1000 units)/sulfatase (100 units), β-glucuronidase (1000 units) and sulfatase (100 units) dissolved in 1 mL of 1 M ammonium acetate solution are added to 0.5 mL aliquots of plasma. The respective enzyme solutions are incubated at 37 °C for 3 h. After the reaction is

Table 13.1 Representative assay methods and their conditions used for isoflavone conjugated metabolites.

Analyte	Plasma or urine	Column	Mobile phase or carrier gas	Detection	LOQ or LOD	References
HPLC						
• daidzein genistein etc.	urine	Supelco Discovery Rpamide-C16	A: 25% MeOH containing 10 mM NH₄OAc and 71 mM Et₃N B: 95% MeOH containing 10 mM NH₄OAc and 71 mM Et₃N gradient: 65–95% B, 0–10 min; 95% B, 10–11 min flow rate: 1 mL/min	MS	LOQs > 4 ng/mL LODs 1 ng/mL	Cimino et al. 1999
	urine					Shelnutt et al. 2000
	plasma					Shelnutt et al. 2002
• daidzein genistein etc.	plasma	Luna Phenyl-Hexyl (150 mm × 4.6 mm, 5 μm) guard: Zorbax Eclipse XDB-phenyl (12.5 mm × 4.6 mm, 5 μm)	A: 0.05 M HCOONH₄ B: MeOH/MeCN (50 : 50, v/v) gradient: 100%–40% B, 0–0.5 min; 40% B, 0.5–11.5 min; 40–80% B, 11.5–12.5 min; 80% B, 12.5–15.5 min flow rate: 2 mL/min	UV 259 nm	LOQ for daidzein 1.76 ± 0.07 ng/mL LOQ for genistein 2.11 ± 0.22 ng/mL	Thomas et al. 2001
	urine		A: 0.05 M HCOONH₄ B: MeOH gradient: 10% B, 0–0.5 min; 45% B, 0.5–1 min; 45% B, 10.5 min; 45–80% B, 10.5–11.5 min; 80% B, 11.5 min–14.5 min flow rate: 2 mL/min			

Table 13.1 (Continued)

Analyte	Plasma or urine	Column	Mobile phase or carrier gas	Detection	LOQ or LOD	References
• genistein	plasma and urine (rat)	C18 column (250 mm × 4.6 mm, 5 μm)	MeOH : 0.1M NH$_4$OAc : 25 mM EDTA soln. (55 : 50 : 1, v/v) flow rate: 1 mL/min	ECD	LOQ for genistein 0.2 μmol/L (54 ng/mL)	King et al. 1996
• daidzein genistein	plasma and urine		MeOH : 0.1 M NH$_4$OAc : 25 mM EDTA soln. (50: 50 : 1, v/v) flow rate: 1mL/min			King and Bursil 1998
• daidzein genistein etc.	urine	Targa C18 column (150 mm × 2.1 mm, 3 μm)	A: MeOH : 0.1M NH$_4$OAc (40:60, v/v) B: MeOH gradient: 0–25% B, 0–1 min; 25–60% B, 1–8 min 100% B, 8–9 min; 100% A, 9–10 min flow rate: 200 μL/min (0–8 min, 9–10 min) 250 μL/min (8–9 min)	MS/MS (ion trap and triple quadrupole)	LOQs 0.1 ng/mL (Both MS/MS systems)	Grace et al. 2007
• 4′-, 7-mono glucuronides, and sulfate of daidzein and genistein	plasma	Prodigy ODS-3 (250 mm × 4.6 mm 5 μm)	A: 0.1% HCOOH in H$_2$O B: 0.1% HCOOH in MeCN gradient: 5% B, 0–3 min; 5–50% B, 3–18 min; 50% B, 18–23 min flow rate: 1mL/min	UV 260 nm	LODs 1 pmol on column (for daidzein 0.43 ng, for genistein 0.45 ng)	Doerge et al. 2000

Analyte	Sample	Column	Mobile phase / Conditions	Detection	LOD / LOQ	Reference
• 4′-, 7-mono glucuronides, and sulfates of daidzein and genistein *etc.*	urine	YMC-ODS AM (250 mm × 3 mm, 5 μm)	A: H$_2$O contained 0.1% HCOOH B: MeCN contained 0.1% HCOOH gradient: 10% B, 0–5 min; 10–22% B, 5–20 min; 22–30% B, 20–35 min; 30–46%, 35–36 min; 46–80% B, 36–56 min; 80–85% B, 56–60 min; 85–10% B, 60–65 min; 10% B, 65–75 min. flow rate: 0.5 mL/min	isotope dilution ESI MS/MS	LODs 50 ng/mL	Clarke *et al.* 2002
• various 4′-, 7-conjugated metabolies of daidzein and genistein	plasma	YMC-Hydrosphere C18 (100 mm × 4.6 mm 3 μm) guard: same material (23 mm × 4.0 mm)	A: 10 mM NH$_4$OAc B: MeCN gradient: 95.3% A, 0–1.5 min; 95.3–68.0% A, 1.5–27 min; 68.0–95.3% A, 27–34 min, and held for 10 min flow rate: 1.5 mL/min at 45 °C	UV 250 nm	LOQs 25 ng/mL LODs 8 ng/mL	Hosoda *et al.* 2010
GC						
• daidzein genistein *etc.*	plasma	bonded phase BP-1 (SGE) vitreous silica colimn (12.5 m × 0.2 mm)	Helium gas oven temperature; 100 °C for 1 min, increased by 30 °C/min to 280 °C	isotope dilution MS	LODs 1 μmol/mL (for daidzein: 254 ng/mL genistein: 270 ng/mL)	Adlercreutz *et al.* 1994
• daidzein genistein *etc.*	plasma				LODs 0.01 μmol/mL (for daidzein: 2.54 ng/mL genistein: 2.70 ng/mL)	Watanabe *et al.* 1998

completed, the respective enzyme solutions are twice extracted with 5 mL of diethyl ether. The extracted solutions are then evaporated to dryness at 55 °C. The residue containing aglycones librated from the conjugated metabolites is dissolved in a moderate solvent as sample preparation for liquid chromatography–mass spectrometry (LC-MS). Typical chromatographic conditions are as follows:

LC column:	Supelco Discovery RPamide-C16 high performance LC column [250 mm × 4.6 mm internal diameter (i.d.), 5 μm particle size; Supelco, Bellafonte, PA, USA]
Flow rate:	1.0 mL/min
Mobile phase:	Linear gradient mode (65% B to 95% B in 10 min, and the held 95% B for 1 min) (A) 25% MeOH solution containing 10 mM NH_4OAc and 71 mM triethylamine (B) 95% MeOH solution containing 10 mM NH_4OAc and 71 mM triethylamine
Instrument:	PE Sciex API 100
Ion mode:	Negative single-ion monitoring
Ionization mode:	Atmospheric pressure chemical ionization (APCI)
Stock solution:	2–5 mg of daidzein and genistein per 5 mL of MeOH
Monitored:	Daidzein, mass-to-charge ratio (m/z) 253.2 [retention time (RT) 5.7 min]; genistein, m/z 269.2 (RT 7.4 min)
Calibration:	1–80 ng

The interassay coefficient of variation (CV) for daidzein and genistein ranges from 10% to 17% for this method. The limit of detection (LOD) and limit of quantification (LOQ) are 1 ng and 4 ng, respectively.

After enzymatic hydrolysis, the liberated aglycones can also be analyzed by HPLC-UV, HPLC–electrochemical detector (HPLC-ECD), LC-MS and LC–tandem mass spectrometry (MS/MS). For example, Thomas et al. (2001) measured the concentrations of daidzein and genistein in human plasma after enzymatic hydrolysis of conjugated metabolites, followed by HPLC-UV. The samples were eluted from a Luna Phenyl-Hexyl HPLC column (150 mm × 4.6 mm i.d., 5 μm particle size; Waters) with a gradient of solvent A (0.05 M ammonium formate, pH 4) to solvent B (methanol/acetonitrile, 50:50, v/v). The LOQ of plasma daidzein and genistein obtained by this method was 1.76 ± 0.07 ng/mL (CV 3.9%, accuracy 99%) and 2.11 ± 0.22 ng/mL (CV 4.4%, accuracy 81%), respectively.

In addition, King and Bursill (1998) determined plasma levels of daidzein and genistein after digestion of a soy meal by HPLC-ECD using an octa decyl silyl (ODS) column (250 mm × 4.6 mm i.d., 5 μm particle size; SGE Australia, Ringwood, Victoria, Australia). After enzymatic hydrolysis, the ethereal extracts were eluted by using a solvent system composed of methanol, 0.1 mol/L ammonium acetate solution (pH 4.6) and 25 mmol/L ethylenediamine-tetra-acetic acid (EDTA) solution (55:50:1, v/v), and a flow rate of 1.0 mL/min.

The operating potential of ECD was 0.75 V. The LOD of genistein was determined as 0.2 μmol/L (54 ng/mL). When this method was first applied to the assay of daidzein and genistein in rat plasma (King et al. 1996), daidzein could not be quantified due to an interfering peak. Subsequently, however, this method was used successfully for the assay of daidzein and genistein in human plasma (King and Bursill 1998).

Lastly, Grace et al. (2007) developed a method for the quantification of daidzein and genistein in human urine and serum by using solid-phase extraction (SPE) and LC-MS/MS. Two different LC-MS/MS systems, a Quattro Premier triple quadrupole mass spectrometer (Waters, Manchester, UK) and a Sciex 4000 QTrap (Applied Biosystems, Warrington, UK), were examined with the following instrument set-up. The ion mode was negative ion electrospray ionization (ESI); the capillary voltage was 2.5 kV for Quattro Premier, 4.3 kV for 4000 QTrap; and the selected reaction monitoring (SRM) mode was used with m/z 253→224 for daidzein, m/z 256→227 for [$^{13}C_3$]daidzein, m/z 269→133 for genistein, m/z 272→183 for [$^{13}C_3$]genistein. For the analytical column, a Targa ODS column (150 mm × 2.1 mm i.d., 3 μm particle size; Higgins Analytical, CA, USA) was used with a gradient of solvent A (methanol: 0.1% ammonium acetate solution, 40:60, v/v) to solvent B (methanol). The LOQ was determined to be 0.1 ng/mL for various aglycones by both instruments. The interassay CV was 6.8% for both instruments. The mean interassay CV was 5.5%.

13.2.2 Fractionation by Ion-exchange Chromatography

Adlercreutz et al. (1994) developed a method for the determination of plasma lignans and isoflavones by isotope-dilution gas chromatography–MS (GC-MS). Plasma samples containing conjugated metabolites of isoflavones were extracted by a Sep-Pak C18 cartridge (Waters Associates Inc., Milford, MA). This method involves the multi-step separation of various conjugates into free and sulfate fractions (mono- and disulfates), and glucuronide fractions (mono- and diglucuronides and sulfoglucuronides) by ion exchange chromatography using DEAE- and QAE-Sephadex in acetate form. Watanabe et al. (1998) previously purified sulfates and glucuronides in plasma by incorporating several clean-up steps as continuous stages of the above method before hydrolysis, and the aglycones obtained were converted into trimethylsilyl derivatives for GC-MS analysis. The isoflavones daidzein and genistein were determined by isotope-dilution GC-MS-selected ion monitoring (SIM) using deuterated internal standards of daidzein and genistein. For the analytical column, a bonded phase BP-1 (SGE) capillary vitreous silica column (12.5 m × 0.2 mm) was used with a helium carrier gas. The following ions were monitored: m/z 398 (M+) for daidzein (RT 9.5 min), m/z 402 (M+) for deuterated daidzein, m/z 471 (M-15) for genistein (RT 9.7 min), m/z 473–475 (M-15, M-16, M-17) for deuterated genistein. The inter-assay CVs ranged from 8.7% to 13.6% at an excess concentration of 0.01 μmol (2.54 ng for daidzein, 2.70 ng for genistein) per mL.

13.2.3 Direct Assay

Doerge *et al.* (2000) analyzed glucuronides and sulfates (4'- and 7-glucuronides, and 4'- and 7-sulfates) of daidzein and genistein in human serum after the ingestion of a supplement by using LC-UV without enzymatic hydrolysis. For the separation of these conjugates, a Prodigy ODS-3 column (250 mm × 4.6 mm i.d., 5 μm particle size; Phenomenex Co., Torrance, CA, USA) was used with a gradient composed of 0.1% aqueous formic acid (solvent A) and 0.1 % formic acid in acetonitrile (solvent B). The UV detection wavelength was set at 260 nm. The LOD value was approximately 1 pmol (0.43 ng for daidzein glucuronides, 0.45 ng for genistein glucuronides). Both 4'- and 7-glucuronides and 4'- and 7-sulfates were eluted within 20 min. The presence of daidzein glucuronide and genistein glucuronide in serum was confirmed by using positive-ion LC-MS/MS with multiple reaction monitoring (MRM) of the following positive ions: daidzein glucuronides, m/z 431 → 255; genistein glucuronides, m/z 447 → 271. For the confirmation of sulfates, negative-ion LC-MS/MS with MRM was performed by monitoring the following negative ions: daidzein sulfates, m/z 333 → 253; genistein sulfates, m/z 349 → 269. Isoflavone conjugates in serum were evaluated as equivalents of aglycone by using LC-ESI/MS after selective enzymatic hydrolysis. The aglycones were determined by using a SIM method to detect $(M + 1)^+$ ions: m/z 255 for daidzein, m/z 271 for genistein. Deuterated daidzein (6,3',5'-^{2}H$_3$) and genistein (6,8,3',5'-^{2}H$_4$) were used as the internal standard.

Clarke *et al.* (2002) determined intact sulfate and glucuronide conjugates in human urine, although not in plasma, by using isotope-dilution LC-ESI-MS/MS with stable-isotope-labeled internal standard [^{13}C$_3$]isofavones. Their report is the only one in the literature to describe a method for the assay of intact conjugated metabolites, except for a recently developed HPLC-UV method (Hosoda *et al.* 2010). Clarke *et al.* (2002) synthesized daidzein-7-glucuronide (D-7-G), daidzein-4',7-diglucuronide (D-4',7-diG), daidzein-4'-sulfate (D-4'-S), daidzein-7-sulfate (D-7-S), daidzein-4',7-disulfate (D-4',7-diS), genistein-7-glucuronide (G-7-G), genistein-4'-sulfate (G-4'-S) and genistein-4',7-disulfate (G-4',7-diS) for use as analytical standards, because most conjugated metabolites of daidzein and genistein are not commercially available. Because they found that the aglycones, the glucuronides and their positional isomers had similar response factors, they proposed a method for the semi-quantification of isoflavone-conjugated metabolites by stable-isotope dilution LC-MS with synthesized [3,4,1'-^{13}C$_3$]genistein and [3,4,8-^{13}C$_3$]daidzein as the internal standards. Because it was not possible to separate these isomers on the base line of sulfates (4'- and 7-isomers of daidzein), these isomers were evaluated as a group sum in urine. The mass spectrometer was operated in negative ESI mode. A YMC-ODS AM analytical column (250 mm × 3.0 mm i.d., 5 μm particle size) was used with a gradient of water and acetonitrile. The LODs were *ca.* 50 ng/mL, with interassay CVs of less than 10% for the conjugates.

13.3 A Simple Direct Assay of Conjugated Metabolites Using HPLC-UV

Numerous studies on the bioavailability and pharmacokinetics of the isoflavones daidzein and genistein have been reported. As described above, the concentrations of glucuronic acid- and sulfuric acid-conjugated metabolites of daidzein and genistein in human plasma are generally estimated by using HPLC to measure free aglycones after enzymatic hydrolysis with β-glucuronidase and sulfatase. This method, however, does not provide detailed information with respect to the type of conjugates and the position(s) of conjugation on the isoflavone rings (daidzein and genistein). A direct assay of intact glucuronic acid- and sulfuric acid-conjugated isoflavones in plasma and urine is indispensable for quantitative estimation of isoflavones in their biological active form, which is necessary to study the mechanisms of pharmacological action and transport of isoflavones, and so on.

One problem in the direct determination of isoflavone-conjugated metabolites containing free-form aglycones is the difficulty in obtaining authentic standards of various conjugated metabolites and their positional isomers: most of these compounds are not commercially available. A second difficulty lies in developing a clean-up method for the desired components, aglycones and their conjugated metabolites, which possess various polarities ranging from low to high. As mentioned above (Section 13.2), reversed-phase HPLC has been widely used for the determination of isoflavones in biological fluids with detection by UV, electrochemical and MS instruments. In particular, because of its simple and commonly available chromatographic instruments, the HPLC-UV method has been frequently used. The disadvantage of this method is its poor specificity as compared with other detecting instruments. However, effective removal of endogenous interference compounds, coupled with effective extraction of the required compounds during HPLC, will compensate for the lack of specificity of the detection method. In other words, a good clean-up method is a prerequisite for the successful analysis of biological fluids, especially those containing highly polar compounds. A third difficulty lies in choosing the reversed-phase column and the conditions for separating the highly polar metabolites from one another on the column.

This Section describes a simple method for the direct assay of intact conjugates of daidzein and genistein in human plasma by HPLC-UV using SPE. Various isoflavone conjugates, such as glucuronides, sulfates and sulfoglucuronides, were chemically synthesized as standards for analysis. Using this method, it has been possible to detect the existing conjugates and forms of isoflavones in human plasma after the ingestion of soybean products. This HPLC-UV assay method, combined with SPE using an Oasis HLB cartridge, has enabled the direct and simultaneous determination of 16 intact metabolites of daidzein and genistein in plasma. This method will contribute to elucidation of the pharmacokinetics, bioavailability and metabolism of isoflavones.

13.3.1 Reference Compounds

Figure 13.2 shows the possible conjugated metabolites and metabolic pathways of daidzein [7-hydroxy-3-(4-hydroxyphenyl)-4H-1-benzopyran-4-one, 4′,7-dihydroxyisoflavone] and genistein [5,7-dihydroxy-3-(4-hydroxyphenyl)-4H-1-benzopyran-4-one, 4′,5,7-trihydroxyisoflavone]. Daidzein, genistein and luteolin-3′,7-di-O-glucoside are commercially available. D-7-G, G-7-G, D-4′,7-diG, genistein-4′,7-diglucuronide (G-4′,7-diG), daidzein-7-glucuronide-4′-sulfate (D-7G-4′S), genistein-7-glucuronide-4′-sulfate (G-7G-4′S), daidzein-7-sulfate (D-7-S), D-4′-S, D-4′,7-diS, G-7-S, G-4′-S and G-4′,7-diS were synthesized (Hosoda *et al*. 2008; 2010) according to previous reports (Nakano *et al*. 2004; Needs and Williamson 2001; Soidinsalo and Wähälä 2007). Daidzein-4′-glucuronide (D-4′-G) and genistein-4'-glucuronide (G-4'-G) were isolated from human urine (Hosoda *et al*. 2008). The structures of the synthesized compounds were confirmed by LC-ESI-MS (Table 13.2) and/or ^{1}H-nuclear magnetic resonance (NMR) analyses (Table 13.3).

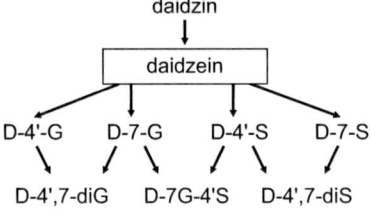

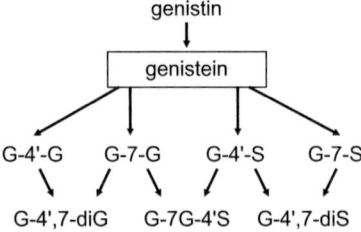

Figure 13.2 Possible metabolic pathways of daidzin and genistin in humans. Absorbed daidzein and genistein are metabolized to various glucuronides, sulfates and sulfoglucuronides. D-4′-G, daidzein-4′-glucuronide; D-7-G, daidzein-7-glucuronide; D-4′-S, daidzein-4'-sulfate; D-7-S, daidzein-7-sulfate; D-4′,7-diG, daidzein-4′,7-diglucuronide; D-7G-4′S, daidzein-7-glucuronide-4′-sulfate; D-4′,7-diS, daidzein-4′,7-disulfate; G-4′-G, genistein-4'-glucuronide; G-7-G, genistein-7-glucuronide; G-4′-S, genistein-4′-sulfate; G-7-S, genistein-7-sulfate; G-4′,7-diG, genistein-4′,7-diglucuronide; G-7G-4′S, genistein-7-glucuronide-4′-sulfate; G-4′,7-diS, genistein-4′,7-disulfate.

Table 13.2 Mass-to-charge ratio (m/z) value of $[M+H]^+$ molecular ions of isoflavone-conjugated metabolites. Positive ESI analysis was carried out under the following conditions: capillary temperature, 350 °C; capillary voltage, 30 V; tube lens voltage, 105 V. Chromatographic separation was performed on a Hydrosphere C18 (100 mm × 3 mm i.d., 3 µm particle size; YMC Co. Ltd., Kyoto, Japan) and a guard cartridge (10 mm × 2.0 mm i.d.). Samples were eluted using a solvent system comprising 10 mM ammonium acetate (solvent A) and acetonitrile mixed using a linear gradient, held at 95.3% solvent A at 0–1.5 min; 68.0% solvent A at 27 min; 95.3% solvent A at 34 min, and then held for 10 min. The flow rate was 0.5 mL min at 45 °C.

	$[M+H]^+$ (m/z)		
Compound	MS	MS/MS	MS/MS/MS
D-4′,7-diG	607	431	255
D-7G-4′S	511	335	255
D-7-G	431	255	
D-4′-G	431	255	
D-4′,7diG	623	447	271
G-7G-4′S	527	351	271
G-7-G	447	271	
G-4′-G	447	271	

13.3.2 SPE

For the direct and simultaneous determination of conjugated metabolites and aglycones, the method must not involve any analytical process that might affect the actual ratio of these metabolites and aglycones in plasma. As described above, selective extraction of conjugated metabolites possessing high polarity in plasma and urine is essential for the HPLC-UV method. The combined solid-phase cartridge, Oasis HLB, which consists of lipophilic divinylbenzene polymer and hydrophilic N-vinylpyrrolidone polymer, is more effective for the extraction of isoflavone conjugates possessing both a lipophilic isoflavone ring and hydrophilic conjugated groups. For sufficient binding of the plasma sample to the solid phase, the ionized isoflavone-conjugated compounds are acidified with 50 mM phosphoric acid solution (pH 2.0; 2.0 mL to plasma 1.0 mL) for deionization. After sample loading, the cartridge is washed with 50 mM phosphoric acid solution containing methanol maintaining a low pH. The last process, the extraction process, is to elute the aglycones and conjugated metabolites as their ionized form from the cartridge using methanol solution containing concentrated ammonia solution (pH 10.0) (Figure 13.3). The recoveries of the 16 isoflavone metabolites from human plasma by this method are over 80%.

Table 13.3 ^{1}H-NMR spectra of aglycones and glucuronides (δ[ppm] in DMSO-d_6). NMR spectra were recorded using the 600 MHz ^{1}H-NMR spectrometer. Chemical shifts are given in δ values downfield from tetramethylsilane.

Compound	2-H δ/ppm	5-H δ/ppm	5-H J/Hz	6-H δ/ppm	6-H J/Hz	7-H δ/ppm	8-H δ/ppm	8-H J/Hz	2', 6'-H δ/ppm	2', 6'-H J/Hz	3', 5'-H δ/ppm	3', 5'-H J/Hz	4'-H δ/ppm	1''-H δ/ppm	5''-H δ/ppm
Dein	8.28 (s)	7.96 (d)	8.75	6.93 (dd)	8.77, 2.17, 2.18		6.86 (d)	2.17	7.38 (d)	8.52	6.81 (d)	8.56			
D-7G-4'S	8.45 (s)	8.07 (d)	8.89	7.16 (dd)	8.90, 2.24, 2.30		7.27 (d)	2.24	7.49 (d)	9.01	7.22 (d)	9.02			
D-7-G	8.38 (s)	8.05 (d)	8.72	7.15 (d)	8.35		7.27 (s)		7.40 (d)	8.04	6.82 (d)	8.05	9.57 (s)	5.56 (s)	4.01 (s)
D-4'-G	8.35 (s)	7.97 (d)	8.74	6.94 (d)	8.76		6.88 (s)		7.49 (d)	8.54	7.08 (d)	8.54	9.60 (s)		
Gein	8.32 (s)	12.96 (s)		6.22 (d)	2.04	10.89 (s)	6.38 (d)	2.06	7.37 (d)	8.57	6.82 (d)	8.56			
G-4',7-diG	8.49 (s)			6.73 (d)	1.99		6.48 (d)	1.93	7.52 (d)	8.64	7.09 (d)	8.75			
G-7G-4'S	8.51 (s)	12.87 (s)							7.42 (broad s)		7.20 (d)	7.28		5.00 (s)	
G-7-G	8.43 (s)	12.95 (s)		6.75 (d)	1.93		6.50 (d)	1.99	7.40 (d)	8.43	6.83 (d)	8.47	9.62 (s)	5.55 (s)	4.05 (s)
G-4'-G	8.39 (s)	12.95 (s)		6.24 (s)			6.41 (s)		7.49 (d)	8.41	7.09 (d)	8.63			

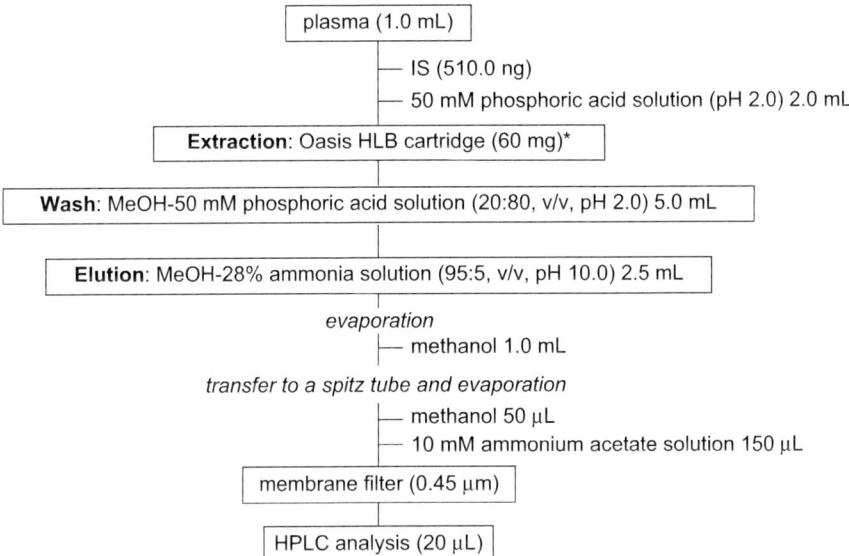

Figure 13.3 Flow diagram of the method for SPE. *The SPE cartridge is activated with 2.5 mL of methanol and 2.5 mL of 50 mM phosphoric acid solution prior to the sample loading.

13.3.3 Chromatographic Conditions

Isoflavones are easily separated by HPLC using reversed-phase C8 or C18 columns and acetonitrile/water and methanol/water systems as the mobile phase. The solvents usually contain trace amounts of acid, such as formic acid, acetic acid and torifluoroacetic acid (~0.1–1%). Ten mM ammonium formate and ammonium acetate can also be used instead of water. These solvent systems make it possible to simultaneously analyze aglycones and their highly polar conjugated metabolites in a short time.

For example, Zhang *et al.* (1999) synthesized daidzein glucuronides and genistein glucuronides by using uridine 5′-diphosphate (UDP)-glucuronosyl-transferase. The reaction products were separated on a YMC C18 AM 303 column (250 mm × 4.6 mm i.d., 5 μm particle size; YMC Co. Ltd., Wilminton, NC, USA) at flow rate of 0.8 mL/min using the solvents 0.1% acetic acid in water (solvent A) and 0.1% acetic acid in acetonitrile (solvent B). The gradient program was as follows: (1) 20% solvent B for 5 min; (2) increasing linearly to 30% solvent B at 30 min; and then (3) 50% solvent B at 45 min; (4) back to 20% solvent B. The peaks of D-7-G, G-7-G, daidzein and genistein appeared on the chromatogram at retention times of approximately 14 min, 22 min, 30 min and 38 min, respectively. Clarke *et al.* (2002) used a YMC-ODS AM column (250 mm × 3 mm i.d., 5 μm particle size) to estimate sulfates and glucuronides of isoflavones in human urine using stable-isotope dilution LC-MS/MS with [$^{13}C_3$]isoflavone internal standards. The eluting solvents were water containing 0.1% formic acid and acetonitrile containing 0.1% formic acid. Although the

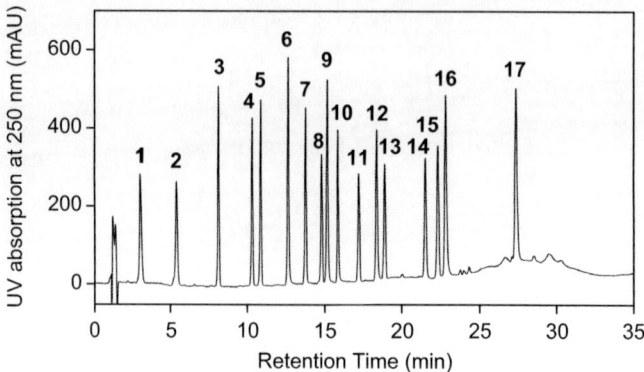

Figure 13.4 HPLC chromatogram of a standard mixture. Numbers correspond to those given in Table 13.4. Binary linear gradient elution was used with a Hydrosphere C18 (100 mm × 4.6 mm i.d., 3 μm particle size; YMC Co. Ltd., Kyoto, Japan) and a guard cartridge (23 mm × 4.0 mm i.d.), and a flow rate of 1.5 mL/min. Solvent A, 10 mM ammonium acetate; solvent B, acetonitrile. Gradient profile: 95.3% A at 0–1.5 min; 68.0% A at 27 min; 95.3% A at 34 min, and then held for 10 min. The UV detection wavelength was set at 250 nm. HPLC analysis was performed at 45 °C.

baseline separation of D-7-S and D-4′-S was not accomplished, other metabolites were separated on the column within *ca.* 50 min.

Recently, an HPLC Hydrosphere C18 column, which strongly retains highly polar compounds, has been developed. Hosoda *et al.* (2010) separated 16 conjugated isoflavone metabolites containing daidzein and genistein on this column (Figure 13.4). As can be seen in Figure 13.4 and Table 13.4, the conjugated metabolites of daidzein and genistein were well resolved within 28 min on this HPLC column under these chromatographic conditions; furthermore, baseline resolution was obtained between the D-4′-S and G-7-S isomers with respective retention times of 18.39 min and 18.91 min.

The correlation coefficient of calibration graphs for the daidzein and genistein conjugated metabolites (*ca.* 5–5000 ng/mL plasma) ranged from 0.9976 to 0.9999. The relative errors for these metabolites determination were ±20%, and the inter-assay CVs were less than 10%. The LOQ and LOD of this method were found to be 25 ng/mL and 8 ng/mL, respectively (at a signal-to-noise ratio of approximately 3).

13.3.4 Application

To test the assay of daidzein and genistein in human plasma, 10 g of kinako, a baked soybean powder, was ingested by a healthy volunteer. Ten grams of kinako includes 5.5 mg of Din, 9.2 mg of Gin, 6.1 mg of daidzein and 11.1 mg of genistein. Figure 13.5 shows a typical HPLC chromatogram of the plasma extract at 8 h after ingestion. Figure 13.6 shows time courses of the

Table 13.4 HPLC retention times of isoflavone conjugated metabolites. See Figure 13.4.

Peak	Compound	Retention time (min)
1	D-4',7-diG	3.04
2	G-4',7-diG	5.38
3	D-7G-4'S	8.12
4	G-7G-4'S	10.33
5	D-7-G	10.88
6	D-4'-G	12.65
7	G-7-G	13.78
8	D-4',7-diS	14.82
9	G-4'-G	15.18
10	IS[a]	15.86
11	G-4',7-diS	17.23
12	D-4'-S	18.39
13	D-7-S	18.91
14	G-4'-S	21.51
15	G-7-S	22.32
16	daidzein	22.82
17	genistein	27.36

[a]IS, internal standard; luteolin-3',7-di-O-glucoside.

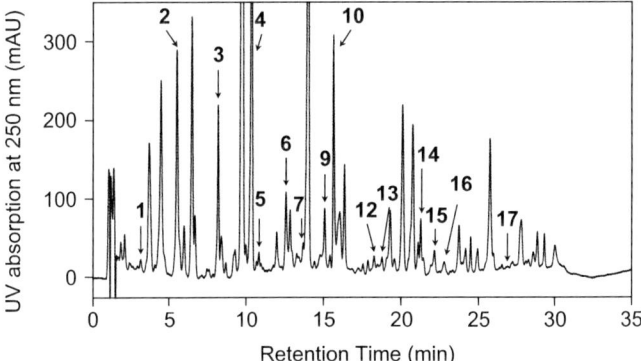

Figure 13.5 HPLC chromatogram of the plasma extract obtained at 8 h after ingestion of kinako (10 g) by a healthy volunteer. Numbers in Figure 13.5 correspond to those given in Table 13.4. Chromatographic conditions are the same as in Figure 13.4.

plasma concentration of two isoflavone metabolites: daidzein metabolites (Figure 13.6A) and genistein metabolites (Figure 13.6B).

Most investigators have attempted to describe the analysis of isoflavones, especially total daidzein and genistein concentrations, after the ingestion of soybean products containing primarily isoflavonoid glycosides such as daidzin and genistin. Recently, it has been reported that these conjugates may have biological activity themselves, or they may be precursors of biologically active

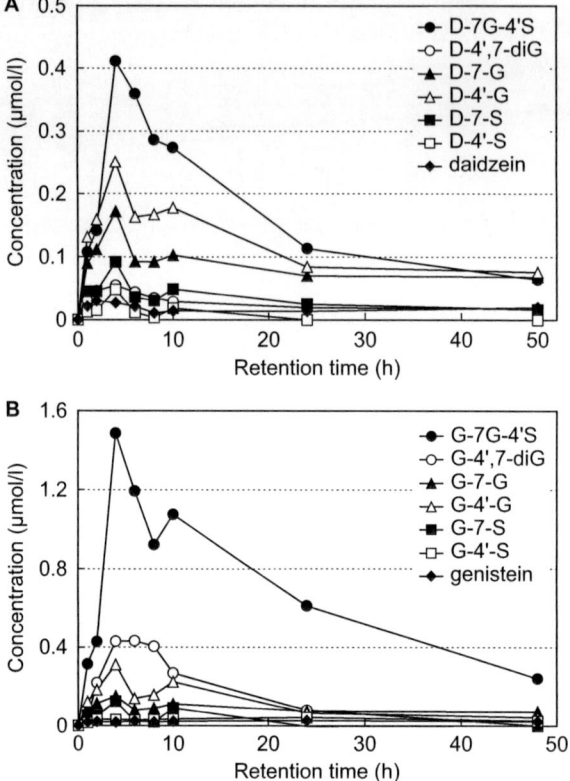

Figure 13.6 Time courses of the plasma concentration of isoflavone metabolites. (A) Daidzein metabolites and (B) genistein metabolites after ingestion of kinako (10 g) by a healthy volunteer.

compounds at or within target cells. The results based on the HPLC-UV analysis indicated that the double conjugated metabolites D-7G-4′S, G-7G-4′S and G-4′,7-diG were the most abundant compounds in plasma after the ingestion of kinako (Figure 13.6). The HPLC-UV analysis should provide a useful method for successful investigation of the pharmacological and medicinal properties, as well as the metabolic pathways, of isoflavones in humans.

Summary Points

- This Chapter focuses on the methods for the assay of glucuronic acid- and sulfuric acid-conjugated metabolites of daidzein and genistein in human plasma and urine.
- These metabolites can be evaluated as: (1) the aglycones obtained after selective enzymatic hydrolysis, (2) the aglycones obtained after hydrolysis of fractions separated according to conjugation type, and (3) the intact conjugated metabolites.

- For the determination of daidzein and genistein, high-performance liquid chromatography (HPLC) is frequently used in combination with ultraviolet (UV), electrochemical detector, mass spectrometry (MS) and tandem mass spectrometry (MS/MS) instruments.
- For the elucidation, particularly structural characterization, of isoflavone conjugates in plasma, the direct assay is indispensable. The limit of quantification of the direct assay of glucuronides and sulfates using liquid chromatography–tandem mass spectrometry (LC-MS/MS) is more than 50 ng/mL plasma.
- Among the available detectors, the UV detector is simple and relatively inexpensive; however, its specificity towards the target compounds is low.
- The low specificity of the UV detector for highly polar conjugates in biofluids can be, to some extent, compensated by a successful clean-up method.
- By using an Oasis HLB cartridge that is able to retain compounds possessing hydrophilic and lipophilic moieties, conjugated metabolite samples can be cleaned up.
- The recently developed high-performance liquid chromatography–ultraviolet (HPLC-UV) method, in combination with solid-phase extraction (SPE), enables the direct and simultaneous determination of 16 intact metabolites of daidzein and genistein in human plasma.
- The limit of quantification and limit of detection of this method are 25 ng/mL and *ca.* 8 ng/mL plasma, respectively.

Key Facts

- Isoflavones are mainly present as phenolic glycosides in soy and soy products.
- After ingestion, isoflavone glycosides are converted into aglycones, such as daidzein and genistein, and sugar moieties by enteral micro-organisms; the aglycones are then absorbed from the gastrointestinal tract.
- Absorbed aglycones are conjugated with either glucuronic acid or sulfuric acid inboth liver and intestine.
- Absorbed daidzein and genistein circulate as glucuronides and sulfates in blood and are then excreted in urine.
- It is thought that aglycones may have prevention effects against hormone-dependent cancers such as breast and prostate cancers.
- Conjugated metabolites themselves also have weak estrogenic activity *in vitro*; this effect is dependent on the type of conjugate (glucuronide or sulfate) and the position of the conjugate on the isoflavone ring.

Definitions of Words and Terms

Aglycone: This is a simple phenolic structure, with the non-sugar moiety of glycoside and the non-sulfuric acid-moiety of sulfate. For example, daidzein is a compound formed either by removing the sugar moiety from daidzin, or by removing the sulfuric acid moiety from daidzein sulfates.

Atmospheric pressure chemical ionization (APCI): This is an ionization method for MS analysis based on the corona discharge of organic compounds in a sample solution eluted from LC. The APCI-MS method is applicable to the analysis of the organic compounds possessing a range of low and middle polarities.

Bioavailability: The quantity and speed by which drugs and biological active compounds enter into systemic circulation after administration.

Conjugated metabolites: Metabolites possessing carboxylic and/or phenolic hydroxy groups are often conjugated with highly polar compounds such as glucuronic acid and sulfuric acid, which are present in the body. The chemical reaction, which occurs in the liver, is called phase II metabolism, and the combined compounds are called phase II metabolites or so-called "glucuronides and sulfates".

Electrospray ionization (ESI): This is a "soft" ionization method. The sample solution eluted from an LC column is introduced to a high-voltage capillary. Charged minute drops containing ions in water, which are formed by spraying the sample from the high-voltage capillary, are evaporated and the collected ions are introduced into the mass spectrometer. The ESI-MS method is applicable to the analysis of highly polar organic compounds.

High-performance liquid chromatography (HPLC): Chromatography using liquid solvents as the mobile phase is called liquid chromatography (LC). In particular, LC with a high-pressure pump and high-performance loading materials for separation is called high-performance liquid chromatography (HPLC). HPLC methods are suitable for the analysis of nonvolatile, so-called polar organic compounds in food, environmental and biological fluids.

Isoflavone glycoside: Most isoflavones are present as glycosides such as daidzin and genistin, among others, in the plant kingdom. The representative sugar moiety is glucose.

Limit of detection (LOD): In general, the LOD is defined as a signal-to-noise ratio of approximately 3.

Limit of quantification (LOQ): The LOQ is the lowest concentration within the standard curve generated with 5–8 concentrations (excluding the blank). The LOQ is defined with acceptable accuracy and precision. From international recommendations, at least five samples of the same concentration should be determined. The mean value obtained should be less than $\pm 15\%$ of the actual concentration and less than $\pm 20\%$ at the LOQ. Furthermore, the CV value should also be less than $\pm 15\%$ and less than $\pm 20\%$ at LOQ.

Liquid chromatography–mass spectrometry (LC-MS): The technique of LC-MS combines LC for the separation of various target components in a sample and MS analysis, and is used, in particular, for structural analyses.

Pharmacokinetics: This is the branch of pharmacology and biopharmaceutics that deals quantitatively with the movement of drugs or exogenous substances within the body of humans and animals, and is kinetically concerned with the absorption, distribution, metabolism and excretion of drugs.

Solid-phase extraction (SPE): This is frequently used as a method of sample pre-treatment before analysis. In general, the sample solution is passed through the solid phase, which is then washed with a moderate solvent to exclude non-target components. The group of target compounds binding to the solid phase can be eluted using a suitable solvent.

List of Abbreviations

APCI	atmospheric pressure chemical ionization
CV	coefficient of variation
D-4′,7-diG	daidzein-4′,7-diglucuronide
D-4′,7-diS	daidzein-4′,7-disulfate
D-4′-G	daidzein-4′-glucuronide
D-7-G	daidzein-7-glucuronide
D-7G-4′S	daidzein-7-glucuronide-4′-sulfate
D-4′-S	daidzein-4′-sulfate
D-7-S	daidzein-7-sulfate
ECD	electrochemical detector
ESI	electrospray ionization
G-4′,7-diG	genistein-4′,7-diglucuronide
G-4′,7-diS	genistein-4′,7-disulfate
G-4′-G	genistein-4′-glucuronide
G-7-G	genistein-7-glucuronide
G-7G-4′S	genistein-7-glucuronide-4′-sulfate
G-4′-S	genistein-4′-sulfate
G-7-S	genistein-7-sulfate
GC	gas chromatography
HPLC	high-performance liquid chromatography
i.d.	internal diameter
LC	liquid chromatography
LOD	limit of detection
LOQ	limit of quantification
m/z	mass-to-charge ratio
MRM	multiple reaction monitoring
MS/MS	tandem mass spectrometry
MS	mass spectrometry
NMR	nuclear magnetic resonance
ODS	octa decyl silyl
RT	retention time
SIM	selected ion monitoring
SPE	solid-phase extraction
UV	ultraviolet

References

Adlercreutz, H., 2002. Phyto-oestrogens and cancer. *The Lancet Oncology.* 3: 364–373.

Adlercreutz, H., Fotsis, T., Watanabe, S., Lampe, J., Wähälä, K., Mäkelä, T., and Hase, T., 1994. Determination of lignans and isoflavonoids in plasma by isotope dilution gas chromatography-mass spectrometry. *Cancer Detection and Prevention.* 18: 259–271.

Cimino, C.O., Shelnutt, S.R., Ronis, M.J.J., and Badger, T.M., 1999. An LC-MS method to determine concentrations of isoflavones and their sulfate and glucuronide conjugated in urine. *Clinica Chimica Acta.* 287: 69–82.

Clarke, D.B., Lloyd, A.S., Botting, N.P., Oldfield, M.F., Needs, P.W., and Wiseman, H., 2002. Measurement of intact sulfate and glucuronide phytoestrogen conjugates in human urine using isotope dilution liquid chromatography-tandem mass spectrometry with [13C3]isoflavone internal standards. *Analytical Biochemistry.* 309: 158–172.

Doerge, D.R., Chang, H.C., Churchwell, M.I., and Holder, C.L., 2000. Analysis of soy isoflavone conjugation in vitro and in human blood using liquid chromatography-mass spectrometry. *Drug Metabolism and Disposition.* 28: 298–307.

Grace, P.B., Mistry, N.S., Carter, M.H., Leathem, A.J.C., and Teale, P., 2007. High throughput quantification of phytoestrogens in human urine and serum using liquid chromatography/tandem mass spectrometry (LC-MS/MS). *Journal of Chromatography B.* 853: 138–146.

Hosoda, K., Furuta, T., Yokokawa, A., Ogura, K., Hiratsuka, A., and Ishii, K., 2008. Plasma profiling of intact isoflavone metabolites by high-performance liquid chromatography and mass spectrometric identification of flavone glycosides daidzin and genistin in human plasma after administration of kinako. *Drug Metabolism and Disposition.* 36: 1485–1495.

Hosoda, K., Furuta, T., and Ishii, K., 2010. Simultaneous determination of glucuronic acid and sulfuric acid conjugated metabolites of daidzein and genistein in human plasma by high-performance liquid chromatography. *Journal of Chromatography B. Analytical Technologies in the Biomedical and Life Sciences.* 878: 628–636.

King, R.A., and Bursill, D.B., 1998. Plasma and urinary kinetics of the isoflavones daidzein and genistein after a single soy meal in humans. *The American Journal of Clinical Nutrition.* 67: 867–872.

King, R.A., Broadbent, J., and Head, R.J., 1996. Absorption and excretion of the soy isoflavone genistein in rats. *The Journal of Nutrition.* 126: 176–182.

Nakano, H., Ogura, K., Takahashi, E., Harada, T., Nishiyama, T., Muro, K., Hiratsuka, A., Kadota, S., and Watabe, T., 2004. Regioselective monosulfation and disulfation of the phytoestrogens daidzein and genistein by human liver sulfotransferases. *Drug metabolism and Pharmacokinetics.* 19: 216–226.

Needs, P.W., and Williamson, G., 2001. Syntheses of daidzein-7-yl-D-glucopyranosiduronic acid and daidzein-4′,7-yl di-β-D-glucopyranosiduronic acid. *Carbohydrate Research.* 330: 511–515.

Pugazhendhi, D., Watson, K.A., Mills, S., Botting, N., Pope, G.S., and Darbre, P.D., 2008. Effect of sulphation on the oestrogen agonist activity of the phytoestrogens genistein and daidzein in MCF-7 human breast cancer cells. *Journal of Endcrinology*. 197: 503–515.

Shelnutt, S.R., Cimino, C.O., Wiggins, P.A., and Badger, T.M., 2000. Urinary pharmacokinetics of the glucuronide and sulfate conjugates of genistein and daidzein. *Cancer Epidemiology, Biomarkers and Prevention*. 9: 413–419.

Shelnutt, S.R., Cimino, C.O., Wiggins, P.A., Ronis, M.J., and Badger, T.M., 2002. Pharmacokinetics of the glucuronide and sulfate conjugates of genistein and daidzein in men and women after consumption of a soy beverage. *The American Journal of Clinical Nutrition*. 76: 588–594.

Soidinsalo, O., and Wähälä, K., 2007. Synthesis of daidzein 7-O-β-D-glucuronide-4′-O-sulfate. *Steroids*. 72: 851–854.

Thomas, B.F., Zeisel, S.H., Busy, M.G., Hill, J.M., Mitchell, R.A., Scheffler, N.M., Brown, S.S., Bloeden, L.T., Dix, K.J., and Jeffcoat, A.R., 2001. Quantitative analysis of the principle soy isoflavones genistein, daidzein and glycitein, and their primary conjugated metabolites in human plasma and urine using reversed-phase high-performance liquid chromatography with ultraviolet detection. *Journal of Chromatography B. Biomedical Sciences and Applications*. 760: 191–205.

Watanabe, S., Yamaguchi, M., Sobue, T., Takahashi, T., Miura, T., Arai, Y., Mazur, W., Wähälä, K., and Adlercreutz, H., 1998. Pharmacokinetics of soybean isoflavones in plasma, urine and faces of men after ingestion of 60 g baked soybean powder (kinako). *The Journal of Nutrition*. 128: 1710–1715.

Zhang, Y., Song, T.T., Cunnick, J.E., Murphy, P.A., and Hendrich, S., 1999. Daidzein and genistein glucuronides in vitro are weakly estrogenic and activate human natural killer cells at nutritionally relevant concentrations. *The Journal of Nutrition*. 129: 399–405.

CHAPTER 14

High-throughput Quantification of Pharmacologically Active Isoflavones using LC-UV/PDA and LC-MS/MS

WAHAJUDDIN*[a] AND SUMIT ARORA[b]

[a] Pharmacokinetics and Metabolism Division, CSIR – Central Drug Research Institute, Post Box No. 173, Chhattar Manzil Palace, Lucknow – 226001 (U. P.), India; [b] Department of Pharmaceutics, National Institute of Pharmaceutical Education and Research, Rae Bareli–229001 (U. P.), India
*Email: wahajuddin@cdri.res.in

14.1 Introduction

It has always been seen that people belonging to certain geographical regions are less susceptible to some diseases than the people native for other regions. For instance, epidemiological studies reveal that the incidence of breast cancer in Japanese and Chinese women is lower than those women in the west (Adlercreutz *et al.* 2000). What might be a plausible reason for such a difference? The reason could be attributed to varied dietary patterns that prevail in different parts of world. Dietary components are known to exert remarkable effect on our health. In fact, nowadays people are moving towards natural therapy to circumvent the side effects resulting from conventional therapy.

These natural dietary components are increasing being investigated by researchers all around the globe to yield potent and safe drug molecules. One such category of dietary components which have shown potential health benefits in humans is isoflavones, a sub-class of natural anti-oxidants, flavanoids.

Isoflavones are commonly known as dietary phytoestrogens because of their structural and functional similarity to 17β-estradiol (a female hormone). Figure 14.1 represents the chemical structures of most pharmacologically active isoflavones. These isoflavones have been associated with a variety of preventive and curative health-related effects. These include inhibition of cancer cell proliferation (Lo *et al.* 2007), improved blood lipid profile (Lovati *et al.* 1987), lowered incidence of coronary artery disease (Lissin and Cooke 2000), anti-oxidative (Arora *et al.* 1998) and anti-inflammatory potential (Kao *et al.* 2007), prevention of osteoporosis (Yamaguchi 2002), anti-estrogenic as well as estrogenic activity (Kuiper and Gustafsson 1997) and anti-viral activity (Kaul

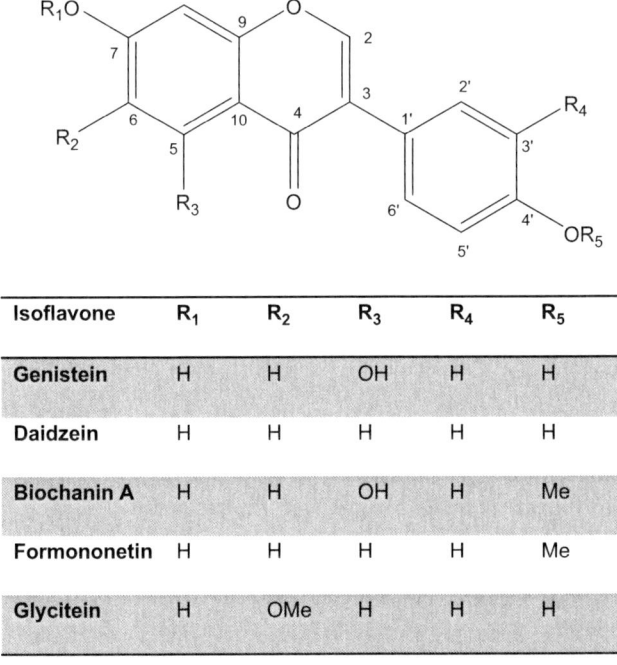

Isoflavone	R_1	R_2	R_3	R_4	R_5
Genistein	H	H	OH	H	H
Daidzein	H	H	H	H	H
Biochanin A	H	H	OH	H	Me
Formononetin	H	H	H	H	Me
Glycitein	H	OMe	H	H	H

Figure 14.1 Chemical structures of pharmacologically active isoflavones Chemical structures of most pharmacologically active isoflavone phytoestrogenic compounds present in soy and red clover, the richest sources of isoflavones. Bicochanin A (BCA), formononetin and their malonyl derivatives are the major components present in red clover, whereas genistein, daidzein, glycitein and their glycosides conjugates are the major isoflavones present in soy.

et al. 1985). Owing to their multifaceted therapeutic benefits, these dietary isoflavones have generated much interest. Among the most studied isoflavones are the soy-derived isoflavones genistein and daidzein. These isoflavones are accredited to exercise more profound pharmacological effect compared to other isoflavones.

Genistein, daidzein and glycitein are the major isoflavones present in soy and soy-derived products. The soyabean is a member of the leguminosae family and is one of the richest sources of dietary isoflavones. The concentration of isoflavones in soyabean varies with habitat conditions, variety and time of harvesting, but the average content ranges from 0.2–0.4% (w/w). Genistein and daidzein owing to their weak estrogenic activity are being considered as the potential alternatives for hormone replacement therapy (HRT) in postmenopausal women with concomitant prevention of hormone-dependent cancers, prostate cancer and breast cancer (Adlercreutz *et al.* 1992). *In vitro* studies revealed that these phytoestrogens possess anti-neoplastic activity (Barnes 1995). Various mechanisms may contribute to their anti-tumor activity such as the inhibition of tyrosine kinase activity (Fotsis *et al.* 1995), suppression of angiogenesis (Schnaper *et al.* 1996) or partial antagonism of estrogen receptors (Adlercreutz *et al.* 1992).

Another leguminosae plant, red clover (*Trifolium pratense*) is reported to contain significantly higher concentrations of isoflavones. Formononetin (FMN) and biochanin A (BCA) are the major isoflavones present in red clover. FMN is reported to aid fracture healing through angiogenesis activation in the early stage of fracture repair and osteogenesis acceleration in the later stages (Gautam *et al.* 2011). BCA, a 4-*O*-methyl derivative of genistein, is the major isoflavone in red clover but is not present in soy foods. BCA, like genistein, has been shown to inhibit chemical-induced tumor carcinogenesis and prevent tumor growth after implantation in animal models (Gotoh *et al.* 1998). Preclinical and clinical studies have shown that BCA, a major component in commercially available red clover extract normalizes the low-density lipoprotein (LDL) level in men and reduces the loss of lumbar spine bone mineral content and bone mineral density in women (Nestel *et al.* 2004). Red clover is found in many herbal formulas and its extract is commercially available as tablets (*e.g.* Promensil and Trinovin from Novogen), capsules, tea (*e.g.* Alvita, Botanic Choice and TerraVita) and liquid preparations. These red clover-derived isoflavones are used for relieving postmenopausal symptoms, such as hot flushes, treatment of bone loss and helps maintain prostate and urinary health. BCA is regarded as prodrug of genistein and is reported to be metabolized into the isoflavone genistein *in vitro* and *in vivo* (Tolleson *et al.* 2002) under the catalytic effect of cytochrome (CYP) isoforms particularly CYP1A2 (Hu *et al.* 2003). Some reports have shown that BCA exhibits a different biological potential than GEN.

In light of the widespread uses of these dietary isoflavones and their impact on human health, it has become necessary to develop specific, sensitive, accurate and precise analytical methods for the separation, identification and quantitation of these phytoestrogens in food products as well as biological matrices. Several analytical methods have been reported that quantify isoflavones in food supplements and in various biometrics, *e.g.* plasma, serum and urine. In 1980s, the primary analytical technique employed for analysis of phytoestrogens was gas chromatography (GC). However, the technique suffered from a number of disadvantages like lengthy procedures, chemical derivatization of the phytoestrogens, significant losses during analysis *etc.* However, with the advancement of technology and popularization of the concept of "high-throughput", other techniques have emerged which not only provides accurate and precise analysis of these phytoestrogens in low volume of bio-matrices but also drastically reduces the time taken for analysis. The most common analytical techniques include high-performance liquid chromatography (HPLC) coupled with detection by ultraviolet (UV) absorption, fluorescence detection, electrochemical detection, mass spectrometry (MS) and nuclear magnetic resonance (NMR) spectroscopy. Capillary electrochromatography and immunoassays are some of other analytical techniques used to identify and quantify phytoestrogens in biological matrices. Several sensitive immunoassays have been developed to analyse various isoflavones in foods and biological fluids. However, HPLC-MS continues to hold its prime most position in pharmaceutical industry as the technique of choice for determining the concentrations of drugs/dietary constituents in biological matrices. This technique serves as an important tool of analysis both at the early stages of drug discovery [determination of absorption, distribution, metabolism and excretion (ADME) characteristics of new chemical entities/new molecular entities (NCEs)/NMEs)], as well as late stages of clinical analysis. This technique offers several advantages, such as improved sensitivity and specificity, making it ideal for trace level quantification of target analytes, relatively minimal method development and faster sample analysis.

Due to a lower number of drug molecules reaching the market, there is tremendous pressure on pharmaceutical industries to yield potent, safe and efficacious drug molecules. With the advancement of combinatorial chemistry and parallel growth of high-throughput organic synthesis, thousands of compounds are being synthesized which require early screening of their ADME characteristics for lead selection and optimization. Faster analytical methods with the requisite sensitivity and selectivity are the present need. High-throughput HPLC-MS is increasingly being used to cater for the ever increasing needs of the drug discovery process.

Several strategies are being employed to achieve high throughput HPLC-MS bioanalysis in the pharmaceutical industry. These includes high liquid chromatography (LC) flow rates, fast gradient elution, monolithic columns, column switching, programmed multiple extraction, restricted access media, multiple electrospray ionization (ESI) channels and automated 96-well sample preparations. Entire automation of the analytical technique starting from sample preparation, processing to the interpretation of data has enabled faster sample analysis and data acquisition (Berna *et al.* 2004; Kyranos *et al.* 2001).

The present focus of this Chapter is to provide an overview of high-throughput analytical and bioanalytical methods developed for the identification and quantification of the most studied isoflavones using HPLC along with UV/MS detection.

14.2 High-throughput HPLC-UV/MS Analytical/Bioanalytical Methods

HPLC offers ease in applicability and less labour-intensive sample preparation compared with other chromatographic techniques such as GC. However, some disadvantages of HPLC include poor chromatographic resolution, non-specificity and decreased sensitivity when coupled with UV detection. These restrictions limit the use of this technique for identification and quantitative analysis of phytoestrogens in plant-derived products and biological fluids. Nevertheless, when this technique is coupled to MS as a detection unit, these disadvantages hold no significance. The combination of HPLC-MS offers exclusive selectivity and a dramatic increase in sensitivity with an overall reduction in the time required for analysis.

Since most of the drugs are either weak acids or weak bases, reverse-phase (RP)-HPLC is the preferred analytical tool. Generally, the polar solvents such as acetonitrile (ACN) and/or methanol in combination with water containing small amount of acid as modifiers are used as mobile phases for the separation of phytoestrogens utilizing RP-HPLC columns. Phytoestrogens and their metabolites contain phenolic hydroxyl groups and/or carboxyl groups which exhibit weakly acidic behaviour. The presence of acidic modifiers, such as formic acid (HCOOH), acetic acid (AcOH), trifluoroacetic acid (TFA) and phosphoric acid (H_3PO_4), can dramatically affect the degree of ionization of isoflavones, facilitating their dissociation in mobile phase, thus ensuring better chromatographic resolution, separation and improved peak shape. An additional benefit is that, in presence of acidic modifiers, MS detection sensitivity of these isoflavones significantly improves due to enhanced positive ion formation, thus enabling quantitation of even ng mL^{-1} concentrations of isoflavones in biological matrices with greater accuracy and precision.

Table 14.1 at the end of this Chapter documents some of the high throughput analytical/bioanalytical methods developed for the identification and quantitation of most studied isoflavones in food products, as well as in preclinical/clinical samples.

14.2.1 High-throughput HPLC Methods using UV Detection

HPLC coupled with UV or UV/diode array detector (DAD) is often used for the identification of the phytoestrogens and their metabolites. As all phytoestrogens and their metabolites contain atleast one aromatic ring, they show UV absorption with maximum absorption wavelength (λ_{max}) lying in the range

230–280 nm. As such, the usefulness of this technique is remarkable for the analysis of phytoestrogens. High-throughput bioanalytical methods utilizing advanced chromatographic techniques such as ultrafast-HPLC (U-HPLC) or rapid resolution LC systems using UV- photodiode array (PDA) as a detection system are increasing being developed and validated. Such methods enable analysis of a large number of samples, making them suitable for epidemiological studies.

Klejdus and co-workers have contributed significantly to the faster analysis of phytoestrogens using RP-U-HPLC (Klejdus et al. 2005b; 2005c; 2007; 2008). The results obtained in these studies are very promising in reducing the conventional analysis time for phytoestrogens from 30–60 min to 6 min or even less, with acceptable accuracy, precision, specificity and sensitivity. They optimized a rapid resolution HPLC technique coupled on-line with UV-visible (VIS) detection at 254 nm that allowed the determination of fmol amounts of isoflavones aglycones as well as their glucoside conjugates (Klejdus et al. 2007). In this method, soy parts and other soy preparations were dried at 60 °C, milled and extracted with 90% aqueous methanol using modified Soxhlet technique. The isolated isoflavones were separated and identified in less than 1 min by elution on a RP Zorbax SB C18 (particle size 1.8 µm) column employing a linear gradient profile with 0.2% acetic acid and methanol as mobile phase. They demonstrated that the combination of the column filled with the sorbent of particle size 1.8 µm and the Rapid Resolution HPLC system is well-suited for the high-throughput separation of isoflavones. In another approach, this group has utilized the fast column HPLC-UV-PDA technique and separated up to 10 isoflavones within 8 min without losing its separation efficiency and sensitivity (Klejdus et al. 2005c).

Rostango et al. (2007) separated 12 selected isoflavones (genistein, daidzein, glycitein and their β-glucosides and/or acetyl- and malonyl-β-glucosides) with retention time of <10 min using two linked monolithic column. As samples were extracted using solid–liquid extraction, which is, in general, non-selective, they showed the applicability of the developed method by determining the isoflavone content of extracts of soybean flour, texturized soy protein, soy fibre, powdered soy milk and liquid soy drink. The developed method demonstrated high chromatographic resolution (>1.06), high reproducibility [relative standard deviation (RSD)$<0.9\%$] and low limit of quantitation (LOQ) (0.80–1.96 mg L^{-1}).

Since phytoestrogens are reported to exert favourable cardiovascular effects and are increasingly being prescribed along with β-blockers, Baranowska et al. (2011) developed a rapid resolution U-HPLC method using absorbance detector for the simultaneous determination of β-blockers (milrinone, sotalol, metoprolol, propranolol and carvedilol), isoflavones (genistein, daidzein, glycitin, glycitein, puerarin and BCA) and their metabolites in human urine with run time no longer than 8 min. In order to evaluate the applicability of the developed method, urine samples of patients on a diet rich in an isoflavone content and with metoprolol (50 mg) and propranolol (80 mg) twice a day, was analysed. Sample preparation involved enzymatic hydrolysis with β-glucuronidase/sulphatase followed by solid phase extraction (SPE) using Oasis HLB cartridges. The analytes were eluted with 5 mL of methanol/acetone/formic acid (4.5:4.5:1; v/v/v). This bioanalytical method can successfully be applied to pharmacokinetic studies after administering β-blockers and isoflavones in humans. The method is apt for routine application in clinical laboratories because of the speed of an analysis and simple extraction procedure.

14.2.2 High-throughput HPLC Methods using MS

The application of HPLC-MS has been tremendous in pharmaceutical research. In the current scenario, quadrupole or triple quadrupole and ion trap MS analyzers equipped with ESI and/or atmospheric pressure chemical ionization (APCI) interfaces have been the mainstay methodology for phytoestrogen analysis, particularly in complex biological fluids, due to their relatively low cost, physically small size, ideal mass range and reasonable scanning speed. ESI is the primary ionization technique for isoflavone analysis because of its high sensitivity and greater ionization stability, as well as compatibility with the solvent system components, *e.g.* ACN and/or methanol and water with small amount of acid modifiers. Triple quadrupole multiple MS instrument operating in selected ion monitoring (SIM) or multiple reaction monitoring (MRM) achieves high sensitivity and selectivity for qualitative and quantitative analysis. In this, the first quadrupole isolates analyte the molecular ion, which is then subjected to collision-induced dissociation followed by analysis of daughter fragment ions by the third quadrupole. This produces a tandem MS (MS/MS) spectrum providing useful information confirming the identity of target analyte. Monitoring the transition of precursor molecular ion to unique daughter fragment ions in MRM provides a highly sensitive, selective, accurate and precise method for the determination of phytoestrogens in food matrices and complex biological samples.

A simple, reliable, reproducible, specific and high-throughput LC–ESI-MS/MS method for quantifying the isoflavones BCA, genistein and their metabolite conjugates in female rat plasma was developed and validated by our group (Singh *et al.* 2010a). In this study, sample preparation involved simple liquid–liquid extraction of BCA, genistein and their metabolites from female rat plasma with diethyl ether with 4-hydroxy mephenytoin added as internal standard. This is followed by separation of fixed volume of supernatant (1.6 mL) which is then evaporated to dryness and finally reconstituted in 120 µL of the mobile phase. 10 µL of the resulting samples were injected on to the LC–ESI-MS/MS system (under negative ion mode) for detection using MRM operation with overall run time of 3 min. Figure 14.2 shows the product ion mass spectrum for BCA and its metabolite genistein and internal standard (IS). In order to determine the BCA and genistein conjugates, 0.1 mL of plasma was incubated with 2000 units of glucuronidase/sulfatase at 37 °C for 4 h and then

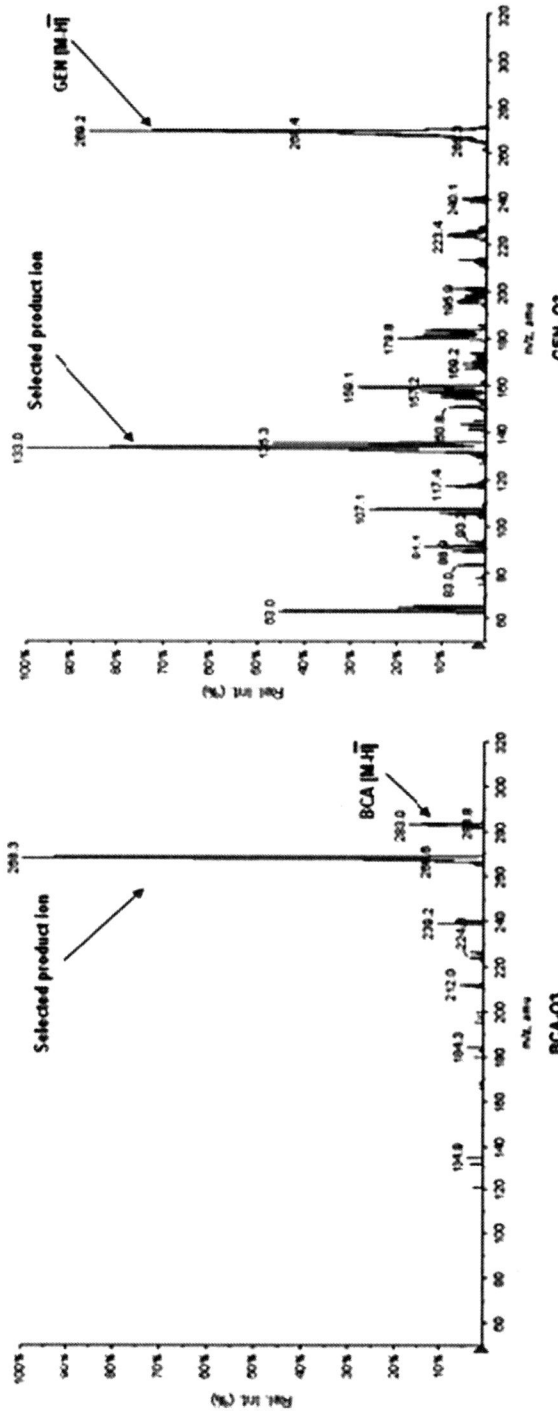

Figure 14.2 MS/MS spectra of BCA and genistein (GEN) showing prominent precursor to product ion transitions. The Figure represents the MS/MS spectra of BCA and GEN. The mass spectrometer was operated in ESI negative ion mode and the detection of the ions was performed in MRM mode. Following the detailed optimization of MS conditions [instrument parameters: ion-spray voltage: −4500 V, nebulizer gas: 30, curtain gas: 10, auxillary gas: 30 and collision gas: 10; compound parameters: declustering potential: −92 (BCA); −93 (GEN), collision energy: −30 (BCA); −42 (GEN), entrance potential: −8 (BCA); −8 (GEN) and collision exit potential: −10 V (BCA); −10 V (GEN)], m/z 283 precursor ion [M−H]$^-$ to the m/z 268 product ion for BCA and m/z 269 precursor ion [M−H]$^-$ to the m/z 133 product ion for GEN was used for the quantitation purpose.

processed as described above. Enzymatic hydrolysis prevents degradation of aglycones which occurs in case of acidic hydrolysis and provides simplicity in work up. Validation of this method was carried out by repetitive analysis of spiked blank female rat plasma. The intra- and inter-day assay precision ranged from 2.66 to 8.34% and 4.40 to 8.10% (RSD %), respectively, and intra- and inter-day assay accuracy was between 90.67–109.25% and 95.86–106.32%, respectively, for both the analytes. The lowest quantitation limit for BCA and genistein was 0.5 ng mL^{-1} in 0.1 mL of rat plasma. The developed and validated LC-ESI-MS/MS method for quantification of BCA and genistein in female rat plasma was successfully applied for pharmacokinetic study in female rats. Similarly, we have also developed a high-throughput LC-MS/MS method with 3 min run time for the quantitation of formononetin and its metabolite daidzein in male rat plasma after intravenous bolus administration of formononetin at dose of 10 mg kg^{-1} (Singh *et al.* 2010b). Figure 14.3 shows the product ion mass spectrum for formononetin and its metabolite daidzein. Sample preparation and analysis were carried out in similar fashion as reported for BCA above

Mallis *et al.* (2003) developed and validated a rapid on-line extraction/ quantitation methodology for cassette dosing analysis of five soy-derived phytoestrogens (genistein, daidzein, BCA, coumestrol, and zearalenone) in male Sprague–Dawley rats. These strategies coupled with the use of dual column setup were applied to achieve high-throughput analysis of phytoestrogens in rat plasma. The sample preparation involved SPE using an Oasis trapping column. Using the MRM technique, all the phytoestrogens were evaluated simultaneously. The LOQ for each compound was 1–1000 ng mL^{-1} with each plasma sample analysis taking less than 2 min. The percentage oral bioavailability was determined to be between 11 and 28%.

Yang *et al.* (2010) reported a sensitive, rapid, reproducible and robust U-HPLC–MS/MS method to simultaneously quantify genistein, genistein-7-*O*-glucuronide (G-7-G), genistein-4′-*O*-glucuronide (G-4′-G), genistein-4′-*O*-sulfate (G-4′-S) and genistein-7-*O*-sulfate (G-7-S) in mouse blood samples. They reported the first LC-MS/MS method that directly determines individual conjugates of genistein without enzymatic hydrolysis. The method was fully validated and applied to the pharmacokinetic study of genistein in FVB mice after intravenously (i.v.) and oral administration at the same dose (20 mg kg^{-1}). As individual standards of genistein metabolites were difficult to purify, they determined the concentration of the metabolites in plasma samples using a genistein standard curve and conversion factors of extinction coefficients for each of the metabolites. Daidzein was used as the internal standard. The total analysis time for each sample was 4.5 min. The standard curves were linear in the concentration range of 19.5–10 000 nM for genistein, 12.5–3200 nM for G-7-G, 20–1280 nM for G-4′-G, 1.95–2000 nM for G-4′-S and 1.56–3200 nM for G-7-S, respectively. The lower limit of quantification (LLOQ) was 4.88, 6.25, 5, 0.98 and 0.78 nM for genistein, G-7-G, G-4′-G, G-4′-S and G-7-S, respectively. Due to high sensitivity of the method, only 20 μL of the blood sample was needed for analysis, so one mouse was used for complete

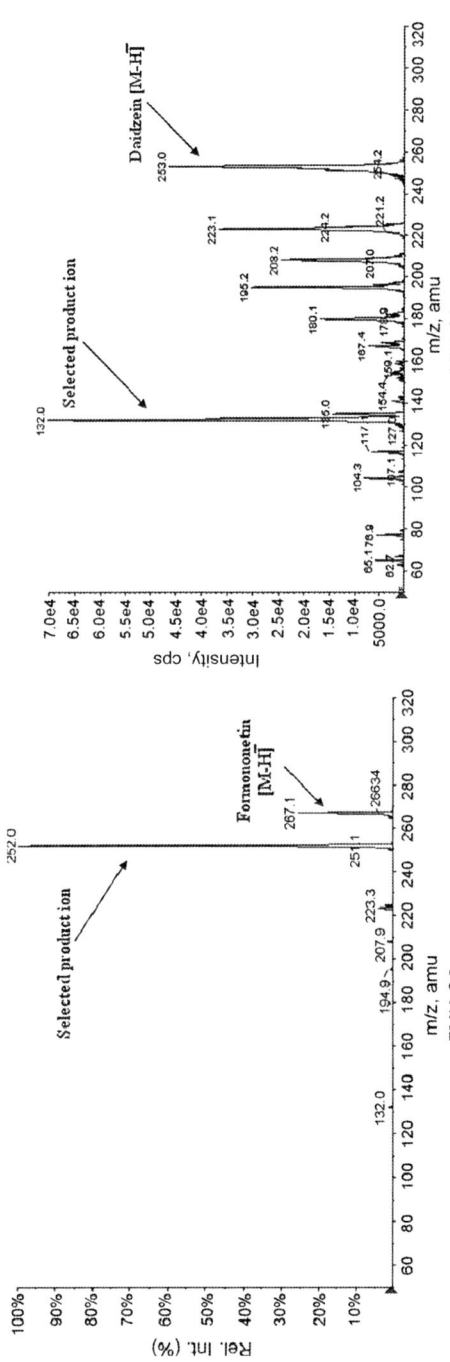

Figure 14.3 MS/MS spectra of formononetin (FMN) and daidzein (DZD) showing prominent precursor to product ion transitions. The Figure represents the MS/MS spectra of FMN and DZD. The mass spectrometer was operated at ESI negative ion mode and detection of the ions was performed in MRM mode. Following the detailed optimization of MS conditions [instrument parameters: ion-spray voltage: –4000 V, nebulizer gas: 50, curtain gas: 15, auxillary gas: 40 and collision gas: 10; compound parameters: declustering potential: –80 (FMN); –102 (DZD), collision energy: –28 (FMN); –51 (DZD), entrance potential: –8 (FMN); –8 (DZD) and collision exit potential: –10 V (FMN); –10 V (DZD)]. m/z 267 precursor ion [M–H]⁻ to the m/z 252 product ion for FMN and m/z 253 precursor ion [M–H]⁻ to the m/z 132 product ion for DZD was used for the quantitation purpose.

pharmacokinetic study. The accuracy and precision data were well within the 15% acceptable range.

Prasain *et al.* (2010) developed and validated a simple, sensitive, reproducible, reliable and a rapid 2 min LC-MS/MS method operating under MRM mode that allows the characterization of 11 phytoestrogens within a single analytical run. The method involves simple liquid–liquid extraction with diethyl ether. All the analytes showed a linear response over the concentration range of 1–5000 ng mL^{-1}, with the exception of dihydrodaidzein whose LLOQ is 2 ng mL^{-1}. The validation of this method achieved appreciable recoveries (>75%) for all the analytes, except for genistein whose mean recovery was 57.48%. Accuracy and precision were well within their acceptable range. The authors have utilized this method for the analysis of over 2000 serum samples, making it well-suited for carrying epidemiological studies.

An analytical method for the determination of the free phytoestrogens and their conjugates in human urine and serum using LC–ESI-MS/MS was developed and validated by Grace *et al.* (2007). Various natural phytoestrogen isoflavones and their major metabolites, mammalian lignans and naringenin were included in this assay and seven stable isotopically labelled [^{13}C$_3$] internal standards in methanol (1 µg mL^{-1}) were used for isotope dilution MS. In order to achieve high throughput, samples were extracted by SPE on 96-well Strata-X plates with the help of Multiprobe II HT EX robotic liquid handling system and analysed using LC-MS/MS incorporating column switching, thus making the assay suitable for analysing a large number of samples as in epidemiological studies. The validation assay revealed that the LC-MS/MS system was capable of quantifying the lowest point of calibration curve 0.1 ng mL^{-1}, setting it as the LLOQ for all the analytes. All the validation parameters were reported to be in accordance with the US Food and Drug Administration (FDA) guidelines for validation of bioanalytical methods.

In order to analyse clinical urine and serum samples collected in a dietary intervention trial with food supplements combining different classes of phytoestrogens, Wyns *et al.* (2008) developed a new chromatographic high-throughput LC/APCI-MS quantitative assay for the simultaneous determination of 13 phytoestrogen, including their gut microbial metabolites, in human urine and serum (Wang *et al.* 2008). The method employed a simple and low-cost sample preparation procedure. Samples were subjected to enzymatic deconjugation followed by liquid–liquid extraction or SPE for urine or serum samples, respectively. The samples were separated with a linear gradient on a Water XBridge C18 RP column with the mobile phases water and methanol/MeCN (80:20, w/w) both acidified with 0.025% (v/v) formic acid. MS analysis was performed with a single quadrupole mass spectrometer using APCI, operating both in the positive and negative ionization mode. Owing to the high sensitivity in the lower parts-per-billion (ppb) range, the developed and validated method is extremely useful for the accurate and precise analysis of various clinical samples obtained from patients fed on diet rich in phytoestrogens. Table 14.1 presents a comprehensive list of the high-throughput HPLC

Table 14.1 High-throughput analytical/bioanalytical methods for the analysis of phytoestrogens. The Table lists some of the high-throughput analytical/bioanalytical methods which have been developed and validated for the estimation of isoflavones in food products and biological matrices. The ratio of mobile phase constituents is expressed as v/v. APCI, atmospheric pressure chemical ionization; CV, coefficient of variation; DAD, diode array detector; DZD, daidzein; ESI, electrospray ionization; GEN, genistein; HPLC, high-performance liquid chromatography; LOD, limit of detection; LOQ, limit of quantitation; MS, mass spectrometry; NI, negative ionization; PDA, photodiode array; PI, positive ionization; RSD, relative standard deviation; SPE, solid phase extraction; SQ, single quadrupole; SRM, selected reaction monitoring; TQ, triple quadrupole; UPLC, ultra-performance liquid chromatography; UV-VIS, ultraviolet-visible.

Sample	Isoflavones analysed	Sample preparation	Internal standard	Analytical technique	Instrument conditions	Detector conditions	Chromatographic run time	Validation	Published report
Male rat plasma	Genistein, daidzein, biochanin A, coumestrol, vearalenone	Rapid on-column solid-phase extraction using Oasis trapping column (2.1 × 20; 25 μm particle size; ambient temperature)	n-(3-Chloro-4-fluoro-phenyl)-2-morpholin-4-yl-kacetamide	LC-MS/MS	XTerra C18 3 mm × 30 mm (3.5 μm particle size; 40 °C; A: 0.02% TEA in water, B: 0.02% TEA in ACN; gradient: 50% B; 0–0.2 min, 50-100 % B; 0.2–1.2 min, 100–50% B; 1.2–1.5 min, flow rate: 1 mL min^{-1}	Micromass Quattro Ultima TQ-MS, ESI, NI, MRM	2 min	Calibration curve range: 1–1000 ng mL^{-1}, LOD 100 pg mL^{-1}	Mallis et al. (2003)
Female rat plasma	Biochanin A, genistein and their conjugates	Liquid–liquid extraction with diethyl ether, organic layer separated, evaporated to dryness and reconstituted with mobile phase	4-hydroxy mephenytoin	LC-MS/MS	Supelco Discovery C18 (4.6 × 50 mm, 5.0 mm); isocratic: mobile phase A: ACN/methanol (50:50, v/v), mobile phase B: acetic acid (0.1%): ratio A/B, 90:10, v/v, flow rate: 0.7 mL min^{-1}	API 4000 mass spectrometer, ESI, NI, MRM	3 min	Calibration curve range: 0.5–200 ng mL^{-1}, intra-day precision: 2.66 to 8.34%; inter-day precision: 4.40 to 8.10%; intra-day accuracy: 90.67–109.25%; inter-day accuracy: 95.86–106.32%; LOQ: 0.5 ng mL^{-1} in 0.1 mL of rat plasma	Singh et al. (2010a)

Table 14.1 (*Continued*)

Sample	Isoflavones analysed	Sample preparation	Internal standard	Analytical technique	Instrument conditions	Detector conditions	Chromatographic run time	Validation	Published report
Male rat plasma	Formononetin, daidzein, and their conjugates	Same as described in Singh et al. (2010a)	4-Hydroxy mephenytoin	LC-MS/MS	Supelco Discovery C18 (4.6 × 50 mm, 5.0 mm); isocratic: mobile phase A: ACN/methanol (50:50, v/v), mobile phase B: acetic acid (0.1%); ratio A/B, 90:10, v/v, flow rate: 0.7 mL min^{-1}	API 4000 mass spectrometer, ESI, NI, MRM	3 min	Calibration curve range: 5–100 ng mL^{-1}, intra-day precision: 1.66–6.82%; inter-day precision: 1.87–6.75%; intra-day accuracy: 89.98–107.56%; inter-day accuracy: 90.54–105.63%; LOQ: 5 ng mL^{-1} in 0.1 mL of rat plasma	Singh et al. (2010b)
Dried leaves of red clover	Formononetin, genistein, daidzein, biochanin A	Dried leaves (10 mg) were extracted with 4 mL of 6 M HCl at 100 °C for 15 min		HPLC-UV	C18 reversed-phase column (Nova-Pak, 4 μm, 3.9 × 150 mm) with guard-column (C18), mobile phase A: acetonitrile/water/trifluoroacetic acid (20:80:0.01; v/v/v); mobile phase B: acetonitrile/trifluoroacetic acid (100:0.1; v/v); gradient: 0–40% B; 0–10 min, 40% B; 10–11 min, 40–100% B; 11–12 min. Flow rate: 0.7 mL min^{-1}	UV and DAD, 260 nm	12 min	Recovery: 85.6 ± 2.06% (DZD), 100.0 ± 0.0% (GEN), 101.0 ± 2.4% (FMN), 97.7 ± 2.9% (BCA), intra-day precision: 0.84–2.54%; inter-day precision: 4.99–7.22%; LOD: 0.0003 μg mL^{-1} (DZD), 0.0015 μg mL^{-1} (GEN), 0.0505 μg mL^{-1} (FMN) and	Ramos et al. (2008)

Sample	Analytes	Extraction	Technique	Instrument/conditions	Detection	Run time	Results	Reference
Bits of soy and methanolic extracts of red clover	Genistin, genistein, daidzein, daidzin, glycitin, glycitein, ononin, formononetin, sissotrin and biochanin A	Extracted with 90% aqueous methanol	HPLC	Agilent 1200 Series Rapid Resolution LC system, Zorbax SB C18 column (1.8 μm particle size), linear gradient: mobile phase A: 0.2% (v/v) acetic acid, mobile phase B: methanol; 0 min 22% B, 1.0 min 80% B, 1.4 min 100% B, 1.8 min 22% B, flow rate: 1.4 mL min^{-1} at temperature 80 °C	UV-VIS DAD	1 min	RSD: 0.24–0.62% for all analytes, LOD and LOQ in fmol range in sub-μL quantities of samples 0.0215 μg mL^{-1} (BCA) LOQ: 0.0009 μg mL^{-1} (DZD), 0.0050 μg mL^{-1} (GEN), 0.1683 μg mL^{-1} (FMN) and 0.0715 μg mL^{-1} (BCA)	Klejdus et al. (2007)
Mouse blood	Daidzein, Genistein, genistein-7-O-glucuronide (G-7-G), Genistein-4'-O-glucuronide (G-4'-G), genistein-4'-O-sulfate (G-4'-S) and genistein-7-O-sulfate (G-7-S)	Extracted with ACN, supernatant separated, evaporated to dryness, reconstituted with 100 μl of 15% acetonitrile aqueous solution	U-HPLC–MS/MS	Acquity UPLC BEH C18 column (50 mm × 2.1 mm i.d., 1.7 μm), mobile phase A: 100% aqueous buffer (2.5 mM ammonium acetate, pH 7.4), mobile phase B: 100% acetonitrile, gradient, initial, 5% B, 0–0.5 min, 5–19% B, 0.5–2 min, 19% B,	API 3200-Qtrap TQ-MS, ESI, NI, MRM	4.5 min	Accuracy: within 85–115%, intra- and inter-day precision: <15%, LLOQ: 4.88 nM (GEN), 6.25 nM (G-7-G), 5 nM (G-4'-G), 0.98 nM (G-4'-S) and 0.78 nM (G-7-S). Recovery: 82.1–114.2% for all analytes	Yang et al. (2010)

Table 14.1 (*Continued*)

Sample	Isoflavones analysed	Sample preparation	Internal standard	Analytical technique	Instrument conditions	Detector conditions	Chromatographic run time	Validation	Published report
					2–2.5 min, 19–40% B, 2.5–3.1 min, 40–52% B, 3.1–3.5 min, 52–80% B, 3.5–4 min, 80–5%, 4–4.5 min, 5% B; flow rate, 0.45 mL min^{-1}; column temperature, 45 °C; sample temperature, 20 °C; and injection volume, 10 μl				
Soy bits, roots, aerial parts, beans	Genistin, genistein, daidzein, daidzin, glycitin, glycitein, ononin, formononetin, sissotrin and biochanin A	Extracted with 90% aqueous methanol	Flavone	HPLC	Atlantis dC18, 20 mm × 2.1 mm, 3 μm, mobile phase A: 0.1% (v/v) acetic acid at pH 3.75 and mobile phase B: methanol, linear gradient:13–22% B (v/v) from 0–2.5 min, up to 30% B to 3.21 min, up to 35% B to 4 min, up to 40% B to 4.5 min, up to 50% B to 5.14 min and followed by negative gradient up 13% B to 7.71 min, flow rate: 0.35 mL min^{-1}, column temperature: 36 °C	PDA	8 min	LOD = 1.1 ng mL^{-1} (GEN), 9.4 ng mL^{-1} (BCA), recoveries: 96–106%	Klejdus *et al.* (2005c)

Sample	Analytes	Sample preparation	Internal standard	Technique	Column and conditions	Detection	Time	LOD/LOQ	Reference
Soy foods	Daidzein, glycitein, daidzein, glycitin, genistin, acetyl-daidzin, acetyl-glycitin, acetyl-genistin malonyl-daidzin, malonyl-glycitin, malonyl-genistin	Extracted under sonication with 25 mL of 50% ethanol (in water) for 20 min at 60 °C	2,5-Dihydroxy-benzaldehyde	HPLC	Two linked monolithic column, chromolith RP-18e (100 mm × 4.6 mm, particle size: 5 μm), mobile phase A: water with 0.1% acetic acid (v/v); mobile phase B: methanol with 0.1% acetic acid (v/v), gradient: 0 min (0% B), 2.0 min (31% B), 4.0 min (31% B), 5.0 min (35% B), 8.0 min (35% B), 9.5 min (100% B), flow rate: 5 mL min^{-1}, column temperature: 35 °C	PDA, 200–400 nm	10 min	LOQ = 800–1960 ng mL^{-1}, RSD: <0.9%	Rostagno et al. (2007)
Trifolium pratense, Ononis spinosa, Pisum sativum, Glycine max	Daidzein, glycitein, daidzin, glycitin, genistin, ononin, formononetin, sissotrin and biochanin A	Sample preparation involves extraction with 90% methanol		U-HPLC	Waters BEH C18 and BEH Phenyl, 50 mm × 2.1 mm, 1.7 μm; Zorbax SB-CN, 50 mm × 2.1 mm, 1.8 μm, mobile phase A: 0.3% aqueous acetic acid, mobile phase B: methanol	UV-VIS-PDA, 200–400 nm	1.8–2.1 min	LOD = 0.2–0.4 ng mL^{-1}	Klejdus et al (2008)
Isoflavones in sea water and fresh water algae	Daidzein, glycitein, daidzin, glycitin, ononin, formononetin	Extraction with supercritical CO_2 modified by 3% (v/v) of methanol/water mixture		LC-MS/MS	Agilent 1200 Series Rapid Resolution Liquid Chromatography, Zorbax SB-CN chromatographic column	MS detector Agilent Technologies 6460 TQ-LC/MS, ESI, NI, MRM	10 min		(Klejdus et al. (2010)

Table 14.1 (Continued)

Sample	Isoflavones analysed	Sample preparation	Internal standard	Analytical technique	Instrument conditions	Detector conditions	Chromatographic run time	Validation	Published report
	sissotrin and biochanin A	(9:1, v/v) at 35 MPa and 40 °C for 60 min			(100 mm × 2.1 sp;mm, particle size 3.5 μm), mobile phase A: 0.2% (v/v) acetic acid in water, mobile phase B: acetonitrile linear gradient: 30% B at 0 min, from 0 min to 3 min up to 50% B, from 3 to 6 min up to 80% B and from 6 to 10 min down to 30% B, flow rate: 0.4 mL min^{-1}, column temperature: 35 °C				
Human urine	β-Blockers, isoflavones and their metabolites	Extracted by means of SPE using Oasis HLB columns	Hesperetin and Paracetamol	UHPLC	Hypersil GOLD™ (50 mm × 2.1 mm, 1.9 μm)	UV	8 min	Intra- and inter-day precision: <4.48%, intra-day and inter-day accuracy: <4.74%, mean recoveries: 70.14% to 99.85%	Baranowska et al. (2011)
Human urine and serum	Daidzein, genistein, glycitein, equol and O-desmethylangolensin, three lignans and one	Aglycones were extracted from the hydrosylate via solid phase extraction (SPE) on 96-well Strata-X	[^{13}C$_3$]-labelled internal standards	LC-MS/MS	Targa C18 columns (2.1 mm × 150 mm, 3 μm), mobile phase A: 60% of 0.1% ammonium acetate in water (pH 4.8) plus 40% methanol;	Quattro Premier TQ-MS, ESI, NI, SRM	10 min	Calibration curve range: 0.1 to 2000 ng mL^{-1}, mean inter-assay Precision: 5.5% (range 0.7–9.8%), mean intra-assay	Grace et al. (2007)

Sample	Analytes	Sample preparation	Internal standard	Technique	Column and mobile phase	MS detection	Run time	Validation	Reference
	flavanone (naringenin)	plates (30 mg, Phenomenex) using a Multiprobe II HT EX robotic liquid handling system			mobile phase B methanol, gradient program: solvent B 0–25% for 1 min; flow rate: 200 μL min^{-1}, 25 to 60% for 7 min; flow rate: 200 μL min^{-1}, 100% solvent B for 1 min; flow rate: 250 μL min^{-1},100% solvent A for 1 min; flow rate:200 μL min^{-1}, during 10 min gradient, auxiliary pump equilibrating second column by pumping 100% A at flow rate: 200 μL min^{-1}			precision: 3.6% (range 1.2–6.5%); accuracy: within 85–115%, recoveries: 82–103% for all the analytes	
Human and mouse serum	Genistein, diadzein, equol	Mixed with acetonitrile to eliminate protein binding, Extracted by means of parallel solid phase extraction (SPE) processing 96 samples simultaneously	Deuterated internal standards (d4-genistein, d3-daidzein, d4-equol)	LC-MS LC-MS/MS	Ultracarb ODS column (235150 mm, 3 μm particle size), isocratic elution: 65% 0.1% formic acid –35% acetonitrile, flow rate: 0.2 mL min^{-1}	Quattro Ultima TQ-MS, ES, PI, SIM and MRM	7 min	Intra-day and inter-day accuracy: within 88–99%, Precision (RSD %): 3–13%, LOD: 0.001 μM for all isoflavones, LLOQ for all isoflavones: 0.005 μM except GEN and DZD with LLOQ: 0.03 μM	Twaddle et al. (2002)
Human serum	Equol, daidzein, dihydrodaidzein, O-desmethylangolesin, genistein,	Serum samples are incubated with β-glucuronidase/	Chrysin	LC-MS/MS	Synergi Polar-RP 2.5 micron (50 mm × 2.0 mm i.d.), mobile phase A:	MRM	2 min	Calibration curve: 1–5000 ng nl^{-1}, except dihydrodaidzein with LLOQ:	Prasain et al. (2010)

Table 14.1 (Continued)

Sample	Isoflavones analysed	Sample preparation	Internal standard	Analytical technique	Instrument conditions	Detector conditions	Chromatographic run time	Validation	Published report
	glycitein, formononetin, biochanin A, coumestrol, enterodiol and enterolactone	sulfatase at 37 °C overnight, extracted with diethyl ether, organic layer evaporated to dryness, reconstituted in methanol/water (80:20, v/v) prior to analysis			10mM ammonium acetate, mobile Phase B: acetonitrile containing 10 mM ammonium acetate, gradient: 25–70% B at 0.75 min and return to 25% B at 1.0 min, flow rate: 0.75 mL min^{-1}, column temperature: 50 °C			2ng mL^{-1}, accuracy ranges from 87.53 to 114.83% and precision ranges from 0.83 to 14.85%,	
Human urine and serum	Genistein, daidzein, equol, dihydrodaidzein, O-desmethylangolensin, coumestrol, secoisolariciresinol, matairesinol, enterodiol, enterolactone, isoxanthohumol, xanthohumol and 8-prenylnaringenin	Enzymatic hydrolysis using helix pomatia β-glucuronidase/sulfatase, followed by extraction with diethyl ether for urine samples and solid phase extraction for serum samples	4-Hydroxy-benzophe-none	HPLC–MS	Waters XBridge C18 reversed phase (3.5 μm) column (3.0mm i.d. × 150 mm column) connected to a 3.0 mm i.d. × 20 mm C18 guard column (3.5 μm), mobile phase A: H$_2$O and moblie phase B: MeOH/MeCN (80:20, w/w), both acidified with 0.025% (v/v) formic acid, gradient: 0–3.5 min:35–45% B (flow: 0.6 mL min^{-1}), 3.5–5.5 min: isocratic 45% B (flow:	MS detector, SQ-MS, APCI, PI, NI, SIM	20 min	Accuracy : 0.2% to 14.6% (in urine) and 0.01% to 18.7% (in serum), %CV for the inter-assay precision: 4.3% to 18.8% (in urine) and 4.4% and 15.7% (in serum), intra-assay precision : ~10% (in urine) and ~15% (in serum), LOD: 0.2 ng mL^{-1} to 7.7 ng mL^{-1} (in urine), and from 1.4 ng mL^{-1} to 20.4 ng mL^{-1} (in serum),	Wyns et al. (2008)

					0.6 mL min^{-1}), 5.5–13 min: 45–100% **B** (flow: 0.6 mL min^{-1}), 13–16 min: isocratic 100% **B** (flow: 0.8 mL min^{-1}), 16–16.10 min: 100–35% **B** (flow: 0.6 mL min^{-1}) and 3.9 min for re-equilibration of the column			recovery: > 67% (in urine), > 83% (in serum)	
Dried leaves of red clover or soy bits	Daidzin, genistin, glycitin, ononin, sissotrin, daidzein, glycitein, genistein, formononetinand biochanin A	Supercritical fluid extraction	Flavone	HPLC-UV/MS	Atlantis^{TM} dC18 high-speed reversed phase chromatographic column (2062.1 mm, 3 lm particle size), mobile phase A: 0.3% (v/v) acetic acid and mobile Phase **B**: acetonitrile, linear gradient: 13– 20% **B** (v/v) from 0–1.25 min, up to 30% **B** to 1.61 min, up to 35% **B** to 2.0 min, up to 40% **B** to 2.25 min, up to 50% **B** to 2.57 min, and followed by negative gradient to 13% **B** to 3.86 min, flow rate: 0.68 mL min^{-1}, column temperature: 36 °C	UV-PDA, MSD detector, SQ-MS, ESI, NI, SIM	4 min	Accuracy : within 5%, precision: 4–7%, LOD for isoflavone glycosides :1.3–3.6 fmol and 0.2–1.0 fmol for aglycones, recoveries: 96.64–102.34% for all analytes	Klejdus *et al.* (2005a)

methods coupled with various MS detectors employed for the rapid and efficient analysis of phytoestrogens.

14.3 Conclusions

A large amount of literature has been published confirming the health-related benefits of dietary isoflavones. These phytoestrogenic compounds have shown to exhibit potent protective effects against cardiovascular diseases, hormone-dependent breast and prostate cancers, and conditions associated with osteoporosis and menopause. They are increasingly being investigated as a safe and effective alternative for HRT. Large-scale clinical trials are required to be conducted in order to establish the potential usefulness of phytoestrogens in the treatment of various disease conditions. This requires the development of rapid, validated assays for the characterization and quantification of the phytoestrogenic precursors and their metabolites in biological matrices.

HPLC coupled with UV and MS detection has emerged as the technique of choice to achieve high-throughput bioanalysis of these phytoestrogens in food products as well as in clinical samples. LC-MS/MS offers high sensitivity, specificity, simple sample preparation and rapid analysis of the targeted analytes. Additionally, with the development of innovative techniques such as on-line and off-line 96-well format sample preparation, use of high flow rates and fast gradients, development and implementation of monolithic columns, application of column switching to MS, *etc.* have significantly reduced the overall analysis time. Development and proper validation of such high-throughput analytical methods will surely strengthen the therapeutic role of phytoestrogens in current medical practices.

Summary Points

- This Chapter focuses on various high-throughput analytical/bioanalytical HPLC-UV/MS methods reported in literature for the separation, identification and quantitation of dietary isoflavones in food products as well in biological matrices.
- Isoflavones are plant secondary metabolites exhibiting a variety of health benefits in humans.
- HPLC coupled with UV or MS offers rapid, reliable and reproducible determination of isoflavones in biological fluids due to their high sensitivity and selectivity.
- The recent high-throughput analytical/bioanalytical methods developed for the quantitation of dietary phytoestrogens are accurate, efficient, precise, reliable and sensitive.
- These developed and validated methods are well suited for analyzing large number of samples from epidemiological studies which will help in establishing the therapeutic potential of these phytoestrogenic compounds.

Key Facts

- Mass spectrometry has effectively being utilized in every aspect of research and development of medicines.
- The mass spectrometer consists of five components: sample inlet, ion source, mass analyser, detector and data system.
- The sample inlet introduces sample to the low pressure region of the mass spectrometer, *i.e.* to the ion source.
- The ion source transforms the sample molecules into gas phase ions. Most commonly used ionization techniques are: electron ionization, chemical ionization, electrospray ionization (ESI), atmospheric pressure chemical ionization (APCI), matrix-assisted laser desorption ionization (MALDI) and fast atom bombardment.
- The mass analyser separates the sample ions based on their mass-to-charge (m/z) ratio. Commonly employed mass analysers are: magnetic sector mass analyser, quadrupole mass analysers and time-of-flight (TOF) mass analysers.
- The detector counts the ions of interest and the signal is recorded and processed by the data system. The output is in the form of a mass spectrum, which is a graph of the number of ions detected as a function of their m/z ratio.

Definitions of Words and Terms

Accuracy: This is defined as how close the determined value is to the normal or true value under the given set of conditions.

Biological matrix: Any material of biological origin whose sample can be collected and processed in a reproducible manner, *e.g.* blood, urine, faeces, saliva, tears, sputum and various tissues.

Calibration range: This is a range of concentrations including the upper limit of quantification as well as the lower limit of quantification (LLOQ), which can be reliably quantified with acceptable accuracy and precision.

Internal standard: This a chemical moiety similar in structure to the analyte of interest or stable labelled compound that is included in the analysis to account for the loss of the analyte of interest during sample preparation *etc.* and minimize the errors.

Limit of detection (LOD): The lowest concentration of the analyte that can be clearly distinguished from the instrument base line noise.

Lower limit of quantitation (LLOQ): This is defined as the lowest amount of analyte which can be quantified with acceptable accuracy and precision.

Method: A detailed and comprehensive explanation of all the procedures used in the analysis of target analyte.

Precision: This determines the degree of closeness of a set of values obtained from homogenous sample.

Selectivity: This defines the ability of the developed method to identify and measure analytes of interest in presence of other analytes such as impurities *etc.*
Validation: The establishment of all validation parameters so that target analytes can be identified and quantified with greater reliability.

List of Abbreviations

ACN	acetonitrile
ADME	absorption, distribution, metabolism and excretion
APCI	atmospheric pressure chemical ionization
BCA	biochanin A
CYP	cytochrome
DAD	diode array detector
ESI	electrospray ionization
FVB	Friend leukemia virus B
GC	gas chromatography
HRT	hormone replacement therapy
HPLC	high-performance liquid chromatography
LC	liquid chromatography
LOQ	limit of quantitation
LLOQ	lower limit of quantitation
MRM	multiple reaction monitoring
MS	mass spectrometry
MS/MS	tandem mass spectrometry
PDA	photodiode array
RP	reverse phase
RSD	relative standard deviation
SIM	selected ion monitoring
SPE	solid phase extraction
U-HPLC	ultrafast-HPLC
UV	ultraviolet
VIS	visible

References

Adlercreutz, H., Mousavi, Y., Clark, J., Höckerstedt, K., Hämäläinen, E., Wähälä, K., Mäkelä, T., and Hase, R., 1992. Dietary phytoestrogens and cancer: *in vitro* and *in vivo* studies. *The Journal of Steroid Biochemistry and Molecular Biology*. 41: 331–337.

Adlercreutz, H., Mazur, W., Stumpf, K., Kilkkinen, A., Pietinen, P., Hulten, K., and Hallmans, G., 2000. Food containing phytoestrogens, and breast cancer. *Biofactors*. 12: 89–93.

Arora, A., Nair, M.G., and Strasburg, G.M., 1998. Antioxidant activities of isoflavones and their biological metabolites in a liposomal system. *Archives of Biochemistry and Biophysics*. 356: 133–141.

Baranowska, I., Magiera, S., and Baranowski, J., 2011. UHPLC method for the simultaneous determination of beta-blockers, isoflavones and their metabolites in human urine. *Journal of Chromatography B. Analytical Technologies in the Biomedical and Life Sciences.* 879: 615–626.

Barnes, S., 1995. Effect of genistein on *in vitro* and *in vivo* models of cancer. *Journal of Nutrition.* 125: 777S–783S.

Berna, M.J., Ackermann, B.L., and Murphy, A.T., 2004. High-throughput chromatographic approaches to liquid chromatographic/tandem mass spectrometric bioanalysis to support drug discovery and development. *Analytica Chimica Acta.* 509: 1–9.

Fotsis, T., Pepper, M., Adlercreutz, H., Hase, T., Montesano, R., and Schweigerer, L., 1995. Genistein, a dietary ingested isoflavonoid, inhibits cell proliferation and *in vitro* angiogenesis. *Journal of Nutrition.* 125: 790S–797S.

Gautam, A.K., Bhargavan, B., Tyagi, A.M., Srivastava, K., Yadav, D.K., Kumar, M., Singh, A., Mishra, J.S., Singh, A.B., Sanyal, S., Maurya, R., Manickavasagam, L., Singh, S.P., Wahajuddin, W., Jain, G.K., Chattopadhyay, N., and Singh, D., 2011. Differential effects of formononetin and cladrin on osteoblast function, peak bone mass achievement and bio-availability in rats. *Journal of Nutrional Biochemistry.* 22: 318–327.

Gotoh, T., Yamada, K., Yin, H., Ito, A., Kataoka, T., and Dohi, K., 1998. Chemoprevention of N-nitroso-N-methylurea-induced rat mammary carcinogenesis by soy foods or biochanin A. *Cancer Science.* 89: 137–142.

Grace, P.B., Mistry, N.S., Carter, M.H., Leathem, A.J.C., and Teale, P., 2007. High throughput quantification of phytoestrogens in human urine and serum using liquid chromatography/tandem mass spectrometry (LC-MS/MS). *Journal of Chromatography B.* 853: 138–146.

Hu, M., Krausz, K., Chen, J., Ge, X., Li, J., Gelboin, H.L., and Gonzalez, F. J., 2003. Identification of CYP1A2 as the main isoform for the phase I hydroxylated metabolism of genistein and a prodrug converting enzyme of methylated isoflavones. *Drug Metabolism and Disposition.* 31: 924.

Kao, T.H., Wu, W.M., Hung, C.F., Wu, W.B., and Chen, B.H., 2007. Anti-inflammatory effects of isoflavone powder produced from soybean cake. *Journal of Agricultural and Food Chemistry.* 55: 11068–11079.

Kaul, T.N., Middleton, Jr, E., and Ogra, P.L., 1985. Antiviral effect of flavonoids on human viruses. *Journal of Medical Virology.* 15: 71–79.

Klejdus, B., Lojková, L., Lapčik, O., Koblovská, R., Moravcová, J., and Kubáň, V., 2005a. Supercritical fluid extraction of isoflavones from biological samples with ultra-fast high-performance liquid chromatography/mass spectrometry. *Journal of Separation Science.* 28: 1334–1346.

Klejdus, B., Mikelová, R., Petrlova, J., Potesil, D., Adam, V., Stiborova, M., Hodek, P., Vacek, J., Kizek, R., and Kubán, V., 2005b. Evaluation of isoflavone aglycon and glycoside distribution in soy plants and soybeans by fast column high-performance liquid chromatography coupled with a diode-array detector. *Journal of Agricultural and Food Chemistry.* 53: 5848–5852.

Klejdus, B., Mikelova, R., Petrlova, J., Potesil, D., Adam, V., Stiborová, M., Hodek, P., Vacek, J., Kizek, R., and Kuban, V., 2005c. Determination of isoflavones in soy bits by fast column high-performance liquid chromatography coupled with UV-visible diode-array detection. *Journal of Chromatography A.* 1084: 71–79.

Klejdus, B., Vacek, J., Benesova, L., Kopecky, J., Lapcik, O., and Kuban, V., 2007. Rapid-resolution HPLC with spectrometric detection for the determination and identification of isoflavones in soy preparations and plant extracts. *Analytical and Bioanalytical Chemistry.* 389: 2277–2285.

Klejdus, B., Vacek, J., Lojková, L., Benesova, L., and Kuban, V., 2008. Ultrahigh-pressure liquid chromatography of isoflavones and phenolic acids on different stationary phases. *Journal of Chromatography A.* 1195: 52–59.

Klejdus, B., Lojková, L., Plaza, M., Snoblova, M., and Sterbova, D., 2010. Hyphenated technique for the extraction and determination of isoflavones in algae: ultrasound-assisted supercritical fluid extraction followed by fast chromatography with tandem mass spectrometry. *Journal of Chromatography A.* 1217: 7956–7965.

Kuiper, G.G.J.M., and Gustafsson, J.A., 1997. The novel estrogen receptor-β subtype: potential role in the cell-and promoter-specific actions of estrogens and anti-estrogens. *FEBS Letters.* 410: 87–90.

Kyranos, J.N., Cai, H., Wei, D., and Goetzinger, W.K., 2001. High-throughput high-performance liquid chromatography/mass spectrometry for modern drug discovery. *Current Opinion in Biotechnology.* 12: 105–111.

Lissin, L.W., and Cooke, J.P., 2000. Phytoestrogens and cardiovascular health. *Journal of the American College of Cardiology.* 35: 1403–1410.

Lo, F.H., Mak, N.K., and Leung, K.N., 2007. Studies on the anti-tumor activities of the soy isoflavone daidzein on murine neuroblastoma cells. *Biomedicine and Pharmacotherapy.* 61: 591–595.

Lovati, M.R., Manzoni, C., Canavesi, A., Sirtori, M., Vaccarino, V., Marchi, M., Gaddi, G., and Sirtori, C.R., 1987. Soybean protein diet increases low density lipoprotein receptor activity in mononuclear cells from hypercholesterolemic patients. *Journal of Clinical Investation.* 80: 1498–502.

Mallis, L.M., Sarkahian, A.B., Harris, H.A., Zhang, M.Y., and McConnell, O.J., 2003. Determination of rat oral bioavailability of soy-derived phytoestrogens using an automated on-column extraction procedure and electrospray tandem mass spectrometry. *Journal of Chromatography B.* 796: 71–86.

Nestel, P., Cehun, M., Chronopoulos, A., DaSilva, L., Teede, H., and McGrath, B., 2004. A biochanin-enriched isoflavone from red clover lowers LDL cholesterol in men. *European Journal of Clinical Nutrition.* 58: 403–408.

Prasain, J.K., Arabshahi, A., Moore, D., Greendale, G.A., Wyss, J.M., and Barnes, S., 2010. Simultaneous determination of 11 phytoestrogens in human serum using a 2min liquid chromatography/tandem mass spectrometry method. *Journal of Chromatography B.* 878: 994–1002.

Ramos, G.P., Dias, P.M.B., Morais, C.B., Fröehlich, P.E., Dall'Agnol, M., and Zuanazzi, J.A.S., 2008. LC determination of four isoflavone aglycones in red clover (*Trifolium pratense* L.). *Chromatographia.* 67: 125–129.

Rostagno, M.A., Palma, M., and Barroso, C.G., 2007. Fast analysis of soy isoflavones by high-performance liquid chromatography with monolithic columns. *Analytica Chimica Acta*. 582: 243–249.

Schnaper, H.W., McGowan, K.A., Kim-Schulze, S., and Cid, M.C., 1996. Oestrogen and endothelial cell angiogenic activity. *Clinical and Experimental Pharmacology and Physiology*. 23: 247–250.

Singh, S.P., Wahajuddin Ali, M.M., and Jain, G.K., 2010a. High-throughput quantification of isoflavones, biochanin A and genistein, and their conjugates in female rat plasma using LC-ESI-MS/MS: application in pharmacokinetic study. *Journal of Separation Science*. 33: 3326–3334.

Singh, S.P., Wahajuddin, D.K., Yadav, P., Maurya, R., and Jain, G.K., 2010b. Quantitative determination of formononetin and its metabolite in rat plasma after intravenous bolus administration by HPLC coupled with tandem mass spectrometry. *Journal of Chromatography B. Analytical Technologies in the Biomedical and Life Sciences*. 878: 391–397.

Tolleson, W.H., Doerge, D.R., Churchwell, M.I., Marques, M.M., and Roberts, D.W., 2002. Metabolism of biochanin A and formononetin by human liver microsomes *in vitro*. *Journal of Agricultural and Food Chemistry*. 50: 4783–4790.

Twaddle, N.C., Churchwell, M.I., and Doerge, D.R., 2002. High-throughput quantification of soy isoflavones in human and rodent blood using liquid chromatography with electrospray mass spectrometry and tandem mass spectrometry detection. *Journal of Chromatography B*. 777: 139–145.

Wang, S.W.J., Chen, Y., Joseph, T., and Hu, M., 2008. Variable isoflavone content of red clover products affects intestinal disposition of biochanin A, formononetin, genistein, and daidzein. *The Journal of Alternative and Complementary Medicine*. 14: 287–297.

Wyns, C., Bolca, S., De Keukeleire, D., and Heyerick, A., 2008. Development of a high-throughput LC-APCI-MS method for the determination of 13 phytoestrogens (including gut microbial metabolites) in human urine and serum. *Planta Medica*. 74: 1100.

Yamaguchi, M., 2002. Isoflavone and bone metabolism: Its cellular mechanism and preventive role in bone loss. *Journal of Health Science*. 48: 209–222.

Yang, Z., Zhu, W., Gao, S., Xu, H., Wu, B., Kulkarni, K., Singh, R., Tang, L., and Hu, M., 2010. Simultaneous determination of genistein and its four phase II metabolites in blood by a sensitive and robust UPLC-MS/MS method: Application to an oral bioavailability study of genistein in mice. *Journal of Pharmaceutical and Biomedical Analysis*. 53: 81–89.

CHAPTER 15
Methods for Isoflavones: A Focus on Beverage Analysis

RITA C. ALVES*[a,b] AND M. BEATRIZ P. P. OLIVEIRA[b]

[a] REQUIMTE, Instituto Superior de Engenharia do Porto, Instituto Politécnico do Porto, Rua Dr. António Bernardino de Almeida, n°431, 4200-072 Porto, Portugal; [b] REQUIMTE, Departamento de Ciências Químicas, Faculdade de Farmácia, Universidade do Porto, Rua de Jorge Viterbo Ferreira n.° 228, 4050-313 Porto, Portugal
*Email: rita.c.alves@gmail.com

15.1 Introduction

Isoflavones have been a focus of special attention, mainly concerning their antioxidant and weak estrogenic properties. They are estrogen-like molecules with a common 3-phenylchromen-4-one core structure and different substituents (methoxy or hydroxyl groups) (Delmonte et al. 2006; Mahesha et al. 2006). Genistein [5,7-dihydroxy-3-(4-hydroxyphenyl], daidzein [7-hydroxy-3-(4-hydroxyphenyl)chromen-4-one], glycitein [7-hydroxy-3-(4-hydroxyphenyl)-6-methoxychromen-4-one], biochanin A [5,7-dihydroxy-3-(4-methoxyphenyl)chromen-4-one] and formononetin [7-hydroxy-3-(4-methoxyphenyl)chromen-4-one] are the most abundant and studied isoflavones.

In plants, the 6″-O-malonyl-7-O-β-D-glycosides are the prevailing forms. These might be converted during food processing or sample analysis into 6″-O-acetyl-7-O-β-D-glycosides, along with 7-O-β-D-glycosides and free isoflavones (or aglycones), in this order (Delmonte et al. 2006).

Food and Nutritional Components in Focus No. 5
Isoflavones: Chemistry, Analysis, Function and Effects
Edited by Victor R Preedy
© The Royal Society of Chemistry 2013
Published by the Royal Society of Chemistry, www.rsc.org

This complex group of compounds is described mainly in soy and soy products (Setchell and Cassidy 1999), these being one of the major sources of isoflavones, especially in Eastern populations, where their consumption is high. However, isoflavones are also present (although in lower amounts) in other vegetables, such as broccoli, cauliflower and apple (Mazur 1998). Beverages are also an important source of isoflavones, especially the soy-based ones. Coffee, milk or fruit juices, if moderately consumed, can also contribute significantly to the intake of these compounds, although to a lower extent due to their comparatively small contents (Alves et al. 2010; Kuhnle et al. 2008a; Thompson et al. 2006).

Isoflavone contents may vary naturally (with species and harvesting conditions) or as a result of food processing. Therefore, the use of accurate and precise analytical methods is very important in order to transmit correctly labelled information about the levels of isoflavones in each final product to consumers.

The aim of this Chapter is to present a review about the different methodologies published in literature for isoflavone analysis in beverages.

A wide range of methods, with distinct extraction procedures or even different chromatographic techniques, are described in literature. Therefore, the comparison of results for similar samples between different laboratories is a hard challenge.

Table 15.1 summarizes the methods used in the analysis of isoflavone content in beverages to the date.

15.2 Methods of Extraction

The extraction is a challenging step to obtain an accurate determination of isoflavones. The chosen process will depend on several factors, namely, the aim of the study, type of matrix and isoflavone forms naturally occurring on it, availability of standards, among others. Indeed, according to the extraction method, the analysed compounds might vary. If the aim is to study the native forms of isoflavones in a matrix, a direct extraction has to be performed. In this case, the extract should mimic the original composition and profile as much as possible (Rostagno et al. 2009). If the preference is to analyse the total aglycones (the bioactive form), hydrolysis of the conjugates is necessary. The differences between these procedures will be discussed below.

15.2.1 Direct Extraction

The most commonly used direct extraction methods consist on stirring or shaking a beverage aliquot (liquid or lyophilized) together with an organic solvent (acidified or mixed with water) at room or higher temperatures, followed by centrifugation/filtration and analysis of the supernatant/filtrate. Several solvents have been employed for this procedure, namely, 80% aqueous methanol (Coward et al. 1998; Nakamura et al. 2000), 50% and 80% aqueous acetonitrile (ACN) (Schwartz and Sontag 2009), ACN/HCl (Jackson et al. 2002; Wang et al. 1990), 80% aqueous ethanol (Genovese and Lajolo 2002), 50% aqueous ethanol (Rostagno et al. 2007) and acetone/HCl (Kao et al. 2004).

Coward et al. (1998) compared different temperatures of extraction and present the extraction with 80% methanol at 4 °C for 2–4 h as the best method to preserve original isoflavone forms of soy and soy products (including soy milk).

Wang et al. (1990) found that the system using ACN/HCl/water (80:0.5:19.5) resulted in a fast settling of suspending sample particles and there were less interfering impurities in the extract when compared with other mixtures such as 80% ethanol/HCl/water or 80% methanol/HCl/water. Samples of soy drink were, then, shaken for 1 min with 1 M HCl/ACN (1:4), settled for a few minutes and filtered for high-performance liquid chromatography (HPLC) analysis.

In some situations, the use of sonication instead of stirring or shaking is a preferred process, since the result is usually a faster extraction. That is the example of the method used by Rostagno et al. (2007) for native isoflavones analysis in soy beverages blended with fruit juices: 50% ethanol in an ultrasound bath at 45 °C for 20 min. Also, in a previous study, Nakamura et al. (2000) used 80% methanol and sonication, followed by 24 h at room temperature, to extract isoflavones from soy milk samples.

With regards to the influence of sample/solvent ratio in isoflavone extraction, Rostagno et al. (2007) reported very interesting results: gradually reducing the ratios from 5:1 to 0.2:1 (v/v) increased the amount of all of the isoflavones detected in the sample. Moreover, the authors found that for the smaller ratio, ethanol extracted approximately 15% more of isoflavones than methanol. Also, 45 °C was described as the best temperature of extraction, with higher ones leading to degradation of malonyl glycosides and lower temperatures producing an incomplete extraction of the compounds. Additionally, an extended extraction time could be a source of variation and malonyl glycoside degradation (Rostagno et al. 2007).

Kao et al. (2004) used an extraction system of acetone/0.1M HCl (5:1), shaking a soy milk sample and the solvent for 2 h at room temperature. However, with this method, slightly lower recoveries were obtained when comparing with the remaining ones (Table 15.1).

Genovese and Lajolo (2002) used a solid-phase extraction method to clean-up and concentrate the soy beverage aliquots. Polyamide columns (1 g/6 ml, CC 6, Macherey-Nagel, Germany) were previously conditioned with methanol and water, and, afterwards, liquid samples were directly apllied (or after centrifugation if they contained suspended particles). Interferents were washed out with water and isoflavones eluted with methanol/ammonia at 99.5:0.5.

Schwartz and Sontag (2009) compared different sample preparations for isoflavones analysis in foodstuffs, including soy drinks. With regards to direct extraction for quantification of native isoflavones, the best conditions obtained by these authors were as follows: use of an additive (zinc sulfate heptahydrate) to obtain clear extracts, sonication (60 °C, 20 min), first extraction with 50% aqueous ACN and, then, with 80% aqueous ACN.

Table 15.1 Methods for isoflavones quantification in beverages. 6OAcDaidzin, 6″-O-acetyl-daidzin; 6OAcGenistin, 6″-O-acetyl-genistin; 6OAcGlycitin, 6″-O-acetyl-glycitin; 6OMalDaidzin, 6″-O-malonyl-daidzin; 6OMalGenistin, 6″-O-malonyl-genistin; 6OMalGlycitin, 6″-O-malonyl-glycitin; ACN, acetonitrile; BHT, butylated hydroxytoluene; CV, coefficient of variation; ESI-MS, electrospray ionization mass spectrometry; GC, gas chromatography; HPLC, high-performance liquid chromatography; MS, mass spectrometry; MS/MS, tandem MS; PAD, photodiode array detector; SPE, solid-phase extraction; TFA, trifluoroacetic acid.

Beverage	Isoflavones	Isoflavones extraction	Internal standard	Chromatographic separation and detection	Validation data	Reference
Soy drink	Free: Daidzein Genistein Total: Daidzein Genistein	Free isoflavones: Organic extraction (ACN) + HCl Total isoflavones: 1) Acid hydrolysis (HCl, 98–100 °C, 2 h) 2) Organic extraction (ACN)	No	HPLC Column: μ-Bondapak C18, 10 μm, 3.9 mm i.d. ×30 cm from Waters Associates (Milford, MA) Mobile phase: methanol/1 mM ammonium acetate (6:4) Flow rate: 1 mL min^{-1} Detector: UV/FL; λ = 254 nm; λ$_{exc}$ = 365 nm and λ$_{em}$ = 418 nm	(with other soy products) Recoveries: 79–112% Precision: Free daidzein: 2–9% Limit of detection: 2 μg ml^{-1}	Wang et al. (1990)
Soy milk	Daidzein Genistein Glycitein Daidzin Genistin Glycitin 6OAcDaidzin 6OAcGenistin 6OAcGlycitin 6OMalDaidzin 6OMalGenistin 6OMalGlycitin	1) Extraction with 80% methanol	Fluorescein	1) HPLC Column: Brownlee Aquapore C8, 4.6 mm i.d. ×22 cm, from Rainin (Woburn, MA) Mobile phases: A) 5% ACN with 0.1% TFA; B) 95% ACN with 0.1% TFA Gradient of 0–30% B over 30 min. Flow rate: 1.5 ml min^{-1} Detection: UV; λ = 262 nm 2) HPLC Column: Aquapore C8, 4.6 mm i.d. ×10 cm	(Data not shown)	Coward et al. (1998)

Table 15.1 (Continued)

Beverage	Isoflavones	Isoflavones extraction	Internal standard	Chromatographic separation and detection	Validation data	Reference
Tea Coffee	Formononetin Biochanin A Daidzein Genistein	1) Enzymatic hydrolysis (Helix pomatia) 2) Ether extraction	Deuterium-labelled isoflavones	Mobile phases: ACN with ammonium acetate; linear gradient (0–50%) Flow rate: 1.0 ml min^{-1} Detection: MS, $[M+H]^+$ 1) Ion-exchange chromatography: -DEAE-Sephadex OH$^-$ -QAE-Sephadex Ac$^-$ 2) Derivatization (to tri-methylsilyl ethers) 3) Isotope dilution capillary GC-MS	Recoveries: ~100% Sensitivity limit: 0.02–0.03 mg kg^{-1} CV (method): 3.1–9.6%	Mazur et al. (1998)
Soy milk	Free: Daidzein Genistein Glycitein Daidzin Genistin Glycitin Total: Daidzein Genistein Glycitein	1) Free isoflavones: 1.1) sonication in 80% methanol (30 min) 1.2) extraction for 24 h (room temperature) Total isoflavones: Acid hydrolysis (HCl) in presence of BHT: 1.1) sonication (30 min) 1.2) hydrolysis in a boiling water bath by reflux (3 h) 2) Free and total isoflavones: SPE: Sep-pak plus C18 from Waters Corp. (Milford, MA, USA)	Flavone	HPLC Column (at 35 °C): STR ODS-II, 5 μm, 4.6 mm i.d. ×30 cm from Shinwa-Kako, Inc. (Kyoto, Japan) Mobile phase: (A) water/phosphoric acid, 1000:1 (v/v) (B) ACN/water/phosphoric acid, 800:200:1 (v/v/v); Gradient program: (B) 10% (0 min) → 80% (50 min), linear gradient Detector: PDA λ = 260 nm for daidzein, daidzin, genistein, genistin, glycitein, and glycitin λ = 280 nm for flavone	Mean recoveries: Free: 89–104% Total: 88–106% CV (retention time) = 0.22–0.43% CV (peak area) = 0.90–3.28% Detection limit: 0.109–0.660 μg ml^{-1}	Nakamura et al. (2000)

Sample	Analytes	Internal standard	Extraction	Analysis	Performance	Reference
Soy milk Coffee Licorice tea	Daidzein Genistein Formononetin Biochanin A	4-Methylumbelliferone	1) Extraction with 80% aqueous methanol 2) Enzymatic hydrolysis with β-glycosidase (pH 5.0, 37°C, overnight) 3) Addition of glacial acetic acid 4) Extraction with ether	HPLC Column: C8 HPLC column, 300 Å, 4.6 mm i.d. × 10 cm from Rainin Instruments (Woburn, MA, USA) Mobile phase: 30% ACN in 10 mM ammonium acetate (isocratic) Flow rate: 1.0 ml min^{-1} Detection: MS m/z, [M-H]$^-$: Daidzein: 253, 132 Genistein: 269, 133 Formononetin: 267, 208 Biochanin A: 283, 239	Limit of detection: 25 μg/100 g of sample	Horn-Ross et al. (2000)
Soy beverage	Daidzein Genistein Glycitein Daidzin Genistin Glycitin 6OAcDaidzin 6OAcGenistin 6OAcGlycitin 6OMalDaidzin 6OMalGenistin 6OMalGlycitin	No	Extraction with ACN/ 0.1 M HCl (5:1), overnight, room temperature	HPLC Column: YMC-pack ODS-AM-303 column, 5 μm, 4.6 mm i.d. × 25 cm, from YMC Inc. (Wilmington, NC, USA) Mobile phases: (A) water with 0.1% of acetic acid (B) ACN with 0.1% of acetic acid Solvent gradient: 15 to 25% B (35 min), to 26.5% B (12 min), to 50% (50 s), held for 14.50 min. Flow rate: 1 ml/min up to 48 min, increased to 1.3 ml min^{-1} from 48.5 till 63 min Detector: PAD λ = 254 nm	Recoveries (using a biochanin A standard): 80–100% Limit of detection: Daidzein: 185 ng ml^{-1} Genistein: 100 ng ml^{-1}	Jackson et al. (2002)

Table 15.1 (Continued)

Beverage	Isoflavones	Isoflavones extraction	Internal standard	Chromatographic separation and detection	Validation data	Reference
Soy-based Beverages	Daidzein Genistein Glycitein Daidzin Genistin Glycitin 6OAcDaidzin 6OAcGenistin 6OAcGlycitin 6OMalDaidzin 6OMalGenistin 6OMalGlycitin	SPE: polyamide columns (1 g/6 ml, CC 6, Macherey-Nagel, Germany)	No	HPLC Column: C18 Nova Pak (4.6 mm i.d. ×30 cm) from Waters (Milford, MA, USA) Detection: PDA	(Data not shown)	Genovese and Lajolo (2002)
Soy milk	Daidzein Genistein Glycitein Daidzin Genistin Glycitin 6OAcDaidzin 6OAcGenistin 6OAcGlycitin 6OMalDaidzin 6OMalGenistin 6OMalGlycitin	1) Extraction with acetone: 0.1 M HCl (5:1, v/v), at room temperature for 2 h	Formononetin	HPLC Column: Vydac 201TP54 C18 column, 5 μm (4.6 mm i.d. ×25 cm) was from Vydac Co. (Hesperia, CA, USA) Mobile phases: (A) ACN (B) water Gradient: 0 min, 8% A; 10% A in 2 min, 12% A in 3 min, 22% A in 10 min, 23% A in 11 min, 35% A in 12 min, 50% A in 13 min, maintained for 3 min and returned to the starting condition of 8% A in 20 min Flow rate: 2.0 ml min^{-1} Detection: UV, λ = 262 nm	Recoveries: 74–90%	Kao et al. (2004)

Sample	Analytes	Internal standard	Sample preparation	Analytical method	Validation	Reference
Nutritional beverage with soy	Daidzein Glycitein Genistein	4-Methylumbelliferone sulfate, formononetin, and flavone	1) Extraction with 80% aqueous methanol 2) Enzymatic hydrolysis (overnight, 39 °C): β-glycosidase (pH 5) and arylsulfatase 3) Extraction with ether	HPLC Column: Hydrobond PS C18 analytical column, 5 μm, (3.0 mm i.d.×100 cm) from MacMod Analytical Inc (Chadds Ford, PA) Mobile phases: (A) methanol (B) water Solvent gradient: 25–95% A, linearly in 30 min, at 95% A for 4.5 min, 95–25% A linearly in 0.5 min, and 25% for 5 min Flow rate: 0.2 ml min^{-1} Detector: PAD (for >0.04 mg/100g), λ = 260 nm, and ESI-MS (for <0.04 mg/100g)	Limits of detection: Daidzein: 0.014 mg/100 g Glycitein: 0.014 mg/100 g Genistein: 0.015 mg/100 g Inter-assay CV: 3.0–12%	Umphress et al. (2005)
Soy milk Soy yogurt Coffee Tea Fruit juices Alcoholic beverages	Daidzein Genistein Glycitein Formononetin	5α-Androstane-3 β, 17β-diol	1) Extraction with 70% methanol (60–70 °C, 2 h) 2) SPE: Octadecyl C18/14%, 200 mg/3 ml from Applied Separations (Allentown, PA, USA) 3) Enzymatic hydrolysis with *Helix pomatia* (37 °C, overnight) 4) SPE again	GC Column: (5%-phenyl)-methylpolysiloxane capillary column, 0.25 μm (0.12 mm i.d. × 25 m) from Agilent Technologies (Wilmington, DE) -injection port temperature: 250 °C -carrier gas: He (1 ml min^{-1}) -oven program temperature: 100 °C held initially for 1 min, increased at 15 °C min^{-1} to 280 °C and held for 17 min at 280 °C Derivatization: Tri-Sil Reagent (0.5 h, 60 °C) Detection: MS ionization by electron ionization (70 eV) m/z	(Data not shown)	Thompson et al. (2006)

Table 15.1 (Continued)

Beverage	Isoflavones	Isoflavones extraction	Internal standard	Chromatographic separation and detection	Validation data	Reference
Soy beverages blended with fruit juices	Daidzein Genistein Glycitein Daidzin Genistin Glycitin 6OAcDaidzin 6OAcGenistin 6OAcGlycitin 6OMalDaidzin 6OMalGenistin 6OMalGlycitin	Extraction with 50% ethanol in an ultrasound bath at 45 °C, 20 min	2,5-Dihydroxy-benzaldehyde	Formononetin: 340.1, 325.1; daidzein: 398.1, 383.1 genistein: 414.1, 399.1; glycitein: 428.1, 398.1 HPLC Column: Chromolith TH Performance RP-18e (monolithic type column) 4.6 mm × 10 cm from Merck (Darm-stadt, Germany) Mobile phase: (A) acidified water (0.1% acetic acid) (B) acidified methanol (0.1% acetic acid) Flow rate: 3.0 ml min^{-1} Gradient: 0 min, 20% B; 3 min, 35% B; 8 min, 35% B; 11 min, 40% B and 15 min, 100% B. Detection: UV, $\lambda = 254$ nm	Intra-day and inter-day precision: <3% Limits of detection: 0.36–0.62 mg L^{-1} Limits of quantification: 1.48–2.05 mg L^{-1} Recoveries: 98.1–102.6%	Rostagno et al. (2007)
Goat's milk Cow's milk Soy milk Coffee Tea Alcoholic beverages Lemon juice	Daidzein Genistein Formononetin Biochanin A Glycitein	1) Extraction with 10% methanol in sodium acetate (0.1%, pH 5) 2) Enzymatic hydrolysis with purified *Helix pomatia* juice (β-glucuronidase), cellulase, and β-glycosidase 3) SPE: Strata C18E cartridges (50 mg ml^{-1}) from Phenomenex (Macclesfield, Cheshire, UK)	$^{13}C_3$-Daidzein $^{13}C_3$-Genistein $^{13}C_3$-Formononetin $^{13}C_3$-Biochanin A $^{13}C_3$-Glycitein	LC Column (at 50 °C): Diphenyl column, 3 μm, 2.0 mm i.d. × 15 cm from Varian Pursuit (Varian, Oxford, Oxfordshire, UK) Mobile phase: (A) 40% ammonium acetate (0.1%, pH 4.8) in methanol (B) 100% methanol Gradient program: 0 min: 100% A, 2 min: 75% A, 9 min: 42% A, 9.01 min: 0% A, 10 min: 0% A, 10.01 min: 100% A, 15 min: 100% A	Intra-day CV: 3–14%, Inter-batch CV: 1–6%. Limit of detection: 1.5 μg/100 g (d.w.)	Kuhnle et al. (2007) Kuhnle et al. (2008a) Kuhnle et al. (2008b)

| Soy drink | Daidzein
Genistein
Glycitein
Daidzin
Genistin
Glycitin
6OAcDaidzin
6OAcGenistin
6OAcGlycitin
6OMalDaidzin
6OMalGenistin
6OMalGlycitin | A) Extraction
  -50% aqueous ACN + 80% aqueous ACN
  -additive: zinc sulfate heptahydrate
  -sonication (60 °C, 20 min)
B) Basic hydrolysis
  -NaOH
C) Enzymatic hydrolysis
  -*Helix pomatia* juice
D) Acid hydrolysis
  -HCl | No | Flow rate: 250 µl min^{-1}
Detection: MS/MS
m/z, [M-H]$^{-}$
Biochanin A: 283, 268; Daidzein: 253, 224; Formononetin: 267,252; Genistein: 269, 133; Glycitein: 283, 268
HPLC
Column (at 30 °C):
ACE 3 C18 column, 3 µm, 2.1 mm i.d. × 10 cm from Advanced Chromatography Technologies (Aberdeen, Scotland, UK)
Mobile phase:
(A) 0.1% glacial acetic acid in water
(B) 0.1% glacial acetic acid in ACN.
Solvent gradient:
For aglycones and glycosides: 0 min: 12% B; 0–12 min: linear increase to 32% B; 12–18 min: linear increase to 60% B; 18–19 min: decrease to 12% B; 19–28 min: re-equilibration.
For all native forms:
0 min: 10% B; 0–15 min: linear increase to 30% B; 15–25 min: linear increase to 60% B; 25–26 min: decrease to 12% B; 26–35 min: reequilibration
Flow rate: 0.25 ml min^{-1}
Detection: UV, λ = 260 nm, and ESI-MS | Limits of quantification: 0.8–1.4 mg/L
Precision: ≤3.8%
Recoveries (for textured vegetable protein): 92–110% | Schwartz and Sontag (2009) |

Table 15.1 (Continued)

Beverage	Isoflavones	Isoflavones extraction	Internal standard	Chromatographic separation and detection	Validation data	Reference
Coffee beverages	Daidzein Genistein Formononetin	Methanolic acid (HCl) hydrolysis (75 °C, 150 min)	2-Methoxyflavone	HPLC Column (at 40 °C): Mediterranea Sea 18 column (5 μm, 4.6 mm i.d. × 15 cm) from Teknokroma (Barcelona, Spain) Mobile phases: (A) formic acid 0.1% (B) ACN Solvent gradient: 0 min 95% A, 15 min 68% A, 18 min 63% A, 26 min 63% A, 26.5 min 50% and 30 min 95%. Flow rate: 1 ml min^{-1} Detector: PDA, $\lambda = 254$ nm	Limits of detection: Daidzein: 4.7 ng ml^{-1} Genistein: 4.5 ng ml^{-1} Formononetin: 8.2 ng ml^{-1} Limits of quantification: Daidzein: 14.1 ng ml^{-1} Genistein: 13.7 ng ml^{-1} Formononetin: 25.0 ng ml^{-1} Intra-day precision: 3.8–4.9% Inter-day precision: 6.7–8.2% Recoveries: 91–94%	Alves et al. (2010)

In a general way, although all these methods were applied in order to evaluate native forms of isoflavones, namely malonyl glycosides, acetyl glycosides, β-glycosides and aglycones (if all present), differences in the procedures will certainly lead to different results and variations on the extraction yields. Moreover, for the same method, performance can vary with the sample being studied. The instability of esterified glycosides in standard and sample solutions (occuring isomerization and deconjugation to glycosides) could also be a serious limitation to this type of extraction (Schwartz and Sontag 2009).

15.2.2 Hydrolysis

To surpass the limitations of direct extraction, several authors opted for a hydrolytic method to convert the unstable malonyl and acetyl derivatives into more stable forms such as β-glycosides and/or aglycones, which are more suitable for analysis. On the other hand, the addition of an additional hydrolytic step can be seen as a disadvantage due to an increase of analysis time and/or possible degradation of compounds. In general, hydrolysis (basic, enzymatic or acid) might be carried out before, during or after the extraction, using different conditions and reagents (Rostagno *et al.* 2009).

15.2.2.1 Basic Hydrolysis

Basic hydrolysis consists in treating the sample with a NaOH solution. It allows the determination of both total glycosides and free aglycones, since it breaks the ester bond of the malonyl and acetyl groups esterified to the alcoholic function of the sugar moieties of isoflavones (Delmonte *et al.* 2006; Schwartz and Sontag 2009). Comparatively to other types of hydrolysis, such as enzymatic or acid, it has not been very used for isoflavones analysis in beverages.

From all the studies depicted in Table 15.1, only Schwartz and Sontag (2009) used basic hydrolysis in a comparative study with other hydrolytic methods. The authors stated that the stability of daizin and glycitin (β-glycosides of daidzein and genistein, respectively) upon basic hydrolysis depended highly on the type and percentage of the organic modifier used ($\sim 100\%$ stability was guaranteed with 80% aqueous methanol). However, Schwartz and Sontag (2009) also showed that total glycosides concentration in soy drinks, obtained by basic hydrolysis after direct extraction by sonication (Section 15.2.1) was 3.3% lower than that determined only by direct extraction. This could be caused by the partial conversion of glycosides into aglycones or by the formation of glucosidic water adducts (observed for $\leq 50\%$ aqueous methanol as organic modifier), due to the high water content of beverages, compared with other solid matrices.

15.2.2.2 Enzymatic Hydrolysis

Enzymatic hydrolysis is the mildest hydrolysis technique (Schwartz and Sontag 2009). It involves the use of an enzyme or a combination of enzymes to release free aglycones. Different enzymes show dissimilar activities, therefore, tests for each matrix should be performed before analysis in order to evaluate if cleavage of glycosides and esterified glycosides is actually efficient.

With regards to beverage analysis, *Helix pomatia* digestive juice (β-glucuronidase/arylsulfatase) (Kuhnle *et al.* 2007; 2008a; 2008b; Mazur *et al.* 1998; Thompson *et al.* 2006; Schwartz and Sontag 2009), arylsulfatase (Umphress *et al.* 2005), β-glycosidase (Horn-Ross *et al.* 2000; Kuhnle *et al.* 2007; 2008a; 2008b; Umphress *et al.* 2005) and cellulase (Kuhnle *et al.* 2007; 2008a; 2008b; Schwartz and Sontag 2009) have been widely used for enzymatic hydrolysis (Table 15.1).

Mazur *et al.* (1998) performed an extraction with ethyl ether after enzymatic hydrolysis to separate the free isoflavones of coffee and tea from lignin glycosides.

Both Horn-Ross *et al.* (2000) and Umpress *et al.* (2005) carried out an extraction with 80% aqueous methanol, prior to enzymatic hydrolysis. After removal of the organic fraction by evaporation, the residual solution was incubated with the enzymes referred to above (overnight, 37 °C). Afterwards, an extraction with ethyl ether was performed to recover free isoflavones; Horn-Ross *et al.* (2000) added a small amount of glacial acetic acid before the ether extraction.

Thompson *et al.* (2006) used 70% methanol to extract native isoflavones followed by a solid-phase extraction in a C18 column. After subjecting the extract to the enzymatic hydrolysis (overnight, 37 °C), a new solid-phase extraction was performed.

Kuhnle *et al.* (2007) compared the activity of several enzymes, individually and in different combinations, on different foods. Higher yields of extraction of free isoflavones were obtained with a mixture of all. Briefly, the authors extracted native isoflavones with 10% methanol in sodium acetate (0.1%, pH 5) by sonication (30 min), and the extract was incubated with purified *H. pomatia* juice, cellulase and β-glycosidase, for 16 h at 37 °C. It is advisable to purify the *H. pomatia* digestive juice through solid-phase extraction before using it to remove natural amounts of isoflavones contained in it, which can affect the sample quantification (Grace and Teale 2006). After enzymatic hydrolysis, Kuhnle *et al.* (2007) also carried out a solid-phase extraction, in a reversed-phase cartridge, for clean-up of the incubated extracts before chromatographic analysis.

Schwartz and Sontag (2009) performed a complete study regarding the efficiency of enzymatic hydrolysis under the following conditions: after extraction with an aqueous organic solvent, after extraction and basic hydrolysis (in this case, only isoflavones glycosides had to be cleaved) and in the presence of the matrix (before any extraction). The results showed that enzymatic hydrolysis after basic hydrolysis was the worst option: it yielded lower concentrations of free isoflavones. On the contrary, enzymatic hydrolysis before or after an extraction step provided similar results, but the last one was significantly faster (there was no need of an overnight incubation), included fewer sample preparation steps and required only one enzyme preparation (*H. pomatia*).

15.2.2.3 Acid Hydrolysis

Acid hydrolysis has been widely used for isoflavone quantification in several matrices, as it is a fast and simple method. It is usually performed by refluxing the sample for 1–3 h in aqueous methanolic or ethanolic solution containing 1–3 M HCl. This process breaks the carbon–oxygen bond between the isoflavone ring and the sugar, converting all of the isoflavones forms into their aglycones. The advantage is to quantify isoflavones linked to sugars other than glucose and also glycosides that are not commercially available. As disadvantages, one might refer the fact that isoflavone derivatives could be incompletely cleaved in mild conditions (≤ 1 M HCl, refluxing for ≤ 2 h) and, in certain circumstances (for example, ≥ 2 M HCl for ≥ 1 h, at 99–100 °C), genistein has been shown to be unstable (Delmonte *et al.* 2006; Schwartz and Sontag 2009; Wang *et al.* 1990).

For total isoflavone quantification in soy drink, Wang *et al.* (1990) used 1 M HCl heated inside a steam bath (98–100 °C) for 2 h, followed by the addition of ACN for a faster settling of suspended sample particles and cleaner extracts.

Nakamura *et al.* (2000) performed an acid hydrolysis in soy milk using 5 ml of sample, 10 ml of 10 M HCl solution and 40 ml of 0.05% butylated hydroxytoluene (BHT) solution as antioxidant. After a 30 min sonication, the mixture was refluxed for 3 h in a boiling water bath. After, the authors used solid-phase extraction (C18 cartridge) to purify the extract.

According to Schwartz and Sontag (2009), for soy drink, aglycone concentrations determined by hydrolysis with 1 M HCl for 2 h were significantly higher than for 1 h.

For coffee beverage analysis, Alves *et al.* (2010) compared enzymatic hydrolysis (with a mixture of cellulose, β-glycosidase and *H. pomatia* juice incubated overnight at 37 °C) with acid hydrolysis (methanol/3.4 M HCl, 1:1, at ~ 45 °C for ~ 3.5 h); aglycone recoveries were 2–4-fold higher in the latter case. This acid hydrolysis, originally applied to soybean hypocotyls by Chiang *et al.* (2001), was then optimized for aglycone analysis in coffee beverages and the optimum conditions described by Alves *et al.* (2010) were as follows: for 1.7 ml of beverage, 1.7 ml of methanol, 670 µl of 10.2 M HCl and 1% BHT (refluxing at 75 °C for 150 min). The replacement of methanol by ethanol was also tested, but methanol provided chromatograms [HPLC/photodiode array detector (PAD)] with less interferents.

15.2.3 Internal Standards

The great majority of the methods described in Table 15.1 include the use of an internal standard. In a general way, an internal standard is used to account for losses during extraction. It should be chemically similar to the coumpounds in study, should not react with any component of the matrix nor originally exist in the sample. Moreover, it should appear separated from the remaining coumpounds in the chromatogram (Rostagno *et al.* 2009).

The internal standards used by the different authors (Table 15.1) varied widely. For liquid chromatography analysis, fluorescein (Coward *et al.* 1998), flavone (Nakamura *et al.* 2000; Umphress *et al.* 2005), 4-methylumbelliferone

(Horn-Ross et al. 2000; Umphress et al. 2005), formononetin (Kao et al. 2004; Umphress et al. 2005), 2,5-dihydroxybenzaldehyde (Rostagno et al. 2007), $^{13}C_3$-isoflavones (Kuhnle et al. 2007; 2008a; 2008b) and 2-methoxyflavone (Alves et al. 2010) were the compounds used. For gas chromatography (GC) analysis, Mazur et al. (1998) employed deuterium-labelled isoflavones as an internal standard, whereas Thompson et al. (2006) used 5-α-androstane-3β,17β-diol.

15.3 Chromatographic Methods

As can be observed in Table 15.1, liquid and gas chromatography are the methods of choice to separate and analyse isoflavones in beverages. From these, HPLC is undoubtly the most used one.

According to Table 15.1 isoflavones can be separated by HPLC using reversed-phase columns and mixtures of water/methanol, water/ACN or only methanol as eluents, sometimes together with aqueous acids or buffers as modifiers, namely, acetic acid (Rostagno et al. 2007), ammonium acetate (Horn-Ross et al. 2000; Kuhnle et al. 2007; 2008a; 2008b; Wang et al. 1990), trifluoroacetic acid (TFA) (Coward et al. 1998), phosphoric acid (Nakamura et al. 2000), acetic acid (Jackson et al. 2002; Schwartz and Sontag 2009) and formic acid (Alves et al. 2010). In the majority of the cases, a gradient program was needed, especially when the aim was to separate all the native isoflavones (malonyl and acetyl glycosides, glycosides and aglycones) (Table 15.1). With regards to detectors, several kinds were employed, including ultraviolet (UV) or PDA (Alves et al. 2010; Coward et al. 1998; Genovese and Lajolo 2002; Jackson et al. 2002; Kao et al. 2004; Nakamura et al. 2002; Rostagno et al. 2007; Schwartz and Sontag 2009; Umphress et al. 2005; Wang et al. 1990), mass spectrometry (MS) or tandem MS (MS/MS) (Coward et al. 1998; Horn-Ross et al. 2000; Kuhnle et al. 2007; 2008a; 2008b; Schwartz and Sontag 2009; Umphress et al. 2005) and fluorescence (Wang et al. 1990).

Some authors (Mazur et al. 1998; Thompson et al. 2006) used GC instead of HPLC. Although very sensitive, GC is a more labourious method, since it requires a derivatization step in order to convert isoflavones into trimethylsilyl ethers (Mazur et al. 1998; Thompson et al. 2006).

In a general way, the methodologies presented in Table 15.1 show good precisions and recoveries, as well as good detection and quantification limits. However, sometimes not all of these validation parameters were described for the beverages analysed. This is of importance, because for different samples the same methodology might have not the same efficiency. Due to the highly variable profile of native isoflavones and the influence of the matrix, in particular, a correct validation procedure is always required.

15.4 Conclusion

A wide range of methods with distinct extraction procedures and different chromatographic techniques are described in literature. Some were adapted from methods previously used for solid samples analysis, whereas other were developed and optimized specially for the study of a specific beverage. Differences consist mainly of the solvents used for extraction, presence or absence of hydrolysis, type of hydrolysis and associated reagents, and chromatographic conditions. Due to these variations, sometimes, the comparison of results obtained by different investigation groups might be complex. Moreover, although essential, it was observed that a correct validation procedure sometimes is lacking, which could lead to the erroneous measurements of isoflavones.

Summary Points

- Isoflavones are antioxidant and weak estrogenic-like compounds.
- Beverages, especially the soy-based ones, can be significant sources of isoflavones.
- In plants, isoflavones can be found as malonyl- and acetyl-glycosides, β-glycosides and aglycones.
- A direct extraction, usually with a mixture of water and an organic solvent (ethanol, methanol or ACN), followed by chromatographic analysis allows the study of the native isoflavones in a sample.
- Hydrolysis converts native forms into aglycones (enzymatic and acid) or β-glycosides and aglycones (basic).
- β-Glycosides and aglycones are more stable than the malonyl and acetyl forms.
- *Helix pomatia* juice (β-glucuronidase/arylsulfatase), cellulase and β-glucosidase, alone or combined, are usually used for enzymatic hydrolysis.
- Acid hydrolysis is fast and simple, but in certain circumstances might lead to genistein degradation.
- HPLC is the most used chromatographic method for isoflavone evaluation in beverages.
- Although not always performed, a complete validation of the method is recommended.

Key Facts

- Sample preparation is a key procedure in modern chemical analysis.
- It is estimated that up to 80% of the work and operating costs in an analytical lab is spent preparing samples for subsequent analysis.
- Sample preparation allows the concentration of the analyte to adequate levels for measurement.
- Sample preparation permits the removal of contaminants and interferring compounds to yield cleaner chromatograms.
- Solid-phase extraction is one of the simplest and most effective methods of sample preparation.

Definitions of Words and Terms

Acid hydrolysis: The process of macromolecular breakdown based on the use of a strong acid (*e.g.* HCl), usually at high temperatures.
Basic hydrolysis: The process of macromolecular breakdown of based on the use of a strong base (*e.g.* NaOH), usually at high temperatures.
Bioactive form: The chemical structure that act on living organisms.
Derivatization: The conversion of a chemical compound into a derivative in order to modify its functionality.
Enzymatic hydrolysis: The process of macromolecular breakdown based on the use of enzymes (*e.g.* cellulase, β-glycosidase, *etc.*).
Lyophilization: The process in which samples are frozen below $-30\,°C$ and the ice formed is removed by sublimation in a temperature-controlled chamber under vaccum.
Method validation: The process used to confirm if an analytical methodology is suitable for its intended use.
Reversed-phase: A chromatographic method with a non-polar stationary phase.
Solid-phase extraction (SPE): A sample preparation technique that uses a solid phase (packed in a cartridge) and a liquid phase to purify and/or concentrate samples.
Sonication: This is the application of ultrasound energy to samples.

List of Abbreviations

ACN	acetonitrile
BHT	butylated hydroxytoluene
GC	gas chromatography
HPLC	high-performance liquid chromatography
MS	mass spectrometry
UV	ultraviolet

Acknowledgements

Rita C. Alves is grateful to the Fundação para a Ciência e a Tecnologia for a post-doctoral research grant (SFRH/BPD/68883/2010) financed by POPH–QREN–Tipologia 4.1–Formação Avançada, subsidized by the Fundo Social Europeu and Ministério da Ciência, Tecnologia e Ensino Superior. This work has been supported by Fundação para a Ciência e a Tecnologia through grant no. PEst-C/EQB/LA0006/2011.

References

Alves, R.C., Almeida, I.M.C., Casal, S., and Oliveira, M.B.P.P., 2010. Method development and validation for isoflavones quantification in coffee. *Food Chemistry*. 122: 914–919.

Chiang, W.-D., Shih, C.-J., and Chu, Y.-H., 2001. Optimization of acid hydrolysis conditions for total isoflavones analysis in soybean hypocotyls by using RSM. *Food Chemistry*. 72: 499–503.

Coward, L., Smith, M., Kirk, M., and Barne, S., 1998. Chemical modification of soy foods during cooking and processing. *American Journal of Clinical Nutrition*. 68: 1486S–1491S.

Delmonte, P., Perry, J., and Rader, J.I., 2006. Determination of isoflavones in dietary supplements containing soy, red clover and kudzu: extraction followed by basic or acid hydrolysis. *Journal of Chromatography A*. 1107: 59–69.

Genovese, M.I., and Lajolo, F.M., 2002. Isoflavones in soy-based foods consumed in Brazil: levels, distribution, and estimated intake. *Journal of Agricultural and Food Chemistry*. 50: 5987–5993.

Grace, P.B., and Teale, P., 2006. Purification of the crude solution from *Helix pomatia* for use as β-glucoronidase and arylsulfatase in phytoestrogen assays. *Journal of Chromatography B*. 832: 158–161.

Horn-Ross, P.L., Barnes, S., Lee, M., Coward, L., Mandel, J.E., John, E.M., Smith, M., Koo, J., John, E.M., and Smith, M., 2000. Assessing phytoestrogen exposure in epidemiologic studies: development of a database (United States). *Cancer Causes and Control*. 11: 289–298.

Jackson, C.-J. C., Dini, J.P., Lavandier, C., Rupasinghe, H.P.V., Faulkner, H., Poysa, V., Buzzell, D., and DeGrandis, S., 2002. Effects of processing on the content and composition of isoflavones during manufacturing of soy beverage and tofu. *Process Biochemistry*. 37: 1117–1123.

Kao, T.H., Lu, Y.F., Hsieh, H.C., and Chen, B.H., 2004. Stability of isoflavone glucosides during processing of soymilk and tofu. *Food Research International*. 37: 891–900.

Kuhnle, G.G., Dell'aquila, C., Low, Y.L., Kussmaul, M., and Bingham, S.A., 2007. Extraction and quantification of phytoestrogens in foods using automated solid-phase extraction and LC/MS/MS. *Analytical Chemistry*. 79: 9234–9239.

Kuhnle, G.G., Dell'Aquila, C., Aspinall, S.M., Runswick, S.A., Mulligan, A.A., and Bingham, S.A., 2008a. Phytoestrogen content of foods of animal origin: dairy products, eggs, meat, fish, and seafood. *Journal of Agricultural and Food Chemistry*. 56: 10099–10104.

Kuhnle, G.G.C., Dell'Aquila, C., Aspinall, S.M., Runswick, S.A., Mulligan, A.A., and Bingham, S.A., 2008b. Phytoestrogen content of beverages, nuts, seeds, and oils. *Journal of Agricultural and Food Chemistry*. 56: 7311–7315.

Mahesha, H.G., Singh, S.A., Srinivasan, N., and Appu Rao, A.G., 2006. A spectroscopic study of the interaction of isoflavones with human serum albumin. *Federation of European Biochemical Societies Journal*. 273: 451–467.

Mazur, W., 1998. Phytoestrogen content in foods. *Baillière Clinical Endocrinology and Metabolism*. 12: 729–742.

Mazur, W.M., Wahala, K., Rasku, S., Salakka, A., Hase, T., and Adlercreutz, H., 1998. Lignan and isoflavonoid concentrations in tea and coffee. *British Journal of Nutrition*. 79: 37–45.

Nakamura, Y., Tsuji, S., and Tonogai, Y., 2000. Determination of the levels of isoflavonoids in soybeans and soy-derived foods and estimation of isoflavonoids in the Japanese daily intake. *Journal of AOAC International*. 83: 635–650.

Rostagno, M.A., Palma, M., and Barroso, C.G., 2007. Ultrasound-assisted extraction of isoflavones from soy beverages blended with fruit juices. *Analytica Chimica Acta*. 597: 265–272.

Rostagno, M.A., Villares, A., Guillamón, E., García-Lafuente, A., and Martínez, J.A., 2009. Sample preparation for the analysis of isoflavones from soybeans and soy foods. *Journal of Chromatography A*. 1216: 2–29.

Schwartz, H., and Sontag, G., 2009. Comparison of sample preparation methods for analysis of isoflavones in foodstuffes. *Analytica Chimica Acta*. 633: 204–215.

Setchell, K.D., and Cassidy, A., 1999. Dietary isoflavones: biological effects and relevance to human health. *The Journal of Nutrition*. 129: 758S–767S.

Thompson, L.U., Boucher, B.A., Liu, Z., Cotterchio, M., and Kreiger, N., 2006. Phytoestrogen content of food consumed in Canada, including isoflavones, lignans and coumestan. *Nutrition and Cancer*. 54: 184–201.

Umphress, S.T., Murphy, S.P., Franke, A.A., Custer, L.J., and Blitz, C.L., 2005. Isoflavone content of foods with soy additives. *Journal of Food Composition and Analysis*. 18: 533–550.

Wang, G., Kuan, S.S., Francis, O.J., Ware, G.M., and Carman, A.S., 1990. A simplified HPLC method for the determination of phytoestrogens in soybean and its processed products. *Journal of Agricultural and Food Chemistry*. 38: 185–190.

CHAPTER 16
The Determination of Isoflavones in Supplemented Foods: An Overview

ALBERTO ZAFRA-GÓMEZ,* SONIA CAPEL-CUEVAS AND NOEMÍ I. DORIVAL-GARCÍA

Department of Analytical Chemistry, University of Granada, Av. Fuentenueva S/N, Granada, Spain, E-18071
*Email: azafra@ugr.es

16.1 Introduction

Over the past few years, the consumption of soybean products has been widely promoted in the media. One main reason that explains this phenomenon is that soybeans, as well as legumes and beans, are important sources of isoflavones, compounds that belong to the family of phytoestrogens. Phytoestrogens show several important health benefits to humans, such as hormonal effects, providing protection against several hormone-dependent diseases, especially different types of cancer (breast, endometrial, prostate, colon, rectum, stomach and lung). Similarly, these substances have important effects on physiological functions related to estrogenic activity, including menopausal symptoms and osteoporosis (Rostagno *et al*. 2009).

The potential benefits of phytoestrogens are giving rise to the marketing of various "health" food supplements and drinks, including tablets of isoflavone extracts sold as a "natural" hormone replacement therapy (Bingham *et al*.

1998). Soy- and soybean-based foodstuffs have become flagship products because they are one of the most important sources of isoflavones for human intake. In the United States, approximately 25% of infant formula sold is based on soy extracts (Cao *et al.* 2009).

Isoflavones belong to the group of flavonoids, a very important member of the family of phytoestrogens as it represents one of largest classes of higher plants with approximately 5000 compounds (Xiao *et al.* 2009), of which 1600 are isoflavonoids (Andersen and Jordheim 2006). Isoflavones are usually found in plants such as glycosides, *i.e.* provided with sugar substituents such as galactose, rhamnose or glucose, or glycoside malonate. Malonates are of great biological interest because plants use this conjugated form to store the less soluble aglicone compounds.

From a chemical point of view, isoflavones are a wide group of structurally related compounds with a chromane skeleton that has a phenyl substituent in the C_3 position as depicted in Figure 16.1. The basic structure of the isoflavone nucleus is composed of two benzene rings linked to a heterocyclic ring. As it is shown in Table 16.1, the principal isoflavone phytoestrogens are daidzein and genistein, in the form of their glucosides, daidzin and genistin, as well as their methyl ether precursors, formononetin and biochanin A (Price and Fenwick 1985).

Although isoflavones have been widely described as health products, some severe criticism has been levelled in relation to their structural and functional

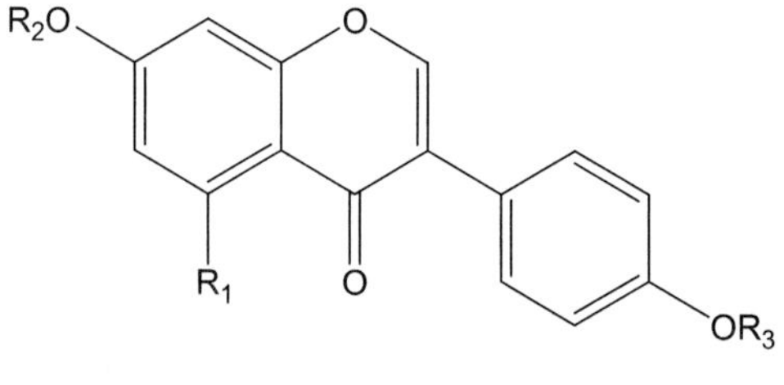

Figure 16.1 Chemical structures of isoflavones. R1, R2 and R3; see Table 16.1. Unpublished.

Table 16.1 Principal isoflavones in supplemented foods. glc, glucoside; MW, molecular weight; R, subtituents of the different isoflavones showed in Figure 16.1.

Name	R_1	R_2	R_3	MW
Daidzein	H	H	H	254
Daidzin	H	7-*O*-β-D-glc	H	416
Genistein	OH	H	H	270
Genistin	OH	7-*O*-β-D-glc	H	432
Formononetin	H	H	CH_3	268
Biochanin A	OH	H	CH_3	284

similarity to 17-β-estradiol and their ability to trigger estrogenic and/or antiestrogenic effects in animals. It has been demonstrated that, depending on the type of estrogen receptor on the cells, isoflavones may reduce or increase the activity of the estrogens. Isoflavones can compete for the same receptor sites, thereby decreasing the health risks related to excess estrogen. If during menopause the body's natural level of estrogen drops, isoflavones can offset this effect by binding to the same receptor and therefore alleviate the symptoms of menopause.

The biological effects of phytoestrogens on reproductive functions have been researched in many animal species. The findings reported seem relatively consistent and explain decreased fertility, and the effects seem to be more prominent in females than males. Potential sites of action include the genital tract, ovaries, pituitary and the central nervous system. However, there is considerable epidemiological evidence to suggest that the consumption of foodstuffs containing phytoestrogens may exert beneficial rather than harmful effects in humans, at least in adults.

According to the literature, the consumption of isoflavones is a controversial issue. For this reason, it is important to develop analytical methodology for a sensitive and accurate identification and quantification of isoflavones in foodstuffs, especially when intended for human consumption. In recent years, a large number of methodologies have been developed for the determination of isoflavones in different matrices (Figure 16.2).

In this Chapter, we present a brief summary of the techniques proposed in the literature for the isolation and determination of these compounds in foods.

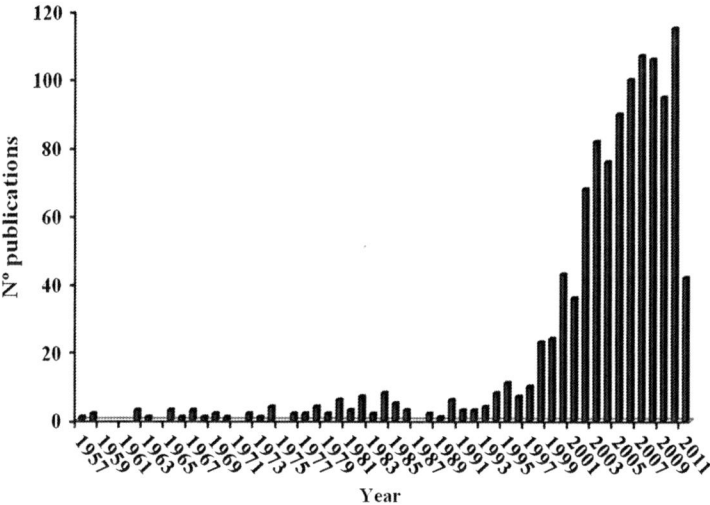

Figure 16.2 Analytical methodologies proposed for isoflavones determination. Years from 1957 to 2011.
Unpublished.

16.2 Isoflavone Isolation

The determination of isoflavones in a complex matrix, as is a foodstuff, usually requires a prior step to extract and concentrate the analytes, sometimes including a clean-up procedure to reduce the matrix effect. The isolation of isoflavones from foods is a complex task as there are several aspects that can influence the quantitative extraction of compounds from the original sample. The extraction technique, the extraction solvent, time and temperature, sample to solvent ratio, sample characteristics and isoflavone stability during extraction are crucial factors that may strongly affect extraction efficiency, isoflavone profile and recovery (Rostagno *et al.* 2010). The elimination of undesired matrix components and the enrichment of samples are two main objectives of the sample treatment. With this purpose, the use of purification/clean-up procedures, such as liquid–liquid extraction (LLE) and solid phase extraction (SPE), has been suggested. The use of SPE, both directly/alone or coupled in-line/on-line with other extraction techniques, for the extraction of isoflavones from liquid foods has been reported. These procedures are aimed at eliminating interferences or serving as a post-extraction step to increase the concentration of isoflavones and as a clean-up procedure (Corradini *et al.* 2011; Delmonte and Rader 2006; Kurie Mitani *et al.* 2003; Rostagno *et al.* 2005). The use of SPE or LLE extractions includes a final step of solvent removal by rotary evaporation and re-dissolution of the dry residue on the mobile phase that is used for the chromatographic analysis (Rostagno *et al.* 2010). The elimination of the solvent increases detection and quantification levels, which is an attractive approach for the analysis of samples with low isoflavone concentration. Finally, it is important to note that, after extraction, the removal of insoluble and solid materials by centrifugation and/or filtration is used as a post-extraction procedure before analysis.

It is also worth noting that during these procedures and storage, the compounds of interest in the samples may undergo degradation, and the isoflavone profiles and sample composition may be altered as a result of sample handling and factors such as temperature, relative humidity and light or product type (Kim *et al.* 2005; Lee *et al.* 2003). Maintaining the representativeness of the sample throughout the determination procedure in order to avoid significant errors and erroneous conclusions in the identification and quantification of the referred analytes is of crucial importance.

There are several techniques available for the extraction of isoflavones from supplemented foods, including the *classical* soxhlet or stirring/agitation technique, and *modern extraction techniques*, such as ultrasound extraction (UE), microwave-assisted extraction (MAE), pressurised liquid extraction (PLE), supercritical fluid extraction (SFE) and SPE techniques. In general, classical methods are not recommended as they are time-consuming (2 to 24 h at extraction temperatures ranging from 4 °C to 80 °C) and require large amounts of solvent. In contrast, modern techniques show important advantages, such as a significant reduction in extraction time and solvent consumption, as well as the ability to achieve highly reproducible quantitative recoveries, without changing the isoflavone profile (Rostagno et al. 2007a, 2009; Vacek et al. 2008). However, the choice of technique depends on practical aspects, such as instrument availability and cost, level of automation and the possibility of on-line coupling with an analysis technique. The choice of sample treatment and extraction procedure also depends on the physical state of samples. For the routine analysis of solid samples, with an approximate knowledge of concentration and distribution of isoflavones, using ultrasound, including sequential extractions, can be an efficient and low-cost technique. Since the sample is not expected to have great variability, its influence on extraction efficiency is reduced to a minimum. In other contexts where a large variety of samples with unknown isoflavone concentration and distribution is present, more efficient techniques, such as PLE, may be more appropriate (Rostagno et al. 2009).

Once the extraction technique is chosen, it is necessary to optimise some extraction parameters, especially those of crucial importance (extraction solvent, volume, extraction time and temperature). For isoflavone extraction, the use of pure solvents is not recommended. The presence of water (40–60%) notably improves recovery rates, although this parameter depends on the food matrix. It has also been reported that, depending on the characteristics of the sample, some solvents cannot quantitatively extract some individual isoflavones. In that case, a sequential extraction is recommended with a solvent or various solvents to maximise the extraction efficiency of all chemical forms of isoflavones present in the sample (Schwartz and Sontag 2009). Another important parameter that requires optimisation is extraction time. A short extraction time is desirable, as long extraction times may affect isoflavone stability. Some studies found in the literature have reported that most isoflavones (80–90%) present in food samples are extracted in no more than 15 min (Rostagno et al. 2010). Lastly, the optimisation of extraction temperature is also important, as high temperatures increase the solubility of the compounds, thus improving the extraction efficiency and reducing the analysis time. However, time and temperature should be taken into account to ensure stability during the extraction.

When samples are liquid or dissolvable, a very useful alternative to avoid tedious extraction procedures is the precipitation of proteins, fats and other interferences from the matrix. To this end, the Association of Official Analytical Chemists (AOAC) proposed a methodology for the determination of isoflavones in soy and selected foods based on the extraction of isoflavones using a mixture of methanol/water (1:1, v/v), followed by a neutralisation step by addition of a sodium hydroxide solution and then glacial acetic acid. The sample is analysed after dilution (1:50) with methanol/water (AOAC 2001). Recently, a very simple and accurate method has been proposed for the isolation of isoflavones in infant formulae supplemented milk or milk-based juices. The method uses a precipitation solution containing zinc and wolframium salts in an acidic medium and acetonitrile (Zafra-Gómez et al. 2010). In this methodology, solid samples are suspended in slightly hot water before precipitation. Samples are placed in 15 mL Falcon flasks and 2 mL of the precipitation

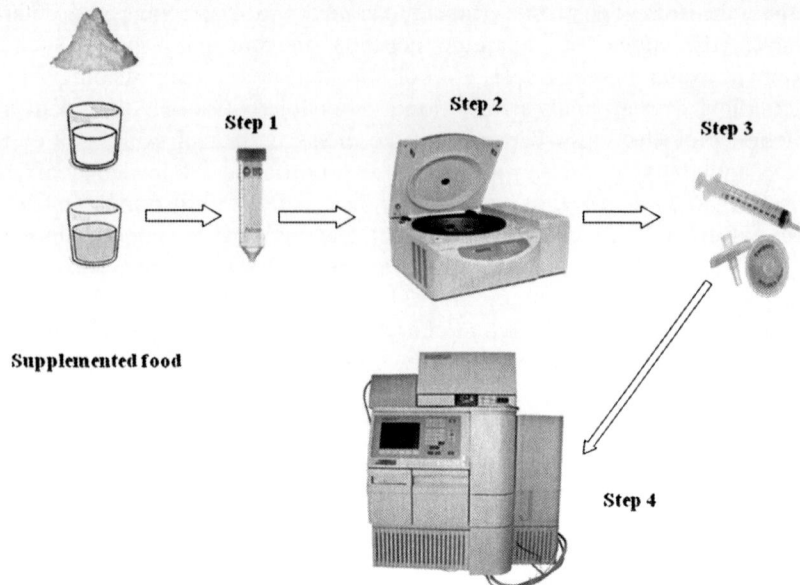

Figure 16.3 Schedule of a supplemented food treatment method based on precipitation. Step 1, addition of the precipitation solution, acetonitrile; Step 2, centrifugation; Step 3, filtration and dilution; Step 4, LC analysis. Unpublished.

solution and 2 mL of acetonitrile are added. Lastly, the samples are diluted to 10 mL with Milli-Q water and centrifuged and the supernatant filtered through 0.22 μm pore membrane before injection into the liquid chromatography (LC) unit (Figure 16.3).

In addition, some authors have proposed using a freeze–drying procedure for liquid or wet samples to remove water. However, this procedure has two main disadvantages: the process is very time-consuming and could affect the stability of isoflavones. Direct extraction from the sample without any pre-treatment or interference precipitation could be appropriate alternatives to freeze–drying.

16.3 Techniques for the Analysis of Isoflavones in Supplemented Food

Several kinds of techniques are currently available for the analytical determination of isoflavones in supplemented foods, including chromatographic methods, as well as CE (capillary electrophoresis) and biological assays (immunoassay). Biological methods, specifically immunoassay, are the most sensitive and specific analytical methods. However, they are limited by the availability of specific antisera, which are quite expensive, as well as being subject to cross-reactivity. Chromatographic methods, although not as sensitive as biological methods, make the simultaneous screening and determination of isoflavones and other related compounds of interest possible. Moreover, LC-mass spectrometry (MS) has gained popularity in recent years because of its high sensitivity and specificity.

Unlike GC, LC has the ability to detect isoflavone-related compounds (aglycones, glucosides and glucoside malonates), which can be determined without a derivatisation process, despite their known non-volatility and high molecular weight. Recent reviews in the literature summarise the different techniques and methods developed (Delmonte and Rader 2006).

16.3.1 Liquid Chromatography (LC)

Nowadays, LC is one of the most used techniques for the determination of isoflavones in foods. It allows for an appropriate separation and accurate quantification of the analytes, high efficiency and reproducibility, as well as the measurement of all of the chemical forms of isoflavones. In addition, one very important advantage of this technique is its simple sample preparation requirements, considering that the main difficulty in the determination of compounds is generally related to the treatment of the samples beforehand, such as cleaning and reducing the matrix effect. Although LC separation can be combined with a variety of detectors, the most common, besides MS coupling, for many years, have been classical ultraviolet (UV), ultraviolet-visible (UV-Vis) and photodiode array detection, as well as fluorescence detection, considering the native fluorescence of some isoflavones (Table 16.2). Chemiluminiscence techniques can also be used for the sensitive analyses of isoflavones. UV and photodiode array detectors are widely used in the range of 200–400 nm, which is the region of the spectrum where all isoflavones show absorption (Zafra-Gómez *et al.* 2010). This wavelength range shortens to 255–260 nm for quantification purposes in order to obtain the highest signal for all chemical forms, as there are differences in the maximum absorption peaks of the derivatives. The main limitation for LC-UV application is its lack of high

Table 16.2 UV-Vis, fluorescence wavelengths and MS data of isoflavones. λ, wavelength (excitation and emission); for MS: capillary voltage, 2 kV; cone voltage, 35 V; desolvation temperature, 350 °C; source temperature, 130 °C; [–15], loss of a -CH3 group; [–56], loss of a -CO-CO group; [–162], loss of a glucoside group.

Name	UV λ (nm)	FL (λ_{exc}, λ_{em}) (nm)	MS spectra (ESI+)
Daidzein	255	250, 305	255 [M+H-142]$^+$, 199 [M+H-56]$^+$
Daidzin	255	250, 305	417 [M+H]$^+$, 255 [M+H-162]$^+$, 199 [M+H-56]$^+$
Genistein	260	225, 260	271 [M+H]$^+$, 215 [M+H-56]$^+$
Genistin	260	225, 260	433 [M+H]$^+$, 271 [M+H-]$^+$, 215 [M+H-56]$^+$
Formononetin	260	250, 305	269 [M+H]$^+$, 254 [M+H-15]$^+$, 213 [M+H-56]$^+$
Biochanin A	260	250, 305	285 [M+H]$^+$, 270 [M+H-15]$^+$, 229 [M+H-56]$^+$

selectivity and sensitivity. This disadvantage is successfully overcome by using MS detection instead, which is characterised by its exclusive selectivity and excellent sensitivity. Because of this, over the past few years, LC coupled with various MS detectors has become the most commonly used technique (Griffith and Collison 2001).

One of the most important issues in an LC analysis is the selection of mobile phases and additives. Methanol and acetonitrile as organic solvents, and water containing small amounts of acids as modifiers (formic, acetic, phosphoric and trifluoracetic acid) are frequently used. The presence of these acids facilitates the dissociation of the analytes in the solvent system, as their structures mostly contain phenolic hydroxyl groups, which show a weak acidity improving the separation and peak shape. In addition, in MS, the sensitivity of the analytes improves notably by enhancing positive ion formation when the mobile phase is acidified (Wu *et al.* 2004).

Other crucial issue in LC is column selection. The separation of isoflavones is generally performed with a conventional reversed-phase column with C18 or C8 stationary phases and a 5 µm particle size. These columns require extended periods of time for complete separation of all isoflavones (Peñalvo *et al.* 2004). In recent years, an important technological advance, known as ultra-performance liquid chromatography (UPLC), has represented a major improvement in LC, as it offers significant practical advantages in resolution, speed and sensitivity for analytical determination, especially when coupled with MS detection, being capable of high-speed acquisitions (Churchwell *et al.* 2005). UPLC techniques allow columns with particles <2 µm in size to be used in a chromatographic system able to withstand pressures close to 15 000 psi (1000 bar) that are equipped with detectors with higher sampling rates and the ability to integrate accurately an analyte peak. All these characteristics result in reduced analysis times, with no loss in separation efficiencies, which were even better in some cases. A recent work (Klejdus *et al.* 2008) reported the optimal conditions for the determination of isoflavones (daidzein, genistein and glycitein, their glucosides and other derivatives) using octadecyl-, phenyl-, and cyanopropyl-modified stationary phases for the simultaneous separation of these substances, which occurred in less than 1 min.

Monolithic columns have also been used in recent years (Rostagno *et al.* 2007b; Unger *et al.* 2008). These allow for fast separations due to the low-pressure drop across the column, which allows for higher flow rates and thereby reduces the chromatogram time without losing efficiency. Many methods proposed in the literature use these stationary phases. Rostagno *et al.* (2007b) validated a fast method (10 min) for the determination of 12 isoflavones that uses two monolithic columns with an elution gradient of acidified water (0.1% acetic acid) and methanol (0.1% acetic acid) at a flow rate of $5\,\text{mL}\,\text{min}^{-1}$.

Finally, another simple strategy to obtain faster chromatographic separations is with the use of high column temperatures. This drastically reduces mobile phase viscosity without affecting column efficiency, thereby allowing the same resolution to be achieved in less time and with no changes to the pressure system (Vanhoenacker and Sandra 2008).

16.3.2 Gas Ghromatography (GC)

In the case of analysing phytoestrogenic compounds and their metabolites in different matrices (including food samples), GC, namely GC-MS, was the preferred technique before the use of LC for a routine analysis of all isoflavones in foods and vegetables, with good specificity, sensitivity and very low detection limits (Liggins et al. 1998; Mazur et al. 1996; Morton et al. 1999). However, there are some major disadvantages associated with GC that should be taken into account, such as the requirement of large sample sizes, the complex sample preparation, which includes derivatisation and purification steps that may cause variability in the results, as well as an increase in the cost of the analysis. GC-MS methods have been developed with SIM mode detection for the determination of daidzein and genistein in soy flour. GC with a flame ionisation detector (FID) has been used for analysis of aglycone in extracts of soybean with a low-temperature treatment (Hsu et al. 2010).

The sample preparation for GC methods always requires a minimum of three steps: hydrolysis, Sep-Pak purification and derivatisation. This procedure is usually time-consuming. The need for hydrolysis constitutes an important limitation of this technique when analysing isoflavones, as all isoflavones are determined based on daidzein and genistein, since applying the method to all soy isoflavones is not possible.

16.3.3 Capillary Electrophoresis (CE)

These separation techniques are based on differences in the electromigration of compounds in an electric field. Capillary zone electrophoresis, capillary gel electrophoresis, micellar electrokinetic chromatography, capillary electrochromatography, capillary isoelectric focusing and capillary isotachophoresis are the main types. UV-Vis detectors, mainly diode array detectors, are the detectors of choice. However, the spectra of many isoflavones are very similar and the applicability of this kind of detection in their identification is very limited. In this sense, one of the most important advantages of CE techniques is how easy it is to couple them with sensitive electrochemical detectors, as the application of electrochemical detectors when using chromatographic techniques is limited due to the presence of organic modifiers in mobile phases. In addition to these detection techniques, electromigration column separations can also be directly coupled with fluorescence or mass selective detectors.

Procedures for a highly effective separation of isoflavones using CE techniques were first introduced in the 1990s, when a large number of protocols for the determination of these compounds in foods and other natural matrices were proposed (Herrero et al. 2005). Recently, Xiao et al. (2011) has proposed the use of electrokinetic capillary chromatography for the determination of isoflavones in food. This method proved to be appropriate accurate, reliable, inexpensive and easy to operate (Xiao et al. 2011).

16.3.4 Mass Spectrometry (MS)

In recent years, MS has become the leading technique in the identification and quantification of isoflavones in matrices of any kind, including foods. MS is usually used coupled mainly with LC and GC, and, to a lesser extent, CE (Vacek *et al.* 2008). Its importance and popularity in analytical laboratories is due to the fact that MS is one of the most sensitive and selective techniques for the determination of unknown compounds (especially in the absence of reference standards). A mass spectrometer can detect individual aglycones, glucosides and other conjugates, and it is the most effective tool for the identification of the separated substances (Figure 16.4).

Mass spectrometers consist of three fundamental parts: the ionisation source, the analyser and the detector. The ionisation source is a key component for its coupling with LC or GC. In the case of GC, the most common ionisation techniques are electron impact and chemical ionisation. In the case of LC, the coupling process is more complicated. There are several ionisation techniques, the most important of which include atmospheric pressure chemical ionisation (APCI), electrospray ionisation (ESI), thermospray ionisation (TI), matrix-assisted laser desorption ionisation (MALDI), among others. ESI, both in positive and negative modes, is the most frequently used ionisation method for the analysis of isoflavones (Griffith *et al.* 2001). This type of ionisation can be coupled with both LC and CE.

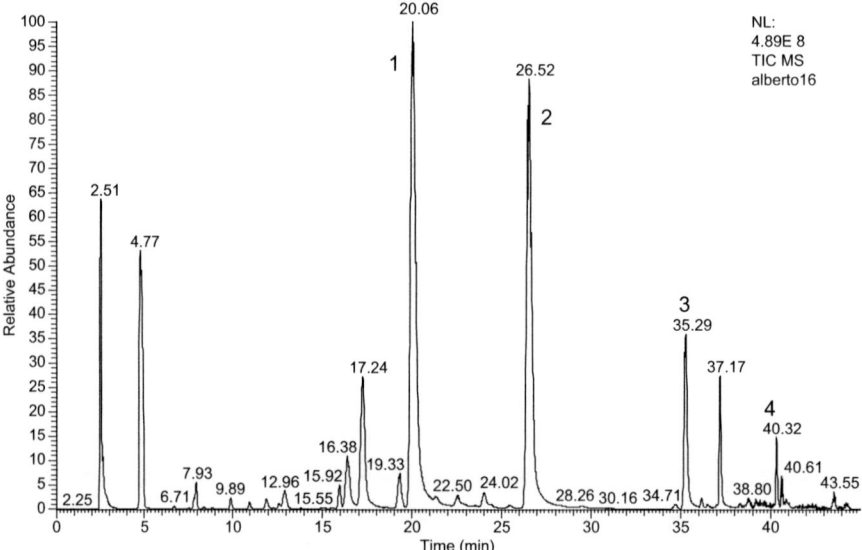

Figure 16.4 Example of isoflavone profile of supplemented cow milk (SCAN mode). Conditions: column, Zorbax SB-C18, 250 mm × 4.6 mm I.D., 5 μm; mobile phase, gradient of methanol and aqueous 10 mM ammonium formate buffer, pH 4; flow rate, 1 mL min^{-1}; injection volume, 10 μL. Peak assignment: 1, daidzin; 2, genistin; 3, daidzein and 4, genistein. Unpublished.

The importance of MS as an isoflavone determination technique is due to the fact that each isoflavone can be characterised not only by its molecular ion, but also by its specific fragmentation products (Table 16.2).

It is important to take into account how the sample was treated beforehand, in addition to the fact that as MS is a highly sensitive technique, a suitable sample preparation method is essential to make data analysis easier. Many works have obtained very good results using ESI-quadrupole MS in combination with LC after purification techniques, such as SPE or SFE, were used (Vacek *et al.* 2008).

In addition to selective analyses, MS can be used for structural analysis. The combination of column separations with nuclear magnetic resonance (NMR) and MS and/or tandem MS can be used in the determination of the structure of isoflavones.

16.3.5 Immunoanalysis

Immunochemical methods are the fastest and most accurate ways to analyse isoflavones (Vacek *et al.* 2008). Radioimmunoassay was one of the first techniques to be applied. This method has since been replaced by enzyme-linked immunosorbent assay (ELISA) and its modifications. These methods are commonly applied, mainly due to their simplicity and the speed of the analyses. The ELISA procedure is based on the reaction between the isoflavones or their conjugates (antigens) with specific antibodies (monoclonal or polyclonal) that need to be prepared in advance. The reaction takes place in a microtitration plate, where antibodies bind to antigens present in the sample. After washing procedures, the number of bound antibodies is quantified using secondary antibodies that are usually marked by peroxidase. Before the addition of a peroxidase substrate, suitable conditions are required to perform the enzymatic reaction. Once the final product is detected, its concentration corresponds to the amount of isoflavone present in the sample. ELISA has been used for the determination of biochanin A, daidzein and genistein, and has shown limits of detection (LODs) that ranged from 1.1 to 5.3 pg of isoflavones per well (Vacek *et al.* 2008).

16.4 Conclusions

This work is an overview of the main advances made in the determination of isoflavones in foods supplemented with phytoestrogens. The most important sample treatments and the most common instrumental techniques, such as LC, GC and MS, and immunological techniques, are discussed. GC methods provide not only a specific and sensitive analysis of isoflavones, but also reduce solvent cost and solvent disposal. However, most GC methods can only separate a small number of soy isoflavones and require a time-consuming sample preparation step. Similarly, LC methods provide rapid and sensitive analysis for all isoflavones simultaneously. Neither GC nor LC is suitable for the analysis of a large number of samples at the same time. Conversely, immunoassay provides not only high efficiency, but also a fast and appropriate analysis of soy isoflavones at very low concentrations, even in complex samples. Nevertheless, only a few isoflavones can be determined by immunoassay methods and quantification accuracy may decrease.

Summary Points

- This Chapter presents an overview of the analytical methodologies for the determination of isoflavones in food.
- Isoflavones are phytoestrogens that have been widely described as health products due to their health benefits and effects on physiological functions.
- The health benefits have an effect on the marketing of several 'health' food supplements and drinks.
- Isoflavones have received severe criticism due to their structural and functional similarity to 17-β-estradiol.
- Isoflavones are mentioned in more than 2000 scientific publications and this fact provides clear evidence of the importance currently given to these substances.
- The determination of isoflavones in foods requires a sample treatment step to extract, isolate and concentrate the analytes.
- The most used techniques for isoflavone isolation are the classical soxhlet or stirring/agitation and the modern techniques such as UE, MAE, PLE, SPE or SFE.
- For liquid samples, a simple precipitation of fat and proteins and filtration procedures is proposed in the literature.
- The extracts are analysed usually by chromatographic (LC, MS or CE) or immunological techniques, being LC coupled with MS the most often.
- Maintaining the representativeness of the sample throughout the determination procedure is of crucial importance.

Key Facts

Key Facts for Isoflavones

- Isoflavones are phytoestrogens.
- Isoflavones are described as health products.
- Isoflavones are added into the processing of several "functional foods".

Key Facts for Isolation Procedures

- Soxhlet is a classical extraction technique used to isolate isoflavones from supplemented foods.
- UE, MAE, PLE, SPE and SFE are modern extraction techniques used to isolate isoflavones from supplemented foods.

- Extraction techniques allow for the isolation, concentration and clean-up of the sample.
- Modern extraction techniques not only shorten the total analysis time but also minimise the use of solvents.

Key Facts for Chromatographic Techniques

- LC, GC and CE are analytical chemistry techniques for the separation of mixtures.
- The equipment consists of a reservoir of solvent (LC), a buffer (CE) or a gas stream (GC), a pump (LC or CE), an injector, a separation column/ capillary and a detector.
- The different analytes pass through the stationary phase at different rates due to different partition behaviour between the mobile and stationary phases.
- MS is the most common detector coupled with LC, CG and CE units.
- MS uses the mass-to-charge ratio (m/z) of ionised compounds to separate analytes.
- MS allows for quantitative and qualitative (chemical and structural) information to be gathered about molecules based on their distinctive fragmentation patterns.

Key Facts for Immunochemical Techniques

- They are one of the most important types of techniques for the fast and accurate identification of isoflavones.
- Radioimmunoassay was the technique initially used. ELISA is now the most used technique.

Definition of Words and Terms

Enzyme-linked immunosorbent assay (ELISA): A biochemical technique used to detect the presence of an antibody or an antigen in a sample. This antibody is linked to an enzyme, and in a final step a substance is added so that the enzyme can it convert into some detectable signal, generally the production of a coloured product in the chemical substrate.

17-β-Estradiol (1,3,5-estratriene-3,17β-diol): An estrogenic steroid. In humans, it is synthesised by the ovary, from which it is secreted directly into the blood stream. It is synthesised from testosterone that is converted by the aromatase complex, which is located in ovarian granulosa cells, and is stimulated by follitropin; lutropin, in turn, stimulates production of testosterone–the aromatase substrate–by ovarian theca foliculli cells.

Monolithic columns: Chromatographic columns where the stationary phase consists of a single piece of solid made of either a porous cross-linked polymer or porous silica with interconnected skeletons and interconnected flow paths that pass through these skeletons. They are more permeable than particle-packed columns and the number of theoretical plates is notably higher. They can operate at high flow rates because the separation efficiency diminishes only slightly due to the small equivalent diameter.

Phytoestrogen: A naturally occurring, plant-derived chemical that functions as the primary female sex hormone, although it is not generated by the endocrine system but rather obtained through the consumption of phytoestrogonic plants. Because of their structural similarity to 17-β-estradiol, phytoestrogens can have estrogenic or/and anti-estrogenic effects.

Supplemented food: Food that has a substance or substances added to it or that has been modified in some way to perform a physiological role beyond the provision of a simple nutritive requirement.

Soxhlet: A classical method applied for the extraction of compounds from solid samples. The sample is placed in a porous cellulose thimble inside the extraction chamber, suspended above a flask containing the solvent and below a condenser. The flask is heated and the solvent evaporates and moves up into the condenser where it is converted into a liquid that trickles into the extraction chamber containing the sample. The extraction chamber is designed so that when the solvent surrounding the sample exceeds a certain level it overflows and trickles back down into the boiling flask. At the end of the extraction process, which lasts a few hours (12–24 h), the flask containing the solvent and analyte is removed. The solvent in the flask is then evaporated and the extract analysed.

List of Abbreviations

CE	capillary electrophoresis
ELISA	enzyme-linked immunosorbent assay
ESI	electrospray ionisation
GC	gas chromatography
LC	liquid chromatography
LLE	liquid–liquid extraction
MAE	microwave-assisted extraction
MS	mass spectrometry
PLE	pressurised liquid extraction
SFE	supercritical fluid extraction
SIM	Selected Ion Monitoring
SPE	solid phase extraction
UE	ultrasound extraction
UPLC	ultra-performance liquid chromatography
UV	ultraviolet
UV-Vis	ultraviolet-visible

References

Andersen, O.M., and Jordheim, M., 2006. In: Andersen, O. and, Markham K. (ed.). 2006. Flavonoids. Chemistry, Biochemistry and Applications. CRC Press, Boca Ratón, FL, USA, p. 471.

AOAC Official Method, 2001.10. Determination of isoflavones in soy and selected foods containing soy. In: Official Methods of Analysis, 17th edn. Chapter 45.4.14. AOAC International, Gaithersburg, MD, USA.

Bingham, S.A., Atkinson, C., Liggins, J., Bluck, L., and Coward, A., 1998. Plant oestrogens: where are we now? *British Journal of Nutrition.* 79: 393–406.

Cao, Y., Calafat, A.M., Doerge, D.R., Umbach, D.M., Bernbaum, J.C., Twaddle, N.C., Ye, X., and Rogan, W.J., 2009. Isoflavones in urine, saliva and blood of infants – data from a pilot study on the estrogenic activity of soy formula. *Journal of Exposure Science and Environmental Epidemiology.* 19(2): 223–234.

Corradini, E., Foglia, P., Giansanti, P., Gubbiotti, R., Samperi, R., and Lagan, A., 2011. Flavonoids: chemical properties and analytical methodologies of identification and quantification in foods and plants. *Natural Product Research.* 25(5): 469–495.

Churchwell, M.I., Twaddle, N.C., Meeker, L. R, and Doerge, D.R., 2005. Improving LC-MS sensitivity through increases in chromatographic performance: Comparisons comparisons of UPLC–ES/MS/MS to LC–ES/MS/MS. *Journal of Chromatography B.* 825(2): 134–143.

Delmonte, P., and Rader, J.I., 2006. Analysis of isoflavones in foods and dietary supplements. *Journal of AOAC International.* 89(4): 1138–1146.

Griffith, A.P., and Collison, M.W., 2001. Improved methods for the extraction and analysis of isoflavones from soy-containing foods and nutritional supplements by reversed-phase high-performance liquid chromatography and liquid chromatography–mass spectrometry. *Journal of Chromatography A.* 913: 397–413.

Herrero, M., Elena Ibáñez, E., and Cifuentes, A., 2005. Analysis of natural antioxidants by capillary electromigration methods. *Journal of Separation Sciences.* 28: 883–897.

Hsu, B.Y., Inbaraj, B.S., and Chen, B.H., 2010. Analysis of soy isoflavones in foods and biological fluids: an overview. *Journal of Food and Drug Analysis.* 18(3): 141–154.

Klejdus, B., Vacek, J., Lojková, L., Benešová, L., and Kubáň, V., 2008. Ultra-high-pressure liquid chromatography of isoflavones and phenolic acids on different stationary phases. *Journal of Chromatography A.* 1195: 52–59.

Kim, J.J., Kim, S.H., Hahn, S.J., and Chung, I.M., 2005. Changing soybean isoflavone composition and concentrations under two different storage conditions over three years. *Food Research International.* 38: 435–444.

Kurie Mitani, K., Narimatsu, S., and Kataoka, H., 2003. Determination of daidzein and genistein in soybean foods by automated on-line in-tube solid-phase microextraction coupled to high-performance liquid chromatography. *Journal of Chromatography A.* 986(2): 169–177.

Lee, S.J., Chung, I.M., Ahn, J.K., Kim, J.T., Kim, S.H., and Hahn, S.J., 2003. Variation in isoflavones of soybean cultivars with location and storage duration. *Journal of Agricultural and Food Chemistry.* 51: 3382–3389.

Liggins, J., Bluck, L.J.C., Coward, W.A., and Bingham, S.A., 1998. Extraction and quantification of daidzein and genistein in food. *Analytical Biochemistry*. 264: 1–7.

Mazur, W., Fotsis, T., Wähälä, K., Ojala, S., Salakka, A., and Adlercreutz, H., 1996. Isotope dilution gas chromatographic-mass spectrometric method for the determination of isoflavonoids, coumestrol, and lignans in food samples. *Analytical Biochemistry*. 233: 169–180.

Morton, M., Arisaka, O., Miyake, A., and Evans, B., 1999. Analysis of phyto-oestrogens by gas chromatography-mass spectrometry. *Environmental Toxicology and Pharmacology*. 7(3): 221–225.

Peñalvo, J.L., Nurmi, T., and Adlercreutz, H., 2004. A simplified LC method for total isoflavones in soy products. *Food Chemistry*. 87(2): 297–305.

Price, K.R., and Fenwick, G.R., 1985. Naturally occurring oestrogens in foods – a review. *Food Additives and Contaminants*. 2(2): 73–106.

Rostagno, M.A., Palma, M., and Barroso, C.G., 2005. Solid-phase extraction of soy isoflavones. *Journal of Chromatography A*. 1076: 110–117.

Rostagno, M.A., Palma, M., and Barroso, C.G., 2007a. Ultrasound-assisted extraction of isoflavones from soy beverages blended with fruit juices. *Analytica Chimica Acta*. 597: 265–272.

Rostagno, M.A., Palma, M., and Barroso, C.G., 2007b. Fast analysis of soy isoflavones by high-performance liquid chromatography with monolithic columns. *Analytica Chimica Acta*. 582: 243–249.

Rostagno, M.A., Manchón, N., Guillamón, E., García- Lafuente, A., Villares, A. and Martínez, J.A., 2010. Methods and techniques for the analysis of isoflavones in foods. In: Quintin, T.J. (ed.) *Chromatography Types, Techniques and Methods*. Nova Science Publishers Inc, Hauppauge, New York, pp. 157–198.

Rostagno, M.A., Villares, A., Guillamón, E., García-Lafuente, A., and Martínez, J.A., 2009. Sample preparation for the analysis of isoflavones from soybeans and soy foods. *Journal of Chromatography A*. 1216: 2–29.

Schwartz, H., and Sontag, G., 2009. Comparison of sample preparation methods for analysis of isoflavones in foodstuffs. *Analytica Chimica Acta*. 633: 204–215.

Unger, K.K., Skudas, R., and Schulte, M.M., 2008. Particle packed columns and monolithic columns in high-performance liquid chromatography-comparison and critical appraisal. *Journal of Chromatography A*. 1184: 393–415.

Vacek, J., Klejdus, B., Lojková, L., and Kubán, V., 2008. Current trends in isolation, separation, determination, and identification of isoflavones: a review. *Journal of Separation Science*. 31: 2054–2067.

Vanhoenacker, G., and Sandra, P., 2008. High temperature and temperature programmed LC: possibilities and limitations. *Analytical and Bioanalytical Chemistry*. 390: 245–248.

Wu, Q., Wang, M., and Simon, J.E., 2004. Analytical methods to determine phytoestrogenic compounds: Review. *Journal of Chromatography B*. 812: 325–355.

Xiao, O.S., Ying, Z., Hui, C., Kai, G., Zhi, C., Wei, Z., and Wei, L., 2009. Soy food intake and breast cancer survival. *Journal of the American Medical Association*. 302(22): 2437–2443.

Xiao, M., Ye, J., Tang, X., and Huang, Y., 2011. Determination of soybean isoflavones in soybean meal and fermented soybean meal by micellar electrokinetic capillary chromatography (MECC). 126: 1488–1492.

Zafra-Gómez, A., Garballo, A., García-Ayuso, L.E., and Morales, J.C., 2010. Improved sample treatment and chromatographic method for the determination of isoflavones in supplemented foods. *Food Chemistry*. 123: 872–877.

CHAPTER 17

Isoflavones: LC-MS/MS Profiling of Isoflavone Glycosides and Other Conjugates

PIOTR KACHLICKI*[a] AND MACIEJ STOBIECKI[b]

[a] Laboratory of Metabolomics, Institute of Plant Genetics, Polish Academy of Sciences, Strzeszyńska 34, 60-479 Poznań, Poland; [b] Laboratory of Metabolomics and Proteomics, Institute of Bioorganic Chemistry, Polish Academy of Sciences, Noskowskiego 12/14, 61-704 Poznań, Poland
*Email: pkac@igr.poznan.pl

17.1 Introduction

Plants are very complex organisms built from different types of tissues and are subjected to various changes of environmental conditions. Extreme temperatures and an excess or deficiency of nutritive compounds, water or light, as well as infestation by herbivores or infection by pathogenic fungi or bacteria, often happen during the plant lifespan. All these environmental factors influence greatly the quantitative and qualitative composition of low molecular weight compounds present in tissues. In many cases plants increase the biosynthesis of secondary metabolites as defence factors against the biotic or abiotic threats or signal compounds. On the other hand, the ontogenetic development of plants also alters the sets of natural products present in their tissues. Our knowledge of all these changes is supported by results of metabolic profiling. Metabolic profiling provides information on quantitative and qualitative composition of the chosen class of compounds in the studied plant organ or tissue. Liquid chromatography combined with mass spectrometry (LC-MS) is the analytical method which offers many advantages and is increasingly used in these studies. The profiling of isoflavonoids using the LC-MS methods is described in this Chapter.

17.2   Chromatographic Separation of Flavonoids

Isoflavonoids, similarly to other polyphenolic compounds present in plant tissues, occur in a form of complex conjugates with glycosidic, acyl or alkyl moieties. Aromatic aglycones of these compounds are slightly polar compounds and the polarity of the conjugates depends on the number and characteristics of their substituents. In general, the polarity increases with the number of sugar units and decreases when some hydroxyls of the conjugate are esterified or alkylated. Reversed phase LC (RP-LC) and capillary electrophoresis are methods of choice for effective separations of naturally occurring mixtures of such compounds (for the overview of these methods see Valls *et al.* 2009; and Chapter 13 and 16 of this book). The connection of LC or capillary electrophoresis instruments on-line to mass spectrometers sometimes requires modification of methods efficiently used for separation of compounds in these systems alone. Often the liquid phase in high-performance liquid chromatography (HPLC) is acidified by an addition of 0.1–0.5% (v/v) formic, acetic or trifluoroacetic acid which increases its resolving power. However, trifluoroacetic acid should be avoided in the LC-MS experiments as it decreases the efficiency of ionisation in most of ion sources applied in mass spectrometers. The use of the improved ultra-high performance instruments [such as ultra-performance liquid chromatography (UPLC)] instead of the standard HPLC is recommended for profiling the isoflavonoids in plant tissues. The application of UPLC-MS to the analysis of flavone and isoflavone glycosides in narrow leaf lupin (*Lupinus angustifolius* L.) leaves resulted in the detection of 38 compounds, which was a much higher number than that found using the standard HPLC-MS (23 compounds; Muth *et al.* 2008). Despite the less frequent use of the capillary electrophoresis instruments hyphenated to mass spectrometers in the analyses of flavonoids in plants, this method may give very satisfactory results. For example, its application to the analysis of flavonoids from *Genista tenera* allowed separation and identification of 18 compounds, whereas only five could be found using the HPLC-based methods (Edwards *et al.* 2006).

17.3   Mass Spectrometry (MS) as a Tool for Identification of Isoflavone Glycoconjugates

MS is a technique used for the qualitative and quantitative analysis of compounds on the basis of a measurement of the mass-to-charge ratio (m/z) of ions formed from these substances, as well as of their fragments. This measurement is performed in a high vacuum, therefore solutions of the analysed compounds have to be deprived of the solvent and ionised prior to the analysis. These functions are performed in the ion source of mass spectrometers. Ion sources most frequently applied in mass spectrometers hyphenated to LC instruments operate on the basis of the electrospray ionisation concept. The analyte solution (the chromatographic column effluent or a directly injected sample) is sprayed through a steel capillary in an electric field of several kV. In these conditions ambient molecules of organic compounds are readily ionised, in most cases giving rise to protonated or deprotonated molecules, designed as $[M+H]^+$ or $[M-H]^-$ ions, respectively. If other ions (*e.g.* inorganic salts) are present in the analysed solution, then metal adduct ions such as $[M+Na]^+$ or $[M+K]^+$ may also be created. Similarly, $[M+HCOO^-]^-$ ions may be registered in analyses in which formic acid has been used as the acidifier of the chromatographic eluents. The outlet of the capillary is placed in a stream of drying gas (usually nitrogen) at elevated temperature. The size of droplets of the sprayed solution rapidly diminishes due to the solvent evaporation and the Coulomb forces of the ion repulsion cause the droplet explosion and the ion liberation to the gas phase. This "soft" ionisation procedure does not supply excessive energy to the analyte molecules and, in general, does not cause their fragmentation in the ion source.

Ions generated in the ion source of the spectrometer have to be transported to its analyser. The role of the analyser is the sorting of the ions and the measurement of their m/z values. This measurement may be performed with different precision: quadrupole (Q) and most of ion trap (IT) instruments measure them up to the unit or the first decimal point values, whereas time-of-flight (ToF) and Fourier transform (FT) instruments are capable of high-resolution measurement up to the fourth or fifth decimal point of an atomic mass unit (amu) performed with a high accuracy (for the best instruments exceeding one part per million). The high-resolution instruments are particularly useful as they allow the calculation of the most probable atomic composition of the analysed compounds. As it was mentioned previously, the important benefit of MS is the possibility of structural and quantitative conclusions according to considerations based on the m/z values of fragments of the analysed compounds registered in the spectra. As almost no fragmentation occurs in the ion sources of modern spectrometers hyphenated to liquid chromatographs, these instruments have to apply another way of obtaining fragments appropriate for such analysis. The most frequently employed mechanism of fragmentation is called collision-induced dissociation (CID). Briefly, a very small amount of inert gas (helium or argon) is added to the MS analyser working in a high vacuum. The $[M+H]^+/[M-H]^-$ ions entering this analyser are accelerated by the internal electric field and thereafter they collide with the gas atoms. The most fragile bonds within these molecules are eventually cleaved and fragments are formed. Some of them carry on the charge and their m/z values may be measured, whereas the chemical nature of the neutral fragments may be inferred from their masses. Some spectrometers (*e.g.* IT instruments) have the function of collecting the charged fragments and submitting them to further CID events, allowing for the registration of the multistage fragmentation MS (MS^n) spectra corresponding to a sequential release of different uncharged parts of the original molecule. In most cases, however, the measurements of the [M + H]$^+$/[M − H]$^−$ along with the CID ions is performed in so-called tandem spectrometers, which consist of two or more analysers, one of which is the CID chamber and the other(s) register the mass spectra. It should be emphasized that only the CID mass spectrometers are useful for the natural products research, as most information about these compounds comes from the analysis of their fragmentation patterns.

Processes of ionisation of the analysed compounds, transport of ions to the analyser, the efficiency of the CID fragmentation and the possibility of quantitative determinations depend on optimization of multiple parameters of the mass spectrometer. It should be kept on mind that only a precise control of the most important settings allows a comparison of results obtained in different laboratories. The voltage applied at the electrospray ionization (ESI) ion source, along with the flow rate and temperature of the drying gas, are the crucial factors influencing the efficiency of the sample ionization. Typically, the ESI voltage is set to 2.5 to 4 kV and usually higher voltages promote a more effective in-source ion fragmentation. A spontaneous decomposition of the [M + H]$^+$/[M − H]$^−$ ions may occur due to this phenomenon, which may lead to errors in identification of the analysed compounds. The drying gas settings (the flow rate, pressure and temperature) should be sufficient for the total evaporation of liquids from the sample. As the ESI ion source operates at atmospheric pressure and a high vacuum is present in the mass analyser, the residual particles of the solvent (especially water) may be sucked into the analyser and drastically reduce its performance.

Ions generated in the source are transferred to the mass analyser and this process is assisted by different elements of so-called ion optics that may discriminate which ions are analysed and which are discarded. The ion optics depends on the spectrometer make, and patents used by individual producers and setting of its components must be optimised for each group of compounds. The collision energy set for the CID process is the most important factor influencing the yield of different fragments of the [M + H]$^+$/[M − H]$^−$ ions corresponding to analysed molecules. In general, the higher is the collision energy, the more abundant are small fragments of the analysed compounds. The influence of the collision energy on fragmentation of genistein-7-O,8-C-diglucoside is shown on Figure 17.1.

17.4 Sources of Structural Variability of Flavonoid Compounds

Isoflavones occur in plants mostly substituted with a various number of different sugar moieties and are then called aglycones, whereas the whole differently substituted compounds are referred to as glycoconjugates. Free isoflavone aglycones are not abundant in plant tissues, but their concentration may be increased as a result of different abiotic and biotic stresses. The naturally

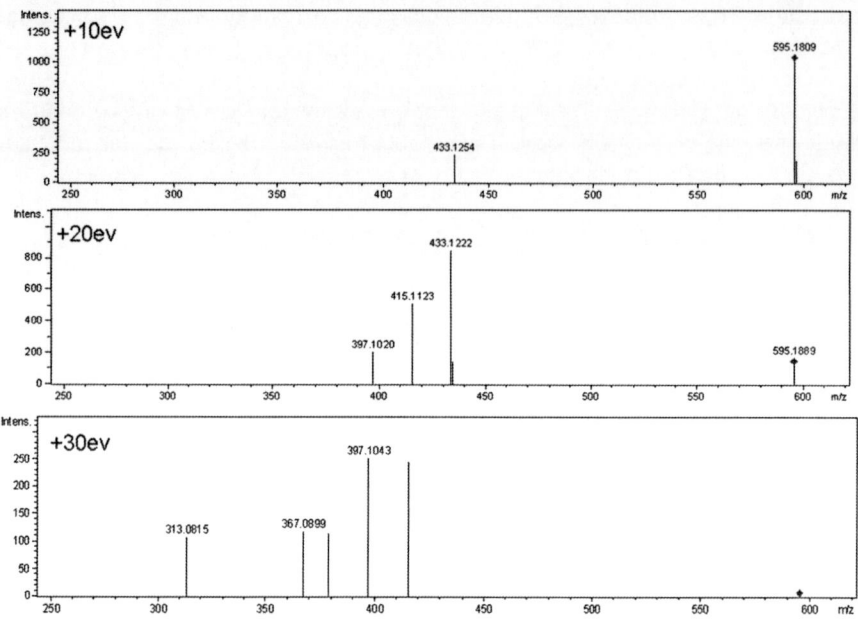

Figure 17.1 Dependence of intensity of product ions obtained in MS/MS spectra using micro-ToFQ spectrometer (Bruker Daltonics, Germany) at different collision energy settings. 7-O, 8-C diglucoside of genistein from *Lupinus rotundiflorus* was analysed with HPLC-MS in the positive ion mode with the collision energy set to 10 eV, 20 eV and 30 eV. (M. Stobiecki, unpublished.)

occurring isoflavonoids are present in plant tissues in complex mixtures, most often along with glycoconjugates of other flavonoids such as flavones and flavonols. There are several other sources of the observed structural variability among the flavonoids. Various types of hydroxylation, methoxylation or prenylation of the isoflavonoid core leads to the presence of such compounds such as daidzein, genistein, 2′-hydroxygenistein, biochanin, wighteone or luteone. Glycosylation of isoflavonoids leading to the formation of O-glycosides may be observed at different hydroxyl groups, most frequently at position 7 and/or 4′. Additionally, a direct glycosylation of carbon atoms of the ring A, at position 6 or 8, or both 6 and 8, is often observed and such compounds are called C-glycosides. The saccharide part of both types of glycosides consists most frequently of one or more glucose moieties, however, other hexoses (galactose), deoxyhexoses (rhamnose), pentoses (arabinose, xylose) or acidic sugars (glucuronic acid) are also common. These sugar moieties may be present as mono-, di-, or tri-glycosides attached at one or more positions of the isoflavonoid aglycone, giving rise to different O- or C-glycosylglucosides, di-O-; C-, O- or di-C-glycosides *etc*. Finally, esterification of either the sugar or the aglycone moiety with different aliphatic (malonic, malic) or aromatic acids (caffeic, ferulic) may also occur. All these various possibilities of substitution lead to hundreds of isoflavonoid compounds and frequently tens of them are simultaneously present in extracts from a plant tissue. For this reason, the analytical method designed for analysis and characterisation of the isoflavonoid glycoconjugates has to be capable of the resolution, differentiation and description of many very similar compounds present in the analysed sample. HPLC-MS methods are particularly well suited to this purpose.

17.5 Differentiation of Isoflavone and Flavone Glycoconjugates with Instrumental Methods

The structural feature that distinguishes isoflavones and flavones is the position of the ring B at which the ring C is attached and that is carbon 3 and 2, respectively. This structural feature notably influences the characteristic absorption of the ultraviolet (UV) light by molecules of both classes of secondary metabolites. In the case of isoflavones and their glycoconjugates one maximum of absorption is observed at a wavelength of approximately 260 nm with a shoulder at higher wavelength values; on the other hand, the UV absorption spectra of flavones and flavonols are characterised by the presence of two maxima at 280 nm and between 320 and 370 nm (Markham and Mabry 1975). This property has been used for the distinction of both classes of compounds analysed using HPLC instruments recording UV spectra of consecutive chromatographic peaks. However, it frequently happens that overlapping UV spectra of compounds belonging to different classes and co-eluting from the chromatographic column make this differentiation very difficult. The application of MS combined on-line with HPLC instruments provides additional possibilities of identification of the analysed compounds. Generally, the flavonoid glycoconjugates may be observed both in positive and negative ion modes, however, the mechanisms of fragmentation of $[M+H]^+$ and $[M-H]^-$ ions are different and give supplementary information. For this reason it is advised to carry on the metabolic profiling of these compounds in two consecutive chromatographic experiments using each mode of ionisation in separate runs. The m/z values of protonated and deprotonated molecule ions of flavonoid O-glycosides are observed in the first order mass spectra (MS) spectra and the tandem MS (MS/MS) spectra provide fragments obtained in result of the CID process. In most cases an abundant ion corresponding to the aglycone (denoted as Y_0^+ or Y_0^- depending on the ionisation mode) is observed in these spectra (for the notation of ions produced in result of fragmentation of flavonoids, see Domon and Costello 1988). Products of further fragmentation of Y_0^+ and Y_0^- of flavones and isoflavones are usually different (Figure 17.2.) and allow for identification of the individual compounds. As it may be observed in Figure 17.2, two types of Y_0^- ions differing in their m/z value by 1 amu are created that correspond to the even electron ($[M-H-sugar]^- = Y_0^-$) and odd electron ($[M-H-sugar-H^\bullet]^{-\bullet} = [Y_0^- - H^\bullet]$) product ions (March et al. 2004). However, since the relative intensity of ions resulting from the CID depends strongly on the setting of the spectrometer (especially the collision energy and

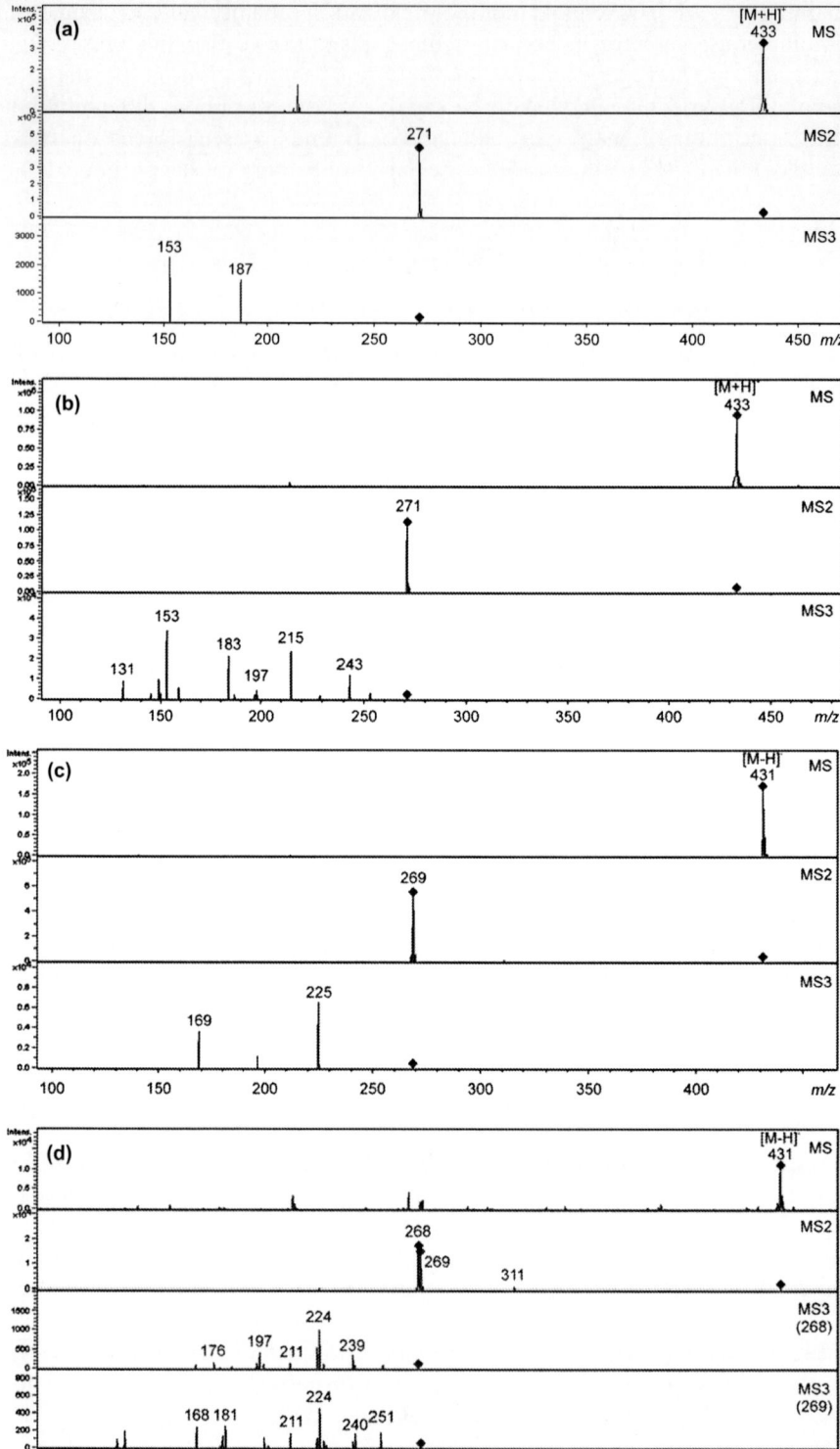

ionisation potential), several standard compounds should be analysed using the same analysis parameters for a proper identification of compounds present in plant material. Nevertheless, MS/MS spectra registered as a result of the fragmentation of the $[M \pm H]^{+/-}$ ions of the aglycone standards are comparable to spectra showing the decomposition of the $Y_0^{+/-}$ product ions obtained in result of the MS^n or pseudo MS^3 experiments of the corresponding flavonoid glycoconjugates (Kachlicki et al. 2005; van der Hooft et al. 2011).

17.6 Differentiation of *C*-Glycosides and *O*-Glycosides of Isoflavonoids

As it has been shown in the previous Section (Figure 17.2), the CID fragmentation of the *O*-glycosides of flavonoids, in general, leads to the charged $Y_0^{+/-}$ fragments with the release of uncharged fragments of the glycosidic moieties. In case of the isoflavonoid glycoconjugates containing more than one glycosidic units, a sequential release of fragments corresponding to monosugars is observed that leads to the formation of the $Y_1^{+/-}$ or $Y_2^{+/-}$ ions (Kachlicki et al. 2005). However, the C–C bond between the glycosidic moiety and the aglycone present in the *C*-glycosides is much more stable than the C–O bond of the *O*-glycosides. For this reason the CID reaction of the former group of compounds causes a sequential decomposition of the glycosidic moiety rather than its release as a whole. As it can be seen in Figure 17.3, many different fragments are formed as result of the CID of the positively charged genistein 6,8-di-*C*-glucoside. The MS^n spectra of this compound obtained in the negative ion mode are dominated by the release of the 120 amu fragments coming from the disruption of the C_1–O and C_2–C_3 bonds of each glucose moiety. Plants containing isoflavonoids frequently synthesise, in addition to simple *C*- or *O*-substituted glycosides, mixed *O,C*-diglycosides or *C*-glycosylglycosides. An exemplary spectrum of such a compound is shown in Figure 17.1. However, a precise differentiation of the last two groups of isoflavone *C*-glycoconjugates is a very difficult task and seldom may be achieved exclusively using the MS methods. It must be also noted that as the fragmentation of the *C*-glycosides of all types of flavonoids occurs at the glycosidic moiety rather than at the aglycone, it is not possible to distinguish glycoconjugates of flavones and isoflavones using the mass spectra. So apigenin-6,8-di-*C*-glucoside and genistein-6,8-di-*C*-glucoside that both are present in leaves of different lupin species may be identified only on the basis of their retention time and the corresponding UV spectra.

Figure 17.2 Sequential mass spectra of apigenin 7-*O*-glucoside (a, c) and genistein 7-*O*-glucoside (b, d) obtained in positive (a, b) and negative (c, d) ion modes using an Esquire 3000 IT spectrometer (Bruker Daltonics, Germany).
(P. Kachlicki, unpublished.)

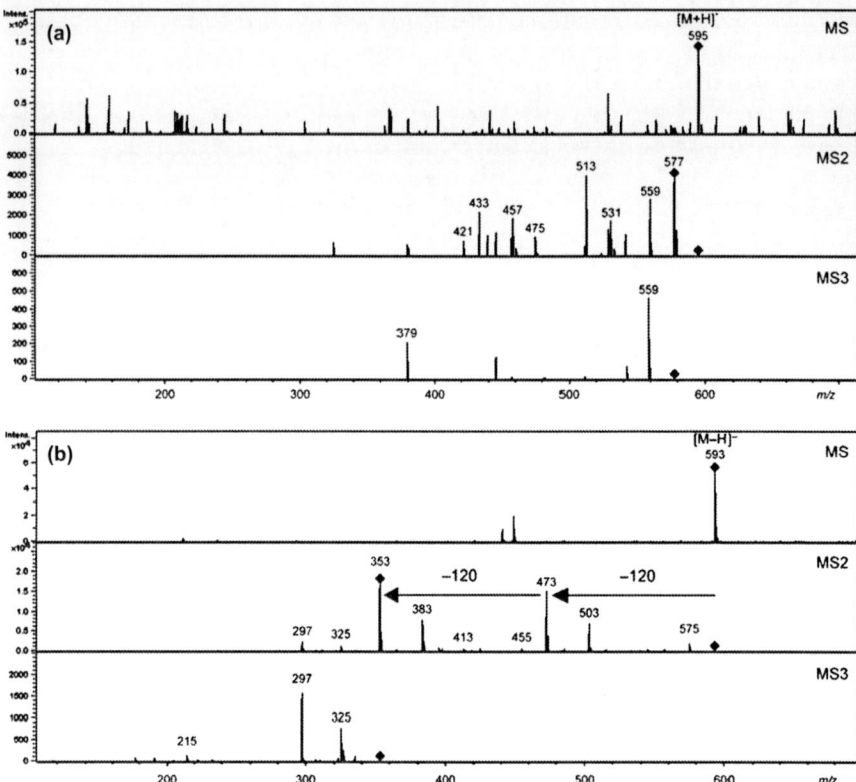

Figure 17.3 Sequential mass spectra of genistein 6,8-di-*C*-glucoside from *Lupinus hintoni* obtained in positive (a) and negative (b) ion modes using an Esquire 3000 IT spectrometer (Bruker Daltonics, Germany).
(P. Kachlicki, unpublished.)

17.7 Differentiation of Acylated Glycoconjugates of Isoflavonoids

Acylation of flavonoid glycoconjugates is a ubiquitous phenomenon and compounds esterified with aliphatic as well as aromatic acids are observed in many monocotyledonous and dicotyledonous plant species. The detection of such flavonoid derivatives using the LC-MS methods is not difficult as the acid moiety is easily ruptured from the molecule during the CID experiments, especially in the positive ion mode (Figure 17.4). In case of common malonylated derivatives such as genistein 7-*O*-malonylglucoside (Figure 17.4) analysed in the negative ion mode some problems may be expected as the $[M-CO_2-H]^-$ ions are extremely easy obtained instead of the $[M-H]^-$ ions. Another problem that frequently occurs during the identification of malonylated isoflavone glycosides is connected with the existence of multiple positional isomers in which malonic acid is attached at different glycosidic

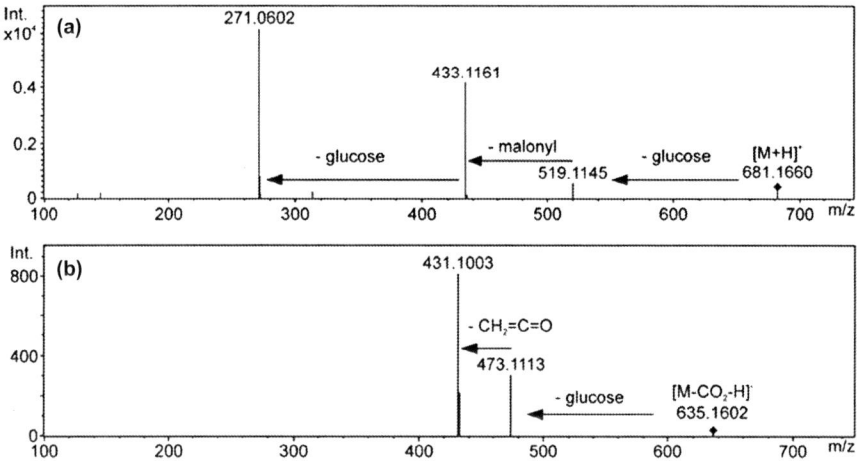

Figure 17.4 Mass spectra of malonylated genistein 4′,7-di-O-glucoside registered in (a) positive and (b) negative ion modes.
(Reprinted with permission from Stobiecki et al. 2010. Copyright 2010 American Chemical Society.)

moieties or at different positions of the same sugar unit. For example, both malonylated and dimalonylated genistein 4′,7-O-diglucosides present in narrow leaf lupin (*L. angustifolius* L.) have been found to have three isomeric forms (Muth et al. 2008). These compounds could be separated using by UPLC, but their MS/MS spectra were quite similar and did not reveal the positions of the esterification. de Rijke et al. (2004) solved a similar problem with the identification of isomeric malonylated isoflavonoids in clover (*Trifolium pratense* L.) using LC-nuclear magnetic resonance (NMR) analysis. The analysis of one dimensional ^{1}H and two dimensional correlation NMR spectra revealed the presence of two formononetin 7-O-glucosides that are malonylated at positions 4″ and 6″. However, such analytical approach may be applied in the case of well separated compounds occurring in relatively large amounts, as the sensitivity and selectivity of NMR spectrometric methods are lower than those of the MS techniques. Some structural information regarding the pattern of glycosylation and/or esterification may be obtained from the analysis of flavonoids ionised with metal ions added to the sample or to the column effluent. Systematic investigation conducted after application of double- or triple-charged metal cations (copper, zinc, cobalt, nickel, aluminium and iron) for the ionisation of standard compounds has been described (March and Brodbelt 2008). Post-column addition of sodium salts to LC-MS analysed narrow leaf lupin phenolic compounds allowing the location of the malonyl group at each sugar unit of isomeric flavone chrysoeriol 4′-glucosyl-7-xylosylglucosides (Kachlicki et al. 2008).

Malonylation of the glycosidic moiety of the isoflavone glycoconjugates is not the only possibility of esterification of these natural products. Analyses of

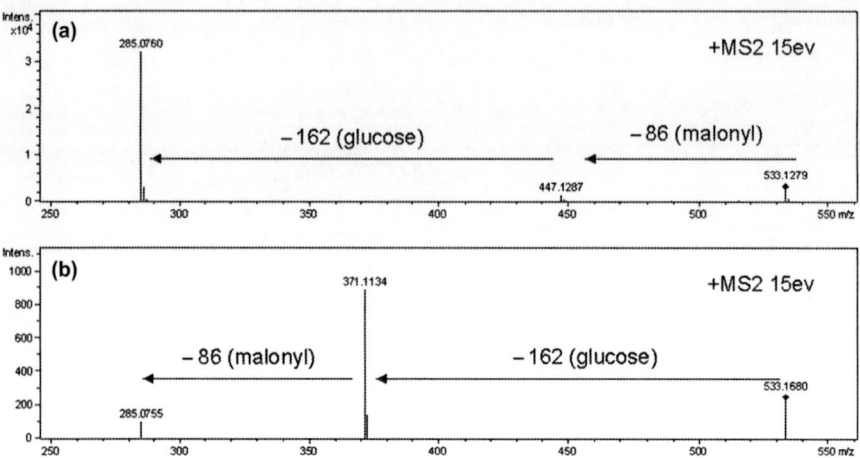

Figure 17.5 Mass spectra of malonylated biochanin A 7-*O* monoglucosides from *Lupinus stipulatus*. The malonyl moiety is attached to the sugar moiety (a) or to the aglycone (b).
(M. Stobiecki, unpublished.)

leaf extracts from different Mexican lupin species revealed the presence of isoflavone glycosides substituted with other aliphatic acids such as malic acid as well as methyl esters of both malic and malonic acids (Stobiecki *et al.* 2010). These acids may be attached either at the sugar part or at the aglycone of the analysed compounds and such substituted isoflavonoids can be distinguished according to their fragmentation pattern (Figure 17.5). The MS/MS spectrum of a malonylated glucoside of biochanin B esterified at the aglycone (Figure 17.5b) clearly reveals the ion at 371 m/z corresponding to the malonylated biochanin fragment. Other ways of esterification of isoflavonoids are also observed. Glucuronidated and acylated with *p*-coumaric or ferulic acid isomeric derivatives of genistein and daidzein were detected in roots of *Medicago truncatula* using the HPLC-MS profiling (Staszków *et al.* 2011). Partial structural characterisation of these natural products with a high-resolution ToF analyser was possible. It is noteworthy that the nominal masses (measured with unit accuracy) of uncharged fragments corresponding to a release of different groups may be equal. For example, fragments corresponding to deoxyhexose (*e.g.* rhamnose) and *p*-coumaric acid have mass equal to 146 amu and similar holds for hexose and caffeic acid (162 amu) or glucuronic acid and ferulic acid (176 amu). These compounds may be distinguished according to the elucidation of their elemental composition done according to the mass measurement with the accuracy to the fourth decimal point of amu with maximum 5 ppm difference. Only on this basis it is possible to differentiate substitution with acyl (coumaroyl and feruloyl) or glucose and glucuronic acid moieties, respectively. It is interesting that leaves of *M. truncatula* contain mainly acylated glucuronides of flavone apigenin (Marczak *et al.* 2010).

Summary Points

- LC-MS techniques are superior methods for profiling isoflavonoid glycoconjugates in plant tissues and other biological materials.
- Especially powerful in this respect are modern high-resolution liquid chromatographs (UPLC and equivalent instruments) hyphenated to fast high-resolution mass spectrometers.
- Separation, identification and at least partial structural characterisation of multiple isomeric isoflavonoid glycoconjugates on the basis of LC-MS experiments are possible.
- Different patterns of substitution of the isoflavones such as glycosylation, alkylation or acylation may be revealed.
- Special care must be paid to standardise analytical conditions to achieve full repeatability of results in order for the comparison of results obtained in different laboratories.

Key Facts

- Fragmentation patterns of protonated or deprotonated molecules of isoflavonoid glycoconjugates registered during LC-MS experiments allow to differentiate numerous compounds present in plant tissues and at least partially identify their structures.
- Registration of exact masses (with the accuracy to the fourth decimal point) of the analyzed compounds allows distinguishing isobaric conjugates with different atomic compositions.
- The application of tandem mass spectrometers capable for registration of high resolution MS/MS spectra connected to ultra-high performance liquid chromatographs is recommended for profiling of isoflavonoid glycoconjugates in plant tissues.

Definitions of Word and Terms

Acylation: Isoflavones are frequently esterified with aliphatic or aromatic acids (mainly with malonic or hydroxycinnamic acids). The acylation takes place either on the aglycone or sugar moiety.

Aglycone: The core part of the isoflavone conjugate, consisting of three aromatic rings with a characteristic pattern of hydroxylation, alkylation or acylation.

Chromatography-mass spectrometry: The hyphenation of different chromatographic techniques (gas chromatography, liquid chromatography or capillary electrophoresis) with mass spectrometry or NMR spectroscopy.

Collision-induced dissociation (CID): The rupture of an ion after collisions with atoms of atoms of noble gas (helium or argon).

Glycoconjugates: The native form of isoflavonoids that occur in plant tissues mainly in a form of glycosides containing one or more sugar moieties consisting of one to four saccharide units.

Mass-to-charge ratio (m/z): After ionization in the ion source, ions are created with a single charge, m/z value corresponds to the mass of a defined ion.

Metabolite profiling: The separation, identification and quantification of secondary metabolites present in the samples of biological material from organisms grown in different environmental conditions (biotic and abiotic).

Metabolomics: The systematic study of the unique chemical status of a cell, organ/tissue or whole organism.

Productions: The fragment ions obtained during CID MS/MS experiments with all types of MS/MS analysers.

Protonated/deprotonated molecule: The positive/negative ion created after proton attachment or elimination to/from an organic compound molecule during electrospray ionization (ESI).

Tandem mass spectrometry (MS/MS): The application of a double analyser in mass spectrometer for the analysis of product ions created after collisions.

Systems biology: This describes the complex interactions occurring in an organism on different molecular and organizational levels; the analysis and integration of data from: genome, transcriptome, proteome, metabolome, lipidome or interactomics and phenomics studies.

List of Abbreviations

amu	atomic mass unit ($1.66053886 \times 10^{-24}$ g)
CID	collision-induced dissociation
ESI	electrospray ionization
FT	Fourier transform
HPLC	high-performance liquid chromatography
IT	ion trap
LC	liquid chromatography
$[M + H]^+$	protonated molecule
$[M - H]^-$	deprotonated molecule
MS	mass spectrometry
MS/MS	tandem mass spectrometry
MS^n	multistage fragmentation mass spectrometry
m/z	mass-to-charge ratio
NMR	nuclear magnetic resonance
ToF	time-of-flight
UPLC	ultra-performance liquid chromatography
UV	ultraviolet

References

de Rijke, E., de Kanter, F., Ariese, F., Brinkman, U.A.T., and Gooijer, C., 2004. Liquid chromatography coupled to nuclear magnetic resonance spectroscopy for the identification of isoflavone glucoside malonates in *T. pratense* leaves. *Journal of Separation Science*. 27: 1061–1070.

Domon, B., and Costello, C.E., 1988. A systematic nomenclature for carbohydrate fragmentations in FAB MS/MS spectra of glycoconjugates. *Glycoconjugates Journal*. 5: 397–409.

Edwards, E.L., Rodrigues, J.A., Ferreira, J., Goodall, D.M., Rauter, A.P., Justino, J., and Thomas-Oates, J., 2006. Capillary electrophoresis-mass spectrometry characterisation of secondary metabolites from the antihyperglycaemic plant *Genista tenera*. *Electrophoresis*. 27: 2164–2170.

Kachlicki, P., Marczak, Ł., Kerhoas, L., Einhorn, J., and Stobiecki, M., 2005. Profiling isoflavone conjugates in root extracts of lupine species with LC/ESI/MS^n systems. *Journal of Mass Spectrometry*. 40: 1088–1103.

Kachlicki, P., Einhorn, J., Muth, D., Kerhoas, L., and Stobiecki, M., 2008. Evaluation of glycosylation and malonylation patterns in flavonoid glycosides during LC/MS/MS metabolite profiling. *Journal of Mass Spectrometry*. 43: 572–586.

March, R.E., and Brodbelt, J., 2008. Analysis of flavonoids: tandem mass spectrometry, computational methods, and NMR. *Journal of Mass Spectrometry*. 43: 1581–1617.

March, R.E., Miao, X.S., Metcalfe, C.D., Stobiecki, M., and Marczak, Ł., 2004. A fragmentation study of an isoflavone glycoside, genistein-7-O-glucoside, using quadrupole time of flight mass spectrometry at high mass resolution. *International Journal of Mass Spectrometry*. 232: 171–183.

Marczak, Ł., Stobiecki, M., Jasiński, M., Oleszek, W., and Kachlicki, P., 2010. Fragmentation pathways of acylated flavonoid diglucuronides from leaves of *Medicago truncatula*. *Phytochemical Analysis*. 21: 224–233.

Markham, K.R. and Mabry, T.J., 1975. Ultraviolet-visible and proton magnetic resonance spectroscopy of flavonoids. In: Harborne, J. B., Mabry, T.J. and Mabry, H. (ed.) The Flavonoids. Chapman and Hall Ltd., London, UK, pp. 45–77.

Muth, D., Marsden-Edwards, E., Kachlicki, P., and Stobiecki, M., 2008. Differentiation of isomeric malonylated flavonoid glyconjugates in plant extracts with UPLC-ESI/MS/MS. *Phytochemical Analysis*. 19: 444–452.

Staszków, A., Swarcewicz, B., Muth, D., Jasiński, M., and Stobiecki, M., 2011. LC-MS profiling of flavonoid glycoconjugates isolated from hairy roots, suspension root cell cultures and seedling roots of *Medicago truncatula*. *Metabolomics*. 7: 604–613.

Stobiecki, M., Staszków, A., Piasecka, A., Garcia-Lopez, P.M., Zamora-Natera, F., and Kachlicki, P., 2010. LC-MSMS Profiling of flavonoid conjugates in wild Mexican lupine *Lupinus reflexus*. *Journal of Natural Products*. 73: 1254–1260.

Valls, J., Millan, S., Marti, M.P., Borras, E., and Arola, L., 2009. Advanced separation methods of food anthocyanins, isoflavones and flavonols. *Journal of Chromatography A*. 1216: 7143–7172.

van der Hooft, J.J.J., Vervoort, J., Bino, R.J. and de Vos, R.C.H., 2011. Spectral trees as a robust annotation tool in LC-MS based metabolomics. *Metabolomics*. doi:10.1007/s11306-011-0363-7 (in press).

CHAPTER 18
Puerariae radix *Isoflavones*

LEI WAN*[a] AND CHIA-HUNG LIN[b]

[a] School of Chinese Medicine, China Medical University, No. 91 Hsueh-Shih Road, Taichung, Taiwan, and Department of Health and Nutrition Biotechnology, College of Health Science, No. 500, Lioufeng Road, Wufeng, Taichung, Taiwan; [b] Institute of Molecular Medicine, National Tsing Hua University, 101, Section 2, Kuang Fu Road, Hsinchu, Taiwan
*Email: leiwan@mail.cmu.edu.tw

18.1 Introduction

Puerariae radix (PR), the root of the vine species *Pueraria lobata*, or kudzu, is used widely as a remedy in Chinese clinical healthcare and as a food in many east Asian countries (Chen *et al.* 2001) (Figure 18.1). In traditional practice, it is used to relax muscles, soothe burns and rashes, treat diarrhea and dysentery, alleviate fevers, relieve thirst by promoting salivation, and remove bacteria and viruses (Zhang *et al.* 2010).

A class of chemical compounds known as phytoestrogens accounts for many of the disease-fighting properties of PR (Cherdshewasart *et al.* 2008). The phytoestrogens found in PR are isoflavones, a type of flavonoid (Yan *et al.* 2006a). In edible plants and Chinese herbs, flavonoids constitute the largest and most important group of polyphenolic compounds—chemical compounds that affect the color and taste of the vegetables in which they are found (Chen *et al.* 2010). The molecular structure of flavonoids consists of two benzene rings, joined together by a linear three-carbon chain that forms an oxygenated heterocycle (Prasain *et al.* 2007) (Figure 18.2). Flavonoids can be divided into

Food and Nutritional Components in Focus No. 5
Isoflavones: Chemistry, Analysis, Function and Effects
Edited by Victor R Preedy
© The Royal Society of Chemistry 2013
Published by the Royal Society of Chemistry, www.rsc.org

Figure 18.1 Image of the *Puerariae radix* (PR) root. Radices Puerariae is a national plant of the People's Republic of China. The primary active constituents—and the effective components—of *Radix Puerariae lobatae* are puerarin and flavone.

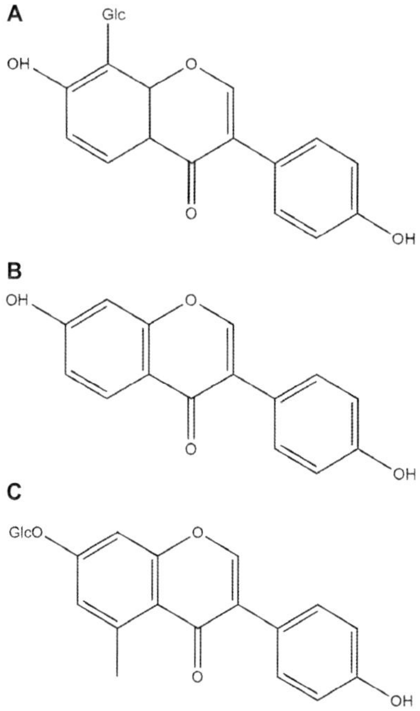

Figure 18.2 Chemical structure of isoflavonoids from *Radix Puerariae*. (A) Puerarin, (B) daidzein, and (C) daidzin are the major isoflavones present in PR, which is widely used in Chinese clinics as a medicinal therapy to relax muscles, soothe burns and rashes, treat diarrhea and dysentery, alleviate fevers, relieve thirst by promoting salivation, and remove bacteria and viruses.

six subclasses: flavones, flavanones, flavanols, flavonols, anthocyanidins, and isoflavones (Yang et al. 2001). These compounds confer various biological properties, including anti-allergic (Shimosaki et al. 2011), anti-inflammatory (Korkina et al. 2011), anti-viral (Sanchez et al. 2000), anti-oxidative (Bebrevska et al. 2010), and anti-tumor activities (Le Bail et al. 1998). Over the past few years, flavonoids have been shown to be capable of suppressing cancer cell proliferation and inducing apoptosis (Valachovicova et al. 2004) (Figure 18.3).

The isoflavones such as puerarin, daidzin, and daidzein are regarded as phytoestrogens because of their estrogenic activity in certain animals (Cheng et al. 1955; Kallela 1975; Miksicek 1995). Phytoestrogens exert dose-dependent estrogenic effects on the reproductive system of ovariectomized rats, cyclic female monkeys, and aged menopausal monkeys. In a study of male rats, a diet high in phytoestrogens disrupted the male sex organs, including the epididymis and seminal vesicles, as well as affecting sperm motility and viability (Glover and Assinder 2006). However, such a diet protected against osteoporosis in orchidectomized male rats (Zheng et al. 2002a; Zou et al. 2009). In addition to this estrogenic activity, phytoestrogens are known to relieve or to cure several different disorders (Bagheri et al. 2011). In this Chapter, we present the medicinal effects of phytoestrogenic isoflavones of PR.

18.2 Use of Resources

P. lobata, or kudzu, is a perennial vine that reaches a total length of approximately 10 m. There are 18 different kudzu species worldwide, most of which are found in the subtropical and temperate regions (Liu et al. 2011) (Table 18.1). *P. lobata* are widely distributed in China, except in Tibet, Xinjiang, and Qinghai. In Taiwan, *P. lobata* is mostly cultivated. China, Yunnan, and its neighboring provinces are the major production centers of the plant. Nine species and two varieties of kudzu grow in this region. The *P. lobata* varieties that are most widely used in food or medicine are those that contain the most flavonoids.

18.3 Chemical Composition

PR is abundant in isoflavones, including 3-hydroxypuerarin, puerarin, 3′-methoxypuerarin, daidzin, daidzein, genistein, and formononetin-7-glucoside (Jungbauer et al. 2003; Liang et al. 2005). The major isoflavones in PR are puerarin and daidzin (Zhao et al. 2011). The total content of isoflavones in PR may reach more than 30 mg (g of dry weight)$^{-1}$, of which puerarin comprises 50%. Daidzin content may reach 8 mg (g of dry weight)$^{-1}$.

The content and concentration of isoflavones in PR are thought to correlate with its quality and therapeutic efficacy. The concentrations of isoflavones vary with the age and harvest season of PR, and these concentrations significantly affect the therapeutic efficacy of the roots and that of the medicines derived

Table 18.1 Individual isoflavonoid and total isoflavonoid contents in mg/100 g of *Pueraria mirifica* tuberous powder collected from 28 provinces of Thailand, in comparison with *P. lobata* from China. *P. mirifica* tubers from 28 provinces showed interprovincial diversity for puerarin, daidzin, genistin, daidzein, genistein, and total isoflavonoid, as calculated from HPLC fingerprints. Data are from Cherdshewasart *et al.* (2007), with permission from the Publisher.

No.	Province	Puerarin	Daidzin	Genistin	Daidzein	Genistein	Total
1	Kanchanaburi	45.25 ± 1.11*	50.24 ± 3.23*	85.69 ± 1.23*	13.92 ± 1.26*	3.19 ± 0.29*	198.29 ± 4.61*
2	Lamphnn	33.18 ± 0.92	28.35 ± 0.68*	84.13 ± 0.54*	8.59 ± 0.09	0.76 ± 0.36	155.00 ± 1.42*
3	Chiang Mai	35.55 ± 3.57	27.39 ± 5.32*	58.00 ± 0.71*	8.38 ± 0.22	1.93 ± 0.54	131.25 ± 9.01*
4	Siikon Nakhon	87.05 ± 0.79*	11.48 ± 0.21	14.83 ± 0.22†	4.78 ± 0.37	1.42 ± 0.14	119.57 ± 1.39*
5	Mae Hong Son	36.99 ± 2.07	17.63 ± 1.74	55.44 ± 3.43*	7.52 ± 1.27	1.54 ± 0.08	119.12 ± 6.54*
6	Ulhai Thani	10.85 ± 1.01	21.70 ± 0.84	50.17 ± 3.57	16.48 ± 1.35*	3.66 ± 0.16*	102.86 ± 6.53
7	Sukhothai	14.12 ± 0.94	25.09 ± 1.50*	51.43 ± 2.40	11.16 ± 0.85*	0.73 ± 0.23	102.52 ± 5.35
8	Lampang	34.65 ± 1.34	16.59 ± 0.08	33.30 ± 0.08	5.72 ± 0.09	1.54 ± 0.12	91.80 ± 1.72
9	Tak	29.06 ± 2.07	8.97 ± 0.99	43.86 ± 1.91	4.56 ± 0.32	1.15 ± 0.19	87.60 ± 4.87
10	Saraburi	23.42 ± 1.21	17.92 ± 0.59	37.94 ± 142	4.86 ± 0.95	0.87 ± 0.46	85.01 ± 4.04
11	Ratcliaburi	8.85 ± 0.36	15.39 ± 0.79	51.15 ± 1.75	6.84 ± 0.53	2.54 ± 0.15*	84.77 ± 2.67
12	Phitsanulok	35.24 ± 1.06	12.26 ± 0.13	26.53 ± 0.57	8.36 ± 0.23	1.63 ± 0.05	84.02 ± 1.91
13	Phelcliaburi	13.19 ± 0.45	20.82 ± 1.78	37.56 ± 1.33	6.00 ± 0.24	1.13 ± 0.04	78.71 ± 3.15
14	Plnac	25.20 ± 1.54	10.55 ± 1.18	30.61 ± 0.81	5.45 ± 0.56	1.34 ± 0.14	73.16 ± 4.21
15	Lop Buri	19.50 ± 1.44	6.84 ± 0.09	39.47 ± 1.65	2.42 ± 0.79†	0.98 ± 0.09	69.21 ± 3.77
16	Chaiyaphum	15.83 ± 2.43	12.91 ± 1.44	29.48 ± 2.33	7.02 ± 0.89	1.89 ± 0.42	67.13 ± 5.47
17	Uttraradith	30.25 ± 0.44	13.69 ± 0.21	10.27 ± 0.19†	7.88 ± 0.18	0.00†	62.96 ± 1.03
18	Nakhon Sawan	13.34 ± 1.46	16.28 ± 1.64	27.71 ± 0.75	4.70 ± 0.37	0.72 ± 0.32	62.75 ± 3.24
19	Chiang Raj	20.02 ± 1.42	8.61 ± 1.12	29.58 ± 2.43	2.16 ± 0.08†	0.50 ± 0.28	60.87 ± 3.30
20	Nong Bua Lam Phu	12.65 ± 2.42	11.91 ± 3.02	23.65 ± 2.14	7.46 ± 0.96	1.91 ± 0.25	57.58 ± 3.61
21	Phayao	12.91 ± 0.99	8.46 ± 0.62	32.43 ± 1.35	3.03 ± 0.36	0.73 ± 0.30	57.56 ± 3.02
22	Prachuap Kbiri Khan	10.42 ± 1.03	9.62 ± 0.44	30.31 ± 1.17	2.11 ± 0.36†	0.59 ± 0.07	53.05 ± 2.57
23	Chumphon	8.45 ± 0.22	7.38 ± 1.11	34.17 ± 4.81	2.64 ± 0.26	0.07 ± 0.06†	52.70 ± 5.46
24	Prachin Buri	12.42 ± 0.26	13.05 ± 0.65	16.69 ± 0.78†	4.28 ± 0.56	0.51 ± 0.09	46.94 ± 1.12
25	Phetchabun	9.40 ± 0.46	10.48 ± 0.67	15.54 ± 1.61*	8.11 ± 0.05	1.29 ± 0.02	44.83 ± 1.73†
26	Nakhon Ratchasima	13.09 ± 0.77	5.61 ± 0.07†	24.15 ± 1.42	1.20 ± 0.37†	0.21 ± 0.19†	44.27 ± 1.27†
27	Kainphaeng Phet	15.44 ± 1.14	7.01 ± 1.10	18.50 ± 4.45	2.31 ± 0.11*	0.46 ± 0.08	43.71 ± 4.02†
28	Nan	5.32 ± 0.44†	2.36 ± 0.22†	7.62 ± 1.36†	3.31 ± 0.31	0.00 ± 0.00†	18.61 ± 1.11†
	Mean ± S.E.M.	23.01 ± 1.80	14.94 ± 1.07	35.39 ± 2.09	6.12 ± 0.40	1.19 ± 0.10	80.67 ± 4.11
	Pueraha lobafa	32.85 ± 0.72	21.9 ± 0.74	25.63 ± 0.86	10.34 ± 0.79*	0.81 ± 0.08	91.57 ± 3.18

*Greater than mean at $P < 0.05$.
†Less than mean at $P < 0.05$.

from them (Sibao et al. 2007). Traditionally, however, PR is harvested according to the size of the root, not according to its chemical composition (Chen et al. 2010). Developed a high-performance liquid chromatography (HPLC) method to determine the concentration of the seven major isoflavones (puerarin, 3'-methoxypuerarin, daidzin, genistein, formononetin-7-glucoside, and daidzein) present in PR. They used these measurements to determine the optimum time to harvest PR. Their research found that isoflavone content increased with age from the second to the third year, when it reached its maximum concentration, before gradually decreasing. Regarding season, isoflavone concentrations reach their highest levels in winter (December to March) and their lowest in spring and summer (April to August) (Chen et al. 2010). Therefore, concluded that harvesting PR in January of its third year will produce the highest yields of isoflavone compounds (Beecher, 2003; Chen et al. 2010).

18.4 Pharmacology and Potential Applications

PR extracts containing puerarin and other isoflavones have shown potential for supporting various types of biological activity *in vivo* and *in vitro*. Puerarin and daidzin account for approximately 75% of the total isoflavones in PR (Lau et al. 2009). Thus, we will focus on these compounds—along with their aglycone form daidzein—briefly discussing their biological activity and their potential applications.

18.4.1 Anti-tumor Activity

Cancer cells are characterized by uncontrolled proliferation. Therefore, cancer is usually treated by inhibiting abnormal cell growth or by inducing programmed cell death (apoptosis). PR, as well as the crude extracts and the single compounds derived from it, is known to induce apoptosis in several types of cancer cells (Figure 18.3).

18.4.1.1 Puerarin

Puerarin can prevent the well-known cancer cell lines Hs578T, MDA-MB-231, and MCF-7 from entering the G0/G1 phase of the cell cycle and can induce apoptosis of these cells (Lin et al. 2009) (Figure 18.4). Puerarin inhibits the expression of the drug resistance gene *MDR1* in adriamycin-resistant MCF-7 cells (MCF-7/adr), hence reversing the multidrug resistance (MDR) phenotype of this cell line. The inhibition of MDR1 expression is mediated through the inhibition of nuclear factor κB (NF-κB) activity and through the degradation of inhibitor of κB (IκB), as well as through the stimulation of AMP-activated protein kinase (AMPK) and acetyl-CoA carboxylase and through glycogen synthase kinase 3-b phosphorylation. Puerarin also induces apoptosis of HT-29 colon cancer cells by inhibiting the expression of *bcl-2* and c-*myc* and by up-regulating bax proteins (Hien et al. 2010).

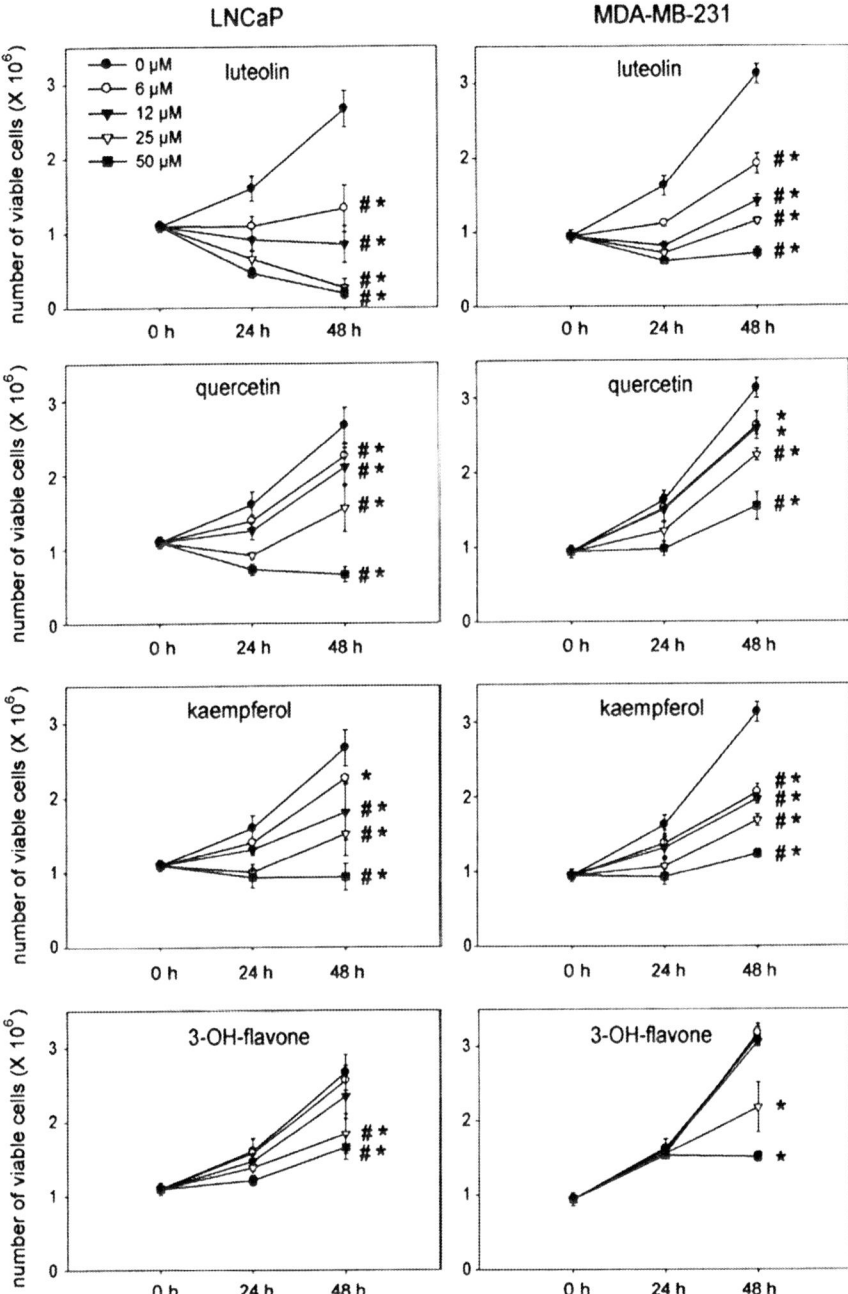

Figure 18.3 Flavonoids inhibit cancer cell proliferation. LNCaP cells and MDA-MB-231 cells were treated with different concentrations of flavonoids. Administering flavonoids reduced cancer cell growth over time, in a dose-dependent manner.
Data are from Brusselmans *et al.* (2005), with permission from the Publisher.

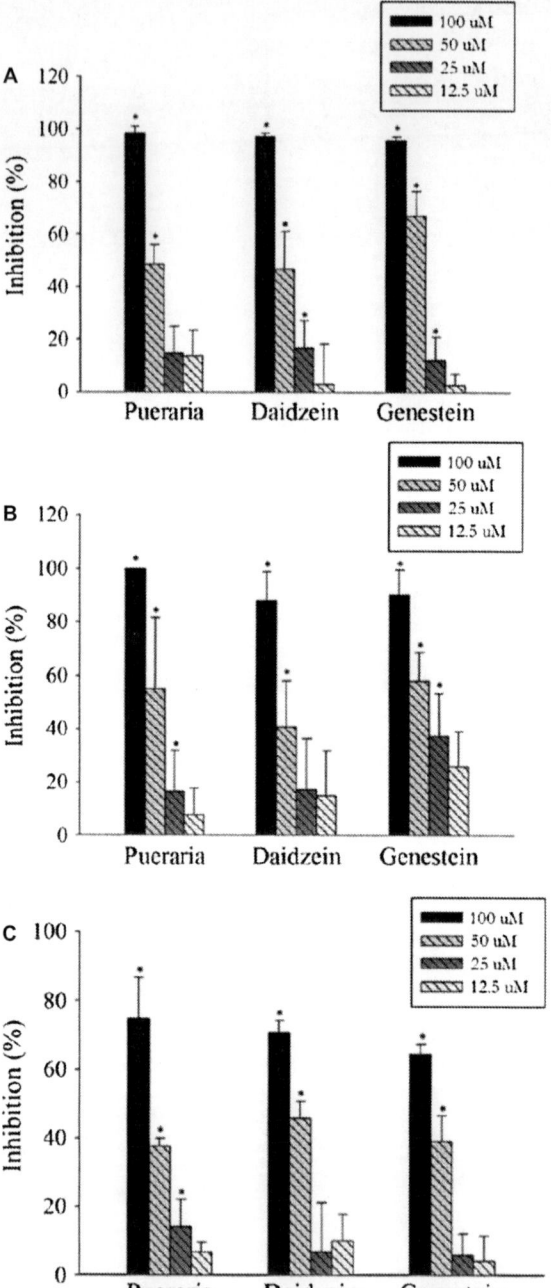

Figure 18.4 Effect of puerarin, daidzein, and genistein on cell viability of Hs578T(A), MDA-MB-231(B), and MCF-7(C) cells. Puerarin, daidzein and genestein inhibit cell growth dose-dependently. Cells are treated with different concentrations of puerarin, daidzein and genestein for 72 h. The viability of the cells is determined by the MTT assay.
Data are from Lin *et al.* (2009), with permission from the Publisher.

18.4.1.2 Daidzin

Daidzin has been shown to reduce the migration of MDA-MB-231 cells (Valachovicova et al. 2004). However, apoptosis-inducing activity has seldom been reported. Daidzin is relatively non-toxic, compared with its aglycone form, daidzein. Daidzin confers strong chemopreventive benefits when included in the diet of rats at a concentration of 0.1% for 40 weeks. At these levels, it will reduce the number of prostate tumors induced by the injection of 3',2'-dimethyl-4-aminobiphenyl (Kato et al. 2000). Daidzin also confers strong chemopreventive properties against heterocyclic-amine-induced colon cancer, mammary cancer, and hepatoma (Yu and Li, 2006).

18.4.1.3 Daidzein

Daidzein, the aglycone form of daidzin, inhibits several types of tumor cells, including those found in breast cancer, prostate cancer, glioma, neuroblastoma, colon cancer, and ovarian cancer (Yu and Li, 2006). It functions by inhibiting precancerous lesions, inducing apoptosis, regulating the immune response, and inhibiting angiogenesis and metastasis. The induction of apoptosis is mediated through the activation of caspases 3, 7, and 9; through up-regulation of bax, p21, and p53; and through down-regulation of bcl-2 and bcl-XL (Jin et al. 2010). Daidzein arrests the cell cycle in the G2/M phase and induces apoptosis in breast cancer cells (Lin et al. 2009). Cell cycle regulatory gene analysis has shown that daidzein increases cyclin-dependent kinase 1 (CDK1) levels but decreases CDK2 and CDK4 levels (Casagrande and Darbon 2001).

Daidzein (20 mg kg^{-1}) also stimulates immune responses, as shown by increases in the phagocytic activity of peritoneal macrophages (Zhang et al. 1997). Activation of the humoral immune response was shown by an increase in the number of IgM-producing cells compared with the number of sheep red blood cells. Stimulation of the cell-mediated immune response was shown by increases in peripheral blood T-cells as compared with a saline control. Daidzein inhibited 49% of angiogenesis compared with negative controls, as observed in a chicken chorioallantoic membrane assay. Daidzein also inhibited the expression of several angiogenesis-related genes in prostate cancer cells, including alanyl aminopeptidase (ANPEP), cadherin 5 (CDH5), chemokine (C-X-C motif) ligand 10 (CXCL10), platelet-derived endothelial cell growth factor (ECGF1), endoglin (Osler–Rendu–Weber syndrome 1) (ENG), fibroblast growth factor 1 (FGF1), insulin-like growth factor 1 (IGF1), midkine (MDK), matrix metallopeptidase 9 (MMP9), epidermal growth factor (EGF), epiregulin (EREG), neuropilin 1 (NRP1), plasminogen activator urokinase (PLAU), prostaglandin-endoperoxide synthase 1 (PTGS1), and tumor necrosis factor α-induced protein 2 (TNFAIP2). Finally, daidzein inhibited the metastasis of MDA-MB-231 cells through inhibition of NF-κB and activator protein 1 (AP-1), as well as suppressed the expression of urokinase-type plasminogen activator (uPA) (Valachovicova et al. 2004).

18.4.2 Anti-inflammation

18.4.2.1 Puerarin

Puerarin inhibits the expression of C-reactive protein (CRP) in lipopolysaccharide (LPS)-induced peripheral blood mononuclear cells (PBMCs) isolated from patients with unstable angina pectoris (Yang et al. 2010). The inhibition of CRP was mediated through inhibition of phosphorylation and through degradation of IκB, which reduces NF-κB transactivation activity. Puerarin also increases the phosphorylation of protein kinase Cδ (PKCδ), in turn increasing the expression of heme oxygenase-1 (HQ-1), which inhibits the inflammatory responses mediated by N-carboxymethyl-lysine in mouse mesangial cells (Kim et al. 2010).

18.4.2.2 Daidzein

Treatment with daidzein-rich isoflavone aglycones (DRIAs) activates the expression of interleukin-6 (IL-6) and IL-8 in monocytes activated by LPS or Pam3CSK4 (PAM) in a dose-dependent manner (Morimoto et al. 2009). DRIAs also reduce the severity of dextran sodium sulfate (DSS)-induced colitis (an animal model for inflammatory bowel disease) via the reduction of IFN-γ, IL-12p40, and IL-6 expression. It additionally increases the level of the anti-inflammatory cytokine IL-10 in mesenteric lymph node cells. Finally, daidzein inhibits the production of NO and of the prostaglandin E2 in LPS- and LPS-plus-IFN-γ-activated murine macrophages (Sheu et al. 2001).

18.4.3 Anti-hypertension

18.4.3.1 Puerarin

Puerarin relieves a Sprague–Dawley SD rat thoracic aorta constricted with norepinephrine (Dong et al. 2004). This flavonoid can cause vasodilation of up to 30%, suggesting it as a potential treatment for hypertension (Jin et al. 2009). In a model of pulmonary artery hypertension induced by monocrotaline, treatment with puerarin lowered the mean pulmonary arterial pressure, mean right ventricular pressure, mean carotid arterial pressure, and weight ratio of the right ventricle to left ventricle plus septum. Puerarin also down-regulates the expression of fractalkine, a cytokine associated with pulmonary hypertension (Li et al. 2008).

18.4.3.2 Daidzein

Daidzein enhances the relaxation effects of acetylcholine treatment in the aortic rings isolated from spontaneously hypertensive rats (SHRs) (Song et al. 1988). Daidzein treatment increases NO synthase activity and inhibits the production of vascular superoxide and prostaglandins (Chang et al. 2009). In addition, daidzein relieves heart-muscle contraction induced by phenylephrine, KCl, and $CaCl_2$ by inhibiting receptor-mediated Ca^{2+} influx (Dong et al. 2004).

Smooth muscle cells participate in cardiovascular disease by proliferating and migrating to form neointima, as well as by depositing extracellular matrices to induce vascular remodeling. Daidzein inhibits the fetal-calf-serum-induced proliferation of human aortic smooth muscle cells and inhibits the migration of these cells induced by platelet-derived growth factor-BB (PDGF-BB). These anti-proliferation and migration effects are mediated by inhibition of mitogen-activated protein kinase (MAPK) activity (Dubey et al. 1999). Daidzein also inhibits the proliferation of cardiac fibroblasts.

18.4.4 Anti-atherosclerosis

18.4.4.1 Puerarin

The anti-hypertensive properties of puerarin are demonstrated by its effects on CRP (Yang et al. 2010). Originating in coronary atheromatous lesions, the concentration of CRP correlates with the progression of cardiovascular disease and atherosclerosis. It is therefore considered a marker for these conditions. As mentioned previously, puerarin can inhibit the expression of LPS-activated PBMCs. Puerarin may act as a chemopreventive compound for atherosclerosis (Liu et al. 2010).

This flavonoid also reduces hypercholesterolemia caused by feeding SD rats a hypercholesterolemic diet. Puerarin increases the expression levels of 3-hydroxy-3-methylglutaryl-CoA reductase (HMGR), an enzyme in the cholesterol biosynthesis pathway, and cholesterol 7α-hydroxylase (CYP7A1), an enzyme in the bile acid biosynthesis pathway. Moreover, puerarin increases endothelial NO synthase (eNOS) expression in rat thoracic aortas impaired by a hypercholesterolemic diet (Yan et al. 2006b).

18.4.4.2 Daidzin

The anti-hypertensive effects of daidzin are shown through its effects on peroxynitrite. This potent oxidant—produced by activated macrophages, neutrophils, and endothelial cells—oxidizes low-density proteins, which are associated with atherosclerosis. Daidzin reduces peroxynitrite-mediated low-density lipoprotein (LDL) oxidation (by 41%), suggesting it as a potential treatment to prevent atherosclerosis (Lai and Yen 2002).

18.4.4.3 Daidzein

Daidzein also reduces peroxynitrite-mediated LDL oxidation. The potencies of daidzin and daidzein are approximately the same. Compared with a control diet, a high-isoflavone diet (21.2 mg of daidzein and 34.8 mg of genistein per day) lowers lipid peroxidation and LDL oxidation. Further, the plasma concentration of the lipid oxidation marker 8-epi-prostaglandin F2a (8-epi-PGF2a) was lower, and the rate of LDL oxidation induced by copper was slower with a high-isoflavone diet. This evidence suggests that consumption of high levels of isoflavones prevents atherosclerosis (Wiseman et al. 2000).

18.4.5 Neuroprotection

18.4.5.1 Puerarin

Puerarin exerts a potent neuroprotective effect against middle cerebral artery occlusion (MCAO). Ten min before MCAO, Wistar rats were treated with puerarin (25 or 50 mg kg^{-1}). The results revealed a dose-dependent protection against focal cerebral ischemia. The protective effects occur through inhibition of the expression of hypoxia-inducible factor-1α (HIF-1α), of inducible nitric oxide synthase (iNOS), and of active caspase 3 (Chang et al. 2009). Puerarin decreased these expression levels by inhibiting the pro-inflammatory cytokine tumor necrosis factor-α (TNF-α).

Puerarin treatment also protected a hippocampal neuron culture from oxygen/glucose-deprivation-induced cell death. Intracellular Ca^{2+} and the level of NO were down-regulated after treatment with puerarin (Xu and Zheng 2007). The nigrostriatal neurotoxin 1-methyl-4-phenylpyridinium (MMP+) causes Parkinson's disease in humans and in laboratory animals. Puerarin reduces toxicity induced by MMP+ in rat pheochromocytoma PC12 cells by inhibiting the phosphorylation of MAPK kinase 7 (MKK7), c-Jun N-terminal kinase (JNK), and c-Jun, and by reducing the activation of caspases 3 and 9 (Wang et al. 2011). In a rat model of Alzheimer's disease, puerarin dissipated β-amyloid$_{1-42}$-induced cognitive dysfunction by inhibiting apoptosis. This was accomplished through the activation of Akt and the subsequent phosphorylation of Bad and activation of caspase 9. Finally, puerarin can improve motor function that has been impaired by a spinal cord ischemia–reperfusion injury when administered within 4 h of the injury (Li et al. 2010).

18.4.5.2 Daidzin

Daidzin protects primary hippocampal neurons from excitotoxic glutamate and from plasma membrane damage induced by β-amyloid$_{25-35}$, as indicated by lower levels of lactate dehydrogenase (LDH). The protective action is effective at a concentration of 10 ng ml^{-1} for glutamate-induced neurotoxicity and at 1000 ng ml^{-1} for neurotoxicity induced by β-amyloid$_{25-35}$ (Zhao et al. 2002).

18.4.5.3 Daidzein

Similar to daidzin, daidzein protects primary hippocampal neurons from plasma membrane damage induced by excitotoxic glutamate and by β-amyloid$_{25-35}$, as indicated by lower levels of LDH. The protective action occurs at a concentration of 10 ng ml^{-1} for glutamate-induced neurotoxicity and at 100 ng ml^{-1} for neurotoxicity induced by β-amyloid$_{25-35}$. A combination of daidzein, genistein, and equol enhances this neuroprotective activity. In general, this combination will enhance the protective effect against plasma membrane damage induced by excitotoxic glutamate and β-amyloid$_{25-35}$ by up to 20% compared with the use of daidzein alone (Zhao et al. 2002). This protective effect is mediated through increases in the expression level of the anti-apoptotic proteins bcl2 and bcl-XL.

Arginase 1 (Arg1), an enzyme regulated by cAMP, is known to overcome neurite outgrowth inhibition by myelin. It also protects neurons from trophic-factor-deprivation-induced neural cell apoptosis. Arg1 activity increases with increasing cAMP concentration. However, daidzein can enhance Arg1 activity through a cAMP-independent pathway and can prevent neuronal death and damage due to myelin-inhibited neurite outgrowth *in vivo*. Daidzein also promotes regeneration of optic nerves damaged mechanically by surgery (Ma *et al.* 2010).

18.4.6 Anti-allergy

18.4.6.1 Puerarin

The anti-allergic properties of puerarin are demonstrated by its effects on PBMCs. In asthma patients, the expression levels of the pro-inflammatory cytokine TNF-α in PBMCs were 116% higher than in the control. The number of NF-κB-positive cells was also higher (321%) in asthma patients. Treatment of PBMCs in asthma patients using puerarin reduced TNF-α expression by 30% and reduced the number of NF-κB-positive cells by 41%. This suggests that puerarin reduces the severity of inflammatory responses in asthma patients (Liu *et al.* 2010).

18.4.6.2 Daidzin

The anti-allergy effects of daidzin were shown in a guinea pig model of asthma. Guinea pigs were challenged with aerosol ovalbumin (OA), following 2 weeks on a diet enriched with daidzin. The total number of white blood cells, neutrophils, eosinophils, and red blood cells was lower in the bronchoalveolar lavage (BAL) of mice fed the high-isoflavone diet than in the BAL of those fed a normal diet. Moreover, the protein concentration in BAL was higher in the mice fed a normal diet. This indicates that a high-isoflavone diet can prevent asthma both by reducing inflammation and by protecting against protein leaks into the air spaces (Regal *et al.* 2000).

18.4.6.3 Daidzein

A diet with a high concentration of isoflavones (daidzein and genistein) reduced the allergic reaction to peanuts in C3H/HeJ mice. Mice fed this diet also showed a lower incidence of anaphylactic reaction, along with decreased mast cell degranulation in the murine ear and the intestinal mucosa [determined by the mouse mast cell protease 1 (MMCP-1) levels] and reductions in the sera expression of peanut-specific antibodies (IgG1, IgG2, and IgE). The anti-allergic activity of daidzein may be mediated through the inhibition of dendritic cell maturation, as suggested by lower expression of the mature dendritic cell markers CD80, CD86, and CD83. Isoflavones can also dose-dependently inhibit the maturation of human monocyte-derived dendritic cells (MDDCs), as indicated by reductions in the expression levels of CD80 and CD83 (but not of CD86). Isoflavones also inhibit the secretion of the pro-inflammatory cytokines IL-6 and IL-8 in MDDCs treated with cholera toxin. Moreover, isoflavone-treated dendritic cells inhibit the secretion of IL-13, IL-5, and IL-9, as well as all Th2 cytokines in CD4+ T-cells. Moreover, a cross-sectional study (1002 pregnant women) showed that consumption of an isoflavone-rich product between episodes of allergic rhinitis reduced the number of subsequent episodes. These results suggest that isoflavone supplements to a normal diet may prevent allergic diseases (Masilamani *et al.* 2011).

18.4.7 Anti-hyperglycemic Effect

18.4.7.1 Puerarin

Intravenous injection of puerarin decreased blood glucose in mice with streptozotocin (STZ)-induced diabetes. This injection also increased glucose uptake in soleus muscle through an increase in the expression of glucose transporter subtype 4 (GLUT4). Another possible mechanism was meditated through an increase in β-endorphin-like immunoreactivity (BER). The glucose-lowering effect was mediated through μ-opioid receptors, as shown by the fact that the μ-opioid receptor antagonists naloxone and naloxonazine inhibited the effect of puerarin. Moreover, puerarin could not lower plasma glucose in diabetic mice that were also μ-opioid receptor knockouts (Hsu *et al.* 2003) (Figure 18.5).

Puerarin pretreatment also protected pancreatic islets from oxidative damage induced by hydrogen peroxide. In type I diabetes, reactive oxygen species generated by macrophages kill β-islet cells. Reactive oxygen species are generated in excess amounts in chronic hyperglycemia and cause cellular injury in type II diabetes (Fu *et al.* 2010).

18.4.7.2 Daidzein

When administered at a concentration of $0.2\,g\,kg^{-1}$, daidzein lowers blood glucose levels (40% below the control value) in non-obese diabetic mice. It also lowers fasting glucose levels but not postprandial glucose levels. The plasma glucose-lowering effect was mediated through increases in insulin (50% higher), C-peptide levels (6-fold higher), and the number of insulin-expressing pancreatic islet β-cells. Gluconeogenesis was also down-regulated through a reduction in the expression of hepatic glucose-6-phosphatase and phosphoenolpyruvate carboxykinase and through an increase in glucokinase activity (Choi*et al.* 2008). Daidzein lowers the plasma free fatty acid and triglyceride concentrations and increases the ratio of high-density lipoprotein (HDL)-C to total cholesterol (Zheng *et al.* 2002b).

In addition to its effect in type I diabetes, daidzein improves hepatic glucose in type II diabetes, as shown in C67BL/KsJ-db/db mice. The HbA1c

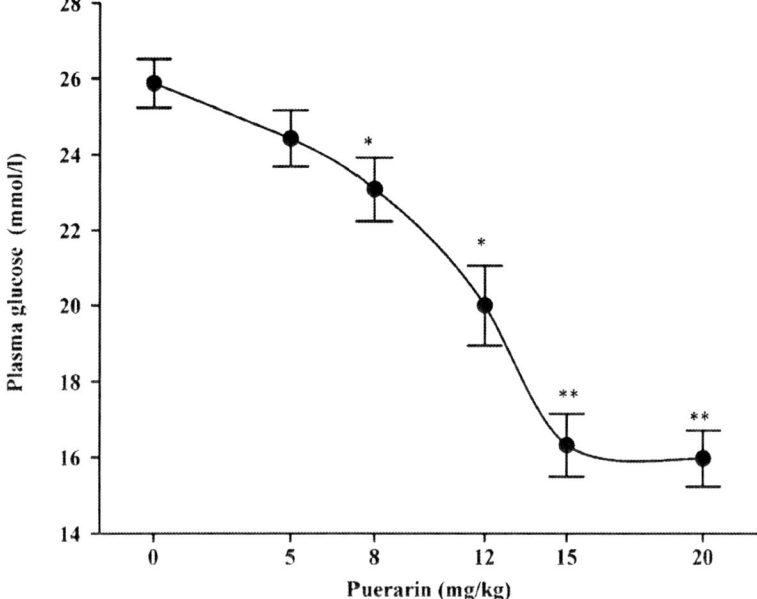

Figure 18.5 Effect of puerarin on plasma glucose concentration in STZ diabetic rats. The basal plasma glucose concentration in STZ diabetic rats was 25.6 mmol L^{-1}. A dose-dependent reduction of plasma glucose was observed in STZ diabetic rats after intravenous injection of puerarin at a dose of 5.0–20.0 mg kg^{-1}.
Data are from Hsu et al. (2003), with permission from the Publisher.

concentrations were lower in mice fed 0.2 g kg^{-1} daidzein. Gluconeogenesis was also down-regulated through a reduction in the expression of hepatic glucose-6-phosphatase and phosphoenolpyruvate carboxykinase and through an increase in glucokinase activity. It should be noted that the positive effects of daidzein on type I and II diabetes mice were not significant with less than 4 weeks of ingestion, suggesting that long-term exposure to daidzein is essential (Choi et al. 2008).

Daidzein also helps in the treatment of diabetes in humans. An epidemiological study (25 872 men and 33 919 women) found that consumption of isoflavone lowers the risk of type II diabetes in overweight women. In men with type II diabetes and nephropathy, isoflavone consumption improves albumin excretion (–9.5%) and increases HDL concentration, HDL/LDL ratio, and HDL/total cholesterol ratio (Teixeira et al. 2004).

Conclusions

The PR root has been used as a traditional remedy to treat various diseases and is effective due to its high isoflavone content. However, progress in the development of effective PR therapies for clinical use is slow because of the complex nature of traditional Chinese medicine (TCM) compounds and the difficulties in standardizing treatments. Several effective applications use either purified PR compounds or mixtures of several PR isoflavones. However, most of the single compounds do not show the same efficacy as more complex extracts. Moreover, these compounds are usually found in only small amounts in the herbal extracts and generally have low bioavailability. It is sometimes impossible for these compounds to reach therapeutic concentrations in the sera.

Thus, more studies on the absorption, distribution, metabolism, and excretion of these extracts should be conducted before application of PR in patient treatment. Such studies will provide more information on the actual active components in these extracts, as well as on the potential compound–compound interactions in the herbal extracts. Most clinical studies involving PR and PR isoflavones have been conducted in China. These studies have involved various diseases and have, almost invariably, reported positive therapeutic effects. However, such clinical studies are consistently labeled as "low quality", compromising their claims regarding the beneficial effects of PR and PR isoflavones. For better basic and clinical science studies on TCM, rigorously designed and stringent criteria are required.

Summary Points

- This Chapter focuses on *Puerariae radix*.
- *Puerariae radix* is rich in isoflavones, with puerarin being the most abundant.
- Puerarin, daidzin, and daidzein exhibit anti-tumor activity in breast cancer, prostate cancer, glioma, neuroblastoma, colon cancer, ovarian cancer, and hepatoma.
- Puerarin, daidzin, and daidzein reduce inflammatory responses.
- Puerarin, daidzin, and daidzein lower blood pressure.
- Puerarin, daidzin, and daidzein prevent atherosclerosis.
- Puerarin, daidzin, and daidzein have a neuroprotective effect from toxic agents that induce neuronal cell death.
- Puerarin, daidzin, and daidzein relieve allergic reactions.
- Puerarin, daidzin, and daidzein down-regulate blood glucose and have the potential to treat type I and II diabetes.

Key Facts

- Puerarin is the most abundant isoflavone in kudzu root.
- It is the major active ingredient in *Puerariae radix*.
- Puerarin reduces tissue damage caused by free radicals.
- Puerarin inhibits the growth of various cancers to prevent or cure cancer.
- Puerarin inhibits inflammation by inhibiting nuclear factor κ-light-chain-enhancer of activated B cells (NF-κB) activity.

- Puerarin relieves hypertension.
- It is a chemopreventive compound for atherosclerosis.
- Puerarin exerts a potent neuroprotective effect, caused by middle cerebral artery occlusion (MCAO).
- Puerarin has a positive effect in relieving the inflammatory response in asthma patients. It lowers blood glucose in diabetic mice.

Definitions of Words and Terms

Amyloid: When the polypeptides are being synthesized, the polypeptide chain starts to form a native three-dimensional structure to execute the biological activity. However, some proteins may misfolded or unfolded. Those partially misfolded or unfolded proteins can aggregate to form amyloid fibrils. Amyloid fibrils are associated with various diseases such as Alzheimer's, Parkinson's and Huntington's diseases.

Angina pectoris: Angina is a clinical manifestation of myocardial hypoxia. Mostly due to the accumulation of cholesterol in the arteries which results to coronary artery disease. If the situation continues to deteriorate, it will reduce arterial oxygen and nutrients to the heart, which will lead to the occurrence of angina.

Angiogenesis: This occurs from existing capillaries through a complex process in embryonic development, tissue repair, and the female reproductive cycle. Because tumor growth and metastasis must occur by means of angiogenesis, the inhibition of angiogenesis is a new strategy for treating cancer.

Apoptosis: A cell undergoes apoptosis when it is affected by environmental stimulation that results in gene regulation to undergo natural mortality, called "apoptosis" or "programmed cell death".

Cancer chemoprevention: This is now generally defined as the use of single or multiple compounds of natural or synthetic origin to achieve prevention, blockage, inhibition, or reversal of the effects of cancer. Alternatively, chemoprevention is the process of reducing risk factors for cancer or reducing the opportunity for recurrence.

Diabetes: Diabetes is a chronic metabolic disorder, mainly caused by insulin deficiency or dysfunction, which can reduce the use of sugar or even make it impossible to use, causing the blood sugar to rise.

Inflammation: In response to catastrophic stimuli such as bacterium or virus infections, allergens, or damaged cell debris, the immune cells respond to those irritants and produce large amount of cytokines, chemokines, and various inflammatory mediators (prostaglandins, histamines) to remove those stimuli and initiate the healing process.

Metastasis: Cancer metastasis is the major cause of death in patients. Cancer cell metastasis by local invasion into surrounding normal tissue, an even through the body's circulation and lymphatic systems to other parts of the body, greatly enhance the difficulty of treatment. Therefore, for cases in which the formation of cancer cannot be prevented, attempting to block the metastasis of cancer cells may be a potential direction for growth.

Multidrug resistance: Patients become resistant to drugs, so that the drugs lose their effectiveness.

List of Abbreviations

Arg1	arginase 1
BAL	bronchoalveolar lavage
CDK	cyclin-dependent kinase
CRP	C-reactive protein
DRIA	daidzein-rich isoflavone aglycones
DSS	dextran sodium sulfate
HDL	high-density lipoprotein
HPLC	high-performance liquid chromatography
IFN-γ	interferon γ
IκB	Inhibitor of κB
IL	interleukin
LDH	lactate dehydrogenase
LDL	low-density lipoprotein
LPS	lipopolysaccharide
MAPK	mitogen-activated protein kinase
MCAO	middle cerebral artery occlusion
MDDC	monocyte-derived dendritic cell
MDR	multidrug resistance
MMP+	1-methyl-4-phenylpyridinium
NF-κB	nuclear factor κB
PBMC	peripheral blood mononuclear cell
PR	*Puerariae radix*
SD	Sprague–Dawley
STZ	streptozotocin
TCM	traditional Chinese medicine
TNF-α	tumor necrosis α

References

Bagheri, M., Joghataei, M.T., Mohseni, S., and Roghani, M., 2011. Genistein ameliorates learning and memory deficits in amyloid β(1–40) rat model of Alzheimer's disease. *Neurobiology of Learning and Memory*. 95(3): 270–276.

Bebrevska, L., Foubert, K., Hermans, N., Chatterjee, S., Van Marck, E., De Meyer, G., Vlietinck, A., Pieters, L., and Apers, S., 2010. In vivo antioxidative activity of a quantified Pueraria lobata root extract. *Journal of Ethnopharmacology*. 127(1): 112–117.

Beecher, G.R., 2003. Overview of dietary flavonoids: nomenclature, occurrence and intake. *Journal of Nutrition.* 133(10): 3248S–3254S.

Brusselmans, K., Vrolix, R., Verhoeven, G., and Swinnen, J.V., 2005. Induction of cancer cell apoptosis by flavonoids is associated with their ability to inhibit fatty acid synthase activity. *The Journal of Biological Chemistry.* 280(7): 5636–5645.

Chang, Y., Hsieh, C.Y., Peng, Z.A., Yen, T.L., Hsiao, G., Chou, D.S., Chen, C.M., and Sheu, J.R., 2009. Neuroprotective mechanisms of puerarin in middle cerebral artery occlusion-induced brain infarction in rats. *Journal of Biomedical Science.* 16: 9.

Chen, G., Zhang, J., and Jiannong, Y., 2001. Determination of puerarin, daidzein and rutin in *Pueraria lobata* (Wild.) Ohwi by capillary electrophoresis with electrochemical detection. *Journal of Chromatography A.* 923(1–2): 255–262.

Chen, T.R., Chen, L.A., and Wei, Q.K., 2010. Evaluation of quality of *Radix Puerariae* herbal medicine by isoflavonoids. *Journal of Pharmacy and Pharmacology.* 62(5): 644–650.

Cheng, E.W., Yoder, L., Story, C.D., and Burroughs, W., 1955. Estrogenic activity of some naturally occurring isoflavones. *Annals of the New York Academy of Sciences.* 61(3): 652–658; Discussion. 658–659.

Cherdshewasart, W., Subtang, S., and Dahlan, W., 2007. Major isoflavonoid contents of the phytoestrogen rich-herb Pueraria mirifica in comparison with Pueraria lobata. *J. Pharm. Biomed. Anal.* 43(2): 428–434.

Cherdshewasart, W., Traisup, V., and Picha, P., 2008. Determination of the estrogenic activity of wild phytoestrogen-rich Pueraria mirifica by MCF-7 proliferation assay. *Journal of Reproduction and Development.* 54(1): 63–67.

Choi, M.S., Jung, U.J., Yeo, J., Kim, M.J., and Lee, M.K., 2008. Genistein and daidzein prevent diabetes onset by elevating insulin level and altering hepatic gluconeogenic and lipogenic enzyme activities in non-obese diabetic (NOD) mice. *Diabetes Metabolism Research and Reviews.* 24(1): 74–81.

De Groot, J.F., Fuller, G., Kumar, A.J., Piao, Y., Eterovic, K., Ji, Y., and Conrad, C.A., 2010. Tumor invasion after treatment of glioblastoma with bevacizumab: radiographic and pathologic correlation in humans and mice. *Neuro-oncology.* 12(3): 233–242.

Dong, K., Tao, Q.M., Shan, Q.X., Jin, H.F., Pan, G.B., Chen, J.Z., Zhu, J.H., and Xia, Q., 2004. Endothelium-independent vasorelaxant effect of puerarin on rat thoracic aorta. *Conference Proceedings: IEEE Engineering in Medicine and Biology Society.* 5: 3757–3760.

Dubey, R.K., Gillespie, D.G., Imthurn, B., Rosselli, M., Jackson, E.K., and Keller, P.J., 1999. Phytoestrogens inhibit growth and MAP kinase activity in human aortic smooth muscle cells. *Hypertension.* 33(1): 177–182.

Fu, J., Woods, C.G., Yehuda-Shnaidman, E., Zhang, Q., Wong, V., Collins, S., Sun, G., Andersen, M.E., and Pi, J., 2010. Low-level arsenic impairs glucose-stimulated insulin secretion in pancreatic beta cells: involvement of cellular adaptive response to oxidative stress.. *Environmental Health Perspectives.* 118(6): 864–870.

Glover, A., and Assinder, S.J., 2006. Acute exposure of adult male rats to dietary phytoestrogens reduces fecundity and alters epididymal steroid hormone receptor expression. *Journal of Endocrinology*. 189(3): 565–573.

Hien, T.T., Kim, H.G., Han, E.H., Kang, K.W., and Jeong, H.G., 2010. Molecular mechanism of suppression of MDR1 by puerarin from *Pueraria lobata* via NF-κB pathway and cAMP-responsive element transcriptional activity-dependent up-regulation of AMP-activated protein kinase in breast cancer MCF-7/adr cells. *Molecular Nutrition and Food Research*. 54(7): 918–928.

Hsu, F.L., Liu, I.M., Kuo, D.H., Chen, W.C., Su, H.C., and Cheng, J.T., 2003. Antihyperglycemic effect of puerarin in streptozotocin-induced diabetic rats. *Journal of Natural Products*. 66(6): 788–792.

Jin, G., Yang, P., Gong, Y., Fan, X., Tang, J., and Lin, J., 2009. [Effects of puerarin on expression of apelin and its receptor of 2K1C renal hypertension rats]. *Zhongguo Zhong Yao Za Zhi*. 34(24): 3263–3267.

Kallela, K., 1975. The effect of storage on the estrogenic effect of red clover silage. *Nordisk Veterinaermedicin*. 27(11): 562–569.

Kato, K., Takahashi, S., Cui, L., Toda, T., Suzuki, S., Futakuchi, M., Sugiura, S., and Shirai, T., 2000. Suppressive effects of dietary genistin and daidzin on rat prostate carcinogenesis. *Japanese Journal of Cancer Research*. 91(8): 786–791.

Kim, K.M., Jung, D.H., Jang, D.S., Kim, Y.S., Kim, J.M., Kim, H.N., Surh, Y.J., and Kim, J.S., 2010. Puerarin suppresses AGEs-induced inflammation in mouse mesangial cells: a possible pathway through the induction of heme oxygenase-1 expression. *Toxicology and Applied Pharmacology*. 244(2): 106–113.

Korkina, L., Kostyuk, V., De Luca, C. and Pastore, S., 2011. Plant phenylpropanoids as emerging anti-inflammatory agents. *Mini-reviews in Medicinal Chemistry*. 11(10): 823–835.

Lai, H.H., and Yen, G.C., 2002. Inhibitory effect of isoflavones on peroxynitrite-mediated low-density lipoprotein oxidation. *Bioscience, Biotechnology and Biochemistry*. 66(1): 22–28.

Lau, C.C., Chan, C.O., Chau, F.T., and Mok, D.K., 2009. Rapid analysis of *Radix puerariae* by near-infrared spectroscopy. *Journal of Chromatography A*. 1216(11): 2130–2135.

Le Bail, J.C., Varnat, F., Nicolas, J.C., and Habrioux, G., 1998. Estrogenic and antiproliferative activities on MCF-7 human breast cancer cells by flavonoids. *Cancer Letters*. 130(1–2): 209–216.

Li, J.W., Chen, P., Guan, X.Q., Gong, Y.S., and Yang, P.L., 2008. [Inhibition of puerarin on pulmonary hypertension in rats with hypoxia and hypercapnia]. *Zhongguo Zhong Yao Za Zhi*. 33(5): 544–549.

Li, J., Wang, G., Liu, J., Zhou, L., Dong, M., Wang, R., Li, X., Lin, C., and Niu, Y., 2010. Puerarin attenuates amyloid-β-induced cognitive impairment through suppression of apoptosis in rat hippocampus *in vivo*. *European Journal of Pharmacology*. 649(1–3): 195–201.

Liang, X.M., Zhang, Y., Chen, J.P., Zhang, C.Z., and Wu, W.Z., 2005. Analysis of the estrogenic components in kudzu root by bioassay and high performance liquid chromatography. *Journal of Steroid Biochemistry and Molecular Biology*. 94(4): 375–381.

Lin, Y.J., Hou, Y.C., Lin, C.H., Hsu, Y.A., Sheu, J.J., Lai, C.H., Chen, B.H., Lee Chao, P.D., Wan, L., and Tsai, F.J., 2009. Puerariae radix isoflavones and their metabolites inhibit growth and induce apoptosis in breast cancer cells. *Biochem. Biophys. Res. Commun*. 378(4): 683–688.

Liu, X.J., Zhao, J., and Gu, X.Y., 2010. The effects of genistein and puerarin on the activation of nuclear factor-κB and the production of tumor necrosis factor-alpha in asthma patients. *Pharmazie*. 65(2): 127–131.

Liu, D., Yu, Z., Liu, C., and Wang, X., 2011. [Classification of Pueraria lobata in different geographical regions]. *Zhongguo Zhong Yao Za Zhi*. 36(3): 299–301.

Lo, F.H., Mak, N.K., and Leung, K.N., 2007. Studies on the anti-tumor activities of the soy isoflavone daidzein on murine neuroblastoma cells. *Biomedicine and Pharmacotherapy*. 61(9): 591–595.

Ma, T.C., Campana, A., Lange, P.S., Lee, H.H., Banerjee, K., Bryson, J.B., Mahishi, L., Alam, S., Giger, R.J., Barnes, S., Morris, S.M., Willis, D.E., Twiss, J.L., Filbin, M.T., and Ratan, R.R., 2010. A large-scale chemical screen for regulators of the arginase 1 promoter identifies the soy isoflavone daidzeinas a clinically approved small molecule that can promote neuronal protection or regeneration via a cAMP-independent pathway. *Journal of Neuroscience*. 30(2): 739–748.

Markou, A., Androulakis, II, Mourmouris, C., Tsikkini, A., Samara, C., Sougioultzis, S., Piaditis, G., and Kaltsas, G., 2010. Hepatic steatosis in young lean insulin resistant women with polycystic ovary syndrome. *Fertility and Sterility*. 93(4): 1220–1226.

Masilamani, M., Wei, J., Bhatt, S., Paul, M., Yakir, S. and Sampson, H.A., 2011. Soybean isoflavones regulate dendritic cell function and suppress allergic sensitization to peanut. *Journal of Allergy Clinical Immunology*. 128(6): 1242–1250.

Miksicek, R.J., 1995. Estrogenic flavonoids: structural requirements for biological activity. *Proceedings of the Society for Experimental Biology and Medicine*. 208(1): 44–50.

Morimoto, M., Watanabe, T., Yamori, M., Takebe, M., and Wakatsuki, Y., 2009. Isoflavones regulate innate immunity and inhibit experimental colitis. *Journal of Gastroenterology and Hepatology*. 24(6): 1123–1129.

Prasain, J.K., Reppert, A., Jones, K., Moore, D.R., Barnes, S., and Lila, M.A., 2007. Identification of isoflavone glycosides in *Pueraria lobata* cultures by tandem mass spectrometry. *Phytochemical Analysis*. 18(1): 50–59.

Regal, J.F., Fraser, D.G., Weeks, C.E. and Greenberg, N.A., 2000. Dietary phytoestrogens have anti-inflammatory activity in a guinea pig model of asthma. *Proceedings of the Society for Experimental Biology and Medicine*. 223(4): 372–378.

Sanchez, I., Gomez-Garibay, F., Taboada, J., and Ruiz, B.H., 2000. Antiviral effect of flavonoids on the dengue virus. *Phytotherapy Research*. 14(2): 89–92.

Sheu, F., Lai, H.H., and Yen, G.C., 2001. Suppression effect of soy isoflavones on nitric oxide production in RAW 264.7 macrophages. *Journal of Agricultural and Food Chemistry*. 49(4): 1767–1772.

Shimosaki, S., Tsurunaga, Y., Itamura, H., and Nakamura, M., 2011. Antiallergic effect of the flavonoid myricitrin from *Myrica rubra* leaf extracts in vitro and in vivo. *Natural Product Research*. 25(4): 374–380.

Sibao, C., Dajian, Y., Shilin, C., Hongx, X., and Chan, A.S., 2007. Seasonal variations in the isoflavonoids of *Radix puerariae*. *Phytochemical Analysis*. 18(3): 245–250.

Song, X.P., Chen, P.P., and Chai, X.S., 1988. [Effects of puerarin on blood pressure and plasma renin activity in spontaneously hypertensive rats]. *Zhongguo Yao Li Xue Bao*. 9(1): 55–58.

Teixeira, S.R., Tappenden, K.A., Carson, L., Jones, R., Prabhudesai, M., Marshall, W.P., and Erdman, Jr, J.W., 2004. Isolated soy protein consumption reduces urinary albumin excretion and improves the serum lipid profile in men with type 2 diabetes mellitus and nephropathy. *Journal of Nutrition*. 134(8): 1874–1880.

Valachovicova, T., Slivova, V., Bergman, H., Shuherk, J., and Sliva, D., 2004. Soy isoflavones suppress invasiveness of breast cancer cells by the inhibition of NF-κB/AP-1-dependent and -independent pathways. *International Journal of Oncology*. 25(5): 1389–1395.

Wang, G., Zhou, L., Zhang, Y., Dong, M., Li, X., Liu, J., and Niu, Y., 2011. Implication of the c-Jun-NH2-terminal kinase pathway in the neuroprotective effect of puerarin against 1-methyl-4-phenylpyridinium (MPP+)-induced apoptosis in PC-12 cells. *Neuroscience Letters*. 487(1): 88–93.

Wiseman, H., O'Reilly, J.D., Adlercreutz, H., Mallet, A.I., Bowey, E.A., Rowland, I.R., and Sanders, T.A., 2000. Isoflavone phytoestrogens consumed in soy decrease F2-isoprostane concentrations and increase resistance of low-density lipoprotein to oxidation in humans. *American Journal of Clinical Nutrition*. 72(2): 395–400.

Xu, X., and Zheng, X., 2007. Potential involvement of calcium and nitric oxide in protective effects of puerarin on oxygen-glucose deprivation in cultured hippocampal neurons. *Journal of Ethnopharmacology*. 113(3): 421–426.

Yan, B., Wang, W., Zhang, L., Xing, D., Wang, D., and Du, L., 2006a. Determination of puerarin in rat cortex by high-performance liquid chromatography after intravenous administration of Puerariae flavonoids. *Biomedical Chromatography*. 20(2): 180–184.

Yan, L.P., Chan, S.W., Chan, A.S., Chen, S.L., Ma, X.J., and Xu, H.X., 2006b. Puerarin decreases serum total cholesterol and enhances thoracic aorta endothelial nitric oxide synthase expression in diet-induced hypercholesterolemic rats. *Life Science*. 79(4): 324–330.

Yang, C.S., Landau, J.M., Huang, M.T., and Newmark, H.L., 2001. Inhibition of carcinogenesis by dietary polyphenolic compounds. *Annual Reviews in Nutrition*. 21: 381–406.

Zhang, W., Liu, C.Q., Wang, P.W., Sun, S.Y., Su, W.J., Zhang, H.J., Li, X.J., and Yang, S.Y., 2010. Puerarin improves insulin resistance and modulates adipokine expression in rats fed a high-fat diet. *European Journal of Pharmacology*. 649(1–3): 398–402.

Zhao, L., Chen, Q., and Diaz Brinton, R., 2002. Neuroprotective and neurotrophic efficacy of phytoestrogens in cultured hippocampal neurons. *Experimental and Biological Medicine*. 227(7): 509–519.

Zhao, C., Chan, H.Y., Yuan, D., Liang, Y., Lau, T.Y. and Chau, F.T., 2011. Pueraria lobata. *Phytochemical Analysis* (in press).

Zheng, G., Zhang, X., Meng, Q., Gong, W., Wen, X., and Xie, H., 2002a. [Protective effect of total isoflavones from *Pueraria lobata* on secondary osteoporosis induced by dexamethasone in rats]. *Zhong Yao Cai*. 25(9): 643–646.

Zheng, G., Zhang, X., Zheng, J., Gong, W., Zheng, X., and Chen, A., 2002b. [Hypocholesterolemic effect of total isoflavones from *Pueraria lobata* in ovariectomized rats]. *Zhong Yao Cai*. 25(4): 273–275.

Zou, S.E., Zhang, S.F., Zhang, R., and Zhang, J., 2009. [Role of the cross-talk between estrogen receptors and peroxisome proliferator-activated receptor gamma in daidzein's prevention and treatment of osteoporosis in ovariectomized rats]. *Zhonghua Yi Xue Za Zhi*. 89(42): 2972–2975.

CHAPTER 19

Pattern Profiling and Quantitative Determination of Isoflavones in Herbal Chemotypes using Liquid Chromatography Tandem Mass Spectrometry

LAKSHMI MANICKAVASAGAM, SMRITI MISHRA AND GIRISH KUMAR JAIN*

Division of Pharmacokinetics and Metabolism, Central Drug Research Institute, PO Box-173, Lucknow-226001, India
*Email: gk_jain@cdri.res.in

19.1 Introduction

Osteoporosis is defined as "a skeletal disorder characterized by compromised bone strength predisposing a person to an increased risk of fracture. Bone strength reflects the integration of 2 main features: bone density and bone quality" (Anonymous 2000). Most of the pharmacological therapy available for osteoporosis such as calcium/vitamin D supplements, calcitonin, bisphosphonates, raloxifene and hormone replacement therapy act by decreasing the rate of bone resorption by slowing the rate of bone loss and thereby reducing bone turnover. Parathyroid hormone supplementation is currently the only clinically efficacious therapy in enhancing bone formation.

Anti-resorptive agents are associated with side effects such as endometrial cancer and venous thromboembolism, resulting in mid-therapy withdrawal. Therefore, there is a great need for better therapies that focuses on bone formation with a good safety profile. This has led to a widespread research for developing alternative and effective therapeutic agents for prevention and treatment of osteoporosis.

19.2 Herbal Preparations for Osteoporosis

Extensive investigations for new therapies for osteoporosis have shown that phytoestrogens, especially isoflavones, are effective for the prevention of bone loss and new bone formation. Soy and red clover are the two important sources of isoflavones. There are many soy and red clover-based products being marketed as food supplements for postmenopausal osteoporosis (Setchell et al. 2001). There are also other natural sources that contain isoflavones which have been reported to exhibit anti-osteoporotic activity. Some of them are *Dioscorea spongiosa* (Yin et al. 2010), *Cuminum cyminum* (Shirke et al. 2008), *Drynaria fortune* (Wang et al. 2008), *Cissus quadrangularis* (Potu et al. 2009) and *Ulmus wallichiana* (Sharan et al. 2010).

19.2.1 Pattern Profiling and Quantitative Analysis of Herbal Fractions

With the increase in the availability of herbal products or traditional medicines for the management of postmenopausal osteoporosis, it is necessary to ensure their quality and regulate their use. Therefore, the quality control (QC), quality assurance (QA) and pharmacokinetic studies of herbal preparations are essential to generate scientific data to support their safety and efficacy (Lee 2000). Moreover, QC and standardization of the herbal chemotype is a mandatory requirement before it can be approved for human consumption.

Quantitative methods and/or qualitative pattern profiling (fingerprint analysis) are used for the QC and QA of herbal medicines. In quantitative methods, few active components are identified and used as reference standards to evaluate the quality of herbal preparations. Pattern profiling provides the comprehensive and accurate assessment of a large number of active constituents in herbal extracts. The U.S. FDA and the European Medicine Agency (EMEA) also emphasize the role of chromatographic fingerprints to ensure the quality of herbal products (Anonymous 2001; 2004).

19.2.2 Osteogenic Herbal Fractions and Compounds of *Butea monosperma*

In our thrust to discover new and safer osteogenic and osteoprotective agents from natural sources, a group of researchers at the Central Drug Research

Institute (CDRI), Lucknow, India, observed that the activity-guided acetone soluble fraction F147 from stem bark of *B. monosperma* possesses potential osteogenic properties (Pandey *et al.* 2010). *B. monosperma* is a medium-sized tree distributed in the mountainous regions of India and Burma. In India, this tree is used in traditional medicine for the treatment of dyspepsia, diarrhea, ulcers, diabetes and snake bites (Varier 1993). The phytochemical investigation of crude extract of *B. monosperma* revealed that it contains isoflavones, coumestans, pterocarpans, flavanol, triterpenes and fatty esters. On further investigation it was found that the isoflavones daidzein (DZN), genistein (GEN), formononetin (FMN), isoformononetin (IFMN), cajanin (CJN), cladrin (CLN), prunetin (PRN) and the pterocarpan medicarpin (MCN) exhibit positive effects on bone formation (Figure 19.1) (Maurya *et al.* 2009).

Although the extracts from *B. monosperma* have been used in traditional medicine for various activities, there are limited reports on the analytical methods for the identification of active constituents in these extracts.

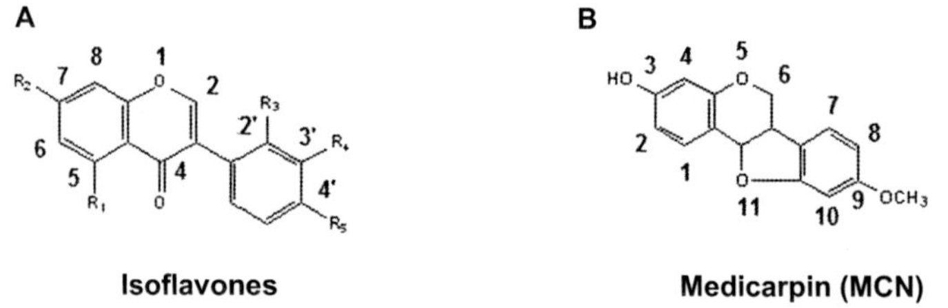

Markers	Chemical name	R_1	R_2	R_3	R_4	R_5
Daidzein	7,4'-Dihydroxy isoflavone	H	OH	H	H	OH
Cajanin	5,2',4'-Trihydroxy,7-methoxy isoflavone	OH	OCH_3	OH	H	OH
Isoformononetin	7-Methoxy,4'-hydroxy isoflavone	H	OCH_3	H	H	OH
Genistein	5,7,4'-Trihydroxy isoflavone	OH	OH	H	H	OH
Cladrin	7-Hydroxy,3',4'-dimethoxy isoflavone	H	OH	H	OCH_3	OCH_3
Formononetin	7-Hydroxy,4'-methoxy isoflavone	H	OH	H	H	OCH_3
Prunetin	5,4'-Dihydroxy,7'-methoxy isoflavone	OH	OCH_3	H	H	OH
IS	7-Hydroxy isoflavone	H	OH	H	H	H

Figure 19.1 Chemical structure of (A) isoflavones and (B) medicarpin. Chemical structures of active markers present in herbal fraction F147 and the internal standard (IS) used in LC-MS/MS analysis.

Thin-layer chromatography (TLC) fingerprinting and gas chromatography have been reported for the extracts of *B. monosperma* flowers (Iqbal *et al.* 2010) and seeds respectively (Sengupta and Basu 1978).

19.3 LC-MS/MS Method for Qualitative and Quantitative Analysis of Herbal Fractions

Liquid chromatography coupled with tandem mass spectrometry (LC-MS/MS) is an effective tool for quantitative analysis of active constituents of herbal fractions because of its high sensitivity, high selectivity and rapid screening ability. The application of the LC-MS/MS technique for the analysis of flavonoids in herbal extracts and dietary supplements has been reported in numerous publications (Table 19.1).

The development of LC-MS/MS analytical method requires optimization of both chromatographic and mass spectrometric conditions. Once the method is developed, it has to be validated for selectivity, sensitivity, linearity, accuracy and precision (Anonymous 1996). Finally, the validated LC-MS/MS method should be applied for the analysis of herbal fractions. The following Section reviews the experimental considerations in the development and validation of LC-MS/MS method for the analysis of the herbal fraction F147.

19.3.1 Chromatographic Conditions

The separation and elution of compounds by liquid chromatography depends upon the selection of suitable mobile phase and column. Generally, the development of the LC-MS/MS method does not require complete chromatographic separation of the analytes. The multiple reaction monitoring (MRM)

Table 19.1 Analytical methods for isoflavones. List of LC-MS methods reported for the analysis of isoflavones in dietary supplements and herbal extracts. APCI, atmospheric pressure chemical ionization; LC-MS, liquid chromatography-mass spectrometry; UV, ultraviolet.

S. No	Analysis	Analytical methods	Reference
1.	Isoflavonoids in kudzu dietary supplements	LC-ESI-MS/MS	Prasain *et al.* (2003)
2.	Isoflavones in edamame and tofu soybeans	LC/UV/ESI-MS	Wu *et al.* (2004)
3.	Isoflavones in soybean seeds	LC-PDA and LC-MS	Heimler *et al.* (2004)
4.	Isoflavones in soy supplements	LC-APCI-MS	Chen *et al.* (2005)
5.	Flavones and isoflavones in red clover	LC-ESI-MS	Nancy *et al.* (2006)
6.	Isoflavones in soymilk	LC-ESI-MS	Otieno *et al.* (2007)
7.	Isoflavonoids and astragalosides in *Astragalus* sp.	LC-APCI-MS/MS	Zhang *et al.* (2007)

mode of mass spectrometry allows for the selective and simultaneous analysis of compounds. In MRM, the precursor and the product ions are selected from their MS/MS spectra for each analyte to obtain a selective and sensitive signal. However, if the co-eluting components are structurally similar, they may undergo similar fragmentation and produce the same precursor and product ions, resulting in cross interferences between them. The probability of cross interference is high in the herbal fractions due to the presence of structurally similar compounds (Lee *et al.* 2005). Therefore, the chromatographic separation of compounds is essential for the qualitative and quantitative analysis in herbal extracts (Robards 2003).

F147 contains the isoflavones FMN and IFMN, which are functional isomers with similar chemical structures (Figure 19.1) and fragmentation pattern (Figure 19.2). Thus the marker FMN produces a signal in the MRM condition (269/226) of IFMN, affecting the selectivity of the method.

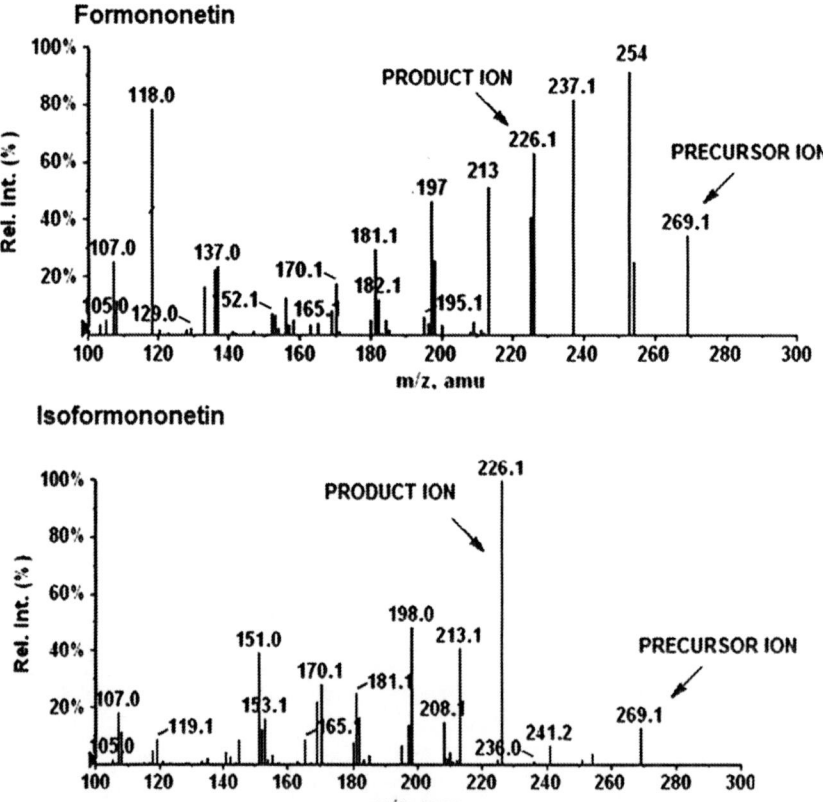

Figure 19.2 MS/MS spectra of FMN and IFMN showing precursor ion (269.1) and product ion (226.1). MS/MS spectra of FMN and IFMN shows similar fragmentation pattern. The prominent precursor ion 269 and product ion 226 of IFMN is present in MS/MS spectra of FMN.

Table 19.2 Gradient condition for LC-MS/MS analysis of F147. The Table shows the linear change in the percentage of mobile phase with time for gradient elution of markers and compounds C1–C13 of F147; mobile phases: A, acetonitrile/methanol (4:6); B, 0.1% acetic acid.

Time (min)	A (%)	B (%)
0.1	40	60
35	45	55
37	45	55
39	50	50
43	50	50
44	90	10
49	90	10
50	40	60
60	40	60

The complete resolution of these two isomers is important for their selective analysis by LC-MS/MS method. A high-performance liquid chromatography-photodiode array detector (HPLC-PDA) method has been reported for F147, wherein these two active constituents were separated using phosphate buffer (Gupta *et al.* 2010). However, for the LC-MS/MS method, phosphate buffer cannot be used as it will suppress the ionization of analytes. Therefore the LC-MS/MS method had to be developed with a mobile phase composition that is different from that of HPLC-PDA method reported earlier. The aqueous modifier with 0.1% acetic acid and a mixture of methanol and acetonitrile as organic modifier were tried in various proportions. The best resolution was achieved with 4:6 (v/v) acetonitrile/methanol as an organic modifier. Different gradient conditions were applied and the attempts to shorten the run time affected the resolution between FMN and IFMN. Finally, the gradient condition with run time of 60 min and the mobile phase composition given in Table 19.2 were selected. The elution of isoflavones was carried out in Phenomenex C18 column (150 × 4.6 mm, 5 µm) using the selected gradient for proper separation of all the eight active constituents (markers) and fingerprinting of isoflavones of F147 in a single LC-MS/MS run.

19.3.2 Mass Spectrometric Conditions

The selectivity and sensitivity of LC-MS/MS method relies on a suitable ionization method. Electro-spray ionization (ESI) with a quadrupole analyser was used for the mass spectrometric analysis of isoflavones of F147. ESI can be operated in either positive or negative mode. Reference standards were available for all eight markers of F147. Therefore, the selection of MRM (precursor-product ion pair) for each marker was done by analyzing the mass spectra (MS/MS spectra) of corresponding reference standard compounds. In addition to these eight markers, F147 contains other isoflavones, for which standards were not available. For screening of such unidentified compounds, an information-dependent acquisition (IDA) approach was applied (Marquet et al. 2003). IDA is a powerful tool of mass spectrometry, which is useful for identifying any unknown compounds in the multi-component mixture (Suryawanshi et al. 2006). To identify other constituents of F147, an IDA experiment was performed in both positive and negative mode of ionization by applying the optimized gradient elution. Initially, enhanced mass scan (EMS) to enhanced product ion scan (EPI) and EMS to enhanced resolution scan was performed for F147 in a single LC-MS/MS run. The presence of eight markers in F147 was confirmed by comparison of MS/MS spectra of reference standards with the EPI spectra of the IDA experiment and their retention time (t_r) in the LC-MS/MS chromatogram. The chromatogram of the IDA experiment showed that F147 contains 13 compounds (compounds C1–C13) in addition to the eight active markers. The MRMs for compounds C1–C13 were selected from their EPI scan. Generally, isoflavones give intense signals when analysed in negative mode of ESI (Rijke et al. 2003). However, among the markers of F147, IFMN gave intense signal in positive mode. Similarly, the compounds C7, C11 to C13 also gave prominent signals in positive mode of ionization, whereas the negative mode of ESI was suitable for other markers and compounds (Table 19.3). Therefore, to analyze all markers and compounds of F147 in both negative and positive mode of ESI simultaneously, polarity switching approach was applied (Lee et al. 2005). Thus, by applying all the selected MRMs of eight markers and 13 compounds, the chromatographic fingerprint was obtained from a single LC-MS/MS run (Figure 19.3). 7-Hydroxy isoflavone was used as an internal standard (IS) for simultaneous analysis of the constituents of F147 in both negative and positive ionization. 7-Hydroxy isoflavone is a synthetic isoflavone with a good signal intensity in both modes of ionization. The MRMs used for the IS in negative and positive mode were selective with no interferences from any of the natural isoflavones and pterocarpan present in the herbal fraction F147.

19.3.3 Validation of the LC-MS/MS Method

The validation process of developed analytical method ensures robustness and reproducibility of the method. The developed method should be validated for selectivity, sensitivity, linearity, accuracy and precision using calibration standards and QC samples. The validation of LC-MS/MS method for analysis of F147 is discussed below. The LC-MS/MS method was validated for both qualitative and quantitative analysis.

To perform validation for qualitative analysis, the reference standards of compounds C1–C13, were not available. Hence, the working solutions (WS) of herbal fraction F147 in methanol ($n = 6$) was used as QC samples, after adding IS. The herbal extract F147 was dissolved completely in methanol to obtain the

Table 19.3 Mass spectrometric conditions of markers and compounds C1–C13 of F147. This Table contains the MRM (precursor–product ion pair) and the ionization mode, declustering potential (DP) and collision energy (CE) selected for each marker and compounds C1–C13 for analysis in F147.

Markers	Molecular weight	MRM transitions (ionization mode)	DP	CE
DZN	254	253/132 (−)	−95	−55
CJN	300	299/164.7 (−)	−90	−30
IFMN	268	269/226 (+)	100	44
GEN	270	269/133 (−)	−90	−44
CLN	298	297/238.9 (−)	−90	−47
FMN	268	267/252 (−)	−75	−28
MCN	270	269/254 (−)	−80	−25
PRN	284	283/239 (−)	−90	−40
C1	268	267/252 (−)	81	44
C2	268	267/252 (−)	−90	−46
C3	282	281/253 (−)	−100	−35
C4	282	281/253 (−)	−100	−35
C5	282	281/253 (−)	−100	−35
C6	282	281/253 (−)	−100	−35
C7	282	283/240 (+)	100	35
C8	284	283/268 (−)	−100	−35
C9	284	283/268 (−)	−100	−35
C10	284	283/268 (−)	−100	−35
C11	284	285/215 (+)	100	35
C12/C13	296	297/239 (+)	100	35
IS	238	239/137 (+)	100	35
IS	238	237/117 (−)	−100	−35

stock solution of 1 mg ml^{-1}. This stock solution was diluted serially in dilution solvent (DS; methanol + acetonitrile/0.1% acetic acid in 1:1 ratio) to obtain the concentrations of 500 and 50 µg ml^{-1}. These WS were taken as QC samples. The validation of the LC-MS/MS method for pattern profiling (qualitative analysis) was assessed by determining the % C.V. (coefficient of variation) of the peak area ratio of each compound to IS. The values for the % C.V. calculated for the compounds C1–C13 were in the range of 5.62 to 9.93, which were within the acceptable limit of ±15%.

The LC-MS/MS method was also validated for sensitivity, linearity, intra-day, inter-day accuracy (% bias) and precision (% C.V.) for quantitative analysis of eight markers in F147. The sensitivity and linearity of the method was determined from the composite WS consisting of eight active markers: DZN, CJN, IFMN, GEN, CLN, FMN, PRN and MCN in methanol. The WS was prepared in the concentration range of 100 µg ml^{-1} to 0.0975 µg mL^{-1} in methanol. 10 µl of these WS were spiked in 990 µl of DS to get the calibration standards in the concentration range from 1000 ng ml^{-1} to 0.975 ng ml^{-1}. For the preparation of QC samples, the composite WS of concentration 80, 40, 10, 0.5 and 0.25 µg ml^{-1} were prepared in methanol. 10 µl of these WS were spiked in 990 µL of DS to obtain concentrations of 800, 400, 100, 10, 5 and 2.5 ng ml^{-1}

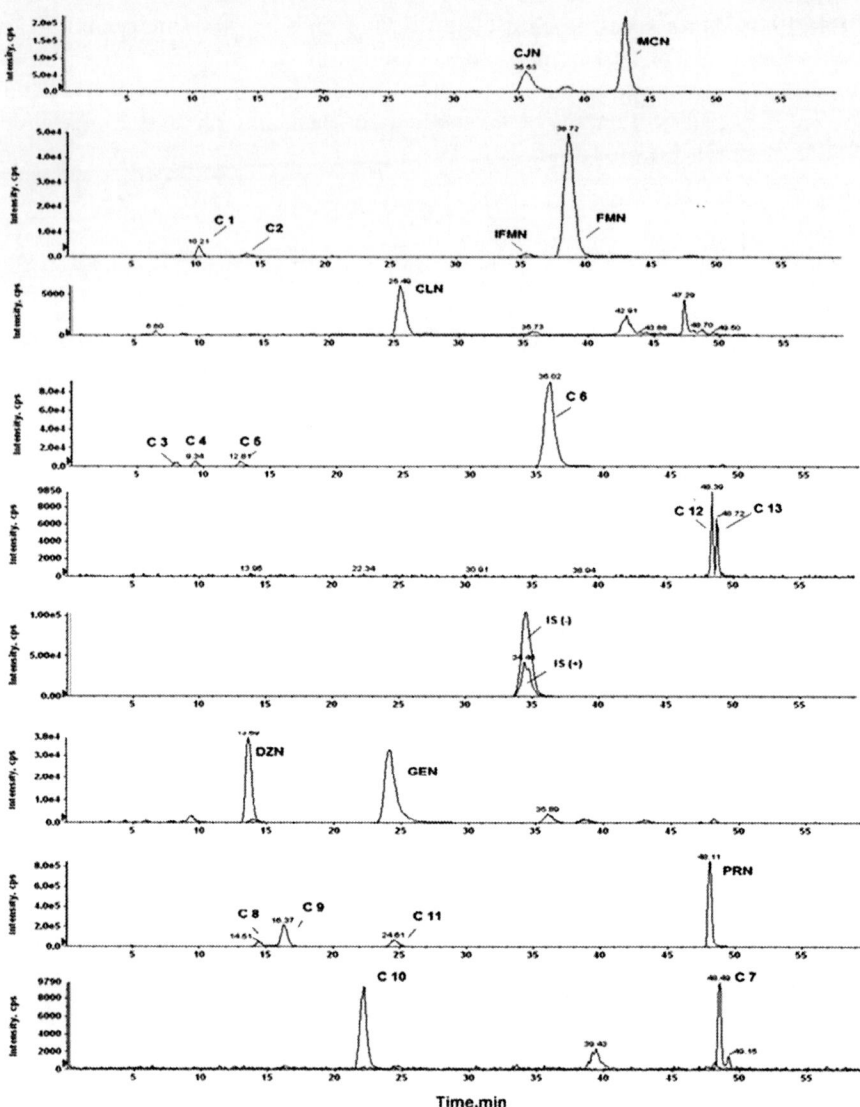

Figure 19.3 LC-MS/MS chromatographic fingerprint of herbal fraction F147. LC-MS/MS chromatogram showing the profile of markers and compounds C1–C13 of F147 obtained in a single LC-MS/MS run.

respectively. The validation parameters intra-day and inter-day accuracy (% bias) and precision (% C.V.) were assessed using the QC samples prepared at low, intermediate and high concentrations for each marker. The stock solution of 7-hydroxy isoflavone (IS) of $1\,\text{mg}\,\text{mL}^{-1}$ was also prepared in methanol. The WS of concentration $100\,\mu\text{g}\,\text{mL}^{-1}$ was prepared from this stock solution in methanol. $5\,\mu\text{L}$ of IS WS was spiked in calibration standard and in QC samples of markers.

Table 19.4 Sensitivity and linearity of markers of F147. The Table lists the validation data of LC-MS/MS method that indicates sensitivity and linearity of eight markers of F147. LLOQ, lower limit of detection; r, regression coefficient.

Markers	LOD (ng mL^{-1})	LLOQ (ng mL^{-1})	Linearity (ng mL^{-1})	r
DZN	3.9	7.8	7.8–1000	0.997 ± 0.002
CJN	1.95	3.9	3.9–1000	0.998 ± 0.001
IFMN	1.95	3.9	3.9–500	0.994 ± 0.002
GEN	3.9	7.8	7.8–500	0.998 ± 0.001
CLN	0.975	3.9	3.9–1000	0.998 ± 0.001
FMN	0.487	0.975	0.975–500	0.998 ± 0.001
MCN	1.95	3.9	3.9–1000	0.999 ± 0.0003
PRN	3.9	7.8	7.8–1000	0.997 ± 0.002

19.3.3.1 Sensitivity and Linearity

The method was sensitive for all the eight markers with the limit of detections (LODs) being in the range of 0.975 ng/ml^{-1} to 3.9 ng ml^{-1}. The linearity was determined by applying the weighting scheme of 1/x or 1/x^2. The regression coefficient (r) obtained for all the markers were greater than 0.99. All the data for sensitivity and linearity are given in Table 19.4.

19.3.3.2 Accuracy and Precision

The % bias and % C.V. values calculated for both intra-day and inter-day analysis shows that the method was linear and precise for all the markers. The values of % bias were in the range of –13.28 to 11.47, whereas the values of % C.V. were in the range of 1.49 to 13.67. All these values were within the acceptable limit of ±15%.

19.3.4 Pattern Profiling and Quantitative Analysis of Isoflavones

The validated LC-MS/MS was applied for both pattern profiling and quantitative analysis of isoflavones in F147. The percentage content of all markers was less than 0.5%. Among the marker compounds, FMN was present in higher proportion (0.3%). MCN and PRN were found to be in almost same quantity (0.2%). Similarly, the percentage content of markers CJN, GEN and DZN were in equal proportions (0.75 to 0.1%). CLN and IFMN (an isomer of FMN), were present in very low quantities in the range 0.002% to 0.004%.

19.4 Conclusion

Natural sources are being increasingly explored in the drug discovery process and traditional medicine proves to be a good source in the identification of new leads. The chemical profiling of herbal extracts helps to understand the role of its active constituents for their biological activity and to control the quality of such preparation to maintain their content uniformity. Thus, pattern profiling and quantitative analysis of active constituents in herbal fractions and traditional medicines is gaining importance. Various instrumental methods are available for the analysis of herbal fractions. Among them, LC-MS/MS is being widely used to analyze herbal fractions. The multi-component nature of herbal preparation poses many challenges in developing selective and sensitive analytical methods for their qualitative and quantitative analysis. However, advanced technology, such as IDA associated with mass spectrometry coupled with liquid chromatography, proves to be an efficient analytical tool to meet these challenges. The use of LC-MS/MS for analysis of herbal fraction has been reviewed in this Chapter by discussing the development, validation and application of the LC-MS/MS method for the analysis of F147.

Summary Points

- Phytoestrogens are effective in preventing bone loss and promoting bone formation.
- Soy and red clover are the major sources of phytoestrogens, especially isoflavones.
- Herbal chemotypes from *Dioscorea spongiosa, Cuminum cyminum, Drynaria fortune, Cissus quadrangularis, Ulmus wallichiana* and *Dalbergia sisso* have also been reported to exhibit promising osteogenic properties, which are associated with the flavonoids constituents present in them.
- In our drive to discover new and safer osteogenic and osteoprotective agents from natural sources, a group of researchers at the Central Drug Research Institute (CDRI), Lucknow, India, observed that the activity-guided fraction F147 from *Butea monosperma* show potential osteogenic properties.
- The extracts from *Butea monosperma* contain active osteogenic constituents, which include the isoflavones DZN, GEN, FMN, IFMN, CJN, CLN and PRN, and the pterocarpan MCN.
- A highly sensitive, selective, accurate and precise method for pattern profiling of isoflavones and simultaneous quantification of DZN, GEN, FMN, IFMN, CJN, CLN, PRN and MCN has been developed and validated using LC-MS/MS.
- The validated LC-MS/MS method was successfully applied for chromatographic fingerprinting and determination of DZN, GEN, FMN, IFMN, CJN, CLN, PRN and MCN in osteogenic herbal chemotype F147 from *Butea monosperma*.
- The method will find wide spread use in quantifying these isoflavones and MCN in herbal extracts from other plants and dietary supplements.

Key Facts

Key Facts for Osteoporosis

- United States National Institutes of Health consensus conference in 2000, defined osteoporosis as "a skeletal disorder characterized by compromised bone strength predisposing a person to an increased risk of fracture. Bone strength reflects the integration of 2 main features: bone density and bone quality."
- According to a World Health Organization (WHO) report, more than 270 million people in India and China are likely to be affected by osteoporosis by the year 2020. The studies from India indicate that 50% of women and 36% of men above the age of 50 have a low bone mass index.
- In women, there is an accelerated rate of bone loss following menopause.
- Osteoporosis significantly increases the risk of vertebral and hip fractures. Fractures cause physical impairment, functional loss, disability and substantial increase in health cost.

Key Facts for Traditional Medicines in the Management of Osteoporosis

- Traditional medical practice, which existed in human societies before the application of modern science to health, was first officially recognized by the World Health Organization (WHO) in 1976 by globally addressing its Traditional Medicine Programme.
- In traditional medicine, there are many natural crude drugs that have the potential to treat bone diseases. The extracts of *Butea monosperma* have been used in traditional medicine for various disorders and has recently been shown to possess osteogenic and osteoprotective activity.
- Qualitative pattern profiling and quantitative analysis of herbal extracts is essential for the comprehensive and accurate assessment of numerous active constituents present in them.

Definition of Words and Terms

Accuracy: It is defined as the nearness of the calculated concentration to the nominal concentration. Intra-batch and inter-batch accuracy is determined by calculating the % bias from the theoretical concentration using the following equation:

$$\% \text{ Bias} = \frac{\text{Observed concentration} - \text{Nominal concentration}}{\text{Nominal concentration}} \times 100$$

Collision Energy: The amount of energy the precursor ion receives as it is accelerated into the collision cell, where they collide with gas molecules and fragment.

Declustering potential: The potential difference between the quadrupole Q0 and the orifice plate to minimize solvent cluster ions, which may attach to the sample.

Electro-spray ionization (ESI): A technique for producing ions from large molecules to produce fragment ions in electron-impact ionization sources.

Enhanced mass scan (EMS): In this scan, ions are transferred directly from the ion source and orifice region to the third quadrupole (Q3) where they are collected, scanned out of Q3 to produce enhanced, single MS-type spectra.

Enhanced product ion (EPI): In this scan, product ions are generated in the Q2 collision cell by the precursor ions from Q1 colliding with the collision (CAD) gas in Q2. These characteristic product ions are transmitted and collected in Q3. These ions are scanned out of Q3 to produce enhanced product ion spectra to achieve good resolution and intensity.

Enhanced resolution: This scan is similar to the EPI scan, except that the Q1 precursor ions pass gently through the Q2 collision cell without fragmenting. A small range about the precursor mass is scanned out of Q3 at the slowest scan rate to produce a narrow window of the best-resolved spectra.

Information-dependent acquisition (IDA): This refers to the real-time acquisition of MS/MS spectra during a chromatographic run through the combination of specific survey scans, high-resolution scans and sensitive product ion scans.

Limit of Detection (LOD): According to ICH, it is the lowest concentration of analyte in a sample which can be detected but not necessarily quantified as an exact value.

Linearity: The linearity of an analytical procedure is its ability (within a given range) to obtain test results which are directly proportional to the concentration of analyte(s) in the sample.

Lower Limit of Quantification (LLOQ): This is the lowest concentration of a sample that can be quantified with acceptable precision and accuracy (bias). The acceptance criteria for these two parameters at LLOQ are $\pm 20\%$ C.V. for precision and $\pm 20\%$ for bias.

Multiple reaction monitoring (MRM): The mode of operating a triple quadrupole instrument so that an ion of given mass (Q1) must fragment or dissociate to give a product ion of specific mass (Q3) in order for a response to be detected. It is used for very specific target compound analysis.

Pattern profiling/chromatographic fingerprinting: An analytical technique by which the chromatographic pattern of various compounds present in complex mixtures of herbal preparations can be obtained.

Precision: This is defined as the repetitiveness of the calculated concentration. Intra-batch and inter-batch precision in terms of coefficient of variation (% C.V.) is obtained by subjecting the data to one-way analysis of variance (ANOVA).

Selectivity: This is the ability of the analytical method to measure unequivocally and to differentiate the analyte(s) in the presence of other components.

List of Abbreviations

CJN	cajanin
CLN	cladrin
% C.V.	coefficient of variation
DS	dilution solvent
DZN	daidzein
EMS	enhanced mass scan
EPI	enhanced product ion
ESI	electro-spray ionization
FMN	formononetin
GEN	genistein
HPLC	high-performance liquid chromatography
IDA	information-dependent acquisition
IFMN	isoformononetin
IS	internal standard
LC-MS/MS	liquid chromatography tandem mass spectrometry
LOD	limit of detection
MCN	medicarpin
MRM	multiple reactions monitoring
PDA	photodiode array detector
PRN	prunetin
QA	quality assurance
QC	quality control
WS	working solution(s)

Acknowledgements

The authors are thankful to the Director CSIR-CDRI for his keen interest and encouragement; the Department of Science and Technology, Government of India, for providing equipment facility at the National Centre for Metabolic and Pharmacokinetic Studies; the Council of Scientific and Industrial Research, New Delhi, India, and the Indian Council of Medical Research, New Delhi, India, for providing research scholarships to Dr M. Lakshmi and Dr Smriti Mishra; the Ministry of Health and Family Welfare, Government of India, for providing financial support to carry out this work.

References

Anonymous, 1996. ICH, Q2B, Harmonized tripartite guideline, validation of analytical procedures: Methodology, IFPMA. In Proceedings of the International Conference on Harmonization. Nov 6, 1996, Geneva, Switzerland.

Anonymous, 2000. NIH Consensus Statement. *Osteoporosis Prevention, Diagnosis, and Therapy*. 17: 1–45.

Anonymous, 2001. The European agency for the evaluation of medicinal products. Note for Guidance on Quality of Herbal Medicinal Products, London, UK.

Anonymous, 2004. U.S. Food and Drug Administration. Guidance for Industry Botanical Drug Product, Rockville, MD, USA.

Chen, L.J., Zhao, X., Plummer, S., Tang, J., and Games, D.E., 2005. Quantitative determination and structural characterization of isoflavones in nutrition supplements by liquid chromatography–mass spectrometry. *Journal of Chromatography A*. 1082: 60–70.

Gupta, V., Dwivedi, A.K., Yadav, D.K., Kumar, M., and Maurya, R., 2010. Reverse phase-HPLC method for determination of marker compounds in NP-1, an anti-osteoporotic plant product from *Butea monosperma*. *Natural Product Communication*. 5: 47–50.

Heimler, D., Vignolini, P., Galardi, C., Pinelli, P., and Romani, A., 2004. Simple extraction and rapid quantitative analysis of isoflavones in soybean seeds. *Chromatographia*. 59: 361–365.

Iqbal, D., Pawar, R.K., and Sharma, R.K., 2010. Physico-chemical standardization of *Butea monosperma* (Lam.) Kuntze (Palasha): An ayurvedic drug. *International Journal of Pharmaceutical Quality Assurance*. 2: 49–51.

Lee, K.H., 2000. Research and future trends in the pharmaceutical development of medicinal herbs from Chinese medicine. *Public Health Nutrition*. 3: 515–522.

Lee, J.S., Kim, D.H., Liu, K.H., Oh, T.K., and Lee, C.H., 2005. Identification of flavonoids using liquid chromatography with electrospray ionization and ion trap tandem mass spectrometry with an MS/MS library. *Rapid Communications in Mass Spectrometry*. 19: 3539–3548.

Marquet, P., Marcoux, F.S., Gamble, T.N., and Leblanc, J.C.Y., 2003. Comparison of a preliminary procedure for the general unknown screening of drugs and toxic compounds using a quadrupole-linear ion-trap mass spectrometer with a liquid chromatography–mass spectrometry reference technique. *Journal of Chromatography B*. 789: 9–18.

Maurya, R., Yadav, D.K., Singh, G., Bhargavan, B., Murthy, P.S.N., Sahai, M., and Singh, P.S.N., 2009. Osteogenic activity of constituents from *Butea monosperma*. *Bioorganic and Medicinal Chemistry Letters*. 19: 610–613.

Nancy, L., Booth, N.L., Overk, C.R., Yao, P., Burdette, J.E., Nikolic, D., Chen, S.N., Bolton, J.L., Breemen, R.B., Pauli, G.F., and Farnsworth, N.R., 2006. The chemical and biological profile of a red clover (*Trifolium pratense*) phase II clinical extract. *Journal of Alternative and Complementary Medicine*. 12: 133–139.

Otieno, D.O., Rose, H.H., and Shah, N.P., 2007. Profiling and quantification of isoflavones in soymilk from soy protein isolate using extracted ion chromatography and positive ion fragmentation techniques. *Food Chemistry*. 105: 1642–1651.

Pandey, R., Gautam, A.K., Bhargavan, B., Trivedi, R., Swarnkar, G., Nagar, G.K., Yadav, D.K., Kumar, M., Rawat, P., Manickavasagam, L.,

Kumar, A., Maurya, R., Goel, A., Jain, G.K., Chattopadhyay, N., and Singh, D., 2010. Total extract and standardized fraction from the stem bark of *Butea monosperma* have osteoprotective action: evidence for the non-estrogenic osteogenic effect of the standardized fraction. *Menopause*. 17: 602–610.

Potu, B.K., Rao, M.S., Nampurath, G.K., Chamallamudi, M.R., Prasad, K., Nayak, S.R., Dharmavarapu, P.K., Kedage, V., and Bhat, K.M.R., 2009. Evidence-based assessment of antiosteoporotic activity of petroleum-ether extract of *Cissus quadrangularis* Linn. on ovariectomy-induced osteoporosis. *Upsala Journal of Medical Science*. 114: 140–148.

Prasain, J.K., Jones, K., Kirk, M., Wilson, L., Johnson, S.M., Weaver, C., and Barnes, S., 2003. Profiling and quantification of isoflavonoids in kudzu dietary supplements by high-performance liquid chromatography and electrospray ionization tandem mass spectrometry. *Journal of Agricultural and Food Chemistry*. 51: 4213–4218.

Rijke, E., Zappey, H., Ariese, F., Gooijer, C., Udo, A., and Brinkman, T., 2003. Liquid chromatography with atmospheric pressure chemical ionization and electrospray ionization mass spectrometry of flavonoids with triple-quadrupole and ion-trap instruments. *Journal of Chromatography A*. 984: 45–58.

Robards, K., 2003. Strategies for the determination of bioactive phenols in plants, fruit and vegetables. *Journal of Chromatography A*. 1000: 657–691.

Sengupta, A., and Basu, S.P., 1978. The triglyceride composition of *Butea monosperma* seed oil. *Journal of the American Oil Chemists Society*. 55: 533–535.

Setchell, K.D.R., Brown, N.M., Desai, P.B., Nechemias, Z.L., Wolfe, B.E., Brashear, W.T., Kirschner, A.S., Cassidy, A., and Heubi, J.E., 2001. Bioavailability of pure isoflavones in healthy humans and analysis of commercial soy isoflavone supplements. *Journal of Nutrition*. 131: 1362S–1375S.

Sharan, K., Siddiqui, J.A., Swarnkar, G., Tyagi, A.M., Kumar, A., Rawat, P., Kumar, M., Nagar, G.K., Arya, K.R., Lakshmi, M., Jain, G.K., Maurya, R., and Chattopadhyay, N., 2010. Extract and fraction from *Ulmus wallichiana* Planchon promote peak bone achievement and have a non-estrogenic osteoprotective effect. *Menopause*. 17: 393–402.

Shirke, S.S., Jadhav, S.R., and Jagtap, A.G., 2008. Methanolic extract of *Cuminum cyminum* inhibits ovariectomy-induced bone loss in rats. *Experimental Biology and Medicine*. 233: 1403–1410.

Suryawanshi, S., Mehrotra, N., Asthana, R.K., and Gupta, R.C., 2006. Liquid chromatography/tandem mass spectrometric study and analysis of xanthone and secoiridoid glycoside composition of *Swertia chirata*, a potent antidiabetic. *Rapid Communications in Mass Spectrometry*. 20: 3761–3768.

Varier, S.P.V., 1993. *Indian Medicinal Plants*. Orient Longman Limited, Madras, India. *284.*

Wang, X.L., Wang, N.L., Zhang, Y., Gao, H., Pang, W.Y., Wong, M.S., Zhang, G., Qin, L., and Yao, X.S., 2008. Effects of eleven flavonoids from the osteoprotective fraction of *Drynaria fortunei* (Kunze) J. SM. on osteoblastic proliferation using an osteoblast-like cell line. *Chemical and Pharmaceutical Bulletin.* 56: 46–51.

Wu, Q., Wang, M., Sciarappa, W.J., and Simon, J.E., 2004. LC/UV/ESI-MS analysis of isoflavones in edamame and tofu soybeans. *Journal of Agricultural and Food Chemistry.* 52: 2763–2769.

Yin, J., Han, N., Liu, Z., Song, S., and Kadota, S., 2010. The *in vitro* antiosteoporotic activity of some glycosides in *Dioscorea spongiosa. Biological and Pharmaceutical Bulletin.* 33: 316–320.

Zhang, X., Xiao, H.B., Xue, X.Y., Sun, Y.G., and Liang, X.M., 2007. Simultaneous characterization of isoflavonoids and astragalosides in two *Astragalus* species by high performance liquid chromatography coupled with atmospheric pressure chemical ionization tandem mass spectrometry. *Journal of Separation Science.* 30: 2059–2069.

CHAPTER 20
Analysis of Novel Isoflavone Digycoside in Nuts

KAZUHIRO NARA

Department of Life Design, Faculty of Home Economics,
Tokyo Kasei-Gakuin University, Machida, Tokyo 194-0292, Japan
Email: k_nara@kasei-gakuin.ac.jp

20.1 Introduction

The groundnut (*Apios americana* Medik), a leguminous perennial vine native to North America, is said to have been brought to Aomori Prefecture, Japan, when apple seedlings were introduced there. The roots of groundnut grow to a depth of 1 m or more, with tubers that are attached 5–10 cm apart on the fibrous roots and grow larger, as the plant grows, to mature bead-like tubers. There are several reports regarding the nutritional value of groundnut tubers (hereafter designated as groundnuts), as well as their fatty acid, amino acid (Okubo *et al.* 1994; Iwai and Matsue 2007; Wilson *et al.* 1986; 1987), and carbohydrate (such as mono- and oligosaccharides) (Ogasawara *et al.* 2006) compositions. Iwai and co-workers recently reported the protective effect of groundnuts against blood pressure elevation (Iwai and Matsue, 2007; Iwai *et al.* 2008). In light of these observations, groundnuts have gained increasing attention as a health food with a high nutritional value and functional components.

However, despite belonging to the legume family, groundnuts have not yet been fully studied regarding their isoflavone profile. Although Krishnan (1998)

Food and Nutritional Components in Focus No. 5
Isoflavones: Chemistry, Analysis, Function and Effects
Edited by Victor R Preedy
© The Royal Society of Chemistry 2013
Published by the Royal Society of Chemistry, www.rsc.org

identified the aglycone genistein in the plant, no information is available with regards to its glycoside content or other groundnut isoflavones.

20.2 Isoflavones

Soybeans have attracted considerable attention worldwide for their health-promoting, disease-preventing effects, which are mainly attributed to isoflavones, a group of flavonoids widely found in plants, predominantly in the legume family, particularly soybeans. Soybeans are thus considered to be a food source that can be incorporated into normal meals to provide a sufficient amount of isoflavones to achieve their potential physiological functions. To date, 12 types of isoflavone, including three aglycones (daidzein, genistein, and glycitein) and their glycosides, have been identified in soybeans (Kudou et al. 1991). Recent reports have shown that soybean isoflavones possess estrogen-like activity and reduce osteoporosis risks (Arjmandi et al. 1996; Zhang et al. 2007). Additionally, isoflavones have been proven to exert various pharmacological actions (Chung et al. 2008; Liu et al. 2005; Wang et al. 2009) such as carcinostatic (anti-carcinogenic and anti-proliferative), anti-hypertensive, anti-oxidative, and anti-allergic (Liu et al. 2005) activities, gaining increasing interest as functional components of soybeans.

In this Chapter I will focus on groundnuts, a leguminous crop, as a source of isoflavones, of which the potential health benefits (*i.e.*, health-promoting, disease-preventing effects) have been the focus of interest, and have identified a novel isoflavone component (Nara et al. 2011).

20.3 Isoflavones in Groundnuts

20.3.1 Plant Material and Preparation of Groundnut Extract

Freeze-dried groundnuts, tubers of *A. americana* Medik, were pulverized in a blender and stored in a freezer until use. The powdered sample, approximately ($\sim$0.1 g), was extracted with 10 ml of 80% ethanol by shaking for 1 h and centrifuged (3000 rpm, 10 min). The resulting supernatant was collected and used for the high-performance liquid chromatography (HPLC) analysis of isoflavones after filtration through a 0.45 μm membrane filter.

To evaluate the influence of the extraction temperature on the isoflavone profile, the powdered sample ($\sim$0.1 g) was extracted at 4 or 45 °C by shaking with 10 mL of distilled water and centrifuged to obtain the supernatant for isoflavone analysis.

20.3.2 HPLC Analysis of Isoflavones

The isoflavone profile was analyzed under the following HPLC conditions: detector, a diode array detector (Agilent Technologies, Inc., USA); column, an Inertsil ODS-3 column (4.6 × 250 mm, GL Sciences, Inc., Japan); column temperature, 40 °C; mobile phase, 0.1% acetic acid (A) and acetonitrile (B); solvent gradient, 90–60% A to 10–40% B in 0–50 min, 60–30% A to 40–70% B in 50–55 min, and 30–90% A to 70–10% B in 55–60 min; detection wavelength, 260 nm; and flow rate, 1.0 mL min^{-1}.

20.3.3 Isolation and Purification of a Novel Isoflavone from Groundnut Extract

Raw groundnuts with the skin were added to three volumes of ethanol, and homogenized in a blender. After vacuum filtration, the ethanol-insoluble fraction was then mixed with 80% ethanol in a blender and vacuum-filtered (this step was repeated twice). All of the filtrate was collected, evaporated, and freeze-dried to obtain an alcohol-soluble solid fraction (ASS). The ASS was dissolved in MQ water and centrifuged (3000 rpm, 10 min) to remove insoluble materials. The supernatant was fractionated through an HP-20 column (2.5 × 25 cm) with an elution gradient from MQ water to ethanol (20, 40, 60, and 80%). The 40% ethanol-eluted fraction, which contained a major peak (peak-1), was further fractionated by HPLC on an Inertsil PREP-ODS column (20 × 250 mm, GL Sciences), which was eluted with a mobile phase of 0.1% acetic acid/acetonitrile (20:80, v/v) at a flow rate of 4.0 mL min^{-1}. The eluates were monitored by measuring the absorbance at 260 nm. The subfraction containing peak-1 was subjected to a preparative HPLC to obtain genistin 7-O-gentiobioside as a colorless solid. The chemical structure was characterized by ^{1}H- and ^{13}C-nuclear magnetic resonance (NMR), and two-dimensional (2D)-NMR spectroscopy (JEOL EX-400, Akishima, Japan), infrared (IR) spectroscopy (Horiba FT-720, Kyoto, Japan), polarimetry (Horiba SEPA-300, Kyoto, Japan) and liquid chromatography-mass spectrometry (LC/MS). Liquid chromatography electrospray ionization (ESI) tandem mass spectrometry (LC/ESI-MS/MS) analysis was performed using a triple-quadrupole TSQ Quantum mass spectrometer (Thermo Fisher Scientific, Inc., USA) equipped with the semi-micro HPLC system SI-1 (Shiseido Co., Ltd., Japan) and an ESI ion source. The sample solution was ionized by ESI and analyzed in a positive ionization mode under the following conditions: column, CAPCELL-PAK C18 MG (3 × 150 mm, Shiseido Co., Ltd.); column temperature, 40 °C; flow rate, 0.4 mL min^{-1}; and mobile phase, 0.1% acetic acid (A) and acetonitrile (B) with a linear gradient increasing from 10 to 20% B over 30 min. The LC/ESI-MS/MS operating conditions were as follows: capillary voltage, 3 kV; cone voltage, 25 V; nebulizer gas flow, 17 L h^{-1}; and ion source temperature, 180 °C.

20.3.4 Content of a Novel Isoflavone in Groundnuts

After the isolation and purification processes, the major component of the groundnut extract (purified peak-1) was dissolved in acetonitrile, diluted to appropriate concentrations, and subjected to HPLC analysis. A standard curve

($R^2 > 0.999$) for purified peak-1 was created based on the HPLC peak areas. Groundnuts from seven different cultivation areas (A–G) were evaluated. The groundnuts were respectively freeze-dried, extracted with 80% ethanol by shaking for 1 h, and centrifuged to collect supernatants. For each supernatant, purified peak-1 content was analyzed by HPLC under the same conditions.

Next, the different parts of tuber, such as peel and flesh (without peel), were determined for novel isoflavone.

20.4 Analytical Profiles of Groundnuts

20.4.1 Isoflavone Profile of Groundnuts

An aliquot of the 80% ethanol extract of groundnuts showed a peak at 260 nm in the UV spectrum (data not shown). We also measured the UV spectrum of soybeans, a leguminous crop, in a similar manner and confirmed the spectrum of isoflavones with a peak at approximately 260 nm, indicating that groundnuts, another leguminous plant, contained isoflavones as well. The groundnut extract was then subjected to HPLC analysis to determine the isoflavone composition in comparison with commercially obtained isoflavones. HPLC analysis showed a major peak (peak-1), along with several other peaks (Figure 20.1), all of which, except for one identified as genistin, were not consistent with the commercially obtained standards. Genistin, a glycoside that contains

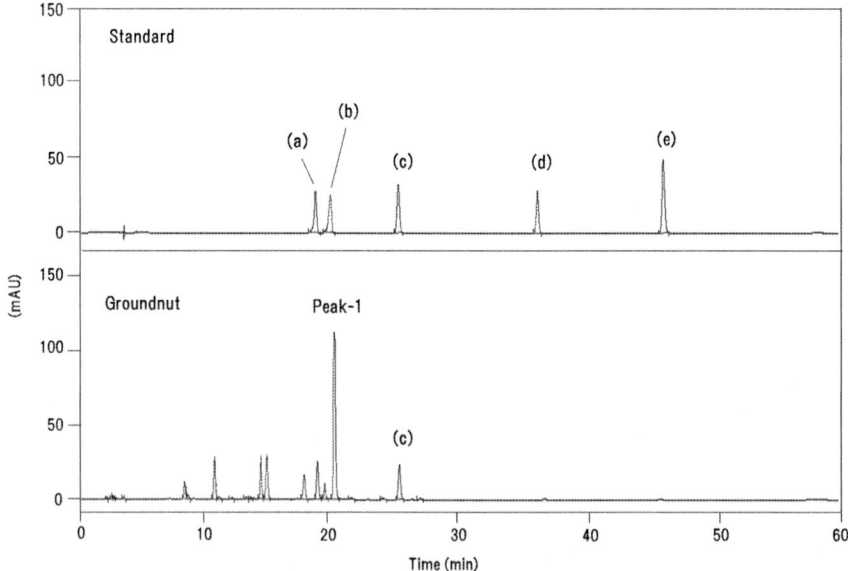

Figure 20.1 HPLC chromatogram of isoflavones from the groundnut extract. (a) Daidzin, (b) glycitein, (c) genistin, (d) daidzein, and (e) genistein. Data are from Nara et al. (2011).

genistein as its aglycone skeleton, is hydrolyzed by enzymes in saliva and the intestinal mucosa, as well as by enterobacterial β-glucosidase to produce genistein (Day et al. 2000; Izumi et al. 2000; Ismail and Hayes 2005). Several reports have shown that genistein serves a wide variety of functions including anti-oxidative, anti-carcinogenic, and anti-allergic activities (Chung et al. 2008; Liu et al. 2005; Wang et al. 2009), suggesting that the groundnut is a beneficial crop with potential health effects. These results indicate that groundnuts can be an important source of isoflavones, with comparable significance to soybeans, which are now recognized as the primary dietary isoflavone source.

20.4.2 Influence of the Extraction Temperature on the Isoflavone Composition of Groundnut Extract

During soaking, the glycoside genistin in soybeans is converted into the aglycone genistein by endogenous β-glucosidase (Day et al. 2000; Izumi et al. 2000; Ismail and Hayes 2005), suggesting that a similar conversion may occur in groundnut isoflavones. In the present study, groundnut extracts were prepared from the dry powdered material either at 4 °C (4 °C extract) or at 45 °C (45 °C extract), and the isoflavone content of each extract was serially measured to evaluate the influence of the extraction temperature. The HPLC elution patterns were clearly different between the two extracts after 24 h of extraction (data not shown), demonstrating a strong influence of the extraction temperature on the composition of groundnut isoflavones. The 4 °C extract exhibited peak-1 and a peak for genistin, whereas the 45 °C extract did not show these peaks but rather revealed a new peak which corresponded to that of authentic genistein. Serial changes in the genistin and genistein contents of groundnut extracts are shown in Figure 20.2. After 1 h of extraction, the genistin content was higher in the 45 °C than in the 4 °C extract; the value for the 4 °C extract was only approximately 74% of that for the 45 °C extract. In the 45 °C extract, the genistin content was increased during the first 4 h of extraction and markedly decreased thereafter, whereas the value for the 4 °C extract was increased with time of extraction. The genistein content in the 45 °C extract increased over time, showing a ~6-fold increase after 24 h compared with 1 h of extraction, whereas genistein was not detected in the 4 °C extract. Extraction at 4 °C yielded a limited amount of the aglycone genistein, whereas the amount was markedly increased with extraction at 45 °C, indicating that groundnuts contained β-glucosidase, by which genistein was produced during the extraction process at 45 °C.

The influence of the extraction temperature was also examined for the major peak (peak-1) content. As the peak did not correspond to any peaks for commercially obtained isoflavones, the influence of the extraction temperature was evaluated in terms of changes in the peak area. Figure 20.3 shows serial changes in the peak-1 content of the extracts. The values are expressed as a percentage relative to the peak-1 content of the 4 °C extract measured after 1 h

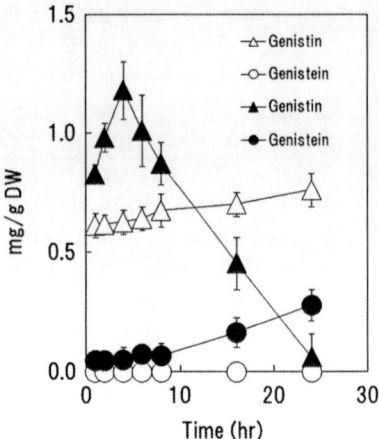

Figure 20.2 Serial changes in the isoflavone content of groundnut extracts prepared at different temperatures.
Data are from Nara *et al.* (2011).

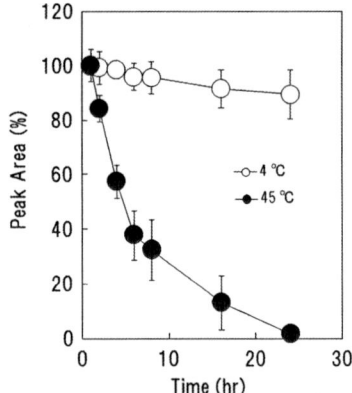

Figure 20.3 Serial changes in the peak-1 content of groundnut extracts. The peak-1 area of the 4 °C extract measured after 1 h of extraction was regarded as 100%.
Data are from Nara *et al.* (2011).

of extraction (regarded as 100%). The peak-1 content in the 4 °C extract tended to be decreased over time to ∼90% after 24 h of extraction, whereas that in the 45 °C extract was markedly decreased with time to ∼57 and ∼44% after 4 and 8 h, respectively, and to an undetectable level after 24 h of extraction.

In the 45 °C extract, both genistin and the peak-1 contents were decreased, whereas the genistein content was increased, suggesting that β-glucosidase in groundnuts not only converted genistin into genistein but also possibly acted on peak-1. Furthermore, an increase in the content of the aglycone genistein in the 45 °C extract indicated that peak-1 had the same aglycone skeleton (genistein) as genistin.

20.4.3 Purification of a Novel Isoflavone (Peak-1) from the Groundnut Extract

The size of peak-1 of the water extract of groundnut was markedly affected by the extraction temperature, whereas that of the alcohol extract remained stable regardless of the temperature conditions, providing a basis for the use of the alcohol extract for the isolation and purification of peak-1.

Groundnut extract prepared using 80% ethanol was evaporated and freeze-dried to obtain ASS, which was then dissolved in MQ water and centrifuged. The supernatant was subjected to an HP-20 column (2.5 × 25 cm) and eluted with MQ water (for a non-absorbed fraction) followed by 20, 40, 60, and 80% ethanol (for absorbed fractions). The absorbance at 260 nm was measured for each fraction. The value for the 40% ethanol-eluted fraction was clearly high, a measurement which was consistent with the results of HPLC analysis in which the peak-1 was only found in the 40% ethanol fraction. After evaporation, the 40% ethanol fraction was separated by HPLC into subfractions of 4 mL per tube, displaying an elution pattern with several peaks. The subfractions, each of which was numbered Fr. (i) to (vii) in the order of elution, were respectively subjected to an Inertsil ODS-3 column, confirming that Fr. (vi) contained peak-1. The Fr. (vi) subfraction was further purified until it showed a single peak (purified peak-1), which was subsequently analyzed using an ESI-MS system.

The mass spectrum showed a molecular ion peak [M + H]+ at m/z 595, and the MS/MS spectrum revealed additional fragment peaks at m/z 433 and 271 (Figure 20.4). m/z 271 (270 + 1) corresponded to the ion signal that originated from genistein, which is also found in soybeans, and, with an m/z difference of 162 from m/z 271, m/z 433 (270 + 162 + 1) was assumed to be its glycoside, genistin. The difference between m/z 595 (433 + 162 + 1) and 447 was 162,

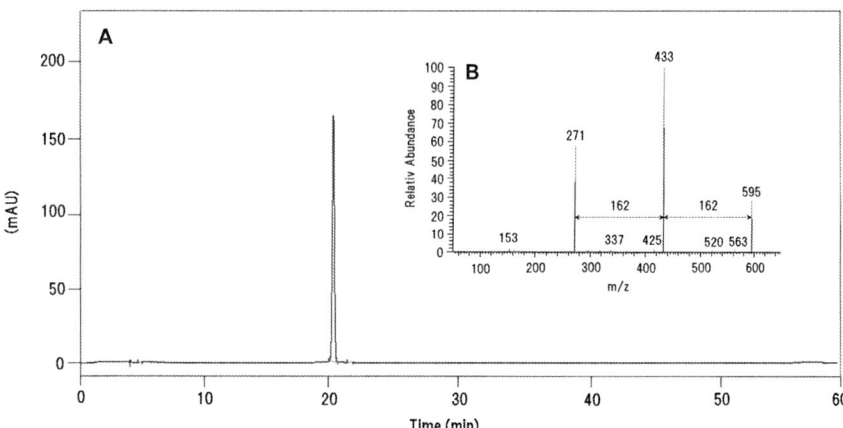

Figure 20.4 HPLC chromatogram (A) and mass spectrum (B) of purified peak-1 from groundnut (*Apios americana* Medik).
Data are from Nara et al. 2011.

suggesting the addition of one sugar molecule to m/z 477 to yield m/z 595; purified peak-1 was deduced to be a glucoside derivative consisting of a genistein and two monosaccharide molecules attached to it.

20.4.4 Structural Analysis of a Novel Isoflavone from the Groundnut Extract

A characteristic singlet signal for the isoflavone ring was observed at δ 8.08 (H-2) in the ^{1}H-NMR spectrum. In addition, a long-range coupling was detected between two aromatic methine signals at δ 6.87 (H-6) and 7.09 (H-8) in the double-quantum filtered–correlated spectroscopy (DQF-COSY) spectrum. Hence, the aglycone of the compound was identified as genistein. By the heteronuclear multiple quantum coherence (HMQC) experiment, two methine signals at δ 5.75 (H-1″) and 101.3 (C-1″), and δ 5.08 (H-1″) and 105.1 (C-1″) were assigned to be anomeric positions so that the compound contained two sugar moieties. An oxygenated aromatic quaternary carbon at δ 163.4 (C-7) possessed three heteronuclear multiple bond correlation (HMBC) correlations with δ 6.87 (H-6), 7.09 (H-8) and 5.75 (H-1″), indicating that the glycoside (Glc I) was attached at C-7. The methylene proton at δ 4.81 (H-6″) and 4.38 (H-6″) in Glc I was correlated to the anomeric carbon (δ 105.1) of Glc II in the HMBC spectrum. Considering *trans*-vicinal coupling constants (7.3–8.3 Hz) from all the oxymethine proton signals of sugar moieties, the glycoside was identical to gentiobioside. From the results of ^{1}H- and ^{13}C-NMR spectroscopic analyses, the structure of purified peak-1 was deduced to be genistein-7-O-genitiobioside (Figure 20.5), which is comprised of genistein attached to gentiobiose (6-O-β-D-glucopyranosyl-D-glucose). Unlike genistin (genistein-7-O-glucoside), which is found in soybeans as a conjugate of genistein and one glucose molecule, genistein-7-O-genitiobioside, which was isolated and purified from groundnuts in this study, has not been found in soybeans nor has it been reported to date. This component was isolated and purified for the first time in the present study. In the 45 °C extract of groundnuts, the amount of genistein (genistein-7-O-glucoside) was increased with time during the first 4 h of extraction. This is probably due to β-glucosidase acting on peak-1 (genistein-7-O-genitiobioside) to release one sugar molecule, leading to a temporal increase in the genistein

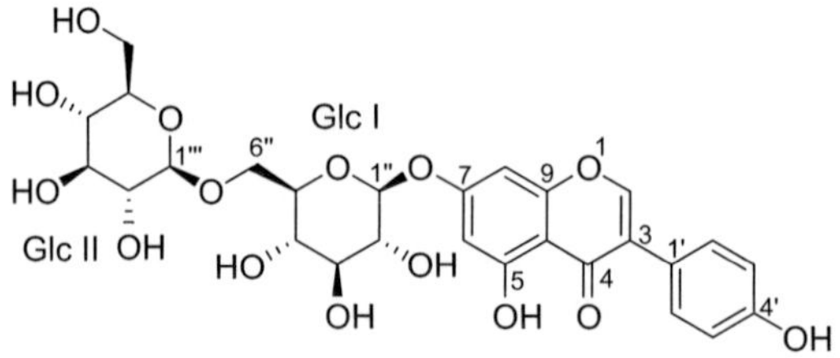

Figure 20.5 Structure of isoflavone diglicoside isolated in this study from groundnut (*Apios americana* Medik).
Data are from Nara *et al.* (2011).

content. Previous studies have shown that the conversion of isoflavone glycosides into aglycones by enterobacterial β-glucosidase has a significant impact on the absorption efficiency and biological activities of the isoflavones (Han *et al.* 2009; Kulling *et al.* 2001; Miura *et al.* 2002). In the present study, β-glucosidase treatment of the groundnut extract resulted in the degradation of genistein-7-*O*-genitiobioside and the formation of genistein, suggesting that genistein-7-*O*-genitiobioside is converted into the aglycone and, thereafter, exerts various biological functions such as anti-oxidative, carcinostatic (anti-carcinogenic and anti-proliferative), and anti-cholesterolemic activities. Generally, isoflavones are relatively insoluble in water, and there is a significant difference in the degree of water solubility between the glycoside and aglycone forms; glycosides have a ∼1000- to 10 000-fold higher water solubility compared with aglycones. The feasibility of adding a sugar molecule(s) to genistein has recently been explored using synthesized enzymes, and changes in the bioactivity associated with the chemical modification have begun to be examined. Having more sugar molecules than genistein and genistin, genistein-7-*O*-genitiobioside in groundnuts is likely to show increased water solubility. Along with this, the degree of water solubility and bioactivity of this novel component, in comparison with genistein and genistin, merit further investigation.

20.4.5 Content of a Novel Isoflavone from the Groundnut Extract

Groundnuts from seven different cultivation areas (A–G) were evaluated for their genistein-7-*O*-genitiobioside content (Table 20.1). There were differences among these groundnuts in their unit weight (8.5–28.6 g) and genistein-7-*O*-genitiobioside content per 1 g of dried groundnuts [2.6–7.9 mg (g of dried groundnuts)$^{-1}$] depending on the cultivation area. No association was found between the unit weight and genistein-7-*O*-genitiobioside content. Even when measured per 100 g of fresh weight, the genistein-7-*O*-genitiobioside content differed among the samples from seven areas, with a mean genistein-7-*O*-genitiobioside content of 190.8 mg (100 g of fresh groundnuts)$^{-1}$. Several reports

Table 20.1 Genistein-7-*O*-genitiobioside in alcohol soluble fraction of groundnuts.

Sampling point	Weight (g FW)	Dry weight (gDW/100 g FW)	Genistein-7-O-gentiobioside	
			(mg/g DW)	(mg/100 g FW)
A	8.5	42.8	4.2	178.2
B	8.7	41.1	5.7	233.6
C	8.8	41.8	2.6	106.5
D	8.8	39.2	5.4	212.5
E	10.8	43.1	2.9	125.4
F	10.8	44.8	7.9	352.5
G	28.6	45.0	2.8	127.3

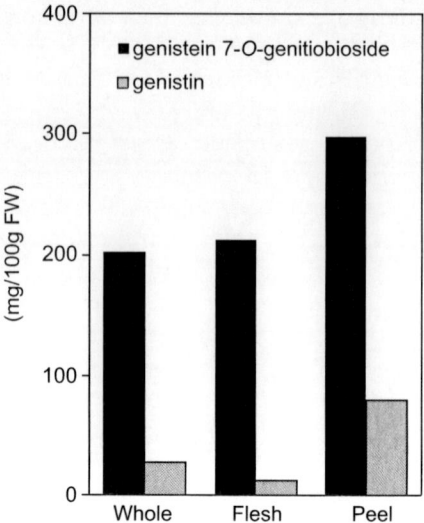

Figure 20.6 Content of genistein-7-O-genitiobioside and genistein from groundnut (*Apios americana* Medik).

have demonstrated that the isoflavone content of soybeans is affected not only by cultivar or growing location, but most significantly by crop year, possibly due to the effects of environmental conditions, particularly ambient temperature during the grain-filling period (Dhaubhadel *et al.* 2003; Kim *et al.* 2005), indicating that the isoflavone content of groundnuts may vary in a similar fashion.

Interestingly, different parts of tuber showed significantly different concentrations of isoflavone. The amount of genistein-7-O-genitiobioside in peel had more 1.5-fold than it in flesh (Figure 20.6). In addition, the amount of genistin showed 6-fold. Therefore, the amount of isoflavone was affected by the existence of the peel.

20.5 Conclusion

The present study demonstrated that groundnuts (tubers of leguminous *A. americana* Medik) contain a novel isoflavone, genistein-7-O-genitiobioside, which has not been reported in any other plant. Similarly to soybean genistin, genistein-7-O-genitiobioside is likely to be converted into genistein by enterobacterial β-glucosidase, a process by which the absorption efficiency and bioactivities of the isoflavone can be strongly affected. We consider that these techniques and methods can be applied to other foods in an endeavor to identify putative neutraceuticals. Currently, soybeans are used as the dominant dietary source of isoflavones in our diets; however, this study demonstrated that groundnuts can provide sufficient amounts of isoflavones. Further studies are needed to elucidate the nature of groundnut isoflavones, including changes in the content of genistein-7-*O*-genitiobioside in groundnuts during crop growth. Given the previous findings in soybeans revealing that their isoflavone composition varies depending on the processing methods (Riedl *et al.* 2005; Yin *et al.* 2005), along with the potential of groundnuts as an important isoflavone source, the effects of processing methods and storage conditions on the groundnut isoflavone composition also merit further research.

Summary Points

- This Chapter focuses on novel isoflavone in nuts.
- In the present HPLC-based analysis of the isoflavone profile of groundnut (*Apios americana* Medik) tubers, we identified a major peak that did not correspond to any known isoflavones.
- The major peak of groundnuts was then isolated and purified through columns, and the chemical structure of the resultant component was analyzed and identified.
- The structure of purified major peak was deduced to be genistein-7-*O*-genitiobioside.
- There were differences among these groundnuts in their unit weight (8.5–28.6 g) and genistein-7-*O*-genitiobioside content per 1 g of dried groundnuts [2.6–7.9 mg (g dried groundnuts)$^{-1}$] depending on the cultivation area.
- We consider that these techniques and methods can be applied to other foods in an endeavor to identify putative neutraceuticals.

Key Facts

Key Facts for Groundnuts

- Groundnut (*Apios americana* Medik), a leguminous perennial vine native to North America, is said to have been brought to Aomori Prefecture, Japan, when apple seedlings were introduced there.
- The roots of groundnut grow to a depth of 1 m or more, with tubers that are attached 5–10 cm apart on the fibrous roots and grow larger, as the plant grows, to mature bead-like tubers.
- It is not clear about functional components in the groundnuts.

Key Facts for Genistein-7-*O*-gentiobiside

- The present study demonstrated that groundnuts contain a novel isoflavone, genistein-7-*O*-genitiobioside, which has not been reported in any other plant.
- There novel isoflavone is major compound.

- The major peak was deduced to be a glucoside derivative consisting of a genistein and two monosaccharide molecules attached to it.
- The size of major peak of the water extract of groundnut was markedly affected by the extraction temperature.
- During soaking, the glycoside genistin in groundnuts is converted into the aglycone genistein by endogenous β-glucosidase.

Definitions of Words and Terms

Aglycon form: An aglycone is the non-sugar compound remaining after replacement of the glycosyl group, such as genistein, daidzin, and glycitin.

Chromatographic methods: Usually a purification protocol contains one or more chromatographic steps. The basic procedure in chromatography is to flow the solution containing the useful component through a column packed with various materials.

Genistein: Genistein is an isoflavonoid derived from soy and soy products. A phytoestrogen found in soy, having anti-oxidant and anti-cancer properties. Moreover, genistein is a well-known inhibitor of protein tyrosine kinases.

Genistin: Genistein is an isoflavonoid derived from soy and soy products. It is hydrolyzed by removing the covalently bound glucose to form genistein.

Gentiobiose: Gentiobiose is a disaccharide composed of two units of D-glucose, O-β-D-glucopyranosyl-$(1 \rightarrow 6)$-D-glucose.

β-Glucosidase: The enzyme β-glucosidase hydrolyzes the isoflavone glucosides form developing aglycones form, which are compounds with anti-cancer effects which are compounds with anti-cancer effects.

Glycosidic form: Soybean isoflavone occur primarily as β-glycoside with a small content as the principal bioactive. Three glycosides form can exist in three forms: malonyl-, acetyl- and non-conjugated glycoside. Essentially, a glycosidic form has a sugar moiety attached to the aglycon structure.

Groundnut (Apios americana Medik): Groundnut (*Apios americana* Medik), a leguminous perennial vine native to North America, is said to have been brought to Aomori Prefecture, Japan. The roots of groundnut grow to a depth of 1 m or more, with tubers that are attached 5–10 cm apart on the fibrous roots and grow larger, as the plant grows, to mature bead-like tubers.

Isoflavones: These are biphenolic compounds mainly found in members of the legume family. Isoflavones are natural compounds, which are found in a variety of plant, predominantly soybean. These are the most common phytoestrogenic compounds found in plants.

Liquid chromatography–mass spectrometry (LC-MS): This is an analytical chemistry technique that combines the physical separation capabilities of high-performance liquid chromatography (HPLC) with the mass analysis capabilities of mass spectrometry (MS).

Purification: This separates the useful components from complex mixture by the several separation steps.

List of Abbreviations

ASS   alcohol-soluble solid fraction
ESI   electrospray ionization
HMBC   heteronuclear multiple bond correlation
HPLC   high-performance liquid chromatography
LC   liquid chromatography
MS/MS   tandem mass spectrometry
MS   mass spectrometry
NMR   nuclear magnetic resonance

References

Arjmandi, B.H., Alekel, L., Hollis, B.W., Amin, D., Stacewicz-Sapuntazakis, M., Guo, P., and Kukreja, S.C., 1996. Dietary soybean protein prevents bone loss in an ovariectomized rat model of osteoporosis. *Journal of Nutrition*. 126: 161–167.

Chung, H., Hogan, S., Zhang, L., Rainey, K., and Zhou, K., 2008. Characterization and comparison of antioxidant properties and bioactive components of Virginia soybeans. *Journal of Agricultural of Food Chemistry*. 56: 11515–11519.

Day, A.J., Canada, F.J., Diaz, J.C., Kroon, P.A., McLauchlan, R., Faulds, C.B., Morgan, M.R., and Williamson, G., 2000. Dietary flavonoid and isoflavone glycosides are hydrolyzed by the lactase site and lactase phlorizin hydrolase. *FEBS Letters*. 486: 166–170.

Dhaubhadel, S., McGarvey, B.D., Williams, R., and Gijizen, M., 2003. Isoflavonoid synthesis and accumulation in developing soybeans seeds. *Plant Molecular Biology*. 53: 733–743.

Han, R.M., Tian, Y.X., Liu, Y., Chen, C.H., Ai, X.C., Zhang, J.P., and Skibsted, L.H., 2009. Comparison of flavonoids and isoflavonoids as antioxidants. *Journal of Agricultural of Food Chemistry*. 57: 3780–3785.

Ismail, B., and Hayes, K., 2005. β-Glycosidase activity toward different glycosidic forms of isoflavone. *Journal of Agricultural of Food Chemistry*. 53: 4918–4924.

Iwai, K., and Matsue, H., 2007. Ingestion of *Apios americana* Medikus tuber suppresses blood pressure and improves plasma lipids in spontaneously hypertensive rats. *Nutrition Research*. 27: 218–224.

Iwai, K., Kuramoto, S., and Matsue, H., 2008. Suppressing effect of *Apios americana* on blood pressure in SHR and its active peptide. *Journal of Clinical Biochemistry and Nutrition*. 43: 315–318.

Izumi, T., Piskula, M., K.Osawa, S., Obata, A., Tobe, K., Saito, M., Kataoka, S., Kubota, Y., and Kikuchi, M., 2000. Soy isoflavone aglycones are absorbed faster and in higher amounts than their glucosides in human. *Journal of Nutrition*. 130: 1695–1699.

Kim, J.J., Kim, S.H., Hahn, S.J., and Chung, I.M., 2005. Cahnging soybean isoflavone composition and concentrations under two different storage conditions over three years. *Food Research International*. 38: 435–444.

Kudou, S., Fleury, Y., Welti, D., Magnolato, D., Uchida, T., Kitamura, K., and Okubo, K., 1991. Malonyl isoflavone glycosides in soybean seeds. (*Glycine max* Merrill), *Agricultural and Biological Chemistry*. 55: 2227–2233.

Krishnan, H.B., 1998. Identification of genistein, an anticarcinogenic compound, in the edible tubers of the American groundnut (*Apios americana* Medikus). *Crop Science*. 38: 1052–1056.

Kulling, S.E., Honig, D.M., and Metzler, M., 2001. Oxidative metabolism of the soy isoflavones daidzein and genistein in humans *in vitro* and *in vivo*. *Journal of Agricultural of Food Chemistry*. 49: 3024–3033.

Liu, J., Chang, S.K., and Wiesenborn, D., 2005. Antioxidant properties of soybean isoflavone extract and tofu *in vitro* and *in vivo*. *Journal of Agricultural and Food Chemistry*. 53: 2333–2340.

Miura, T., Yuan, L., Sun, B., Fujii, H., Yoshida, M., Wakame, K., and Kosuna, K., 2002. Isoflavone aglycon produced by culture of soybean extracts with basidiomycetes and its anti-angiogenic activity. *Bioscience, Biotechnology, and Biochemistry*. 66: 2626–2631.

Nara, K., Nihei, K., Ogasawara, Y., Koga, H., and Kato, Y., 2011. Novel isoflavone diglycoside in groundnut (*Apios americana* Medik). *Food Chemistry*. 124: 703–710.

Ogasawara, Y., Hidano, Y., and Kato, Y., 2006. Study on carbohydrate composition of Apios (*Apios Americana* Medikus) flowers and tubers. *Nippon Shokuhin Kagaku Kogaku Kaishi*. 53: 130–136.

Okubo, K., Yoshiki, Y., Sugihara, T., Tsukamoto, C., and Hoshikawa, K., 1994. DDMP-conjugated saponin (soyasaponin βg) isolated from American groundnut (*Apios americana*). *Bioscience, Biotechnology, and Biochemistry*. 58: 2248–2250.

Riedl, K.M., Zhang, Y.C., Schwartz, S.J., and Vodovotz, Y., 2005. Optimizing dough proofing conditions to enhance isoflavone aglycons in soy bread. *Journal of Agricultural of Food Chemistry*. 53: 8253–8258.

Wang, B.F., Wang, J.S., Lu, J.F., Kao, T.H., and Chen, B.H., 2009. Antiproliferation effect and mechanism of prostate cancer cell lines as affected by isoflavones from soybean cake. *Journal of Agricultural of Food Chemistry*. 57: 2221–2232.

Wilson, P.W., Gorny, J.R., Blackmon, W.J., and Rwynolds, B.D., 1986. Fatty acids on the American gruoundnut (*Apios Americana*). *Journal of Food Science*. 51: 1387–1388.

Wilson, P.W., Pichardo, F.J., Liuzzo, J.A., Blackmon, W.J., and Rwynolds, B.D., 1987. Amino acids on the American groundnut (*Apios Americana*). *Journal of Food Science*. 52: 224–225.

Yin, L., Li, L., Liu, H., Saito, M., and Tatsumi, E., 2005. Effects of fermentation temperature on the content and composition of isoflavones and β-activity in sufu. *Bioscience, Biotechnology, and Biochemistry*. 69: 267–272.

Zhang, E.J., Ng, K.M., and Luo, K.Q., 2007. Extraction and purification of isoflavones from soybeans and characterization of their estrogenic activities. *Journal of Agricultural of Food Chemistry*. 55: 6940–6950.

Function and Effects

CHAPTER 21

Isoflavone Ingestion by Multiethnic Populations: Implications for Health

BASKARAN STEPHEN INBARAJ AND BING HUEI CHEN*

Department of Food Science, Fu Jen University, Taipei 242, Taiwan
*Email: 002622@mail.fju.edu.tw

21.1 Introduction

Isoflavones are a class of natural heterocyclic phenols which act as phytoestrogen compounds due to their estrogenic activity similar to estradiol hormones (Hsu *et al.* 2010). Soybean is the major dietary source of isoflavones consisting of 12 chemical forms grouped into four categories, namely aglycone (genistein, daidzein and glycitein), β-glucosides (genistin, daidzin and glycitin), acetyl-glucosides (acetylgenistin, acetyldaidzin and acetylglycitin) and malonyl-glucosides (malonylgenistin, malonyldaidzin and malonylglycitin). Among them, the aglycone isoflavones have drawn considerable attention due to their higher biological activity (Kao and Chen 2006). Several possible health-promoting effects of soy isoflavones include estrogenic or anti-estrogenic activity, anti-proliferation of cancer cells, anti-oxidant activity, anti-inflammatory effect, regulation of immune system, alteration in cellular signaling, prevention of cardiovascular disease and osteoporosis, as well as alleviation of postmenopausal syndrome (Albulescu and Popovici 2007; Barnes 2010; Kao and Chen 2006; Kao *et al.* 2007; Pan *et al.* 2010; Wang *et al.* 2009). Soybeans are usually processed into a wide variety of soy or soy-based products. However, the stability of different chemical forms of isoflavones can be affected during processing, particularly under high temperature and long heating time (Chien *et al.* 2005). Besides, the difference in processing/cooking methods can result in soy foods with different nutritional values (Coward *et al.* 1993; Fukutake *et al.* 1996). The Key Facts at the end of this Chapter highlights the difference in soy foods consumed by Asian and Western population. The objectives of this Chapter are to review the isoflavone ingestion by different ethnic populations and its implication on health.

21.2 Isoflavone Ingestion by Multiethnic Populations

21.2.1 Australia and the UK

Hutabarat *et al.* (2001) determined total daidzein plus genistein content in four soybean commodities including soybean, non-fermented soybean products, fermented soybean products and second-generation soyfoods, which were purchased from retail outlets in Australia. Their contents on wet weight basis were 27.4, 43.9 and 137.4 mg $100\,g^{-1}$ for fresh, canned and dried soybeans, respectively, with moisture content being 67, 72 and 9.1%. Whereas in non-fermented soybean products, various soymilk and tofu products showed a lower level, ranging from 4.3–9.5 and 10.2–28.4 mg $100\,g^{-1}$, respectively, with moisture content being 86.4–93.8% and 71.5–89.4%. Fermented soybean products like Nutrisoy tempeh contained daidzein plus genistein at 13.7 mg $100\,g^{-1}$, whereas the second-generation soyfoods such as soy sausage rolls, tofu dessert, tofu burger and soycheese had 2.7, 12.1, 27.8 and 2.3 mg $100\,g^{-1}$ with moisture content being 50.8, 81.4, 56.5 and 49.5%, respectively. Additionally, the mean isoflavone content in 85 commercial soymilks from 40 different brands in Australia and USA was 63.6 ± 21.9 mg L^{-1} for whole soybean milks and 30.2 ± 5.8 mg L^{-1} for soy protein isolated milk, with the ratio of genistein to daidzein being 1.7-fold higher for the latter (Figure 21.1) (Setchell and Cole 2003).

In the UK, both commercial- and home-prepared high- and low-soy foods were measured, with soybean (774 mg kg^{-1}) and soybean-containing foods (14–407 mg kg^{-1}) showing the highest isoflavone content among the foods examined (Table 21.1) (Wiseman *et al.* 2002). Table 21.2 shows the high-soy rotating diet developed for 11 days to provide an average of 100.0 mg of total isoflavone per day. Liggins *et al.* (2002) quantified total daidzein plus genistein (mg kg^{-1}) in a variety of cereals commonly consumed in Europe, of which soy flours showed the largest value (1639–2117), followed by different varieties of bread (0.03–11.87), rice (0.06–1.33), breakfast cereal (0.04–1.13), biscuits (0.06–0.55), flour (0.11–0.36), noodles (0.05–0.23) and pasta (0.04–0.22). Using the gas chromatography–mass spectrometry (GC-MS) data of total daidzein plus genistein for 300 foods available in the UK and extensive recipe calculations, a new database was constructed by Ritchie *et al.* (2006). Eventually, the

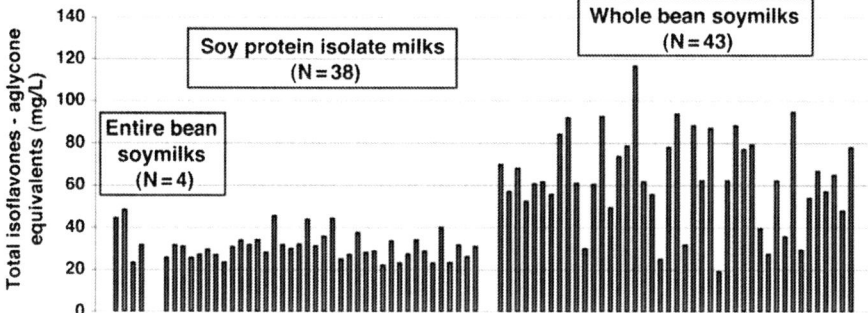

Figure 21.1 Total isoflavone content of 85 samples from 40 different brands of commercial soy milks. The histogram shows the levels of total isoflavone in commercial soy milks grouped according to milk type. Each bar represents the mean values from 3 or 4 replicate analyses.
Data are from Setchell and Cole (2003), with permission from the publishers.

isoflavone intake by vegetarian and omnivorous groups of healthy volunteers was estimated by analyzing 7 day weighed food diaries and the mean value was 7.4 ± 3.1 and 1.2 ± 0.4 mg day^{-1}, respectively (Figure 21.2).

21.2.2 Brazil

Isoflavone content in processed soy-based products consumed in Brazil varied significantly from 2 to 100 mg 100 g^{-1} (wet basis) for soy-based infant formula, soy beverage, textured soy protein, oral/enteral diet and traditional Asian soy food (Genovese and Lajolo 2002). The isoflavone distribution in seven infant formulas revealed 24–42% of daidzein, 6–16% of glycitein and 52–62% of genistein, which were similar (29, 12 and 59%) to that reported for six soy-based infant formulas in the USA (Murphy *et al.* 1997). Depending on the product and infant age, the daily intake of isoflavone varied between 5.9 and 35 mg day^{-1}, with infants aged 0–2 weeks being 2.0–6.1 mg kg^{-1}; 2–8 weeks, 1.7–6.6 mg kg^{-1}; 2–3 months, 1.6–5.2 mg kg^{-1}; 3–6 months, 1.4–5.4 mg kg^{-1}. The daily intake of four oral/enteral diets, including soyac, soya diet, ensure and diet shake, ranged from 0.91–12.3 mg (Genovese and Lajolo 2002). For nine soy-based Tonyu, Mupy and Ades beverages, the total isoflavone content varied significantly from 17.8–82.9 mg L^{-1} and an intake of 250 mL soy beverage could result in 20 mg ingestion of isoflavone, which is equivalent to the Korean daily intake (21 mg day^{-1}) (Choi *et al.* 2000), revealing soy beverage is an important source of isoflavones in Brazilian diet. Analysis of commercially available defatted soy flour, soy isolate and soy concentrate produced by thermoplastic extrusion showed a high isoflavone content of 87–100 mg 100 g^{-1}, which was only slightly lower than reported for industrial samples (149 mg 100 g^{-1}) in Brazil. Textured soy proteins are largely consumed by

Table 21.1 Total isoflavone equivalents (mg kg^{-1}) and isoflavones per serving (g) of high- and low-soy foods in the UK. The commercially prepared high- and low-soy foods are purchased from supermarkets in London, UK, and analyzed for individual isoflavones for the estimation of total isoflavone equivalents and its amount per serving. nd, not detected. Data are from Wiseman et al. (2002), with permission from the publishers.

Meal no.	Food item	Total isoflavone equiv.	Portion size (g)	Total isoflavones per serving (mg)
High-soy foods				
1	Spaghetti bolognaise	78	305	24
2	Lamb stew	118	610	72
3	Turkey curry	106	250	26
4	Turkey chilli (soybeans and red kidney beans)	231	191	44
5	Mixed soybeans and baked beans	407	80	32
6	Mixed soybeans, red kidney beans and vegetables	249	270	67
7	Soy sausage and batter dish "toad in the hole"	81	190	15
8	Soy meat balls and spaghetti	nd	350	nd
9	Turkey and soybean casserole	159	435	69
10	Hazelnut soymilk pudding	117	125	15
11	Chocolate soymilk pudding	69	125	9
12	Chocolate milk drink	96	250	24
13	Strawberry milk drink	209	250	52
14	Banana milk drink	105	250	26
15	Plain milk	108	280	30
16	Vanilla soymilk yogurt	183	125	23
17	Cherry soymilk yogurt	159	125	20
18	Vanilla soymilk pudding	88	125	11
19	Custard	120	125	15
20	Banana cake	152	77	12
21	Soy sausages (1)	176	60	11
22	Soy burgers	14	60	0.8
23	Soybeans	774	104	80
24	Soy sausages (2)	233	60	14
Low-soy foods				
31	Pork sausage and batter dish "toad in the hole"	8	120	1.0
34	Wholemeal bread	3	70	0.2
35	White bread	7	70	0.5
37	Sponge cake	2	40	0.1
39	Currant bun	4	34	0.1

vegetarians as a meat substitute and added as ingredients in food systems, with an intake of 100 g corresponding to 9.7–11.2 mg ingestion of isoflavone (Genovese and Lajolo 2002). Three frequently consumed traditional Asian soy foods, shoyu (soy sauce produced by fermentation of soybeans, rice and other

Table 21.2 High-soy rotating diet for the UK-based population with daily menu and daily total isoflavone equivalents intake (mg day^{-1}) for 11 days. Based on the meal item no. and total isoflavone equivalents provided in Table 21.1, the daily menu and daily ingestion of high-soy foods are constructed to provide an average of 100 ± 21.4 mg of mean isoflavone intake over a period of 11 days. Data are from Wiseman et al. (2002), with permission from the publishers.

	Meal position in 11 day cycle										
	1	2	3	4	5	6	7	8	9	10	11
Meal item no.	16	10	11	11	13	11	11	10	11	18	10
	12	12	13	10	4	12	6	12	13	13	13
	1	2	9	5	19	3	18	7	5	7	5
	11	11		14	20	10	12	5	10		
					11						
Isoflavones per serving (mg)	23	15	9	9	52	9	9	15	9	11	15
	24	24	52	15	44	24	67	24	52	52	52
	24	72	69	32	15	26	11	15	32	15	32
	9	9		26	12	15	24	32	15		
					9						
Daily total (mg)	80	120	130	82	132	74	111	86	108	78	99

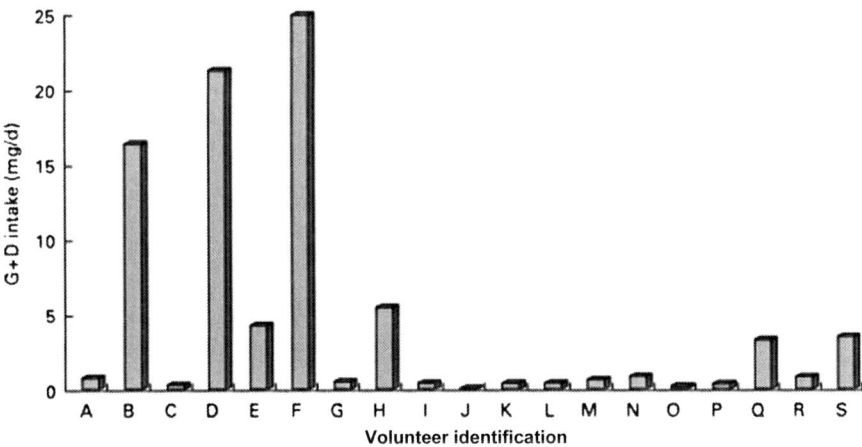

Figure 21.2 Genistein plus daidzein intake by vegetarians A–J and omnivorous K–S volunteers. This histogram is based on a 7 day weighed food dairies and isoflavone database. The main food sources consumed by vegetarians are soya milk, soya mince, wholemeal bread, white bread and others, whereas the omnivorous volunteers ingested soya yogurts, wholemeal bread and toast, white bread and toast and others.
Data are from Ritchie et al. (2006), with permission from the publishers.

cereals), tofu (precipitation of a curd from soymilk with calcium salts) and miso (fermented rice and soybeans) in Brazilian dishes account for a total isoflavone content of 5.7 mg L^{-1}, 6.8 mg 100 g^{-1} and 20.0 mg 100 g^{-1}, respectively. In all the soy foods, the largest amount of isoflavones corresponded to β-glucosides, with genistein glucosides dominating over daidzein and glycitein glucosides, but for fermented products miso and shoyu, the aglycones were present in largest amount (Genovese and Lajolo 2002).

21.2.3 Canada

A database of nine phytoestrogens in 121 foods consumed in Western diets revealed that soy products, especially soybean and soy nut, contained the highest amount of isoflavone (primarily daidzein and genistein), followed by legumes, meat products and other processed foods, cereal and bread, nut and oil seed, vegetables, alcoholic beverages, fruit and non-alcoholic beverages (Thompson et al. 2006). Among the soybean and soy products, the total isoflavone content (mg) ingested per 100 g wet weight decreased in the order: soybean (103.7)>soy nut (68.6)>tofu (27.1)>tempeh (18.3)>textured vegetable protein (16.1)>miso paste (11.1)>soy yoghurt (10.2)>soy protein powder (8.8)>bacon bits (6.0)>veggie burger (1.7)>miso soup (1.5)>soybean sprouts (0.79)>soy sauce (0.14). Non-soy products also contained a moderate amount of isoflavones with the following order: black bean sauce (5.3)>doughnuts (2.9)>protein bar (2.7)>flax bread (0.30), which may arise from soy addition during processing. For non-soy beverages, red wine had the highest level (0.017 mg 100 g^{-1}) of isoflavones (Thompson et al. 2006). Based on the data from 21 non-mineral, non-vitamin dietary supplements commonly consumed by Canadian women, the supplements containing soy or red clover showed the highest total isoflavones (0.73–35.4 mg g^{-1}), followed by both licorice and licorice-containing supplements (0.04–0.36 mg g^{-1}) (Thompson et al. 2007).

21.2.4 China and Hong Kong

Liu et al. (2004) evaluated the intake of soy foods and soy isoflavones by rural adult women and their average intake by 1139 subjects from two different provinces of China (Gansu and Hebei) was 38.7 and 0.018 g day^{-1}, respectively, with tofu contributing mainly to intake of isoflavones. Nevertheless, 89.2% of the subjects consumed 0–0.035 g day^{-1} of soy isoflavones, and the women aged 41–50 years consumed less soy foods and isoflavones than those aged 20–30 and 31–40 years. Also, intake by women having secondary education or above was significantly higher than illiterate women. The average isoflavone intake of 0.018 g day^{-1} by Chinese rural women population was much higher than that by Western population, which was estimated to be 0.001 g day^{-1} (Liu et al. 2004). In a later study, the amount of isoflavones in soybean and soybean products (chungkukjang and doenjang) from Chinese and Korean origin were shown to be 319.4 and 202.5 mg 100 g^{-1} for soybean, 260.7 and 247.2 mg 100 g^{-1} for chungkukjang and 3.5 and 7.7 mg 100 g^{-1} for doenjang, respectively

(Lee et al. 2007). To estimate the dietary soy isoflavone intake in Hong Kong Chinese population, Chan et al. (2009) determined the isoflavone level and distribution in 47 soy-based foods, with soybeans containing the highest total isoflavone content (8–80 mg 100 g^{-1}), followed by different varieties of beancurd skin (5–76 mg 100 g^{-1}), fermented soybean (2–62 mg 100 g^{-1}), soybean milk (6–29 mg 100 g^{-1}), fermented tofu (8–29 mg 100 g^{-1}), tofu (1–28 mg 100 g^{-1}) and deep-fried tofu (8–9 mg 100 g^{-1}), with the distribution of conjugated isoflavones varying within and between food groups depending on soybean types, processing or cooking techniques used. In another study, the daily intake of soy isoflavones by 141 Hong Kong Chinese women aged 50–61 years was estimated to be 7.8 mg from 22 dietary sources, of which tofu products accounted for 1.0–20.8% of intake, followed by beancurd (1.5–7.1%) and soybean milk (1.6–6.3%) (Chan et al. 2007).

21.2.5 Italy

In view of the impact of isoflavones on human health, the Italian Health Authority has issued a recommended limit (RL) of 80 mg day^{-1} for isoflavones in dietary supplements (Gazzetta Ufficiale Italiana 2002). The level of total isoflavones fall in the range of 5–20 mg per serving for most of the soy-based foods consumed by an Italian, but the imitation meat products, mainly soy burger and soy meat ball, exceeded this value reaching 27–33 mg per serving (Morandi et al. 2005). On an average a minimum of 4–5 servings per day is necessary to exceed the RL value (80 mg day^{-1}). Based on the RL value, the theoretical maximum daily intake (TMDI) for infants and children was estimated to be 5–7 mg day^{-1} for 6-month-old with a weight of 5–6 kg, 10–12 mg day^{-1} for 12-month-old of 9–10 kg, 14 mg day^{-1} for 18-month-old of 12 kg, 23 mg day^{-1} for 5–6-year old of 20 kg, 29 mg day^{-1} for 7–8-year old of 25 kg and 37 mg day^{-1} for 10–11 years of 32 kg (Morandi et al. 2005). Consumption of one soy burger or 250 mL of a soy drink plus pudding or yoghurt by a child aged 10–11 years with an average weight of 32 kg or intake of 250 mL of a soy drink by a child aged 5–6 year with a weight of 12 kg is sufficient to exceed the TMDI (Morandi et al. 2005). Isoflavone contents ranging 77–220 µg g^{-1} were found in gluten-free products including soy biscuit, raw pasta, cooked pasta and toasted bread. Each serving of these products could provide an isoflavone level of 2–16 mg, and a gluten-intolerant children aged 5–6 years far exceeds the TMDI value by consuming only two servings. In contrast to some other countries such as USA, a lower number of soy-based infant foods are sold on the Italian market because of concern over possible contamination by genetically modified soybeans (Mendez et al. 2002). With one initial and three follow-on formulas containing a total isoflavone content of 121–427 µg g^{-1} by dry weight, a child may consume approximately 14.7–25.6 mg day^{-1}, which is 3-fold higher than the TMDI probably caused by efficient absorption of isoflavones in babies. In common soy foods, especially soy gluten-free foods and infant formulas, the conjugated forms of genistein were present in the largest amounts, followed by conjugated forms of daidzein and glycitein.

21.2.6 Japan

The daily intake of soybean and related bean products by a Japanese person was estimated based on the amounts of genistein and genistin, which ranged from 4.6–18.2 and 200.6–968.1 $\mu g\,g^{-1}$, respectively, in soybean, soynut and soy powder, 1.9–13.9 and 94.8–137.7 $\mu g\,g^{-1}$ in both soymilk and tofu, 38.5–229.1 and 71.7–492.8 $\mu g\,g^{-1}$ in fermented soybean products (miso and natto) (Fukutake et al. 1996). The genistein level in the fermented soybean products was much higher than in soybeans and non-fermented soybean products, suggesting that the β-glycosyl bond of genistin is cleaved to produce genistein by microbes during fermentation of miso and natto. Based on the average data for annual consumption of soybean and related products, the daily intake of genistein and genistin in Japan was calculated to be 1.5–4.1 and 6.3–8.3 mg person^{-1}, respectively (Fukutake et al. 1996). In a later study, Nakamura et al. (2000) estimated the daily intake of total isoflavones from 11 domestic and imported soybeans, as well as 12 soybean-based processed foods, and the highest content was found in roasted soybean powder, whereas the lowest in soy sauce. The amount of aglycone was larger in miso and soy sauce as they were processed through heating and fermentation. The Japanese daily intake of isoflavones was estimated to be 27.80 mg day^{-1}, with the daidzein, glycitein and genistein distribution being 12.02, 2.30 and 13.48 mg, respectively. These levels are much higher than those ingested by Americans or Western Europeans, which may explain why a lower mortality rate of breast, colon and prostate cancer was observed in Japan.

21.2.7 Korea

Choi et al. (2000) determined the isoflavone contents in 83 soybean varieties including 20 modern cultivars, 53 Korean landraces and 10 wild soybeans collected from Korean farmers and the Crop Research Station, Rural Development and Administration, Korea. Wild soybeans showed a higher content than cultivars and landraces, with total isoflavone ranging from 976–2223, 560.3–1979 and 623.8–2330 mg kg^{-1}, respectively, in which the dominant chemical form was glucosides. In Korea, soybeans are consumed in the form of soymilk, soybean curd, soy sauce, sprout, chungkukjang (fermented soybeans), kochujang and doenjang (soybean paste), raw soybean and soybean sprout. Of the various soybean products analyzed, the total isoflavone level was largest in soybean sprouts (2935–3498 mg kg^{-1} dry basis), whereas kochujang (hot soybean paste with red pepper and rice or wheat) registered the lowest level (112–287 mg kg^{-1}) (Choi et al. 2000). The average daily intake per person based on the Korean National Nutrition Survey was 3.6 g of soybean, 25.7 g of soybean curd, 20.8 g of soybean sprout, 8.1 g of soy sauce, 7.0 g of denjang, 1.3 g of chungkukjang, 2.6 g of kochujang and 2.5 g of soymilk (Choi et al. 2000). Based on these data, the daily average intake of total genistein and daidzein by a Korean was estimated to be 21.0 mg person^{-1}. Also, the average intake of isoflavone by 178 Korean middle-aged women (35–60 years) was

27.2 mg day^{-1} (Lee et al. 1999). Obviously, the isoflavone intake can be varied with different sex and age groups (Park et al. 2007). Out of 791 subjects (426 boys and 365 girls aged 8–11 years) assessed for dietary intake, 71% of boys and 83% of girls had an isoflavone intake <10 mg, with >50% of girls consuming 2.5–7.5 mg. The daily intake ranged from 0.002–43 mg for girls and 1.8–41 mg for boys. Koreans aged 7–12 years have lower isoflavone intake (19 g day^{-1}) when compared with adults aged >30 years (35 g day^{-1}). However, between women of different age groups, the daily intake by a college-going woman and middle-aged woman was only 17 and 24 mg, respectively. Apparently in Korea, children consume a lesser amount of isoflavones than middle-aged (24 mg) and menopausal women (27 mg). The isoflavone database in 142 common Korean foods was also reported in the same article by Park et al. (2007).

21.2.8 Singapore and Indonesia

The isoflavone contents of raw and cooked soy foods obtained from open markets and local stores in Singapore were investigated by Franke et al. (1999). The total isoflavone level varied from 295–2661 mg kg^{-1} for raw soy foods and 162–1282 mg kg^{-1} for cooked ones (boiling for 5–10 min), with the following order: malonylglucoside > β-glucoside > aglycone > acetylglucosides. In all soy foods except for tau kwa (pressed tofu), the total isoflavone was higher in raw foods than in cooked ones, with their levels being 297 and 259 mg kg^{-1} for tofu, 369 and 162 mg kg^{-1} for tau pok (fried tau kwa), 2661 and 554 mg kg^{-1} for foo jook (skimmed dried supernatant of boiled soybean curd sticks) and 1789 and 1282 mg kg^{-1} for soybean, respectively. Fermented tofu contained only aglycones in an abundant amount (587 mg kg^{-1}), which is mainly due to hydrolysis of isoflavone conjugates by bacteria during fermentation process. For isoflavone levels in soybean and soybean products in Indonesia, Hutabarat et al. (2001) analyzed daidzein and genistein in soybean, non-fermented and fermented soybean products. The daidzein plus genistein levels in fresh and dried soybean on wet weight basis were 27.4 and 211.1 mg 100 g^{-1} with the moisture content being 67 and 9%, respectively. Similarly, in Mony Susu Kedelai (Mony), Soya Bean milk (Yeo's) and traditional Susu Kedelai soymilk products, their levels were 5, 5.1 and 4.5 mg 100 g^{-1} with moisture content being 90, 86 and 87%, respectively, whereas for traditional tofu, silken tofu (Kong Kee), tahu tau kwa (Miko Sejati), tofu sakake (Mitra Boga Segar), silken tofu sakura (Harum Sari food) and street fried tofu, the levels were 22.5, 24.6, 30.4, 17.3, 12.1 and 37.1 mg 100 g^{-1} with the moisture content being 77, 84, 75, 83, 82 and 67%. In fermented products such as traditional tempeh, fresh oncom, taucho (soy paste), asli no. 1 taucho, taucho medan mekar, taucho medan harum sedap and taucho no. 1 macan, the daidzein plus genistein levels were 46.0, 9.7, 51.0, 31.1, 40.8, 38.3 and 22.9 mg 100 g^{-1} with the moisture content being 71, 74, 64, 60, 60 and 60%, respectively. Indonesians consume the most frequently served soy foods tofu, tempeh and oncom as fried products. In a similar study, Prabhakaran et al. (2006) evaluated the isoflavone content in various commercial soy-based dietary supplements, soy health products and infant formulas collected from South East Asian countries (Singapore, Malaysia, Thailand and Indonesia), which ranged from 405–57570, 46.32–1333.80 and 59.54–226.84 $\mu g\,g^{-1}$, respectively.

21.2.9 Taiwan

In nine different soybean cultivar varieties grown in Taiwan, namely, CH 1 (regular soybean), VS-KS 2 (vegetable soybean), HBS (black soybean), Kaohsuing 10, Tainan 1, Tainan 4, commercial, Tainan 5 and a genetically modified variety, total isoflavone contents of 224.3, 295.7, 170.9, 346.8, 146.3, 142.5, 169.7, 59.0 and 130.4 mg $100\,g^{-1}$ were found, respectively (Tsai et al. 2007; Wei et al. 2004). The isoflavone glucoside (mainly daidzin and genistin) was the major isoflavone present in soybeans, accounting for 93–96% of total isoflavones. The isoflavone contents (mg $100\,g^{-1}$) in soybeans, non-fermented (tofu, soymilk, soybean meal) and fermented soybean products (natto, miso and sufu) were shown to vary from 77.0–534.8 mg $100\,g^{-1}$ with soybean meal containing the highest level (264.0–534.8 mg $100\,g^{-1}$) and soymilk the lowest level (77.0–161.3 mg $100\,g^{-1}$) (Chen and Wei 2008). Both soybean and tofu showed a similar isoflavone content ranging from 126.0–332.4 and 119.9–311.0 mg $100\,g^{-1}$, respectively. Likewise, the distribution of various isoflavones showed a similar trend in both soybeans and non-fermented soybean products, with daidzin ranging from 23.1–214.0 mg $100\,g^{-1}$, genistin from 26.9–261.5 mg $100\,g^{-1}$ and glycitin from 14.1–44.9 mg $100\,g^{-1}$. Nonetheless, daidzin outweighed genistin in soybeans, whereas the reverse was true in non-fermented soybean products. The total isoflavone in fermented soybean products including natto, miso and sufu ranged from 364.7–574.9, 83.3–154.8 and 259–363 mg $100\,g^{-1}$, respectively, with the aglycone form dominating in both sufu and miso (Chen and Wei 2008). The average consumption of isoflavones by a Taiwanese was reported to be 35 g day^{-1} per capita (Coward et al. 1993).

21.2.10 USA

In three different studies, the isoflavone level and distribution in 29 commercial soybean (CS) foods, different brands of retail and institutional (RI) soy foods and six soy-based infant formulas (IFs) were determined (Murphy et al. 1997; 1999; Wang and Murphy 1994). Among the CS foods, Vinton 81 soybean variety showed the highest total isoflavone (277.6 mg $100\,g^{-1}$), followed by roasted soybean (266.1), soy granule (240.4), green soybean (239.8), textured vegetable protein (229.5), soy flour (201.4), instant beverage (191.8), soy isolate (98.9), tempeh (86.5), bean paste (64.7), tofu (53.2), honzukuri miso and fermented bean curd (38.9), tempeh burger (38.6), tofu yogurt (28.2), soy hot dog (23.6), Cheddar cheese (19.7), soy bacon (14.4), flat noodle (12.7), mozzarella cheese (12.3), soy Parmesan (8.8) and soy concentrate (7.2) (Wang and Murphy 1994). Isoflavone levels in five different-branded RI foods including aseptically processed soymilk ($n=7$), pasteurized soymilk ($n=5$), tofu ($n=12$), raw ($n=5$)

and cooked soy/beef hamburgers ($n = 5$) ranged from 9.6–16.5, 8.0–9.7, 20.2–34.7, 0.6–1.8 and 0.9–3.0 mg 100 g^{-1}, respectively (Murphy et al. 1999). In six commercially available soy-based IFs, namely, Alsoy, Prosobee, Enfamil, Gerber, Isomil and Nursoy, the isoflavone content ranged from 21.4–28.5 mg 100 g^{-1} (Murphy et al. 1997). Umphress et al. (2005) analyzed isoflavone level in 167 processed and fast foods with soy additives, which were ranged from 0–93.9 mg 100 g^{-1} for bread and grain products, 0–2.7 mg 100 g^{-1} for gravies and sauces, 0–15.9 mg 100 g^{-1} for meat and poultry products, 0.3–149.9 mg 100 g^{-1} for meat substitutes, 0.4–36.5 mg 100 g^{-1} for nutritional products, 0.1–4.0 mg 100 g^{-1} for reduced fat peanut butters, 0–1.6 mg 100 g^{-1} for seafood products and 2.9–109.3 mg 100 g^{-1} for soybean products. In another study, Franke et al. (1999) assayed the isoflavone level in Western-style soy foods from Hawaii, with the mean values in soy supplements, soy flours, soybean seeds, soy protein drinks, soy yogurts, soy cheeses, burgers and soymilk being 745.1, 257.0, 197.3, 81.6, 60.7, 10.7, 4.6 and 3.5 mg 100 g^{-1}, respectively.

Overall, Asian fermented soy foods contain predominantly aglycone forms of isoflavone, whereas, in non-fermented soy foods of both American and Asian origin, isoflavones are present mainly as β-glycoside conjugates. Daily consumption of isoflavone by an Asian was estimated to be 25–100 mg, which is comparable on a body weight basis with the isoflavone amount in powdered soybean chip-containing diets, which were shown to inhibit mammary tumorigenesis, such as breast cancer, in rats (Coward et al. 1993). On the contrary, the average American or Western European ingests only a few mg of isoflavones per day. Additionally, the urinary levels of equol are generally low in subjects consuming a Western-style diet, but are higher in vegetarians consuming soy-based products (Fukutake et al. 1996). In British diet, the isoflavone level was reported to be lower than that is necessary to have a physiological effect (Coward et al. 1993). Thus, the lower incidence and mortality due to cancer (especially breast cancer) in Asians may be attributed to the high isoflavone level in their daily diets.

Summary Points

- Isoflavone ingestion varies depending on species, food processing/cooking methods, diet-style, cultural differences, literacy, age, sex and health awareness.
- Isoflavone intake by vegetarians, literate women, girls, middle-aged women and menopausal women is significantly higher than non-vegetarians, illiterate women, boys, college-going women and middle-aged women, respectively.
- Isoflavone contents in commercial soymilks prepared from whole soybean milk exceeds that from soy protein isolate, and the daily isoflavone intake by an infant varies between 5.9 and 35 mg day^{-1} depending on food product and age.
- In unprocessed/non-fermented foods, the isoflavones in glucosidic form are present in largest amount with genistein glucosides dominating over

daidzein and glycitein glucosides, whereas aglycones outweighed conjugated forms in fermented foods.
- In South East Asian countries, the isoflavone contents in various commercial soy-based supplements, soy health products and infant formulas range from 40.5–5757.0, 4.6–133.4 and 6.0–22.7 mg $100\,g^{-1}$, respectively.
- The daily isoflavone ingestion by an Asian is much higher than by an American or European, implicating a lower mortality rate of breast, colon and prostate cancers in Asia.

Key Facts

Asian *versus* Western Soy Foods

- Soybean-containing foods consumed in Asia are different from the forms ingested by Western population.
- Unlike American soy foods, Asian soy foods are often fermented by micro-organisms to convert conjugated isoflavone forms into biologically active aglycones.
- Some fermented soy-based foods in Asia include miso (Japan), soy paste (Korea) and tempeh (Indonesia).
- In the USA, soybeans are grown as a source of edible oil. The defatted soy flour is converted into soy protein concentrate through either washing with water to remove soluble carbohydrates or extracted with hot aqueous 65% ethanol to remove carbohydrates, lipids and isoflavones.
- Soy flour is also toasted at 250 °C and added to many bread and cake products including doughnuts.
- Soy protein isolate (>92% protein) are prepared by first solubilizing the proteins in soy flour with a mild alkaline extraction and precipitating proteins by lowering the pH to 4.5. It is used to make low-fat soymilk and tofu, as well as fermented isoflavone-protein-enriched products.
- In Asia, soymilk and tofu are manufactured directly from soybeans as full-fat soymilk and tofu.
- Regardless of ethnic origin, soy milk is recommended as an important alternative to mother's milk in countries with a high incidence of lactase insufficiency.

Definitions of Words and Terms

Daidzin, glycitin and genistin: The glucosides of biologically active isoflavone aglycones daidzein, glycitein and genistein which are present in abundant quantities in unprocessed foods/plant materials, but are hydrolyzed by acid, base or enzyme into respective aglycone.

Fermented bean curd: Also called as sufu, fermented tofu, tofu cheese or preserved tofu, it is widely used in East Asian cuisine. The ingredients typically are soybeans, salt, rice wine and sesame oil or vinegar.

Gluten: A protein composite found in foods processed from wheat and related species. It gives elasticity to dough and chewy texture to the final product. A gluten-free diet is a diet that excludes foods containing gluten.

Ingestion: This is the process of consuming something through the mouth and taking it into the body *via* gastrointestinal tract.

Licorice: A flavorful herb (*Glycyrrhiza glabra*) used for food and medicinal remedies. It is a perennial growing wild in parts of Europe and Asia. Licorice supplements are made from roots and underground stems for treating a variety of illness, ranging from common cold to liver disease.

Oncom: A traditional staple food of Sudanese cuisine served as red and black oncom. It is closely related to tempeh and made from the by-products of other food preparations.

Phytoestrogens: These are plant-derived estrogens functioning as the primary female sex hormone not generated within the endocrine system but consumed by eating phytoestrogonic plants.

Red clover: This is a wild plant (*Trifolium pratense*) belonging to the legume family. Because of the presence of isoflavones, it is used to treat several medical conditions including cardiovascular disease, menopause, cancer, osteoporosis, psoriasis, eczema and rashes.

Thermoplastic extrusion: A modern food processing technology of combining heat, moisture and mechanical stress to modify the raw material into new forms and structures with improved functional and nutritional characteristics.

Tofu: A food made by coagulating soy milk and then pressing the resulting curds into soft white blocks.

List of Abbreviations

CS   commercial soybean
IF   infant formula
RI   retail and institutional
RL   recommended limit
TMDI   theoretical maximum daily intake

References

Albulescu, M., and Popovici, M., 2007. Isoflavones – biochemistry, pharmacology and therapeutic use. *Revue Roumaine de Chimie*. 52: 537–550.

Barnes, S., 2010. The biochemistry, chemistry and physiology of the isoflavones in soybeans and their food products. *Lymphatic Research and Biology*. 8: 89–98.

Chan, S., Ho, S. C., Kreiger, N., Darlington, G., So, K. F., and Chong, P.Y.Y., 2007. Dietary sources and determinants of soy isoflavone intake among midlife Chinese women in Hong Kong. *The Journal of Nutrition*. 137: 2451–2455.

Chan, S.G., Murphy, P.A., Ho, S.C., Kreiger, N., Darlington, G., So, E.K.F., and Chong, Y.Y., 2009. Isoflavonoid content of Hong Kong soy foods. *Journal of Agricultural and Food Chemistry*. 57: 5386–5390.

Chen, T.R., and Wei, Q.K., 2008. Analysis of bioactive aglycone isoflavones in soybean and soybean products. *Nutrition and Food Science*. 38: 540–547.

Chien, J.T., Hsieh, H.C., Kao, T.H. and Chen, B.H., 2005. Kinetic model for studying the conversion and degradation of isoflavones during heating. *Food Chemistry*. 91: 425–434.

Choi, Y.S., Lee, B.H., Kim, J.H., and Kim, N.S., 2000. Concentration of phytoestrogens in soybeans and soybean products in Korea. *Journal of the Science of Food and Agriculture*. 80: 1709–1712.

Coward, L., Barnes, N.C., Setchell, K.D.R., and Barnes, S., 1993. Genistein, daidzein and their β-glycoside conjugates: antitumor isoflavones in soybean foods from American and Asian diets. *Journal of Agricultural and Food Chemistry*. 41: 1961–1967.

Franke, A.A., Hankin, J.H., Yu, M.C., Maskarinec, G., Low, S.H., and Custer, L.J., 1999. Isoflavone levels in soy foods consumed by multiethnic populations in Singapore and Hawaii. *Journal of Agricultural and Food Chemistry*. 47: 977–986.

Fukutake, M., Takahashi, M., Ishida, K., Kawamura, H., Sugimura, T., and Wakabayashi, K., 1996. Quantification of genistein and genistin in soybeans and soybean products. *Food and Chemical Toxicology*. 34: 457–461.

Gazzetta Ufficiale Italiana, 2002. Annexure 2. No. 188.

Genovese, M.I., and Lajolo, F.M., 2002. Isoflavones in soy-based food consumed in Brazil: levels, distribution and estimated intake. *Journal of Agricultural and Food Chemistry*. 50: 5987–5993.

Hsu, B.Y., Inbaraj, B.S. and Chen, B.H., 2010. Analysis of soy isoflavones in foods and biological fluids: An Overview. *Journal of Food and Drug Analysis*. 18: 141–154.

Hutabarat, L.S., Greenfield, H., and Mulholland, M., 2001. Isoflavones and coumestrol in soybeans and soybean products from Australia and Indonesia. *Journal of Food Composition and Analysis*. 14: 43–58.

Kao, T.H. and Chen, B.H., 2006. Functional components in soybean cake and their effects on antioxidant activity. *Journal of Agricultural and Food Chemistry*. 54: 7544–7555.

Kao, T.H., Wu, W.M., Hung, C.F., Wu, W.B. and Chen, B.H., 2007. Anti-inflammatory effects of isoflavone powder produced from soybean cake. *Journal of Agricultural and Food Chemistry*. 55: 11068–11079.

Lee, S., Yoon, S., and Lee, M. J., 1999. Estimated isoflavone intake from soy products in Korean middle-aged women. *8th Asian Congress of Nutrition Abstract, Federation of Asian Nutritional Societies,* Seoul, p. 309.

Lee, Y.W., Kim, J.D., Zheng, J., and Row, K.H., 2007. Comparisons of isoflavones from Korean and Chinese soybean and processed products. *Biochemical Engineering Journal*. 36: 49–53.

Liggins, J., Mulligan, A., Runswick, S., and Bingham, S.A., 2002. Daidzein and genistein content of cereals. *European Journal of Clinical Nutrition.* 56: 961–966.

Liu, Z., Li, W., Sun, J., Liu, C., Zeng, Q., Huang, J., Yu, B., and Huo, J., 2004. Intake of soy foods and soy isoflavones by rural adult women in China. *Asia Pacific Journal of Clinical Nutrition.* 13: 204–209.

Mendez, M.A., Anthony, M.S., and Arab, L., 2002. Soy-based formulae and infant growth and development: a review. *Journal of Nutrition.* 132: 2127–2130.

Morandi, S., D'Agostina, A., Ferrario, F., and Arnoldi, A., 2005. Isoflavone content of Italian soy food products and daily intakes of some specific classes of consumers. *European Food Research and Technology.* 221: 84–91.

Murphy, P.A., Song, T., Buseman, G., and Barua, K., 1997. Isoflavones in soy-based infant formulas. *Journal of Agricultural and Food Chemistry.* 45: 4635–4638.

Murphy, P.A., Song, T., Buseman, G., Barua, K., Beecher, G.R., Trainer, D., and Holden, J., 1999. Isoflavones in retail and institutional soy foods. *Journal of Agricultural and Food Chemistry.* 47: 2697–2704.

Nakamura, Y., Tsuji, S., and Tonogai, Y., 2000. Determination of the levels of isoflavonoids in soybeans and soy-derived foods and estimation of isoflavonoids in the Japanese daily intake. *Journal of AOAC International.* 83: 635–644.

Pan, M.H., Lai, C.S. and Ho, C.T., 2010. Anti-inflammatory activity of natural dietary flavonoids. *Food and Function.* 1: 15–31.

Park, M.K., Song, Y.J., Joung, H., Li, S., and Paik, H.Y., 2007. Establishment of an isoflavone database for usual Korean foods and evaluation of isoflavone intake among Korean children. *Asian Pacific Journal of Clinical Nutrition.* 16: 129–139.

Prabhakaran, M.P., Hui, L.S., and Perara, C.O., 2006. Evaluation of the composition and concentration of isoflavones in soy based supplements, health products and infant formula. *Food Research International.* 39: 730–738.

Ritchie, M.R., Cummings, J.H., Morton, M.S., Steel, C.M., Bolton-Smith, C., and Riches, A.C., 2006. A newly constructed and validated isoflavone database for the assessment of total genistein and daidzein intake. *British Journal of Nutrition.* 95: 204–213.

Setchell, K.D.R. and Cole, S.J., 2003. Variations in isoflavone levels in soy foods and soy protein isolates and issues related to isoflavone databases and food labeling. *Journal of Agricultural and food Chemistry.* 51: 4146–4155.

Thompson, L.U., Boucher, B.A., Liu, Z., Cotterchio, M., and Kreiger, N., 2006. Phytoestrogen content of foods consumed in Canada, including isoflavones, lignans and coumestan. *Nutrition and Cancer.* 54: 184–201.

Thompson, L.U., Boucher, B.A., Cotterchio, M., Kreiger, N., and Liu, Z., 2007. Dietary phytoestrogens, including isoflavones, lignans, and coumestrol, in nonvitamin, nonmineral supplements commonly consumed by women in Canada. *Nutrition and Cancer.* 59: 176–184.

Tsai, H.S., Huang, L.J., Lai, Y.H., Chang, J.C., Lee, R.S., and Chiou, R.Y.Y., 2007. Solvent effects on extraction and HPLC analysis of soybean isoflavones and variations of isoflavone compositions as affected by crop season. *Journal of Agricultural and Food Chemistry*. 55: 7712–7715.

Umphress, S.T., Murphy, S.P., Franke, A.A., Custer, L.J., and Blitz, C.L., 2005. Isoflavone content of foods with soy additives. *Journal of Food Composition and Analysis*. 18: 533–550.

Wang, B.F., Wang, J.S., Lu, J.F., Kao, T.H., and Chen, B.H., 2009. Antiproliferation effect and mechanism of prostate cancer cell lines as affected by isoflavones from soybean cake. *Journal of Agricultural and Food Chemistry*. 57: 2221–2232.

Wang, H.J., and Murphy, P.A., 1994. Isoflavone content in commercial soybean foods. *Journal of Agricultural and Food Chemistry*. 42: 1666–1673.

Wei, Q.K., Jone, W.W., and Fang, T.J., 2004. Study on isoflavones isomers contents in Taiwan's soybean and GM soybean. *Journal of Food and Drug Analysis*. 12: 324–331.

Wiseman, H., Casey, K., Clarke, D.B., Barnes, K.A., and Bowey, E., 2002. Isoflavone aglycon and glucoconjugate content of high- and low-soy U.K. foods used in nutritional studies. *Journal of Agricultural and Food Chemistry*. 50: 1404–1410.

CHAPTER 22
Isoflavones in Beverages

RITA C. ALVES*[a,b] AND M. BEATRIZ P. P. OLIVEIRA[b]

[a] REQUIMTE, Instituto Superior de Engenharia do Porto, Instituto Politécnico do Porto, Rua Dr. António Bernardino de Almeida, n° 431, 4200-072 Porto, Portugal; [b] REQUIMTE, Departamento de Ciências Químicas, Faculdade de Farmácia, Universidade do Porto, Rua de Jorge Viterbo Ferreira n° 228, 4050-313 Porto, Portugal
*Email: rita.c.alves@gmail.com

22.1 Introduction

Isoflavones are phenolic and phytoestrogenic compounds. They are structurally and functionally similar to 17β-estradiol and show weak estrogenic activity (being able to act as estrogen agonists or anti-estrogens, according to the situation), as well as antioxidant properties. These compounds have been receiving increased attention, as some epidemiological studies have associated their consumption with potential beneficial health effects. Their protective action might involve the etiology of hormone-dependent cancers, cardiovascular diseases, osteoporosis and hot flushes of menopause (Xiao 2008).

The most abundant and the best characterized isoflavones are daidzein [7-hydroxy-3-(4-hydroxyphenyl)chromen-4-one)], genistein [5,7-dihydroxy-3-(4-hydroxyphenyl)chromen-4-one], glycitein [7-hydroxy-3-(4-hydroxyphenyl)-6-methoxychromen-4-one], formononetin [7-hydroxy-3-(4-methoxyphenyl)chromen-4-one] and biochanin A [5,7-dihydroxy-3-(4-methoxyphenyl)chromen-4-one)], the last two being methylated precursors of daidzein and genistein, respectively. Chemically, they all present a common 3-phenylchromen-4-one core structure and differ by substituents such as methoxy or hydroxyl functions (Delmonte *et al.* 2006; Mahesha *et al.* 2006).

In plants, these isoflavones (also known as aglycones or free isoflavones) are usually linked to glucose moieties (7-O-β-D-glucosides), and acetylated or malonylated (6″-O-acetyl-7-O-β-D-glucosides or 6″-O-malonyl-7-O-β-D-glucoside derivatives, respectively). Usually, malonylglucosides are the prevailing forms in plants. Acetylglucosides, glucosides and free aglycones are produced through food processing, modifying the original profile of compounds present in the matrix (Delmonte et al. 2006).

Also, when consumed through diet, native forms of isoflavones are hydrolyzed by gastric acid and by bacterial glucosidases of the intestinal microflora. Sugar groups are cleaved and aglycones are released. Formononetin can be further metabolized to daidzein by losing its methoxy group, and biochanin A can be converted into genistein. It is generally assumed that aglycones are absorbed faster and, consequently, are more bioavailable compared with the glucoside forms. However, instead of being promptly absorbed, aglycones can subsequently be biotransformed by gastrointestinal bacteria to specific metabolites such as equol, O-desmethylangolensin (metabolites of daidzein) and p-ethylphenol (metabolite of genistein). Equol and O-desmethylangolensin are physiologically important, as they are more estrogenic than daidzein. On the contrary, p-ethylphenol does not show estrogenic action. As the conversion into these metabolites is mainly due to gastrointestinal flora, it is understandable that dietary intake, antibiotic use, bowel disease, gender and other factors could be factors interfering in this metabolism. Indeed, intestinal microflora strongly influence isoflavone bioavailability (Barnes et al. 2011; Kano et al. 2006; Yuan et al. 2007).

Nevertheless, it has been reported that human saliva is able to convert glucosides into aglycones, which could also improve their bioavailability (Allred et al. 2001).

22.2 Isoflavone Sources

The Fabaceae (also known as Leguminosae) family includes several important plants with regards to isoflavone contents. Some examples of well-known plants from this family are: soy (*Glycine max* L.), alfalfa (*Medicago sativa* L.), red clover (*Trifolium pretense* L.), peas (*Pisum sativum* L.), bean plant (*Phaseolus vulgaris* L.) and chickpeas (*Cicer arietinum* L.).

Soy and soy-based products have been recognized as the major sources of isoflavones in the diet, especially genistein and daidzein (Mateos-Aparicio et al. 2008).

Although soy-based products are not traditional foods in Western countries as they are in Asian ones, its consumption has been increasing. Nowadays, it is possible to find in the Western markets several dietary supplements, products and beverages containing extracts of soy as the main ingredient, with alleged beneficial health properties, for example, the reduction of serum levels of total and low-density lipoprotein (LDL) cholesterol and the relief of menopausal symptoms. They are considered a good source of high-quality proteins being a good alternative not only for vegetarians, but also for lactose-intolerant and cow's milk allergic people, as well as for all consumers that appreciate these kind of products (Amigo-Benavent et al. 2007).

While a major importance is given to soy and soy-based products due to their high isoflavone contents, other foodstuffs without soy in their composition have also been recognized as relevant sources of these compounds, although they have comparative lower amounts. They should be considered important dietary contributors of isoflavones, especially for those populations that do not consume soy-containing products. Some examples are alfalfa sprouts (43 µg g^{-1}), peanuts (1.5 µg g^{-1}), barley (0.2 µg g^{-1}), apple (0.1 µg g^{-1}), broccoli (0.1 µg g^{-1}), cauliflower (0.1 µg g^{-1}) (Mazur 1998) and several beverages (cow's milk, coffee and tea) (Alves et al. 2010; Kuhnle et al. 2008a; Mazur et al. 1998).

It is of major importance to know the accurate levels of isoflavones (as well as their forms) in foods and beverages in order to correctly use these data in epidemiological studies regarding isoflavone exposure and properly interpret results.

Investigations that study the effects of only a few foods or only traditional soy-based foods on health may suffer from uncontrolled confounding factors by unmeasured sources of exposure, which results in misclassification of exposure and misinterpretation of the real effects. This is of importance especially in Western populations or in Eastern populations with a large proportion of individuals consuming a "western" diet (Horn-Ross et al. 2000).

Investigators have been using different methodologies to assess isoflavone contents, which sometimes makes the comparison of different studies focusing similar matrices difficult. Some of them use a direct extraction with a polar organic solvent to isolate isoflavones, followed or not by a clean-up step to eliminate interfering compounds, and by a chromatographic technique, identifying and quantifying the native isoflavone forms present in the matrix, namely, malonylglycosides, acetylglycosides, glycosides and aglycones. Other researchers opt for hydrolysis, most frequently enzymatic or acidic, which releases isoflavones from their sugar moieties, making it possible to quantify total aglycones. In fact, the diversity of techniques that exist in literature presents not only different procedure steps, but also differences in precision, accuracy and the limits of detection. Moreover, the nature of the matrix should also be considered, and one method that is applied to one food or beverage may not be adequate to another foodstuff, making method validation required. However, correct validation is not always performed.

Therefore, it is difficult to clearly know if differences in isoflavone contents are due to distinct methodologies or to factors inherent in samples such as environmental factors (growth, harvesting and processing seasonality), food processing and storage, among others.

22.2.1 Beverages as Isoflavone Sources

Beverages can be a significant dietary source of isoflavones. In literature, there are several studies concerning isoflavone contents in these types of products, some of which are summarized in Table 22.1, and will be discussed below.

Table 22.1 Isoflavone contents of several beverages. 6″AcGlc, 6″-O-acetyl-glucoside conjugate; B, biochanin; D, daizein; d.w.b., dry weight basis; F, formononetin; G, genistein; Gl, glycitein; 6″MalGlc, 6″-O-malonyl-glucoside conjugate; tr, traces; –, not considered or referred in the study.

Sample	Unit	Hydrolysis	Aglycones					β-Glucosides	Others	References
			Daidzein	Genistein	Glycitein	Formononetin	Biochanin A			
Soy milk and soy-based beverages										
Soy milk	µg/100 g (d.w.b.)	Yes/No	Free: 2210 Total: 13030	Free: 2870 Total: 9180	–	n.d.	n.d.	–	–	Wang et al. 1990
Soy milk	µg/100 g	Yes	5023	6299	–	n.d.	n.d.	–	–	Horn-Ross et al. 2000
Soy milk, sweetened, fortified, unflavored	µg/100 g	Yes	Total (D, G, Gl, F, B): 9307					–	–	Kuhnle et al. 2008b
Soy milk, unsweetened	µg/100 g	Yes	Total (D, G, Gl, F, B): 6018					–	–	Kuhnle et al. 2008b
Soy milk (n = 3)	µg/100 ml	Yes	2640–12600	5240–16800	151–1630	–	–	–	–	Nakamura et al. 2000
Regular soy milk	µg/100 ml	No	n.d.	90	n.d.	–	–	Daidzin: 3430 Genistin: 4250 Glycitin: 140	6″MalGlcD: 460 6″MalGlcG: 890 6″AcGlcD: 250 6″AcGlcG: 100	Coward et al. 1998
Low-fat soy milk	µg/100 ml	No	n.d.	n.d.	n.d.	–	–	Daidzin: 1540 Genistin: 2110 Glycitin: 60	6″MalGlcD: 170 6″MalGlcG: 310	Coward et al. 1998
Non-fat soy milk	µg/100 ml	No	n.d.	n.d.	n.d.	–	–	Daidzin: 430 Genistin: 540 Glycitin: n.d.	6″MalGlcD: 70 6″MalGlcG: 110	Coward et al. 1998
Soy milk	µg/100 g	Yes	921	1852	170	1	–	–	–	Thompson et al. 2006
Soy yogurt	µg/100 g	Yes	3364	6565	297	2	–	–	–	Thompson et al. 2006

Beverage	Units									Reference
Soy beverages blended with fruit juices ($n=11$)	μg/100 ml	No	n.d.–70	40–90	n.d.	—	—	Daidzin: 200–2040; Genistin: 270–2270; Glycitin: n.d.–160	6″MalGlcD: n.d.–610; 6″MalGlcGl: n.d.–50; 6″MalGlcG: n.d.–740; 6″AcGlcD: n.d.–90; 6″AcGlcGl: n.d.; 6″AcGlcG: 110–160	Rostagno et al. 2007
Nutritional beverages with soy ($n=9$)	μg/100 g	Yes	100–12900	200–21100	0–2500	—	—	—	—	Umphress et al. 2005
Soy-based beverages with fruits or chocolate ($n=9$)	μg/100 ml	Yes	Total aglycones (D, G, Gl): 1210–8290	—	—	—	—	—	—	Genovese et al. 2002
Cow's and goat's milk beverages										
Cow's milk, semi-skimmed	μg/100 g	Yes	Total (D, G, Gl, F, B): 4			—	—	—	—	Kuhnle et al. 2008a
Cow's milk, skimmed	μg/100 g	Yes	Total (D, G, Gl, F, B): 14			—	—	—	—	Kuhnle et al. 2008a
Cow's milk, whole	μg/100 g	Yes	Total (D, G, Gl, F, B): 6			—	—	—	—	Kuhnle et al. 2008a
Goat's milk	μg/100 g	Yes	Total (D, G, Gl, F, B): 1			—	—	—	—	Kuhnle et al. 2008a
Milk, cow	μg/100 g	Yes	0.0	0.1	0.0	0.1	—	—	—	Thompson et al. 2006
Milkshake, instant	μg/100 g	Yes	0.2	0.6	0.1	0.5	—	—	—	Thompson et al. 2006
Coffee beverages										
Coffee	μg/100 g	Yes	50	<25	—	n.d.	n.d.	—	—	Horn-Ross et al. 2000

Table 22.1 (Continued)

Sample	Unit	Hydrolysis	Aglycones					β-Glucosides	Others	References
			Daidzein	Genistein	Glycitein	Formononetin	Biochanin A			
Coffee, infusion	μg/100 g	Yes	n.d.	<1	n.d.	<1	<1	—	—	Kuhnle et al. 2008b
Coffee, decaffeinated infusion	μg/100 g	Yes	n.d	<1	n.d.	<1	<1	—	—	Kuhnle et al. 2008a
Coffee, decaffeinated	μg/100 g	Yes	0.1	0.1	0.2	0.2	—	—	—	Thompson et al. 2006
Coffee, regular	μg/100 g	Yes	0.1	0.1	0.0	0.2	—	—	—	Thompson et al. 2006
Espresso, regular ($n=10$)	μg/30 ml	Yes	11–26	0.8–16	—	34–215	n.d.	—	—	Alves et al. 2010
Espresso, decaffeinated ($n=3$)	μg/30 ml	Yes	18–32	2–14	—	31–54	n.d	—	—	Alves et al. 2010
Espresso, 100% arabica, medium roast ($n=2$)	μg/30 ml	Yes	13–22	2–4	—	14–37	n.d	—	—	Alves et al. 2010
Espresso, 100% robusta, medium roasted ($n=2$)	μg/30 ml	Yes	26–33	5–8	—	151–255	n.d.	—	—	Alves et al. 2010
Mocha coffee[a]	μg/100 ml	Yes	94	25	—	478	n.d.	—	—	Alves et al. 2010
Press-pot coffee[a]	μg/100 ml	Yes	46	8	—	164	n.d.	—	—	Alves et al. 2010
Filter coffee[a]	μg/100 ml	Yes	16	n.d.	—	136	n.d.	—	—	Alves et al. 2010
Coffee ($n=6$)	μg/100 g grain	Yes	n.d.–70	n.d.–30	—	n.d.–80	n.d.	—	—	Mazur et al. 1998

Tea beverages

Beverage	Unit	Analyzed							Reference
Licorice tea	µg/100 g	Yes	tr	—	—	100	n.d.	—	Horn-Ross et al. 2000
Tea, strong (15 g L^{-1}), from tea leaves	µg/100 g	Yes	<1	<1	n.d	<1	<1	—	Kuhnle et al. 2008b
Tea, standard, from tea bag (4 bags L^{-1})	µg/100 g	Yes	<1	<1	<1	<1	<1	—	Kuhnle et al. 2008b
Tea, decaffeinated, from tea bag	µg/100 g	Yes	<1	<1	<1	<1	<1	—	Kuhnle et al. 2008b
Tea, chamomile	µg/100 g	Yes	<1	3	<1	<1	3	—	Kuhnle et al. 2008b
Tea, black	µg/100 g	Yes	0.4	0.1	0.1	0.1	—	—	Thompson et al. 2006
Tea, green	µg/100 g	Yes	0.4	0.2	0.0	0.1	—	—	Thompson et al. 2006
Tea ($n=19$)	µg/100 g leaves	Yes	n.d.–70	n.d.–30	—	n.d.	n.d.–20	—	Mazur et al. 1998

Alcoholic beverages

Beverage	Unit	Analyzed							Reference
Beer ($n=4$)	µg/100 g	Yes	<1	<1–2	<1–4	<1–2	<1–6	—	Kuhnle et al. 2008b
Beer	µg/100 g	Yes	1.3	0.0	0.0	0.3	—	—	Thompson et al. 2006
Wine, red	µg/100 g	Yes	<1	<1	<1	<1	n.d.	—	Kuhnle et al. 2008b
Wine, red	µg/100 g	Yes	2	3	0.3	12	—	—	Thompson et al. 2006
Wine, white, dry	µg/100 g	Yes	<1	<1	<1	<1	<1	—	Kuhnle et al. 2008b
Wine, white	µg/100 g	Yes	0.7	2	0.1	2	—	—	Thompson et al. 2006
Cider, dry	µg/100 g	Yes	<1	n.d.	4	<1	1	—	Kuhnle et al. 2008b
Sherry, dry	µg/100 g	Yes	n.d.	n.d.	<1	2	1	—	Kuhnle et al. 2008b

Table 22.1 (Continued)

Sample	Unit	Hydrolysis	Aglycones					β-Glucosides	Others	References
			Daidzein	Genistein	Glycitein	Formononetin	Biochanin A			
Sherry, cream	μg/100 g	Yes	<1	<1	<1	<1	<1	—	—	Kuhnle et al. 2008b
Gin	μg/100 g	Yes	<1	<1	<1	<1	n.d.	—	—	Kuhnle et al. 2008b
Whiskey	μg/100 g	Yes	<1	<1	<1	<1	n.d.	—	—	Kuhnle et al. 2008b
Cranberry cocktail	μg/100 g	Yes	0.2	0.3	0.0	1.0	—	—	—	Thompson et al. 2006

[a]Coffee beverages prepared with the same coffee blend.

22.2.1.1 Soy Milk and Soy-based Beverages

Table 22.1 shows a high variability with regards to the isoflavone content of soy milk, a beverage prepared by extracting ground soybeans with water.

The differences observed might be due not only to intrinsic factors of soybeans, but also to the way soy milk is prepared, and the methodologies involved in sample analysis.

Coward *et al.* (1998) analyzed the original isoflavone forms present in soy milk. The β-glucoside derivatives daidzin and genistin were the main compounds present in this beverage, followed by malonyl- and acetylglucosides, and the aglycone genistein. A similar profile was described by Rostagno *et al.* (2007) in soy beverages blended with fruit juices.

Comparatively with soy beans, which contain malonylglucoside derivatives as the main isoflavones (Wang and Murphy 1994), soy milk shows a higher content of glucosides. This is due to the conversion of malonyl and acetyl forms into glucosides during soy milk preparation, as the former formss are very unstable at high temperatures (Jackson *et al.* 2002). Contrary to these results, Wang *et al.* (1990) reported a higher content of free aglycones in soy milk, namely daidzein and genistein. However, this could be due to the methodology used for compounds extraction, as the authors used fast extraction with HCl and acetonitrile, and the presence of the acid might be responsible for some release of aglycones.

Coward *et al.* (1998) showed that low-fat and non-fat soy milk contained predominantly β-glucoside conjugates and were depleted of aglycones. Moreover, when comparing regular, low-fat and non-fat soy milks, a decrease in isoflavones content was found in a similar manner.

The other researchers listed in Table 22.1 opted for assessing total aglycones of soy milk by using hydrolytic methods (Horn-Ross *et al.* 2000; Kuhnle *et al.* 2008b; Nakamura *et al.* 2000; Thompson *et al.* 2006). In general, total values ranged from ~ 3000 to $\sim 30\,000\,\mu g\,100\,g^{-1}$ (or 100 ml). The main aglycone found was genistein, followed by daidzein and glycitein.

Thompson *et al.* (2006) reported small amounts of formononetin (1.5 μg $100\,g^{-1}$) in soy yogurt. Its presence could be due to soy milk, as Thompson *et al.* (2006), contrary to Wang *et al.* (1990) and Horn-Ross *et al.* (2000), also found this compound in a sample of soy milk. However, it could also result of, for example, the addition of cow's milk that is sometimes used to fortify soy yogurt (Yazici *et al.* 1997).

Soy beverages mixed with fruit juices or flavours are becoming very popular in some countries, being a convenient way to include soy in diet (Rostagno *et al.* 2007). Genovese and Lajolo (2002) analyzed several samples of soy-based beverages showing that, generally, protein contents were derived almost exclusively from soy (0.6 to 2.5 g 100 ml^{-1}. The authors also evaluated the isoflavone distribution (%) in these beverages. The percentage of malonylglucosides varied from 1.6 to 38.9, and the proportion of aglycones was low, ranging from 0.3 to 5.6. Acetylglucosides derivatives were generally not present, with the most abundant compounds being the β-glucosides ($\sim 90\%$).

Also, some beverages showed a higher percentage of malonylglucosides, indicating a less severe heat treatment. Genistein conjugates were the main isoflavones of the products analyzed (43–68%), followed by daidzein conjugates (29–42%), similar to what was observed in soy protein products.

Rostagno et al. (2007) also evaluated soy beverages blended with fruit juices. It was shown that higher amounts of soy used in the product do not necessarily result in a higher isoflavone concentration. Indeed, samples with the same soy source and amount (3.0% soybeans) from different manufacturers contained different isoflavone levels. This could be due to differences in the fruit components, such as enzymes and pH that may act during storage affecting the stability of the isoflavone glucoside derivatives (Rostagno et al. 2007).

Besides, fruit juices blended with soy beverages could also contribute, although in minor extent, to the isoflavone content of the product, as several fruits contain measurable amounts of daidzein and genistein, namely, clementines, mango, melon, passion fruit, pears, figs and strawberry (Liggins et al. 2000). Thompson et al. (2006) also reported the presence of daidzein, genistein and formononetin in orange juice (0.1 µg 100 g of each^{-1}).

22.2.1.2 Cow's and Goat's Milk Beverages

Table 22.1 shows the isoflavone contents found in cow's milk by Thompson et al. (2006) and Kuhnle et al. (2008a). As expected, the amounts are very low compared with soy milk or soy-based beverages. Isoflavones are secondary plant metabolites normally ingested by animals through diet, as rations, fodder or pasture, and subsequently excreted in urine and milk. Therefore, isoflavones content in milk may depend from season. Indeed, King et al. (1998) reported 0.2 µg 100 ml^{-1} genistein in samples collected in summer and 2–3 µg 100 ml^{-1} for samples collected during spring when clover is dominant in pasture. Moreover, the mean concentrations of genistein in milk collected in the spring were approximately the double from those from milk collected in the autumn. The authors also revealed that pasteurization did not have any effect on isoflavone concentration (King et al. 1998).

According to data of Kuhnle et al. (2008a), the fat content might not be correlated with the total aglycone content of milk, as a sample of whole cow's milk contained more isoflavones (6 µg 100 ml^{-1}) than a semi-skimmed milk (4 µg 100 ml^{-1}), but less than a skimmed (14 µg 100 ml^{-1}). Goat's milk was observed to have 1 µg 100 ml^{-1} total aglycones.

Thompson et al. (2006) detected essentially genistein and formononetin in cow's milk, and reported even lower values than Kuhnle et al. (2008a), namely 0.1 µg 100 ml^{-1} for each compound.

22.2.1.3 Coffee Beverages

Coffee beverages are increasingly popular worldwide due to its organoleptic characteristics and stimulating effect. Coffee can be prepared by many procedures according to consumers' preferences. Variations in the coffee/water amount, methods of percolation, coffee species involved and respective roast degrees directly influence the composition of the brew. Table 22.1 shows the isoflavone contents of coffee beverages reported by several authors. The values clearly differ, which could be due to the factors mentioned above. In a previous study by our group (Alves et al. 2010), we explored the isoflavone concentrations of espresso coffee, a very concentrated brew usually prepared with both arabica and robusta coffee species that is widely consumed in Portugal. We also ascertained the influence of the technological parameters in the composition of the final beverage. A standard Portuguese espresso is usually prepared with 6.5 g of ground coffee per 30–40 ml. In our study, we observed that isoflavone levels decreased by approximately 30–40% during the roasting of coffee beans. However, this degradation was not visible when analyzing brews of different roast degrees, as, generally, practically no variation or a small variation in total isoflavone content was observed. This was due to a significant increase in the extractability of the compounds from light- to medium-dark-roasted degrees. However, this increase was less marked for daidzein and genistein in comparison with formononetin. In fact, formononetin was more extensively extracted than other isoflavones (40–48%), especially at dark roast degrees, followed by genistein (23–37%) and daidzein (18–25%).

Independently of the roast degree, robusta coffees always contained superior amounts of isoflavones. The main difference between both species was formonotenin content (~5–6-fold higher in robusta coffees). Also, robusta coffee contained ~2-fold higher daidzein content than arabica one (Alves et al. 2010).

The final volume of espresso coffee may vary according to the consumers' preferences, ranging from a short brew (20 ml) to a long one (70 ml). We observed that a longer percolation resulted on a better extraction of isoflavones from ground coffee to espresso beverage. Indeed, extraction percentages varied between 19, 16 and 31% in 20 ml espressos and 61, 50 and 72% in 70 ml ones for genistein, daidzein, and formononetin, respectively. Thus, a long espresso (70 ml) contains double the amount of isoflavones compared with a short one (20 ml) (Alves et al. 2010).

As for other authors (Kuhnle et al. 2008b; Thompson et al. 2006), we also found no significant differences between caffeinated beverages and decaffeinated ones.

Generally, our values are higher than those presented by other authors. This is probably a result of the methods of beverage preparation (coffee/water ratio), as well as the use of robusta coffees and dark roasts in the blends, as usual in Portugal. The influence of the percolation method in the brews' composition can be observed in Table 22.1 by analyzing the data for mocha, press-pot and filter coffees, as the same commercial blend (comprising a mixture of arabica and robusta coffees) was used to prepare these beverages. Differences between these samples are exclusively due to the method of brewing (Alves et al. 2010).

22.2.1.4 Tea Beverages

Tea (*Camellia sinensis*) infusion is one of the most known and appreciated beverages all over the world. Its richness in phenolic compounds, especially catechins, has been related to nutritional and pharmacological health benefits when consumed moderately (Cabrera *et al.* 2006). However, with regards to isoflavones, its content is very low (Table 22.1) (Kuhnle *et al.* 2008b; Thompson *et al.* 2010). Kuhnle *et al.* (2008b) found slightly higher levels of genistein and biochanin A in chamomile tea. Mazur *et al.* (1998) analyzed a large range of plant infusions and found a high variability in isoflavone contents of samples. The highest levels of isoflavones, especially genistein and formononetin, were described for licorice tea (26 and 100 µg 100 g^{-1}, respectively) due to the richness of licorice in these compounds (Horn-Ross *et al.* 2000).

22.2.1.5 Alcoholic Beverages

Among all the alcoholic beverages shown in Table 22.1, a sample of red wine contained the highest levels of isoflavones, especially of formononetin (Thompson *et al.* 2006). The same authors report levels ~3.5-fold lower for a white wine.

Kunhle *et al.* (2008b) also analyzed a red and a white wine, but the compounds were not detected or were below 1 µg 100 g^{-1}.

Generally, isoflavone content may vary with the type of beer (Kuhnle *et al.* 2008b), but total contents are still very low, although slightly higher than those for other alcoholic beverages such as cider, sherry, gin or whiskies.

22.3 Conclusion

Among all beverages, soy milk is the most important dietary source of isoflavones. It could be consumed pure or blended with fruit juices, which will improve the flavor of the product and consumers' acceptability.

Although having substantially lower amounts of isoflavones, coffee beverages could also be a relevant source of these compounds, especially in Western populations, where they are moderately consumed. It is important to notice, however, that the isoflavone content of a coffee brew vary widely with the percolation method and the type of coffee (species and roast) used to prepare the brew.

Although other beverages such as cow's milk, plant infusions, fruit juices, red wine or beer are minor sources of isoflavones, they might also contribute to the daily isoflavones intake, especially by populations that usually do not consume soy-based products.

Summary Points

- Isoflavones are phenolic and phytoestrogenic compounds.
- Isoflavones show weak estrogenic activity and antioxidant properties.

- Malonylglucosides are the prevailing forms of isoflavones in plants.
- Food processing converts malonylglucosides derivates into acetylglucosides, glucosides and free aglycones, in this order.
- Soy milk is the most important beverage source of isoflavones.
- Robusta coffee contains significantly higher amounts of isoflavones than arabica coffee.
- The percolation method highly influences the isoflavone content of coffee beverages.
- Cow's milk contains higher isoflavone content than goat's milk.
- Chamomile tea presents higher isoflavone levels than green tea.
- Cow's milk, plant infusions, fruit juices, red wine or beer are minor sources of isoflavones.
- Beverages can be a significant dietary source of isoflavones.

Key Facts

Key Facts for Coffee

- Approximately 70% of Portuguese people between 20 and 65 years old drink at least one espresso coffee per day.
- Brazil is the largest producer of arabica coffee and the second one of robusta.
- Robusta coffee is mainly used in countries such as Italy, Portugal or France to increase the body and the foam of espresso coffee.
- Robusta coffee contains more caffeine than arabica coffee.
- Arabica coffee is more aromatic and flavored than robusta coffee.

Definitions of Words and Terms

Acid hydrolysis: The process of macromolecular breakdown based on the use of a strong acid, usually at high temperatures.

Antioxidant: A chemical compound that protects the cells of the body against the effects of free radicals.

Bioavailability: The capacity of a chemical compound to be absorbed and reach the target site.

Catechins: A subgroup of flavan-3-ols (also known as flavanols), from the flavonoids family.

Chromatography: An analytical technique based on differences in partitioning behavior between a flowing mobile phase and a stationary phase to separate components in a mixture.

Enzymatic hydrolysis: The process of macromolecular breakdown based on the use of enzymes (*e.g.* cellulase, β-glucosidase, ...).

Method validation: The process used to confirm that an analytical method is adequate for its intended use.

Pasteurization: The process of heating to a specific temperature for a specific period of time in order to kill unwanted micro-organisms.
Percolation: The passage of a liquid through a permeable substance.
Phytoestrogen: A chemical compound found in plants that can act (to a certain extent) as endogenous estrogens.

Acknowledgements

Rita C. Alves is grateful to the Fundação para a Ciência e a Tecnologia for a post-doctoral research grant (SFRH/BPD/68883/2010) financed by POPH-QREN and subsidized by ESF and MCTES.

References

Allred, C. D., Ju, Y. H., Allred, K. F., Chang, J., and Helferich, W. G., 2001. Dietary genistin stimulates growth of estrogen-dependent breast cancer tumors similar to that observed with genistein. *Carcinogenesis*. 22: 1667–1673.

Alves, R.C., Almeida, I.M.C., Casal, S., and Oliveira, M.B.P.P., 2010. Isoflavones in coffee: influence of species, roast degree, and brewing method. *Journal of Agricultural and Food Chemistry*. 58: 3002–3007.

Amigo-Benavent, M., Villamiel, M., and del Castillo, M.D., 2007. Chromatographic and electrophoretic approaches for the analysis of protein quality of soy beverages. *Journal of Separation Science*. 30: 502–507.

Barnes, S., Prasain, J., D'Alessandro, T., Arabshahi, A., Botting, N., Lila, M.A., Jackson, G., Janle, E.M., and Weaver, C.M., 2011. The metabolism and analysis of isoflavones and other dietary polyphenols in foods and biological systems. *Food and Function*. 2: 235–244.

Cabrera, C., Artacho, R., and Giménez, R., 2006. Beneficial effects of green tea – a review. *Journal of the American College of Nutrition*. 25: 79–99.

Coward, L., Smith, M., Kirk, M., and Barne, S., 1998. Chemical modification of soy foods during cooking and processing. *American Journal of Clinical Nutrition*. 68: 1486S–1491S.

Delmonte, P., Perry, J., and Rader, J.I., 2006. Determination of isoflavones in dietary supplements containing soy, red clover and kudzu: extraction followed by basic or acid hydrolysis. *Journal of Chromatography A*. 1107: 59–69.

Genovese, M.I., and Lajolo, F.M., 2002. Isoflavones in soy-based foods consumed in Brazil: Levels, distribution, and estimated intake. *Journal of Agricultural and Food Chemistry*. 50: 5987–5993.

Horn-Ross, P.L., Barnes, S., Lee, M., Coward, L., Mandel, J.E., John, E.M., Smith, M., Koo, J., John, E.M., and Smith, M., 2000. Assessing phytoestrogen exposure in epidemiologic studies: development of a database (United States). *Cancer Causes and Control*. 11: 289–298.

Jackson, C.-J. C., Dini, J.P., Lavandier, C., Rupasinghe, H.P.V., Faulkner, H., Poysa, V., Buzzell, D., and DeGrandis, S., 2002. Effects of processing on the content and composition of isoflavones during manufacturing of soy beverage and tofu. *Process Biochemistry.* 37: 1117–1123.

Kano, M., Takayanagi, T., Harada, K., Sawada, S., and Ishikawa, F., 2006. Bioavailability of isoflavones after ingestion of soy beverages in healthy adults. *The Journal of Nutrition.* 136: 2291–2296.

King, R.A., Mano, M.M., and Head, R.J., 1998. Assessment of isoflavonoid concentrations in Australian bovine milk samples. *Journal of Dairy Research.* 65: 479–489.

Kuhnle, G.G.C., Runswick, S.A., Dell'Aquila, C., Mulligan, A.A., Aspinall, S.M., and Bingham, S.A., 2008a. Phytoestrogen content of foods of animal origin: dairy products, eggs, meat, fish, and seafood. *Journal of Agricultural and Food Chemistry.* 56: 10099–10104.

Kuhnle, G.G.C., Dell'Aquila, C., Aspinall, S.M., Runswick, S.A., Mulligan, A.A., and Bingham, S.A., 2008b. Phytoestrogen content of beverages, nuts, seeds, and oils. *Journal of Agricultural and Food Chemistry.* 56: 7311–7315.

Liggins, J., Bluck, L.J.C., Runswick, S.A., Atkinson, C., Coward, W.A., and Bingham, S.A., 2000. Daidzein and genistein content of fruits and nuts. *Journal of Nutritional Biochemistry.* 11: 326–331.

Mahesha, H.G., Singh, S.A., Srinivasan, N., and Appu Rao, A.G., 2006. A spectroscopic study of the interaction of isoflavones with human serum albumin. *Federation of European Biochemical Societies Journal.* 273: 451–467.

Mateos-Aparicio, I., Cuenca, A.R., Villanueva-Suárez, M.J., and Zapata-Revilla, M.A., 2008. *Nutrición Hospitalaria.* 23: 305–312.

Mazur, W., 1998. Phytoestrogen content in foods. *Baillière Clinical Endocrinology and Metabolism.* 12: 729–742.

Mazur, W.M., Wahala, K., Rasku, S., Salakka, A., Hase, T., and Adlercreutz, H., 1998. Lignan and isoflavonoid concentrations in tea and coffee. *British Journal of Nutrition.* 79: 37–45.

Nakamura, Y., Tsuji, S., and Tonogai, Y., 2000. Determination of the levels of isoflavonoids in soybeans and soy-derived foods and estimation of isoflavonoids in the Japanese daily intake. *Journal of AOAC International.* 83: 635–650.

Rostagno, M. A., Palma, M., and Barroso, C.G., 2007. Ultrasound-assisted extraction of isoflavones from soy beverages blended with fruit juices. *Analytica Chimica Acta.* 597: 265–272.

Thompson, L.U., Boucher, B.A., Liu, Z., Cotterchio, M., and Kreiger, N., 2006. Phytoestrogen content of food consumed in Canada, including isoflavones, lignans and coumestan. *Nutrition and Cancer.* 54: 184–201.

Umphress, S.T., Murphy, S.P., Franke, A.A., Custer, L.J., and Blitz, C.L., 2005. Isoflavone content of foods with soy additives. *Journal of Food Composition and Analysis.* 18: 533–550.

Wang, G., Kuan, S.S., Francis, O.J., Ware, G.M., and Carman, A.S., 1990. A simplified HPLC method for the determination of phytoestrogens in soybean and its processed products. *Journal of Agricultural and Food Chemistry.* 38: 185–190.

Wang, H.-J., and Murphy, P.A., 1994. Isoflavone composition of American and Japanese soybeans in Iowa: effects of variety, crop year, and location. *Journal of Agricultural and Food Chemistry*. 42: 1674–1677.

Xiao, C.W., 2008. Health effects of soy protein and isoflavones in humans. *The Journal of Nutrition*. 138: 1244S–1249S.

Yazici, F., Alvarez, V.B., and Hansen, P.M.T., 1997. Fermentation and properties of calcium-fortified soy milk yogurt. *Journal of Food Science*. 62: 457–461.

Yuan, J.-P., Wang, J.-H., and Liu, X., 2007. Metabolism of dietary soy isoflavones to equol by human intestinal microflora – implications for health. *Molecular Nutrition and Food Research*. 51: 765–781.

CHAPTER 23

Use of Isoflavones in Inherited Metabolic Diseases: A Focus on Mucopolysaccharidoses

DANIEL SCHERMAN,* AUDREY ARFI AND
CORINNE MARIE

CNRS UMR 8151; INSERM U1022, Ecole Nationale Supérieure de Chimie de Paris, Chimie ParisTech, Université Paris Descartes, Sorbonne Paris Cité, Faculté de Pharmacie, Chemical and Genetic Pharmacology Laboratory, 4 avenue de l'Observatoire 75006, Paris, France
*Email: Daniel.Scherman@parisdescartes.fr

23.1 General Introduction

Mucopolysaccharidoses (MPS) belong to the family of lysosomal storage diseases (LSDs), which are inherited metabolic diseases resulting from a mutation in a gene encoding either a lysosomal enzyme or a lysosome-related protein. Lysosomes are small intracellular organelles present in each cell of the organism. They contain more than 50 different acid hydrolases involved in the catabolism of macromolecules such as glycosphingolipids, glycogen, mucopolysaccharides, oligosaccharides, cholesterol, peptides and glycoproteins. The deficiency of one of the lysosomal enzymes generates the accumulation of their respective substrate in all cells, as lysosomal enzymes are ubiquitously expressed (Pan 2011).

LSDs are generally classified according to the major storage materials. Clinical syndromes can vary, depending on the types of diseases considered and of the genetic mutations, but often include hepatosplenomegaly, cardiac symptoms and abnormal skeletal growth, as well as severe impairment in the central nervous system (CNS).

To date, over 50 different LSDs have been identified with a global incidence that varies from 1:5000 to 1:7000 live births. Several types of therapies have been proposed and are currently being evaluated (Beck 2010; de Ruijter *et al.* 2011; Gritti 2011; Lachmann 2010; Pan 2011). They include: (a) gene therapy that consists of introducing a functional gene into the deficient cells; (b) enzyme replacement therapy (ERT) that involves administering the missing enzyme into a patient; (c) pharmacological chaperone therapy (PCT) that involves the use of small molecules that stabilize the mutated protein, thus reducing its premature degradation and restoring its functionality; and (d) substrate reduction therapy (SRT) that aims to limit the accumulation of the deficient enzyme's substrate. Each therapy has its own pros and cons that have been reviewed in details in several recent comprehensive articles (Hemsley and Hopwood 2011).

The two last ones, PCT and SRT, involve small molecules, which can be of real benefit when taking into consideration the production cost and the possibility of passing through biological barriers such as the blood–brain barrier (BBB). This last property is of prime importance when neurological symptoms need to be addressed. In 2006, the isoflavone genistein was identified as a promising drug medicine that limits the accumulation of mucopolysaccharides in lysosomes of patient cells. Since then, several molecules of the same family have been identified as playing similar roles. The aim of this Chapter is to review our current knowledge and discuss the use of isoflavones for the treatment of MPS.

23.2 Introduction on Mucopolysaccharidoses (MPS)

23.2.1 Clinical Information, Genetics and Biochemistry

MPS represent a subgroup of LSDs that are characterized by the impaired catabolism of mucopolysaccharides, more often designated as glycosaminoglycans (GAGs). The MPS diseases result from the deficiency of an enzyme involved in the degradation of at least one of the five GAGs which are classified as heparan sulfate (HS), dermatan sulfate (DS), chondroitin sulfate (CS), keratan sulfate (KS) or hyaluronan (see Table 23.2).

To date, eleven enzymatic deficiencies resulting in GAG accumulation have been identified. These are classified into seven clinical MPS types: I, II, III (with four different subtypes collectively referred as Sanfilippo disease), IV (subtypes A and B), VI, VII and IX (Table 23.1). Most MPS diseases are transmitted in an autosomal and recessive manner. However, an exception to this rule is the inheritance of MPS II disease, which only affects boys as the deficient gene is located on chromosome X (Beck 2010).

Table 23.1 Classification of MPS diseases. MPS diseases can be divided into four classes, depending on the relative involvement of somatic *versus* neurological symptoms that are directly linked to the types of accumulating GAGs (CS, chondroitin sulfate; DS, dermatan sulfate; HS, heparan sulfate; KS, keratan sulfate). All patients accumulating HS present neurological symptoms (Pan 2011; Wegrzyn *et al.* 2010; Les cahiers d'Orphanet 2011).

Symptoms	MPS types	Accumulating GAGs	Prevalence (patients per 100 000 births)
Somatic	MPS IVA	CS, KS	0.4 for MPS IVA and IVB
	MPS IVB	KS	0.4 for MPS IVA and IVB
	MPS VI	DS	0.16
	MPS IX	Hyaluronan	1 patient
Somatic > neurological	MPS VII	CS, DS, HS	< 40 cases
Somatic < neurological	MPS IIIA	HS	1.1 for all subtypes
	MPS IIIB	HS	
	MPS IIIC	HS	
	MPS IIID	HS	
Somatic = neurological	MPS I	DS, HS	1.3
	MPS II	DS, HS	0.6

Depending on its role in GAG catabolism, the deficiency of a specific enzyme determines the type(s) of accumulating GAG in lysosomes, resulting in various symptoms. The clinical features in patients with MPS disorders are associated with organ dysfunction such as hepatosplenomegaly or corneal clouding, skeletal abnormalities and/or CNS manifestations characterized by progressive and severe behavioural disturbances. Recently, Pan (2011) and Wegrzyn *et al.* (2010) have published comprehensive reviews showing a correlation between the type(s) of accumulating GAGs and the relative involvement of neurological and visceral manifestations in MPS patients, pointing to a direct link between the accumulation of heparan sulfate and severe neurological disorders and the deterioration of brain functions. For the most severe forms of MPS diseases, a patient's death occurs in childhood or early adulthood.

23.2.2 Pathophysiology

A lysosomal defect leads to accumulation of undegraded material, causing cell and organ dysfunction. In order to explain how substrate accumulation results in disease, many investigations in a great number of tissues and organs have been carried out over the years. From these studies it has been realized that there are many factors that play a role in the pathophysiology of lysosomal storage disorders, such as MPS. Some studies have provided evidence that inflammatory processes, such as microglial activation or astrocytosis, contribute significantly to neurodegeneration in several LSDs. This microglial activation is a characteristic shared by numerous neurodegenerative diseases, including MPS I, MPS IIIB (Ohmi *et al.* 2003), MPS VII (Richard *et al.* 2008) and MPS IIIA (Arfi et al. 2011). Astrocytosis, like microglial activation, has also been widely described in neurodegenerative disorders (Arfi et al. 2011; Wu and Proia 2004).

Furthermore, a clear oxidative stress was observed in several neurological LSDs. Other studies previously highlighted reactive oxygen species (ROS) release and/or increased expression of components of the phagocyte nicotinamide adenine dinucleotide phosphate (NADPH) oxidase in MPS IIIB (Villani et al. 2007), MPS I (Reolon et al. 2009) or MPS IIIA (Arfi et al. 2011). It is still not defined whether the previously reported and presently observed oxidative imbalance is triggered by lysosomal dysfunction (Butler and Bahr 2006), by autophagy impairment as recently described in MPS IIIA and multiple sulfatase deficiency mouse models (Settembre et al. 2008), or is a secondary result from microglial activation. Nevertheless, it seems clear that these oxidative alterations may enhance damage to the brain cells, and may contribute to cognitive impairment as well as neurodegeneration through a decline in synaptic integrity.

In summary, although we presently do not completely understand the pathophysiology of lysosomal storage disorders, there is no doubt that the accumulation of storage material is the first pathogenic factor which triggers secondary structural and biochemical alterations, thus leading to disease initiation and progression. Removal of this material should be the first therapeutic goal.

23.3 Introduction on Glycosaminoglycan (GAGs) and Isoflavone Families

23.3.1 GAGs: Structure, Function and Life Cycle

The most abundant heteropolysaccharides in the body are the GAGs. These molecules are long unbranched polysaccharides containing a repeating disaccharide unit. The disaccharide units contain either of two modified sugars, *N*-acetylgalactosamine (GalNAc) or *N*-acetylglucosamine (GlcNAc), and an uronic acid such as glucuronate or iduronate. The GAGs are highly negatively charged molecules, with an extended conformation that imparts high viscosity to the solution. The GAGs are located primarily on the surface of cells or in the extracellular matrix (ECM). Along with the high viscosity of GAGs comes low compressibility, which makes these molecules ideal for a lubricating fluid in the joints. At the same time, their rigidity provides structural integrity to cells and provides passageways between cells, allowing for cell migration. The specific GAGs of physiological significance are hyaluronic acid, dermatan sulfate, chondroitin sulfate, heparin, heparan sulfate and keratan sulfate. Although each of these GAGs has a predominant disaccharide component (Table 23.2), heterogeneity does exist in the sugars present in the make-up of any given class of GAG (Esko et al. 2009; King 2005).

Table 23.2 Main groups of GAGs and their disaccharide unit structure. GAGs molecules are long unbranched polysaccharides containing a repeating disaccharide unit. The disaccharide unit of hyaluronates, dermatan sulfates, chondroitin 4- and 6- sulfates, heparin and heparan sulfates, and keratan sulfates are presented (King 2005; used with permission from themedicalbiochemistrypage, LLC).

D-glucuronate   N-acetyl-D-glucosamine

Hyaluronates: composed of D-glucuronate + GlcNAc linkage is β(1,3)

L-iduronate   N-acetyl-D-galactosamine-4-sulfate

Dermatan sulfates:
composed of L-iduronate (many are sulfated) + GalNAc-4-sulfate linkages is α(1,3)

D-glucuronate   N-acetyl-D-glalactosamine-4-sulfate

Chondroitin 4- and 6-sulfates :
composed of D-glucuronate and GalNAc-4- or 6-sulfate linkage is β(1,3)
(the figure contains GalNAc 4-sulfate)

Table 23.2 (*Continued*)

L-iduronate-2-sulfate   N-sulfo-D-glucosamine-6-sulfate

Heparin and heparan sulfates:
composed of iduronate-2-sulfate (D-glucuronate-2-sulfate) and
N-sulfo-D-glucosamine-6-sulfate linkage is α(1,4)
(heparans have less sulfate than heparins)

D-galactose   N-acetylD-glucosamine-6-sulfate

Keratan sulfates:
composed of galactose + GlcNAc-6-sulfate linkage is β(1,4)

Hyaluronic is unique among the GAGs in that it does not contain any sulfate and is not found covalently attached to proteins as a proteoglycan. It is, however, a component of non-covalently formed complexes with proteoglycans in the ECM. Hyaluronic acid polymers are very large (with molecular weights of 100 000–10 000 000) and can displace a large volume of water. This property makes them excellent lubricators and shock absorbers.

The majority of GAGs in the body are linked to core proteins, forming proteoglycans (also called mucopolysaccharides). The GAGs extend perpendicularly from the core in a brush-like structure. The linkage of GAGs to the protein core involves a specific trisaccharide composed of two galactose residues and a xylose residue (GAG-GalGalXyl-O-CH$_2$-protein). The trisaccharide linker is coupled to the protein core through an *O*-glycosidic bond to a serine residue in the protein. Some forms of keratan sulfates are linked to the protein core through an *N*-asparaginyl bond. The protein cores of proteoglycans are rich in serine and threonine residues, which allow the attachment of multiple GAGs (King 2005).

23.3.2 Isoflavone Family: Structure, Function and Life Cycle

Isoflavones of nutritional interest are substituted derivatives of isoflavone, being related to the parent by the replacement of two or three hydrogen atoms with hydroxyl groups (Figure 23.1). The parent isoflavone is of no nutritional interest. Isoflavones are produced *via* a branch of the general phenylpropanoid pathway that produces flavonoid compounds in higher plants. Soybeans are the most common source of isoflavones in human food; the major isoflavones in soybean are genistein and daidzein. The phenylpropanoid pathway begins from the amino acid phenylalanine, and an intermediate of the pathway, naringenin, is sequentially converted into the isoflavone genistein by two legume-specific enzymes, isoflavone synthase and a dehydratase. Similarly, another intermediate naringenin chalcone is converted into the isoflavone daidzein by sequential action of three legume-specific enzymes: chalcone reductase, type II chalcone isomerase and isoflavone synthase. Plants use isoflavones and their derivatives as phytoalexin compounds to ward off disease-causing pathogenic fungi and other microbes. In addition, soybean uses isoflavones to stimulate soil–microbe rhizobium to form nitrogen-fixing root nodules (Thompson 2011).

23.4 Use of Isoflavones for Substrate Reduction Therapy (SRT)

Certain MPS, such as MPS I, MPS II and MPS VI, with no neurological manifestations can be efficiently treated after intravenous injection of the missing protein (Beck 2010; Lachmann 2010; Pan 2011). Lysosomal enzymes are glycosylated proteins that can be efficiently endocytosed by cells containing mannose 6-phosphate (M6P) receptors, thus correcting peripheral organs. Nevertheless, a similar treatment for MPS that affects the CNS is not appropriate for two main reasons: (1) the BBB restricts the passage, by simple diffusion of most hydrophilic molecules, even that of small molecules that have a molecular weight smaller than 500 Da, and (2) the presence of M6P receptors on the surface on endothelial cells of the BBB seems to be developmentally

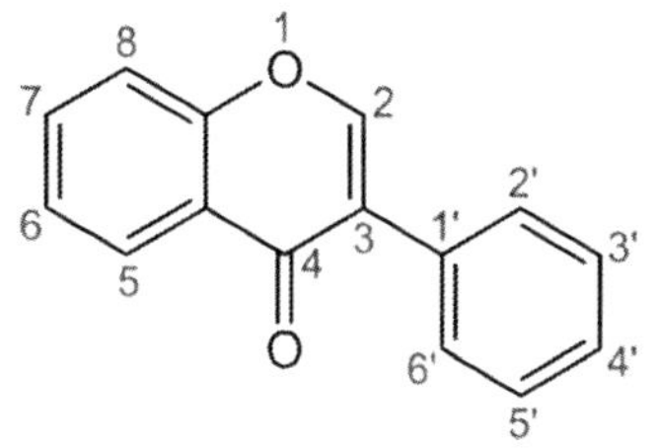

Figure 23.1 Isoflavone structure and numbering. Genistein (5-OH, 7-OH, 4′-OH) or daidzein (7-OH, 4′-OH) are example members of the isoflavone family. Isoflavone differs from flavone (2-phenyl-4*H*-1-benzopyr-4-one) in location of the phenyl group (Thompson 2011; used with permission from Nova Science Publishers).

regulated. In mice, the M6P receptor-dependent process is present only in the first 2 weeks of life (Urayama *et al.* 2008).

Thus, the inefficient ability of lysosomal proteins to gain access to the brain (but also the skeleton) raised the necessity of identifying small molecules that are able to cross biological barriers. Over the last 5 years, isoflavones have been identified as molecules that could be used for the treatment of MPS in an SRT approach.

23.4.1 Substrate Reduction Therapy (SRT): Definition

SRT, also called substrate deprivation therapy (SDT), consists of reducing the amount of storage material instead of enhancing or replacing the activity of the degrading-deficient enzymes (Beck 2010; de Ruijter *et al.* 2011; Lachmann 2010). This approach is based on the utilization of small molecules that may directly or indirectly impair synthesis of macromolecules that are not efficiently degraded in patients. The objective of this therapy is to restore the GAG turnover that is altered due to the deficiency in GAG degradation.

Recently, Hemsley and Hopwood (2011) proposed a research pipeline divided into three main parts: (i) construction/manufacture of therapeutic drugs and *in vitro* testing; (ii) transition into the *in vivo* testing phase; and (iii) a scale-up of therapy for human use once the pathological and clinical efficacy has been observed in animal models.

In the case of SRT, the first phase will consist of the identification of potential molecules that can modulate GAG synthesis *in vitro* by using fibroblasts of patients suffering from MPS diseases. The second phase then requires the use of MPS disease animal models and the translation to larger animals and human while taking into consideration drug administration, potency, side effect and cost issues.

23.4.2 Isoflavones and Analogs are Potent Agents for SRT

Genistein [4′,5,7-trihydroxyisoflavone or 5,7-dihydroxy-3-(4-hydroxyphenyl)-4*H*-1-benzopyran-4-one], an isoflavone extracted from soybean, was first identified for its potential for inhibiting GAG synthesis. Two methods are currently used for assessing the effect of genistein or related compounds on GAG accumulation. One of them consists of quantifying the total GAG concentration (Arfi *et al.* 2010), whereas the alternative one involves the measurement of ^{35}S incorporation into proteoglycans, thus reflecting GAG synthesis (Kloska *et al.* 2011). Using either technique, genistein was found to cause a reduction in GAG levels in fibroblasts of patients suffering from MPS I, II, IIIA, IIIB or MPS VII diseases (Arfi *et al.* 2010; Kloska *et al.* 2011; Piotrowska *et al.* 2006). It has been postulated by Piotrowska *et al.* (2006) that the reduction in GAG accumulation could result from GAG synthesis inhibition and/or from the dilution of GAG amount in dividing fibroblasts. In addition, these authors also showed that genistein treatment reversed abnormal structures (onion skin structures, zebra bodies, flocculent inclusions and complex vacuoles with inclusions) in MPS I fibroblasts. Interestingly, genistein exhibited a similar efficacy in reducing GAG levels in the different MPS models independently of the type of accumulated GAG, suggesting a common mechanism at the signal transduction level (see Section 23.4.3 for more details).

Subsequently, five natural isoflavones have been selected for their structural analogy with genistein: diadzein, glycitein, formononetin, prunetin and biochanin A (Arfi et al. 2010). All of these molecules are natural compounds; the two former ones are detected in significant concentrations in soybeans, whereas the three last ones are extracted from red clover.

In term of toxicity, the addition of genistein, glycitein or biochanin A (at concentrations up to 100 mM) did not affect the viability of MPS IIIA human fibroblasts. Formononetin and daidzein appeared to display a slight toxicity at a concentration higher than 50 µM. In contrast, prunetin (at a concentration as low as 25 µM) had a marked negative impact on cell viability (Arfi et al. 2010), as shown in Figure 23.2.

Within the range of a concentration of 25–100 µM, all of the tested non-toxic isoflavones exhibit an efficacy comparable with or superior to genistein for reducing GAG accumulation in MPS IIIA and MPS VII human fibroblasts. Interestingly, an equimolar mix of these isoflavones (at a total concentration of

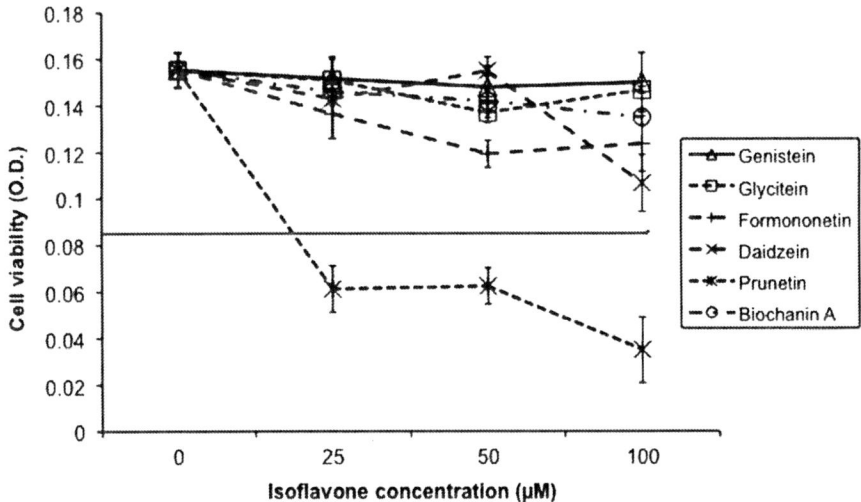

Figure 23.2 Toxicity study in isoflavones-treated MPS IIIA human fibroblasts, using the MTT cell assay. Cells were treated with the selected isoflavones at different concentrations, ranging from 0 to 100 µM, and cell viability was quantified using the MTT assay. Values are means ± S.D. from three independent cultures in which three wells were analysed for each isoflavone concentration. The grey line indicates 50% viability level compared to cells treated with DMSO only (used with permission from Arfi et al. 2010). O.D., optical density.

50 μM) was more efficient than one single molecule in reducing GAG accumulation in both MPS IIIA and MPS VII human fibroblasts, suggesting a synergistic effect (Arfi *et al.* 2010), as highlighted in Figure 23.3.

At this point, we believe that it is worth mentioning that other molecules which are not isoflavones but belong to other subclasses of flavonoids can also reduce the GAG accumulation in fibroblasts of patients suffering from MPS (Kloska *et al.* 2011). Indeed, the following molecules were assessed: apigenin (flavone), daidzein (isoflavone), kaempferol (flavonol) and naringenin (flavanone).

For all compounds, the LC50 (concentration that is lethal to 50% of the cells exposed for 24 h) appears to be superior to 100 μM, indicating the low toxicity of these molecules. The proliferation assay revealed that the IC50 (half maximal inhibitory concentration) that represents the concentration that is required for 50% inhibition of cell proliferation *in vitro* was of 17 μM for genistein and above 100 μM for naringenin (Kloska *et al.* 2011). As observed with isoflavones, a combination of various flavonoids appears to have a synergistic effect on the decrease in GAG accumulation in MPS III fibroblasts.

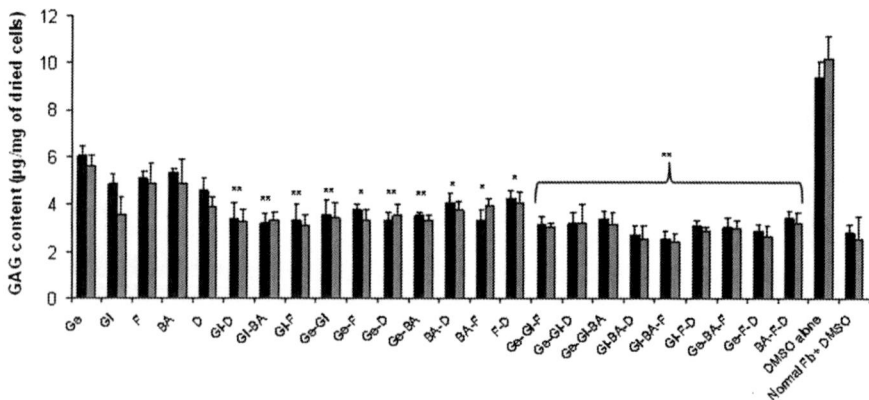

Figure 23.3 GAG-decreasing efficacies of various isoflavone combinations on MPS IIIA and MPS VII human fibroblasts. Fibroblasts from MPS IIIA (black bars) or MPS VII patients (grey bars) were treated with different combinations of DMSO-solubilized genistein (Ge), glycitein (Gl), formononetin (F), biochanin A (BA) or daidzein (D). Each mix was at a 50 μM total concentration: 25 μM for each component when two compounds are used and 16 μM for each component when three compounds are used in combination. GAG contents were measured using the Dimethylene Blue complexation method. Non-treated MPS IIIA or MPS VII fibroblasts (DMSO alone) and normal fibroblasts (Normal Fb + DMSO) were used as negative and positive controls, respectively. The results were expressed as μg of GAG per mg of dried cells. Values are means ± S.D. of three independent cultures, including two flasks per condition. Statistical significance *vs.* genistein at the same concentration was assessed by the Student's *t* test, *$P<0.005$, **$P<0.0005$ (used with permission from Arfi *et al.* 2010)

23.4.3 Isoflavone Action on GAG Synthesis and GAG Accumulation

We expect that other isoflavones and flavonoids will be further tested *in vitro* to assess their potency for a SRT. As potential medicines should display low cytotoxicity and acceptable anti-proliferative activity, any screening of compounds for reduction in GAG accumulation should be combined with tests assessing their toxicity both *in vitro* and *in vivo* (see below). Furthermore, to predict and anticipate putative side effects, it will be important to decipher the mechanism of action of these different compounds.

To the best of our knowledge, data only exist for the isoflavone genistein, which has been shown to inhibit the tyrosine-specific protein kinase activity of epidermal growth factor receptor (EGFR) (Jakobkiewicz-Banecka *et al.* 2009). The EGFR is present on the cell surface. Upon activation by ligands, such as epidermal growth factor (EGF), EGFR dimerization stimulates its autophosphorylation and the activation of downstream proteins involved in several signal transduction cascades, leading to DNA synthesis and cell proliferation. Genistein appears to inhibit EGFR phosphorylation, almost as efficiently as PD168390, a potent EGFR tyrosine kinase inhibitor (Kloska *et al.* 2011). In addition, it has been reported that EGF increased GAG synthesis, leading to levels above those observed in untreated cells (Jakobkiewicz-Banecka *et al.* 2009) and seems to compete with genistein for its effect on GAG synthesis. No such effects were observed in the presence of 17β-estradiol, indicating that genistein-mediated inhibition of GAG synthesis is mediated by EGF-dependent processes unrelated to estrogen activities (Jakobkiewicz-Banecka *et al.* 2009).

Thus, the tyrosine kinase inhibitory effect of genistein most probably affects the expression of genes encoding enzymes necessary for GAG synthesis. Nevertheless, this remains to be clearly established, as well as the mechanism of action of other flavonoid compounds which appearing to mediate their effect *via* an alternative pathway (Kloska *et al.* 2011).

Clearly, considering the key role played by GAGs in the proper functioning of many tissues and organs, it is not medically intended to totally inhibit GAG synthesis, but rather to restore a GAG level comparable with that measured in healthy patients. It is, therefore, relevant to mention that in all studies reported so far, isoflavone addition did not cause any reduction in GAG level to a value significantly lower than that measured in untreated healthy cells (Arfi *et al.* 2010; Piotrowska *et al.* 2006). From this observation, it was postulated that the GAG level in healthy cells might be sensed and that GAGs were degraded only when present in excess (Piotrowska *et al.* 2006).

Since a treatment with genistein appeared to ensure the synthesis of a sufficient amount of GAG required for proper cell and organ functions, several pharmacological studies have been carried out using animal models replicating the MPS diseases and in patients suffering from MPS (see below). Although pure genistein or other analogs are commercially available for laboratory use, they are not currently offered for therapeutic purposes. In contrast, isoflavone-containing diet supplements are marketed. A recent study by Piotrowska *et al.*

(2009) showed that the real content in isoflavones in these products is variable. For example, in seven analysed products, the genistein content was found to vary from 0 to 59% and should therefore be therefore established before carrying out *in vivo* experiments (Piotrowska *et al.* 2009).

23.4.3.1 Effect of Genistein and Related Compounds on GAG Accumulation in Mouse MPS Models

Malinowska *et al.* (2009) recently described significant reductions in accumulated heparan sulphate substrate in liver of a mouse model of MPS IIIB using the tyrosine kinase inhibitor genistein for 8 weeks of treatment.

Subsequent studies tested high doses of the aglycone genistein (160 mg kg^{-1} day^{-1}) in MPS IIIB mice diet for 36 weeks, whereas two different dose regimens, 5 or 25 mg kg^{-1} day^{-1}, were administered for 10 or 20 weeks in MPS II mice. These experiments indicated that genistein may be effective in the treatment of mice suffering from MPS IIIB and MPS II (Figure 23.4), including treating the CNS in short-term treatment of MPS II animals, and a complete correction of behavior in long-term treatment of MPS IIIB animals (Friso *et al.* 2010; Malinowska *et al.* 2009).

23.4.3.2 Effect of Genistein and Related Compounds on GAG Accumulation in MPS Patients

In pilot clinical studies, it has been demonstrated that an oral 12-month treatment with a genistein-rich isoflavone extract (5 mg kg^{-1} day^{-1}) of patients suffering from MPS IIIA and MPS IIIB diseases resulted in statistically important improvement of all tested parameters (when mean values from all patients were compared), including cognitive functions, assessed by using a special psychological test (Piotrowska *et al.* 2008). After a 2-year follow-up of this pilot study, the results showed that, in five children treated for 36 months, this genistein treatment might improve cognitive functions and behavioral symptoms, or at least stop their deterioration over a defined period of time (Piotrowska *et al.* 2011).

23.5 Conclusions and Perspectives for the Clinic

During the last 5 years, various isoflavones and flavonoids molecules have been identified as potential drug medicines to cure patients suffering from MPS that necessitate a life-time treatment. Compared with other proposed therapies, such as ERT which is often prohibitive and requires a hospital setting for enzyme infusion, SRT seems to be indeed an attractive option. Small molecules present the following advantages: (i) they can be synthesised, thus potentially reducing production costs; (ii) they can be orally administrated; and (iii) they can cross more easily biological barriers such as the BBB which is generally considered impermeable to large molecules. Furthermore, either in animal or

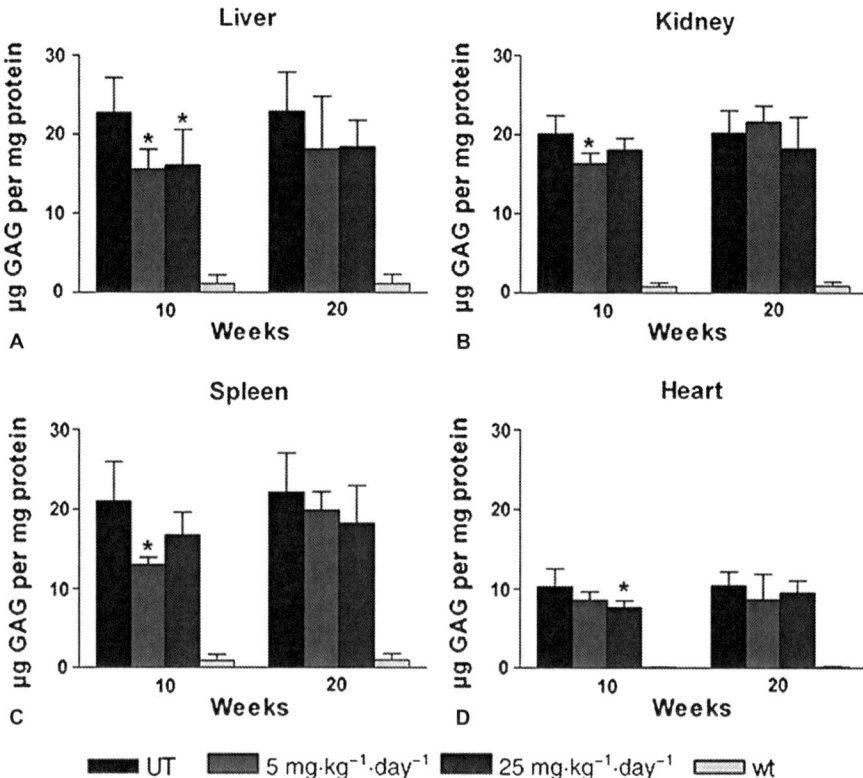

Figure 23.4 Tissue levels of GAGs measured biochemically in genistein-treated MPS II mice. $n = 5$ for untreated (UT) MPS II mice, $5\,\text{mg}\,\text{kg}^{-1}\,\text{day}^{-1}$ treated MPS II mice, $25\,\text{mg}\,\text{kg}^{-1}\,\text{day}^{-1}$ treated MPS II mice; $n = 3$ for untreated wild-type (wt) mice. Values, presented as means ± S.D., significantly decreased compared with untreated MPS II animals (UT), in liver ($P < 0.01$), kidney ($P < 0.01$) and spleen ($P < 0.005$) of MPS II mice treated with $5\,\text{mg}\,\text{kg}^{-1}\,\text{day}^{-1}$ for 10 weeks, in liver ($P < 0.05$) and heart ($P < 0.05$) of MPS II mice treated with $25\,\text{mg}\,\text{kg}^{-1}\,\text{day}^{-1}$ for 10 weeks. *Statistically significant decrease compared with UT (used with permission from Friso et al. 2010).

human patients, genistein treatment appears to be safe and well-tolerated. However, although the *in vitro* and preliminary results are encouraging, some adjustments (regimen dose, treatment periodicity...) are clearly needed in order to increase the therapeutic effects. Furthermore, since only limited amounts of genistein (*i.e.* ~10% of blood levels) can cross the BBB (Tsai 2005), it might be relevant to substitute it by other isoflavones or flavonoids or a combination of several types of molecules, once their toxicity, mechanistic and pharmacological analyse will have been determined.

Due to the complexity of MPS diseases, it is likely that an efficient treatment will require a combination of several therapies. To reach inaccessible tissues, such as skeletal bones and CNS, small molecules such as isoflavones appear to be promising candidates, especially to treat diseases with neurological alterations.

Summary Points

- Mucopolysaccharidoses (MPS) are genetic lysosomal storage diseases (LSDs), characterized by enzymatic deficiency, leading to the accumulation of glycosminoglycans (GAGs) in cell lysosomes.
- Although we presently do not completely understand the pathophysiology of these disorders, there is no doubt that accumulation of storage material is the first pathogenic factor which triggers secondary structural and biochemical alterations, thus leading to disease initiation and progression.
- Substrate reduction therapy (SRT) can be used in the treatment of MPS, and consists of reducing the production of glycosaminoglycans (GAGs) in order to limit lysosomal storage and its damaging consequences.
- Isoflavones recently demonstrated their efficacy in reducing GAG levels in cells from MPS patients, in MPS animal models or in the clinic.
- This therapeutic approach could also be used in combination with other strategies aiming at restoring deficient enzymatic activity.

Key Facts

Key Facts of Mucopolysaccharidoses

- Mucopolysaccharidoses (MPS) belong to the family of lysosomal storage diseases (LSDs), which are inherited metabolic rare diseases resulting from a mutation in a gene encoding either a lysosomal enzyme or a lysosome-related protein.
- MPS result from the deficiency of an enzyme involved in the degradation of at least one of the five glycosaminoglycans (GAGs) which are classified as heparan sulfate, dermatan sulfate, chondroitin sulfate, keratan sulfate or hyaluronan.
- Most MPS diseases are transmitted in an autosomal and recessive manner.
- The clinical features in patients with MPS disorders are associated with organ dysfunction such as hepatosplenomegaly or corneal clouding, skeletal abnormalities and/or central nervous system (CNS) manifestations characterized by progressive and severe behavioural disturbances.
- For the most severe forms of MPS diseases, a patient's death occurs in childhood or early adulthood.

Definitions of Words and Terms

Blood-brain barrier (BBB): This is a poorly permeable barrier that separates the blood from the extracellular fluid in the central nervous system (CNS).

Enzyme Replacement Therapy (ERT): This therapy consists of injecting a recombinant protein that is otherwise missing or deficient in patients.

Glycosaminoglycans (GAGs): These are unbranched polysaccharides of high molecular weight, often associated with proteins and recycled in lysosomes.

IC50: The value that designates the concentration of a compound required for 50% inhibition of cell proliferation.

Isoflavones: A subclass of flavonoid molecules with rings B and C linked at C3.

LC50: The value that designates the concentration of a compound that is lethal to 50% of the cells.

Lysosomal storage diseases (LSDs): These diseases are characterized by the accumulation of macromolecules in lysosomes, resulting in cellular dysfunction.

Mucopolysaccharidosis (MPS): A subclass of lysosomal storage disease (LSD), characterized by the accumulation of mucopolysaccharides (also called GAGs) in lysosomes.

Pharmacological chaperone therapy (PCT): This therapy consists of using a molecule to stabilize a mutated protein to restore its activity.

Substrate Reduction Therapy (SRT): This therapy consists of using a molecule to reduce the accumulation of a substrate resulting from the deficiency of an enzyme.

List of Abbreviations

BBB	blood–brain barrier
CNS	central nervous system
ECM	extracellular matrix
EGF	epidermal growth factor
EGFR	epidermal growth factor receptor
ERT	enzyme replacement therapy
GAG	glycosaminoglycan
GalNAc	*N*-acetylgalactosamine
GlcNAc	*N*-acetylglucosamine
LSD	lysosomal storage disease
M6P	mannose-6-phosphate
MPS	mucopolysaccharidosis
PCT	pharmacological chaperone therapy
SRT	substrate reduction therapy

References

Arfi, A., Richard, M., Gandolphe, C., and Scherman, D., 2010. Storage correction in cells of patients suffering from mucopolysacccharidoses type IIIA and type VII after treatment with genistein and other isoflavones. *Journal of Inherited Metabolic Disease*. 33: 61–67.

Arfi, A., Richard, M., Gandolphe, C., Bonnefont-Rousselot, D., Thérond, P., and Scherman, D., 2011. Neuroinflammatory and oxidative stress phenomena in MPS IIIA mouse model: the positive effect of long-term aspirin treatment. *Journal of Molecular Genetic Metabolism*. 103: 18–25.

Beck, M., 2010. Emerging drugs for lysosomal storage diseases. *Expert Opinion on Emerging Drugs*. 15: 495–507.

Butler, D., and Bahr, B.A., 2006. Oxidative stress and lysosomes: CNS-related consequences and implications for lysosomal enhancement strategies and induction of autophagy. *Antioxidant Redox Signal*. 8: 185–196.

de Ruijter, J., Valstar, M. J. and Wijburg, F. A., 2011. Mucopolysaccharidosis type III (Sanfilippo syndrome): Emerging treatment strategies. *Current Pharmaceutical Biotechnology*. 12(6): 923–930.

Esko, J.D., Kimata, K., and Lindahl, U., 2009. Proteoglycans and Sulfated Glycosaminoglycans. In: Varki, A., Cummings, R.D., Esko, J.D., Freeze, H.H., Stanley, P., Bertozzi, C.R., Hart, G.W., and Etzler, M.E. (ed.) *Essentials of Glycobiology*, 2nd edn. Cold Spring Harbor Laboratory Press, Cold Spring Harbor, NY, USA, Chapter 16.

Friso, A., Tomanin, R., Salvalaio, M., and Scarpa, M., 2010. Genistein reduces glycosaminoglycan levels in a mouse model of mucopolysaccharide type II. *British Journal of Pharmacology*. 159: 1082–1091.

Gritti, A., 2011. Gene therapy for lysosomal storage disorders. *Expert Opinion on Biological Therapy*. 29(10): 1559–1571.

Hemsley, K.M., and Hopwood, J.J., 2011. Emerging therapies for neurodegenerative lysosomal storage disorders – from concept to reality. *Journal of Inherited Metabolic Disease*. 34(5): 1003–1012.

Jakobkiewicz-Banecka, J., Piotrowska, E., Narajczyk, M., Baranska, S., and Wegrzyn, G., 2009. Genistein-mediated inhibition of glycosaminoglycan synthesis, which corrects storage in cells of patients suffering from mucopolysaccharidoses, acts by influencing an epidermal growth factor-dependent pathway. *Journal of Biomedical Science*. 16: 26.

King, M.W., 2005. The Medical Biochemistry Page. Available at: http://themedicalbiochemistrypage.org. Accessed 31 December 2006.

Kloska, A., Jakobkiewicz-Banecka, J., Narajczyk, M., Banecka-Majkutewicz, Z., and Wegrzyn, G., 2011. Effects of flavonoids on glycosaminoglycan synthesis: implications for substrate reduction therapy in Sanfilippo disease and other mucopolysaccharidoses. *Metabolic Brain Disease*. 26: 1–8.

Lachmann, R., 2010. Treatments for lysosomal storage disorders. *Biochemical Society Transactions*. 38: 1465–1468.

Les Cahiers d'Orphanet, 2011. Série Maladies rares. Numéro 2. Available at: http://www.orpha.net/orphacom/cahiers/docs/FR/Prevalence_des_maladies_rares_par_prevalence_decroissante_ou_cas.pdf. Accessed May 2011.

Malinowska, M., Wilkinson, F.L., Bennett, W., Langford-Smith, K.J., O'Leary, H.A., Jakobkiewicz-Banecka, J., Wynn, R., Wraith, J.E., Wegrzyn, G., and Bigger, B.W., 2009. Genistein reduces lysosomal storage in peripheral tissues of mucopolysaccharide IIIB mice. *Molecular Genetics and Metabolism*. 98: 235–242.

Ohmi, K., Greenberg, D.S., Rajavel, K.S., Ryazantsev, S., Li, H.H., and Neufeld, E.F., 2003. Activated microglia in cortex of mouse models of mucopolysaccharidoses I and IIIB. *Proceedings of the National Academy of Sciences of the United States of America.* 100: 1902–1907.

Pan, D., 2011. Recent advances in treatment approaches of lysosomal storage diseases. Cell- and gene-based therapeutic approaches for neurological deficits in mucopolysaccharidoses. *Current Pharmaceutical Biotechnology.* 12: 884–896.

Piotrowska, E., Jakobkiewicz-Banecka, J., Branska, S., Tylki-Szymanska, A., Czartoryska, B., Wegrzyn, A., and Wegrzyn, G., 2006. Genistein-mediated inhibtion of glycosaminoglycan synthesis as a basis for gene expression-targeted isoflavone therapy for mucopolysaccharidoses. *European Journal of Human Genetics.* 14: 846–852.

Piotrowska, E., Jakóbkiewicz-Banecka, J., Tylki-Szymańska, A., Wegrzyn, A., Czartoryska, B., Slominska-Wojewodzka, M., and Wegrzyn, G., 2008. Genistin rich soy isoflavone extract in substrate reduction therapy for Sanfilippo syndrome: an open-label, pilot study in 10 pediatric patients. *Current Therapy Research Clinical Experiment.* 69: 166–179.

Piotrowska, E., Jakobkiewicz-Banecka, J., and Wegrzyn, G., 2009. Different amounts of isoflavones in various commercially available soy extracts in the light of gene expression-targeted isoflavone therapy. *Phytotherapy Research.* Suppl. 1: S109–S113.

Piotrowska, E., Jakobkiewicz-Banecka, J., Maryniak, A., Tylki-Szymanska, A., Puk, E., Liberek, A., Wegrzyn, A., Czartoryska, B., Slominska-Wojewodzka, M., and Wegrzyn, G., 2011. Two-year follow-up of Sanfilippo disease patients treated with a genistein-rich isoflavone extract: Assessment of effects on cognitive functions and general status of patients. *Medical Science Monitor.* 17: 196–202.

Reolon, G.K., Reinke, A., de Oliveira, M.R., Braga, L.M., Camassola, M., Andrades, M.E., Moreira, J.C., Nardi, N.B., Roesler, R., and Dal-Pizzol, F., 2009. Alterations in oxidative markers in the cerebellum and peripheral organs in MPS I mice. *Cellular and Molecular Neurobiology.* 29: 443–448.

Richard, M., Arfi, A., Rhinn, H., Gandolphe, C., and Scherman, D., 2008. Identification of new markers for neurodegeneration process in the mouse model of Sly disease as revealed by expression profiling of selected genes. *Journal of Neuroscience Research.* 86: 3285–3294.

Settembre, C., Fraldi, A., Jahreiss, L., Spampanato, C., Venturi, C., Medina, D., de Pablo, R., Tacchetti, C., Rubinsztein, D.C., and Ballabio, A., 2008. A block of autophagy in lysosomal storage disorders. *Human Molecular Genetics.* 17: 119–129.

Thompson, M., 2011. *Isoflavones: Biosynthesis, Occurrence and Health effects.* Nova Science Publishers Inc, New York, U.S.A.

Tsai, T.H., 2005. Concurrent measurement of unbound genistein in the blood, brain and bile of anesthetized rats using microdialysis and its pharmacokinetic application. *Journal of Chromatography A.* 1073: 317–322.

Urayama, A., Grubb, J.H., Sly, W.S., and Banks, W.A., 2008. Mannose 6-phosphate receptor–mediated transport of sulfamidase across the blood–brain barrier in the newborn mouse. *Proceedings of the National Academy of Sciences of the United States of America*. 16: 1261–1266.

Villani, G.R., Gargiulo, N., Faraonio, R., Castaldo, S., Gonzalez, Y., Reyero, E., and Di Natale, P., 2007. Cytokines, neurotrophins, and oxidative stress in brain disease from mucopolysaccharidosis IIIB. *Journal of Neuroscience Research*. 85: 612–622.

Wegrzyn, G., Jakóbkiewicz-Banecka, J., Narajczyk, M., Wisniewski, A., Piotrowska, E., Gabig-Ciminska, M., Kloska, A., Slominska-Wojewódzka, M., Korzon-Burakowska, A., and Wegrzyn, A., 2010. Why are behaviors of children suffering from various neuronopathic types of mucopolysaccharidoses different? *Medical Hypotheses*. 75: 605–609.

Wu, Y.P., and Proia, R.L., 2004. Deletion of macrophage-inflammatory protein 1α retards neurodegeneration in Sandhoff disease mice. *Proceedings of the National Academy of Sciences of the United States of America*. 101: 8425–8430.

CHAPTER 24

Optimizing Isoflavone-rich Food Delivery Systems for Human Clinical Trials

JENNIFER AHN-JARVIS, STEVEN SCHWARTZ AND YAEL VODOVOTZ*

Department of Food Science and Technology, College of Food, Agricultural, and Environmental Sciences, The Ohio State University, 2015 Fyffe Court, Columbus, Ohio, USA,
*Email: vodovotz.1@osu.edu

24.1 Introduction

Substantial evidence from epidemiological, *in vitro*, and animal studies suggest that a diet rich in soy protein and isoflavones reduce the risk of age-related disease such as coronary heart disease, cancer, and osteoporosis (Messina 2010). In Asian countries where consumption of soy products is considerably more frequent than Western countries, there are relatively low rates of coronary heart disease, hormone-related cancers, and other age-related diseases. Much of the health benefits of soy observed in Asian countries are hypothesized not to have materialized in Western countries because of the low quantity and quality of the soy foods consumed in Westernized diets (Reinwald *et al.* 2010).

24.1.1 Delivery of Isoflavone Dosage without Compromising Palatability

Because of their bioactive potential, functional food design begins with identifying the ingredient containing the lead phytochemical and the challenges associated with delivery of the appropriate dose and the stability of the compound(s) within the food delivery system (Figure 24.1). For soy isoflavones, meta-analyses examining the effects of soy protein with isoflavones on endpoints of cardiovascular disease (Zhan and Ho 2005) and cancer risk (Yan and Spitznagel 2009) suggest that isoflavone doses (50 mg day^{-1}) similar to those found in Asian diets are important for biological response. Therefore, substantial amounts of soy need to be integrated into soy foods appropriate for clinical trials. Traditional Asian soy foods readily meet this benchmark for isoflavones; however, Westernized palates are not accustomed to flavors and textures associated with soy foods. Developing soy foods acceptable to Western palates may aid in increasing soy intake. Soy inclusion at 3% is used to improve baked goods color and batter water absorption (Liu and Limpert 2005), but

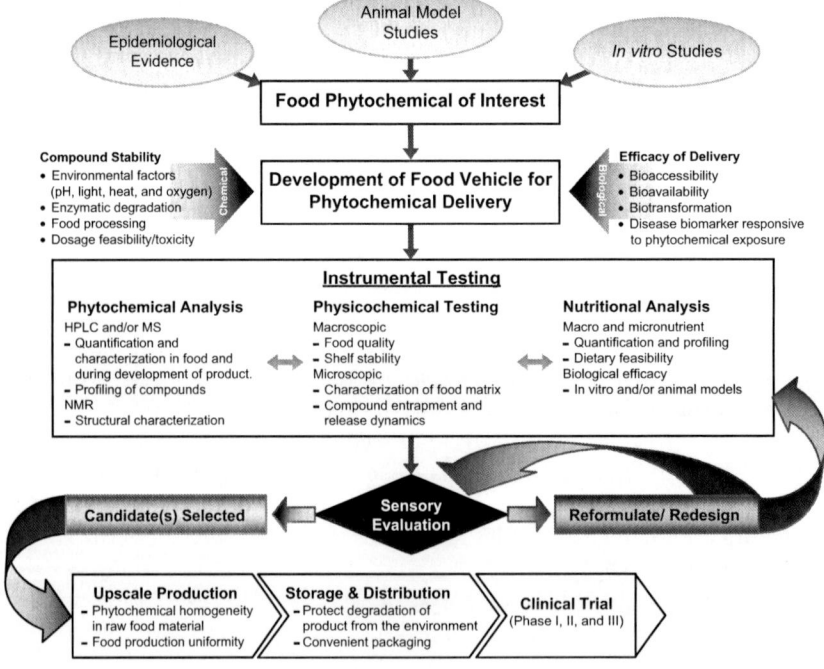

Figure 24.1 Flowchart of the development of phytochemical delivery system for clinical trials. This flowchart details the step-wise process of designing an appropriate vehicle system for focused delivery of the phytochemical of interest and sensory evaluation is the decision tree for selection of the study agent for clinical trials.
Unpublished.

for clinical trials soy quantities need to exceed 30% of the formulation, which can cause detrimental effects on loaf volume, crumb texture, off-flavors, and aromas (Dhingra and Jood 2002). At these high levels of soy in yeasted breads, non-gluten proteins from soy predominate and outcompete gluten for water (Knorr and Betschart 1978); this, combined with the dilution of gluten proteins, results in the loss of gluten strength (Brewer et al. 1992). Consequently, loss of bread quality affects palatability and compliance is compromised. Dietary compliance is critical to ensure that the therapeutic intent of the isoflavone-rich food delivery system is met (Dove 1946).

24.1.2 Processing Effects on Soy Isoflavones

In native soybeans, isoflavones exist as daidzein, genistein, and glycitein and predominately as conjugated β-glucosides. Food processing, such as fermentation, converts glucosides into their respective aglycone form (Coward et al. 1998). Because of sensitivity of isoflavones to chemical conversion from food processing, soy foods can be engineered to deliver a specific soy isoflavone profile. Isoflavone chemical composition can be altered using fermentation and various thermal (steaming and roasting) processing strategies (Figure 24.2). Presumably, these changes in isoflavone composition may affect the bioavailability of soy isoflavones in soy products (Coward et al. 1998). From our previous studies, soy bread crumb and crust was observed to have discrete isoflavone compositions which suggest that differences in heat exposure (crumb, 100 °C; crust, 165 °C) and moisture content (crumb, 44%; crust, 16%) affect isoflavone composition (Zhang et al. 2003). β-Glucosidase is the key enzyme that cleaves the glucose moiety from the isoflavone glucosides resulting in isoflavone aglycones. Endogenous β-glucosidase activity was detected in the

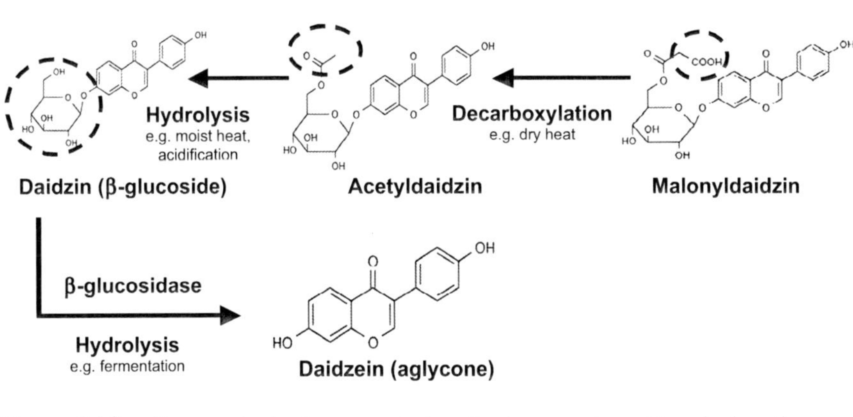

Figure 24.2 Changes in isoflavones during food processing. Chemical pathway of isoflavone glucoside transformation into aglycone as a result of different processing methods and the activity of β-glucosidase enzyme. The Figure illustrates the principles driving the experiments for development of a aglycone-rich soy bread.
Unpublished.

various bread ingredients including wheat flour, soy ingredients (soy flour, soy milk powder), and active dry yeast (ADY, *Saccharomyces cerevisiae*) with the soy ingredients providing the greatest activities (Hsieh and Graham 2001). During the bread-making process, β-glucosidase activity was highest during the fermentation process. Partial conversion of soy isoflavone glucosides into aglycones occurs during dough fermentation through the action of endogenous soy β-glucosidase, thereby producing a soy bread (SB) with 15% of isoflavones in the aglycone form (Figure 24.2). Almond meal, a rich natural source of β-glucosidase activity, when added to the formulation (5%, w/w) over the same fermentation period produced a soy bread with 30% aglycone composition (Vodovotz 2007). Conversion of glucosides into their aglycone forms can be achieved with prolonged fermentation in a SB system; however, extensive exposure to proofing conditions weaken the interaction of the gluten proteins which affect gas entrapment and air cell formation (Schofield *et al.* 1983). The combined effort of pre-processing the soy ingredients and utilizing the β-glucosidase activity from almonds provides a strategy to control isoflavone conversion hence modulate isoflavone composition within a soy bread system.

24.1.3 Relevance of Isoflavone Chemical Composition

Soy isoflavone absorption is assumed to occur first with passive absorption of the aglycones in the small intestines, whereas conjugated β-glucosides have been suggested to be hydrolyzed by β-glucosidase present in the human gut mucosa (Setchell *et al.* 2002), and by the bacterial β-glucosidase in the large intestine (Turner *et al.* 2003). Although aglycones are readily absorbed in the upper intestinal tract, there is ongoing discussion about the bioavailability of isoflavone aglycones compared with isoflavone glucosides. Studies have suggested higher (Izumi *et al.* 2000), lower (Setchell *et al.* 2001), or no difference in absorption (Kano *et al.* 2006) of the aglycone compared with their glucosides. This controversy might be explained by the different forms in which isoflavones have been administered–supplement (Setchell *et al.* 2001), or within a food (Kano *et al.* 2006, Faughnan *et al.* 2004). Moreover, metabolites such as equol, *O*-desmethylangolensin, and dihydrodaidzein are often not included in the analyses and thus overall isoflavone absorption is underestimated. Kano *et al.* (2006) found lower urinary *O*-desmethylangolensin excretion for two aglycone-rich soymilks *versus* a glycoside-rich version and mixed results were found for equol excretion.

Included with the changes in isoflavone composition, physicochemical changes were observed with the finished SB and functional properties of the soy ingredients; hence these differences are expected to affect the quality and acceptability of the resultant breads and may compromise dough machinability and hinder its translation for large-scale production. The effects of pre-processed soy flour on flour functionality, bread quality, and consumer acceptability has not been examined. Therefore the primary objective for this study was to develop an acceptable soy bread that is suitable for clinical trials and where isoflavone composition was carefully controlled.

24.2 Materials and Methods

24.2.1 Preparation of Fermented and Thermally Treated Soy Ingredients

The physicochemical effects of fermented and thermally treated soy ingredients integrated into SB were examined using a patented combination of soy flour and soymilk (Vodovotz and Ballard 2010). Four different processing methods were applied to this mix of soy ingredients. The soy ingredients were formed into a sponge with almond meal and yeast and fermented for 2 h. Thermally treated soy ingredients were examined in a microwave-heated soy mix using 80% formula water combined with soy mix and placed in a commercial microwave (1500 W, Model R995J Sharp) for 10 min. Soy ingredients were steamed using a #30 sieve lined with filter paper (No.1, Whatman Inc.). The internal temperature of the soy ingredients was measured during steaming and was maintained at 98 °C for 2 h. Steamed soy acquired 11% moisture during steaming process. Soy ingredients were oven-roasted for 60 min at 175 °C. Soy was fermented with 80% formula water, covered with polyethylene, and stored at 25 °C for 96 h. Many other variations in fermentation and thermal processing (boiling soy-water slurry and frying) of soy ingredients were examined but produced poor quality bread.

24.2.2 Soy-almond Bread (SAB) Preparation

All ingredients were measured in the proportions described in Table 24.1 and scaled for 1.00 kg loaves. The SB and SABs were produced using a sponge-dough process. For SABs, two separate sponge mixtures were combined during dough formation. One mixture containing high gluten, wheat flour (13%

Table 24.1 Formulation for SAB based upon the percentage dry basis. The Table details ingredients and formulation for SAB. Soy bread formulation uses the same ingredients excludes almonds. This is the optimized formulation for delivery of 31 mg of AE of isoflavones and 6.8 g of soy protein in one slice of bread. Unpublished.

Ingredients	Dry basis (%)
Wheat flour (Bouncer, Bay State Milling)	53.8
Vital Gluten (Hodgson Mill)	3
Soy flour (Baker's soy, ADM Protein Specialties Division) and soymilk (Benesoy, Devansoy)	30
Yeast (SAF-instant, Lesaffre Group)	1
Sugar (Domino Foods Corp.)	3
Shortening (Crisco, Proctor & Gamble)	3
Almonds (Whole Foods)	5
Kosher salt (Morton International Inc.)	1
Dough conditioner (Caravan Products Co.)	0.2

protein), vital gluten, instant active dry yeast, and water mixture was rested for 2 h at 25 °C. The other mixture with soy flour, soymilk, and almond meal was placed in an electric proofing cabinet (CM2000 FlavorView InterMetro Industries Corp.) set at ≥ 90% relative humidity at 40 °C for 2 h. Whole raw almonds were ground into meal prior to dough formation using commercial blender (model 50200MP, Hamilton Beach Brands Inc.) and passed through a No. 20 stainless steel sieve (USA Standard Testing Sieve). The sponge ingredients were mixed using a Kitchen Aid planetary mixer (model KV25GO, Kitchen Aid) with a paddle attachment for 5 min. Soy bread (SB), an almond-free variant, was produced using the wheat ingredient sponge with integration of the dry soy ingredients during dough formation. Dough was developed by mixing dough ingredients with a sponge using the dough hook attachment. Formed loaves were panned and returned to high humidity proofing cabinet at 38 °C for 60 min. Proofed loaves were immediately placed into a preheated convection oven (model: JA14, Doyon, Liniere) set at 150 °C. Total baking time was 50 min and breads were determined to be done once an internal temperature of 95 °C was achieved. Finished loaves were cooled for 4 h and stored in sealed polyethylene bags (Ziploc, SC Johnson & Sons Inc.). Instrumental analysis of soy breads were conducted with stored breads within 24 h after baking. Sensory evaluation and consumer acceptance was conducted on breads that were stored and handled in the same manner anticipated for clinical trials, which was to store cooled breads in sealed polyethylene bags, freeze (–25 °C) for 72 h, and thaw at room temperature (25 °C) for 12 h prior to evaluation.

24.2.3 Isoflavone Analysis

24.2.3.1 Sample Preparation

Samples of SB from each of the eight baking days were freeze-dried and ground to a fine powder and stored at –25 °C (Murphy *et al.* 2002). Using methods adapted from Achouri *et al.* (2005), 500 mg samples were extracted (10 mL × 2 volumes) using aqueous acetonitrile (ACN) (60%, v/v) were sonicated (Fisher Scientific FS30H, Fisher Scientific; 100 W at 42 kHz output) and centrifuged. Extracts were dried under nitrogen and stored at –25 °C until high-performance liquid chromatography (HPLC) analysis was carried out. Each food sample was redissolved in 80% (v/v) methanol, sonicated, and filtered (Alltech Associates Inc.) prior to HPLC analysis. Soy bread samples of each baking day were analyzed in duplicate.

24.2.3.2 Separation and Quantification of Isoflavones in Soy Foods

HPLC-grade solvents and authenticated isoflavone standards were used. Food extracts were analyzed utilizing Waters model 2690 HPLC equipped with a Waters 2996 photodiode array detector (PDA), an autosampler (15 °C), and a column heater at 30 °C (Waters Associates). Reverse-phased separation was performed using a 3.0 × 100 mm, 3 μm particle, Symmetry™ PS C_{18} column (Waters Associates) with a Waters guard column. The binary mobile phase (1% aqueous acetic acid/ACN) gradient began at 95:5 for 2 min, to 65:35 in 24 min, 15:85 by 27 min, and then returned to starting conditions for a total run time of 28 min. The injection volume was 10 μL.

Retention times of standards in HPLC chromatograms and previously published UV–visible spectral signatures were used for identification of unknowns (Murphy et al. 2002). The HPLC peak areas of each isoflavone were analyzed and quantified using Empower Pro (version 5.0 Waters Associates) software. The isoflavone concentration in food is reported as means ± S.D. mg of aglycone equivalents (AEs).

24.2.4 Physicochemical Experiments

24.2.4.1 Water Holding Capacity (WHC)

Assessment of the approximate WHCs of soy flour, soy milk, gluten, wheat flour, and soy mixture variables (untreated, fermented, and thermally treated soy ingredients) was carried out (Quinn and Paton 1979). Wheat flour, gluten, soy flour, soy milk, and pre-processed soy ingredients were freeze-dried (Labconco) for 72 h and stored in a desiccator. Prior to analysis, samples (5.0 g) were combined with distilled water. The slurries were stirred, vortexed, rested, and centrifuged at $3000\,g$ for 30 min (IEC HN-SII, Damon/IEC Division, Damon Corp.). The free liquid present as supernatant was removed and the remaining pellet was weighed. The difference between the mass of the starting material and pellet per g of dry sample was the WHC.

24.2.4.2 Specific Loaf Volume

SAB loaf volume was determined using standard rapeseed displacement methods as detailed in AACC approved method 10-05 (2008). Specific loaf volume was derived from the ratio of loaf volume and loaf mass from eight loaves of each treatment.

24.2.4.3 Crumb and Crust Color

Instrumental analysis of crumb and crust color was measured using a tristimulus colorimeter (Minolta Chroma Meter CR-300, Minolta) equipped with a data processor (DP 301, Minolta). For each bread type L*, a*, and b* values were obtained for six loaves with six different readings from the crumb and crust. Hue angle, chroma, and browness index were calculated using the calculations described by Maskan (2001).

24.2.4.4 Texture

Bread texture was characterized with a texture profile analysis two-bite test on an Instron 5542 Universal Testing Machine (Instron Corp.) equipped with a

100 N load cell. The endpoints measured were crumb firmness, springiness, and chewiness. Texture analysis conditions were those described by Lodi and Vodovotz (2008). The Blue Hill Software for texture profile analysis program (Instron Corp.,) was used to assess texture parameters (hardness, springiness, and chewiness).

24.2.4.5 Thermogravimetric Analysis (TGA)

A thermogravimetric analyzer (Q5000, TA Instruments) equipped with autosampler was used to determine the moisture content of bread crumbs (Schiraldi et al. 1996). Approximately 20 mg of bread crumb was obtained from the loaf center and placed in pre-tared platinum pans. The experiment was performed using a linear ramp of temperature from room temperature ($\sim 20\,^\circ$C) up to $180\,^\circ$C at the rate of $5\,^\circ$C min^{-1}, and held isothermally for 5 min. The moisture content on a wet basis was determined as the ratio between the difference in mass from the start of the experiment to the end, and the starting material mass, expressed as a percentage. The reported moisture content is the mean $\pm$ S.D. of six loaves prepared over six baking days.

24.2.4.6 Differential Scanning Calorimetry (DSC)

In brief, calorimetric measurements were performed on a DSC, model 2920, equipped with a refrigerated cooling system (TA Instruments). Approximately 15.0 mg of bread crumb was obtained from the loaf center, placed in a large volume stainless steel pans, sealed with O-ring and lid (PerkinElmer Life And Analytical Sciences, Inc.), and immediately analyzed. The experimental conditions were those described by Lodi and Vodovotz (2008). The endothermic melting peak ($\sim 0\,^\circ$C) reflects the melting of ice and the enthalpy of this transition is used to calculate "freezable" water (FW). The percentage FW is determined using calculations described by Vittadini and Vodovotz (2003).

24.2.5 Characterization of Organoleptic Properties

24.2.5.1 Consumer Acceptance

After the Ohio State University Institute Review Board exemption was approved, all consumers were prescreened for allergies to study products which included wheat, rye, soy, almonds, and gluten prior to participation. Completed consumer evaluations were received from 60 consumers. Using a 9-point hedonic scale consumers were asked to evaluate the acceptability of the flavor, texture, and color of six different soy breads (one SB and five SABs). Additionally, a 5-point "just-about-right" (JAR) scale (1 = much too weak to 5 = much too strong) assessed the appropriateness of the degree of flavor (saltiness and sweetness), aroma (yeast- and bean-like), and appearance (crumb texture, crumb and crust color) (Lawless and Heymann 1998). Evaluations of soy breads (with and without almonds) were conducted in the Ohio State University sensory facilities. Frozen bread samples were defrosted at room temperature for 12 h. The bread samples were cut into cubes (1 × 1 × 0.5 inches) and stored tightly covered in plastic soufflé cups. Consumer testing was conducted in a temperature-, air flow-, and noise-controlled facility. Ambient fluorescent light and incandescent light illuminated each sensory booth.

24.2.5.2 Descriptive Analysis

Untrained panelists ($n = 60$) were oriented to bread attributes using a 15 min information session which was used to define terminology and provide reference foods for firmness, bean-like flavor, crumb, and crust color (Table 24.2). The reference food and bread attributes utilized a 10-point intensity scale (1 = very weak, 5 = moderate, and 10 = very intense). For the attribute test, breads were defrosted at room temperature for 12 h, sliced (SM302B Bread Slicer Doyon,) and from each slice 1 inch squares were extracted from the center. Each participant received two squares of bread which were placed in air tight plastic soufflé cups and labeled with a random 3-digit code. The visual bread samples were one slice of bread placed in a sealed re-sealable plastic sandwich bag. The sandwich bags were coded with a random 3-digit numeric code. Compusense 4.5 software (Compusense,) was used to generate a balanced randomized presentation order for the six samples and 60 panelists. Using a strict serial monadic protocol, panelists were given a warm-up sample prior to the study samples. Panelists were asked to rinse with distilled water in between each sample. Paper ballots with a horizontal scale from 1 to 10 for each attribute were provided. Relative humidity, temperature (25 °C), air flow, and noise were controlled during descriptive analysis testing. Ambient fluorescent and over head white task lighting were used for illumination. Data from attribute tests were evaluated using ANOVA, and Tukey's post hoc test (IBM SPSS 19.0 version) was used to discern significant differences ($P \leq 0.05$). The means ± S.D. were reported.

24.3 Results and Discussion

24.3.1 Process of Selecting the Optimal Phytochemical Delivery Vehicle

Intrinsically functional foods need to be palatable and deliver a specific threshold of compounds to elicit a biological activity that will improve health. The design and development of functional foods for clinical trials requires a systematic scientifically sound step-wise approach (Figure 24.1). Much of the direction in selecting a phytochemical compound of interest emerges from previous epidemiological, animal model, and *in vitro* mechanistic studies which have discerned biological target(s), pivotal to health maintenance and are sensitive to modulation by food components. Soy contains a variety of

Table 24.2 Definitions and food references of bread attributes. A Table of terms and their corresponding definitions as defined during the training sessions. The reference food description provides insight into the handling and preparation of reference foods, the reference value, and their correlation to instrumental measures. Unpublished.

Term	Definition	Reference food
Firmness	The necessary force for the front teeth to bite through the 0.5 inch thick slice of bread.	Sara Lee® Soft and Smooth Classic White bread = 1: two $\frac{1}{2}$ inch cubes served in tightly closed 3.25 oz soufflé cups. Hardness = 2 N, springiness = 4 mm, and chewiness = 8 N. Mestemachers® pumpernickel bread = 10: two $\frac{1}{2}$ inch cubes served in air tight container in tightly closed 3.25 oz soufflé cups. Hardness = 20 N, springiness = 5 mm, and chewiness = 95 N.
Bean-like flavor	The musty, grassy, nutty flavor or aroma commonly found with cooked beans.	Silk™ soymilk = 1: room temperature for 30 min, 20 ml in tightly covered 1 oz. soufflé cups Sahmyook™ soy drink = 10: room temperature for 30 min, 20 ml in tightly covered 1 oz. soufflé cups.
Crumb color	The intensity of brown color at the center of the slice of soy bread (no brown = 1 to light brown = 10).	Arnold's® Italian white bread = 1: 1 slice placed in clear plastic reclosable sandwich bag. $L^* = 75.6$, $a^* = -2.0$, and $b^* = 31.5$. Bakery at Wal-mart all-natural pumpernickel bread = 10: 1 slice placed in clear plastic reclosable sandwich bag. $L^* = 30.6$, $a^* = 9.0$, and $b^* = 24.2$.
Crust color	The color at the top edge of the slice of soy bread referred to as the bread crust (brown = 1 to dark brown = 10).	Schwebel's® Country Potato Bread = 1: 1 slice placed in clear plastic re-sealable sandwich bag. $L^* = 59.2$, $a^* = 14.7$, and $b^* = 30.4$. Bakery at Wal-mart all-natural pumpernickel bread, = 10: 1 slice placed in clear plastic re-sealable sandwich bag. $L^* = 36.7$, $a^* = 10.7$, and $b^* = 18.6$.
Warm-up sample	Improves reliability of responses in descriptive analysis and evaluates effectiveness of training.	Schwebel's® Sweet harvest wheat bread: 1 slice placed in clear plastic re-sealable sandwich bag. Hardness = 5 N, springiness = 5 mm, and chewiness = 10 N. Crumb: $L^* = 59.9$, $a^* = 4.7$, and $b^* = 23.4$. Crust: $L^* = 49.6$, $a^* = 14.2$, and $b^* = 25.1$.

bioactive components, including proteins and isoflavones, which have been studied for their health-promoting properties (Messina 2010). The most well-established and is the basis for the approved Food and Drug Administration claim, albeit controversial benefit, is the hypocholesterolemic effects of soy protein (Food and Drug Administration 1999). Once a phytochemical has been identified, vehicle selection needs to integrate the chemistry (dose and chemical stability) and biology (bioavailability and bioactivity) yet retain its identity as a acceptable food product. In Asian countries there is an abundance of soy food varieties that can be readily assimilated into Asian diets, however, in Western countries soy foods fashioned to integrate into Western diets is limited (Reinwald et al. 2010). Therefore, in developing a vehicle for soy isoflavone delivery, whole soy ingredients (defatted soy flour and soymilk) were incorporated into a soy bread system. Chemical composition was controlled using various processing strategies within a soy-bread system and sensory evaluation was used to select the ideal candidate for intervention studies.

24.3.2 Isoflavone Analysis

Soy isoflavone composition was examined in soy breads with and without almonds and those with pre-processed soy ingredients. All SAB, regardless of the pre-treatment conditions of soy, had significantly greater conversion of glucosides into aglycones than SB (Figure 24.3). Roasted and steamed SAB provided the greatest amount of conversion into the aglycone forms (63 and 67% aglycones, respectively). Much of the conversion of the simple β-glucoside (greatest in roasted and steamed soy mix) occurred during the 2 h fermentation (proofing) of the soy dough system prior to baking. Increasing the aglycone composition was achieved but because aglycone increases bitter flavors product acceptability was a concern (Matsuura et al. 1989).

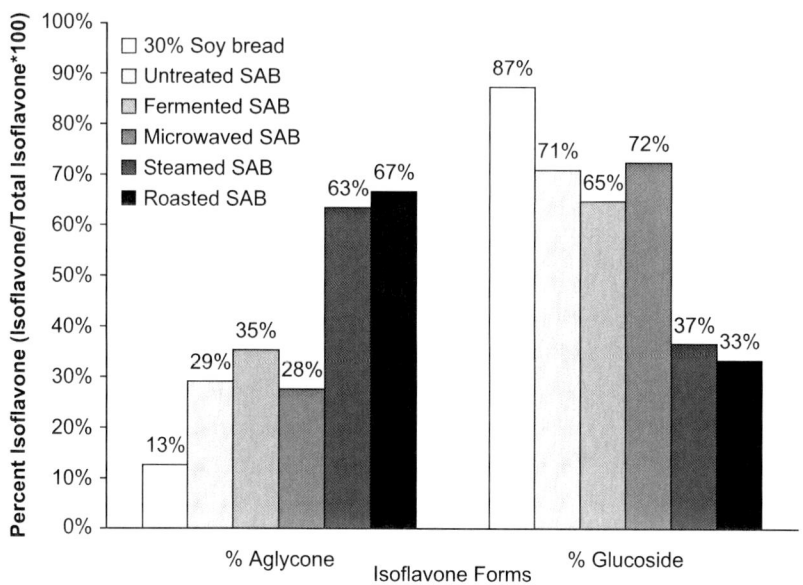

Figure 24.3 Isoflavone composition of aglycones *versus* glycosides in soy breads. Isoflavone composition of aglycones and combined glycosides in soy bread and SAB determined by HPLC analysis ($n = 16$ for each variable). Control is 30% soy bread, whereas all others are SABs with same percentage of soy.
Unpublished.

24.3.3 Physicochemical Properties

24.3.3.1 WHC

Wheat flour had the lowest approximate WHC at 70% and soy flour WHC was almost three times that of wheat flour (Figure 24.4). The WHC in the soy mix was significantly affected by processing. The fermented soy mixture had the lowest WHC (146%) among the soy mixtures, whereas roasted and microwaved soy mix had the greatest WHC. Soy proteins responsible for its water holding functionality may have denatured during thermal processing (Nilufer et al. 2008). Soy mix (steamed and raw) and soy milk had the same WHC and these differences in soy functionality affect soy bread structure.

24.3.3.2 Specific Loaf Volume and Texture

The SB, untreated, microwaved, and fermented SAB had the greatest loaf volume (Table 24.3), whereas the steamed and roasted SAB had the lowest specific loaf volume. The roasted SAB had a 40% reduction in specific loaf volume than the control SB. Both the loss in volume and increase in loaf mass suggest that thermally processed soy ingredients inhibit gas retention. Moreover, the loss of loaf volume can be attributed to the loss of soy ingredient functionality from denaturing of soy proteins with roasting and steaming of the soy ingredients prior to inclusion into soy bread dough. When the WHC is

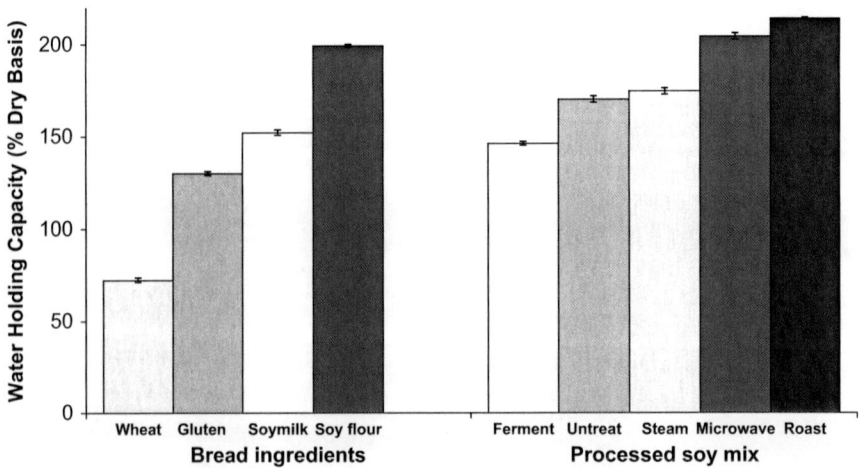

Figure 24.4 WHC of dry ingredients in soy bread and of the processed and unprocessed soy mix. Histogram showing the WHC of dry bread ingredients and of the soy mix (soy flour and soymilk) unprocessed (untreated) and after being fermented and thermally processed (microwaved, steamed, and roasted). Quadruplicate analyses of all variables were significantly different using one-way ANOVA, Tukey's post-hoc ($P \leq 0.05$). Unpublished.

much greater than untreated soy mix (170%), a reduction in specific loaf volume was observed (roasted and steamed SAB).

Bread crumb firmness measured as Instron peak force (N) to compress a 1 inch cube of bread by 40% of its height was lower in SAB (8.7 N) compared with the control SB (12.2 N). The higher specific loaf volume in the untreated SAB (2.5 cm^3 g^{-1}) compared with the same bread without almonds (control SB, 2.2 cm^3 g^{-1}) was previously observed by Lodi and Vodovotz (2008) and attributed to the higher lipid fraction in SAB. The dense crumb structure as shown by the significant decrease in specific loaf volume in the roasted SAB (1.33 cm^3 g^{-1}) compared with the untreated SAB (2.52 cm^3 g^{-1}) or to the control (2.24 cm^3 g^{-1}) may explain the difference in crumb texture (Table 24.3). Moreover, the inclusion of almonds into the soy bread system significantly reduced the springiness of the SAB (5.0 mm) compared to SB (7.3 mm). When the ratio of chewiness to springiness (chewiness/springiness) was evaluated three significantly discernible groups were found. One group having the highest ratio (SB, 9.0, and untreated SAB, 8.9) was the least processed group, essentially the raw soy ingredient. In the second cluster, the soy ingredients were thermally processed (steamed and roasted) then formulated into soy bread. The third group had the lowest ratio (fermented, 5.6, and microwaved, 5.2) and were first hydrated, processed (fermented and microwaved), and then introduced into the soy bread system. The differences in bread texture suggest that processing history affects the functionality of the soy and subsequently the quality of the soy bread.

Table 24.3 Bread quality characteristics of soy breads with and without almonds and pre-processed soy mix. Specific volume analogous to density of soy breads with and without almonds and pre-processed soy mix. Hardness is direct measurement compression force (N = Newtons) applied to the food material, whereas springiness and chewiness are calculated from measures obtained from the second compression cycle during texture profile analysis. Quadruplicate analyses of all variables from eight baking days. Results are means ± S.D., ANOVA analysis performed within crumb and crust soy bread variables. Significantly different samples have different subscripts, Tukey's post-hoc, $P \leq 0.05$. Unpublished.

Soy Bread	Specific volume (cm^3 g^{-1})	Texture profile		
		Hardness (N)	Springiness (mm)	Chewiness (N)
Control	2.24 ± 0.04^a	12.19 ± 1.98^a	7.31 ± 0.56^a	65.75 ± 5.54^a
Untreated SAB*	2.52 ± 0.19^a	8.20 ± 0.72^b	5.86 ± 1.02^b	52.15 ± 6.22^b
Fermented SAB	2.65 ± 0.18^a	8.51 ± 0.81^b	4.48 ± 0.78^b	24.93 ± 5.55^c
Microwaved SAB	2.40 ± 0.13^a	7.37 ± 0.63^b	5.21 ± 0.87^b	26.97 ± 5.98^c
Steamed SAB	1.68 ± 0.18^b	8.99 ± 1.00^{bc}	4.78 ± 0.81^b	34.06 ± 8.75^{cd}
Roasted SAB	1.33 ± 0.07^c	10.34 ± 1.30^c	4.87 ± 1.01^b	38.69 ± 6.82^d

24.3.3.3 Crumb and Crust Color

Crumb color was clearly most affected by differences in processing. The most significant difference was between the control SB and the thermally processed SAB. The SAB containing the thermally processed soy ingredients (steamed and roasted) were significantly darker (L*), redder (a*), yellower (hue angle), and had greater brownness index than the SB or the untreated SAB (Table 24.4). The control SB was much more yellow than any of the SAB. However, significant differences in crust color were found between fermented SAB and SB. Fermented SAB had the darkest, reddest, least yellow color than any of the other soy and SAB variables. Because fermented soy ingredients had the lowest WHC, this difference in crust color maybe associated with difference in WHC among the various breads. The relatively low water content in the fermented SAB bread dough coupled with the high protein content and greater evaporative conditions on the crust may accentuate Maillard browning of the soy bread crust more than in the other soy bread variables. Whereas the temperature in the crumb environment was relatively stable at 95 °C and water loss was minimal, the color was affected by the processing history of the soy ingredient rather than the conditions the soy bread undergoes during baking.

Table 24.4 Colorimeter characteristics of soy bread and SAB pre-processed soy ingredients. L* (lightness), a* (green to red), b* (blue to yellow) color measures of soy bread and SAB were used to calculate hue angle, chroma, and browness index. Quadruplicate analyses of all variables from six baking days. Results are means ± S.D., ANOVA analysis performed within crumb and crust soy bread variables. Significantly different samples have different lettered superscripts, Tukey's post-hoc, $P \leq 0.05$. Unpublished.

Soy Bread	L*	a*	b*	Hue angle	Chroma	Browness Index
Control (30% Soy)						
Crumb	73.6 ± 0.7^a	0.2 ± 0.0^a	22.5 ± 0.5^a	89.5 ± 0.1^a	22.5 ± 0.5^a	35.8 ± 0.9^a
Crust	49.8 ± 0.7^z	18.4 ± 0.3^z	35.9 ± 0.7^z	62.9 ± 0.6^z	40.3 ± 0.6^z	141.9 ± 4.1^a
Untreated SAB						
Crumb	68.5 ± 2.2^b	0.5 ± 0.2^a	24.2 ± 2.9^a	88.8 ± 0.4^a	24.2 ± 2.9^a	43.2 ± 6.1^a
Crust	43.3 ± 7.0^{yz}	18.2 ± 1.2^z	36.4 ± 5.5^z	62.9 ± 4.6^z	40.8 ± 4.7^z	184.5 ± 36.0^b
Fermented SAB						
Crumb	70.0 ± 3.0^b	0.9 ± 0.4^a	24.6 ± 1.9^b	87.9 ± 1.0^a	24.6 ± 1.9^a	43.1 ± 4.0^a
Crust	35.8 ± 2.6^x	18.4 ± 1.3^z	29.0 ± 1.3^y	57.6 ± 1.7^y	34.3 ± 1.5^y	177.2 ± 17.7^{ab}
Microwave SAB						
Crumb	69.6 ± 1.9^b	0.8 ± 0.4^a	24.0 ± 2.6^b	88.2 ± 0.9^a	24.0 ± 2.6^a	42.1 ± 5.0^a
Crust	47.2 ± 9.2^z	15.5 ± 3.5^{xy}	35.8 ± 5.6^z	66.1 ± 6.6^z	39.2 ± 5.0^z	161.7 ± 52.1^{ab}
Steamed SAB						
Crumb	60.2 ± 2.1^c	5.7 ± 0.7^b	28.4 ± 1.9^c	78.6 ± 0.8^b	28.9 ± 2.0^b	69.1 ± 6.7^b
Crust	44.6 ± 6.2^{yz}	17.0 ± 1.0^{yz}	33.3 ± 4.4^{yz}	62.7 ± 3.6^z	37.5 ± 3.8^{yz}	150.6 ± 11.2^{ab}
Roasted SAB						
Crumb	34.5 ± 3.0^d	10.0 ± 0.8^c	27.5 ± 2.0^{bc}	69.9 ± 2.5^c	29.3 ± 1.7^b	161.5 ± 29.8^c
Crust	40.4 ± 1.9^{xy}	13.6 ± 1.2^x	34.3 ± 2.6^{yz}	68.2 ± 2.8^z	37.0 ± 2.2^{yz}	177.2 ± 18.9^{ab}

24.3.3.4 Moisture Content and "Freezable" Water

Because of the sensitivity of isoflavone conversion in water, moisture content, FW, and "un-freezable" water (UFW) were evaluated (Toda *et al.* 2001). Increases in FW permits more chemical conversion of isoflavones from the glucoside forms into their aglycones. Soybeans with a starting aglycone composition of 3% when soaked over a 16 h period at ambient conditions (20 °C) had final composition of 15% aglycones (Matsuura *et al.* 1989). Conceivably, isoflavone conversion in a soy bread system can occur readily, as the system undergoes relatively high temperatures over a 3 h fermentation (42 °C) and baking (150 °C) periods. FW, normalized to moisture content, was examined among the six soy bread variables. In SAB with thermally treated soy mixtures the FW increased with longer exposure to heat (Figure 24.5). The roasted SAB was similar to control SB and had the highest percentage of FW. Moreover, the steamed and roasted SAB had the greatest level of glycoside to aglycone conversion.

24.3.4 Characterization of Organoleptic Properties

The soy ingredients having been processed prior to their inclusion into soy bread dough altered the organoleptic properties of the finished soy breads (Figure 24.6). Consumer acceptability and descriptive analysis of soy bread flavor, texture, and color were evaluated. Evaluating the acceptability and characterizing the organoleptic qualities of the study agent is essential for

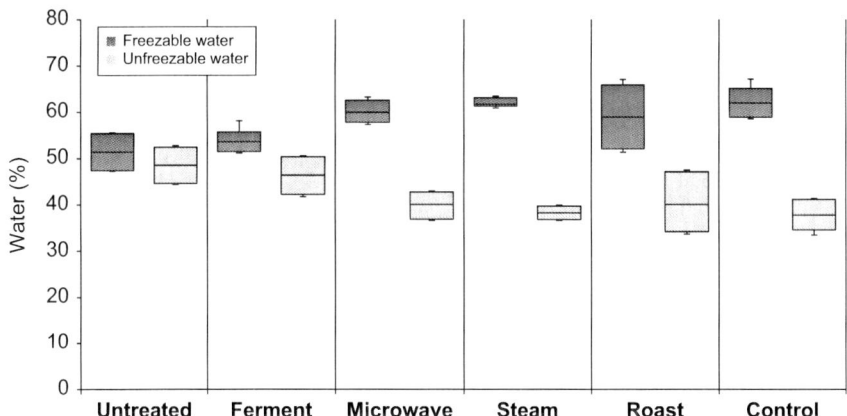

Figure 24.5 Percentages of freezable and unfreezable water in finished fresh soy breads. Box plot of percentages of freezable and unfreezable water in finished fresh soy breads. Values normalized to moisture content. The control is 30% soy bread and all others are SABs with either untreated (raw) or pre-processed soy mix. No significant differences in percent FW or UFW among the soy bread variables were observed. Unpublished.

Figure 24.6 Photo of control soy bread and SABs. Photos of control soy bread (A), untreated SAB (B), steamed SAB (C), fermented SAB (D), and roasted SAB (E). The photographs depict the bread quality (slice height and crumb morphology) and color of the soy and SAB variables. Unpublished.

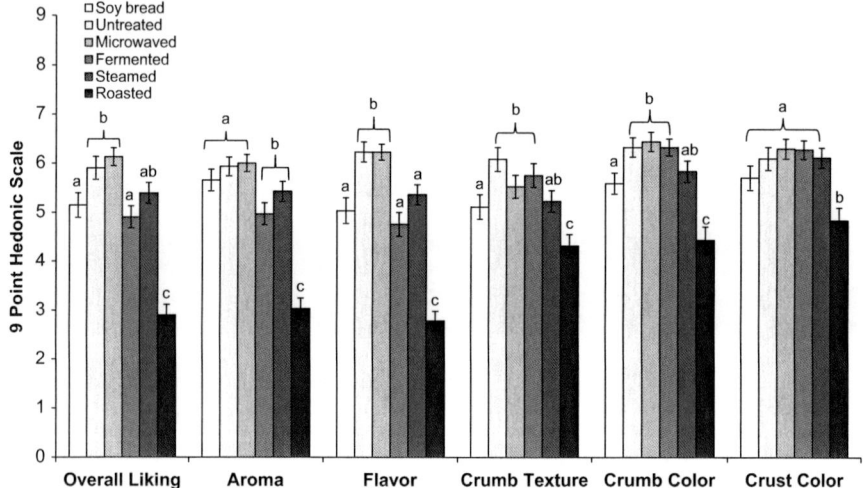

Figure 24.7 Consumer acceptance scores of soy bread and SAB among five attributes evaluated. Consumer acceptance of soy bread (control) and SAB with processed soy ingredients, $n = 60$. Nine-point hedonic scale, dislike extremely $= 1$ to like extremely $= 9$. Significant difference represented in samples with differing letters; ANOVA, Tukey's ($P \leq 0.05$). Unpublished.

selection of a palatable study product hence critical for study compliance in clinical trials and market success of the functional food (Figure 24.1).

24.3.4.1 Consumer Acceptance

The most acceptable soy breads were untreated SAB and microwaved SAB, scoring above 6 (like slightly on a 9 point hedonic scale where $1 =$ dislike extremely to $9 =$ like extremely) for all six parameters evaluated (Figure 24.7). From the JAR score, these breads had the greatest percentage (greater than 80% of the panelist) JAR responses for flavor, texture, and color of the soy breads. However, the least desirable was the roasted SAB. Roasted SAB was rated unacceptable because panelists found that there was much too much bitter flavor (48%), yeast flavor (38%), bean-like flavor (53%), dark brown crumb color (66%), and crust color (54%). The roasted SAB provided the greatest degree of glucoside into aglycone conversion but because of its poor consumer acceptability, it was not selected as a possible candidate for clinical trials. In the remaining SAB and SB, the control SB, fermented, and steamed had similar acceptability scores for overall liking, texture, and color; however, fermented SAB had the lowest acceptability for aroma, second to roasted SAB. The JAR scores indicated that the fermented SAB deviated from the ideal by having too many yeast-like aromas (33%). SB and steamed SAB did not have any significant differences in any of the parameters of acceptability. Therefore these two soy breads were selected as final candidates for clinical trials because they deliver equivalent amounts of total isoflavones with significantly different chemical compositions. With regards to soy bread as a convenient study food, 76% of the 60 panelists preferred soy bread over conventional Asian soy foods and 59% preferred soy bread over soy beverage. Moreover, 72% of panelists could consume two slices/day of control soy bread for 1 month.

24.3.4.2 Descriptive Analysis

Although consumer acceptability provides important information on product palatability, assessing the intensity of specific attributes provides information on how ingredients or processing differences affect product characteristics and useful in identifying key sensory attributes that drive product acceptance. Quantities of soy necessary to achieve levels appropriate for clinical trials resulted in firmer crumb texture, emergence of musty bean flavor, yellow crumb and darker crust colors. Combined, these attributes have detrimental effects towards the overall quality and palatability of the soy bread (Brewer et al. 1992).

The values of firmness assigned by the panelists were analogous to findings obtained from the Instron values for hardness. Roasted SAB had greatest intensity of firmness (8), whereas fermented SAB had the least firm texture (5) among the six breads (Figure 24.8). The hardest texture was associated with the bread with the low specific loaf volume and high WHC, whereas the softer bread had the greatest specific loaf volume and lower WHC. Among SAB, the intensity of firmness correlated strongly with the Instron values (Pearson, $R = 0.91$). Bean-like flavors are associated with poor bread quality; fermented SAB had the least bean-like flavor, whereas roasted SAB had the greatest. When these observations were coupled with the JAR scale evaluation of yeast aromas, fermented SAB had too many yeast aromas. Yeast, along with other bread additives such as ascorbic acid, have been found to mask bean flavors (Shogren et al. 2003), therefore the yeast aroma may have masked the flavor of bean-like flavors in the fermented SAB. Conversely, roasting the soy mixture might have facilitated the oxidation of soy lipids known to accentuate bean-like flavors (Liu and Limpert 2005). Because of the high saturation of brown color with the soy bread crust, panelists, like the colorimeter, were not able to

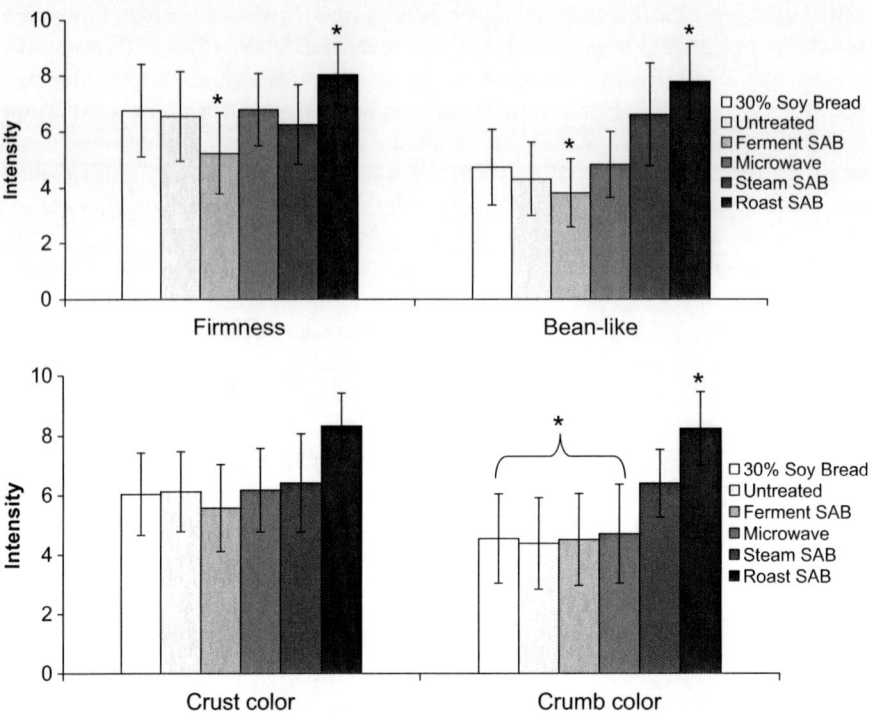

Figure 24.8 Crust and crumb color evaluated by descriptive analysis in soy bread and SAB. Histograms of SB and SAB crust and crumb color, $n = 60$. Attribute sensory intensity scores were none = 0 to very intense = 10. *Significantly different samples, one-way ANOVA, Tukey's ($P \leq 0.05$). Unpublished.

discriminate differences in brown intensity in the soy bread variables. However, panelists were able to distinguish differences in soy bread crumb. Significant differences were found between roasted and the SB and untreated, microwaved, and fermented SAB. Roasted SAB was the darkest followed by steamed SAB. The panelist scores for brown intensity had a strong positive correlation between a* ($R = 0.99$), chroma ($R = 0.96$), and brownness index ($R = 0.96$) and a negative correlation was found with L* ($R = -0.97$) and hue angle ($R = -0.99$). Color intensity was an important indicator of the crumb color acceptability ($R = -0.96$) and overall acceptability ($R = -0.88$). Darker breads were less acceptable than lighter breads.

24.4 Conclusions

The approach outlined in this Chapter describes a systematic method for development of functional foods for use in human clinical trials. An emphasis is placed on the vehicle needing to be palatable, and convenient to consume, with the phytochemicals remaining stable under domestic storage conditions.

Moreover, because dietary intervention studies are typically longer in duration than studies that evaluate therapeutic agents, compliance has a very critical role in directing clinical outcomes. From the processes described in this Chapter, a palatable and convenient glycoside-rich (SB) and an aglycone-rich SAB has been developed for use in clinical trials designed to assess the effects of dietary intervention on isoflavone bioavailability. Hypothesis-driven human intervention studies are necessary for evaluating the effectiveness of the food delivery system. The results presented in these series of experiments have demonstrated that isoflavone composition can indeed be manipulated within a food system. The food delivery system is capable of delivering the appropriate dose and has been designed to be convenient for use in clinical trials.

Summary Points

- Key variables for optimizing isoflavone delivery are identified throughout the process of development of the food vehicle and selection for clinical trials.
- There is ongoing discussion regarding the bioavailability being preferential in the aglycone form or the glucoside form.
- Although isoflavones in native soybeans exist predominantly in their conjugated 6″-O-malonyl-β-glucoside form, food processing methods such as fermentation combined with β-glucosidase from soy or almonds convert glucosides into their respective aglycone form.
- A critical impediment to creating an optimal delivery system for isoflavones is to include substantial amounts of soy ingredients appropriate for clinical trials without compromising the quality of the soy food product.
- Instrumental analysis is essential to monitor isoflavone stability and quality in the food product during the development of the delivery vehicle.
- Sensory evaluation, which includes consumer acceptance and descriptive analysis, provides an opportunity to assess whether the delivery vehicle is appropriate for use in clinical trials or needs reformulation/redesigning.
- A successful delivery vehicle is one that delivers the intended dose (phytochemical stability during production and storage), is palatable and convenient to use for the study participant, and is designed to address specific questions of the clinical trial.

Key Facts

Key Facts for Isoflavone Conversion by β-Glucosidase Enzyme

- Isoflavones in native soybeans exist predominantly conjugated to β-glucoside.
- β-Glucosidase is an enzyme that facilitates the cleavage of the β-glucosides from the isoflavones, thereby transforming the isoflavone from their glucoside form into their aglycone form.

- β-Glucosidase enzymes are found abundantly in sweet almond (*Prunus dulcis*) and to a lesser extent in soy bread ingredients such as yeast, soy, and wheat flour.
- The bread-making process, particularly one involving a sponge, provides a prolonged fermentation period that supports β-glucosidase activity.
- The co-fermentation of the almonds and soy bread ingredients during the production of soy bread facilitates rapid isoflavone conversion (over 2 h) into the aglycone at the levels achieved in this Chapter.

Key Facts for Consumer Acceptance Testing

- Developed by Peryam and colleagues in the 1950s, the hedonic scale is commonly used to quantify consumer opinion of food products.
- The information obtained can be easily misused or misinterpreted, therefore sensory tests rely on a careful study design and scientifically sound methods to remove biasness during consumer testing.
- Sensory scientist rely on many strategies to remove bias such as blinding the test foods with numerical codes, randomizing the presentation order of food samples, controlling physical properties of the study food (temperature, uniform shape and size), and of the environment (red lighting).
- The environment of the sensory testing facility is under constant, controlled conditions where distractions from noise, poor lighting, ambient temperature, and odors are minimized, since these factors can alter their judgment and impact the outcome of selected candidates.
- This is typically used at the final phase of food product development for selection to market or for reformulation.

Key Facts of Physical Properties of Functional Food

- Macronutrients and micronutrients of foods exist in various physical states such as a solid (amorphous and crystalline), liquid or gas.
- The physical state of food molecules is an important factor in the formulation and delivery of food phytochemicals.
- The physical state of phytochemicals within the food system, as well as the overall food matrix, can affect the solubility of the phytochemical and the rate of their dissolution.
- Physical states can be manipulated to facilitate delivery of phytochemicals to biological targets or, conversely, the physical states themselves can deter phytochemical delivery to biological targets.

Definitions of Words and Terms

Chewiness: A calculated measure of food texture which is derived by multiplying gumminess by springiness.

Descriptive analysis: Carefully calibrated human senses are used to quantify the intensity of various food attributes. These tests are important to distinguish the effect of formulation and food processing on food products.

Differential scanning calorimetry: An instrument used in thermal analysis of food materials to identify and monitor phase transitions.

β-Glucosidases: Enzymes that facilitate the cleavage of the β_1–4 glycosidic ester linkage of a sugar to either produce smaller sugar monomers or an aglycone form of a glycosylated bioactive compound. These enzymes are found in pro- and eukaryotes and are essential for nutrient utilization and play a critical role in energy utilization.

Hardness: A measure of force (Newton) that is needed to compress a one inch cube of material. It is the maximum force achieved during the first compression.

Hedonic scale: A testing method commonly used in food science and market research to quantify the "liking" of a given product. The classic hedonic scale involves 9 points of liking, but modified 7 and 5 point hedonic scales are also used.

Just-about-right (JAR) scale: A tool to measure consumer expectations of specific food attributes by quantifying if the intensity of the attribute is appropriate or needs improvement to achieve an ideal level for a specific product.

Springiness: A calculated measure of the material recovery after the two compression cycles of texture profile analysis. It is the ratio of the distance traveled for first compression cycle to the distance traveled during the second compression cycle.

Sponge: Bread ingredients undergo a prolonged fermentation period before it is incorporated into the finished dough.

Texture profile analysis: A tensile test where the food material is subjected to two cycles of deformation (compression) and information regarding the response of the material to the deformation is analyzed.

Thermogravimetric analysis: Thermal analysis quantifies the water content of food material using the difference in mass before and after heating.

List of Abbreviations

ACN	acetonitrile
AE	aglycone equivalent
FW	freezable water
HPLC	high-performance liquid chromatography
JAR scale	just-about-right scale
SAB	soy-almond bread
SB	soy bread (control)
UFW	un-freezable water
WHC	water holding capacity

References

Achouri, A., Boye, I.J., and Belanger, D., 2005. Soybean isoflavones: efficacy of extraction conditions and effect of food type on extractability. *Food Research International.* 38: 1199–1204.

Brewer, M.S., Potter, S.M., Sprouls, G., and Reinhard, M., 1992. Effect of soy protein isolate and soy fiber on color, physical and sensory characteristics of baked products. *Journal of Food Quality.* 15: 245–262.

Coward, L., Smith, M., Kirk, M., and Barnes, S., 1998. Chemical modification of isoflavones in soyfoods during cooking and processing. *The American Journal of Clinical Nutrition.* 68: 1486S–1491S.

Dhingra, S., and Jood, S., 2002. Organoleptic and nutritional evaluation of wheat breads supplemented with soybean and barley flour. *Food Chemistry.* 77: 479–488.

Dove, W.F., 1946. Developing food acceptance research. *Science.* 103: 187–190.

Faughnan, M.S., Hawdon, A., Ah-Singh, E., Brown, J., Millward, D.J., and Cassidy, A., 2004. Urinary isoflavone kinetics: the effect of age, gender, food matrix and chemical composition. *British Journal of Nutrition.* 91: 567–574.

Food and Drug Administration, 1999. Food Labeling: health claims; soy protein and coronary heart disease. *Federal Register.* 64: 57699–57733.

Hsieh, M.-C., and Graham, T.L., 2001. Partial purification and characterization of a soybean β-glucosidase with high specific activity towards isoflavone conjugates. *Phytochemistry.* 58: 995–1005.

Izumi, T., Piskula, M.K., Osawa, S., Obata, A., Tobe, K., Saito, M., Kataoka, S., Kubota, Y., and Kikuchi, M., 2000. Soy isoflavone aglycones are absorbed faster and in higher amounts than their glucosides in humans. *The Journal of Nutrition.* 130: 1695–1699.

Kano, M., Takayanagi, T., Harada, K., Sawada, S., and Ishikawa, F., 2006. Bioavailability of isoflavones after ingestion of soy beverages in healthy adults. *The Journal of Nutrition.* 136: 2291–2296.

Knorr, D., and Betschart, A.A., 1978. The relative effect of an inert substance and protein concentrations upon leaf volume of breads. *Lebensmittel-Wissenschaft und-Technologie.* 11: 198–204.

Lawless, H.T., and Heymann, H., 1998. *Sensory Evaluation of Food: Principles and Practices.* Chapman and Hall, New York, NY, USA.. 430–475.

Liu, K., and Limpert, W.F., 2005. Soy flour: varieties, processing, properties, and applications. In: Liu, K. (ed.) *Soybeans as Functional Foods and Ingredients.* AOCS Press, Champaign, IL, USA, pp. 101–121.

Lodi, A., and Vodovotz, Y., 2008. Physical properties and water state changes during storage in soy bread with and without almond. *Food Chemistry.* 110: 554–561.

Maskan, M., 2001. Kinetics of colour change of kiwifruits during hot air and microwave drying. *Journal of Food Engineering.* 48: 169–175.

Matsuura, N., Obata, A., and Fukushima, D., 1989. Objectionable flavor of soymilk developed during the soaking of soybeans and its control. *Journal of Food Science*. 54: 602–605.

Messina, M.J., 2010. A brief historical overview of the past two decades of soy and isoflavone research. *Journal of Nutrition*. 140: 1350S–1354S.

Murphy, P.A., Barua, K., and Hauck, C.C., 2002. Solvent extraction selection in the determination of isoflavones in soy foods. *Journal of Chromatography B. Analytical Technologies in the Biomedical and Life Sciences*. 777: 129–138.

Nilufer, D., Boyacioglu, D., and Vodovotz, Y., 2008. Functionality of soymilk powder and its components in fresh soy bread. *Journal of Food Science*. 73: 275–281.

Quinn, J.R., and Paton, D., 1979. A practical measurement of water hydration capacity of protein materials. *Cereal Chemistry*. 56: 38–40.

Reinwald, S., Akabas, S.R., and Weaver, C., 2010. Whole versus the piecemeal approach to evaluating soy. *The Journal of Nutrition*. 140: 2335S–2343S.

Setchell, K.D.R., Brown, N.M., Desai, P., Zimmer-Nechemias, L., Wolfe, B. E., Brashear, W.T., Kirschner, A.S., Cassidy, A., and Heubi, J.E., 2001. Bioavailability of pure isoflavones in healthy humans and analysis of commercial soy isoflavone supplements. *The Journal of Nutrition*. 131: 1362S–1375S.

Setchell, K.D., Brown, N.M., Zimmer-Nechemias, L., Brashear, W.T., Wolfe, B.E., Kirschner, A.S., and Heubi, J.E., 2002. Evidence for lack of absorption of soy isoflavone glycosides in humans, supporting the crucial role of intestinal metabolism for bioavailability. *The American Journal of Clinical Nutrition*. 76: 447–453.

Schiraldi, A., Piazza, L., and Riva, M., 1996. Bread staling: a calorimetric approach. *Cereal Chemistry*. 73: 32–39.

Schofield, J.D., Bottomley, R.C., Timms, M.F., and Booth, M.R., 1983. The effect of heat on wheat gluten and the involvement of sulphydryl–disulphide interchange reactions. *Journal of Cereal Science*. 1: 241–253.

Shogren, R.L., Mohamed, A.A., and Carriere, C.J., 2003. Sensory analysis of whole wheat/soy flour breads. *Journal of Food Science*. 68: 2141–2145.

Toda, T., Sakamoto, A., Takayanagi, T., and Yokotsuka, K., 2001. Changes in isoflavone compositions of soybeans during the soaking process. *Food Science Technology Research*. 6: 314–319.

Turner, N.J., Thomson, B.M., and Shaw, I.C., 2003. Bioactive isoflavones in functional foods: the importance of gut microflora on bioavailability. *Nutrition Reviews*. 61: 204–213.

Vittadini, E., and Vodovotz, Y., 2003. Changes in the physico-chemical properties of wheat and soy containing breads during storage as studied by thermal analyses. *Journal of Food Science*. 68: 2022–2027.

Vodovotz, Y., 2007. Soy enriched bread. In: Hamaker, B. (ed.) *Technology of Functional Cereal Products*. Woodhead Publishing Ltd., Cambridge, UK, pp. 388–408.

Vodovotz, Y., and Ballard, C., 2010. Composition and processes for making high soy protein containing bakery products. Patent #7,592,028.

Yan, L., and Spitznagel, E.L., 2009. Soy consumption and prostate cancer risk in men: a revisit of a meta-analysis. *The American Journal of Clinical Nutrition.* 89: 1155–1163.

Zhan, S., and Ho, S., 2005. Meta-analysis of the effects of soy protein containing isoflavones on the lipid profile. *The American Journal of Clinical Nutrition.* 81: 397–408.

Zhang, Y.C., Albrecht, D., Bomser, J., Schwartz, S.J., and Vodovotz, Y., 2003. Isoflavone profile and biological activity of soy bread. *Journal of Agricultural Food Chemistry.* 51: 7611–7616.

CHAPTER 25
Isoflavones and Thyroid Function: An Overview

FRANCESCO SQUADRITO* AND ALESSANDRA BITTO

Department of Clinical and Experimental Medicine and Pharmacology, University of Messina, Messina, Italy
*Email: Francesco.Squadrito@unime.it

25.1 Introduction

The thyroid is a gland located in the lower part of the neck enfolded around the front of the trachea. As it primary function the thyroid gland produces thyroid hormones, which primarily function to regulate basic metabolic rate. The thyroid gland produces three hormones: calcitonin, which is involved in calcium regulation, and 3,5,3′-tri-iodothyroine (T3) and 3,5,3′,5′-tetra-iodothyronine (T4, thyroxine). Iodine is an essential component of both T4 and T3, contributing 65% and 59% of their molecular weights, respectively.

In physiological conditions the human thyroid weighs between 15 and 20 g. However, in response to prolonged iodine deficiency, the thyroid gland can increase approximately 5-fold to the size of an apple, a condition recognized as goiter. If serum iodine levels are low or the thyroid is not able to incorporate enough iodine, the anterior pituitary secretes a hormone called thyroid stimulating hormone (TSH). When in excess, the secretion of TSH compensates by increasing the size of the thyroid and thereby allowing the gland to work more efficiently. For this reason, the primary cause of goiter and hypothyroidism around the world is inadequate iodine intake. The adult iodine recommended daily dose (RDA) is 150 µg, so one teaspoon (approximately

4–5 g) of iodine is all a person requires in a lifetime. However, approximately one billion people are considered to be iodine deficient worldwide. Iodine deficiency is a huge problem in many world regions, especially in the Himalayas, the Andes, the South of Italy, and in the central part of Africa. In the US approximately 50% of the population uses iodized salt, which is why the iodine intake of the US population is considered adequate. Some subsets of the population, such as women of child-bearing age, still have marginal iodine intakes, however, and thyroid problems can also develop in populations consuming adequate amounts of iodine. Thyroid abnormalities in these situations are often due to the presence of dietary goitrogens, such as glucosinolates in cruciferous vegetables, which interfere with utilization of iodine or functioning of the thyroid gland. This could be the case of flavonoids which are abundant in green vegetables and seasonal fruits and, due to their widespread occurrence, are regularly ingested by humans. Flavonoids are polyphenolic compounds that are ubiquitous in nature and are categorized, according to their chemical structure, into flavonols, flavones, flavanones, isoflavones, catechins, anthocyanidins and chalcones. Isoflavones are a group of flavonoids, which can act as antioxidants, due to the ability of trap a singlet oxygen, and as estrogens with protective functions.

Isoflavones are transformed by bacteria in the intestinal flora during digestion. It is only once this transformation has been completed that the isoflavones can be absorbed and exert their beneficial effects in the body. The individual intake of flavonoids, however, varies considerably depending on the type of diet consumed. Asian individuals have based their dietary style on rice, vegetables and beans for centuries, whereas Western countries have a greater intake of cereals, meat and derivatives; this explains the differences in flavonoid plasma levels detected in different populations.

There are considerable regional differences in soy protein and isoflavone consumption, with Asian populations consuming an estimated 25–50 mg of isoflavone aglycone equivalents daily, with 10% of the population consuming more than 100 mg isoflavones daily. Isoflavone intake in the US is several fold lower than in Asia, with intake estimates ranging from 0.15–3 mg per day. In addition to dietary patterns, there is currently an extensive range of flavonoid supplements on the market. Suppliers of such supplements recommend daily flavonoid intakes in amounts that are many times higher than those doses which can normally be achieved from a flavonoid-rich diet. Another example is the isoflavone-based nutraceuticals (*e.g.* pills, tablets, extracts), which are one of the most widely tested and used flavonoid supplements so far. The declared content of isoflavones is extremely variable, ranging from 50 to 500 mg, and different daily doses are recommended. However, at present, no specific dosage of isoflavones has been established to exert a beneficial effect.

As mentioned previously, isoflavones are scarce in Western diets compared with Asian diets, because the main isoflavone sources are soy-derived products. The question arises whether supplements containing such supra-physiological flavonoid levels may induce adverse events. In addition, it is likely that a large proportion of individuals taking dietary flavonoid supplements are also taking conventional drugs. The concomitant intake of "supra-nutritional" flavonoid doses together with conventional drugs may lead to flavonoid–drug interactions. This, additionally, raises concerns about the safe use of dietary flavonoids.

25.1.1 Goitrogenic Effects of Flavonoids

As for other naturally derived substances, flavonoids may exhibit anti-thyroid and goitrogenic activity. Excessive soy intake has been reported as a risk factor for the development of goiter, in both iodine-deficient rodents and infants fed soy-flour-based formula without iodine fortification. Animals fed soy diet require almost twice as much iodine compared with animals not fed soy. Indeed, several experimental studies demonstrate that isoflavones can inhibit the catalytic activity of thyroid peroxidase (TPO). Inhibition of TPO-catalyzed reactions results in decreased levels of circulating thyroid hormones, which lead to increased secretion of TSH by the anterior pituitary. The increased levels of TSH provide a growth stimulus to the thyroid, and it has been proposed that a prolonged stimulus can select for clones of follicular cells with the potential for transformation.

TPO is the heme-containing enzyme found on the apical membrane of thyroid follicular cells that catalyze the two reactions required for thyroid hormone synthesis: iodination of tyrosyl residues on thyroglubulin and subsequent oxidative coupling to yield T3 and T4. Ferreira *et al.* (2002) studied the *in vitro* effects of various flavonoids on thyroid type 1 iodothyronine deiodinase activity in a murine thyroid microsome fraction and found that type 1 iodothyronine deiodinase activity was significantly inhibited by isoflavones, quercetin and catechins.

In a recent study in rats (Chandra and De 2010), a decreased activity of TPO and 5′-deiodinase was reported in response to dietary green tea extracts. Serum T3 and T4 levels were found to be significantly reduced and associated with a significant increase of serum TSH levels. The authors concluded that green tea extracts at high doses could impair thyroid function.

Furthermore, isoflavones have been reported to inhibit thyroid hormone biosynthesis and may exert, at high concentrations, goitrogenic effects in humans. A significant correlation between circulating isoflavone concentrations in blood and thyroid function was found by Milerová and co-workers (2006) in a subgroup of 36 children who ate soy food in the previous 24 h. By contrast, a recent 3-year study by our group involving almost 400 osteopenic, postmenopausal women found that intake of the aglycone genistein does not significantly increase the risk of clinical or subclinical hypothyroidism at a dose of 54 mg day^{-1} (Bitto *et al.* 2010).

25.2 The Case of Genistein

Genistein (4′,5,7-trihydroxyflavone) is a phytoestrogen belonging to the class of soy isoflavones. Among soy isoflavones, genistein is the most widely studied in experimental and clinical studies. The aglycone genistein is an isoflavone found in low concentrations in soybeans and elevated amounts in certain soy-derived food, whereas genistin, the glucoside form of the aglycone genistein, is much more abundant in the unprocessed soybean. Genistein structurally closely resembles 17β-estradiol and it can directly bind to estrogen receptors (ERs) with a stronger affinity for ERβ (Kuiper *et al.* 1998). Acting as a natural selective estrogen receptor modulator (SERM), genistein exerts its estrogen agonist or antagonist action in a tissue- and dose-dependent manner (Setchell 2001). Moreover, the aglycone genistein shows anti-neoplastic effects in several *in vitro* and *in vivo* studies *via*: (a) modulation of cell growth and proliferation throughout tyrosine kinases and topoisomerase II inhibition, (b) stimulation of the immune system, (c) anti-angiogenic effects, and (d) potent antioxidant capacity (Polkowski and Mazurek 2000). Additionally, the anti-cancer action of genistein may be likely due to an epigenetic modulation of gene transcription activity, such as DNA methylation and/or chromatin modification (Li and Tollefsbol 2009). This is in agreement with epidemiological studies suggesting that high dietary intake of soy is associated with lower incidence rates of hormonally dependent and independent cancers (Nedrow *et al.* 2006; Setchell 1998).

The aglycone genistein has been previously shown, at a dose of 54 mg day^{-1}, to improve bone mineral density and markers of bone turnover without harmful estrogenic activity in the breast and uterus in postmenopausal women with low bone mass (Marini *et al.* 2007; 2008). In our studies, pure genistein administration avoided possible interferences with other isoflavones and resulted in documented and stable increases of the serum level of the isoflavone. The same population revealed a reduced exchange is sister chromatids, which associates with DNA stability; and a preserved expression of two breast cancer-associated genes, *BRCA1* and *BRCA2*, suggesting no proliferative effect on the mammary gland (Atteritano *et al.* 2008; Marini *et al.* 2008). The promising safety profile of the aglycone genistein may be a direct consequence of greater genistein affinity for the ERβ found to be particularly abundant in trabecular bone during the mineralization phase and endothelial tissue of arteries, whereas ERα exists at higher densities in the reproductive tissues (Nilsson *et al.* 2011).

25.2.1 Genistein and Thyroid Function

It is known that estrogens exert several effects on thyroid cells by modulation of cell cycle progression, proliferation and function, and then potentially contributing to the pathogenesis of thyroid cancer or thyroid hyperplasia. These effects are mediated by the binding to ERs, as well as through non-genomic distinct molecular pathways (Ben-Rafael *et al.* 1987; Santin and Furlanetto 2011). However, controversial data regarding the expression of ERs in normal and neoplastic human thyroid tissues, probably due to the methods chosen for detection of their expression, have been reported; moreover, an important concern regards the different action of ERα and ERβ in the neoplastic thyroid (Chen *et al.* 2008; Santin *et al.* 2011). Thyroid hormones regulate several metabolic processes and are crucial in normal growth and development and maturation of the central nervous system, the cardiovascular system and the skeleton in humans; additionally, an altered thyroid hormones synthesis and action is related to human health problems such as impaired neurocognitive function and an increased risk for cardiovascular diseases (Biondi and Cooper 2008).

Since 1930 researchers have examined the relationship between soybean intake and thyroid function (Divi *et al.* 1997; Doerge *et al.* 2002a; 2002b; Fitzpatrick 2000); for this purpose, many studies suggested that soy consumption and, specifically, soy isoflavones as genistein, exert goitrogenic action and alter thyroid function.

The goitrogenic effects of genistein seems to be exerted through a direct interaction of this isoflavone with key pathways involved in thyroid hormones synthesis, metabolism and thyroid hormone transport proteins (Radović *et al.* 2006). In the presence of iodide ions, genistein (and its parent compound daidzein) block TPO-catalyzed tyrosine iodination by acting as alternate substrates, yielding mono-, di- and tri-iodoisoflavones (Figure 25.1). Genistein also

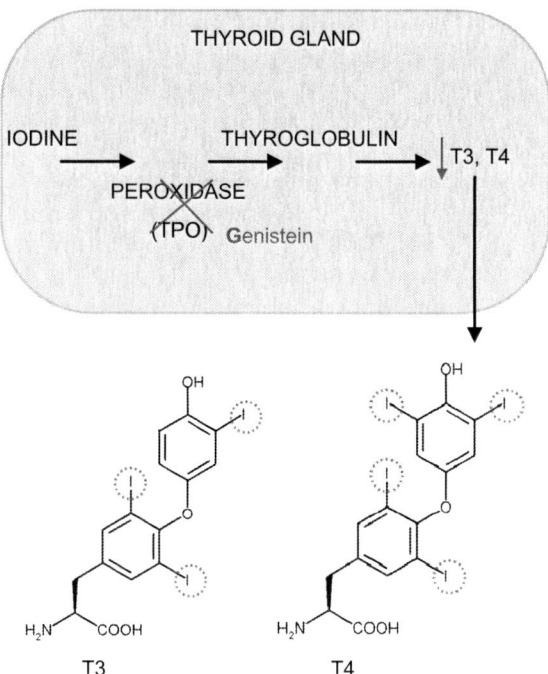

Figure 25.1 A schematic representing the effects of genistein on thyroid hormone production.

inhibits thyroxine synthesis using iodinated casein or human goiter thyroglobulin as substrates for the coupling reaction. In the absence of iodide, both genistein and daidzein are suicide substrates of the TPO enzyme, blocking both iodinating and coupling activities of this enzyme (Chang et al. 1996; Divi and Doerge 1996; Divi et al. 1997; Doerge et al. 2002b).

Moreover, genistein also affect the metabolism of thyroid hormones and iodide re-utilization by inhibition of sulfotransferase enzymes (Ebmeier and Anderson 2004). Indeed, TPO catalyzes the iodination of thyroglobulin and oxidative coupling of di-iodothyronine resulting in the formation of thyroxine (T4), all essential steps involved in thyroid hormones synthesis. Thus, inhibition of TPO leads to a reduction in thyroid hormones levels, with a subsequent increment of TSH release, which, in turn, provides a strong growth stimulus to the thyroid gland.

Divi and co-workers (1997) demonstrated that the TPO inhibitory properties of genistein can be evident by using different doses of genistein, indicating that this isoflavone inhibits TPO-catalyzed iodination of tyrosine with a half maximal inhibitory concentration (IC50) value of 7.6 µM. However, this effect is reversed in the presence of sufficient amount of iodide in the incubation mixtures. Similar inhibitory results were observed in rat microsomal TPO and in rats fed a diet fortified with different doses of genistein, whereas blood thyroid hormone and TSH levels were not affected (Chang and Doerge 2000). Genistein is also known as a potent inhibitor of tyrosine kinase in a dose-dependent manner (Ravindranath et al. 2004), and it is even capable of affecting thyroid hormone deiodination by type I 5′-deiodinase (Mori et al. 1996; White et al. 2004).

More recently, Sosić-Jurjević et al. (2010) showed that a subcutaneous injection of 10 mg kg^{-1} of genistein or daidzein can alter structure and function of the pituitary–thyroid axis and induce hypothyroidism in orchidectomized middle-aged rats. However, data presenting *in vitro* and *in vivo* animal studies are not easy to compare with studies involving humans, as demonstrated by a recent paper (Setchell et al. 2011).

In humans, early studies showed that feeding infants with soy milk caused goiter in those with inadequate iodine intake, even if this effect was reverted by iodine supplementation (Chorazy et al. 1995; Jabbar et al. 1997). A more recent human trial reports the effects of short-term soy eating on thyroid parameters in relation to isoflavone levels in male and female healthy subjects (Hampl et al. 2008). The serum levels of both of the isoflavones genistein and daidzein increased significantly after 7 days of soy consumption. Correlating the levels of isoflavones with thyroid parameters, the authors found a significant correlation only between basal levels of daidzein and thyrotropin, daidzein and anti-thyroglobulin at the end of soy consumption in males, and between daidzein and free thyroxine at the end of the soy ingestion in females, but no correlation between genistein and the above-considered thyroid parameters was found. This discrepancy is probably due to the fact that daidzein can be further converted into equol, which has greater biological activity than its parent compound. These results are in agreement with a previous research investigating whether serum levels of genistein and daidzein could influence thyroid hormone function in a population of 268 children. This study showed only a modest association between isoflavones serum levels and parameters of thyroid function (Milerová et al. 2006).

So far, it is well recognized that thyroid diseases are most common in women and especially during perimenopause and menopause, as consequence of an altered balance between estrogen and progesterone. Therefore, the effects of genistein on thyroid function were also analyzed in postmenopausal women. Results from a 3-month study in postmenopausal women consuming an isoflavone-rich diet (containing 58% of total genistein) showed no significant effect of isoflavones on thyroid hormone levels (Duncan et al. 1999).

Bruce et al. (2003) investigated thyroid function in 38 iodine-replete postmenopausal women at baseline and after 90 and 180 days following supplementation with 90 mg (aglycone weight) of total isoflavones day^{-1}. The results of this study found no difference in thyroid parameters between the placebo and treatment arms (Bruce et al. 2003).

Another study enrolled 77 postmenopausal women randomized to receive, for 16 weeks, cow's milk and a placebo supplement, soy milk and placebo supplement or cow's milk and isoflavone supplement (Ryan-Borchers, et al. 2008). In detail, the daily amount of genistein after ingestion of 706 ml of soy milk was approximately 37 mg day^{-1}, whereas the isoflavone content of the supplements was 15 mg of daidzein, 17 mg of genistein and 3.5 mg of glycitein. The results of this study are congruent with those of Bruce et al. (2003) and provide evidence that the levels of isoflavone intake does not adversely affect thyroid function and not influence cognitive functioning in healthy postmenopausal women, as indicated by serum TSH levels that remained within normal range following the intervention period in all women. A more recent clinical trial evaluated the effects of a 3-year administration of pure genistein aglycone (Figure 25.2) (54 mg day^{-1}) on thyroid-related markers, in postmenopausal women (Bitto et al. 2010). Specifically, changes in thyroid hormone receptors and thyroid hormone enzymes, blood levels of thyroid hormones and thyroid auto-antibodies were assessed. The results of this research showed that daily consumption of genistein aglycone did not modified circulating free T4 (fT4), free T3 (fT3) and TSH levels (Figure 25.3); furthermore, administration of the aglycone genistein over 3 years did not affect the enzymes involved in thyroid hormone production, the thyroid hormone auto-antibodies and the expression of thyroid hormone receptors, therefore confirming that genistein does not appear to alter thyroid function in postmenopausal women (Figure 25.4). Finally, the safety of genistein was also analyzed in males in a recent paper involving 47 men with localized prostate cancer (Lazarevic et al. 2011). Specifically, in this study, genistein was administered at a dose of 30 mg day^{-1} for 3 to 6 weeks prior to prostatectomy, and thyroid hormones levels were analyzed as secondary outcome. The results showed no significant effects on thyroid hormones.

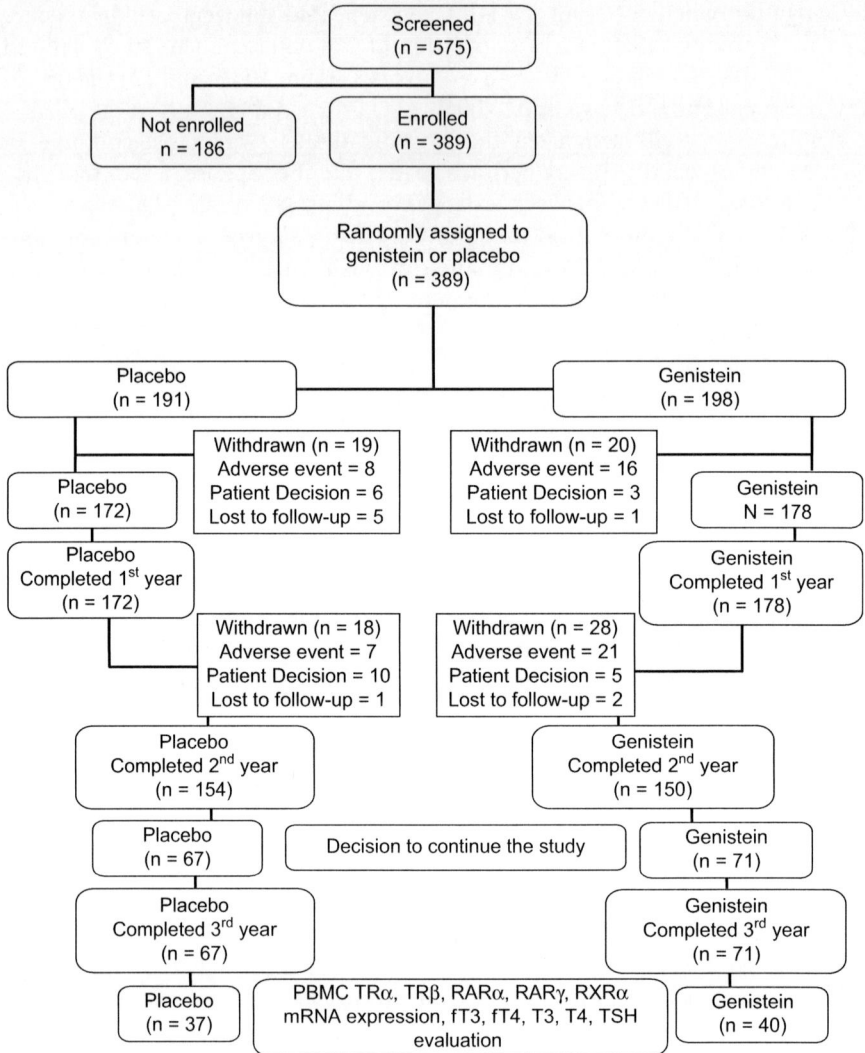

Figure 25.2 Flow chart of the randomized clinical trial carried out to investigate the effects of genistein on thyroid function.

25.3 Conclusions

Overall, there is a lack of information about the effect of isolated isoflavones, such as genistein, on thyroid safety in humans and the results of intervention trials are not easily comparable because the researchers have used: (i) mixed isoflavones or isoflavone and protein mixtures with different dosage regimens, soy foods or supplements as the active treatment; (ii) the quality and amount of genistein varied widely in all of these previous studies; and (iii) the trials were of different duration. Although the overall evidence suggests that the isoflavone

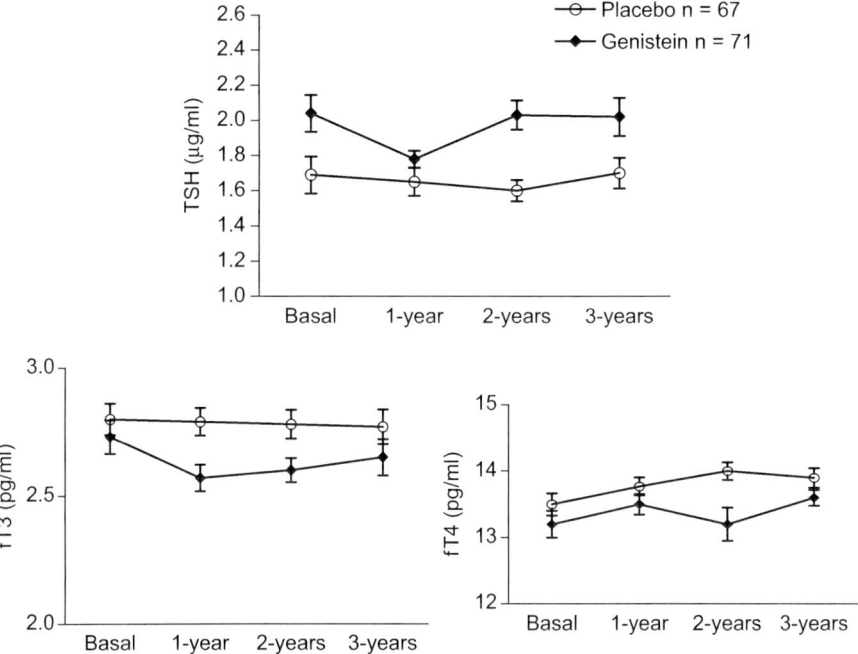

Figure 25.3 Effects of genistein or placebo administration on TSH, free T3 (fT3) and free T4 (fT4) over 3 years in postmenopausal women. TSH and thyroid hormones were measured with routine laboratory methods in serum samples. Data are expressed as means ± S.D.; no significant differences were observed between groups over time.

genistein do not adversely affect thyroid function in iodine-replete euthyroid individuals, further studies are warranted to better define the relationship between genistein and the thyroid gland.

Summary Points

- Isoflavones have been identified as being among the chemicals that could act as endocrine disruptors.
- Genistein has been shown to affect iodine incorporation into thyroid hormones.
- The available data are mainly related to *in vitro* or animal experiments using doses of isoflavones much higher than those normally found in humans.
- A randomized clinical trial showed that genistein administration for 3 years did not affect serum levels of thyroid hormones. In addition, there were no differences in mRNA expression of thyroid hormone receptors following 3 years of treatment.
- Taken together, these data suggest that isoflavones intake does not significantly increase the risk of clinical hypothyroidism.

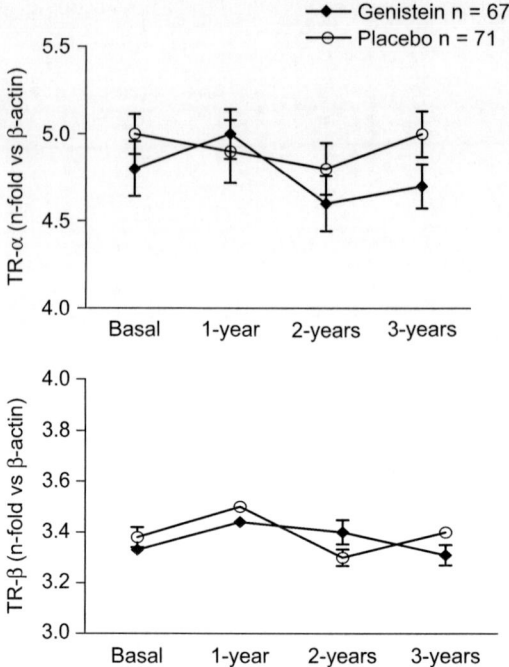

Figure 25.4 Effects of genistein or placebo administration on thyroid hormone receptor (TR-α and TR-β) mRNA expression over 3 years in postmenopausal women. mRNA was extracted from circulating white blood cells and quantified by quantitative real-time PCR. Data are expressed as means ± S.D.; no significant differences were observed between groups over time.

- Clearly, it is important for all individuals, regardless of their soy intake, to consume adequate iodine.

Key Facts

- Isoflavones with estrogen-like activities are used as ingredients in food supplements, dietary foods and medical foods for special medical purposes, advertised primarily to women in menopause.
- Isoflavones have received a great deal of attention over the last few years because of their potentially preventive roles against some of the most prevalent chronic diseases of the present day, namely cardiovascular disease, osteoporosis and hormone-related cancers (breast and prostate cancers).
- Despite the proposed benefits of isoflavones, health concerns regarding the potential harmful effects resulting from long-term use have been reported. One such concern is that isoflavones may adversely affect thyroid function.

- Hypothyroidism, which results when the thyroid gland releases insufficient amounts of thyroid hormones, has an enormous impact on heath. Because thyroid function tends to decline with age and is associated with hormonal transitions, the prevalence is greater in women than men, occurring in 6–10% of postmenopausal women.
- Much of the evidence for the anti-thyroid effects of isoflavones stems from early animal studies conducted in the 1930s in which rats fed soy-based, iodine-deficient diets developed goiters, which are enlargements of the thyroid gland resulted from thyroid hormone deficiency. However, iodine fortification of the diet was shown to ameliorate this problem.

Definitions of Words and Terms

Daidzein: This is the substituted form of genistein. It is a dihydroxyisoflavone, with two hydroxyl groups in the 4′ and 7 positions.

Endocrine disruptors: A class of substances that impair synthesis, secretion, transport, activity and/or elimination of hormones.

Equol: This is an isoflavandiol metabolized by bacterial flora from daidzein.

Genistein: This is an isoflavone which influences different biochemical functions in live cells. It possesses three hydroxilic groups in the 4′, 5 and 7 positions.

Goitrogenic: A substance that interfere with normal uptake of iodine in thyroid gland, causing reduced production of thyroid hormones and then the enlargement of thyroid gland (goitre).

Isoflavones: A class of organic compounds, related to the isoflavonoids. The main compounds (isoflavones) do not have nutritional interest, unlike the derivatives such as genistein and daidzein.

Thyroid hormones (T3 and T4): These are tyrosine-based hormones, produced by follicular cells of thyroid gland, which are primarily responsible for regulation of metabolism. Iodine is essential for synthesis of these two hormones.

Thyroid peroxidase (TPO): The enzyme responsible for the organification of iodine.

Thyroid stimulating hormone (TSH): A peptide hormone, synthesized and secreted by the anterior pituitary gland, which promotes, in turn, the secretion of thyroid hormones. It is controlled by a hypothalamic hormone named thyroid stimulating releasing hormone (TRH).

Tyrosine kinase: An enzyme that play a key role as an "on" or "off" switch in many cellular functions.

List of Abbreviations

BRCA   breast cancer-associated gene
ER   estrogen receptor

T3   3,5,3′-tri-iodothyroine
T4   3,5,3′,5′-tetra-iodothyronine or thyroxine
TPO   thyroid peroxidase
TSH   thyroid stimulating hormone

References

Atteritano, M., Pernice, F., Mazzaferro, S., Mantuano, S., Frisina, A., D'Anna, R., Cannata, M.L., Bitto, A., Squadrito, F., Frisina, N., and Buemi, M., 2008. Effects of phytoestrogen genistein on cytogenetic biomarkers in postmenopausal women: 1 year randomized, placebo-controlled study. *European Journal of Pharmacology*. 589: 22–26.

Ben-Rafael, Z., Struass, 3rd, J.F., Arendash-Durand, B., Mastroianni, Jr, L., and Flickinger, G.L., 1987. Changes in thyroid function tests and sex hormone binding globulin associated with treatment by gonadotropin. *Fertility and Sterility*. 48: 318–320.

Biondi, B., and Cooper, D.S., 2008. The clinical significance of subclinical thyroid dysfunction. *Endocrinology Review*. 29: 76–131.

Bitto, A., Polito, F., Atteritano, M., Altavilla, D., Mazzaferro, S., Marini, H., Adamo, E.B., D'Anna, R., Granese, R., Corrado, F., Russo, S., Minutoli, L., and Squadrito, F., 2010. Genistein aglycone does not affect thyroid function: results from a three-year, randomized, double-blind, placebo controlled trial. *Journal of Clinical Endocrinology and Metabolism*. 95: 3067–3072.

Bruce, B., Messina, M., and Spiller, G.A., 2003. Isoflavone supplements do not affect thyroid function in iodine-replete postmenopausal women. *Journal of Medical Food*. 6: 309–316.

Chandra, A.K., and De, N., 2010. Goitrogenic/antithyroidal potential of green tea extract in relation to catechin in rats. *Food and Chemical Toxicology*. 48: 2304–2311.

Chang, H.C., and Doerge, D.R., 2000. Dietary genistein inactivates rat thyroid peroxidase *in vivo* without an apparent hypothyroid effect. *Toxicol Appl Pharmacol*. 168(3): 244–252.

Chen, G.G., Vlantis, A.C., Zeng, Q., and van Hasselt, C.A., 2008. Regulation of cell growth by estrogen signaling and potential targets in thyroid cancer. *Current Cancer Drug Targets*. 8: 367–377.

Chorazy, P.A., Himelhoch, S., Hopwood, N.J., Greger, N.G., and Postellon, D.C., 1995. Persistent hypothyroidism in an infant receiving a soy formula: case report and review of the literature. *Pediatrics*. 96: 148–150.

Divi, R.L., and Doerge, D.R., 1996. Inhibition of thyroid peroxidase by dietary flavonoids. *Chemical Research Toxicology*. 9: 16–23.

Divi, R.L., Chang, H.C., and Doerge, D.R., 1997. Anti-thyroid isoflavones from soybean: isolation, characterization, and mechanisms of action. *Biochemical Pharmacology*. 54: 1087–1096.

Doerge, D.R., and Chang, H.C., 2002a. Inactivation of thyroid peroxidase by soy isoflavones, *in vitro* and *in vivo*. *Journal of Chromatography. B, Analytical Technologies in the Biomedical Life Sciences*. 777: 269–279.

Doerge, D.R., and Sheehan, D.M., 2002b. Goitrogenic and estrogenic activity of soy isoflavones. *Environmental Health Perspectives*. 110: 349–353.

Duncan, A.M., Underhill, K.E., Xu, X., Lavalleur, J., Phipps, W.R., and Kurzer, M.S., 1999. Modest hormonal effects of soy isoflavones in postmenopausal women. *Journal of Clinical Endocrinology and Metabolism*. 84: 3479–3484.

Ebmeier, C.C., and Anderson, R.J., 2004. Human thyroid phenol sulfotransferase enzymes 1A1 and 1A3: activities in normal and diseased thyroid glands, and inhibition by thyroid hormones and phytoestrogens. *Journal of Clinical Endocrinology and Metabolism*. 89: 5597–5605.

Ferreira, A.C., Lisboa, P.C., Oliveira, K.J., Lima, L.P., Barros, I.A., and Carvalho, D.P., 2002. Inhibition of thyroid type 1 deiodinase activity by flavonoids. *Food and Chemical Toxicology*. 40: 913–917.

Fitzpatrick, M., 2000. Soy formulas and the effects of isoflavones on the thyroid. *New Zealand Medical Journal*. 113: 24–26.

Hampl, R., Ostatnikova, D., Celec, P., Putz, Z., Lapcík, O., and Matucha, P., 2008. Short-term effect of soy consumption on thyroid hormone levels and correlation with phytoestrogen level in healthy subjects. *Endocrinology Regulation*. 42: 53–61.

Jabbar, M.A., Larrea, J., and Shaw, R.A., 1997. Abnormal thyroid function tests in infants with congenital hypothyroidism: the influence of soy-based formula. *Journal of American College of Nutritionists*. 16: 280–282.

Kuiper, G.G., Lemmen, J.G., Carlsson, B., Corton, J.C., Safe, S.H., van der Saag, P.T., van der Burg, B., and Gustafsson, J.A., 1998. Interaction of estrogenic chemicals and phytoestrogens with estrogen receptor β. *Endocrinology*. 139: 4252–4263.

Lazarevic, B., Boezelijn, G., Diep, L.M., Kvernrod, K., Ogren, O., Ramberg, H., Moen, A., Wessel, N., Berg, R.E., Egge-Jacobsen, W., Hammarstrom, C., Svindland, A., Kucuk, O., Saatcioglu, F., Taskèn, K.A., and Karlsen, S.J., 2011. Efficacy and safety of short-term genistein intervention in patients with localized prostate cancer prior to radical prostatectomy: a randomized, placebo controlled, double-blind phase 2 clinical trial. *Nutrition and Cancer*. 63: 889–898.

Li, Y., and Tollefsbol, T.O., 2010. Impact on DNA methylation in cancer prevention and therapy by bioactive dietary components. *Current Medicinal Chemistry*. 17: 2141–2151.

Marini, H., Minutoli, L., Polito, F., Bitto, A., Altavilla, D., Atteritano, M., Gaudio, A., Mazzaferro, S., Frisina, A., Frisina, N., Lubrano, C., Bonaiuto, M., D'Anna, R., Cannata, M.L., Corrado, F., Adamo, E.B., Wilson, S., and Squadrito, F., 2007. Effects of the phytoestrogen genistein on bone metabolism in osteopenic postmenopausal women: a randomized trial. *Annals of Internal Medicine*. 146: 839–847.

Marini, H., Bitto, A., Altavilla, D., Burnett, B.P., Polito, F., Di Stefano, V., Minutoli, L., Atteritano, M., Levy, R.M., D'Anna, R., Frisina, N., Mazzaferro, S., Cancellieri, F., Cannata, M.L., Corrado, F., Frisina, A., Adamo, V., Lubrano, C., Sansotta, C., Marini, R., Adamo, E.B., and Squadrito, F., 2008. Breast safety and efficacy of genistein aglycone for postmenopausal bone loss: a follow-up study. *Journal of Clinical Endocrinology Metabolism*. 93: 4787–4796.

Milerová, J., Cerovska, J., Zamrazil, V., Bilek, R., Lapcik, O., and Hampl, R., 2006. Actual levels of soy phytoestrogens in children correlate with thyroid laboratory parameters. *Clinical and Chemical Laboratory Medicine*. 44: 171–174.

Mori, K., Stone, S., Braverman, L.E., and Devito, W.J., 1996. Involvement of tyrosine phosphorylation in the regulation of 5′-deiodinases in FRTL-5 rat thyroid cells and rat astrocytes. *Endocrinology*. 137: 1313–1318.

Nedrow, A., Miller, J., Walker, M., Nygren, P., Huffman, L.H., and Nelson, D., 2006. Complementary and alternative therapies for the management of menopause-related symptoms: a systematic evidence review. *Archives of Internal Medicine*. 166: 1453–1465.

Nilsson, S., and Gustafsson, J.Å., 2011. Estrogen receptors: therapies targeted to receptor subtypes. *Clin Pharmacol Ther*. 89(1): 44–55.

Polkowski, K., and Mazurek, A.P., 2000. Biological properties of genistein. A review of *in vitro* and *in vivo* data. *Acta Polonica Pharmacologica*. 57: 135–155.

Radović, B., Mentrup, B., and Köhrle, J., 2006. Genistein and other soya isoflavones are potent ligands for transthyretin in serum and cerebrospinal fluid. *Br J Nutr*. 95(6): 1171–1176.

Ravindranath, M.H., Muthugounder, S., Presser, N., and Viswanathan, S., 2004. Anticancer therapeutic potential of soy isoflavone, genistein. *Advanced Experimental Medical Biology*. 546: 121–165.

Ryan-Borchers, T., Chew, B., Soon Park, J., McGuire, M., Fournier, L. and Beerman, K. Effects of dietary and supplemental forms of isoflavones on thyroid function in healthy postmenopausal women. *Top Clin Nutr* 2008; 23 (1): 13–22.

Santin, A.P., and Furlanetto, T.W., 2011. Role of estrogen in thyroid function and growth 268 regulation. *Journal of Thyroid Research*. 2011: 875125, doi:10.4061/2011/875125.

Setchell, K.D., 1998. Phytoestrogens: the biochemistry, physiology, and implications for human health of soy isoflavones. *American Journal of Clinical Nutrition*. 68: 1333S–1346S.

Setchell, K.D., 2001. Soy isoflavones – benefits and risks from nature's selective estrogen receptor modulators (SERMs). *Journal of the American College of Nutritionist*. 20: 354S–362S.

Setchell, K.D., Brown, N.M., Zhao, X., Lindley, S.L., Heubi, J.E., King, E.C., and Messina, M.J., 2011. Soy isoflavone phase II metabolism differs between rodents and humans: implications for the effect on breast cancer risk. *American Journal of Clinical Nutrition*. 94: 1284–1294.

Sosić-Jurjević, B., Filipović, B., Ajdzanović, V., Savin, S., Nestorović, N., Milosević, V., and Sekulić, M., 2010. Suppressive effects of genistein and daidzein on pituitary-thyroid axis in orchidectomized middle-aged rats. *Experimental Biology and Medicine*. 235: 590–598.

White, H.L., Freeman, L.M., Mahony, O., Graham, P.A., Hao, Q., and Court, M.H., 2004. Effect of dietary soy on serum thyroid hormone concentrations in healthy adult cats. *American Journal of Veterinary Research*. 65: 586–591.

CHAPTER 26
Isoflavones against Gastric Cancer: Function and Effects

SUE K. PARK*[a,b,c] AND KWANG-PIL KO[d]

[a] Department of Preventive Medicine, Seoul National University College of Medicine, Seoul, Korea; [b] Department of Biomedical Science, Seoul National University Graduate School, Seoul, Korea; [c] Cancer Research Institute, Seoul National University, Seoul, Korea; [d] Department of Preventive Medicine, Gachon University, Incheon, Korea
*Email: suepark@snu.ac.kr

26.1 Introduction

Soybean and soybean products are gaining public interest due to their positive health benefits in combating many diseases. These benefits are due to the existence of isoflavones in soybeans which have a chemical structure similar to estrogen. Isoflavones are types of phytoestrogens that act as estrogen agonists or antagonists, depending on endocrine estrogenic levels. Soybeans and their products are the richest sources of isoflavones which competitively bind to estrogen receptors (ERs), but have a weak estrogenic effect of nearly 0.1% of that of estradiol. Isoflavone has been observed to produce an anti-estrogenic effect by blocking estradiol and other estrogens from ER binding. In this manner, isoflavone inhibits the growth and proliferation of estrogen-dependent cancers, such as breast cancer. Isoflavones function as estrogen agonists in the cardiovascular system by binding to ER-β at low endocrine estrogen levels, and they can directly relax vessels, possibly by enhancing the promotion of prostacyclin release, and indirectly reducing plaques in vessels by inhibiting collagen-induced aggregation (Cano *et al.* 2010). Isoflavones exhibit estrogen-like effect on bone by binding to ER-β. They are expected to benefit bone health by impeding the effects of osteoporosis by stimulating bone formation and inhibiting bone reabsorption. In postmenopausal syndrome accompanied by hot flushes and night sweats, the prescription of isoflavone can decrease menopause symptoms by inducing an estrogen-like effect (Cassidy et al. 1994; Setchell 1998).

Typically, the stomach has direct contact with food several times each day. Masticated food enters the stomach through the esophagus, and then the stomach releases hydrochloric acid and digestive enzymes such as proteases. Food is churned by the stomach and the bolus (food mass) is converted into partially digested food which slowly passes into the duodenum. Depending on the quantity and contents of the meal, the stomach digests the food into chyme from within 40 min to a few hours. Because the stomach functions in food processing as a reservoir, in digestion and absorption, foods and dietary habit both have a significant influence on stomach health.

Based on the results of various experimental and epidemiologic studies, it is evident that gastric cancer is associated with dietary factors. According to the panel, it is probable that salted or salty foods can increase the risk of gastric cancer, whereas non-starchy vegetables, alliums, and fruits can decrease the risk of gastric cancer. Also, there is evidence that soybean and its products may prevent gastric cancer, but the level of liability is only 'limited-suggestive' which is lower than 'probable' (WCRF Project 2007).

However, there is no obvious evidence that gastric cancer is estrogen-dependent, and the association between gastric cancer and isoflavone remains incompletely understood. This Chapter introduces *in vitro*, *in vivo*, and epidemiological studies conducted specifically to examine possible biological mechanisms associated with the beneficial effects of isoflavone on gastric cancer in humans.

26.2 Absorption, Metabolism, and Excretion

After ingestion, isoflavones, which are conjugated glycosides (genistin, daidzin, and glycitin), are hydrolyzed in the intestinal wall by glucosidases produced either by the body's intestinal enzymes or those from intestinal bacteria. Hydrolysis converts conjugated isoflavones into bioactive aglycones such as daidzein, genistein, and glycitein (Cederroth and Nef 2009).

Genistein and daidzein are the major isoflavones found in soybean and can be produced from their glucosides or from their precursors, biochanin A and formononetin, by the action of intestinal glucosidases. Genistein and daidzein undergo extensive metabolism in the gut and liver. Daidzein undergoes conjugation with glucose to form either the 7-O-glucoside daidzin (in soybean) or the 8-C-glucoside puerarin (Barnes 2010; Kulling et al. 2002; Mortensen et al. 2009). Daidzein is metabolized to dihydrodaidzein, which is further metabolized to desmethylangolensin and equol. Genistein is metabolized by gut bacteria to dihydrogenistein, which is further metabolized to p-ethyl-phenol. Glycitin undergoes minimal metabolic action prior to excretion due to it being resistant to enzymatic hydrolysis by gut bacteria (Barnes 2010; Cederroth and Nef 2009).

Bacterial actions are the main contributors to the metabolism of isoflavones. The conjugated isoflavones undergo enterohepatic circulation, are deconjugated within intestinal bacteria and are then reabsorbed (Barnes 2010) (Figure 26.1). In addition to bacterial metabolism, isoflavones are metabolized by phase I and II isoenzymes in the liver. Genistein and daidzein undergo hydroxylation catalyzed by phase I enzymes (cytochrome P450) and glycetin is metabolized to mono- or dihydrozylated glycetin metabolites (Mortensen et al. 2009), whereas metabolites conjugated by phase II enzymes, such as uridine 5′-diphospho-glucuronosyl transferase and sulfotransferase, are synthesized (Mortensen et al. 2009).

Isoflavone aglycones are passively absorbed in the upper small intestine from blood (Barnes 2010). To date, there is little evidence that isoflavones are actively absorbed in the intestine (Mortensen et al. 2009). Isoflavones are excreted, primarily through urine, by a mechanism of glucuronidation and sulfation (5–35%). A few isoflavones are excreted *via* feces (nearly 1–4%).

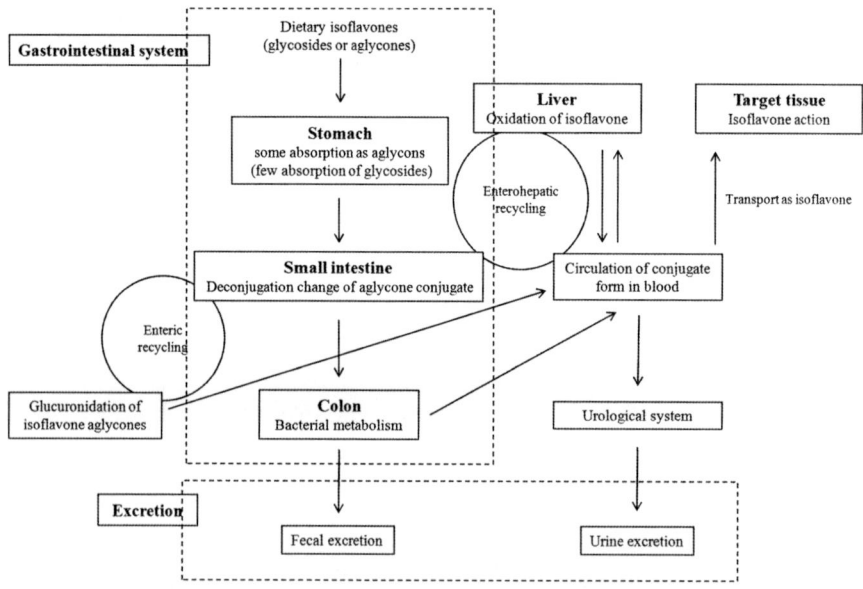

Figure 26.1 Biological pathways of isoflavones in absorption, circulation, metabolism, and excretion. After ingestion, conjugated isoflavones are converted into bioactive aglycones by hydrolysis in the intestinal wall. The conjugated isoflavones undergo enterohepatic circulation, are deconjugated within intestinal bacteria, and are then reabsorbed. In addition to bacterial metabolism, isoflavones are metabolized in the liver. Isoflavone aglycones are passively absorbed in the upper small intestine from blood. Isoflaveones are excreted primarily through urine and some isoflavones are excreted *via* feces.
Modified from Kulling *et al.* (2002), Heinonen (2006), and Barnes (2010).

26.3 Function and Effect of Isoflavones against Gastric Cancer

The general hypothesis describing the development of gastric carcinogenesis is summarized in Figure 26.2. When mild superficial gastritis persists, the normal gastric mucosa becomes atrophic as a result of continuous erosion. After passing through further steps of intestinal cell metaplasia and dysplasia, gastric cancer develops (Correa 1992; Sugiyama and Asaka 2004). In addition, *Helicobacter pylori* infection and lifestyle choices, such as cigarette smoking and high salt food intake, seem to be important contributing factors to development of gastric cancer (Hamajima *et al.* 2006). However, *H. pylori* infection is the primary cause of superficial gastritis and gastric atrophy. In the process of atrophy in the mucosa, apoptosis and cell proliferation accelerate. Vegetable and fruit intake are often identified as foods which prevent gastric cancer. Isoflavones in soybean is also expected to possess anticancer mechanisms.

26.3.1 Anti-bacterial Effects

Gastric cancer is recognized as an infectious disease due to the fact that *H. pylori* infection is a significant risk factor for development of gastric

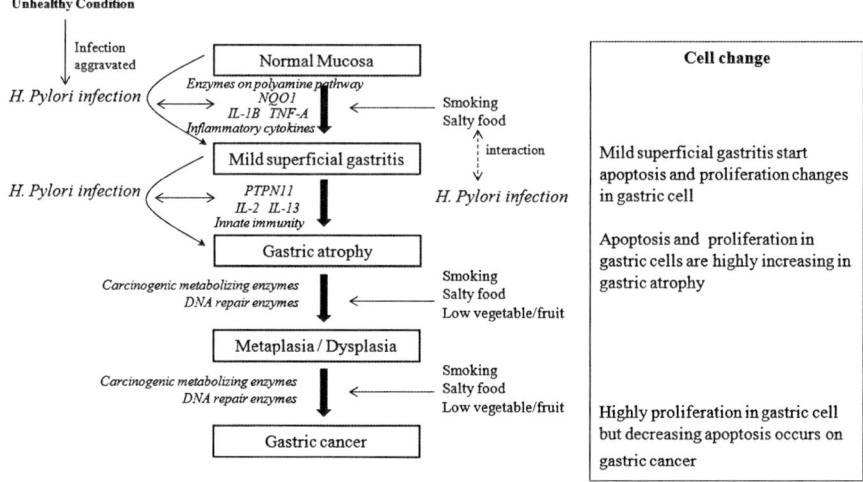

Figure 26.2 Multiple carcinogenetic steps for the development of gastric cancer induced by *H. pylori*. *H. pylori* infection is the primary cause of superficial gastritis and gastric atrophy. When mild superficial gastritis persists, normal gastric mucosa becomes atrophic as a result of continuous erosion. After passing through further steps of intestinal cell metaplasia and dysplasia, gastric cancer develops. Cigarette smoking and high salt food intake seem to be important contributing factors to the development of gastric cancer.
Modified from Sugiyama and Asaka (2004), Correa (1992), and Hamajima *et al.* (2006).

cancer. Chronic *H. pylori* infection of the gastric epithelium can induce chronic inflammation, a critical step in gastric carcinogenesis. In 1994, the International Agency for Research on Cancer classified *H. pylori* as a gastric carcinogen.

As a barrier-sustaining agent, genistein inhibits the extra-intestinal invasion of enteric bacteria and has potential anti-bacterial activities as well (Wells *et al.* 1999). In an *in vitro* study comparing the incubation of *H. pylori* with and without genistein, the growth of *H. pylori* was inhibited by genistein as compared with the control. This inhibition is possibly mediated by the stabilization of the covalent topoisomerase II–DNA cleavage complex. Although there is no obvious epidemiological evidence of an association between isoflavones and *H. pylori*, it is possible that isoflavones may inhibit gastric cancer development by inhibiting the growth of *H. pylori*.

26.3.2 Antioxidant Activity

Oxidative stress occurs in the stomach when it is exposed to exogenous pathogens from food. *H. pylori* is the major cause of oxidative stress. Oxidative stress is a consequence of excessive reactive oxygen species (ROS) build up due to lack of antioxidant capability. Accumulated oxidative stress leads to acute/chronic cell injury and is closely related to the development of gastric cancer (Suzuki *et al.* 2012).

Genistein has antioxidant properties as a scavenger of radicals and performs chelation of metals. This function occurs by promotion of gene expression for enzymes that react with antioxidants, such as catalase and superoxide dismutase, and inhibition of secondary oxidant production associated with hydrogen peroxide or hypochlorous acid. Genistein is a more active antioxidant than daidzein due to its having a third hydroxyl group in the C-5 position. Equol is another antioxidant with improved actions compared with its precursor compounds. This result is due to the absence of the 2,3-double bond in conjunction with the loss of the 4-oxo group, which enhances the antioxidant properties (Barnes 2010; Mortensen *et al.* 2009).

Isoflavone metabolism pathway in the intestine and liver affect the antioxidant properties of isoflavones and their metabolites. Although genistein reacts strongly with superoxide, dismutase, catalase, glutathione peroxidase, isoflavone metabolites such as equol, 8-OH-daidzein, *O*-desmethylangolensin, and 1,3,5-trihydroxybenzene are also potent scavengers. The most potent scavenger for hydroxyl and superoxide anion radicals is 8-hydroxy-daidzein. Genistein and daidzein metabolites are highly chelating when accompanied with ferrous compounds. However, sulfated isoflavones can decrease antioxidant activity (Barnes 2010; Mortensen *et al.* 2009).

26.3.3 Anti-inflammation Effect

As shown in Figure 26.2, the inflammation process resulting from *H. pylori* infection is a critical step in gastric carcinogenesis. *H. pylori* induces pro-inflammatory responses in epithelial cells by two pathways (Figure 26.3). Through a functional type IV secretion system, *H. pylori* delivers effector molecules such as cell wall peptidoglycan (PGN) and the protein CagA to epithelial cells. After recognizing *H. pylori*, PGN, the cytosolic host defense molecule nucleotide-binding oligomerization domain 1 (NOD1) activates the receptor-interacting protein 2 (RIP2)–transforming growth factor-β (TGF-β)-activated kinase 1 (TAK1) pathway, and TAK1 activates nuclear factor-κB (NF-κB) (Fox and Wang 2007; Watanabe *et al.* 2011). Activated NF-κB complexes translocate to the nucleus where they up-regulate expression of genes encoding interleukin (IL)-8. Another pathway is accompanied by CagA translocation by certain *H. pylori* strains. After CagA translocation into epithelial cells, phosphorylated CagA activates Ras–RAF. RAF activates mitogen-activated protein kinase (MAPK) kinase (MEK) and extra-cellular-signal-regulated kinase 1/2 (ERK1/2), which leads to activation of the transcription factor NF-κB and production of IL-8 by host cells (Fox and Wang 2007).

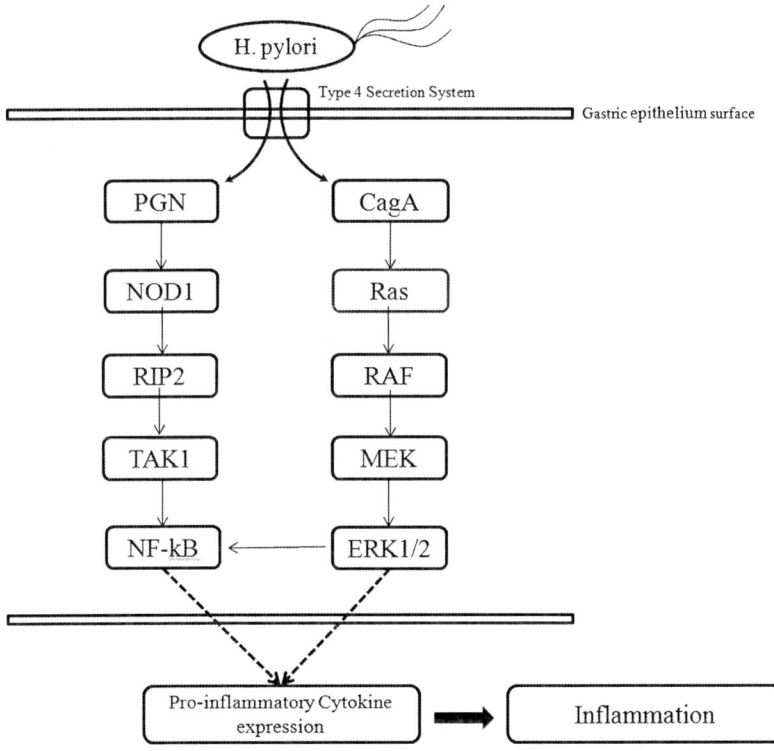

Figure 26.3 Pro-inflammatory responses induced by *H. pylori* in gastric epithelial cells. *H. pylori* induces pro-inflammatory responses in epithelial cells by two pathways, through a functional type 4 secretion system.
Modified from Fox and Wang (2007) and Watanabe *et al.* (2011).

The carcinogenic promotion and progression stages are dependent on activated NF-κB. The promotion stage refers to the proliferative stage of the immortal cancer cell, and the progression stage is settled by a sequence of events such as proliferation, anti-apoptosis, angiogenesis, invasion, and metastasis (Hanahan and Weinberg 2000). During the promotion and progression stages, further mutations can occur. Also, the proliferating cells transport the additional initiating mutations. It is thought that the mechanism involves various components regulated by NF-κB activation (Maeda and Omata 2008).

The isoflavones genistein and daidzein, in particular genistein, inhibit deregulated activation of NF-κB which can suppress inflammation. Suppression in NF-κB activation may occur by inhibition of IκBα kinase, leading to inhibition of phosphorylation and degradation of IκBα, and, consequently, NF-κB DNA binding by p65 nuclear translocation. Daidzein can also inhibit the signal transducer and activator of transcription 1 [transcription factor for inducible nitric oxide synthase (iNOS)], which also suppresses the inflammation pathway. Genistein directly inhibits IL-8. Some isoflavone extracts from plant sources suppress NF-κB and NO activation and may down-regulate several inflammation related genes such as cytochrome oxidase-2, matrix metalloprotein-9 (MMP-9), intercellular adhesion molecule 1 (ICAM1), and iNOS to reduce inflammation (Prasad *et al.* 2010).

26.3.4 Inhibition of Enzymes

Genistein is a protein tyrosine kinase (PTK) inhibitor. PTKs catalyze the process of addition of a tyrosine residue phosphate (PO_4^{3-}) to proteins and other molecules, which in turn activate, or deactivate, broad-spectrum proteins and enzymes, especially in carcinogenesis. PTK inhibition suppresses or slows down the signal transduction pathway which supports carcinogenesis. PTKs are known to be highly expressed in gastric cancers (Lin *et al.* 2000). At high isoflavone doses, PTK actions are suppressed in gastric tissue which results in reduction of carcinogenesis (Piontek *et al.* 1993).

Furthermore, in *in vitro* study (Chacko *et al.* 2005; Okura *et al.* 1988), genistein inhibited DNA replication enzymes such as DNA topoisomerases I/II and MMP-9, which are related to invasion or metastasis of gastric carcinogenesis. Genistein and other isoflavones interact with peroxisome proliferator-activated receptors (PPARs) as transcription factors, and thus isoflavones can act as a regulating factor for gene expression in gastric carcinogenesis.

26.3.5 Inhibition of Angiogenesis

Vascular endothelial growth factor (VEGF) is a key mediator of angiogenesis and plays an important role in the development of neoplasms, including gastric cancer. Isoflavones can also down-regulate the expression of VEGF along with other related growth factor genes. Genistein, in particular, inhibits VEGF-induced endothelial cell activation by inhibiting the action of PTKs, as well as MAPK activation (Yu *et al.* 2010).

26.3.6 Enhancement of Apoptosis

The homeostatic rate between cellular loss and cellular regeneration/proliferation maintains gastric mucosal integrity. Gastric cancer is caused by excessive cell proliferation and correlates with cell cycle regulation disorder which involves numerous complicated factors. Genistein and daidzein directly inhibit growth and proliferation of gastric cancer cells through an apoptosis process by down-regulation of the apoptosis-regulated gene *Bcl2*, up-regulation of apoptosis-regulated gene *Bax*, and cell cycle arrest at the G/early M phases (Zhou *et al.* 2004). *In vitro* data from Cui *et al.* (2005) supported the conclusion that genistein can arrest cell cycle progression of human gastric carcinoma cells at the G2/M phase by promoting the expression of $p21^{waf1/cip1}$ and reducing the degradation of cyclin B1 protein in tumor cells. In this manner, tumor cells cannot process the checkpoint pathway of G2/M and block mitosis (Cui *et al.* 2005).

26.3.7 Evidence from Population Studies

According to the results from *in vitro* and *in vivo* animal studies, isoflavones may act as an anti-gastric cancer agent. Although soybean intake could decrease gastric cancer risk as observed in epidemiological studies, there was a failure to deliver a consistent correlation.

Such inconsistent results are derived from uncontrolled study participant conditions including different genetic attributes, different study designs, and non-validated isoflavone exposure (intake) measurement methods. In a typical animal study, the experimental and control groups are under the same conditions, so there is greater consistency of results. However, since most epidemiological research is performed as an observational study, it is difficult to accurately conclude a causal association between isoflavone intake and gastric cancer.

Depending on the participant racial profile, there is a large difference observed in study results. Mean daily isoflavone intake for Asians is typically about 8–50 mg. Blood genistein levels are generally in the range of $25\,\text{ng}\,\text{mL}^{-1}$ for women in Asia, slightly less for vegetarian women, and under $2\,\text{ng}\,\text{ml}^{-1}$ for women in the United States (Verkasalo *et al.* 2001).

Study results depend on study design. A prospective cohort study is more likely than a case-control study to identify a causal relationship. Case-control studies also are prone to deliver inconsistent results due to factors such as comparability between case and control groups, information bias, *etc.* The amount of isoflavone intake derived from soybeans depends on food preparation recipes, and it is difficult to measure the quantity of individual isoflavone intake. High processed foods contain lower amounts of isoflavones than that found in unprocessed soybeans. For example, soybean products such as tofu contain only 6–20% (5.1–$64\,\text{mg}\,100\,\text{g}^{-1}$) of the isoflavones found in unprocessed soybeans (Verkasalo *et al.* 2001). Foods such as miso, doenjang, and natto contain large amounts (20–$126\,\text{mg}\,100\,\text{g}^{-1}$), and tofu and tempeh contain moderate amounts of isoflavones (5–$64\,\text{mg}\,100\,\text{g}^{-1}$). New generation soy foods such as soy milk, yogurts and cheeses have small amounts of isoflavones (1–33 mg 100 g^{-1}).

In studies, a questionnaire survey is the usual method used to determine the amount of isoflavone intake. However, this method is prone to error due to reliance on participant recalled memory for generating the data. Although serum concentration measurement is a more accurate measurement method than a questionnaire, there are few human studies including isoflavone blood concentration measurements. In a Korean study, isoflavones, including genistein, daidzein and equol, were found to reduce gastric cancer risk, and the higher the concentration of the three isoflavones that were observed, the lower the gastric cancer risk (Ko et al. 2010). The protective effect for gastric cancer can be explained as a result of the anti-inflammatory, anti-tumorigenic, and antioxidative effects of isoflavones.

The effects of isoflavones on established gastric cancer is more complicated and is dependent on food processing. Both Korean and Japanese populations have high incidence rates of gastric cancer, yet frequently eat a wide variety of soy foods. In a recent meta-analyses among Korean and Japanese populations, non-fermented soy foods showed protective effects for gastric cancer, whereas fermented soy foods presented no effect in reducing gastric cancer risk (Kim et al. 2011). Fermented soy foods contain high amounts of salt. These include fermented soy sauce and soybean pastes such as doenjang and miso, which Koreans and Japanese use as food preparation ingredients, which are associated with higher risk for gastric cancer due to the effects of high salt and N-nitroso compounds. Two other meta-analysis studies demonstrated consistent result that conclude that consumption of non-fermented soy foods is a protective factor against gastric cancer, whereas fermented soy foods are risk factors (Tong et al. 2010; Wu et al. 2000) (Table 26.1).

Table 26.1 Meta-analysis or pooled analysis on the association between soybean foods and gastric cancer risk. Non-fermented soy foods were associated with decreased gastric cancer, but fermented soy foods were associated with increased gastric cancer due to salt intake possibly. RR (95% CI), relative risk (95% confidence interval) or odds ratio (95% confidence interval).

			Risk for gastric cancer [RR (95% CI)]	
Reference	Study design	Ethnicity criteria	Non-fermented soy foods	Fermented soy foods
Wu et al. (2000)	Pooled analysis	No restriction	0.72 (0.63–0.82)	1.26 (1.11–1.43)
Kim et al. (2011)	Meta-analysis	Korean/ Japanese	0.64 (0.54–0.77)	1.22 (1.02–1.44)
Tong et al. (2010)	Meta-analysis	No restriction	Soybean products 0.58 (0.52–0.65) Tofu 0.90 (0.80–1.00)	1.18 (1.09–1.28)

Summary Points

- This Chapter focuses on the overview of possible effects and functions of isoflavones against gastric cancer.
- Isoflavones are synthesized as part of the phenylpropanoid pathway, leading to the generation of genistein and daidzein.
- Through both hormonal and non-hormonal functions, isoflavones have beneficial effects on human health such as reducing menopausal symptoms, promoting bone health, reducing cancer risk, and reducing metabolic disease risk.
- Isoflavones possibly inhibit the development of gastric cancer by inhibiting the growth of *H. pylori*, although there is no conclusive epidemiological evidence to support this hypothesis.
- Accumulated oxidative stress is closely related to the development of gastric cancer. Genistein has antioxidant properties.
- The isoflavones genistein and daidzein, in particular genistein, inhibit deregulated activation of NF-κB, which can suppress inflammation.
- Isoflavones can down-regulate the expression of VEGF, which is a key mediator of angiogenesis and plays an important role in gastric carcinogenesis.
- Genistein and daidzein directly inhibit growth and proliferation of gastric cancer cells through the process of apoptosis, and genistein arrests progression of human gastric carcinoma cells at the G2/M cell cycle phase.
- As observed by epidemiological study, non-fermented soy foods demonstrate a protective effect for gastric cancer, whereas fermented soy foods present no effect in reducing gastric cancer risk.

Key Facts

- The US Food and Drug Administration (FDA) issued health claims regarding the association of soy protein and reduced risk of coronary heart disease in 1999.
- Based on the available scientific evidence there is significant agreement that soy protein, included at a level of 25 g day^{-1} and accompanying a diet low in saturated fat and cholesterol, may reduce the risk of coronary heart disease by reducing total and low-density lipoprotein (LDL)-cholesterol levels. Approximately 60–75 mg of isoflavones is contained in 25 g of soy protein.
- The FDA has stated that the recommended daily consumption of soy protein needs to be re-evaluated because many relevant studies containing additional data have been published since the claim.

Definitions of Words and Terms

Antioxidant: A chemical compound or substance that inhibits oxidation or reactions promoted by oxygen.

Apoptosis: A process of programmed cell death in which the cell uses specialized cellular machinery in order to kill itself.

***Bias*:** The systematic error resulting from study subject selection, method of obtaining information, and lack of consideration for confounder which produces a deviated estimation of the effect of exposure on the risk of developing a disease.
***Conjugation*:** The turning of substances into a hydrophilic state in the body.
***Epidemiology*:** A scientific discipline dealing with health events, health characteristics, or health-determinant patterns in a population, and applying it to public health.
***Estrogens*:** Female steroid sex hormones that are secreted chiefly by the ovaries and which control changes in female reproductive organs, promote development of female secondary sexual characteristics, and maintain the female reproductive system.
***Fermentation*:** Enzymatically controlled food processing using yeast or bacteria under anaerobic conditions, which results in the conversion of carbohydrates into alcohols and carbon dioxide or organic acids.
***Helicobacter pylori*:** A Gram-negative bacterium found in the stomach and a major risk factor for gastric cancer.
***Meta-analysis*:** A quantitative statistical analysis for synthesizing research results by using various statistical methods to retrieve, select, and combine results from previous studies which are separate but related.
***Phytoestrogens*:** Naturally-occurring chemical plant compounds which are structurally or functionally similar to estrogen and their active metabolites in the body.

List of Abbreviations

ER	estrogen receptor
ERK	extracellular-signal-regulated kinase
MMP-9	matrix metalloprotein-9
IL	interleukin
iNOS	inducible nitric oxide synthase
MAPK	mitogen-activated protein kinase
MEK	MAPK kinase
NF-κB	nuclear factor-κB
NOD1	nucleotide-binding oligomerization domain 1
PGN	peptidoglycan
PTK	protein tyrosine kinase
RIP2	receptor-interacting protein 2
TAK1	transforming growth factor-β (TGF-β)-activated kinase 1
VEGF	vascular endothelial growth factor

References

Barnes, S., 2010. The biochemistry, chemistry and physiology of the isoflavones in soybeans and their food products. *Lymphatic Research and Biology*. 8: 89–98.

Cano, A., Garcia-Perez, M.A., and Tarin, J.J., 2010. Isoflavones and cardiovascular disease. *Maturitas*. 67: 219–226.

Cassidy, A., Bingham, S., and Setchell, K.D., 1994. Biological effects of a diet of soy protein rich in isoflavones on the menstrual cycle of premenopausal women. *The American journal of Clinical Nutrition*. 60: 333–340.

Cederroth, C.R., and Nef, S., 2009. Soy, phytoestrogens and metabolism: A review. *Molecular and Cellular Endocrinology*. 304: 30–42.

Chacko, B.K., Chandler, R.T., Mundhekar, A., Khoo, N., Pruitt, H.M., Kucik, D.F., Parks, D.A., Kevil, C.G., Barnes, S., and Patel, R.P., 2005. Revealing anti-inflammatory mechanisms of soy isoflavones by flow: modulation of leukocyte-endothelial cell interactions. *Am J Physiol Heart Circ Physiol*. 2: H908–H915.

Correa, P., 1992. Human gastric carcinogenesis: a multistep and multifactorial process--First American Cancer Society Award Lecture on Cancer Epidemiology and Prevention. *Cancer Research*. 52: 6735–6740.

Cui, H.B., Na, X.L., Song, D.F., and Liu, Y., 2005. Blocking effects of genistein on cell proliferation and possible mechanism in human gastric carcinoma. *World Journal of Gastroenterology*. 11: 69–72.

Fox, J.G., and Wang, T.C., 2007. Inflammation, atrophy, and gastric cancer. *The Journal of Clinical Investigation*. 117: 60–69.

Hamajima, N., Naito, M., Kondo, T., and Goto, Y., 2006. Genetic factors involved in the development of *Helicobacter pylori*-related gastric cancer. *Cancer Science*. 97: 1129–1138.

Hanahan, D., and Weinberg, R.A., 2000. The hallmarks of cancer. *Cell*. 100: 57–70.

Heinonen, S., 2006. Academic dssertation: Identification of isoflavonoid metabolites in humans. [Online]. University of Helsinki, Helsinky, Finland. Available: http://ethesis.helsinki.fi/julkaisut/mat/kemia/vk/heinonen/identifi.pdf [Accessed].

Kim, J., Kang, M., Lee, J.S., Inoue, M., Sasazuki, S., and Tsugane, S., 2011. Fermented and non-fermented soy food consumption and gastric cancer in Japanese and Korean populations: a meta-analysis of observational studies. *Cancer Science*. 102: 231–244.

Ko, K.P., Park, S.K., Park, B., Yang, J.J., Cho, L.Y., Kang, C., Kim, C.S., Gwack, J., Shin, A., Kim, Y., Kim, J., Yang, H.K., Kang, D., Chang, S.H., Shin, H.R., and Yoo, K.Y., 2010. Isoflavones from phytoestrogens and gastric cancer risk: a nested case-control study within the Korean Multicenter Cancer Cohort. *Cancer Epidemiology, Biomarkers And Prevention*. 19: 1292–1300.

Kulling, S.E., Lehmann, L., and Metzler, M., 2002. Oxidative metabolism and genotoxic potential of major isoflavone phytoestrogens. *Journal of Chromatography. B, Analytical Technologies in the Biomedical and Life Sciences*. 777: 211–218.

Lin, W., Kao, H.W., Robinson, D., Kung, H.J., Wu, C.W., and Chen, H.C., 2000. Tyrosine kinases and gastric cancer. *Oncogene*. 19: 5680–5689.

Maeda, S., and Omata, M., 2008. Inflammation and cancer: role of nuclear factor-κB activation. *CancerScience*. 99: 836–842.

Mortensen, A., Kulling, S.E., Schwartz, H., Rowland, I., Ruefer, C.E., Rimbach, G., Cassidy, A., Magee, P., Millar, J., Hall, W.L., Kramer Birkved, F., Sorensen, I.K., and Sontag, G., 2009. Analytical and compositional aspects of isoflavones in food and their biological effects. *Molecular Nutrition and Food Research*. 53(Suppl. 2): S266–S309.

Okura, A., Arakawa, H., Oka, H., Yoshinari, T., and Monden, Y., 1988. Effect of genistein on topoisomerase activity and on the growth of [Val 12]Ha-ras-transformed NIH 3T3 cells. *Biochem Biophys Res Commun*. 1: 183–189.

Piontek, M., Hengels, K.J., Porschen, R., and Strohmeyer, G., 1993. Antiproliferative effect of tyrosine kinase inhibitors in epidermal growth factor-stimulated growth of human gastric cancer cells. *Anticancer Research*. 13: 2119–2123.

Prasad, S., Phromnoi, K., Yadav, V.R., Chaturvedi, M.M., and Aggarwal, B.B., 2010. Targeting inflammatory pathways by flavonoids for prevention and treatment of cancer. *Planta Medica*. 76: 1044–1063.

Setchell, K.D., 1998. Phytoestrogens: the biochemistry, physiology, and implications for human health of soy isoflavones. *The American Journal of Clinical Nutrition*. 68: 1333S–1346S.

Sugiyama, T., and Asaka, M., 2004. *Helicobacter pylori* infection and gastric cancer. *Medical Electron Microscopy*. 37: 149–157.

Suzuki, H., Nishizawa, T., Tsugawa, H., Mogami, S., and Hibi, T., 2012. Roles of oxidative stress in stomach disorders. *Journal of Clinical Biochemistry and Nutrition*. 50: 35–39.

Tong, X., Li, W., and Qin, L.Q., 2010. Meta-analysis of the relationship between soybean product consumption and gastric cancer. *Zhonghua Yu Fang Yi Xue Za Zhi*. 44: 215–220.

Verkasalo, P.K., Appleby, P.N., Allen, N.E., Davey, G., Adlercreutz, H., and Key, T.J., 2001. Soya intake and plasma concentrations of daidzein and genistein: validity of dietary assessment among eighty British women (Oxford arm of the European Prospective Investigation into Cancer and Nutrition). *British Journal of Nutrition*. 86: 415–421.

Watanabe, T., Asano, N., Kitani, A., Fuss, I.J., Chiba, T., and Strober, W., 2011. Activation of type I IFN signaling by NOD1 mediates mucosal host defense against *Helicobacter pylori* infection. *Gut Microbes*. 2: 61–65.

WCRF Project. *Food, nutrition, physical activity, and the prevention of cancer: a global perspective*. 2007, American Institute for Cancer Research, Washington DC, USA.

Wells, C.L., Jechorek, R.P., Kinneberg, K.M., Debol, S.M., and Erlandsen, S.L., 1999. The isoflavone genistein inhibits internalization of enteric bacteria by cultured Caco-2 and HT-29 enterocytes. *The Journal of Nutrition*. 129: 634–640.

Wu, A.H., Yang, D., and Pike, M.C., 2000. A meta-analysis of soyfoods and risk of stomach cancer: the problem of potential confounders. *Cancer Epidemiology, Biomarkers and Prevention*. 9: 1051–1058.

Yu, X., Zhu, J., Mi, M., Chen, W., Pan, Q., and Wei, M., 2012. Anti-angiogenic genistein inhibits VEGF-induced endothelial cell activation by decreasing PTK activity and MAPK activation. *Medical Oncology*. 29: 349–357.

Zhou, H.B., Chen, J.J., Wang, W.X., Cai, J.T., and Du, Q., 2004. Apoptosis of human primary gastric carcinoma cells induced by genistein. *World Journal of Gastroenterology*. 10: 1822–1825.

CHAPTER 27

Dietary Isoflavones and Learning and Memory

CRAIGE C. WRENN

Department of Pharmaceutical, Biomedical, and Administrative Sciences, College of Pharmacy and Health Sciences, Drake University, Des Moines, IA 50311, USA
Email: craige.wrenn@drake.edu

27.1 Introduction

The phytoestrogens are plant-derived compounds that have structural similarity to estradiol. Isoflavones comprise one class of phytoestrogen, and are present in several food sources, especially soybeans and other legumes (Kurzer and Xu 1997). In soy the most prominent isoflavones are genistein and daidzein (Patisaul and Jefferson 2010). Both of these compounds have been shown to bind estrogen receptors where they act as either agonists or antagonists, depending on the tissue (Kuiper *et al.* 1998; Patisaul *et al.* 2001). Given that human exposure to these compounds is primarily through the diet and that estrogen receptors mediate a myriad of reproductive and non-reproductive functions in mammals, a number of studies have addressed the question of whether dietary exposure to isoflavones has beneficial or deleterious consequences (Patisaul and Jefferson 2010). Included in this assortment of work are studies that have examined the effects of dietary exposure to isoflavones on learning and memory.

The studies of the effects of dietary isoflavones on learning and memory arise from three distinct ideas regarding the role of estradiol in learning and memory, as outlined below.

(a) One idea springs from the fact that sex hormones, including estradiol, influence learning and memory by modulating brain development during critical periods (Isgor and Sengelaub 1998; 2003). In this context, dietary isoflavones are hypothesized to have deleterious effects by acting as endocrine disruptors that interfere with the role of sex hormones in brain development. A consequence of this altered brain development is impaired learning and memory.
(b) Another idea is inspired by the fact that sex hormones, including estradiol, regulate the neuronal plasticity that is necessary for learning and memory (Foy *et al.* 2008). In this context, dietary isoflavones are hypothesized to reverse cognitive deficits that are associated with aging and diseases of aging (*e.g.* Alzheimer's disease).
(c) A third idea is based on the observation that the cognitive dysfunction associated with aging in women is due in part to the loss of estradiol in menopause (Genazzani *et al.* 2007). In this context, dietary isoflavones are hypothesized to have beneficial effects by serving as a sort of estrogen replacement therapy that enhances cognitive function.

Thus, current thinking regarding the impact of dietary isoflavones on learning memory is that it can be either deleterious or beneficial depending on the physiological context of the dietary intake. This Chapter reviews both pre-clinical studies in animals and clinical studies in humans in support of this nuanced view.

27.2 Pre-clinical Studies of the Deleterious, Endocrine-disrupting Effects of Dietary Isoflavones on Learning and Memory

A number of non-reproductive behaviors are referred to as sexually dimorphic, meaning that they differ between the sexes. Sexual dimorphisms, whether they be anatomical, physiological, or behavioral, are organized by the activity of sex steroids during development. For this reason, agents that can mimic or interfere with endogenous sex steroids are hypothesized to disrupt sexually dimorphic phenotypes. As mentioned above, isoflavones found in the diet have the potential to act in this way because of their ability to stimulate or antagonize estrogen receptors.

Spatial learning and memory is a particularly suitable behavior to employ in testing the hypothesis that dietary isoflavones can negatively impact cognitive function. This suitability stems from an understanding that male-typical patterns of efficient spatial learning and memory (Gresack and Frick 2003) are organized by sex steroids (Isgor and Sengelaub 1998; 2003), and the manipulation of sex steroids alters the morphology of hippocampal neurons and spatial cognition (Roof and Havens 1992).

Whether dietary isoflavones can impact spatial learning and memory was tested by Lund *et al.* (2001). As a means of exposing subjects (rats) to dietary isoflavones, pregnant dams were fed either a phytoestrogen-free diet or a phytoestrogen-containing diet. After birth and into adulthood the offspring of these dams were maintained on their respective diets such that exposure to dietary isoflavone was "life-long". As adults, these subject rats were tested in a food-motivated spatial memory task called the eight-arm radial arm maze. This maze consists of an octagonal hub from which eight arms, or runways, radiate. A food reward is located at the end of four of the runways and rats are allowed to explore the maze for 10 min and eat the rewards. The rewards are not replaced within a trial, so the rat must learn to use spatial working memory to recall which arms have been visited during a trial in order to not waste time in unbaited runways. An entry into a not yet visited runway is recorded as a correct choice, and re-entries into already visited runways are recorded as working memory errors. Since only four of the eight arms are baited, the rat must also use spatial reference memory to recall which arms are never baited and avoid entering those arms. Entrances into never baited runways are recorded as a reference memory error. The authors found that female rats receiving the isoflavone-containing diet learned this maze faster than females that lacked isoflavones in their diet. In males, the effect was in the opposite direction; males on the isoflavone-containing diet had impaired spatial learning.

In a similar experiment Lund *et al.* (2001) allowed rats to learn the radial maze task and then switched some rats to a isoflavone-free diet while keeping others on the isoflavone-containing diet. Females that remained on the isoflavone diet had better spatial memory performance than those on isoflavone-free diet. Conversely, males that remained on the isoflavone diet were impaired relative to those switched to the isoflavone-free diet.

These data from Lund and colleagues (2001) show that dietary isoflavones have a deleterious effect on learning and memory, at least in males. A limitation of this work is that the isoflavone exposure was "life-long" in that the exposure started in the womb and carried through adulthood and cognitive testing. Such an experimental design cannot discriminate between an organizational effect, in which the neurodevelopmental role of sex steroids is disrupted, or an activational effect, in which the neuroplasticity that subserves memory is altered.

As an alternative approach, recent work in my laboratory limited dietary exposure to the isoflavone genistein to either the gestational or lactational periods (Ball *et al.* 2010). We allowed pregnant dams to eat genistein-containing food (5 mg kg^{-1}) only during gestation, only during lactation, or during both periods. At weaning all of the subjects were switched to genistein-free food. At approximately 5 months of age, male rats were tested in the Morris water maze (Morris *et al.* 1982), a task in which the rats must learn the spatial location of a submerged platform in a large tub of water. In this task, rats indicate that they have learned the location of the platform by decreasing their latency to swim to the platform with training (acquisition trials). An additional measure of spatial memory is a bias for the location of the platform even when the platform is not present (probe trials). We found that male rats that were exposed to dietary genistein during *both* gestation and lactation were impaired in acquiring the memory of the platform's location. Rats whose exposures were limited to either the gestational or lactational periods were unaffected. These data show that the isoflavone genistein exerts an organizational effect on spatial learning and memory, but only when present throughout pre- and post-natal development.

27.3 Pre-clinical Studies of the Beneficial Effects of Dietary Isoflavones in Rodent Models of Age-associated Cognitive Deficits

Learning and memory requires neurons to undergo changes in their morphological and electrophysiological properties. Among the many molecules that regulate this plasticity is estradiol. Estradiol has been shown to play a role in altering dendrite density and shape (morphological plasticity) and in long-term potentiation, a form of electrophysiological plasticity that is thought to underlie memory (Foy *et al.* 1999; Woolley and McEwen 1993). For this reason, a few studies have tested the hypothesis that dietary isoflavones, acting as estradiol mimics, can ameliorate aging-associated cognitive deficits.

One approach in this vein has been to produce learning and memory deficits in rats by modeling one feature of Alzheimer's disease pathology in rats. Specifically, investigators have injected β-amyloid peptide into the hippocampus of male rats. The relevance of this approach to Alzheimer's disease is that β-amyloid is the major component of the toxic plaques that are deposited in Alzheimer's brains, and the hippocampus is a major memory center of the brain. Li *et al.* (2010) showed that rats that were injected with intrahippocampal β-amyloid were impaired in the aforementioned Morris water maze. These researchers found that the isoflavone puerarin (administered by oral gavage daily for 28 days) resulted in improvement in both acquisition and probe trials of the water maze. Interestingly, the level of improvement was comparable with that provided in a parallel group of rats by donepezil, an Alzheimer's disease medication. These behavioral data were bolstered by data showing that puerarin's mechanism was to decrease the apoptotic cell death produced by the intrahippocampal β-amyloid.

In a similar study Bagheri *et al.* (2011) treated rats that had been injected with intrahippocampal β-amyloid with oral gavage delivery of the isoflavone genistein (single dose). To assess spatial memory in these subjects, a Y-maze task was used. The Y-maze consists of a three-armed maze in which the animal is allowed to freely explore. Rats with intact spatial memory will explore the arms in an alternating fashion owing to their ability to remember which arm has just been visited. Rats injected with intrahippocampal β-amyloid revisited arms more frequently than controls, but genistein reversed this impairment.

An additional memory test used by Bagheri et al. (2011) was the passive avoidance task. Passive avoidance testing utilizes a two-chambered box connected by a doorway. One chamber is brightly lit; the other is darkened. Rats are placed in the brightly lit chamber. When they move into the preferable darkened chamber, a mild foot shock is administered. In subsequent trials the rat indicates a memory for the foot shock by remaining in the lit chamber. Latency to enter the dark chamber is the measure of the strength of this memory. Intrahippocampal β-amyloid significantly deceased latency showing that the peptide caused memory impairment. Genistein increased latency in β-amyloid treated rats showing that the isoflavone reversed the memory deficit. Mechanistic data from this study indicated that the pro-memory effects of genistein were due to an attenuation of oxidative stress.

Another approach has employed a pharmacological model of cognitive impairment in Alzheimer's disease. Inadequate activity of the neurotransmitter acetylcholine is an important feature of Alzheimer's disease and may explain many of the learning and memory problems. This acetylcholine deficiency can be modeled in animals by administering the acetylcholine antagonist scopolamine and observing the subsequent learning and memory deficits. Investigators can then attempt to reverse those deficits with putative therapies, such as a dietary isoflavone.

Kim et al. (2010) used the scopolamine model in male mice to test the memory-enhancing effects of the isoflavones daidzein and daidzin. The learning and memory tests used were the Morris water maze and passive avoidance tasks. Scopolamine significantly impaired memory in both tasks, but oral administration of either isoflavone ($5\,mg\,kg^{-1}$) ameliorated the scopolamine-induced deficits.

In another scopolamine study Bansal and Parle (2010) used dietary supplementation with soybean to reverse scopolamine-induced memory deficits in mice. The diet was eaten for 60 consecutive days prior to passive avoidance testing and was shown to protect against scopolamine-induced impairment. The authors also showed that the diet diminished acetylcholinesterase activity, increased thiobarbituric acid-reactive substances, and brain glutathione levels. These biochemical findings are consistent with facilitation of acetylcholine signaling, reduced free radical generation, and enhanced scavenging of free radicals, respectively. Each of these phenomena is a feasible mechanism by which dietary isoflavones may enhance cognitive function. However, the study is limited by the use of soybean diet rather than known concentrations of an individual isoflavone.

27.4 Pre-clinical Studies of Beneficial Effects of Dietary Isoflavones in Ovariectomized Rats

Another hypothesis regarding the beneficial effects of dietary isoflavones is that these compounds can act as a cognitive enhancer in menopause. The desire for administering a cognitive enhancer in menopause stems from the observations that menopause is associated with a decrease in cognitive performance and diseases of memory such as Alzheimer's disease (Halbreich et al. 1995; Henderson 1997). An experimental system for testing this hypothesis is the ovariectomized (OVX) rat. The general approach here is to remove endogenous estradiol through the removal of the ovaries and then administer isoflavones through the diet or oral administration. The rationale is that if dietary isoflavones improve performance in various rodent behavioral paradigms of learning and memory in OVX rats, then it becomes reasonable to suggest similar efficacy in menopause.

An early study that took this approach used OVX retired breeder rats (8–10 months old) and fed them diets supplemented with various doses of 17β-estradiol, soy protein isolate with phytoestrogens (SPE), or combinations of 17β-estradiol and SPE for 10 months (Pan et al. 2000). To test the effects of these diets on learning and memory, the authors used the eight-arm radial arm maze. The principal finding was that both estradiol and SPE alone significantly increased the number of correct choices, suggesting an enhancement of spatial working memory in these aged OVX rats. However, a weakness of the study was that the concentrations of specific isoflavones (e.g. genisitein and daidzein) in the SPE were not known which precluded the evaluation of dose–response relationships.

Younger OVX rats have also been used to test the idea that a diet containing isoflavones can enhance learning and memory. Luine et al. (2006) allowed 2-month-old OVX rats to eat a so-called High Phyto Chow, which contained 810 μg g^{-1} of phytoestrogens, or a Low Phyto Chow, which contained no sources of phytoestrogen. After the rats ate their respective diets for 7 weeks, they were tested in a spatial memory task called the object placement task. In this task the rat is allowed to investigate two distinct objects (e.g. a small candlestick and a small statue) in an open field for a few minutes. The rat is then removed from the open field, and one of the objects is moved. After a 4 h delay, the rat is placed back in the arena and allowed to freely investigate the objects. A bias towards investigating the moved object is taken as an indicator of memory for the objects original location. Rats that had eaten the High Phyto Chow explored the moved object significantly more than rats that had eaten the Low Phyto Chow, suggesting better object placement memory in rats that had eaten isoflavones. A strength of this study was that the researchers were able to correlate this behavioral finding with increased density of dendritic spines in the hippocampus and prefrontal cortex, areas of the brain that are critical to object placement memory (Ennaceur and Aggleton 1994; 1997). Thus, the dietary isoflavones appeared to improve memory by altering neuronal morphology. Unfortunately, like the study of Pan et al. (2000), the dose of specific isoflavones consumed by the rats was not known.

Rather than rely only on post-ovariectomy administration of dietary isoflavones, Monteiro et al. (2008) allowed young rats to eat a soy diet for 60 days before OVX (pre-treatment group) or for 30 days after OVX (post-treatment group). Learning and memory was assessed in these rats using the Morris water maze. This study showed that ovariectomy significantly impaired both acquisition and probe performance in the water maze. Remarkably, both pre- and post-OVX soy diet improved acquisition trial performance. Only post-OVX treatment improved probe trial performance. Unfortunately, data that indicates mechanism were not provided, but a strength of the study was that isoflavone concentrations of the soy diet were known (genistein, $0.23\,\text{mg}\,\text{g}^{-1}$; daidzein, $0.25\,\text{mg}\,\text{g}^{-1}$).

Another approach relevant to examining whether dietary isoflavones can replace lost estradiol has been to combine ovariectomy with models of age-associated neurodegenerative diseases. In one study OVX rats received electrolytic lesions of the nucleus basalis magnocellularis as a way of modeling the degeneration of this structure in Alzheimer's disease (Sarkaki *et al.* 2008). In a second study from the same group, OVX rats received chemical lesions of the substantia nigra as a way of modeling the degeneration of this structure in Parkinson's disease (Sarkaki *et al.* 2009). In both studies the rats were fed diets containing soy meal with or without isoflavones. The authors reported that *both* soy meal with and without isoflavones improved spatial learning and memory in the water maze. The fact that even the isoflavone-free soy meal offered a pro-cognitive effect underscores the necessity of studies that use known concentrations of isoflavones in isolate to draw unequivocal conclusions regarding isoflavone action.

Not all studies have found that dietary isoflavones enhance cognitive function. In two well-designed studies by Neese *et al.* (2010; 2012) OVX rats ate known doses of genistein daily and were assessed in various lever-pressing operant tasks. These studies found that genistein exerted impairments in operant tasks that tap into working memory and response inhibition/timing. As pointed out by these authors, the discrepancy between these studies and those using spatially based maze tasks may be related to the different brain structures that modulate these behaviors. The maze tasks require the spatial mapping functions of the hippocampus, whereas the operant tasks require the executive function of the prefrontal cortex. Perhaps, genistein varies in its effect in these two structures. It is also very important to note that the studies showing benefit used treatments that consisted of isoflavone mixtures along with standard rodent diet that is likely to contribute to the isoflavone content of the diet. The lack of firm knowledge of dose in these studies limits interpretation, especially in light of the fact that hormone dose–response curves are often not monotonic. Future work will need to address these issues by administering individual isoflavones over a range of doses.

27.5 Human Clinical Studies of Beneficial Effects of Dietary Isoflavones on Learning and Memory

The studies in OVX rats suggest that dietary isoflavones can serve as an estradiol replacement that improves the cognitive deficits associated with ovariectomy. So, if one accepts the OVX rat as a valid model of menopause, then the question arises whether dietary isoflavones can improve post-menopausal cognition in humans (see below). A few small studies have addressed this question.

In looking at the studies of dietary isoflavones on cognitive effects on humans in total, an interesting trend emerges in which isoflavone-containing supplements have greater benefit to women younger than 65 years of age compare to women older than 65 years. This trend is appealing because of the current hypothesis that there is a "critical window" early in menopause during which estrogen-targeting therapies are beneficial, and beyond which they have neutral or detrimental effects.

Two of the earliest studies (Duffy et al. 2003; File et al. 2005) to examine the effects of isoflavones on cognition in menopause used oral soy supplements in women aged 50–66 years for 6–12 weeks. In both studies there was significant improvement in mental flexibility and planning; however, only one of the studies (Duffy et al. 2003) found an improvement in memory function, specifically verbal and figural memory. These data suggest that the benefits of soy are not on hippocampal-dependent memory *per se*, but are instead on prefrontal-dependent executive function.

A later, somewhat larger, crossover study (Casini et al. 2006) in women less than 65 years old were administered a 60 mg isoflavone tablet or placebo daily for 24 weeks followed by a 4 week washout and then a switch to the opposite treatment. Like the earlier studies, benefits of the isoflavones were seen in tasks that depend on prefrontal function, specifically working memory and psychomotor speed.

A more recent study by Santos-Galduróz et al. (2010) treated women aged 50–65 years with a daily oral dose of 80 mg of isoflavone extract for 16 weeks. The results were consistent with the trend in the aforementioned studies in which the isoflavones reliably enhanced precortical executive function but not memory formation.

Not all studies of menopausal women younger than 65 years have found benefits of isoflavone treatment. Fournier et al. (2007) assessed working memory after administering either soy milk (60–100 mg) or the Novasoy supplement (70 mg) daily for 16 weeks. In contrast to the findings mentioned above, this study found that the soy milk *impaired* working memory, whereas the soy supplement had no effect. A neutral effect of soy supplementation (soy protein powder) on prefrontal functions was also reported by Basaria et al. (2009).

Two studies have directly compared the effects of isoflavone-containing supplements (24 weeks) on menopausal women younger and older than age 65 years. A study by Kritz-Silverstein et al. (2003) reported improvements in the younger women in mental flexibility (a prefrontal cognitive domain) and in verbal memory (a hippocampal cognitive domain). Interestingly, the older women that received the supplement in this study had *impaired* verbal memory, but improved prefrontal executive function. In contrast, Ho et al. (2007) found only neutral effects of isoflavone supplementation in any cognitive domain.

Only one study has examined the cognitive effects of isoflavone supplements exclusively on menopausal women older than age 65 years (Kreijkamp-Kaspers et al. 2004). Here, soy protein was administered for a full year and verbal memory (hippocampal dependent), working memory (prefrontal dependent), and verbal fluency (prefrontal dependent) were assessed. Only neutral effects were obtained.

When one considers the clinical trials of dietary/oral administration of isoflavones in total, two trends emerge. One is that the benefits, when present, seem to tilt towards executive function that is mediated by the prefrontal cortex. Benefits in hippocampally mediated memory function have appeared less reliably. A second trend is that benefits are more consistently seen in women younger than age 65 years, suggesting that isoflavones are more effective in augmenting cognition when they are given closer to transition into menopause. A major caveat to keep in mind as one critically considers these trends in the data is that all of these studies have been relatively small. Clearly, larger scale efforts are needed to unequivocally determine whether isoflavones have cognitive benefits after menopause.

27.6 Summary

Isoflavones are phytoestrogens found in food sources, most notably soy. They are similar in chemical structure to 17β-estradiol and exert either agonist or antagonist activity on estrogen receptors. Given the role of estradiol in neural plasticity, there has been interest in whether dietary isoflavones can impact learning and memory. Data from both rodent and human studies indicate that dietary isoflavones can affect learning and memory, but the direction of the effects depends on physiological context. Dietary isoflavones can be beneficial to cognition when administered as a sort of estrogen replacement therapy in OVX rodents or menopausal women. Dietary isoflavones can be deleterious to cognition when administered as endocrine disruptors during development. In order to better understand the cognitive effects of isoflavones, future animal studies should move away from heterogeneous mixtures of unknown isoflavone concentrations and towards utilizing known doses of individual isoflavones. Future clinical trials could enhance our understanding by relying on larger sample sizes and systematically comparing younger (<age 65 years) and older (>age 65 years) women.

Summary Points

- Isoflavones can act as ligands at estrogen receptors, and thereby impact estradiol-mediated neural plasticity and development.
- The impact of dietary isoflavones on learning and memory can be either beneficial or deleterious depending on the physiological context.
- Exposure of male rats to isoflavones through the maternal diet results in spatial memory impairments in adulthood.
- Administration of dietary isoflavones to ovariectomized rats can augment learning and memory function.
- Oral administration of isoflavones to post-menopausal women can offer cognitive benefits, especially in women younger than age 65 years and in the domain of prefrontal-dependent executive function.

Key Facts

Table 27.1 Key facts regarding rodent behavioral paradigms of learning and memory. The Table identifies the rodent behavioral paradigms of learning and memory that have been used to assess the effects of dietary isoflavones. The behavior required by the animal, the cognitive domain of the behavior, and the brain structure critically involved in regulating the behavior are provided.

Name of paradigm	Behavioral requirement	Cognitive domain	Critical brain structure
Radial arm maze	Enter maze arms for food reward	Spatial reference and working memory	Hippocampus
Morris water maze	Swim to a hidden platform	Spatial reference memory	Hippocampus
Y-Maze	Explore maze arms in an alternating fashion	Spatial working memory	Hippocampus
Passive avoidance	Avoid chamber in which foot shock occurred	Fear conditioning	Hippocampus and amygdala
Object placement task	Investigate a moved object more than an unmoved object	Spatial memory	Hippocampus
Operant conditioning	Regulate lever responses for food rewards	Executive function	Prefrontal cortex

Definition of Words and Terms

Dendrite: Branch-like neuronal structure usually responsible for receiving information from other neurons. Contains dendritic spines which are knob-like structures that undergo morphological and biochemical changes associated with learning and memory.

Endocrine disruptor: An exogenous agent that interferes with the synthesis, secretion, transport, metabolism, binding action, or elimination of endogenous hormones.

Hippocampus: Structure of the brain found in the temporal lobe. It is critically involved in learning and memory, especially spatial learning and memory.

Ovariectomy: Removal of the ovaries. It is used in rodents to model hypoestrogenic states such as menopause.

Prefrontal cortex: Structure found in the frontal lobe of the brain. It is responsible for executive functions of the brain such as planning, strategizing, and working memory.

Sexual dimorphism: A trait that differs between males and females.

Working memory: A form of short term memory in which information is briefly held and manipulated for the purpose of performing some cognitive task.

List of Abbreviations

OVX ovariectomized
SPE soy protein isolate with phytoestrogens

References

Bagheri, M., Joghataei, M.-T., Mohseni, S., and Roghani, M., 2011. Genistein ameliorates learning and memory deficits in amyloid β(1–40) rat model of Alzheimer's disease. *Neurobiology of Learning and Memory*. 95: 270–276.

Ball, E.R., Caniglia, M.K., Wilcox, J.L., Overton, K.A., Burr, M.J., Wolfe, B.D., Sanders, B.J., Wisniewski, A.B., and Wrenn, C.C., 2010. Effects of genistein in the maternal diet on reproductive development and spatial learning in male rats. *Hormones and Behavior*. 57: 313–322.

Bansal, N., and Parle, M., 2010. Soybean supplementation helps reverse age- and scopolamine-induced memory deficits in mice. *Journal of Medicinal Food*. 13: 1293–1300.

Basaria, S., Wisniewski, A.B., Dupree, K., Bruno, T., Song, M.Y., Yao, F., Ojumu, A., John, M., and Dobs, A.S., 2009. Effect of high-dose isoflavones on cognition, quality of life, androgens, and lipoprotein in post-menopausal women. *Journal of Endocrinology Investigation*. 32: 150–155.

Casini, M.L., Marelli, G., Papaleo, E., Ferrari, A., D'Ambrosio, F., and Unfer, V., 2006. Psychological assessment of the effects of treatment with phytoestrogens on postmenopausal women: A randomized, double-blind, crossover, placebo-controlled study. *Fertility and Sterility*. 85: 972–978.

Duffy, R., Wiseman, H., and File, S.E., 2003. Improved cognitive function in postmenopausal women after 12 weeks of consumption of a soya extract containing isoflavones. *Pharmacology Biochemistry and Behavior*. 75: 721–729.

Ennaceur, A., and Aggleton, J.P., 1994. Spontaneous recognition of object configurations in rats: Effects of fornix lesions. *Experimental Brain Research*. 100: 85–92.

Ennaceur, A., and Aggleton, J.P., 1997. The effects of neurotoxic lesions of the perirhinal cortex combined to fornix transection on object recognition memory in the rat. *Behavioural Brain Research*. 88: 181–193.

File, S.E., Hartley, D.E., Elsabagh, S., Duffy, R., and Wiseman, H., 2005. Cognitive improvement after 6 weeks of soy supplements in postmenopausal women is limited to frontal lobe function. *Menopause*. 12: 193–201.

Fournier, L.R., Ryan Borchers, T.A., Robison, L.M., Wiediger, M., Park, J.S., Chew, B.P., McGuire, M.K., Sclar, D.A., Skaer, T.L., and Beerman, K.A., 2007. The effects of soy milk and isoflavone supplements on cognitive performance in healthy, postmenopausal women. *Journal of Nutrition Health and Aging*. 11: 155–164.

Foy, M.R., Xu, J., Xie, X., Brinton, R.D., Thompson, R.F., and Berger, T.W., 1999. 17β-Estradiol enhances NMDA receptor-mediated EPSPs and long-term potentiation. *Journal of Neurophysiology*. 81: 925–929.

Foy, M.R., Baudry, M., Foy, J.G., and Thompson, R.F., 2008. 17β-Estradiol modifies stress-induced and age-related changes in hippocampal synaptic plasticity. *Behavioral Neuroscience*. 122: 301–309.

Genazzani, A.R., Pluchino, N., Luisi, S., and Luisi, M., 2007. Estrogen, cognition and female ageing. *Human Reproduction Update*. 13: 175–187.

Gresack, J.E., and Frick, K.M., 2003. Male mice exhibit better spatial working and reference memory than females in a water-escape radial arm maze task. *Brain Research*. 982: 98–107.

Halbreich, U., Lumley, L.A., Palter, S., Manning, C., Gengo, F., and Joe, S.-H., 1995. Possible acceleration of age effects on cognition following menopause.. *Journal of Psychiatric Research*. 29: 153–163.

Henderson, V.W., 1997. Estrogen, cognition, and a woman's risk of Alzheimer's disease. *The American Journal of Medicine*. 103: 11S–18S.

Ho, S.C., Chan, A.S., Ho, Y.P., So, E.K., Sham, A., Zee, B., and Woo, J.L., 2007. Effects of soy isoflavone supplementation on cognitive function in chinese postmenopausal women: A double-blind, randomized, controlled trial. *Menopause*. 14: 489–499.

Isgor, C., and Sengelaub, D.R., 1998. Prenatal gonadal steroids affect adult spatial behavior, CA1 and CA3 pyramidal cell morphology in rats. *Hormones and Behavior*. 34: 183–198.

Isgor, C., and Sengelaub, D.R., 2003. Effects of neonatal gonadal steroids on adult CA3 pyramidal neuron dendritic morphology and spatial memory in rats. *Journal of Neurobiology*. 55: 179–190.

Kim, D., Jung, H., Park, S., Kim, J., Lee, S., Choi, J., Cheong, J., Ko, K., and Ryu, J., 2010. The effects of daidzin and its aglycon, daidzein, on the scopolamine induced memory impairment in male mice. *Archives of Pharmacy Research, Pharmaceutical Society of Korea*. 33: 1685–1690.

Kreijkamp-Kaspers Sanne, K.L., Grobbee, D.E., de Haan Edward, H.F., Aleman, A., Lampe, J.W., and van der Schouw, Y.T., 2004. Effect of soy protein containing isoflavones on cognitive function, bone mineral density, and plasma lipids in postmenopausal women: A randomized controlled trial. *JAMA: Journal of the American Medical Association*. 292: 65–74.

Kritz-Silverstein, D., Mühlen, D.V., Barrett-Connor, E., and Bressel, M.A.B., 2003. Isoflavones and cognitive function in older women: The soy and postmenopausal health in aging (SOPHIA) study. *Menopause*. 10: 196–202.

Kuiper, G.G.J.M., Lemmen, J.G., Carlsson, B., Corton, J.C., Safe, S.H., van der Saag, P.T., van der Burg, B., and Gustafsson, J.-A., 1998. Interaction of estrogenic chemicals and phytoestrogens with estrogen receptor β. *Endocrinology*. 139: 4252–4263.

Kurzer, M.S., and Xu, X., 1997. Dietary phytoestrogens. *Annual Review of Nutrition*. 17: 353–381.

Li, J., Wang, G., Liu, J., Zhou, L., Dong, M., Wang, R., Li, X., Li, X., Lin, C., and Niu, Y., 2010. Puerarin attenuates amyloid-β-induced cognitive impairment through suppression of apoptosis in rat hippocampus in vivo. *European Journal of Pharmacology*. 649: 195–201.

Luine, V., Attalla, S., Mohan, G., Costa, A., and Frankfurt, M., 2006. Dietary phytoestrogens enhance spatial memory and spine density in the hippocampus and prefrontal cortex of ovariectomized rats.. *Brain Research*. 1126: 183–187.

Lund, T.D., West, T.W., Tian, L.Y., Bu, L.H., Simmons, D.L., Setchell, K.D.R., Adlercreutz, H., and Lephart, E.D., 2001. Visual spatial memory is enhanced in female rats (but inhibited in males) by dietary soy phytoestrogens. *BMC Neuroscience*. 2: 20.

Monteiro, S., de Mattos, C., Ben, J., Netto, C., and Wyse, A., 2008. Ovariectomy impairs spatial memory: Prevention and reversal by a soy isoflavone diet. *Metabolic Brain Disease*. 23: 243–253.

Morris, R.G.M., Garrud, P., Rawlins, J.N.P., and O'Keefe, J., 1982. Place navigation impaired in rats with hippocampal lesions. *Nature*. 297: 681–683.

Neese, S.L., Wang, V.C., Doerge, D.R., Woodling, K.A., Andrade, J.E., Helferich, W.G., Korol, D.L., and Schantz, S.L., 2010. Impact of dietary genistein and aging on executive function in rats. *Neurotoxicology and Teratology*. 32: 200–211.

Neese, S.L., Bandara, S.B., Doerge, D.R., Helferich, W.G., Korol, D.L., and Schantz, S.L., 2012. Effects of multiple daily genistein treatments on delayed alternation and a differential reinforcement of low rates of responding task in middle-aged rats. *Neurotoxicology and Teratology*. 34: 187–195.

Pan, Y., Anthony, M., Watson, S., and Clarkson, T.B., 2000. Soy phytoestrogens improve radial arm maze performance in ovariectomized retired breeder rats and do not attentuate benefits of 17β-estradiol treatment. *Menopause*. 7: 230–235.

Patisaul, H.B., Dindo, M., Whitten, P.L., and Young, L.J., 2001. Soy isoflavone supplements antagonize reproductive behavior and estrogen receptor α- and β-dependent gene expression in the brain. *Endocrinology*. 142: 2946–2952.

Patisaul, H.B., and Jefferson, W., 2010. The pros and cons of phytoestrogens. *Frontiers in Neuroendocrinology*. 31: 400–419.

Roof, R.L., and Havens, M.D., 1992. Testosterone improves maze performance and induces development of a male hippocampus in females. *Brain Research*. 572: 310–313.

Santos-Galduróz, R.F., Galduroz, J.C.F., Facco, R.L., Hachul, H., and Tufik, S., 2010. Effects of iosflavone on the learning and memory of women in menopause: A double-blind placebo-controlled study. *Brazilian Journal of Medical and Biological Research*. 43: 1123–1126.

Sarkaki, A., Badavi, M., Aligholi, H., and Moghaddam, A.Z., 2009. Preventive effects of soy meal (+/−) isoflavone on spatial cognitive deficiency and body weight in an ovariectomized animal model of Parkinson's disease. *Pakistan Journal of Biological Sciences*. 12: 1338–1345.

Sarkaki, A.R.A., Badavi, M., Moghaddam, A.Z., Aligholi, H., Safahani, M., and Haghighizadeh, M.H., 2008. Pre-treatment effect of different doses of soy isoflavones on spatial learning and memory in an ovariectomized animal model of Alzheimer's disease. *Pakistan Journal of Biological Sciences.* 11: 1114–1119.

Woolley, C.S., and McEwen, B.S., 1993. Roles of estradiol and progesterone in regulation of hippocampal dendritic spine density during the estrous cycle in the rat. *Journal of Comparative Neurology.* 336: 293–306.

CHAPTER 28
Glycitein in Health

BRIAN R. STEPHENS AND JOSHUA A. BOMSER*

Department of Human Nutrition, Ohio State University, 1787 Neil Ave., Columbus, Ohio 43210, USA
*Email: bomser.1@osu.edu

28.1 Introduction

Glycitein is the third most abundant isoflavone found in soybeans (Figure 28.1). The abundance of glycitein has high intra- and intercultivar variation depending on seed variety, environment, and processing (Gutierrez-Gonzalez *et al.* 2010). Consequently, heavily processed foods such as soymilk and tofu have 5–10% glycitein content, whereas milled soybeans (soy germ) may have 40%. There is considerable evidence which suggests glycitein may contribute to plant health and have beneficial effects in humans. Studies have shown that glycitein induces a variety of health-promoting benefits which include inhibition of stress-related kinases and inflammation as well as robust antioxidant and estrogenic activity (Kinjo *et al.* 2004; Lee *et al.* 2010). These physiological effects are thought to contribute to the observed reduction of hormone- and inflammation-associated cancers, amelioration of menopausal symptoms, and improved lipid serum profile in individuals who regularly consume soy and soy-derived products (Lee *et al.* 2007). Although glycitein is synthesized from the amino acid phenylalanine through the phenylpropanoid pathway in plants, the enzyme(s) involved in conversion of isoliquiritigenin into glycitein have not been fully elucidated. What is known, however, is that newly synthesized glycitein are stored in plant vacuoles as glucosylated (glycitin), malonylated (malonyl glycitin), and acetylated (acetyl glycitin) conjugates (Carrara *et al.* 2009). These

---

Food and Nutritional Components in Focus No. 5
Isoflavones: Chemistry, Analysis, Function and Effects
Edited by Victor R Preedy
© The Royal Society of Chemistry 2013
Published by the Royal Society of Chemistry, www.rsc.org

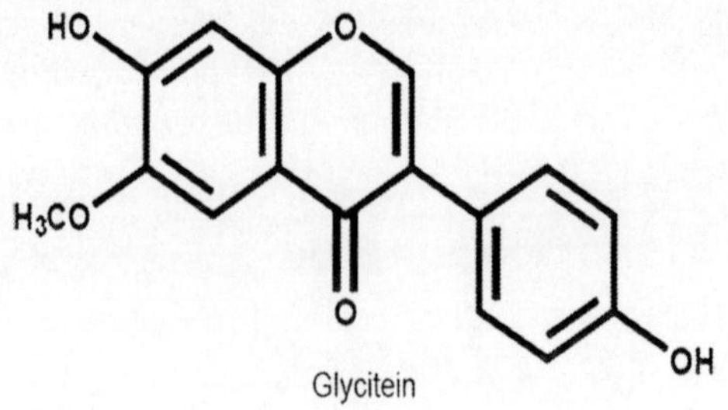

Figure 28.1 Structure of glycitein. The chemical structure of the soy isoflavone glycitein.

conjugates can be released during digestion to glycitein by salivary and pancreatic glucosidases and absorbed from the gut into the blood.

28.2 Absorption

During digestion, glycitin and its acetylated and malonylated conjugates are cleaved by glucosidases present in the saliva, pancreatic juice, small intestinal wall, and intestinal bacteria. Glycitein is then absorbed from the gut into enterocytes either from mixed micelles or through organic anionic transporters. The major metabolic fate of glycitein is glucuronidation by specific isoforms of the uridine diphosphate glucuronosyltransferases (UGTs) family (Tang et al. 2009). UGTs are endoplasmic reticulum-resident proteins enriched in enterocytes and hepatocytes which catalyze the conjugation of glucuronic acid to polyphenols. This addition, or conjugation, of glucuronic acid to polyphenols increases the water-solubility, affinity for apical efflux into the small intestine, and probability for biliary and kidney excretion. Tang et al. (2009) investigated the role of UGTs on the conjugation of isoflavones in human livers and found that isoforms 1A1, 1A8, 1A9, and 1A10 had a preference for glycitein over genistein and daidzein. Despite extensive first-pass metabolism of glycitein, consumption of soy products leads to an increase in glycitein plasma concentrations. Zhang et al. (1999) fed healthy human volunteers a glycitein-poor (soymilk) or glycitein-rich (soy germ) diet and analyzed plasma glycitein concentrations. The results showed glycitein concentrations in plasma peaked within 6 h at 0.2 µM and 0.85 µM after consumption of soymilk and soy germ, respectively. Over a 48 h period, 55% of the ingested glycitein was excreted as glucuronides compared with 28% for genistein (Zhang et al. 1999). The shorter half-life of glycitein (3.4 h) compared with genistein (7.4 h) and daidzen (3.8 h) suggests that glycitein glucuronides may be more water-soluble than genistein or daidzein (Bloedon et al. 2002). The intestine harbors gut bacteria, or microflora, which contribute to polyphenol metabolism through enzymatic modification of substrates. The microflora of the intestine converts glycitein into more estrogenic compounds such as daidzen and hydroxylated equol derivatives (Simons *et al.* 2005); further, studies have shown while wide variety of gut bacteria contribute the glycitein metabolism, *Bacteroides acidifaciens*, *Bacteroides uniformis*, and *Tannerella forsythensis* are the major metabolizers of glycitein (Renouf and Hendrich 2011). More studies are needed to determine the specific strains of gut bacterium that are responsible for glycitein degradation and metabolism in the intestine.

28.3 Glycitein and Bone Health

At any one time, approximately 30% of the mineralized bone is undergoing turnover and remodeling. The precision of this remodeling is controlled by hormones such as estrogen, testosterone, and vitamin D. These hormones differentially stimulate osteoblasts ("bone forming") and suppress osteoclasts ("bone breaking") formation in the skeleton. The precipitous drop in estrogen levels in post-menopausal women leads to enhanced osteoclast activity which increases the likelihood of hip fractures and osteoporosis. While soy supplementation has been shown to suppress bone turnover by 14% in post-menopausal women (Taku *et al.* 2010), the specific effect of glycitein—separate from genistein and daidzein—on bone cells is not well-characterized in the literature. There is evidence that glycitein stimulates osteoblast differentiation (Yoshida *et al.* 2001) and increases osteoclast apoptosis (Winzer *et al.* 2010) in murine cell lines, although the specific mechanism is not known. To determine the effect of glycitein on expression of the vitamin D receptor (VDR), Gilad *et al.* (2006) transfected a vector containing the VDR promoter next to a firefly luciferase reporter into HT29 cells and found glycitein increased promoter activity similar to that of estrogen. The authors also found that glycitein increased transcription and translation of the VDR receptor in the transfected cell line. These results are interesting as osteoblasts constitutively express VDRs, suggesting that osteoblast stimulation by glycitein may be related, in part, to increased responsiveness to serum vitamin D levels. More studies are needed to discern the effect of glycitein on osteoblast and osteoclast activity and its overall impact on bone health.

28.4 Glycitein and Cancer

Chronic inflammation is associated with the development of many types of cancer. Cancer is typically a slow process and often represents a lifetime of accumulated DNA damage and subsequent alteration of gene expression. These genetic alterations induce distinct changes in normal cells that promote uncontrolled cell division, insensitivity to anti-growth signals, apoptotic escape, secretion of angiogenic factors, as well as tissue invasion and metastasis (Hanahan and Weinberg 2000). Constitutive activation of nuclear factor-κB (NF-κB) is a hallmark of inflammation-associated cancers. Active NF-κB translocates from the cytosol to the nucleus and directs synthesis of genes involved in cellular proliferation, apoptotic escape, and angiogenesis (Maeda and Omata 2008). Glycitein has been shown to inhibit NF-κB activity, reduce activator protein 1 (AP-1) binding to its promoter, and decrease the phosphorylation of mitogen-activated protein kinases (MAPKs) including p38, c-Jun N-terminal kinase (JNK), and extracellular-signal-regulated kinase 1/2 (ERK1/2) in glioma cells (Lee et al. 2010). Glycitein was also found to decrease the expression of NF-κB-dependent genes in response to inflammation such as matrix metalloproteinases (MMPs). High MMP expression is correlated with tumor aggressiveness and invasiveness, suggesting that glycitein may reduce the metastatic potential of some cancers. Methanolic fractions of glycitein-containing soybeans have been shown to have peroxisome proliferator-activated receptor α (PPARα) agonist activity (Carrara et al. 2009), which is interesting as PPARα agonists have been shown to have anti-inflammatory (Staels et al. 1998) and anti-cancer (Panigrahy et al. 2008) properties. PPARα is nuclear receptor that upon ligand binding translocates to the nucleus and directs the transcription of genes involved in fatty oxidation and energy homeostasis. Evidence suggests that PPARα activation suppresses tumor growth in many cancer cell lines including breast, colon, and endothelial cells through down-regulation of angiogenic factors such as vascular endothelial growth factor (VEGF) (Panigrahy et al. 2008). Overall, the anti-cancer activity of glycitein is likely attributed, in part, to inhibition of stress-related MAPKs and inflammation.

Epidemiological evidence suggests that early dietary intervention with soy can improve health outcomes in prostate cancer patients. Many studies have focused on the mechanisms by which soy isoflavones can inhibit prostate carcinogenesis. Modulation of MAPK activity has emerged as one target by which glycitein may reduce prostate cancer risk. The ERK1/2–MAPK cascade is involved in prostate cancer progression and its activation may be associated with prevention of this disease. Glycitein was found to induce and sustain ERK1/2 activity to a greater extent than genistein, daidzein, or equol in non-tumorigenic prostate epithelial cells and suppress proliferation of non-tumor-genic RWPE-1 cells through activation of the VEGF receptor (VEGFR) (Clubbs and Bomser 2007) (Figures 28.2 and 28.3). Interestingly, isoflavone-induced ERK1/2 activation was not observed in cancerous prostate (PC-3) cells (Figure 28.2).

Sustained activation of the ERK1/2 signaling cascade is associated with cellular differentiation. Therefore, it is hypothesized that sustained glycitein-induced ERK1/2 activation may alter prostate cell differentiation. This is of particular importance given that the progression of prostate cancer involves a loss of the basal prostate cell differentiation and uncontrolled division of the luminal prostate cell population. Data from our laboratory suggest that glycitein may induce basal cell differentiation and suppress luminal differentiation (Figure 28.4) in RWPE-1 cells (Clubbs and Bomser 2009); interestingly, these results were cancer-stage specific and not recapitulated in a precancerous prostate cell line. Taken together, these data suggest that glycitein has affinity

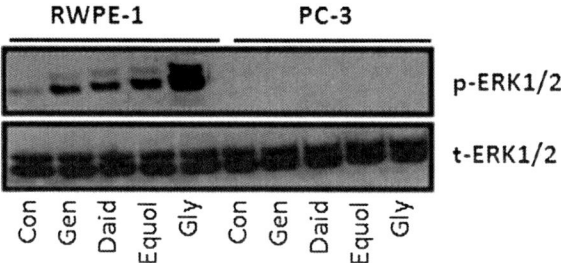

Figure 28.2 Effect of isoflavones on ERK1/2 activity in prostate cells. Activation of ERK1/2 [phospho (p)-ERK1/2] following treatment of non-tumorgenic (RWPE-1) and tumorgenic (PC-3) prostate cells with 50 μM genistein (Gen), diadzein (Daid), equol, and glycitein (Gly) for 2 h. Con, control. Modified from Clubbs and Bomser (2007).

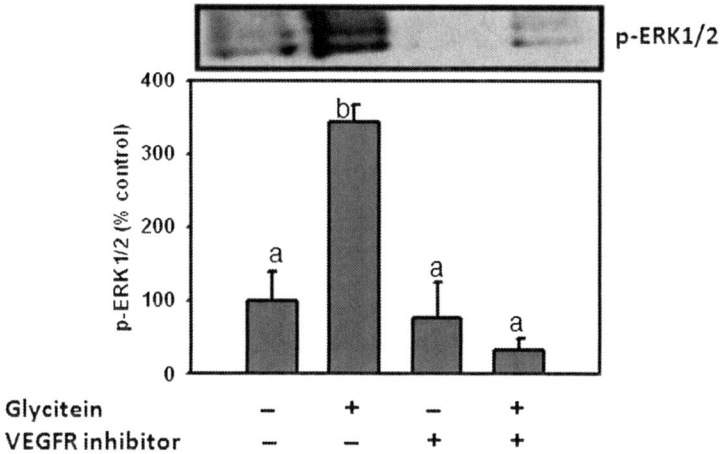

Figure 28.3 Glycitein activates ERK1/2 *via* VEGFR in RWPE-1 cells. Treatment of RWPE-1 cells with glycitein (50 μM, 2 h) activates ERK1/2. Glycitein-induced ERK1/2 activation is blocked using an inhibitor of VEGFR. Treatments with different letters are statistically significant ($P < 0.05$).

for the cell-surface VEGFRs and is unique in the induction of basal cell differentiation in prostate cancer cells. The preservation and induction of basal cell differentiation, as well as attenuated proliferation in the prostate epithelium, by glycitein may represent one mechanism by which soy consumption reduces prostate cancer risk. Isoflavone concentrations in the prostate can reach up toward 50 μM and may serve as a reservoir for the accumulation of these molecules (Hedlund *et al.* 2006). Future studies using glycitein-rich soy germ in the chemoprevention of prostate cancer are warranted.

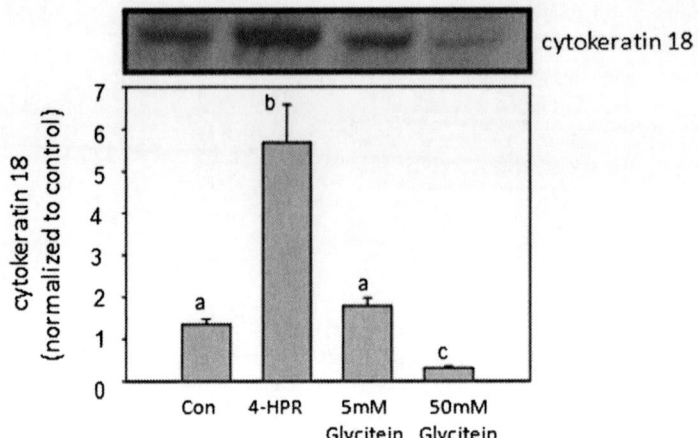

Figure 28.4 Glycitein and cytokeratin 18 expression in RWPE-1 cells. Effects of glycitein on expression of the luminal cell marker cytokeratin 18 in non-tumorgenic RWPE-1 cells. The synthetic retinoid N-(4-hydroxyphenyl)retinamide (4-HPR) was used as a positive control for cytokeratin 18 expression. Treatments with different letters are statistically significant ($P < 0.05$).

28.5 Glycitein and Cardiovascular Health

Blood pressure refers to the pressure exerted on the vasculature and is influenced by age and gender as well as by dietary factors such as consumption of sodium. Chronic high blood pressure, or hypertension, is a risk factor for heart failure, stroke, and cardiovascular disease. Soy consumption is correlated with a reduction in systolic and diastolic blood pressure. In an 8 week crossover trial, normotensive and hypertensive post-menopausal women were fed soy nuts containing 10 mg of glycitein. The results showed that normotensive and hypertensive women had reduced diastolic and systolic blood pressure (Welty et al. 2007). Glycitein-containing nuts lowered systolic blood pressure by 6–15 mm Hg and diastolic blood pressure by 2–6 mm Hg. These results are compelling as: (1) a 2 mm Hg reduction in diastolic blood pressure is correlated with 6% reduction of stroke and 15% reduction in coronary heart disease (Cook et al. 1995), and (2) the reductions in blood pressure were comparable with prescription anti-hypertensive drugs (Welty et al. 2007). In another study, isoflavone preparations containing 1% glycitein reduced systolic and diastolic blood pressure by 7% (Bloedon et al. 2002). Overall, these studies and others suggest that dietary modification alone can improve overall cardiovascular health. It is likely that the vasoprotective properties of glycitein are due its affinity for the estrogen receptor (ER) and subsequent enhancement of nitric oxide-dependent vasodilation (Arnal et al. 2010). Glycitein has been shown to suppress serum cholesterol levels (Lee et al. 2007) and is a thromboxane receptor antagonist (Muñoz et al. 2009). Thromboxane receptors are expressed by platelets; activation by thromboxane, an eicosanoid lipid, induces aggregation and clot formation, suggesting that glycitein may have putative anti-atherogenic and anti-thrombotic properties. More studies, however, are needed to understand the mechanism(s) involved in the vaso- and cardioprotective properties of glycitein alone and in combination of other isoflavones.

28.6   Glycitein as an Estrogen Receptor (ER) Agonist

ERs are cytoplasmic nuclear receptors that upon ligand binding translocate to the nucleus and induce transcription of target genes. There is structural similarity between estrogen and glycitein. Despite glycitein possessing weak binding affinity for the ER (approximately 3600 times less potent than 17β-estradiol) *in vitro*, Song *et al.* (1999) found that glycitein induced a robust enlargement of the mouse uterus compared with genistein and daidzen (Song *et al.* 1999). While the estrogenic activity of possible glycitein metabolites is limited, there are data to support that glycitein glucuronides bind—albeit with less affinity than the aglycone—to the α and β isoforms of the ER (Kinjo *et al.* 2004). When researchers compared transcription of estrogen-responsive genes among isoflavones using DNA microarray technology, glycitein altered more genes than daidzein and genistein; furthermore, among the isoflavones tested glycitein was unique in that treatment down-regulated the activity of estrogen-responsive genes such as *CDKN1A* and *IGFBP4* (Ise *et al.* 2005). Unlike genistein, which is metabolized into non-estrogenic compounds, glycitein is converted into daidzen and hydroxylated equol derivatives by the liver and human gut microflora (Rüfer *et al.* 2007; Simons *et al.* 2005). CYP1A2 has been shown to hydroxylate the methoxy group on glycitein, resulting in daidzen formation. Taken together, these data suggest that glycitein may have greater estrogenic potency *in vivo* compared with other dietary isoflavones. This may be due, in part, to glycitein's higher bioavailability, estrogenic activity and metabolism into more potent estrogenic compounds such as equol.

28.7   Glycitein and Memory and Mood

The hippocampus, the area of the brain involved in short- and long-term storage of memories, possesses neural stem cells that undergo neurogenesis ("birth of new neurons"); indeed, it has become clear that hippocampal neurons are not quiescent and undergo continual turnover throughout life. Hippocampal chandelier cells modulate the maturation of newborn neurons and integration into the hippocampal formation (Hwang *et al.* 2006). When middle-aged ovariectomized mice were fed a soy diet containing 40% glycitein, Hwang *et al.* (2006) found a dose-dependent reduction in parvalbumin-containing chandelier cells. The authors had previously shown this soy diet ameliorated memory deficits in elderly male rats (Lee *et al.* 2004). This reduction of parvalbumin-containing chandelier cells has been shown in the brains of mice in enriched housing and correlates with increased spatial memory performance (Iuvone *et al.* 1996). Since studies have shown that ERs co-localize with hippocampal chandelier cells, this study suggests that soy-induced reduction of chandelier cells may be due to glycitein-induced activation of the ER. Moreover, estrogen has been shown to be both neuroprotective and synthesized *de novo* in the hippocampus, which lends strength to the hypothesis that glycitein may be acting as an ER agonist rather than exerting antioxidant or anti-inflammatory activity.

While there are studies that show anti-anxiety and anti-depressive effects in animals that are administered soy-based diets, limited data exists on the mechanism(s) in which glycitein exerts a neuroprotective effect. Compared with non-neuronal tissues, neurons of the central nervous system (CNS) are especially vulnerable to oxidative stress due to high oxygen consumption and lipid content as well as reduced antioxidant defense. Astrocytes are microglial cells which provide maintenance to neurons through myelination of axons, CNS repair, and nutrient support. Treatment of rat astrocytes in culture with 50 μM glycitein increased the expression of heme oxygenase 1 (HO-1) and nuclear factor erythroid-derived 2 (Nrf2) within 12 h of treatment. HO-1 is an inducible antioxidant defense isoform that cleaves hemoglobin into carbon monoxide, ferrous iron, and biliverdin, which possess free radical scavenging capability (Park *et al.* 2011). The formation of amyloid-β in neurons suppresses synapse formation and induces neuronal death and is thought to underlie the progressive memory impairment observed in Alzheimer's disease patients. Glycitein—not genistein nor daidzein—reduced amyloid-β deposits in *Caenorhabditis elegans*, suggesting that glycitein may be unique in ameliorating amyloid-β-induced toxicity. Since the generation of free radicals precedes the formation of amyloid aggregates, it is hypothesized that the neuroprotective effects of glycitein may be due to its role as an antioxidant (Gutierrez-Zepeda *et al.* 2005). However, the quenching of free radicals did not ameliorate the formation of amyloid-β aggregates in rat hippocampal cultures (Lockhart *et al.* 1994), suggesting that the neuroprotective effect of glycitein may be distinct from its role as an antioxidant.

28.8 Glycitein and Oxidative Stress

Oxidative stress can lead to the production of free radicals that can damage lipid membranes, disrupt protein function, modify purine and pyramidine DNA bases, and elicit single- and double-stranded DNA breaks that pave the way for genomic instability, cellular senescence, and the onset of cancer. Although glycitein possesses low free radical scavenging capability compared with genistein and daidzein, glycitein has been shown to inhibit copper-induced low-density lipoprotein (LDL) oxidation (Rüfer and Kulling 2006) as well as hydrogen peroxide-induced apoptosis (Kang *et al.* 2007). It has been surmised that methylation of the hydroxyl group on the A ring decreases the free radical scavenging capability of glycitein (Rüfer and Kulling 2006), suggesting other mechanism(s) may be involved in antioxidant defense. As polyphenols are common activators of the transcription factor Nrf2 (Surh and Na 2008), one possibility is that the intracellular accumulation of glycitein promotes Nrf2-mediated induction of antioxidant defense genes such as glutathione and quinone reductase. Glutathione is a short peptide consisting of three amino acids (Glu-Gly-Cys) and participates in the scavenging of free radicals in the intracellular environment. Quinone reductase is an intracellular enzyme that participates in the regeneration of vitamin E, vitamin K, and coenzyme Q from an oxidized to a reduced active state. Glutathione and quinone reductase are both Nrf2-inducible and constitutively expressed. Glycitein also suppresses oxidative DNA damage through binding to ERα. Ligand binding ERα enhances aberrant base removal through interaction of proteins involved in DNA repair (Likhite et al. 2004) and decreases oxidant insult through increased expression of superoxide dismutase 1 (SOD1) (Schultz-Norton et al. 2011). SOD1 is found in both the cytoplasm and mitochondria and participates in the conversion of superoxide into molecular oxygen and hydrogen peroxide. Since glycitein and its glucuronide conjugate has been shown to have affinity for both ERα and ERβ (Kinjo et al. 2004) as well as a propensity to be metabolized *in vivo* into more estrogenic compounds (Rüfer et al. 2007; Simons et al. 2005), it is likely that ERα induction of DNA repair and antioxidant genes contribute to the beneficial effects observed with glycitein treatment. Glycitein metabolites, such as equol, have a higher affinity for the ER than the parent compound. It is therefore conceivable that one molecule of glycitein and its multiple estrogenic metabolites could participate in multiple rounds of ER (α and β isoforms) and Nrf2 activation before subsequent inactivation and excretion.

Summary Points

- This Chapter focuses on the recent advances in glycitein and health.
- Glycitein binds to both isoforms (α and β) of the estrogen receptors (ERs). On a mole-for-mole basis, glycitein possesses more estrogenic activity than daidzein or genistein.
- Glycitein is absorbed from the gut and undergoes glucuronidation by the intestine and liver.
- Glycitein can be metabolized *in vivo* into estrogenic compounds such as daidzein and equol.
- Glycitein differentially stimulates osteoblasts and suppresses osteoclast activity. Glycitein has also been shown to increase the responsiveness of osteoblasts to vitamin D.
- Glycitein improves cognitive function through suppression of hippocampal chandelier cell activity.
- Glycitein-containing foods may decrease blood pressure *via* an estrogen receptor-mediated or thromboxane receptor antagonist activity.
- Glycitein attenuates the formation of amyloid-β plaques which have been shown to cause the characteristic memory loss in Alzheimer's disease patients.

- Glycitein is a likely ligand for peroxisome proliferator-activated receptor α (PPARα) which is involved in inflammation and metabolism.
- Although glycitein has less antioxidant activity compared with other isoflavones, glycitein induces the expression of DNA repair enzymes as well as proteins involved in antioxidant defense.
- Glycitein has also been shown to suppress copper-induced oxidation of LDL and hydrogen peroxide-induced apoptosis.
- Glycitein inhibits NF-κB, AP-1, and MAPK activities associated with inflammation and cancer development. Glycitein enhances MAPK activity in prostate epithelial cells through interaction with the vascular endothelial growth factor receptor (VEGFR).

Key Facts

- Glycitein was first isolated and characterized from soy extract at the University of Jerusalem in 1972 (Naim et al. 1973).
- The first synthesis was reported in 1975 at the Hungarian Academy of Sciences in Budapest (Antus et al. 1975).
- The glucosylated form of glycitein (glycitin) was first synthesized in 1996 at the Technical University of Budapest (Nogradi and Szollosy 1996).
- In 2003, a team from Kenyatta University and the University of Iowa utilized a conventional microwave oven to convert 2,4,4'-trihydroxy-5-methoxydeoxybenzoin into glycitein (Lang'at-Thoruwa et al. 2003). This method is amenable to large-scale commercial synthesis.
- In plants, the amino acid phenylalanine is a precursor for glycitein synthesis and is converted into the intermediate, isoliquiritigenin. It is not known how this intermediate is converted into glycitein.
- There is more glycitein in soy germ (40%) than in heavily processed soy products such as tofu (10%). Soy germ is often used in the production of soy-based supplements and contributes to the high glycitein content found in these products. In soy germ and processed soy products, the ratios of genistein/daidzen/glycitein are 1:5:4 and 5:4:1, respectively.

Definitions of Words and Terms

Glycitein: The third most abundant isoflavone from soybeans. The isoflavone has a hydroxyl group on the 7 and 4' position and a unique methoxy group on the 6 carbon (Figure 28.1).

Glutathione: A short peptide consisting of three amino acids (Glu-Gly-Cys) and participates in the scavenging of free radicals in the intracellular environment. Glutathione reductase is responsible for recycling or regeneration of glutathione from an oxidized, inactive form (glutathione disulfide) to a reduced, active state (glutathione).

Nuclear factor-κB (NF-κB): This is a heterodimer comprising of the p50 and p65 subunits that is activated in response to inflammation and/or oxidative stress. It has been referred to as a 'master regulator' of inflammation. Activated NF-κB acts as a transcription factor and translocates to the nucleus where it directs the transcription of genes involved in antioxidant defense, inflammation, and proliferation. Many cancers have constitutive activation of NF-κB, making it a prime target for chemotherapeutic research.

Nuclear factor erythroid-derived 2 (Nrf2): This is a transcription factor involved in the transcription of antioxidant defense genes. It is activated by oxidative stress and by several dietary polyphenols.

Phenylpropanoid pathway: An isoflavone synthetic pathway in plants which utilizes the amino acid phenylalanine as a precursor to synthesize daidzein, genistein, and glycitein.

Peroxisome proliferator-activated **receptor α (PPARα)**: This belongs to a family of nuclear receptors that bind fatty acids and direct the transcription of genes involved in fatty acid oxidation and energy homeostasis.

Quinone reductase: An intracellular enzyme that participates in the regeneration of vitamin E, vitamin K, and coenzyme Q from an oxidized to a reduced active state.

Superoxide dismutase 1 (SOD1): This is found in both the cytoplasm and mitochondria and participates in the conversion of superoxide into molecular oxygen and hydrogen peroxide.

Thromboxane receptor: This receptor is expressed by platelets and binds thromboxane which is secreted by damaged cells. Thromboxane binding induces platelet aggregation and subsequent clot formation.

Uridine diphosphate glucuronosyltransferases (UGTs): This term encompasses a family of phenol-preferring detoxifying enzymes that conjugate glucuronic acid to polyphenols and other xenobiotics. This addition of glucuronic acid increases the water solubility and probability of excretion.

List of Abbreviations

AP-1	activator protein 1
CNS	central nervous system
ER	estrogen receptor
ERK1/2	extracellular-signal-regulated kinase 1/2
HO-1	heme oxygenase 1
LDL	low-density lipoprotein
MAPK	mitogen-activated protein kinase
MMP	matrix metalloproteinase
NF-κB	nuclear factor-κB
Nrf2	nuclear factor erythroid-derived 2
PPARα	peroxisome proliferator-activated receptor α
SOD1	superoxide dismutase 1

VDR	vitamin D receptor
VEGF	vascular endothelial growth factor
VEGFR	vascular endothelial growth factor receptor
UGT	uridine diphosphate glucuronosyltransferase

References

Antus, S., Farkas, L., Kardos-Balogh, Z., and Nogradi, M., 1975. The synthesis of dalpatin, fujikinin, glycitein, and other natural isoflavones. *Chemische Berichte*. 108: 3883.

Arnal, J., Fontaine, C., Billon-Galés, A., Favre, J., Laurell, H., Lenfant, F., and Gourdy, P., 2010. Estrogen receptors and endothelium. *Arteriosclerosis, Thrombosis and Vascular Biology*. 30: 1506–1512.

Bloedon, L., Jeffcoat, A., Lopaczynski, W., Schell, M., Black, T., Dix, K., Thomas, B., Albright, C., Busby, M., Crowell, J., and Zeisel, S., 2002. Safety and pharmacokinetics of purified soy isoflavones: single-dose administration to postmenopausal women. *American Journal of Clinical Nutrition*. 76(5): 1126–1137.

Carrara, V.S., Amato, A.A., Neves, F.A., Bazotte, R.B., Mandarino, J.M., Nakamura, C.V., Filho, B.P., and Cortez, D.A., 2009. Effects of a methanolic fraction of soybean seeds on the transcriptional activity of peroxisome proliferator-activated receptors (PPAR). *Brazilian Journal of Medical and Biological Research*. 42(6): 545–550.

Clubbs, E. A., and Bomser, J. A., 2007. Glycitein activates extracellular signal-regulated kinase via vascular endothelial growth factor receptor signaling in nontumorigeneic (RWPE-1) prostate epithelial cells. *Journal of Nutritional Biochemistry*. 18(8): 525–532.

Clubbs, E. A., and Bomser, J. A., 2009. Basal cell induced differentiation of noncancerous prostate epithelial cells (RWPE-1) by glycitein. *Nutrition and Cancer*. 61(3): 390–396.

Cook, N. R., Cohen, J., Herbert, P. R., Taylor, J. O., and Hennekens, C. H., 1995. Implications of small reductions in diastolic blood pressure for primary prevention. *Archives of Internal Medicine*. 155: 701–709.

Gilad, L. A., Tirosh, O., and Schwartz, B., 2006. Phytoestrogens regulate transcription and translation of vitamin D receptor in colon cancer cells. *Journal of Endocrinology*. 191(2): 387–398.

Gutierrez-Zepeda, A., Santell, R., Wu, Z., Brown, M., Wu, Y., Khan, I., Link, C. D., Zhao, B., and Luo, Y., 2005. Soy isoflavone glycitein protects against β amyloid-induced toxicity and oxidative stress in transgenic. *Caenorhabditis elegans. BMC Neuroscience*. 6: 54.

Gutierrez-Gonzalez, J. J., Wu, X., Gillman, J. D., Lee, J. D., Zhong, R., Yu, O., Shannon, G., Ellersieck, M., Nguyen, H. T., and Sleper, D. A., 2010. Intricate environment-modulated genetic networks control isoflavone accumulation in soybean seeds. *BMC Plant Biology*. 10: 105.

Hanahan, D., and Weinberg, R. A., 2000. The hallmarks of cancer. *Cell*. 100(1): 57–70.

Hedlund, T. E., van Bokhoven, A., Johannes, W. U., Nordeen, S. K., and Ogden, L. G., 2006. Prostatic fluid concentrations of isoflavonoids in soy consumers are sufficient to inhibit growth of benign and malignant prostatic epithelial cells *in vitro*. *Prostate*. 66(5): 557–566.

Hwang, I. K., Lee, Y. B., Yoo, K. Y., Kang, T. C., Lim, S. S., Sohn, H. S., Kim, S. M., Kim, W. J., Shin, H. K., and Won, M. H., 2006. Calbindin D-28k immunoreactivity increases in the hippocampus after long-term treatment of soy isoflavones in middle-aged ovariectomized and male rats. *International Journal of Neuroscience*. 116(8): 991–1003.

Ise, R., Han, D., Takahashi, Y., Terasaka, S., Inoue, A., Tanji, M., and Kiyama, R., 2005. Expression profiling of the estrogen responsive genes in response to phytoestrogens using a customized DNA microarray. *FEBS Letters*. 579(7): 1732–1740.

Iuvone, L., Geloso, M. C., and Dell'Anna, E., 1996. Changes in open field behavior, spatial memory, and hippocampal parvalbumin immunoreactivity following enrichment in rats exposed to neonatal anoxia. *Experimental Neurology*. 139(1): 25–33.

Kang, K. A., Zhang, R., Piao, M. J., Lee, K. H., Kim, B. J., Kim, S. Y., Kim, H. S., Kim, D. H., You, H. J., and Hyun, J. W., 2007. Inhibitory effects of glycitein on hydrogen peroxide induced cell damage by scavenging reactive oxygen species and inhibiting c-Jun N-terminal kinase. *Free Radical Research*. 41(6): 720–729.

Kinjo, J., Tsuchihashi, R., Morito, K., Hirose, T., Aomori, T., Nagao, T., Okabe, H., Nohara, T., and Masamune, Y., 2004. Interactions of phytoestrogens with estrogen receptors α and β (III). Estrogenic activities of soy isoflavone aglycones and their metabolites isolated from human urine. *Biological and Pharmaceutical Bulletin*. 27(2): 185–188.

Lang'at-Thoruwa, C., Song, T. T., Hu, J., Simons, A. L., and Murphy, P. A., 2003. A simple synthesis of 7,4'-dihydroxy-6-methoxyisoflavone, glycitein, the third soybean isoflavone. *Journal of Natural Products*. 66(1): 149–151.

Likhite, V. S., Cass, E. I., Anderson, S. D., Yates, J. R., and Nardulli, A. M., 2004. Interaction of estrogen receptor α with 3-methyladenine DNA glycosylase modulates transcription and DNA repair. *Journal of Biological Chemistry*. 279(16): 16875–16882.

Lee, S. O., Renouf, M., Ye, Z., Murphy, P. A., and Hendrich, S., 2007. Isoflavone glycitein diminished plasma cholesterol in female golden Syrian hamsters. *Journal of Agricultural and Food Chemistry*. 55(26): 11063–11067.

Lee, E. J., Kim, S. Y., Hyun, J. W., Min, S. W., Kim, D. H., and Kim, H. S., 2010. Glycitein inhibits glioma cell invasion through down-regulation of MMP-3 and MMP-9 gene expression. *Chemico-Biological Interactions*. 185(1): 18–24.

Lockhart, B. P., Benicourt, C., Junien, J. L., and Privat, A., 1994. Inhibitors of free radical formation fail to attenuate direct β-amyloid$_{25-35}$ peptide-mediated neurotoxicity in rat hippocampal cultures. *Journal of Neuroscience Research*. 39(4): 494–505.

Maeda, S., and Omata, M., 2008. Inflammation and cancer: role of nuclear factor-κB activation. *Cancer Science*. 99(5): 836–842.

Muñoz, Y., Garrido, A., and Valladares, L., 2009. Equol is more active than soy isoflavone itself to compete for binding to thromboxane A(2) receptor in human platelets. *Thrombosis Research*. 123(5): 740–744.

Naim, M., Gestetner, B., Kirson, I., Birk, Y., and Bondi, A., 1973. A new isoflavone from soya beans. *Phytochemistry*. 12: 169–170.

Nogradi, M., and Szollosy, A., 1996. Synthesis of 4′,7-dihydroxy-6-methoxyisoflavone 7-*O*-β-D-glucopyranoside (glycitin). *Liebigs Annalen der Chemie*. 10: 1651–1652.

Park, J. S., Jung, J. S., Jeong, Y. H., Hyun, J. W., Van Le, T. K., Kim, D. H., Choi, E. C., and Kim, H. S., 2011. Antioxidant mechanism of isoflavone metabolites in hydrogen peroxide-stimulated rat primary astrocytes: critical role of hemeoxygenase-1 and NQO1 expression. *Journal of Neurochemistry*. 119: 909–919.

Panigrahy, D., Kaipainen, A., Huang, S., Butterfield, C. E., Barnés, C. M., Fannon, M., Laforme, A. M., Chaponis, D. M., Folkman, J., and Kieran, M. W., 2008. PPARα agonist fenofibrate suppresses tumor growth through direct and indirect angiogenesis inhibition. *Proceedings of the National Academy of Sciences USA*. 105(3): 985–990.

Renouf, M., and Hendrich, S., 2011. *Bacteroides uniformis* is a putative bacterial species associated with the degradation of the isoflavone genistein in human feces. *Journal of Nutrition*. 141: 1120–1126.

Rüfer, C. E., and Kulling, S. E., 2006. Antioxidant activity of isoflavones and their major metabolites using different in vitro assays. *Journal of Agricultural and Food Chemistry*. 4(8): 2926–2931.

Rüfer, C. E., Maul, R., Donauer, E., Fabian, E. J., and Kulling, S.E., 2007. *In vitro* and *in vivo* metabolism of the soy isoflavone glycitein. *Molecular Nutrition and Food Research*. 51(7): 813–823.

Schultz-Norton, J. R., Ziegler, Y. S., and Nardulli, A. M., 2011. ERα-associated protein networks. *Trends in Endocrinology and Metabolism*. 22(4): 124–129.

Simons, A. L., Renouf, M., Hendrich, S., and Murphy, P. A., 2005. Metabolism of glycitein (7,4′-dihydroxy-6-methoxy-isoflavone) by human gut microflora. *Journal of Agricultural and Food Chemistry*. 53(22): 8519–8525.

Song, T. T., Hendrich, S., and Murphy, P. A., 1999. Estrogenic activity of glycitein, a soy isoflavone. *Journal of Agricultural and Food Chemistry*. 47(4): 1607–1610.

Staels, B., Koenig, W., Habib, A., Merval, R., Lebret, M., Pineda Torra, I., Delerive, P., Fadel, A., Chinetti, G., Fruchart, J., Najib, J., Maclouf, J., and Tedgui, A., 1998. Activation of human aortic smooth-muscle cells is inhibited by PPARα but not PPARγ activators. *Nature*. 393: 790–793.

Surh, Y. J., and Na, II. K., 2008. NF-κB and Nrf2 as prime molecular targets for chemoprevention and cytoprotection with anti-inflammatory and antioxidant phytochemicals. *Genes and Nutrition*. 2(4): 313–317.

Taku, K., Melby, M. K., Kurzer, M. S., Mizuno, S., Watanabe, S., and Ishimi, Y., 2010. Effects of soy isoflavone supplements on bone turnover markers in menopausal women: systematic review and meta-analysis of randomized controlled trials. *Bone*. 47(2): 413–423.

Tang, L., Singh, R., Liu, Z., and Hu, M., 2009. Structure and concentration changes affect characterization of UGT isoform-specific metabolism of isoflavones. *Molecular Pharmacology*. 6(5): 1466–1482.

Welty, F. K., Lee, K. S., Lew, N. S., and Zhou, J. R., 2007. Effect of soy nuts on blood pressure and lipid levels in hypertensive, prehypertensive, and normotensive postmenopausal women. *Archives of Internal Medicine*. 167(10): 1060–1067.

Winzer, M., Rauner, M., and Pietschmann, P., 2010. Glycitein decreases the generation of murine osteoclasts and increases apoptosis. *Wiener Medizinische Wochesnshrift*. 160(17–18): 446–451.

Yoshida, H., Teramoto, T., Ikeda, K., and Yamori, Y., 2001. Glycitein effect on suppressing the proliferation and stimulating the differentiation of osteoblastic MC3T3-E1 cells. *Bioscience, Biotechnology and Biochemistry*. 65(5): 1211–1213.

Zhang, Y., Wang, G. J., Song, T. T., Murphy, P. A., and Hendrich, S., 1999. Urinary disposition of the soybean isoflavones daidzein, genistein and glycitein differs among humans with moderate fecal isoflavone degradation activity. *Journal of Nutrition*. 129(5): 957–962.

CHAPTER 29

Isoflavones and Prenatal Exposure to Equol

EDWIN D. LEPHART

Department of Physiology and Developmental Biology and The Neuroscience Center, College of Life Sciences, 785 WIDB, Brigham Young University, Provo, Utah, USA
Email: Edwin_Lephart@byu.edu

29.1 Background: Polyphenols and Isoflavonoids

Polyphenols represent a wide variety of compounds, which are divided into several classes, *i.e.*, hydroxybenzoic acids, hydroxycinnamic acids, anthocyanins, proanthocyanidins, flavonols, flavones, flavanols, flavanones, isoflavones, stilbenes, and lignans. An example of a stilbene compound is resveratrol that can be extracted from Japanese knotweed, whereas isoflavones are soybean or other plant-derived products. Resveratrol represents a high-profile molecule with many potential health-related benefits (Smoliga *et al.* 2011). Conversely, the isoflavonoid molecules, while not as well-known as resveratrol, have experienced a dramatic increase in research investigative attention, general awareness to their health benefits, especially bioavailability and inclusion in food products.

Polyphenols are abundant micronutrients in the human diet, and evidence for their role in the prevention of certain disorders such as cancer and cardiovascular diseases is emerging. Several thousand molecules having a polyphenol structure (*i.e*, several hydroxyl groups on aromatic rings) have been identified in higher plants, and several hundred are found in edible plants.

Food and Nutritional Components in Focus No. 5
Isoflavones: Chemistry, Analysis, Function and Effects
Edited by Victor R Preedy
© The Royal Society of Chemistry 2013
Published by the Royal Society of Chemistry, www.rsc.org

For example, Kuhnau (1976) calculated that dietary flavonoid intake in the United States was $1\,\text{g}\,\text{day}^{-1}$ and consisted of the following: 16% flavonols, flavones, and flavanones; 17% anthocyanins; 20% catechins; and 45% biflavones. While this is a dated reference, in general, these values are still used today to estimate the daily dietary intake in humans of polyphenolic compounds. However, many analytical studies of polyphenols in foods that have been conducted to date provide a good indication of polyphenol distribution in plants and consumption in human diets. Fruit and beverages, such as tea, red wine, and coffee, constitute the principal sources of polyphenols, but vegetables, leguminous plants, soybeans and cereals are also good sources (USDA 2007). Due to human consumption of plant products throughout the world, isoflavonoid compounds are present in all humans regardless of age, gender, culture or style of dietary intake.

Isoflavones make up the largest group of natural isoflavonoid molecules and include the most investigated compounds with regard to their biological actions. Isoflavonoids can be isolated from most plants tissues including leaves, stems, roots, flowers, seeds, and germs. In germ/seeds and sprouts these compounds occur in abundance and appear to regulate physiological processes important for plant growth.

Finally, in brief, one reason the soybean or soy is often used as the primary example for isoflavonoid molecules is due to the abundance of the precursor molecules contained in these food sources. Notably, soy has been cultivated for over 4000 years and is used widely in food products throughout the world. Published data by the USDA indicate that isoflavones are components in a wide variety of legumes, prepared foods, spices, teas, and, of course, soy food products, including infant formula, tofu, tempeh, cheese, beverages, noodles, sauces, chips, and meat substitutes (USDA 2007). The widespread historical exposure to soy isoflavones in the diet (by either direct consumption or secondary consumption *via* meat and other food products from animals) is without adverse human health effects (Hamilton-Reeves *et al.* 2010) based upon the overall body of available scientific data, to date, which provides compelling evidence of their safety. Several of these topics will be covered in somewhat more detail in the latter sections of this Chapter and postnatal developmental data will be presented as a frame work for the perinatal sections.

29.2 Isoflavonoid Metabolism in Mammals: Postnatal Development

It is important to understand some characteristics and metabolic conversions of isoflavones, especially between models such as rodents *vs.* humans. In foods derived from soy, the isoflavones molecules such a daidzein and genistein (aglycones) occur naturally but are in low abundance (2–4%) (Adlercreutz and Mazur 1997) (Figure 29.1). Soybeans or soy foods are abundant in the precursor molecules, and this is the rationale for using soy as the primary example for the source of isoflavonoids. For example, these sources are

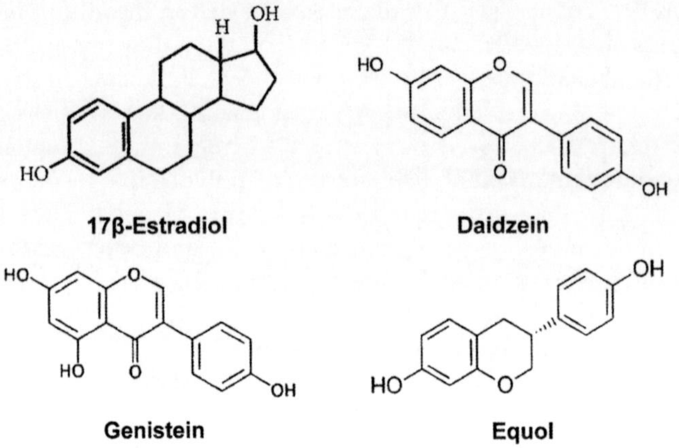

Figure 29.1 Chemical ring structure comparison of the sex steroid hormone 17β-estradiol *versus* the isoflavonoid molecules genistein, daidzein and equol. Equol is a metabolite of daidzein. Additionally, equol contains a chiral carbon shown as the broken elongated triangle between the ring structures, whereas genistein and daidzein do not.

composed mainly of β-glycosides or glucosides (commonly called glycones) that make up over 90% of the isoflavone content which are biologically inactive. After oral consumption and during intestinal metabolism the β-glycoside is removed, and the isoflavones become biologically active. These aglycones have the ability to mimic estrogens and bind to mammalian estrogen receptors [ERβ>ERα, although at a lower affinity) compared with estradiol] (Setchell and Clerici 2010a) (Figure 29.1). Thus, these aglycone isoflavones act like natural selective estrogen receptor modulators (SERMs) at various tissue sites throughout the body. The conversion of the glycones, daidzin, and genistin into their respective aglycones, daidzein, and genistein occurs in every mammal, including all traditional animal models and humans. Previously and up until the early 2000s, genistein was given greater investigative attention and examined at a much higher frequency compared with other isoflavonoids until the equol hypothesis was proposed in the late 1990s (Setchell and Clerici 2010b). Notably, the equol hypothesis suggests that the generation or consumption of equol above arbitrary threshold levels will impart health benefits for various disorders or diseases.

29.3 Equol: Biosynthesis (Plants and Mammals), Antioxidant Activity, and Safety Data in Postnatal Development

In plants, the accumulation of polyphenolic compounds such as flavonoids and phenolic molecules have been demonstrated and linked to pathogen resistance.

There are many polyphenolic molecules that act as antioxidants, including isoflavonoids.

In food, a recent study in 2009 examined the changes in antioxidant compounds in white cabbage during winter storage (Hounsome et al. 2009). Equol was identified in cabbage at high levels similar to other polyphenolic compounds that remained stable for months, and that can serve as an antioxidant during the storage of white cabbage to help prevent oxidative stress (Hounsome et al. 2009). Furthermore, in 2010, equol was identified in lettuces, peas, and beans (in addition to white cabbage) (Hounsome et al. 2010), suggesting the existence of this isoflavonoid in food products.

Comparative studies examining polyphenolic compounds have demonstrated that equol is a superior antioxidant, having greater antioxidant capacity than vitamin C or vitamin E in several *in vitro* tests (Rufer and Kulling 2006). Also, equol has greater antioxidant activity (*i.e.*, preventing oxidative damage to lipid membranes, *etc.*) compared with genistein.

Historically, equol [7-hydroxy-3-(4'-hydroxyphenyl)-chroman] was first discovered in the early 1980s in the urine of adults consuming soy foods (Setchell and Clerici 2010a; 2010b) (Figure 29.1). It was shown to be a key metabolite of daidzein, one of the main isoflavones present in soy foods. Equol, therefore, is naturally produced in the intestine in humans as a metabolite of soy foods, and some individuals are capable of producing higher levels of equol than others. These individuals have been identified as "equol producers" (see below) (Setchell and Clerici 2010b).

Most animal species exclusively and efficiently produce equol at high levels when fed diets containing soy (Setchell and Clerici 2010b). For example, all monkeys, rats, mice, chickens, sheep, cattle, pigs, *etc.* produced exclusively equol from consuming grasses, leguminous plants, clovers, alfalfa, and soy-containing products to increase the protein content of the feed.

Moreover, only approximately 20–40% of humans living in Western countries are "equol producers" after ingesting soy foods, whereas the frequency of "equol producers" in Asian populations that consume soy foods is typically 50–60% (Setchell and Clerici 2010b). It should be noted that the amount of equol production in animals is much higher compared with humans. For example, humans that are considered "equol-producers" display levels at approximately $10–200\,\text{ng ml}^{-1}$ or more, whereas in all other mammals (except pigs) the levels of equol are in the range $800–2500\,\text{ng ml}^{-1}$ or more. Moreover, the term "equol producer" is a descriptive or arbitrary term for humans that maintain equol levels around or above 10 to $20\,\text{ng ml}^{-1}$ after consumption of soy foods that infers protective health benefits (Setchell and Clerici 2010b).

From laboratories that have the ability to measure low levels of equol: all humans whether they are "equol-producers" or not have 1 to approximately $4\,\text{ng ml}^{-1}$ of equol circulating in their bloodstreams. Thus, there are several direct and indirect references of the natural consumption of equol from dietary sources. It is known that equol is contained in meat and milk products of animals consuming grasses, grain, soy-, alfalfa-, or clover-supplemented feeds. Also, equol excretion levels correlate positively with the intake of total fat and meat products in humans. Finally, even if humans do not consume soy or other plant-derived products that contain high levels of the precursor molecule of equol, these molecules are present in low concentrations in foods such as corn or wheat along with hundreds of other plant-food products and in high concentrations in nutritional supplements (Reinli and Block 1996; USDA 2007).

Equol, unlike its precursor daidzein (or genistein), is unique in having a chiral carbon atom at position C-3 of the furan ring (Setchell and Clerici 2010b). It therefore can occur as two distinct isomers as S-equol or R-equol (Figure 29.2). S-Equol, apparently, is the exclusive intestinal metabolite of diadzein in all animals including humans (Setchell and Clerici 2010b). Equol is not only known for its strong antioxidant properties (Rufer and Kulling 2006) but also for its anti-androgenic activities (Lund *et al.* 2004; 2011). For example, it was recently demonstrated that both S-equol and R-equol have unique anti-androgenic properties where they both specifically bind 5α-dihydrotestosterone (5α-DHT) with high affinity, and thereby prevent 5α-DHT from binding the androgen receptor (Lund *et al.* 2004; 2011) (Figure 29.2).

Remarkably, although equol has been shown to possess health benefits, there is some evidence that the R-, rather than the S-, enantiomer is responsible for the *in vivo* chemoprotective properties of equol (Magee *et al.* 2006). Furthermore, a recent clinical study showed that isoflavone supplementation increased the production of serum equol in equol-producers with a corresponding decline in serum 5α-DHT levels in men (Tanaka *et al.* 2009) to provide support for the mechanism of action of equol in health disorders.

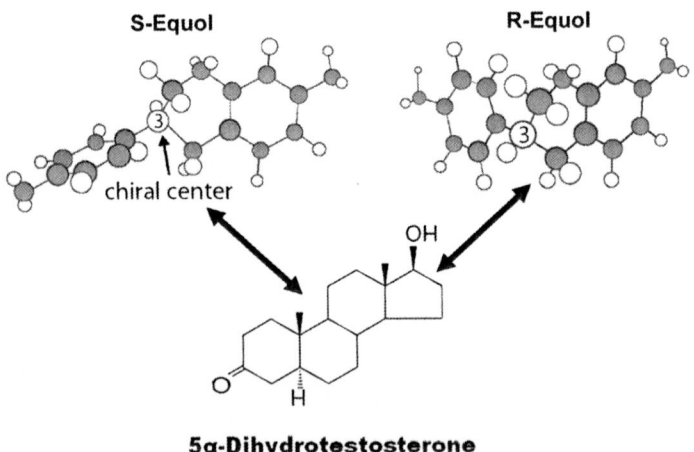

Figure 29.2 The enantiomers of equol (S-equol and R-equol). The chiral carbon is labeled '3' with the single headed-arrow. The bottom of the figure shows the chemical structure of the most potent androgen 5α-DHT. The double-headed arrows indicate that both S-equol and R-equol bind specifically to 5α-DHT with high affinity.

In examining the safety/toxicology studies during postnatal development, equol appears to have either a positive protective or somewhat neutral influence even when administered at relatively high doses (however, other isoflavones are not without controversy in this regard). For example: (a) in an *in vitro* model of cellular cytotoxicity, 5α-DHT enhances cytotoxicity, whereas, if equol is added to the 5α-DHT treatment, it completely reversed the cytotoxic effects on 5α-DHT, suggesting that equol binds 5α-DHT directly and thus inhibits the cytotoxic effects of 5α-DHT (Gopaul *et al.* 2012) [notably: the specificity of equol binding 5α-DHT has been confirmed (Lund *et al.* 2011)]; (b) the estrogenicity of equol on reproductive and non-reproductive organs was examined in mice and although total serum equol levels ranged from 1.4 to 8.1 µM the equol treatments did not significantly increase uterine weight and neither dietary nor injected equol decreased thymic or adipose tissue weights compared with control values (Selavraj *et al.* 2004), although these blood equol levels were 4-to-8-times higher that than seen for human consumption of soy foods; and (c) several studies using cynomolgus monkeys have examined the influence of dietary soy isoflavones on female health parameters such as breast and uterine tissue changes as well as the effects of prior oral contraceptive use in combination with consumption of soy isoflavones. Over 370 monkeys have been studied across several major investigations using soy isoflavones or isolated isoflavonoid molecules from 1 month to 36 months. The general conclusions from these studies indicate that dietary soy isoflavones or isolated isoflavonoid molecules do not alter breast or uterine tissues (meaning, they do not stimulate breast or uterine proliferation) and the high isoflavone dose may contribute to a reduction in the risk of breast cancer. In fact, the authors stated, "high doses of dietary soy isoflavonoids have minimal uterotrophic or mammotrophic effects in an established primate model". (Wood *et al.* 2006).

Finally, human clinical pharmacokinetic studies have been performed showing that the absorption and bioavailablity of equol is higher compared with genistein or daidzein and the half-life of equol or its enantiomers is similar to sex steroid hormones (Setchell and Clerici 2010a).

29.4 Comparison between Human and Rat: Newborn and Adult Parameters

As shown in Table 29.1, select parameters between human and rat models are listed in perinatal and postnatal (adult) development. The obvious differences between the two models involve weight, size or mass. For example, while the rat is an accepted animal model for basic or preclinical research leading to human clinical research, the positive characteristics of this model include its inexpensive costs, ease of experimental manipulation, rapid development, and long-established research database. Although there is some scale of complexity in rat models especially in pharmacology/behavioral studies where different rat strains determine the outcome of applied agents or drugs, the complexity in humans is unmatched with regards to anatomy, physiology, and functional

Table 29.1 Comparison between human and rat: newborn and adult parameters. This table lists some key facts in comparing human to rat parameters. BMI is not a parameter calculated in rats. GD, gestational day.

	Human	Rat
Newborn		
Gestational length	40.5 weeks	20–22 days
Ovarian-dependent pregnancy	No	Yes
Placental development/function	8–12 weeks	GD or E17
Body weight (g)	3000–4000	5.9
Brain weight (g)	350–400	0.280
Blood volume	85 ml kg^{-1} (full term)	<0.6 ml at birth
Alpha-fetoprotein	Present (function?)	Present (function+)
Adult		
Body weight (kg)	70–85	0.300
Body mass index (BMI)	16–32	–
Brain weight (g)	1300–1400	2.0
Blood volume	5–6 L	6.4 ml/100 g
Hematocrit (%)	38–42	43
Heart rate (beats min^{-1})	70	350–450
Cardiac output (volume min^{-1})	5–6 L	74 ml
Respiratory rate (per min)	14–20	100–120
Plasma albumin (g 100 ml^{-1})	4.0–6.0	3.0–3.05
Sex hormone binding globulin (SHBG)	Present	Absent
Surface area (m^2)	1.5–2.0	0.023
Sexual maturation	12–15 years	6–8 weeks
Mean life span (years)	65–85	2.0–3.5

interactions of multiple variables or factors. In humans, functional interactions include levels of compliance, environment, nutritional/supplemental, and pharmacological exposure, and the quantity and quality of higher brain functions such as social, emotional, language, cognition and perception along with age-, sex-, and employment-dependent factors.

Thus, in brief, a few of the parameters listed in Table 29.1 are covered below to provide a perspective for the pre-clinical and clinical studies and the interpretation of the outcomes of the perinatal studies covered below. Also, many authors in journal reports often mix animal and human references so readers are presented with a difficult challenge in following parameters associated with the rodent or human model. It is important to keep in mind what characteristics are shared and which are distinct to each model. For example, if isoflavonoids, other molecules or certain physiological interactions could stimulate the production of alpha-fetoprotein (AFP) or some other estrogenic binding-protein, especially during perinatal development, this might represent a protective mechanism by which estrogens and estrogenic-like molecules can become bound or biologically inactivated. While just a postulation, this notion remains to be determined. Finally, one of the essential biomarkers to be aware of in isoflavonoid research is sex hormone binding globulin (SHBG) which is expressed in humans but not in rats.

29.5 Perinatal Studies and Transplacental Transport of Isoflavonoids

The definition of perinatal refers to the time interval before, at, and shortly after birth. In rats this time interval is usually any time from conception up to approximately 7 days post birth (*i.e.*, for the neonatal period) and sometimes includes the infantile period from 7–20 days post birth. In humans, this period relates to around 20 to 28 week of gestation to approximately 1 month after birth. Most of the studies covered below overlap with these defined intervals, but in some cases the treatments extend into later postnatal development. It should be noted that this coverage is not comprehensive in nature, but provides adequate information for the scope of this Chapter.

29.5.1 Isoflavonoids: Perinatal Rodent Studies

It should be noted in the following studies, and with comparisons to the human investigations, that: (a) treatments, either soy-derived isoflavonoids, separate isoflavonoids or combinations of isoflavoids; (b) single or multiple doses, dosing levels and dosing intervals; (c) mode of treatment administrated [*via* diet (oral) or subcutaneous injection]; (d) basal or stock dietary sources (either known or unknown) with or without phytoestrogens; (e) interactions with mineral, heavy metal, and vitamin levels along with other dietary components or biochemical molecules; (f) rodent strain employed; (g) age and sex; and (h) other parameters (source of animal supplier, housing, lighting, seasonal influences) such as environmental factors may influence the outcome of the results from one study to another.

In general, a larger number of studies have investigated genistein compared with daidzein, while many studies used soy-containing diets as their source of isoflavonoids, and several studies administered isolated phytoestrogens or isoflavonoids molecules. Below is a summary of the rodent perinatal studies. Note: if not stated, dietary sources of phytoestrogens in the stock or basal diets were not reported and blood levels of the circulating isoflavonoids were not measured.

Only one study, in 2002, specifically examined daidzein over perinatal development. Lamartiniere *et al.* (2002) studied the bioavailability, potential reproductive toxicity, and breast cancer chemoprovention *via* diets containing 0 mg, 250 mg or 1000 mg of daidzein kg^{-1} of feed fed to virgin female rats (starting 2 weeks prior to breeding and continuing until the offspring were 50 postbirth). The daidzein and equol (the major metabolite of daidzein) levels in pregnant rats, 7-day-old and 21-day-old post birth female offspring were: 529 and 303 nM, 163 nM and 982 nM and 1188 nM and 1359 nM in the 250 mg daidzein treatment group, whereas in the 1000 mg daidzein group the daidzein and equol levels were 4462 nM and 407 nM, 1013 and 3841 nM and 6472 and 3308 nM, respectively. Even at these supraphysiological concentrations of daidzein, administered *via* the diet, did not cause toxicity to the female offspring reproductive tract or provide protection against chemically induced mammary cancer. However, in another study Lamartiniere et al. (1995) reported that neonatal genistein exposure has a chemoprotective influence against mammary cancer, suggesting the differences in isoflavonoid chemical messenger signaling.

In a series of papers from 1991 to 1995, Faber and Hughes (1991; 1993) and Levy et al. (1995) examined the influence of prenatal, or neonatal exposure to genistein on pituitary function, hypothalamic structures (and presumably function), and sexual differentiation in rats. The main outcomes of this study were: (a) there were no differences in the hypothalamic structure (the volumes of the sexually dimorphic nucleus of the pre-optic area or SDN-POA) by treatment, and (b) exposure to environmental estrogens during neonatal development altered the postpubertal pituitary function or response to gonadotropin-releasing hormone (GnRH).

In the second study, Faber and Hughes (1993) examined similar parameters as were assessed in the first study (pituitary function and hypothalamic structure). The conclusions of this study suggested that increasing exposure to genistein during neonatal development resulted in decreasing Luteinzing hormone (LH) sections, and the higher doses of genistein at 500 or 1000 µg increased SDN-POA volumes in female castrated offspring.

In the third study, Levy et al. (1995) examined the influence of prenatal genistein exposure on sexual differentiation as measured via SDN-POA volumes and pituitary function. No significant differences in pituitary responsiveness to GnRH were observed among any of the treatment groups, and there was a slight decrease in the SDN-POA volumes of animals treated with 5000 µg of genistein. Therefore, these studies suggest that genistein might alter pituitary function and hypothalamic brain structure elements that may be involved in reproduction. This notion that genistein via neonatal exposure may impact hypothalamic structures, GnRH function, and thus reproductive development are supported in a few studies that examined: (a) the hypothalamic kisspeptin signaling pathways in female rats that is involved in the onset of puberty and GnRH expression, and (b) ovarian development and function.

29.5.2 Equol: Perinatal Rodent Studies

Few studies have examined equol during perinatal development before 2000. However, in the last decade there have been several studies covering this topic. For example, Medlock et al. (1995) examined the effects of equol on the developing rat uterus, where postnatal days (PND) 10–14 represent the critical period of uterine gland genesis. Equol administered from PND 1–5 or PND 1–10 did not alter uterine ER levels, and there was no effect on other uterine parameters.

Later mammary gland differentiation by neonatal/infantile exposure to the enantiomers of equol and chemoprotection [using the 7,12-dimethylbenz(a) anthracene (DMBA)-induced breast cancer model] were studied by Brown et al. (2010). Female rats were exposed to an S-equol or R-equol at 250 mg kg^{-1} diet during neonatal/infantile development from birth to PND 21.

Subsequently, histological parameters of mammary gland development and breast cancer protection were assayed. The main outcome of this study was exposure to the equol isomers had no long-term chemoprevention actions against breast cancer, (but neither isomer increased tumor formation in response to DMBA). However, if the equol enantiomers were administered (at the same dose) later during postnatal development (starting at PND 35), S-equol had no chemopreventive action, whereas R-equol reduced palpable tumors by 43%, increased tumor latency, and the tumors were less invasive.

In 2001, perinatal exposure to equol on maternal, fetal and neonatal brain aromatase was determined in rats fed a high soy-containing diet (600 ppm of isoflavonids) or a low soy-containing diet (10 ppm) by Weber et al. (2001). Brain aromatase and isoflavone levels were determined, and the ingested isoflavones/equol cross the placenta and become concentrated in maternal milk as evident from the high infantile plasma concentrations which ranged from 738 ng ml^{-1} in the high soy-fed group to 84 ng ml^{-1} in the low soy-fed neonates. The maternal levels of isoflavones/equol ranged from 1000 to over 1600 ng ml^{-1} in the high soy-fed animals, and in the low soy-fed animals they ranged from 89 to 135 ng ml^{-1}, which is similar to adult male rat values for both treatments. However, brain (medial basal hypothalamic–pre-optic area) aromatase enzyme activity levels were not altered in either the maternal or fetal/neonatal animals by the diet treatments.

In a later study in 2009 by Matulka et al. the toxicity of equol was determined in pregnant Sprague–Dawley rats at doses of 200, 1000, and 2000 mg kg^{-1} day^{-1} by gavage from gestational days 6 through 19 using 22 animals per treatment group by utilizing an equol-rich soy-based ingredient. The male and female offspring of the treated mothers were examined for developmental and reproductive toxicity effects at birth, during various time points of postnatal development, and in subsequent generations of the offspring [first generation (F1), second generation (F2), etc.]. For the reproductive phase of this study, no observed adverse effects were seen in both the male and female offspring during postnatal development, and no adverse effects were observed in the toxicity phase of the study in the embryo–newborns at any of the doses tested. Using the same treatment in adult rats, the toxicity and genotoxicity was determined and an LD$_{50}$ of >4000 mg kg^{-1} was calculated, and additionally, there were no chromosomal aberrations in Chinese hamster ovary cells up to 3000 μg ml^{-1} in a study by Yee et al. in 2008. In brief, both studies provide evidence that equol at very high concentrations does not alter developmental, reproductive parameters or function in rodents.

Finally, two studies examined equol and the equol enantiomers during perinatal development in rats. The first study in 2010 by Blake et al. examined prenatal equol exposure at a high [10.5 mg (kg of body weight)$^{-1}$ day^{-1}] or a supra-pharmacological [63.0 mg (kg of body weight)$^{-1}$ day^{-1}] dose during later gestation [embryonic day 14 (E14)–E20]. Mothers in the supra-pharmacological dose gained less body weight compared with all other treatment groups, and this trend was also seen in the male and female offspring. This result supports postnatal data where soy-containing diets given chronically lead to lower body weights and reduced adipose tissue deposition in rats. Notably, although equol binds 5α-DHT during postnatal development (Lund *et al.* 2004; 2011), this did not occur during perinatal development (see summary below). In this study, during prenatal development in the presence of high pharmacological equol levels maternal circulating equol levels reached approximately 7700 ng ml^{-1} or 32 μM and in the neonates plasma equol concentrations were approximately 1000 ng ml^{-1} or 4 μM. These doses are much higher compared with humans consuming soy-based products in an Asian diet (Adlercreutz and Mazur 1997). For example, total maternal plasma isoflavonoid concentrations of Japanese women during birth range between 19 to 744 nM (mean 232 nM), whereas in cord plasma samples isoflavonoids ranged from 58 to 831 nM (mean 299 nM) (Adlercreutz *et al.* 1999). In the 10.5 mg kg^{-1} equol dose, equol levels (at 894 nM in the maternal and 886 nM in the neonatal blood samples) are more than 3- to 4-times higher compared with the human birth data, suggesting human dietary consumption of soy food products may not approach this dosage. However, these high equol levels did not alter genital development or 5α-DHT levels in (the mothers or) the neonatal rats. This may suggest an unknown protective mechanism by which equol is bound and biologically inactivated in this animal model. Also, in this study by Blake *et al.* (2010) behavioral analysis [*via* the Porsolt forced swim test (PFST), which indexes depressive-like behaviors in rats] was performed on PND 29 in the male and female offspring. Remarkably, depressive-like behaviors indexed by the PFST in the equol-treated offspring were decreased compared with controls.

In the second study by Brown *et al.* (2011), a comprehensive examination of the impact of perinatal exposure to equol enantiomers on the reproductive development in rodents was examined. The experimental design of this study examined three groups (9 rats per group) fed an AIN-93G control diet or an AIN-93G diet containing 250 mg kg^{-1} of S- or R-equol beginning 10 days prior to breeding and continuing until the offspring were 21 days old (postbirth). Neither enantiomer affected the fertility, number of litters produced, number of rat pups per litter, number of male and female pups born, birth weight, anogenital distance, testicular descent or vagina opening. In histological analysis there were no major abnormalities in ovary, testes, prostate or seminal vesicle tissue with the dietary exposure to the S- or R-equol treatments. This comprehensive study in rats displays a lack of overt toxicity on male or female reproductive systems with the exposure to S- or R-equol at 250 mg kg^{-1} during gestation and through the lactation developmental intervals.

29.5.3 Isoflavonoids: Perinatal Human Studies

Although there are millions of people that consume soy food products, there are only a few human studies that have quantified isoflavonoids during pre- or perinatal development compared with adult studies. The first study reported by Adlercreutz *et al.* (1999), examined seven young healthy Japanses women at delivery and quantified isoflavonoid levels in maternal, cord, and amniotic fluid samples. Total maternal plasma isoflavonoid concentrations ranged from 19 to 744 nM (mean 232 nM), cord values ranged from 58 to 831 nM (mean 299 nM), and amniotic fluid values ranged from 52 to 799 nM (mean 223 nM). The conclusions from this study suggest that high levels of isoflavonoid phytoestrogens found in healthy neonatal Japanese infants indicate transfer of isoflavonoid molecules from the maternal to the fetal compartment (Adlercreutz et al. 1999).

In a second study published by Foster et al. (2002) quantified isoflavonoid levels between 15 and 23 weeks of gestation from amniotic fluid samples from 59 healthy women. Their findings showed that the concentrations of daidzein and genistein in amniotic fluid samples were 1.4 and 1.7 ng ml^{-1} (with maximum levels of 5.5 and 6.5 ng ml^{-1}), respectively. These data suggest that during the second trimester quantifiable levels of dietary phytoestrogens are present in the amniotic fluid of fetuses that are transferred from the mother.

In a third study, Todaka et al. (2005) investigated fetal exposure to phytoestrogens (derived from plants) by quantifying serum concentrations of isoflavonoids in maternal and cord blood samples. This study included 51 healthy mothers in Japan who were scheduled for cesarean section. The detection rates of genistein, daidzein, and equol in cord serum were 100%, 80%, and 35%. Levels of genistein and daidzein were higher in cord than in maternal serum (means $= 19.4$ ng ml^{-1} vs. 7.2 ng ml^{-1} and 4.3 ng ml^{-1} vs. 1.8 ng ml^{-1} for genistein and daidzein, respectively). However, a reverse pattern was seen for equol (cord mean $= 0.9$ ng ml^{-1}, whereas maternal mean $= 2.0$ ng ml^{-1}). Thus, this study confirmed the placental transfer of isoflavonoids from mother to fetus (Todaka et al. 2005).

In a fourth study, Nagata et al. (2006) determined whether maternal soy intake may be inversely associated with pregnancy hormone levels. Steroid hormone levels (estradiol, estriol, and testosterone) and isoflavonoid levels (genistein, daidzein, and equol) were quantified in maternal serum and umbilical cord blood samples of 194 women during pregnancy and at delivery. Soy intake was assessed by 5-day diet records at approximately the 29th week of pregnancy. At delivery, the serum cord blood levels were as follows: genistein, 127 nM (range 2–1298 nM); daidzein, 39 nM (range 2–303 nM); and equol, 4 nM (range 2–169 nM); whereas levels in the maternal serum were as follows: genistein, 117 nM (range 2–2223 nM); daidzein, 50 nM (range 2–1182 nM); and equol, 7 nM (range 2–197 nM). There was a high correlation observed for isoflavone levels between the maternal and umbilical cord blood samples, indicating that isoflavonoids can be transferred from the maternal to the fetal compartment. However, none of the steroid hormones quantified in umbilical cord blood was associated with any of the isoflavonoids measured. Thus, there was no indication that soy intake affects steroid hormone levels during human pregnancy (Nagata et al. 2006). These findings are opposite to that found in animal studies by Harrison et al. (1999) in which estradiol levels increased in pregnant rhesus monkeys fed 8 mg (kg of body weight)$^{-1}$ genistein. Although the above studies are few in number, they provide valuable data for: (a) the quantification of isoflavoind levels in the maternal and fetal compartments in humans, (b) circulating isoflavonoids that do not influence steroid hormone production during pregnancy, and (c) result sets to serve as a reference for animal studies that may confirm or display divergent results to the human model.

Exposure to soy-based infant formulas has been investigated by Setchell *et al.* (1997; 1998). It is estimated that a 4-month-old infant fed soy formula would be exposed to approximately 4.5 to 8.0 mg (kg of body weight)$^{-1}$ day^{-1} of total isoflavones. The long-term data for consumption of infant soy formulas do not indicate any abnormalities in reference to growth and development. However, this topic remains a point of controversy based upon one's viewpoint and understanding (or lack of understanding) of the scientific data. One viewpoint is that exposure to these isoflavonoid molecules early in life may have long-term health benefits for hormone-dependent disorders. The historical health record from Asian countries apparently supports this viewpoint for certain types of cancers (Adlercreutz and Mazur 1997). Particularly, it is impossible to remove polyphenolic molecules from one's diet due to the abundance of these molecules in all types of foods. Finally, in general, the data, especially in humans, suggest a level of safety, and isoflavonoids provide a health benefit and protective influence for certain types of age-related diseases (Adlercreutz and Mazur 1997; Gopaul *et al.* 2012; Hamilton-Reeves *et al.* 2010; Lund *et al.* 2011; Rufer and Kulling 2006; Setchell and Clerici 2010a; Tanaka *et al.* 2009).

29.6 Summary and Considerations

Polyphenols are abundant micronutrients found in red wine, coffee, chocolate, fruits, vegetables, cabbage, soybeans, legumes, prepared foods, spices, teas, cereals food products, infant formula, tofu, cheese, beverages, noodles, sauces, chips, and meats. Polyphenolic compounds representing thousands of molecules in the human diet and isoflavonoid compounds fall under this category or biochemical classification. It is impossible to exclude polyphenolic or isoflavonoid compounds from the human diet due to the abundance and widespread distribution of these molecules in plants, meat products, and prepared foods. Interest in isoflavonoids has increased dramatically in the last 30 years, and more recently over the last decade a surge of investigative attention has been focused on equol (a metabolite of daidzein). This is due to the general awareness of their potential health benefits, bioavailability, inclusion into food products, and potential use in health applications (nutraceuticals, cosmetics, and other products).

Isoflavonoid metabolism in mammals is well studied, including in humans. Their biosynthesis in plants and mammals along with their strong antioxidant activity and safety data are presented in this Chapter. It appears that the safety data of isoflavonoids in adults (postnatal development) and presumably during perinatal development in humans has a positive record with a potential for chemopreventive actions against certain age-related diseases and disorders.

A brief comparison between human and rat, newborn and adult parameters is presented to highlight the utilization of the rodent model as a basic science or pre-clinical tool to elucidate the characteristics, biochemical properties, and chemical messenger actions of isoflavonoid compounds. Several items are noted in the differences between the human and rat models as to general characteristics and anatomical/physiological and functional parameters in order to keep in mind the value and utilization of animals *versus* human studies.

Much more investigative information is known about isoflavonoid molecules during postnatal (adult) *versus* perinatal development, and this pattern is similar to that seen in the advances in adult general medicine *versus* neonatal medicine. In this regard, some of the animal studies cited above used regular rat chow to which genistein or other isoflavonoids were added to the stock diets, whereas others used an AIN-93G diet devoid of phytoestrogens, but this diet formulation is high in milk protein (casein) not normally consumed by rodents. Both are problematic in determining the outcomes of the results reported due to not knowing: (a) the circulating levels of isoflavonoids the animals were exposure to, and (b) the impact of changing the stock or other dietary sources on body weight gain/composition, metabolism, brain function, *etc*. Few animal studies report diet, blood and/or tissue levels of isoflavones in a way the readership may determine the true significance of the results reported. In human studies the complexity of the model is much greater compared with animal investigations due to the lack of or level of manipulation to which the subjects can be examined. Keeping these factors in mind may assist in focusing on the utility and purpose for the use of animal studies and implication for human health.

In perinatal studies, it is well documented that isoflavonoid molecules cross the placental barrier during prenatal development and that neonates and infants *via* suckling can be exposed to dietary soy-based formulas or isoflavones in the mother's milk. From perinatal animal studies, there are few studies that have examined daidzein, whereas most investigations have focused on genistein and more recently equol. In human perinatal studies, there are a handful of reports on maternal, prenatal (*via* amniotic fluid samples), and newborn levels of isoflavones from examining Japanese subjects consuming a soy diet. These human studies provide valuable data from which a comparison can be made as to the confirmation or divergent results seen in animal studies.

Summary Points

- This Chapter focuses on polyphenolic compounds and more specifically isoflavonoids that possess health benefits.
- Isoflavonoids are found naturally in plants, teas, and soy food products, including infant formula, tofu, cheese, beverages, and meat products, and are absorbed and/or metabolized in the gastrointestinal tract of mammals.
- The main isoflavonoids molecules covered in this Chapter are: daidzein, genisten, and equol (a metabolite of daidzein).

- The isoflavonoids, particularly equol, are strong antioxidants.
- There is widespread historical exposure in Asian countries to isoflavonoid molecules and the safety and toxicological data, in general, in animals and humans is without adverse health effects.
- Much more research has been performed during postnatal (adult) *versus* pre- or perinatal development in examining isoflavonoids using rodent and human models.
- In comparison between human and rat, newborn and adult parameters are covered to provide a perspective on the benefits and disadvantages of each model.
- Equol is unique in its chemical structure that can be expressed in two different confirmations, S-equol and R-equol or as enantiomers; both S-equol and R-equol can specifically bind 5α-DHT with high affinity, whereas S-equol has a higher affinity for estrogen receptor β (ERβ) compared with R-equol.
- To date, S-equol appears to be the exclusive metabolite of daidzein in the gastrointestinal tract of mammals.
- Prenatal and term studies have shown the transplacental transport of isoflavonoids in mammals.
- In animal studies covering perinatal development most investigations have examined genistein or equol by incorporation of these isoflavonoid molecules into the diet or by injection of a combination of isoflavones or the isolation of single isoflavone compounds in various dosing and time intervals examining a wide range of parameters.
- Human studies in Japanese women during prenatal development or at birth have provided valuable data for the quantification of isoflavonoid levels in the maternal, amniotic, and fetal compartments.
- Animal studies provide valuable basic science or pre-clinical information on the characteristics, biochemical properties, and physiological actions, and serve as a reference. However, the animal data may corroborate or display divergent results compared with that observed in the human model.
- Therefore, further research is warranted to reveal the many benefits of isoflavonoid molecules in human health.

Key Facts

Key Facts for Polyphenols and Isoflavonoids

- Plants, especially soy, has been cultivated for over thousands of years with human consumption occurring for millennia.
- Polyphenolic and isoflavonoid molecules derived from plants are classified as phytoestrogens.
- Dr Herman Adlercreutz is considered the "Father of Phytoestrogens" due to his long and successful research work on this topic. He used the historical information that the incidence of breast, colorectal, and prostate

cancer, along with other disorders, is higher in the Western world compared with countries in Asia and examined the properties of many isoflavonoids with health benefits.
- Dr Kenneth Setchell, also a phytoestrogen expert, discovered, in the 1980s, the production of equol in humans that consumed soy food products; and subsequently has studied numerous aspects of isoflavonoids.
- Dr John Pezzuto discovered the polyphenolic molecule resveratrol in 1997 for its health-promoting benefits, which has expanding since that time to cover several applications for improving human health.
- Due to human consumption of plant products throughout the world, isoflavonoid compounds are present in all humans regardless of age, gender, culture or style of dietary intake.
- The presence of isoflavonoids compounds in all humans is due to either direct consumption of plant or soy food products or secondary consumption *via* meat or other food products from animals.
- Recent scientific reports covering the last 20 years of research indicate that exposure *via* consumption of isoflavonoid molecules is without adverse effects.

Key Facts for Perinatal Isoflavonoids and Equol

- Animal studies, in some cases, suggest alterations in anatomical, developmental, and physiological processes.
- While there are limited human studies, those examining Japanese women show that exposure to isoflavonoids during pregnancy does not alter steroid hormone profiles.
- Animal studies do not always support the results seen in human studies.
- Since the proposed "equol hypothesis" (*i.e.* the production of circulating equol above threshold arbitrary levels imparts health benefits), a dramatic increase in this research topic has occurred in both animals and humans.
- Further investigations examining the health benefits of the isoflavoniods, including equol, will continue to study future applications in nutraceuticals, cosmetics, pharmaceuticals, *etc.*

Definition of Words and Terms

Aglycones: Isoflavonoids without their β-glycoside or glucoside group.
AIN-93G: A rodent diet devoid or very low in soy or isoflavonoid molecules containing casein (milk protein) as the replacement ingredient.
Applied research: This refers to scientific study that seeks to solve practical problems; this may include clinical research.
Basic research: Fundamental research carried out to increase understanding of basic principles. The end results do not often have any direct or immediate commercial benefits.
Chiral carbon: A carbon atom in a molecule that is bonded to four different chemical species, allowing for the formation of isomers.

Clinical research: This refers to scientific study involving human subjects.
Cord blood: This is blood obtained from the umbilical cord at birth in humans.
7,12-Dimethylbenz(a)anthracene (DMBA): A powerful organ-specific laboratory carcinogen.
Equol hypothesis: The generation or consumption of equol above arbitrary threshold levels will impart health benefits for various disorders or diseases.
Equol producer: A descriptive or arbitrary term for humans that maintain equol levels around or above $10–20\,ng\,ml^{-1}$ after consumption of soy foods that infers protective health benefits.
Glycosides: Isoflavonoids molecules with intact carbohydrate or side group portion(s) of the molecule.
Enantiomer: Each of a pair of molecules that are mirror images of each other.
Isomers: Each of two or more compounds with the same formula but a different arrangement of atoms in the molecule.
Isoflavonoids: Polyphenolic compounds derived from plant sources and/or one of a family of phytoestrogens found chiefly in soybeans.
LD_{50}: The lethal dose (or LD) test measures the amount of a substance that will, in a single dose, kill a certain percentage (50%) of animals in a test group.
Mean: The mathematical average of a set of numbers.
Mole: The molecular weight of a compound expressed in grams (g).
Polyphenol: A compound containing more than one phenolic hydroxyl group.
SDN-POA: A brain area in the medial basal hypothalamic region in rats that contains the sexually dimorphic nucleus of the pre-optic area.

List of Abbreviations

5α-DHT	5α-dihydrotestosterone
DMBA	7,12-dimethylbenz(a)anthracene
E14, *etc.*	embryonic day 14 *etc.*
ER	estrogen receptor
GnRH	gonadotropin-releasing hormone
LH	Luteinizing hormone
PFST	Porsolt forced swim test
PND	postnatal day(s)

References

Adlercreutz, H., and Mazur, W., 1997. Phyto-oestrogens and Western diseases. *Annuals Medicine*. 29: 95–120.

Adlercreutz, H., Yamada, T., Wahala, K., and Watanabe, S., 1999. Maternal and neonatal phytoestrogens in Japanese women during birth. *American Journal Obstetrics Gynecology*. 180: 737–743.

Blake, C., Fabick, K.M., Setchell, K.D.R., Lund, T.D., and Lephart, E.D., 2010. Prenatal exposure to equol decreases body weight and depressive-like behaviors in male and female offspring. *Current Topics Nutraceutical Research.* 8: 69–78.

Brown, N.M., Belles, C.A., Lindley, S.L., Zimmer-Nechemias, L., Witte, D.P., Kim, M.O., and Setchell, K.D.R., 2010. Mammary gland differentiation by early life exposure to enantiomers of the soy isoflavone metabolite equol. *Food Chemistry Toxicology.* 48: 3042-3050.

Brown, N.M., Lindely, S.L., Witte, D.P., and Setchell, K.D.R. 2011. Impact of perinatal exposure to equol enantiomers on reproductive development in rodents. *Reproductive Toxicology.* 32: 33–42.

Faber, K.A., and Hughes, C.L. Jr., 1991. The effect of neonatal exposure to diethylstilbestrol, genistein, and zearalenone on pituitary responsiveness and sexually dimorphic nucleus volume in the castrated adult rat. *Biology Reproduction.* 45: 649–653.

Faber, K.A., and Hughes, C.L. Jr., 1993. Dose-response characteristics of neonatal exposure to genistein on pituitary responsiveness to gonadotrophin releasing hormone and volume of the sexually dimorphic nucleus of the preoptic area (SDN-POA) in postpubertal castrated female rats. *Reproductive Toxicology.* 7: 35–39.

Foster, W.G., Chan, S., Platt, L., and Hughes, Jr, C.L., 2002. Detection of phytoestrogens in samples of second trimester human amniotic fluid. *Toxicology Letter.* 129: 199–205.

Gopaul, R., Knaggs, H., and Lephart, E.D., 2012. Biochemical investigation and gene analysis of equol: A plant and soy-derived isoflavonoid with anti-aging and antioxidant properties with potential human skin applications. *Biofactors.* 38: 44–52.

Hamilton-Reeves, J.M., Vasquez, G., Duval, S.J., Phipps, W.R., Kurzer, M. S., and Messina, M.J., 2010. Clinical studies show no effects of soy protein or isoflavones on reproductive hormones in men: results of a meta-analysis. *Fertility Sterility.* 94: 997–1007.

Harrison, R.M., Phillippi, P.P., Swan, K., and Henson, M.C., 1999. Effect of genistein on steroid hormone production in the pregnant rhesus monkey. *Proceedings Experimental Biology Medicine.* 222: 78–84.

Hounsome, N., Hounsome, B., Tomos, D., and Edwards-Jones, G., 2009. Changes in antioxidant compounds in white cabbage during winter storage. *Postharvest Biology Technology.* 52: 173–179.

Hounsome, N., Grail, B., Tomos, A.D., Hounsome, B., and Edwards-Jones, G., 2010. High-throughput antioxidant profiling in vegetables by Fourier-transform ion cyclotron resonance mass spectrometry. *Functional Plant Science and Biotechnology.* 4: (Special Issue 1), 1–10.

Kuhnau, J., 1976. The flavonoids. A class of semi-essential food components: their role in human nutrition. *World Review Nutrition Diet.* 24: 117–191.

Lamartiniere, C.A., Moore, J., Holland, M., and Barnes, S., 1995. Neonatal genistein chemoprevents mammary cancer. *Proceedings Society Experimental Biology Medicine.* 208: 120–123.

Lamartiniere, C.A., Wang, J., Smith-Johnson, M., and Eltoum I.E., 2002. Daidzein: bioavailability, potential for reproductive toxicity, and breast cancer chemoprevention in female rats. *Toxicology Science*. 65: 228–238.

Levy, J.R., Faber, K.A., Ayyash, L., and Hughes, C.L. Jr., 1995. The effects of prenatal exposure to the phytoestrogen genistein on sexual differentiation in rats. *Proceedings Society Experimental Biology Medicine*. 208: 60–66.

Lund, T.D., Munson, D.J., Haldy, M.E., Setchell, K.D.R., Lephart, E.D., and Handa, R.J., 2004. Equol is a novel anti-androgen that inhibits prostate growth and hormone feedback. *Biology Reproduction*. 70: 1188–1195.

Lund, T.D., Blake, C., Bu, L., Hamaker, A.N., and Lephart, E.D., 2011. Equol an isoflavonoid: potential for improved prostate health, *in vitro* and *in vivo* evidence. *Reproductive Biology Endocrinology*. 9: 4.

Magee, P.J., Raschke, M., Steiner, C., Diffin, J.G., Pool-Zobel, B.L., Jokela, T., Wahala, K., and Rowland, I.R., 2006. Equol: a comparison of the effects of the racemic compound with that of the purified S-enantiomer on the growth, invasion and DNA integrity of breast and prostate cells *in vitro*. *Nutrition Cancer*. 54: 232–242.

Matulka, R.A., Matsuura, I., Uesugi, T., Ueno, T., and Burdock, G., 2009. Developmental and Reproductive Effects of SE5-OH: An Equol-Rich Soy-Based Ingredient. *Journal Toxicology*. epub, article ID 307618, 13 pages.

Medlock, K.L, Branham, W.S., and Sheehan, D.W., 1995. The effects of phytoestrogens on neonatal rat uterine growth and development. *Proceedings Society Experimental Biology Medicine*. 208: 307–213.

Nagata, C., Iwasa, S., Shiraki, M., Ueno, T., Uchiyama, S., Urata, K., Sahashi, Y., and Shimizu, H., 2006. Association among maternal soy intake, isoflavone levels in urine and blood samples, and maternal and umbilical hormone concentrations (Japan). *Cancer Causes Control*. 17: 1107–1113.

Reinli, K., and Block, G., 1996. Phytoestrogen content of foods– a compendium of literature values. *Nutrition Cancer*. 26: 123–148.

Rufer, C.E, and Kulling, S.E., 2006. Antioxidant activity of isoflavones and their major metabolites using different in vitro assays. *Journal of Agricultural and Food Chemistry*. 54: 2926–2931.

Selavraj, V., Zakroczymski, M.A., Naaz, A., Mukai, M., Ju, Y.H., Doerge, D.R., Katzenellenbogen, J.A., Helferich, W.G., and Cooke, P.S., 2004. Estrogenicity of the isoflavone metabolite equol on reproductive and non-reproductive organs in mice. *Biology Reproduction*. 71: 966–972.

Setchell, K.D., Zimmer-Nechemias, L., Cai, J., and Heubi, J.E., 1997. Exposure of infants to phyto-estrogens from soy-based infant formula. *Lancet*. 350: 23–27.

Setchell, K.D.R., Zimmer-Nechemias, L., Cai, J., and Heubi, J.E., 1998. Isoflavone content of infant formulas and the metabolic fate of these phytoestrogens in early life. *American Journal Clinical Nutrition*. 68 (suppl): 1453S–1461S.

Setchell, K.D.R., and Clerici, C., 2010a. Equol: pharmacokinetics and biological actions. *Journal of Nutrition*. 140: 1363S–1368S.

Setchell, K. D. R., and Clerici, C., 2010b. Equol: history, chemistry, and formation. *Journal of Nutrition*. 140: 1355S–1362S.

Smoliga, J.M., Baur, J.A., and Hausenblas, H. A., 2011. Resveratrol and health – A comprehensive review of human clinical trials. *Molecular Nutrition and Food Research.* 55: 1–13.

Tanaka, M., Fujimoto, K., Chihara, Y., Torimoto, K., Yoneda, T., Tanaka, N., Hirayama, A., Miyanaga, N., Akaza, H., and Hirao, Y., 2009. Isoflavone supplements stimulated the production of serum equol and decreased the serum dihydrotestosterone levels in healthy male volunteers. *Prostate Cancer and Prostatic Disease.* 12: 247–252.

Todaka, E., Sakurai, K., Fukata, H., Miyagawa, H., Uzuki, M., Omori, M., Osada, H., Ikezuki, Y., Tsutsumi, O., Iguchi, T., and Mori, C., 2005. Fetal exposure to phytoestrogens – the difference in phytoestrogen status between mother and fetus. *Environmental Research.* 99: 195–203.

Weber, K.S., Setchell, K.D.R., and Lephart. E.D., 2001. Maternal and perinatal brain aromatase: effects of dietary soy phytoestrogens. *Developmental Brain Research.* 126: 217–221.

Wood, C.E., Appt, S.E., Clarkson, T.B., Franke, A.A, Lees, C.J., Doerge, D.R., and Cline, J.M., 2006. Biology reproduction. Effect of high-dose soy isoflavones and equol on reproductive tissues in female cynomolgus monkeys. *Biology Reproduction.* 75: 477–486.

Yee, S., Burdock, G.A., Kurata, Y., Enomoto, Y., Narumi, K., Hamada, S., Itoh, T., Shimomura, Y., and Ueno, T., 2008. Acute and subchronic toxicity and genotoxicity of SE5-OH, an equol-rich product produced by Lactococcus garvieae. *Food Chemistry Toxicology.* 46: 2713–2720.

USDA, 2007. USDA – Iowa State University Database on the Isoflavone Content of Foods. *Release.* 1.4. Agricultural Research Service, Beltsville Human Nutrition Research Center, Nutrient Data Laboratory, Beltsville, MD, USA.

CHAPTER 30

Genistein: $GABA_A$ and NMDA Receptors

RENQI HUANG* AND GLENN H. DILLON

Department of Pharmacology and Neuroscience, University of North Texas Health Science Center, Fort Worth, TX, USA
*Email: Ren-qi.huang@unthsc.edu

30.1 Introduction

The phytoestrogen genistein (4′,5′,7-trihydroxyisoflavone) is one of several compounds found in high abundance in several plants. These molecules are so named due to their close similarity in structure to estrogens and thus an ability to bind estrogen receptors (ERs) and in turn influence estrogen-mediated processes. The three main types of phytoestrogens include lignans, which are abundant in flaxseed, coumestans, found in alfalfa and other sprouting plants, and isoflavones. Isoflavones are soy-based phytoestrogens; of the three main classes of phytoestrogens, they are the most widely consumed by humans. Within the isoflavone class of phytoestrogens, a number of specific molecules exist. Among them, genistein, daidzein (4′,7- dihydroxyisoflavone) and glycitein (4′,7-dihydroxy-6-methoxyisoflavone) have been the most studied because they are considered to have the most potent estrogenic effects of the soy phytoestrogens. Several phytoestrogens have a higher affinity for ERβ than for ERα (Kuiper *et al.* 1998). The binding affinity of genistein for ERα is 4% and for ERβ is 87% compared with endogenous estrogen (Kuiper *et al.* 1998). Genistein acts like a natural selective estrogen receptor modulator (SERM) with either agonistic or antagonistic effects, although with considerably reduced potency (Kuiper *et al.* 1998). Besides eliciting ER-mediated effects, experimental evidence suggests that genistein acts through a number of non-ER pathways. These effects include the ability to modulate tyrosine kinases (Akiyama *et al.* 1987), antioxidant activity (Tikkanen *et al.* 1998), cholesterol levels (Anthony *et al.* 1998), cell proliferation and DNA synthesis (Pan *et al.* 2001). Moreover, genistein has been reported to directly inhibit a number of ion channels. These include voltage-gated channels selective for Ca^{2+} (Tao *et al.* 2009), Na^+ (Paillart *et al.* 1997) and K^+ (Smirnov and Aaronson 1995). Genistein inhibits depolarization-induced Ca^{2+} influx and glutamate release from hippocampal synaptosomes, most likely through direct inhibition of voltage-gated Ca^{2+} channels and/or K^+ channels (Pereira *et al.* 2003). It is worthy to note that genistein exerts different effects on its target at largely different concentrations. At nM concentrations, genistein exhibits estrogenic effects such as neuroprotective and neurotrophic effects. At low μM concentrations (<30 μM), genistein displays the effects mediated by inhibition of tyrosine kinase activity and topoisomerase. At higher concentrations (>50 μM), genistein exerts its effects through direct inhibition of channels and receptors.

We attempt herein to summarize the modulatory effects of genistein on γ-aminobutyric acid type A ($GABA_A$) and *N*-methyl-D-aspartate (NMDA) receptors, two major neurotransmitter receptors in the central nervous system (CNS), and the mechanism underlying its effects.

30.2 $GABA_A$ Receptor Function and Structure

30.2.1 $GABA_A$ Receptor Function

The $GABA_A$ receptor is the predominant inhibitory neurotransmitter receptor in the mammalian CNS, and is the site of action of numerous therapeutic agents. This receptor is part of the Cys-loop superfamily of ligand-gated ion channels (LGICs), which also include receptors for acetylcholine (nicotinic), glycine and serotonin ($5-HT_3$ subtype). For a thorough description of the receptor, the reader is referred to the review by Huang *et al.* (2006). Aspects seminal to the present discussion are summarized here.

The binding of GABA to the $GABA_A$ receptors results in opening of a Cl^- channel. In the adult CNS, receptor activation causes hyperpolarization, and neuronal activity is reduced. Alterations in GABAergic neurotransmission have been implicated in many diseases and disorders. For instance, a number of mutations found in $GABA_A$ receptors cause several forms of epilepsy. In addition, work in recent years has established a strong link between GABAergic signaling and schizophrenia (Costa *et al.* 2004). Moreover, the role of $GABA_A$ receptors in anxiety and related disorders is well known. Because of this direct involvement in several disorders, as well as the ubiquitous nature of the $GABA_A$ receptors, $GABA_A$ receptors are targeted by numerous drugs, including sedatives/hypnotics, anxiolytics, muscle relaxants, anesthetics and anticonvulsants.

30.2.2 GABA$_A$ Receptor Structure

The GABA$_A$ receptors are pentamers that are assembled form an array of homologous subunits. To date, 16 subunits (α_{1-6}, β_{1-3}, γ_{1-3}, δ, ϵ, θ and π) have been cloned and sequenced from the mammalian CNS and several of these subunits slice variants have been identified. The most common isoforms are generally thought to compose two α subunits, two β subunits and one γ or δ subunit. The most abundant population of native GABA$_A$ receptors in the mammalian brain is believed to be a configuration of $\alpha1\beta2\gamma2$.

The general structure of the GABA$_A$ receptors is similar to that of the other Cys-loop superfamily members. Each subunit contains a large, extracellular N-terminal domain (NTD), which contains the ligand-binding site and Cys-loop; four hydrophobic transmembrane domains (TM1–TM4); a large intracellular loop between TM3 and TM4, which is a site of regulation by phosphorylation; and a very short extracellular C-terminal domain. TM2 is the most highly conserved domain of the Cys-loop receptor superfamily, and forms the central ion pore (Figure 30.1).

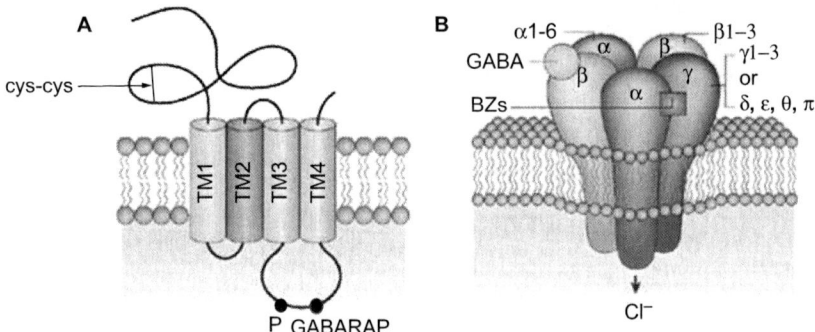

Figure 30.1 A schematic diagram illustrating pertinent features of GABA$_A$ receptors. (**A**) GABA$_A$ receptor subunits consist of the Cys-loop in the extracellular domain, four hydrophobic transmembrane domains (TM1–TM4), with TM2 lining the channel pore, and the cytoplasmic loop localized between TM3 and TM4 which contains the residues for phosphorylation (P) and binding of GABA$_A$ receptor-associated protein (GABARAP). (**B**) Despite the extensive heterogeneity of the GABA$_A$ receptor subunits, most GABA$_A$ receptors expressed in the brain comprise two α (1–6), two β (1–3) and one γ (1–3) subunit with the GABA- and Benzodiazepine (BZD)-binding (BZ) sites located at the α–β and α–γ interfaces, respectively. The γ subunit can be replaced by δ, ϵ, θ or π. The large extracellular N-terminus is the site of GABA binding, and also contains binding sites for psychoactive drugs, such as BZD (BZs). Binding of the neurotransmitter GABA triggers the opening of the channel, allowing the rapid influx of Cl$^-$ into the cell.
This Figure is reproduced and modified, with permission, from Jacob et al. (2008) Nature Publishing Group.

30.2.3 GABA$_A$ Receptor Modulation

The activity of the GABA$_A$ receptors is modulated by numerous pharmacologically and clinically important drugs. For example, enhancing GABA$_A$ receptor function is the key property of several classes of therapeutic agents such as the benzodiazepines (BZDs), barbiturates and several general anesthetics. Conversely, inhibiting GABA$_A$ receptor activity is the basis for the mechanism of action of picrotoxin and other convulsants. Knowledge of particular domains important in the binding and functional effects of many of these ligands to the GABA$_A$ receptors has grown considerably in recent years.

The GABA$_A$ receptor function is susceptible to regulation by endogenously derived agents including trace metals (for example Cu^{2+} and Zn^{2+}), protons, neurosteroids and protein phosphorylation. A number of protein kinases, such as protein kinase C, protein kinase A, protein kinase G, protein tyrosine kinases (PTKs) and Ca^{2+}/calmodulin-dependent kinase II (CaMKII), have been reported to modulate GABA$_A$ receptors by direct phosphorylation of residues at the large intracellular loop between TM3 and TM4. Phosphorylation sites for these kinases have been mapped to subunit major intracellular domains between TM3 and TM4 of the β_{1-3} and γ_2 subunits.

The function of GABA$_A$ receptors is positively regulated by tyrosine phosphorylation. The phosphorylation sites for PTKs have been identified on Y365 and Y367 within the intracellular loop of the γ2 subunits. Phosphorylation of these residues potentiates GABA$_A$ receptor-mediated currents by an increase in channel mean open time and open probability, whereas protein tyrosine phosphatases (PTPs) inhibits GABA response (Huang and Dillon 1998), suggesting that GABA$_A$ receptors are dynamically regulated by a balance between activities of endogenous PTKs and PTPs. Tyrosine phosphorylation is also involved in modulation of the subcellular distribution of GABA$_A$ receptors (Wan *et al.* 1997b) and surface receptor trafficking (Kittler *et al.* 2008), which could also be an important mechanism for synaptic GABA$_A$ receptor plasticity. For example, the residues Y365/Y367 are part of a classical tyrosine-based binding motif for the clathrin adaptor protein 2 (AP2) complex. Enhancing tyrosine phosphorylation of the γ2 subunits reduced clathrin-dependent endocytosis and thus increases synaptic GABA$_A$ receptors (Kittler *et al.* 2008).

30.3 Genistein Modulation of GABA Receptors

30.3.1 Genistein as an Inhibitor of Protein Tyrosine Kinases (PTKs)

One of the mechanisms by which genistein inhibits GABA$_A$ receptors function is *via* its inhibition of protein tyrosine phosphorylation. Genistein was one of the first PTK inhibitors identified (Akiyama *et al.* 1987). It is a potent PTK inhibitor with little effect on serine/threonine kinases (Akiyama *et al.* 1987). Genistein inhibits PTKs with an IC$_{50}$ of 2.6–30 μM *in vitro*, depending on the PTKs inhibited (Akiyama *et al.* 1987; Akiyama and Ogawara 1991; Dean *et al.* 1989).

Genistein is membrane permeable, and is able to inhibit intracellular PTK-dependent pathways. Indeed, for many years genistein has been routinely utilized to probe potential involvement of tyrosine kinase activity in literally dozens of diverse cell signaling events. Daidzein, another isoflavone that is structurally similar to genistein but which poorly inhibits PTKs, is often used as a negative control for study of PTK-dependent events. Genistein has been reported to inhibit $GABA_A$ receptors by inhibiting endogenous PTKs that are required to maintain $GABA_A$ receptor function, modulate receptor surface trafficking and expression as described above. Genistein inhibits $GABA_A$ receptor-mediated currents in recombinant (Huang and Dillon 1998; Valenzuela et al. 1995) and native (Castel et al. 2000; Jassar et al. 1997; Valenzuela et al. 1995; Wan et al. 1997a) tissues, and blocks the activity of autophosphorylation of PTKs associated with the $\gamma 2$ subunits (Jurd et al. 2010). Genistein also causes a reduction of the expression of $GABA_A$ receptors in recombinant preparations (Balduzzi et al. 2001; 2002). However, in most studies, especially for those that genistein was applied extracellularly, the concentrations of genistein were high enough (>50 μM) such that non-PTK-dependent mechanisms (see next section) may also be elicited.

30.3.2 Direct Inhibition of $GABA_A$ Receptors

As described above, genistein had been primarily used as a PTK inhibitor to assess the involvement of tyrosine phosphorylation in $GABA_A$ receptors. However, the effect of genistein is not always attributed to inhibition of tyrosine kinases. At the concentrations commonly used in the aforementioned studies, we showed that genistein is able to directly inhibit the GABA response in recombinant $\alpha 1 \beta 2 \gamma 2$ and $\alpha 1 \beta 2$ receptors expressed in HEK 293 cells (Huang et al. 1999). At a concentration of 100 μM, genistein reduced the GABA current by $\sim 20\%$ when it was co-applied with GABA (Figure 30.2). The inactive analogue daidzien also similarly inhibited $GABA_A$ receptors but with lower potency than genistein. In contrast, lavendustin, another PTK inhibitor that has different structure from genistein, has no such direct action. Similar findings were reported by Dunne et al. (1998) in recombinant $\alpha 1 \beta 2 \gamma 2S$ receptors. This direct inhibition by genistein is not dependent on the presence of the γ subunit or Y365 and Y367 on the $\gamma 2$ subunit which are needed for PTK-dependent pathways (Dunne et al. 1998; Huang et al. 1999). Given the fact that the effect of genistein was rapid in onset (in the ms scale), and unaffected by intracellular pre-treatment with genistein, genistein probably acts on the extracellular sites (Huang et al. 1999). Non-competitive antagonism of genistein is consistent with an allosteric modulation (distinct from GABA binding site) (Dunne et al. 1998; Huang et al. 1999). Genistein inhibition was not influenced by the deletion of the $\gamma 2$ subunit which plays an important role in BZDs and Zn^{2+} modulation (Huang et al. 1999). Genistein had no effect on the pentobarbitone or diazepam response, suggesting no major involvement of the active sites for BZDs or barbiturates in genistein inhibition (Dunne et al. 1998).

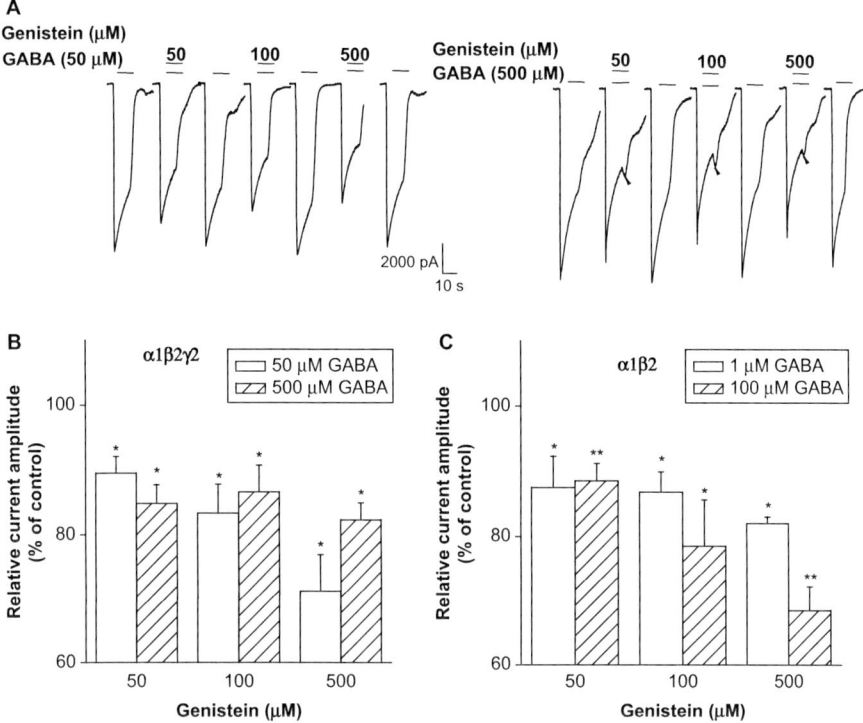

Figure 30.2 Genistein co-applied with GABA decreased the GABA-activated current amplitude and increased the decay rate in α1β2γ2 (C) and α1β2 (D) GABA$_A$ receptors. (**A**) Typical recording traces to 50 and 500 μM GABA recording from the same cell expressing α1β2γ2 subunits show that co-application of genistein with GABA resulted in a decreased current peak and a increased current decay in an reversible and concentration-dependent manner. Current rebound indicated by triangles was not exceptional for effect of genistein on saturating GABA-activated currents. (**B, C**) Histogram summarizing the effect of genistein (50–500 μM) on the GABA response in α1β2γ2 (**B**) and α1β2 (**C**) receptors. Each data point is for 4–13 cells. *$P<0.05$; **$P<0.01$ compared with the control response. This Figure is reproduced in part from our previous publication (Huang et al. 1999) with permission from the publisher Elsevier.

However, whether genistein inhibits GABA$_A$ receptors at a novel sites or other known drug site needs further investigation.

30.4 NMDA Receptor Function, Structure and Modulation

30.4.1 NMDA Receptor Function

The excitatory amino acid glutamate interacts with a family of metabotropic G-protein-coupled receptors and a family of neurotransmitter-gated ion channels. The latter family includes receptors for α-amino-3-hydroxy-5-methyl-4-isoxazole propionic acid (AMPA), kainate and NMDA. The names of each of these receptors arise from the ability of the respective ligands to selectively activate the corresponding receptors. These receptors are also typically categorized as NMDA on non-NMDA (AMPA and kainite) based on the ability to bind (or not bind) this ligand. In the nervous system, the ligand activating each of these receptors is generally glutamate, although aspartic acid and other molecules may serve in a neurotransmitter capacity in some cases. Each of these receptors has important roles in fast excitatory neurotransmission in the CNS; these roles are best understood for AMPA receptors and NMDA receptors, which are often co-localized. NMDA receptors are the subject of this section.

The NMDA receptor possesses a number of unique characteristics that makes it of functional interest when considering signaling within the CNS. In particular, its high permeability to the divalent cation Ca^{2+} confers to it the ability to influence Ca^{2+}-mediated events within the cell. Due to this trait and some additional characteristics described below, the NMDA receptor is well known to play a major role in formation of learning and memory. It is also key in the development of the nervous system, has important influences on the brain during conditions associated with trauma or injury, and is a therapeutic target for a number of conditions and disorders (Lau and Tymianski 2010).

30.4.2 NMDA Receptor Structure

The basic architecture of the NMDA receptor is the same as that of the AMPA and kainite receptors, and has been generally understood for several years. A number of specific structural elements have emerged in recent years. A critical advance came in 2009 with the solving of the crystal structure of an AMPA subtype of the receptor family (Sobolevsky *et al.* 2009). A brief overview of the key structural elements of the NMDA receptor is presented here. For a more detailed description of the structure of this family of the receptor, the reader is referred to recent reviews by Stawski *et al.* (2010) and Traynelis *et al.* (2010).

All ionotropic excitatory amino acid receptors are tetrameric, and are formed by heteromeric combining of a large array of subunits. In the case of the NMDA receptor, seven subunits have been identified (NR1, NR2A–NR2D, NR3A and NR3B). As illustrated in Figure 30.3, the formation of a functional receptor requires NR1 combined with an NR2 type and/or NR3 type of subunit. These subunits are very large, ranging from over 900 to nearly 1500 amino acid residues per subunit. Much of the protein contributes to the extracellular domain, which includes a large N-terminal domain, which appears to be important in receptor oligomerization, trafficking and modulation (Hansen *et al.* 2010) and the ligand-binding domain (LBD). The transmembrane and C-terminal cytosolic domains complete the protein. The LBD is often referred to as being clamshell or venus flytrap-like, as studies have demonstrated that the two components of the "clamshell" move toward one another when ligand binds. The presence of a partial agonist closes the clamshell by approximately 12°, whereas the presence of a full agonist such as glutamate results in 20° of

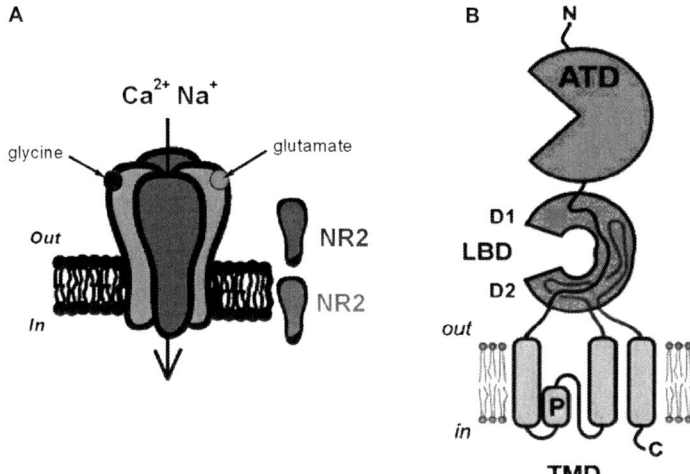

Figure 30.3 A schematic diagram illustrating pertinent features of NMDA neurotransmitter receptors. (**A**) A model of a tetrametric NMDA receptor consisting of two NR1 and two NR2 subunits. Activation of NMDA receptors results in the opening of cation-selective ion channels. Unlike the non-NMDA receptor subtype, NMDA receptor has relatively high permeability to Ca^{2+}, in addition to Na^+. NMDA receptors require co-activation by two ligands, glutamate and glycine. (**B**) Domain structure of ionotropic glutamate receptors. The extracellular domain contains the N-terminal domain (ATD) and the ligand-binding domain (LBD) which consists of upper lobe of the LBD (D1) and lower lobe of the LBD (D2). The transmembrane domain (TMD) consists of three transmembrane segments and a re-entrant pore helix (P). The cytoplasmic domain (C) contains residues for protein phosphorylation, and residues for structural adaptor and scaffolding proteins.
(**A**) is modified from Granger *et al.* (2011) with permission from the publisher John Wiley and Sons. (**B**) is reproduced in part from Stawski *et al.* (2010) with permission from the publisher Elsevier.

channel closure (Armstrong and Gouaux 2000). This closure of the LBD drives opening of the channel, permitting ion passage. The transmembrane domain is made up of M1–M4. Unlike the $GABA_A$ receptor, only three of these domains (M1, M3 and M4) are transmembrane. M2 enters into the membrane from the cytoplasmic side and returns. The M2 domain is highly conserved and forms the ion permeation pathway.

The NMDA receptor is unique among the glutamate family of ion channels in a number of respects in addition to its high Ca^{2+} permeability as noted above. First, it requires presence of a co-agonist, glycine, for channel activation to occur. This binding site for glycine is structurally distinct from the glycine binding site on the inhibitory glycine receptor, and is a potential therapeutic target. The NMDA receptor also has a voltage-dependent component. In the absence of membrane depolarization, Mg^{2+} binds to the channel pore, preventing channel activation. The NMDA receptor is thus considered a coincidence detector because, for it to be activated, both the agonist glutamate and membrane depolarization, often caused by activation of AMPA receptors, must be present. This trait is critical for the role of the NMDA receptor in learning and memory (Traynelis et al. 2010).

30.4.3 NMDA Receptor Modulation

NMDA receptor function is regulated by a variety of protein kinases which mediate serine/threonine and tyrosine phosphorylation. The intracellular C-termini of NMDA receptor subunits contain the phosphorylation sites and a number of the residues have been identified for protein phosphorylation [for reviews, see Chen and Roche (2007) and Salter et al. (2009)]. The phosphorylation can modulate NMDA receptor expression, trafficking, channel opening probability and ultimately affect synaptic strength which underlies synaptic plasticity.

Similar to $GABA_A$ receptors, NMDA receptor function is up-regulated by Src family of tyrosine kinases. NMDA receptor function is maintained by a balance between endogenous PTKs and PTPs (Wang and Salter 1994; Wang et al. 1996). The Src family kinases (SFKs) were found to be components of the NMDA receptor complex (Yu et al. 1997). The NR1 subunit does not appear to be a site for PTKs. Three tyrosine residues on NR2A (Y1292, Y1325 and Y1387) and three residues on NR2B (Y1252, Y1336 and Y1472) have been identified as targets for PTKs. Direct phosphorylation of NMDA receptors by PTKs up-regulates NMDA receptor channel gating *via* a phosphopeptide activator (pYEEI peptide) (Yu et al. 1997). PTKs also modulate surface expression and trafficking of NR2B-containing NMDA receptors. Phosphorylation of NR2B Y1472, which is within a tyrosine-based internalization motif (YEKL), disrupts its binding to AP2 and thus inhibits NR2B-mediated endocytosis. NMDA receptor function is also regulated by endogenous PTPs that are intrinsic to the NMDA receptor complex. One of those PTPs, STEPs (striatal-enriched tyrosine phosphatases) which promote de-phosphorylation of YNR2B, regulate the function of NMDARs in opposition to Src (Chen and Roche 2007).

Up-regulation of NMDA receptors by PTKs is critical for the induction of long-term potentiation (LTP) at Schaffer collateral CA1 synapses in the hippocampus. A model for the role of PTKs and PTPs in the induction of LTP was proposed by Salter et al. (2009).

30.5 Genistein Modulates NMDA Receptor Modulation

30.5.1 Genistein as an Inhibitor of Protein Tyrosine Kinases (PTKs)

The studies on the role of tyrosine phosphorylation in NMDA receptor function partially rely on the use of genistein to inhibit endogenous PTKs. In fact, the first use of genistein as a PTK inhibitor of the channels and receptors in the CNS was in a study of NMDA receptors. Genistein (1–100 µM) resulted in a reversible inhibition of NMDA currents in spinal dorsal horns neurons (Wang and Salter 1994). Similar inhibition by genistein was later reported in recombinant NMDA receptors expressed in HEK 293 cells (Kohr and Seeburg 1996) and in *Xenopus* (Chen and Leonard 1996). Genistein inhibited NMDA receptors and subsequently blocked NMDA receptor-dependent events such as LTP induction in hippocampal CA1 region (Huang and Hsu 1999; O'Dell *et al.* 1991) and dentate gyrus (Casey *et al.* 2002), taste memory formation (Barki-Harrington *et al.* 2009) and NMDA-elicited feeding (Khan *et al.* 2004). Genistein was also shown to block NMDA receptor surface expression regulated by dopamine receptors (Gao and Wolf 2008).

30.5.2 Direct Inhibition of NMDA Receptors

It is notable that the concentrations of genistein used as a pharmacological tool to examine PTK involvement in NMDA receptor-associated processes range from 50 µM up to 500 µM. We first reported that genistein was able to directly inhibit both recombinant and native NMDA receptors at concentrations higher than 30 µM (Huang *et al.* 2010). For example, extracellular application of 100 µM genistein reduced NMDA-activated currents by 20–30% (Figure 30.4). The inhibition of NMDA-activated currents was concentration-dependent but not voltage-dependent. The direct action of NMDA receptors displays the characteristics similar to that of $GABA_A$ receptors. The effect was rapid in onset and reversible. The action site seems to be located on the extracellular sites but the precise details need further investigation. Likewise, daidzein, an inactive analog, also inhibited NMDA-activated current (Figure 30.4). Lavendustin A, a specific PTK inhibitor whose structure is distinct from Isoflavones, had no direct inhibition on NMDA receptors.

30.6 Genistein Modulation of $GABA_A$ and NMDA Receptors *via* the Estrogenic Pathway

Estrogen has been reported to up-regulate NMDA and $GABA_A$ receptors through traditional nuclear ERs and membrane ERs. Estrogen regulates $GABA_A$ receptor mRNA expression at a transcriptional level in the pre-optic area and bed nucleus of the stria terminalis of female rat brain (Herbison and Fenelon 1995) and in NT2-N neurons (Pierson *et al.* 2005). Estradiol increases the number of spines on the apical dendrites of hippocampal CA1 pyramidal neurons and increases the production of mRNA for NMDA receptor subunits and the density of excitatory NMDA receptors on the dendritic spines (Gazzaley *et al.* 1996; Woolley and McEwen 1993). The estrogen-induced effect could result from an increase in receptor synthesis *via* a genomic mechanism. It could also be mediated by a non-genomic mechanism *via* membrane ERs. Estrogen increases tyrosine phosphorylation which is known to up-regulate

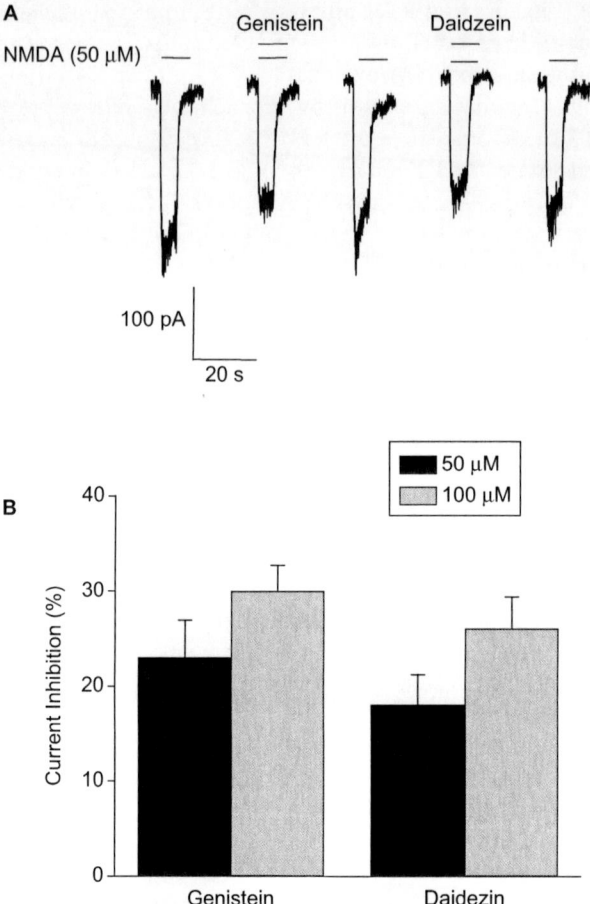

Figure 30.4 Effect of co-applied genistein or daidzein on NMDA-activated currents in mouse hippocampal slices. (**A**) Whole-cell NMDA (50 μM)-activated current was recorded from one hippocampal neuron. Genistein (50 or 100 μM) or daidzein (50 or 100 μM) was co-applied with NMDA to the cell for 10 s. Note that genistein- and daidzein-induced inhibition of the NMDA current was reversible. (**B**) Summary data of the inhibition of the NMDA response by genistein or daidzein. The current amplitude is normalized to the initial NMDA response (assigned as 100%). Each data point represents mean ± S.E.M. from at least 4 cells.
This Figure is reproduced from our previous publication (Huang *et al.* 2010) with permission from the publisher Pergamon.

receptor function and expression (see above). Estrogen may utilize both mechanisms to modulate $GABA_A$ and NMDA receptor function (Finocchi and Ferrari 2011).

Genistein is structurally similar to estrogen and has been shown to bind to both nuclear and membrane ERs (Kuiper *et al.* 1998; Lin *et al.* 2010). Although genistein has higher affinity towards ERβ than ERα (Kuiper et al. 1998), dietary doses of genistein (1–5 µM) activate ERα equally well, like a natural estrogen (Chang et al. 2008). Genistein acts as an estrogen agonist which elicits similar effects to those seen with estrogen. However, it remains to be determined whether genistein regulates $GABA_A$ and NMDA receptors through its effects on ERs.

30.7 Functional Implications

The biological effects of genistein are highly dependent on its concentration *via* multiple mechanisms as summarized by Klein and King (2007). The physiological level of genistein is very low. The plasma level of genistein in people consuming a soy-rich diet was 1–5 µM. As most of genistein is bound to serum proteins, the actual concentration of biologically active genistein in serum is estimated in the nM range (Klein and King 2007). At such low concentration, genistein influences cellular function mainly by acting as a modulator at ERs. Genistein has been shown to be neuroprotective *in vitro* and *in vivo* at a physiologically relevant dose range (Luine et al. 2006). In the case when high dose isoflavones is needed for testing a therapeutic effect such as alternatives to estrogen replacement or as chemopreventive agents, the concentrations of blood genistein (5–20 µM) may exceed physiological levels. At such concentration, genistein may influence cellular function including $GABA_A$ and NMDA receptors *via* tyrosine kinase inhibition. The direct inhibition of the GABA and NMDA receptors is unlikely to occur in humans at dietary relevant concentrations of genistein. However, the biological effects of genistein in *in vitro* cellular activities have been reported to occur over a very wide range of concentration, from less than 1 µM up to 500 µM genistein. It is notable that *in vitro* doses of genistein used in those studies, particularly as a PTK inhibitor, far surpass those with likely physiological relevance for humans. Under those experimental conditions when a high concentration of genistein is used, genistein exerts its modulatory effect on $GABA_A$ receptors and NMDA receptor *via* inhibition PTKs and/or direct action on the receptors.

30.8 Conclusion

We summarize the findings on the regulation of $GABA_A$ and NMDA receptors by genistein, which is found in soy-containing diets. Extensive studies have shown that genistein regulates receptor function and cell surface expression *via* inhibiting the PTK mechanism. In addition, recent studies from our laboratory and others suggest that genistein, at the concentrations commonly used *in vitro* studies, is able to directly inhibit these two receptors in a PTK-independent manner. Thus, great caution should be taken when genistein is used to probe the regulation of target proteins by PTKs. The results reviewed here may be relevant to the study of the neurobehavioral effects of genistein as well as research fields using genistein as a PTK inhibitor.

Summary Points

- This Chapter focuses on the effects of genistein on $GABA_A$ and NMDA receptor function.
- NMDA and $GABA_A$ receptors are two of the major classes of ligand-gated ion channels in the mammalian brain.
- NMDA and $GABA_A$ receptor function is up-regulated by protein tyrosine phosphorylation.
- Genistein displays multi-targeted actions depending on its concentration.
- At dietary relevant concentrations (1–5 µM), genistein influences cellular function, mainly by acting as a modulator of estrogen receptors (ERs).
- At pharmacological concentrations (5–20 µM), genistein inhibits $GABA_A$ and NMDA receptor function by blocking tyrosine kinases.
- Under certain experimental conditions ([genistein] > 30 µM), genistein directly inhibits $GABA_A$ and NMDA receptors *via* a PTK-independent mechanism.

Key Facts

- NMDA and $GABA_A$ receptors are two of the major classes of ligand-gated ion channels in the mammalian brain.
- NMDA and $GABA_A$ receptors are pivotal for normal brain function.
- NMDA and $GABA_A$ receptors are important drug targets.
- Protein tyrosine phosphorylation positively regulates the channel gating, surface expression and trafficking of NMDA and $GABA_A$ receptors.
- Genistein is the first broad PTK inhibitor identified. It has been widely used as a PTK inhibitor to probe PTK-dependent processes. Genistein has been used at concentrations up to 500 µM in those studies.
- Genistein directly inhibits $GABA_A$ and NMDA receptors at high concentrations (> 30 µM) *via* a PTK-independent mechanism.
- Great caution should be taken with the interpretation of results when genistein is used as a PTK inhibitor.

Definitions of Words and Terms

Cys-loop receptor superfamily: A group of receptors that have a characteristic loop formed by a disulfide bond between two cysteine (Cys) residues 13 highly conserved amino acids apart near the N-terminal extracellular domain. It includes nicotinic acetylcholine, $GABA_A$, $GABA_c$, glycine and serotonin (5-HT3 subtype) receptors.

De-phosphorylation: The opposite process to phosphorylation. It is processed by an enzyme called phosphatase.

Endocytosis: The process whereby cells internalize molecules (*e.g.*, proteins).

Inhibitory/excitatory neurotransmitter: The chemicals released by neurons allow the transmission of signals from one neuron to the next across synapses.

A chemical that makes firing of an action potential less likely and thus suppresses the activity of other neurons is called an "inhibitory neurotransmitter". A chemical that increases the probability that the target cell will fire an action potential is called an "excitatory neurotransmitter". GABA is the major inhibitory neurotransmitter, whereas glutamate is the major excitatory neurotransmitter in the brain.

Ligand-gated ion channels (LGICs): A group of ion channels that open after binding to specific molecules (ligands).

Long-term potentiation (LTP): A long-lasting enhancement in synaptic efficacy after high-frequency stimulation of afferent fibers. LTP is the molecular basis for learning and memory.

Phosphorylation: The metabolic process of introducing a phosphate group (PO_4) into a protein of an organic molecule. It is processed by an enzyme called kinase.

Synaptic plasticity: The change in strength of a neuronal connection or synapse.

Voltage-gated ion channels: A group of ion channels that open and close in response to membrane potential.

List of Abbreviations

AMPA	α-amino-3-hydroxy-5-methyl-4-isoxazole propionic acid
AP2	adapter protein 2
BZD	benzodiazepine
CNS	central nervous system
ER	estrogen receptor
$GABA_A$	γ-aminobutyric acid type A
LBD	ligand-binding domain
LTP	long-term potentiation
NMDA	N-methyl-D-aspartate
ATD	N-terminal domain
PTK	protein tyrosine kinase
PTP	protein tyrosine phosphatase
TM (or M)	transmembrane domain (TM used for $GABA_A$ receptors; M used for NMDA receptors)

References

Akiyama, T., and Ogawara, H., 1991. Use and specificity of genistein as inhibitor of protein-tyrosine kinases. *Methods in Enzymology*. 201: 362–370.

Akiyama, T., Ishida, J., Nakagawa, S., Ogawara, H., Watanabe, S., Itoh, N., Shibuya, M., and Fukami, Y., 1987. Genistein, a specific inhibitor of tyrosine-specific protein kinases. *Journal of Biological Chemistry*. 262: 5592–5595.

Anthony, M.S., Clarkson, T.B., and Williams, J.K., 1998. Effects of soy isoflavones on atherosclerosis: potential mechanisms. *American Journal of Clinical Nutrition*. 68: 1390S–1393S.

Armstrong, N., and Gouaux, E., 2000. Mechanisms for activation and antagonism of an AMPA-sensitive glutamate receptor: Crystal structures of the GluR2 ligand binding core. *Neuron.* 28: 165–181.

Balduzzi, R., Cupello, A., Diaspro, A., Ramoino, P., and Robello, M., 2001. Confocal microscopic study of $GABA_A$ receptors in *Xenopus* oocytes after rat brain mRNA injection: modulation by tyrosine kinase activity. *Biochimica et Biophysica Acta.* 1539: 93–100.

Balduzzi, R., Cupello, A., and Robello, M., 2002. Modulation of the expression of $GABA_A$ receptors in rat cerebellar granule cells by protein tyrosine kinases and protein kinase C. *Biochimica et Biophysica Acta.* 1564: 263–270.

Barki-Harrington, L., Elkobi, A., Tzabary, T., and Rosenblum, K., 2009. Tyrosine phosphorylation of the 2B subunit of the NMDA receptor is necessary for taste memory formation. *Journal of Neuroscience.* 29: 9219–9226.

Casey, M., Maguire, C., Kelly, A., Gooney, M.A., and Lynch, M.A., 2002. Analysis of the presynaptic signaling mechanisms underlying the inhibition of LTP in rat dentate gyrus by the tyrosine kinase inhibitor, genistein. *Hippocampus.* 12: 377–385.

Castel, H., Louiset, E., Anouar, Y., Le Foll, F., Cazin, L., and Vaudry, H., 2000. Regulation of $GABA_A$ receptor by protein tyrosine kinases in frog pituitary melanotrophs. *Journal of Neuroendocrinology.* 12: 41–52.

Chang, E.C., Charn, T.H., Park, S.H., Helferich, W.G., Komm, B., Katzenellenbogen, J.A., and Katzenellenbogen, B.S., 2008. Estrogen receptors α and β as determinants of gene expression: influence of ligand, dose, and chromatin binding. *Molecular Endocrinology.* 22: 1032–1043.

Chen, C., and Leonard, J.P., 1996. Protein tyrosine kinase-mediated potentiation of currents from cloned NMDA receptors. *Journal of Neurochemistry.* 67: 194–200.

Chen, B.S., and Roche, K.W., 2007. Regulation of NMDA receptors by phosphorylation. *Neuropharmacology.* 53: 362–368.

Costa, E., Davis, J.M., Dong, E., Grayson, D.R., Guidotti, A., Tremolizzo, L., and Veldic, M., 2004. A GABAergic cortical deficit dominates schizophrenia pathophysiology. *Critical Reviews in Neurobiology.* 16: 1–23.

Dean, N.M., Kanemitsu, M., and Boynton, A.L., 1989. Effects of the tyrosine-kinase inhibitor genistein on DNA synthesis and phospholipid-derived second messenger generation in mouse 10T1/2 fibroblasts and rat liver T51B cells. *Biochemical and Biophysical Research Communications.* 165: 795–801.

Dunne, E.L., Moss, S.J., and Smart, T.G., 1998. Inhibition of $GABA_A$ receptor function by tyrosine kinase inhibitors and their inactive analogues. *Molecular and Cellular Neuroscience.* 12: 300–310.

Finocchi, C., and Ferrari, M., 2011. Female reproductive steroids and neuronal excitability. *Neurological Sciences.* 32(Suppl. 1): S31–S35.

Gao, C., and Wolf, M.E., 2008. Dopamine receptors regulate NMDA receptor surface expression in prefrontal cortex neurons. *Journal of Neurochemistry.* 106: 2489–2501.

Gazzaley, A.H., Weiland, N.G., McEwen, B.S., and Morrison, J.H., 1996. Differential regulation of NMDAR1 mRNA and protein by estradiol in the rat hippocampus. *Journal of Neuroscience.* 16: 6830–6838.

Granger, A.J., Gray, J.A., Lu, W., and Nicoll, R.A., 2011. Genetic analysis of neuronal ionotropic glutamate receptor subunits. *Journal of Physiology.* 17: 4095–4101.

Hansen, K.B., Furukawa, H., and Traynelis, S.F., 2010. Control of assembly and function of glutamate receptors by the amino-terminal domain. *Molecular Pharmacology.* 78: 535–549.

Herbison, A.E., and Fenelon, V.S., 1995. Estrogen regulation of $GABA_A$ receptor subunit mRNA expression in preoptic area and bed nucleus of the stria terminalis of female rat brain. *Journal of Neuroscience.* 15: 2328–2337.

Huang, C.C., and Hsu, K.S., 1999. Protein tyrosine kinase is required for the induction of long-term potentiation in the rat hippocampus. *J. Physiol.* 520: 783–796.

Huang, R.Q., and Dillon, G.H., 1998. Maintenance of recombinant type A γ-aminobutyric acid receptor function: role of protein tyrosine phosphorylation and calcineurin. *Journal of Pharmacology and Experimental Therapeutics.* 286: 243–255.

Huang, R.Q., Fang, M.J., and Dillon, G.H., 1999. The tyrosine kinase inhibitor genistein directly inhibits $GABA_A$ receptors. *Brain Research and Molecular Brain Research.* 67: 177–183.

Huang, R.Q., Gonzales, E.B., and Dillon, G.H., 2006. $GABA_A$ receptors – structure, function and modulation. In: Arlas, H. (ed.) *Biolgical and Biophysical Aspects of Ligand-Gated Ion Channel Receptor Superfamilies.* Research Signpost, Kerala, pp. 171–198.

Huang, R., Singh, M., and Dillon, G.H., 2010. Genistein directly inhibits native and recombinant NMDA receptors. *Neuropharmacology.* 58: 1246–1251.

Jacob, T.C., Moss, S.J., and Jurd, R., 2008. $GABA_A$ receptor trafficking and its role in the dynamic modulation of neuronal inhibition. *Nature Reviews in Neuroscience.* 9: 331–343.

Jassar, B.S., Ostashewski, P.M., and Jhamandas, J.H., 1997. $GABA_A$ receptor modulation by protein tyrosine kinase in the rat diagonal band of Broca. *Brain Research.* 775: 127–133.

Jurd, R., Tretter, V., Walker, J., Brandon, N.J., and Moss, S.J., 2010. Fyn kinase contributes to tyrosine phosphorylation of the $GABA_A$ receptor γ2 subunit. *Molecular and Cellular Neuroscience.* 44: 129–134.

Khan, A.M., Cheung, H.H., Gillard, E.R., Palarca, J.A., Welsbie, D.S., Gurd, J.W., and Stanley, B.G., 2004. Lateral hypothalamic signaling mechanisms underlying feeding stimulation: differential contributions of Src family tyrosine kinases to feeding triggered either by NMDA injection or by food deprivation. *Journal of Neuroscience.* 24: 10603–10615.

Kittler, J.T., Chen, G., Kukhtina, V., Vahedi-Faridi, A., Gu, Z., Tretter, V., Smith, K.R., McAinsh, K., Arancibia-Carcamo, I.L., Saenger, W., Haucke, V., Yan, Z., and Moss, S.J., 2008. Regulation of synaptic inhibition by phospho-dependent binding of the AP2 complex to a YECL motif in the GABA$_A$ receptor γ2 subunit. *Proceedings of the National Academy of Sciences of the United States of America.* 105: 3616–3621.

Klein, C.B., and King, A.A., 2007. Genistein genotoxicity: Critical considerations of *in vitro* exposure dose. *Toxicology and Applied Pharmacology.* 224: 1–11.

Kohr, G., and Seeburg, P.H., 1996. Subtype-specific regulation of recombinant NMDA receptor-channels by protein tyrosine kinases of the Src family. *Journal of Physiology.* 492(2): 445–452.

Kuiper, G.G., Lemmen, J.G., Carlsson, B., Corton, J.C., Safe, S.H., van der Saag, P.T., van der Burg, B., and Gustafsson, J.A., 1998. Interaction of estrogenic chemicals and phytoestrogens with estrogen receptor β. *Endocrinology.* 139: 4252–4263.

Lau, A., and Tymianski, M., 2010. Glutamate receptors, neurotoxicity and neurodegeneration. *Pflügers Archive: European Journal of Physiology.* 460: 525–542.

Lin, A.H., Leung, G.P., Leung, S.W., Vanhoutte, P.M., and Man, R.Y., 2010. Genistein enhances relaxation of the spontaneously hypertensive rat aorta by transactivation of epidermal growth factor receptor following binding to membrane estrogen receptors-α and activation of a G protein-coupled, endothelial nitric oxide synthase-dependent pathway. *Pharmacological Research.* 63: 181–189.

Luine, V., Attalla, S., Mohan, G., Costa, A., and Frankfurt, M., 2006. Dietary phytoestrogens enhance spatial memory and spine density in the hippocampus and prefrontal cortex of ovariectomized rats. *Brain Research.* 1126: 183–187.

O'Dell, T.J., Kandel, E.R., and Grant, S.G., 1991. Long-term potentiation in the hippocampus is blocked by tyrosine kinase inhibitors. *Nature.* 353: 558–560.

Paillart, C., Carlier, E., Guedin, D., Dargent, B., and Couraud, F., 1997. Direct block of voltage-sensitive sodium channels by genistein, a tyrosine kinase inhibitor. *Journal of Pharmacology and Experimental Therapeutics.* 280: 521–526.

Pan, W., Ikeda, K., Takebe, M., and Yamori, Y., 2001. Genistein, daidzein and glycitein inhibit growth and DNA synthesis of aortic smooth muscle cells from stroke-prone spontaneously hypertensive rats. *Journal of Nutrition.* 131: 1154–1158.

Pereira, D.B., Carvalho, A.P., and Duarte, C.B., 2003. Genistein inhibits Ca2+ influx and glutamate release from hippocampal synaptosomes: putative non-specific effects. *Neurochemistry International.* 42: 179–188.

Pierson, R.C., Lyons, A.M., and Greenfield, Jr, L.J., 2005. Gonadal steroids regulate GABA$_A$ receptor subunit mRNA expression in NT2-N neurons. *Brain Research and Molecular Brain Research.* 138: 105–115.

Salter, M. W., Dong, Y., Kalia, L. V., Liu, X. J., and Pitcher, G., 2009. Regulation of NMDA receptors by kinases and phosphatases. In: Van Dongen AM, editor. Biology of the NMDA Receptor. Boca Raton (FL): CRC Press; 2009. Chapter 7.

Smirnov, S.V., and Aaronson, P.I., 1995. Inhibition of vascular smooth muscle cell K^+ currents by tyrosine kinase inhibitors genistein and ST 638. *Circulation Research*. 76: 310–316.

Sobolevsky, A.I., Rosconi, M.P., and Gouaux, E., 2009. X-ray structure, symmetry and mechanism of an AMPA-subtype glutamate receptor. *Nature*. 462: 745–756.

Stawski, P., Janovjak, H., and Trauner, D., 2010. Pharmacology of ionotropic glutamate receptors: A structural perspective. *Bioorganic and Medicinal Chemistry*. 18: 7759–7772.

Tao, J., Zhang, Y., Li, S., Sun, W., and Soong, T.W., 2009. Tyrosine kinase-independent inhibition by genistein on spermatogenic T-type calcium channels attenuates mouse sperm motility and acrosome reaction. *Cell Calcium*. 45: 133–143.

Tikkanen, M.J., Wahala, K., Ojala, S., Vihma, V., and Adlercreutz, H., 1998. Effect of soybean phytoestrogen intake on low density lipoprotein oxidation resistance. *Proceedings of the National Academy of Sciences of the United States of America*. 95: 3106–3110.

Traynelis, S.F., Wollmuth, L.P., McBain, C.J., Menniti, F.S., Vance, K.M., Ogden, K.K., Hansen, K.B., Yuan, H., Myers, S.J., and Dingledine, R., 2010. Glutamate receptor ion channels: structure, regulation, and function. *Pharmacological Reviews*. 62: 405–496.

Valenzuela, C.F., Machu, T.K., McKernan, R.M., Whiting, P., VanRenterghem, B.B., McManaman, J.L., Brozowski, S.J., Smith, G.B., Olsen, R.W., and Harris, R.A., 1995. Tyrosine kinase phosphorylation of $GABA_A$ receptors. *Brain Research and Molecular Brain Research*. 31: 165–172.

Wan, Q., Man, H.Y., Braunton, J., Wang, W., Salter, M.W., Becker, L., and Wang, Y.T., 1997a. Modulation of $GABA_A$ receptor function by tyrosine phosphorylation of β subunits. *Journal of Neuroscience*. 17: 5062–5069.

Wan, Q., Xiong, Z.G., Man, H.Y., Ackerley, C.A., Braunton, J., Lu, W.Y., Becker, L.E., MacDonald, J.F., and Wang, Y.T., 1997b. Recruitment of functional $GABA_A$ receptors to postsynaptic domains by insulin. *Nature*. 388: 686–690.

Wang, Y.T., and Salter, M.W., 1994. Regulation of NMDA receptors by tyrosine kinases and phosphatases. *Nature*. 369: 233–235.

Wang, Y.T., Yu, X.M., and Salter, M.W., 1996. Ca^{2+}-independent reduction of N-methyl-D-aspartate channel activity by protein tyrosine phosphatase. *Proceedings of the National Academy of Sciences of the United States of America*. 93: 1721–1725.

Woolley, C.S., and McEwen, B.S., 1993. Roles of estradiol and progesterone in regulation of hippocampal dendritic spine density during the estrous cycle in the rat. *Journal of Comparative Neurology*. 336: 293–306.

Yu, X.M., Askalan, R., Keil, 2nd, G.J., and Salter, M.W., 1997. NMDA channel regulation by channel-associated protein tyrosine kinase Src. *Science*. 275: 674–678.

CHAPTER 31
Estrogenic Activity and Molecular Mechanisms of Coumestrol-induced Biological Effects

KENNETH NDEBELE, BARBARA GRAHAM AND PAUL TCHOUNWOU*

Department of Biology, College of Science, Engineering and Technology, Jackson State University, 1400 J. R. Lynch Street, Jackson, MS 39217, USA
*Email: paul.b.tchounwou@jsums.edu

31.1 Introduction

Since the beginning of the 21st century a great interest has been placed on compounds that mimic estrogen and induce neoplastic transformation in cells and genotoxicity in animals. Coumestrol is a classical example; it is classified as phytoestrogen as it is a plant-derived non-steroidal compound possessing estrogen-like biological activity. It is an environmentally derived substance that binds to the estrogen receptor (ER) (Kuiper *et al.* 1998; Molteni *et al.* 1995; Shutt and Cox 1972). The chemical shape of coumestrol orients its two hydroxy groups in the same position as the two hydroxy groups in estradiol, allowing it to inhibit the activity of aromatase and hydroxysteroid dehydrogenase (Blomquist *et al.* 2005). These two enzymes are involved in the synthesis of hormones, therefore modification results in the modulation of hormone production (Amin and Buratovich 2007).

The question has been raised whether coumestrol has genotoxic effects in addition to its beneficial activity. It is evident that nutritional factors are important in determining the health status of the human population. It has been suggested that diets high in phytoestrogens, such as coumestrol, may have a chemopreventative action and reduce the risk of breast, colon and prostate cancer (Olen et al. 2001). With consumer awareness of the advantages of diets high in fiber, dietary supplementation with phytoestrogens may increase (Olen et al. 2001). However, recent studies (Kulling and Metzler 1997; Schuler and Eastmond 2000) demonstrating the clastogenicity and mutagenicity of coumestrol in Chinese hamster ovary (CHO) cells have indicated the need for additional studies on the potential deleterious effects of coumestrol. The drawback of studying coumestrol is that its potential chronic health effects are unknown, that is long-term carcinogenic, mutagenic and teratogenic effects are not available. This Chapter presents a review of recent studies on the estrogenic activity and molecular mechanisms of coumestrol-induced biological effects.

31.2 Animal Studies

In the past decades, a great interest has focused on the mechanism of hormone-dependent cell transformation. Certain estrogenic compounds are capable of inducing neoplastic transformation in primary cells and are clearly carcinogenic in animals. Phytoestrogens exist primarily as glycosides (isoflavonoides). Many studies have demonstrated hormonal activities of coumestrol (Olen et al. 2001), but studies on genotoxic activity are still limited. It was shown by Kulling and Metzler (1997) that coumestrol exerted mutagenic and genotoxic effects in V79 cells. At concentrations of 5–50 µM the mutation frequency was increased up to 10-fold compared with the control. Furthermore, DNA single strand breaks (SSBs) were determined in V79 cells after treatment with coumestrol. Induction of SSBs was significant at 25 µM and increased in a dose-dependent manner. When tested for micronucleus formation, coumestrol was also active from 10 to 35 µM. Characterization of the micronuclei showed that the induced micronuclei consisted of acentric chromosomal fragments, indicating clastogenic activity of coumestrol (Kulling and Metzler 1997). Nogowski et al. (1992) studied the effect of coumestrol on carbohydrate metabolism in ovariectomized rats.

Coumestrol had no significant effect on plasma insulin or glucagon concentrations, but it decreased muscle glycogen and inhibited insulin binding to muscle membrane. Thus, the effect of coumestrol on carbohydrate metabolism appears to be *via* changes in insulin receptors (Nogowski et al. 1992). Whether these actions of phytoestrogens on skeletal muscle have an effect on overall glucose disposal *in vivo* is not known. Coumestrol has also been shown to affect lipid metabolism. In chicks, dietary coumestrol decreased plasma cholesterol concentrations in a dose-dependent manner (Beguin and Kincaid 1984). Thus, coumestrol appears to have favorable biological actions on glucose and lipid metabolism, which may explain its potential benefit of obesity and diabetes. Studies on the role of coumestrol in obesity and diabetes are few, and studies of the effect of the coumestrol on obesity and diabetes are needed. Most of the clinical trials that have been conducted have been observational only, have been of relatively short duration, and have involved a small number of subjects. Long-term controlled trials on the safety and effectiveness of coumestrol on the development and progression of diabetes and obesity and their complications in animal and humans with diabetes mellitus and obesity are overdue (Bhathena and Velasquez 2002).

In another study by Draper *et al.* (1997), the phytoestrogen coumestrol prevented or reduced oophorectomy-induced bone loss in skeletally mature rats. This study concluded that it is likely a result of the estrogenic effect of these compounds, because the effect is similar to that of estrogen. Although the mechanisms by which coumestrol reduces bone loss cannot be fully elucidated from these studies, the lower urinary calcium/creatinine ratio in the coumestrol group compared with the controls suggests that coumestrol treatment resulted in a decrease in calcium excretion. This decrease mirrored that observed with estrogen treatment and is consistent with the maintenance of bone mass in both groups. These data suggest a similar mechanism of effect of coumestrol and estrogen on the calcium balance in oophorectomized rats. This study suggests that coumestrol may also be affecting the calcium balance through mechanisms in the bowel. This would agree with studies of the effect of estrogen on calcium balance (Heaney *et al.* 1978) and would suggest that phytoestrogens affect calcium balance in the same way.

31.3 Mechanisms of the Toxic Action

Animal studies have shown that coumestrol exposure induces an increase in uterine weight, a decrease in ovulation rate and an increase in embryo degeneration (Fredricks *et al.* 1981; Tinwell *et al.* 2000). However, there are few studies on the induction of apoptosis by coumestrol. A study done by Moon *et al.* (2009) has shown an increase in apoptosis in the adult rat ovary after lactational exposure to coumestrol. This study examined the effect of coumestrol on the ovary and assessed the level of apoptosis using Western blotting, immunohistochemistry and the terminal deoxynucleotidyl transferase dUTP nick-end labeling (TUNEL) method. It was confirmed that coumestrol significantly reduced the ovarian weight of adult female pups at concentrations of 0.1 and 1.0 mg kg^{-1} and dose-dependently increased the expression of caspases and the number of apoptotic cells in ovarian tissues. In this study the coumestrol-treated group displayed a marked dose-dependent increase in the level of cleaved caspase-3 and caspase-7.

Another study exploring the mechanism action of coumestrol, carried out by Wang *et al.* (2008), demonstrated that coumestrol is a human pregnane X receptor (PXR) and constitutive androstane receptor (CAR) antagonist in

Estrogenic Activity and Molecular Mechanisms

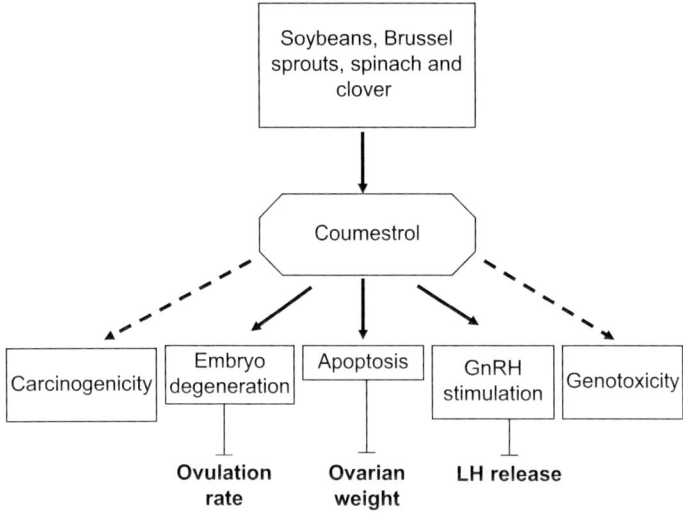

Figure 31.1 The schematic overview of coumestrol proposed mechanisms of action.

in vitro assays and coumestrol can block the effects of an available PXR agonist in primary human hepatocyte cultures. They also demonstrated that coumestrol does not affect the localization of PXR when it is activated by its cognate ligand. These results suggest that the mechanisms of action of coumestrol may involve the PXR and its co-regulatory proteins. Studies carried out by Bulayeva and Watson (2004) also demonstrated that coumestrol regulates intracellular receptors and kinases, which in turn have been shown to activate the PXR.

Studies carried out by McGarvey *et al.* (2004) examined the effects of coumestrol on the activity of the hypothalamic gonadotropin-releasing hormone (GnRH) pulse generator and on pituitary sensitivity to GnRH stimulation *in vivo*. These studies also investigated effects of coumestrol on GnRH-induced pituitary luteinizing hormone (LH) release *in vitro* using primary pituitary cell cultures, and concluded that coumestrol has profound effects on the control of LH secretion by actions at both the pituitary and hypothalamic levels. A pituitary site of action was indicated by the ability of coumestrol to inhibit GnRH-induced LH release *in vivo*. However, the mechanisms of the coumestrol inhibitory effect were not elucidated. These studies, taken together, suggest coumestrol has health risk effect that warrant further investigation (Figure 31.1).

31.4 Human Studies

Endogenous estrogens have dramatic and differential effects on classical endocrine organ and proliferation. Coumestrol is an environmental estrogen that has endocrine impact, acting as both an estrogen agonist and antagonist, but whose effects are not well characterized in humans. Human lymphoblastoid cells exposed to 10–50 μM coumestrol showed a significant increase in micronuclei in a dose-dependent manner (Olen et al. 2001). The highest frequency of micronuclei occurred 2 days after exposure and then decreased. Parallel to the decrease in the micronucleus frequency the percentage of cells undergoing apoptosis increased. This study demonstrated that coumestrol is mutagenic and clastogenic in cultured human lymphoblastoid cells. The genotoxicity of this compound appears to result from its ability to inhibit topoisomerase II function and the subsequent induction of deletions. In addition, the tumor suppressor protein p53 may play an important role in determining the strength of the mutagenic and clastogenic responses through a role in DNA repair, the apoptosis response or a combination of pathways. Other studies (Kulling and Metzler 1997; Schuler and Eastmond 2000) demonstrated the clastogenicity and mutagenicity of coumestrol in CHO cells. In this study, coumestrol proved to be a strong inducer of DNA strand breaks and micronuclei containing acentric fragments. The clastogenicity of coumestrol was shown to act through topoisomerase II inhibition and/or DNA intercalation and inducer of hypoxanthine guanine phosphoribosyl transferase (HPRT) mutations in V79 cells. This study was the first to report on the clastogenicity and mutagenicity of coumestrol in mammalian cells. This suggests the need for additional studies on the potential deleterious effects of coumestrol.

Despite the studies linking the effects of coumestrol to genotoxicity and carcinogenicity, the U.S. Environmental Protection Agency and the International Agency for Research on Cancer do not list it as being a human carcinogen. The data in support of these listings are from animal cancer studies, an extensive literature on the mechanisms of action of coumestrol and human epidemiology findings providing additional support. There is still a lot of work to be done to gather adequate and compelling evidence of basic similarities in the mechanisms by which laboratory animals and humans respond to coumestrol exposure. Another study by Wang et al. (2008) on the human nuclear xenobiotic receptor PXR demonstrated that coumestrol may have important clinical implications in preventing drug–drug interactions and improving therapeutic efficacy. The results showed that coumestrol is an antagonist of the nuclear receptor PXR. In transient transfection assays, coumestrol was able to suppress the agonist effects of SR12813 on human PXR activity. PXR activity was assessed and correlated with effects on the metabolism of the anesthetic tribromoethanol and on gene expression in primary human hepatocytes. Coumestrol suppressed the effects of PXR agonists on the expression of the known PXR target genes, *CYP3A4* and *CYP2B6,* in primary human hepatocytes. Based on the finding that coumestrol reduces PXR activity in *in vitro* and cell-based assays, tests were done to determine whether this activity would be reflected in cellular effects on gene expression. Results showed that coumestrol blocked the induction of *CYP3A4* by PXR agonists. Taken together, these studies by Wang and co-workers provide additional insight into mechanisms of coumestrol-induced PXR activation, while suggesting new avenues of PXR antagonism to prevent harmful drug–drug interactions and improving therapeutic efficacy.

Diet has been implicated as a possible link to the etiology and progression of many diseases, including cancer. Recently, interest has been focused on the increased risk of cancer by several of the hormone-like diphenolic phytoestrogens. Beverly *et al.* (1999) examined the effects of coumestrol on human pancreatic adenocarcinoma cells *in vitro*. Two human adenocarcinoma cell lines, HPAF-11 from a male and Su 86.86 from a female, were used. HPAF-11 cells were exposed for 24 h to these agents at concentrations of 1 and 10 mM. Su 86.86 cells were exposed for 24 h at a concentration of 1 mM. Coumestrol at the higher concentration was toxic to the Su 86.86 cells. Coumestrol displayed marked differences between cell lines in inhibition of growth. It inhibited the growth of the female pancreatic tumor cells by 95% and stimulated the growth of pancreatic tumor cells from the male, suggesting that coumestrol effects are cell specific.

Endogenous estrogens are known to modulate several components of the immune response, including interleukin-2 (IL-2) production. IL-2 is a cytokine that plays an important role in adaptive immune responses. These responses may be modulated by the xenoestrogen coumestrol. In a study examining the effects and potential mechanisms of action of coumestrol on IL-2 production in activated cluster differentiation (CD)4+Jurkat T-cells, Ndebele *et al.* (2004) reported that coumestrol significantly suppressed IL-2 production in activated CD4+Jurkat T-cells at the transcriptional and translational level (Figure 31.2).

The transcriptional suppression of IL-2 was associated with decreased protein levels of nuclear factor κB (NF-κB), an important IL-2 positive transcription factor, without affecting the expression of IκBα protein expression, an important inhibitor of NF-κβ nuclear translocation. Although the direct mechanisms of xenoestrogens modulation of the immune system remain to be elucidated, coumestrol-induced suppression of IL-2 may have ramifications for our understanding of the impact of xenoestrogens on health and disease.

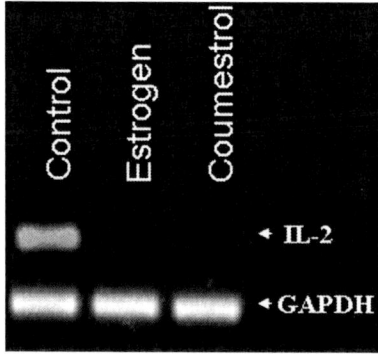

Figure 31.2 Semi-quantitative reverse transcriptase (RT)-PCR of IL-2 mRNA transcripts from activated Jurkat T-cells exposed to estrogen and coumestrol. GAPDH, glyceraldehyde-3-phosphate dehydrogenase (control).

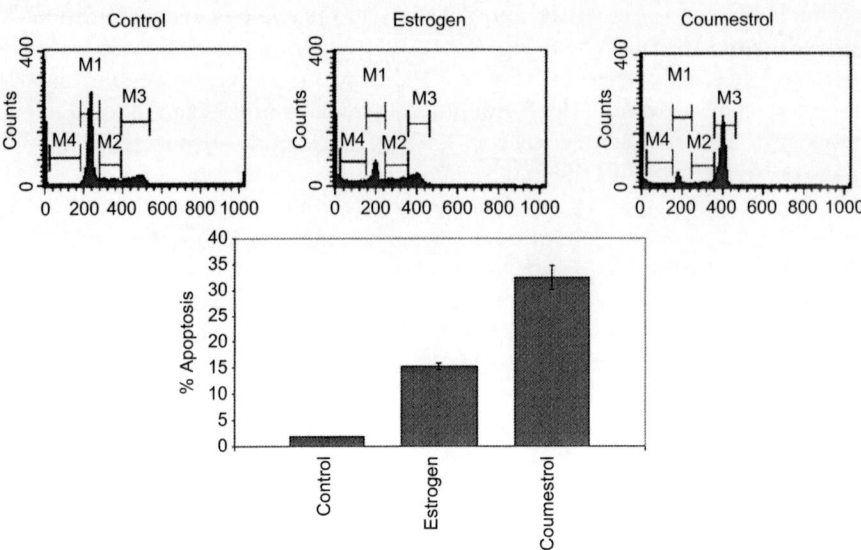

Figure 31.3 Histograms showing apoptotic effect of 17β-estradiol (estrogen) and coumestrol in human cervical cancer.

Another study by Ndebele *et al.* (2010) demonstrated that coumestrol produced accumulation of HeLa cells in G_2/M phase and subsequently induced apoptosis (Figure 31.3).

Results from epidemiological and experimental studies indicate that coumestrol may protect against breast cancer (Hedelin *et al.* 2008). One of the biological effects of coumestrol is estrogenic, therefore it is possible that the preventive effect on breast cancer differs according to the tumor microenvironment. This particular study showed that coumestrol intake was associated with a decreased risk of receptor negative tumors but not positive tumors (Hedelin *et al.* 2008).

31.5 Interpretation of Data in Relation to Human Exposure

Extrapolation of human data is difficult for two reasons. Previous studies clearly showed that coumestrol can produce estrogenic effects in rodents. The human health impacts of these biological effects are still premature due to species differences. Data from animal studies are more useful in terms of risk assessment but are limited in number and scope. For example, human studies specifically examining the potential effects of uterus exposure to coumestrol have not been conducted. It is extremely difficult to examine the effects of coumestrol on human development and reproduction because of practical and ethical reasons. As a result most of the previous studies have been conducted in laboratory animals such as rodents. Extensive data in rodents suggest that coumestrol has estrogenic effects in both male and female rodents but effects may be more pronounced in the female. However, the significance of these effects, such as alterations in sex hormone concentrations, advancement of vaginal opening and mammary gland development by coumestrol to humans, is unclear.

Present data collectively indicate that coumestrol is a functional component of phytoestrogen phenotype and that it could potentially serve as a novel therapeutic and biomarker indicator. However, the information on coumestrol distribution in various tissue types and organs is still inadequate to propose meaningful strategies for these applications. Despite several studies which clearly show coumestrol as a molecule associated with different aspects of malignancy, we are still missing correlation studies that would reveal patterns of coumestrol effect with other cancer-related and signaling proteins. Many definitive aspects of coumestrol presence and specific roles in different cancer types await clarification. Despite this complicated background, some promising directions can be seen already at this stage. For example, the relationship between the hormone activation of the ER and expression of coumestrol could have important implications for cancer biology, because estrogen is frequently used as a component of many chemotherapy regimens.

Summary Points

- This Chapter focuses on coumestrol.
- Coumestrol is a chemical substance found in legumes, soybeans, brussel sprouts and spinach.
- Coumestrol induces neoplastic transformation in cells and genotoxicity in animals.
- Coumestrol may also have a chemopreventative action and reduces the risk of breast, colon and prostate cancers.
- Previous studies have shown the mechanisms of action of coumestrol and estrogen on the calcium balance.
- Coumestrol may have important clinical implications in preventing drug–drug interactions and improving therapeutic efficacy.
- Human studies examining the potential effects of coumestrol have not been conducted.

Key Facts

- Experimental studies with animals have demonstrated that coumestrol is estrogenic.
- Coumestrol exposure at high doses may cause cell death.
- Coumestrol may also be used to prevent cancer.
- The effect of coumestrol is dose-dependent and cell-specific.
- The mechanisms by which coumestrol exerts its functions are not fully understood.
- Additional scientific information is required from human studies.

Definitions of Words and Terms

Adenocarcinoma: A cancer of an epithelium that originates in glandular tissue.
Apoptosis: The process of programmed cell death.
Biomarker: A physical, functional or biochemical indicator of a physiological or disease process.
Carcinogenicity: The potential of a biological, chemical or physical agent to cause cancer.
Clastogenic: The potential of an agent to cause disruption in the number and/or structure of chromosomes.
Diabetes: A medical condition due to high levels of sugar in the blood.
Genotoxicity: The ability of a biological, chemical or physical agent to cause genetic damage.
Mutagenicity: The potential of a biological, chemical or physical agent to cause mutations.
Obesity: A medical condition due to an excess proportion of total body fat.
Phytoestrogen: A foreign estrogen from plants.

List of Abbreviations

CD	cluster differentiation
CHO	Chinese hamster ovary
ER	estrogen receptor
GnRH	gonadotropin-releasing hormone
LH	luteinizing hormone
IL-2	interleukin-2
NF-κB	nuclear factor κB
PXR	pregnane X receptor
SSB	DNA single strand break

Acknowledgements

This project was supported by NCRR (5G12RR013459-15) and NIMHD (8G12MD007581-15) grants from the National Institutes of Health.

References

Amin, A., and Buratovich, M., 2007. The anti-cancer charm of flavonoids: A cup-of-tea will do. *Recent Patents on Anti-Cancer Drug Discovery*. 2(2): 109–117.

Beguin, D.P., and Kincaid, R.L., 1984. 3-Hydroxy-3-methyl-glutaryl coenzyme A reductase activity in chicks fed coumestrol, a phytoestrogen. *Poultry Science*. 63: 686–690.

Beverly, D., Lyn-Cooka, Heather, L., Stottman, B., Yana, Y., Blanna, E., Kadlubara, F.F., and Hammonsa, G.J., 1999. The effects of phytoestrogens on human pancreatic tumor cells *in vitro*. *Cancer Letters*. 142: 111–119.

Bhathena, S.J., and Velasquez, M.T., 2002. Beneficial role of dietary phytoestrogens in obesity and diabetes. *American Journal Clinical Nutrition.* 76: 1191–1201.

Blomquist, C.H., Lima, P.H., and Hotchkiss, J.R., 2005. Inhibition of 3α-hydroxysteroid dehydogenase (3α-HSD) activity of human lung microsomes by genistein, daidzein, coumestrol and C18-, C19- and C21 hydroxysteroids and ketosteroids. *Steroids.* 70(8): 507–514.

Bulayeva, N.N., and Watson, C.S., 2004. Xenoestrogen-induced ERK-1 and ERK-2 activation via multiple membrane-initiated signaling pathways. *Environmental Health Perspectives.* 112: 1481–1487.

Draper, C.R., Edel, M.J., Dick, I.M., Randall, A.G., Martin, G.B., and Prince, R.L., 1997. Phytoestrogens reduce bone loss and bone resorption in oophorectomized rats. *Journal of Nutrition.* 127: 1795–1799.

Fredricks, G.R., Kincaid, R.L., Bondioli, K.R., and Wright, Jr, R.W., 1981. Ovulation rates and embryo degeneracy in female mice fed the phytoestrogen coumestrol. *Proceedings of the Society of Experimental Biology and Medicine.* 167(2): 237–241.

Heaney, R.P., Recker, R.R., and Saville, P.D., 1978. Menopausal changes in calcium balance performance. *Journal of Laboratory and Clinical Medicine.* 92: 953–963.

Hedelin, M., Lo, M., Olsson, M., Adlercreutz, H., Sandin, S., and Weiderpass, E., 2008. Dietary phytoestrogens are not associated with risk of overall breast cancer but diets rich in coumestrol are inversely associated with risk of estrogen receptor and progesterone receptor negative breast tumors in Swedish women. *Journal of Nutrition.* 138: 938–945.

Kuiper, G.G., Shughrue, P.J., Merchenthaler, I., and Gustafsson, J.A., 1998. The estrogen receptor βsubtype: a novel mediator of estrogen action in neuroendocrine systems. *Frontiers in Neuroendocrinology.* 19(4): 253–286.

Kulling, S.E., and Metzler, M., 1997. Induction of micronuclei, DNA strand breaks and HPRT mutations in cultured Chinese hamster V79 cells by the phytoestrogen coumoestrol. *Food Chemistry and Toxicology.* 35: 605–613.

McGarvey, C., Cates, P., Brooks, N., Swanson, I., Milligan, S., cohen, C., and O'Byrne, K., 2004. Phytoestrogens and gonadotropin-releasing hormone pulse generator activity and pituitary luteinizing hormone release in the rat. *Journal of Endocrinology.* 142: 1202–1208.

Molteni, A., Molteni, L., and Persky, V., 1995. *In vitro* hormonal effects of soybean isoflavones. *Journal of Nutrition.* 125(3): 751S–756S.

Moon, S., Seok, J.H., Kim, S.S., Rhee, G.S., Lee, R.D., Yang, J.Y., Chae, S.Y., Kim, S.H., Kim, J.Y., Chung, J., Kim, J.M., and Chung, S.Y., 2009. Lactational coumestrol exposure increases ovarian apoptosis in adult rats. *Archives of Toxicology.* 83: 601–608.

Ndebele, K., Tchounwou, P.B., and McMurray, R.W., 2004. Coumestrol, bisphenol-A, DDT, and TCDD modulation of interleukin-2 expression in activated CD+4 Jurkat T cells. *International Journal of Environmental Research and Public Health.* 1: 3–11.

Ndebele, K., Graham, B., and Tchounwou, P.B., 2010. Estrogenic activity of coumestrol, DDT, and TCDD in human cervical cancer cells. *International Journal of Environmental Research and Public Health.* 7(5): 2045–2056.

Nogowski, L., Nowak, K.W., and Mackowiak, P., 1992. Effect of phytoestrogen coumestrol and oestrone on some aspects of carbohydrate metabolism in ovariectomized female rats. *Archivum Veterinarium Polonicum.* 32: 79–84.

Olen, E., Lynda, D., McGarrity, J., Bishop, M., Yoshioka, M., Chen, J.J., and Morris, S.M., 2001. Evaluation of the genotoxicity of the phytoestrogen, coumestrol, in AHH-1 TKC human lymphoblastoid cells. *Mutation Research.* 474: 129–137.

Schuler, M., and Eastmond, D.A., 2000. Influence of test concentration and treatment protocol on the induction of micronuclei and chromosomal aberrations in cultured human lymphocytes by the phytoestrogen coumestrol. *Environmental and Molecular Mutagenesis.* 35(S31): 54.

Shutt, D.A., and Cox, R.I., 1972. Steroid and phyto-oestrogen binding to sheep uterine receptors *in vitro. Journal of Endocrinology.* 52: 299–310.

Tinwell, H., Soames, A.R., Foster, J.R., and Ashby, J., 2000. Estradiol-type activity of coumestrol in mature, immature ovariectomized rat uterotrophic assays. *Environmental Health Perspectives.* 108: 631–634.

Wang, H., Li, H., Linda, B., Michael, M., Johnson, D.L., Maglich, J.M., Goodwin, B., Ittoop, O.R., Wisely, B., Creech, K., Parks, D. J.., Collins, J.L., Wilson, T.M., Kalpana, G.V., Venkatesh, M., Xie, W., Cho, S.Y., Roboz, J., Redinbo, M., Moore, J.T., and Mani, S., 2008. The phytoestrogen coumestrol is a naturally occurring antagonist of the human pregnane X receptor. *Molecular Endocrinology.* 22: 838–857.

CHAPTER 32
Genistein and Insulin Secretory Function

DONGMIN LIU

Department of Human Nutrition, Foods and Exercise, College of Agriculture and Life Sciences, Virginia Tech, Blacksburg, Virginia 24061, USA
Email: doliu@vt.edu

32.1 Introduction

Recently, some bioactive compounds have drawn wide attention due to their potentially beneficial effects on some human degenerative diseases. Genistein, an isoflavone found in soy and some Chinese herb medicines, such as *Genista tinctoria* Linn and *Sophora subprostrala* Chun et T. Chen, has various biological actions including a weak estrogenic effect by binding to estrogen receptors (ERs) and inhibition of protein tyrosine kinases (PTKs) at pharmacological doses. It is widely used as a dietary supplement for various presumed health benefits, although the research evidence supporting the beneficial effects of genistein consumption on human health is not well established. Genistein consists of two aromatic rings (A and B) linked through a heterocyclic pyrane ring (C) (Figure 32.1). The chemical structure of genistein is similar to that of 17β-estradiol (E2), which is the endogenous estrogen that primarily acts through ERα and ERβ in humans. Unlike E2, however, which binds to both ERα and ERβ with nearly equal affinity, genistein shows much higher affinity to ERβ (87% of E2) than to ERα (4%

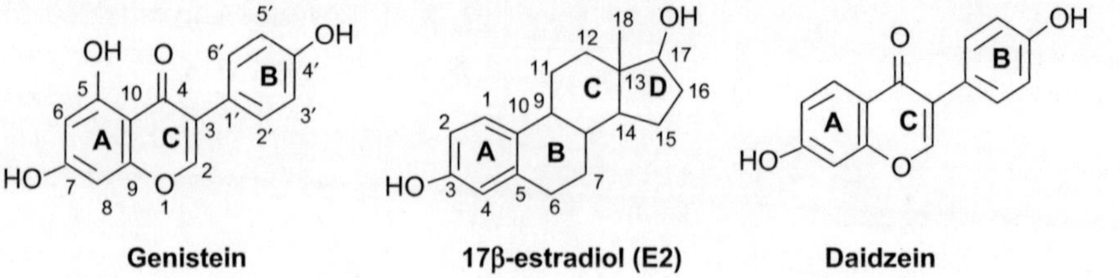

Figure 32.1 Chemical structures of genistein, 17β-estradiol (E2) and daidzein.

of E2) (Kuiper *et al.* 1998). The hydroxyl group at C5 of genistein may primarily determine it affinity to ERs, because daidzein, a genistein analog and another isoflavone which only lacks the hydroxyl group at C5 compared with genistein (Figure 32.1), essentially has no binding activity to either ERα or ERβ (0.1% and 0.5% of E2, respectively) (Kuiper *et al.* 1998). Therefore, genistein can be considered a plant-derived novel selective ERβ agonist. As genistein has weak estrogenic effects in some tissues (Kim *et al.* 1998), there is concern about its potential adverse effects in various estrogen-dependent organs, although research evidence has not been established and the relationship between doses, duration, beneficial or harmful effects are unclear. Genistein has been shown to affect development of the reproductive system and immune functions in young experimental animals (Chen and Rogan 2004), but no such an effect has been reported in infants fed infant formula containing this compound or in various mature animal species fed pharmacological doses of genistein. In addition, a few studies have found that soy isoflavones may have the potential to induce the growth of estrogen-dependent breast and endometrium cancers, whereas epidemiological data from short- or long-term use of soy isoflavones showed a preventive effects on these diseases, and numerous short-term high-dose toxic studies demonstrated that intake of genistein does not appear to cause adverse health effects in humans. The following part of this Chapter summarizes studies exploring the anti-diabetic effects of genistein, with a focus on providing the latest information regarding its insulinotropic effects in pancreatic beta-cells and the underlying mechanism for this action.

32.2 Genistein may have Anti-diabetic Effects

Type 2 diabetes (T2D) is a result of chronic insulin resistance and beta-cell dysfunction (Stoffers 2004). Both in experimental animals and people, obesity is a leading pathogenic factor for developing insulin resistance. However, constant insulin resistance will only progress to T2D when beta-cells are unable to secret adequate amount of insulin to compensate for decreased insulin sensitivity, which is largely due to insulin secretory dysfunction and significant loss of functional beta-cells. As such, the search for novel agents that promote beta-cell function may provide an effective strategy to prevent the onset of diabetes. Genistein has been previously investigated for its potential beneficial effects on cancer treatment, cognitive function, and cardiovascular and skeletal health, with a primary focus on exploring its potential hypolipidemic, anti-oxidative and estrogenic effects. However, studies on whether genistein has an effect on diabetes are limited. Some recent studies performed in animals and humans have shown that ingestion of isoflavone-containing soy protein moderates hyperglycemia (Jayagopal et al. 2002; Lavigne et al. 2000). However, it is not clear whether genistein primarily contributes to this beneficial effect. Emerging studies reported that administration of isoflavones lowered plasma glucose in animals (Ali et al. 2005; Mezei et al. 2003) and in postmenopausal women (Cheng et al. 2004) independent of its effect on food intake or weight gain, suggesting that genistein may be a novel plant-derived anti-diabetic agent, although it is unclear whether this beneficial effect is primarily ascribed to this component. In addition, the underling mechanism for this action of genistein in diabetes is unknown.

Most published trials using isoflavones or genistein have focused largely on elucidating the effect of isoflavones on lipid profiles, and therefore data from recent studies suggest an anti-diabetic effect of genistein presumably by a hypolipidemic effect, thereby increasing insulin sensitivity. However, previous studies investigating the genistein effect on plasma lipid profiles have shown either only a moderate positive effect or a neutral effect. Indeed, a recent report by the American Heart Association summarizing data from 10 randomized trials indicated that soy isoflavones have no effect on human plasma lipids (Sacks et al. 2006). However, recent studies demonstrated that isoflavone administration still lowered plasma glucose, even though lipid profiles or insulin sensitivity were unaffected in obese and diabetic animals (Ali et al. 2005) and in humans (Cheng et al. 2004). These data therefore did not support the idea that isoflavones exert an anti-diabetic effect through lowering plasma lipids or increasing insulin sensitivity. There is increasing evidence showing that oxidative stress and reactive oxygen species (ROS) play a potential role in the initiation of diabetes. Genistein has been reported to exhibit anti-oxidant activity in aqueous phase systems (Ruiz-Larrea et al. 1997; Wei et al. 1993). However, the anti-oxidant effect of genistein is achieved only at concentrations ranging from 25 to 100 µM, suggesting that genistein is not a physiologically effective anti-oxidant, as the achievable levels of total plasma genistein in both humans and rodents through dietary supplementation is usually no more than 5 µM. This result is further consolidated by a recent report indicating that isoflavones have no anti-oxidative effect in healthy postmenopausal women (Steinberg et al. 2003). Consistently, it has been shown that genistein is also a relatively poor ROS scavenger (Chacko et al. 2005; Patel et al. 2001). Therefore, although these data suggest that genistein may have a protective role in diabetes, the mechanisms underlying these beneficial effects are still largely unknown, and there is a possibility that genistein may have a direct effect on the management of diabetes.

32.3 Genistein at Physiological Concentrations Augments Glucose-stimulated Insulin Secretion (GSIS) in Beta-cells and Pancreatic Islets

Although studies are limited and the results are inconsistent, the available data show that genistein may have a direct effect on pancreatic beta-cells. Several earlier studies demonstrated that genistein stimulates insulin secretion from a clonal beta-cell line (Ohno *et al.* 1993) and cultured islets (Jonas *et al.* 1995; Sorenson *et al.* 1994), whereas other studies have found an inhibitory effect on insulin secretion (Jones and Persaud 1994; Persaud *et al.* 1999). These discrepant data may be the result of variations in the experimental conditions and model used. Nevertheless, the concentrations ($>30\,\mu M$) used in most of these studies are well above those physiologically achievable by dietary means ($<5\,\mu M$) as aforementioned. Therefore, it is still unclear from these studies whether genistein at physiologically doses can act directly on pancreatic beta-cells to modulate cellular function.

In contrast to these effects at high concentrations, it was recently reported that genistein increases rapid GSIS in both insulin-secreting cell lines (INS-1 and MIN6) and mouse pancreatic islets (Liu *et al.* 2006). Genistein elicited a significant effect at a concentration as low as 10 nM with a maximal effect at $5\,\mu M$ (Figure 32.2). Consistent with its effect on GSIS, genistein increases intracellular cyclic adenosine monophosphate (cAMP) and activates protein kinase A (PKA) in both cell lines (Figure 32.3) and islets (Figure 32.4). These effects of genistein on GSIS and cAMP signaling were not dependent on ERs and also not related to an inhibition of PTK. The cAMP induced by genistein results primarily from enhanced adenylate cyclase activity. The effect of genistein on the cAMP is as potent as that of glucagon-like peptide-1 (GLP-1), the most potent insulinotropic hormone found so far. Like GLP-1, genistein augments insulin only when glucose is increased to stimulatory levels ($\geq 5.6\,mM$), a characteristic very important for avoiding potential hypoglycemia from an undesired increase in insulin. Pharmacological or molecular intervention of PKA activation indicated that the insulinotropic effect of genistein is primarily mediated through PKA. These findings demonstrate that genistein can directly act on pancreatic beta-cells, leading to activation of the cAMP/PKA signaling cascade to exert an insulinotropic effect, thereby providing a novel role for soy isoflavones in the regulation of insulin secretion. However, it is presently unclear how genistein activates adenylate cyclase which could be differentially regulated by Ca^{2+}, G-proteins and protein kinases (Figure 32.5). In this regard, genistein could directly act on the cell surface to facilitate cAMP production involving a non-genomic mechanism, which is a very interesting question that remains to be investigated.

In addition to inducing rapid insulin secretion in response to elevated glucose, it was found that relative long-term exposure of beta-cells to genistein, at physiologically relevant concentrations through dietary consumption, improves insulin secretory function of pancreatic beta-cells (Fu and Liu 2009).

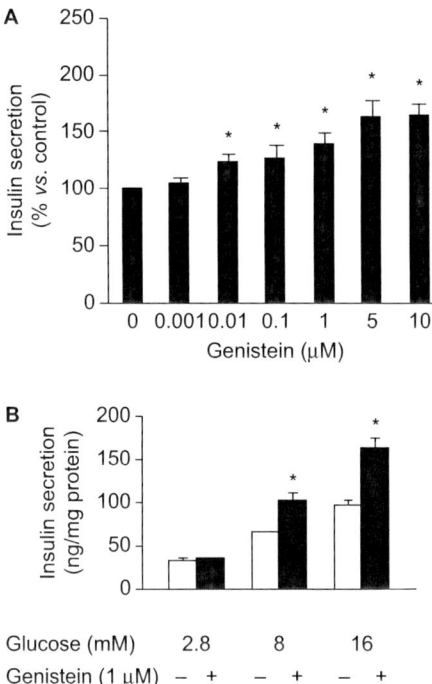

Figure 32.2 Genistein potentiates GSIS. (**A**) INS-1 cells were stimulated with various concentrations of genistein in buffer containing 5.6 mM glucose for 30 min at 37 °C. (**B**) Mouse islets were stimulated with genistein (1 μM) in the presence of indicated concentrations of glucose for 30 min at 37 °C. Insulin secreted in culture buffer was determined by a radioimmunoassay (RIA) kit. *$P < 0.05$ vs. vehicle alone-treated cells.

Specifically, it was reported that exposure of INS-1E cells and mouse and human pancreatic islets to genistein for 48 h enhanced GSIS, showing that this effect of genistein is not species-specific and may be biologically relevant, given that the concentrations used in this study overlap with those of physiologically achievable following dietary consumption of genistein products. However, genistein had no effect on insulin content, suggesting that its effect on insulin secretion is not due to a modulation of insulin synthesis, which could subsequently contribute to the improved GSIS by this compound, given that increased insulin release can result from improved insulin expression. Similar to its rapid effect on insulin secretion, this improvement of the insulin secretory function of beta-cells by chronic exposure to genistein is also PTK-independent as daidzein, an analog of genistein, which does not inhibit PTK, also increased insulin secretion, an effect that is only slightly less potent than that of genistein. It is well characterized that glucose induces insulin secretion through glycolysis and mitochondrial oxidation in the cells, which increase the intracellular ATP/ADP ratio, sequentially leading to the closure of K_{ATP} channels,

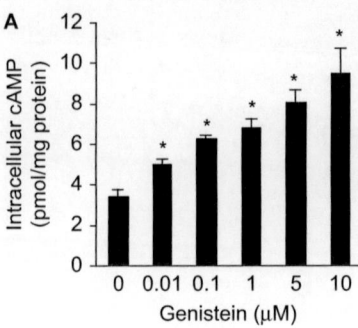

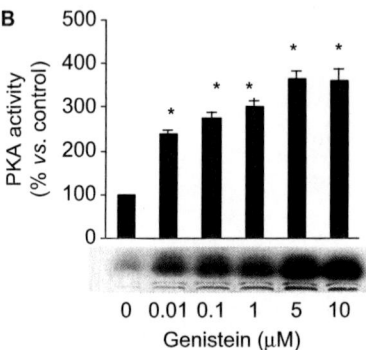

Figure 32.3 Genistein stimulates intracellular cAMP accumulation and PKA activity in INS-1 cells. INS-1 cells were stimulated with various concentrations of genistein in the presence of 5.6 mM glucose in buffer for 20 min (**A**) or 30 min (**B**) at 37 °C. cAMP was measured by EIA (**A**), and PKA activity in cell extracts was determined by measuring phosphorylated kemptide (image panel). $*P<0.05$ vs. vehicle alone-treated cells.

depolarization of voltage-gated L-type Ca^{2+} channels on the plasma membrane, Ca^{2+} influx and the activation of exocytosis of insulin-containing granules (Figure 32.6). Genistein has no effect on K_{ATP} channel, glucose transporter-2 expression or cellular ATP production, but similarly augments pyruvate-stimulated insulin secretion in INS-1E cells (Fu and Liu, 2009), indicating that the improvement of insulin secretory function by long-term genistein exposure is not related to an alternation in glucose uptake or the glycolytic pathway. However, the enhanced insulin secretion by genistein is associated with an elevated intracellular Ca^{2+} concentration, suggesting that the improvement of insulin secretory function by genistein may be at least partially attributable to its modulation of Ca^{2+} signaling in beta-cells. Interestingly, pharmacological inhibition of PKA activity completely abolished improved insulin secretion following long-term exposure to genistein, demonstrating that the chronic effect of genistein on insulin secretion is PKA-dependent (Fu and Liu 2009). However, it is still unclear whether genistein regulates Ca^{2+} influx pathway or release of Ca^{2+} from intracellular stores, and whether PKA is involved in this process.

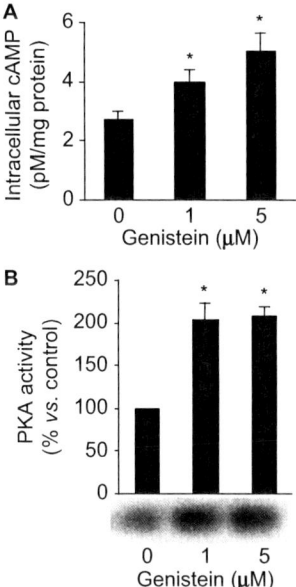

Figure 32.4 Genistein stimulates intracellular cAMP accumulation and PKA activity in mouse islets. The islets were stimulated with indicated concentrations of genistein in buffer containing 8 mM glucose for 20 min (**A**) or 30 min (**B**) at 37 °C. cAMP (**A**) and PKA activity (**B**) in the cell extracts from the islets was determined. *$P<0.05$ vs. vehicle alone-treated cells.

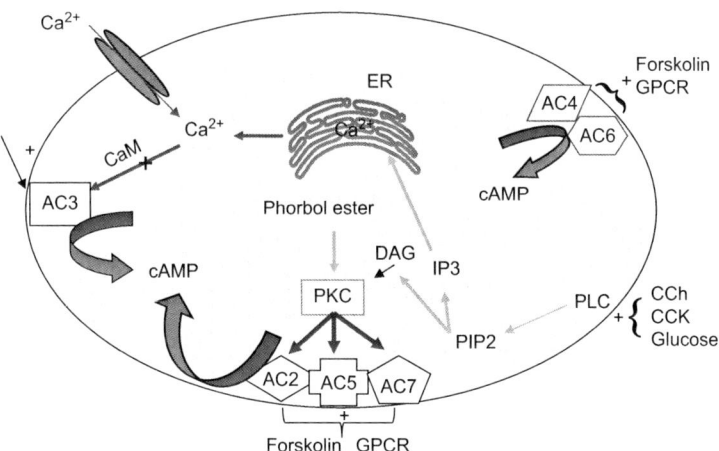

Figure 32.5 Mechanism of adenylate cyclase activation. AC, adenylate cyclase; CaM, Ca^{2+}/calmodulin complex; cAMP, cyclic adenosine monophosphate; CCK, cholecystokinin; CCh, carbamylcholine; DAG, diacylglycerol; ER, endoplasm reticulum; GPCR, G-protein-coupled receptor; IP3, inositol 1,4,5-triphosphate; PIP2, phosphatidylinositol 4,5-bisphosphate; PKC, protein kinase C; PLC, phospholipase C.

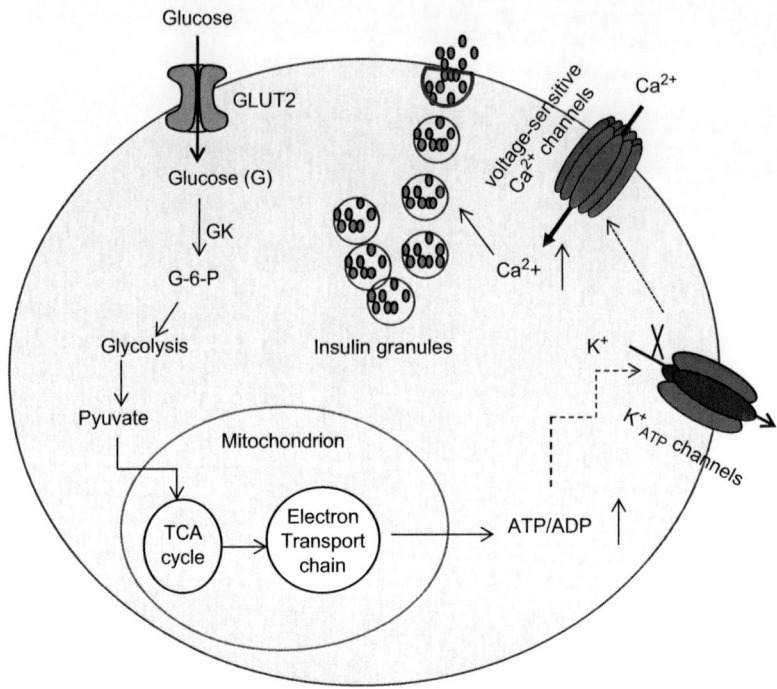

Figure 32.6 Schematic diagram of glucose-stimulated insulin secretion (GSIS). Glucose (G) is transported through glucose transporter-2 (GLUT2) into the cells, where is it phosphorylated by glucokinase (GK) to glucose 6-phosphate (G-6-P). Pyruvate generated from glycolysis enters the mitochondria and is metabolized in the tricarboxylic acid (TCA) cycle, which ultimately leads to the generation of ATP, thereby increasing the ATP/ADP ratio in the cytosol. This leads to the closure of K_{ATP} channels, which then open the voltage-sensitive Ca^{2+} channels, leading to Ca^{2+} influx, which is the main driving force for insulin exocytosis.

While it is unclear how PKA is involved in the effect of long-term genistein exposure on insulin secretion, it is unlikely that this genistein action is mediated through a rapid activation of PKA. Indeed, this effect of genistein on insulin secretion is dependent on new protein synthesis, suggesting that long-term genistein exposure improves beta-cell function *via* a genomic mechanism. Insulin is released from beta-cells through regulated exocytosis, which requires transport and docking of insulin secretory granules to the plasma membrane and subsequent fusion. Studies showed that synaptosomal-associated protein of 25 kDa (SNAP-25), a membrane bound protein, is involved in this process of insulin exocytosis from beta-cells (Blanpied and Augustine 1999; Gonelle-Gispert *et al.* 2000), and its expression is up-regulated by PKA in oocytes and steroidogenic cells (Grosse *et al.* 2000). Therefore, it is interesting to speculate that genistein may improve the insulin secretory function of beta-cells through PKA-mediated up-regulation of SNAP-25 expression.

Summary Points

- Genistein is an isoflavone present in legumes and some Chinese herb medicines.
- It has a weak estrogenic effect by binding to estrogen receptors (ERs) and is a well-known non-specific pharmacological inhibitor of protein tyrosine kinases (PTKs).
- Genistein has been shown to exert an anti-diabetic effect in experimental animals and patients.
- Recent data provide evidence that genistein is a novel insulinotropic agent by targeting the cyclic adenosine monophosphate (cAMP) signaling pathway.
- Loss of insulin secretory function and/or beta-cell mass is the key for the development of both type 1 diabetes (T1D) and type 2 diabetes (T2D). In this context, genistein could potentially be a naturally occurring low-cost agent that can be used as an alternative or complementary treatment for diabetes, if its usefulness in preventing or treating human diabetes or pre-diabetes can be verified.

Key Facts

- Genistein is an isoflavone and a major phytoestrogen.
- Human exposure to genistein occurs primarily through intake of soy foods and/or dietary supplements.
- Genistein is primarily present as a glucoside form genistin.
- Hydrolysis of genistin in the intestine produces the aglycone genistein.
- The majority of genistein exist as conjugates in the blood.

Definitions of Words and Terms

Adenylate cyclase: The enzyme catalyzing the synthesis of cyclic adenosine monophosphate (cAMP) from adenosine triphosphate (ATP).

Cyclic adenosine monophosphate (cAMP): An intracellular signaling molecular critically involved in many biological processes.

Estrogen receptors (ERs): The receptors activated by hormone 17β-estradiol (E2). While ERs are usually referred to ERα and ERβ that belong to superfamily of steroid nuclear receptors, it was recently found that a G-protein-coupled receptor named GPR30 is a plasma membrane ER.

Glucagon-like peptide-1 (GLP-1): An insulinotropic hormone secreted from the intestinal L-cells in response to ingestion of nutrients such as carbohydrates, lipids and proteins.

Plasma membrane: The cell membrane that is made of lipids, particularly phospholipids.

Protein kinase A (PKA): A cAMP-dependent holoenzyme that consists of two regulatory and two catalytic subunits. It is involved in a large spectrum of cellular and metabolic functions such as cell proliferation, apoptosis, insulin secretion, and glucose and lipid metabolism.

Protein tyrosine kinases (PTKs): A large family of enzymes involved in many cellular functions such as cell adhesion, growth, differentiation, migration, etc.

Reactive oxygen species (ROS): Chemically reactive molecules containing oxygen formed during electron transport processes in metabolism involving oxygen.

Synaptosomal-associated protein of 25 kDa (SNAP-25): A plasma membrane-bound protein that is encoded by the SNAP-25 gene. SNAP-25 protein may play a critical role in insulin exocytosis by facilitating fusion of insulin containing vesicles and plasma membranes.

Type 2 diabetes (T2D): The most common form of diabetes due to insulin resistance and pancreatic beta-cell dysfunction.

List of Abbreviations

cAMP	cyclic adenosine monophosphate
ER	estrogen receptor
E2	17β-estradiol
GLP-1	glucagon-like peptide-1
GSIS	glucose-stimulated insulin secretion
PKA	protein kinase A
PTK	protein tyrosine kinase
ROS	reactive oxygen species
SNAP-25	synaptosomal-associated protein of 25 kDa
T2D	type 2 diabetes

References

Ali, A.A., Velasquez, M.T., Hansen, C.T., Mohamed, A.I., and Bhathena, S.J., 2005. Modulation of carbohydrate metabolism and peptide hormones by soybean isoflavones and probiotics in obesity and diabetes. *Journal of Nutritional Biochemistry*. 16: 693–699.

Blanpied, T.A. and Augustine, G.J.1999. Protein kinase A takes center stage in ATP-dependent insulin secretion. *Proceeding of National Academy of Science (USA)*. 96: 329–331.

Chacko, B.K., Chandler, R.T., Mundhekar, A., Khoo, N., Pruitt, H.M., Kucik, D.F., Parks, D.A., Kevil, C.G., Barnes, S., and Patel, R.P., 2005. Revealing anti-inflammatory mechanisms of soy isoflavones by flow: modulation of leukocyte-endothelial cell interactions. *American Journal of Physiology*. 289: H908–H915.

Chen, A., and Rogan, W.J., 2004. Isoflavones in soy infant formula: a review of evidence for endocrine and other activity in infants. *Annual Review of Nutrition*. 24: 33–54.

Cheng, S.Y., Shaw, N.S., Tsai, K.S., and Chen, C.Y., 2004. The hypoglycemic effects of soy isoflavones on postmenopausal women. *Journal of Womens Health (Larchmt)*. 13: 1080–1086.

Fu, Z., and Liu, D., 2009. Long-term exposure to genistein improves insulin secretory function of pancreatic beta-cells. *European Journal of Pharmacology*. 616: 321–327.

Gonelle-Gispert, C., Molinete, M., Halban, P.A., and Sadoul, K., 2000. Membrane localization and biological activity of SNAP-25 cysteine mutants in insulin-secreting cells. *Journal of Cell Science*. 113(18), 3197–3205.

Grosse, J., Bulling, A., Brucker, C., Berg, U., Amsterdam, A., Mayerhofer, A., and Gratzl, M., 2000. Synaptosome-associated protein of 25 kilodaltons in oocytes and steroid-producing cells of rat and human ovary: molecular analysis and regulation by gonadotropins. *Biology of Reproduction*. 63: 643–650.

Jayagopal, V., Albertazzi, P., Kilpatrick, E.S., Howarth, E.M., Jennings, P.E., Hepburn, D.A., and Atkin, S.L., 2002. Beneficial effects of soy phytoestrogen intake in postmenopausal women with type 2 diabetes. *Diabetes Care*. 25: 1709–1714.

Jones, P.M., and Persaud, S.J., 1994. Tyrosine kinase inhibitors inhibit glucose-stimulated insulin secretion. *Biochemical Society Transactions*. 22: 209S.

Jonas, J.C., Plant, T.D., Gilon, P., Detimary, P., Nenquin, M., and Henquin, J.C., 1995. Multiple effects and stimulation of insulin secretion by the tyrosine kinase inhibitor genistein in normal mouse islets. *British Journal of Pharmacology*. 114: 872–880.

Kim, H., Peterson, T.G., and Barnes, S., 1998. Mechanisms of action of the soy isoflavone genistein: emerging role for its effects via transforming growth factor beta signaling pathways. *American Journal of Clinical Nutrition*. 68: 1418S–1425S.

Kuiper, G.G., Lemmen, J.G., Carlsson, B., Corton, J.C., Safe, S.H., van der Saag, P.T., van der Burg, B., and Gustafsson, J.A., 1998. Interaction of estrogenic chemicals and phytoestrogens with estrogen receptor beta. *Endocrinology*. 139: 4252–4263.

Lavigne, C., Marette, A., and Jacques, H., 2000. Cod and soy proteins compared with casein improve glucose tolerance and insulin sensitivity in rats. *American Journal of Physiology*. 278: E491–E500.

Liu, D., Zhen, W., Yang, Z., Carter, J.D., Si, H., and Reynolds, K.A., 2006. Genistein acutely stimulates insulin secretion in pancreatic beta-cells through a cAMP-dependent protein kinase pathway. *Diabetes*. 55: 1043–1050.

Mezei, O., Banz, W.J., Steger, R.W., Peluso, M.R., Winters, T.A., and Shay, N., 2003. Soy isoflavones exert antidiabetic and hypolipidemic effects through the PPAR pathways in obese Zucker rats and murine RAW 264.7 cells. *Journal of Nutrition*. 133: 1238–1243.

Ohno, T., Kato, N., Ishii, C., Shimizu, M., Ito, Y., Tomono, S., and Kawazu, S., 1993. Genistein augments cyclic adenosine $3'5'$-monophosphate(cAMP) accumulation and insulin release in MIN6 cells. *Endocrine Research*. 19: 273–285.

Patel, R.P., Boersma, B.J., Crawford, J.H., Hogg, N., Kirk, M., Kalyanaraman, B., Parks, D.A., Barnes, S., and Darley-Usmar, V., 2001. Antioxidant mechanisms of isoflavones in lipid systems: paradoxical effects of peroxyl radical scavenging. *Free Radical Biology and Medicine*. 31: 1570–1581.

Persaud, S.J., Harris, T.E., Burns, C.J., and Jones, P.M., 1999. Tyrosine kinases play a permissive role in glucose-induced insulin secretion from adult rat islets. *Journal of Molecular Endocrinology*. 22: 19–28.

Ruiz-Larrea, M.B., Mohan, A.R., Paganga, G., Miller, N.J., Bolwell, G. P., and Rice-Evans, C.A., 1997. Antioxidant activity of phytoestrogenic isoflavones. *Free Radical Research*. 26: 63–70.

Sacks, F.M., Lichtenstein, A., Van Horn, L., Harris, W., Kris-Etherton, P., and Winston, M., 2006. Soy protein, isoflavones, and cardiovascular health: an American Heart Association Science Advisory for professionals from the Nutrition Committee. *Circulation*. 113: 1034–1044.

Sorenson, R.L., Brelje, T.C., and Roth, C., 1994. Effect of tyrosine kinase inhibitors on islets of Langerhans: evidence for tyrosine kinases in the regulation of insulin secretion. *Endocrinology*. 134: 1975–1978.

Steinberg, F.M., Guthrie, N.L., Villablanca, A.C., Kumar, K., and Murray, M.J., 2003. Soy protein with isoflavones has favorable effects on endothelial function that are independent of lipid and antioxidant effects in healthy postmenopausal women. *American Journal of Clinical Nutrition*. 78: 123–130.

Stoffers, D.A., 2004. The development of beta-cell mass: recent progress and potential role of GLP-1. *Hormone and Metabolic Research*. 36: 811–821.

Wei, H., Wei, L., Frenkel, K., Bowen, R., and Barnes, S., 1993. Inhibition of tumor promoter-induced hydrogen peroxide formation *in vitro* and *in vivo* by genistein. *Nutrition and Cancer*. 20: 1–12.

CHAPTER 33

Prevention and Management of Obesity by Isoflavones

BARBARA B. DOONAN,[a] ERXI WU[b] AND JOSEPH M. WU*[a]

[a] Department of Biochemistry and Molecular Biology, New York Medical College, Valhalla, NY 10595, USA; [b] Department of Pharmaceutical Sciences, North Dakota State University, Fargo, ND 58105, USA
*Email: Joseph_Wu@nymc.edu

33.1 Introduction

During the last quarter of the 20th century, leading up to the present, a trend towards overweight and obesity, defined as a body mass index (BMI), with a BMI of 30–39 considered as moderately obese and a BMI of >40 as extremely obese, has been observed worldwide both in adults and children; alarmingly, this trend is rapidly reaching epidemic proportions in the United States and elsewhere (World Health Organization 2004; Wang and Beydoun 2007). For instance, 10 years of survey by the National Health and Nutrition Examination Board (NHANES II and III) indicate the prevalence of overweight in the United States alone has approximately doubled from 15.0% of the population in 1980 to 30.9% in 2000, of which 11% belong to the 6–11 year age group (Flegal *et al.* 2002). A recent study in Australia by (Walls *et al.* 2010) has projected a 65% increase in obesity prevalence by 2025 if the current trends are maintained. Notably, obesity is believed to be significantly associated with the increased global burden of chronic disease and disability, and as a risk factor for type 2 diabetes mellitus (T2DM), hypertension, cardiovascular disease, stroke, certain forms of cancer, insulin resistance and abnormalities of lipid metabolism resulting in elevated concentrations of fatty acids, triglycerides, and both low- and very-low-density lipoproteins (LDLs and VLDLs) (Zhan and Ho 2005). Furthermore, obese children emerging as obese adults are more likely to have some of the same associated co-morbidities.

While there is sufficient evidence for a genetic influence for obesity, studies conducted in both animals and in human populations have led to the general recognition that the environmental factors of lifestyle, progressive decreases in physical activity and dietary patterns, are also major contributors. Thus, it may be suggested that the best way to prevent obesity is to maintain a healthy life style, consume healthy food and avoid overeating (Figure 33.1). To these ends, organizations such as the American Heart Association and American Diabetic Association have developed guidelines that can help in maintaining body fat and body weight at an appropriate level. Equally important, awareness of these associations has proved to be a strong driver of research programs to uncover the underlying biochemistry and molecular biology with the goal of identifying the means to prevent and manage this disorder. Current pharmaceutical

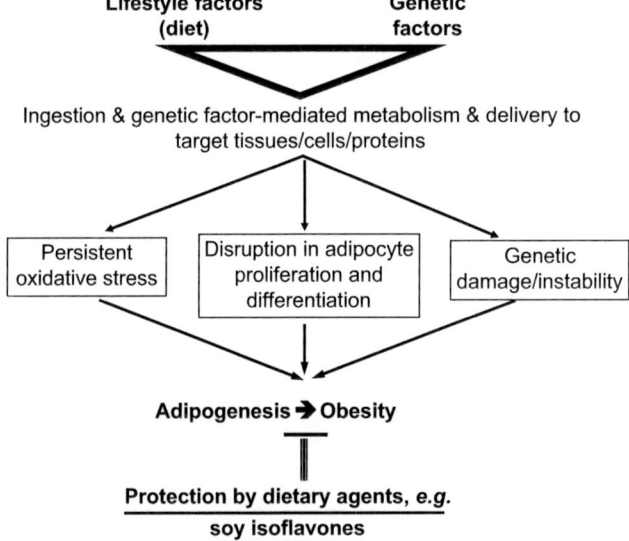

Figure 33.1 A scheme showing that both lifestyle and genetic factors play a role on a host of cellular activities ultimately impinging on adipogenesis and obesity. A scheme that depicts adipogenesis and obesity as the manifestation of a multitude of sustained events encompassing oxidative stress, uncontrolled adipocyte proliferation, and genetic damage and instability, sustained over time, with each subject to adverse control by lifestyle/environmental and genetic factors whose influence are delivered and presented to target tissues, cells and proteins through ingestion in combination with genetic factor-mediated metabolic activities. Protection against lipotoxicity by soy isoflavones may be attributed to their ability to act on the same multiple cellular targets adversely affected by extrinsic and intrinsic risk factors for adipogenesis and obesity.

approaches for treating obesity involve suppression of food intake or induced loss of calories into the stool. Several medications including sibutramine, orlistat, phentermine and amphetamine derivatives are often used to boost metabolism, suppress appetite, elevate mood and reduce absorption of fat from the consumed food. Additionally, a less regimented, more casual approach for countering obesity consists of modification of diet (such as low calorie and low fat), reduction of food intake and routine exercise. Furthermore, it is not uncommon for obese people to seek use of unconventional remedies available widely as dietary supplements from the open market or Internet. These are primarily herbal or plant-derived compounds suggested to have weight control features whose mechanisms of action are largely unknown.

Numerous published studies also suggest that obesity may inversely correlate with intake and dose level of consumption of soy or soy plus individual isoflavones (Orgaard and Jensen 2008). In postmenopausal women (Christie *et al.* 2010) consumption of soy-supplemented diets has been reported to reduce subcutaneous and total abdominal fat, in agreement with *in vivo* results from studies involving ovariectomized rodent models (Kim *et al.* 2006; Rachon *et al.* 2007). The influence of diet on obesity also is supported by the observed reduced prevalence of these diseases in Asian countries where large amounts of soy-based products are consumed on a daily basis (Liao *et al.* 2007). The low percentage of obese individuals in the Asian population may be attributed to their dietary preference for amount and frequency of consumption of soy-based food items, such as soybeans, tofu, tempeh and soy milk, which may contain phytochemicals with anti-obesity properties. Of particular interest are the isoflavones, which are naturally occurring soy components with numerous potential health benefits, based on published scientific studies.

This Chapter will summarize the epidemiological, *in vivo* and *in vitro* evidence describing the involvement of dietary factors in obesity. Emphasis will be directed at the soy isoflavones and their role in regulation of adipocyte differentiation, in the context of adipogenesis and development of obesity. We will present the salient features of an approach aimed at discovering soy isoflavones with anti-obesity potentials. A biospecific affinity chromatography approach will also be introduced that might lead to the identification of molecular targets and mechanistic understanding by which putative soy isoflavone targets counteract the increased prevalence of obesity and mediate prevention of this global problem. Possible interplay between NAD(P)H:quinone oxidoreductase1 (NQO1) and p53 in the control of adipogenesis will also be discussed.

33.2 Evidence that Dietary Factors and Soy Isoflavones Play a Role in Obesity

33.2.1 Introduction to Soy Phytochemicals and Isoflavones

Phytochemicals are present naturally in plants including soy and soy products, which are widely consumed by humans as whole soybeans, tofu, tempeh and soy milk. Isoflavones, which are mainly derived from soy and soy-based foods in the human diet, have been the most extensively studied soy phytochemicals (Xiao *et al.* 2008). The principal soy isoflavones are daidzein, genistein and glycetein and their glycoside conjugates including 7-*O*-glucosides, 6-*O*-acetyl and 6-*O*-malonyl-7-*O*-glucosides. After ingestion of isoflavone-rich foods, the isoflavone glycosides undergo deglycosylation by glucosidases located in the small intestinal brush border, *e.g.* lactose phlorizin hydrolase, and in enterocytes releasing aglycones daidzein and genistein, which are further converted into other metabolites such as equol, or are predominantly conjugated with glucuronic acid and to a lesser extent with sulfate (Coward *et al.* 1998). The conjugates are more readily transported in the blood and excreted in bile or urine than parent aglycones. The absorption of isoflavones by gastrointestinal mucosa may partly depend on their relative hydrophobicity and hydrophilicity. Limited studies have shown that the time to attain peak plasma concentrations after ingesting aglycones is 4–7 h, whereas it is 8–11 h when the glycoside conjugates are ingested. This shift indicates that the rate-limiting step for absorption is the initial hydrolysis of sugar moiety (Setchell *et al.* 2001). Ingestion of 50 mg of isoflavones day^{-1} in human adults yields plasma concentrations ranging from 50–800 ng ml^{-1}. Soy food processing appears to influence isoflavone bioavailability. Urinary isoflavone excretion was similar in 17 males consuming either 112 g of fermented soy tempeh or 125 g of unfermented soybean pieces for 9 days each. Urinary recovery of diadzein and genistein was higher when the subjects consumed fermented tempeh diet ($9.7 \pm 0.6\%$ and $1.9 \pm 0.1\%$) than when they consumed unfermented soy diet ($5.7 \pm 0.6\%$ and $1.3 \pm 0.1\%$), suggesting that isoflavone aglycones in fermented foods are more bioavailable (Hutchins *et al.* 1995). The metabolism of isoflavones is variable among individuals and influenced by other components of the diet.

A major isoflavone present in soy products is genistein. The estimated daily isoflavone intake in the US population is 1–9 mg, which can be increased several fold in heavy soy consumers (Skibola and Smith 2000). Oral administration of genistein has been shown to lower food intake, body weight and fat pad weight, and induce apoptosis in adipose tissue in ovariectomized female mice, suggesting that genistein may offer the prospect of treating or preventing increased adiposity after menopause (Kim *et al.* 2006).

33.2.2 Population-based Studies

The ability of the body to hoard calories *via* storage of excesses as fat during times of plenty was once a critical adaptation in preparation for anticipated leaner times. In our current world such is no longer necessary and this built-in physiological response tends to work against us as an internal clock set in favor of progressive weight gain and obesity. One might therefore surmise that the development and adoption of an appropriate lifestyle and nutritional regimen is necessary and even imperative in order to offset this intrinsic risk and achieve a balance that fits the world we live in. Kohno *et al.* (2006) prepared candies either containing soybean-derived β-conglycinin as the test compound or a placebo to be consumed over a 12 week period. 102 volunteers aged 26 to 69 years with BMIs of 25 to 30 participated in the study. A significant reduction in visceral body fat was observed in the β-conglycinin group. Maesta *et al.* (2007) conducted a controlled trial of 46 overweight postmenopausal women in Brazil testing soy protein or placebo with or without exercise (resistance training 3 times per week). At the end of the 16 week treatment period, an increase in muscle mass and a reduction in abdominal fat correlated with soy and the resistance training approach. Significant decreases in total cholesterol and LDL were observed in the users of soy alone. In a Canadian trial of obese postmenopausal women (Aubertin-Leheudre *et al.* 2007), it was reported that an exercise program combined with isoflavone supplementation had a significant effect on parameters such as body weight, BMI, and total and abdominal fat mass not seen with exercise alone. A study by Sites *et al.* (2007) tested a mixture of 20 g of soy protein plus 160 mg of isoflavones *versus* an isocaloric shake, taken once daily by 15 postmenopausal women over a period of 3 months. Over this period there was a greater, but non-significant, increase of total and subcutaneous fat in the isocaloric group. A later study with 75 postmenopausal women utilizing a similar approach saw no significant differences in BMI after 12 weeks of treatment (Charles *et al.* 2009). Wong *et al.* (2009) reported some positive effects on the lipid profile of hyperlipidemic men and women, with significant increases in high-density lipoprotein (HDL), suggesting a dietary means to improve serum cholesterol. In a recent study of hyperglycemic postmenopausal women consuming soy protein plus isoflavones or placebo (Liu *et al.* 2010), a mild, but significant, effect on body weight, BMI and body fat percentage (−3.74%) was reported following 6 months of soy supplementation. In a study with 33 obese Caucasian and African-American postmenopausal women consuming either a soy-supplemented beverage or a casein placebo beverage (Christie *et al.* 2010), reductions in total and subcutaneous fat in both groups were reported.

33.2.3 Animal Studies

In vivo studies vary in approach from utilization of diets with either soy, individual soy-derived components or soy enhanced with specific isoflavones, *e.g.* added genistein, to be fed in diets to rodents in which obesity has been induced prior to initiating test compound treatments to a rodent model such as the Zucker rat, bred for obesity-related research, or even throughout the gestational period (Cederroth *et al.* 2007). Excellent use has also been made of ovariectomized rodents as models for *in vivo* obesity studies, as the removal of the ovaries closely mimics the changes in fat deposition in women following menopause. A general influence of genistein in the diet of ovariectomized rats on lipid metabolism with reductions in blood serum and muscle triglyceride concentrations and increased free fatty acids in serum has been reported

(Rayalam et al. 2007). Juvenile or adult ovariectomized C57/BL6 female mice were given daily subcutaneous injections of vehicle, genistein (18–200 mg kg^{-1}) or 17β-estradiol (5 µg kg^{-1}), or fed diets containing 0–1500 ppm genistein in a report by Naaz et al. (2003). Genistein injections decreased adipose weight and adipocyte size in both juveniles and adults and dose–response decreases in fat pad weights of 37–57% were seen in juveniles fed 500–1500 ppm genistein. Later, in a report by Kim et al. (2006), genistein at 0, 150 or 1500 mg kg^{-1} was added to the diet of ovariectomized female mice. At the 1500 mg kg^{-1} level genistein was found to reduce food intake, body weight and reduce inguinal, parametrial and retroperitoneal fat pad weights with apoptosis in inguinal fat increased by 290%. Rachon et al. (2007) evaluated dietary equol, a metabolite of diadzein, in an ovariectomized Sprague–Dawley rat model which demonstrated a reduction in weight gain and intra-abdominal fat accumulation, and a decrease in plasma total cholesterol. However, this metabolite also was found to increase uterine mass, although at a level 3.5 times lower as compared with estradiol-3 benzoate. Cederroth et al. (2007) allowed *ad libitum* access to either high-soy or soy-free diets for CD-1 male and female mice from conception to adulthood. Their results indicated significant reductions in body weight and fat content in both sexes at adulthood with greater reductions observed in females. Yao et al. (2010), working with a C57BL/6J male mouse obesity-induced model, reported a 35.29% reduction in white fat mass following a 4 week treatment with a diet containing 100 mg kg^{-1} 4′-methoxy diadzein-7-fatty acid ester. Other positive effects observed were a 12.33% lower body weight, lower triglycerides and increased HDL as compared with control.

Some positive impacts of soy proteins with various levels of isoflavones have been observed in a Zucker rat model focused on obesity-related disease aspects such as development of insulin resistance and metabolic syndrome. However, there appears to be a clear indication that further research needs, especially with respect to the role of phenotype and underlying genetic mutations on isoflavone and phytochemical metabolism in the selected study models, are in order (Davis et al. 2006). This proves to be an important point of consideration for any selected model toward the goal of providing sound evidence for future therapeutic extrapolation to humans.

33.2.4 Laboratory Studies

Evidence suggests that the consumption of foods rich in isoflavones offers an approach for prevention and management of obesity *via* regulation of the adipocyte cycle of division and deposition of fats. The life cycle of adipocytes consists of several stages: differentiation of fibroblasts/mesenchymal stem cells into pre-adipocytes which undergo stages of maturation, followed by an acquisition of the capacity for intracellular accumulation of lipids. Studies have revealed that treatment of mature 3T3-L1 adipocytes and pre-adipocytes with genistein resulted in dose-dependent reduction in cell viability, decreases in lipid accumulation, lipolysis and induction of apoptosis (Rayalam et al. 2007). A $30 \pm 1.7\%$ reduction in lipid accumulation was observed at a dose level of

25 µM. However, when genistein was combined with resveratrol at the same dose level (each at 25 µM), the decrease in accumulation was 77.9 ± 3.4%. A 48 h treatment of maturing adipocytes with genistein alone at a dose level of 100 µM resulted in a 46.44 ± 9.2% increase in apoptosis, which rose to 242 ± 8.7% when combined with resveratrol, each at a 100 µM dose level. Interestingly, it has been observed that genistein can have opposing effects on adipogenesis, behaving as both agonist and antagonist (Relic *et al.* 2009). At low doses (3.25 µM) lipid accumulation and cell viability is increased, whereas at doses >6.25 µM the opposite result has been noted. Utilizing the 3T3-L1 murine cell line, Yang *et al.* (2007) reported the induction of apoptosis in mature adipocytes following treatment with genistein alone, which was enhanced when combined with a plant steroid, guggulsterone. Furthermore, genistein was found to inhibit the accumulation of lipid in maturing adipocytes, and this activity was also potentiated in combination with the plant steroid. In their recent report on the anti-obesity effect of a 4′-methoxy diadzein-7-fatty acid ester, Yao *et al.* (2010) also observed a reduction in viability, inhibition of differentiation and lipid accumulation, and induction of apoptosis, as a result of treatment of 3T3-L1 preadipocytes with this isoflavone fatty acid ester.

The major findings, including the strengths and limitations, from the combination of the aforementioned human, animal and laboratory studies are summarized in Table 33.1.

33.3 Identification and Mechanism of Soy Isoflavones with Anti-obesity Activities

33.3.1 Modulation of Fat Deposition and Adipogenesis by Soy Isoflavones

Epidemiological data and small intervention studies in humans have demonstrated that higher intakes of dietary soy isoflavones are associated with a lower incidence of diabetes and improved insulin sensitivity. In addition, soy isoflavones have also been shown to reduce plasma triglyceride and cholesterol levels, thereby improving cardiovascular disease risk profiles in both diabetic animals and humans. Taken together, soy isoflavones may target insulin resistance and improve the cluster of metabolic abnormalities that occur in T2DM.

Obese individuals have an increase in number and size of adipocytes, reflecting a change in proliferation and hypertrophy leading to increased adipose mass and size, and alterations in the distribution and extent of body fat and metabolic sequelae of obesity such as association with insulin resistance and T2DM. For instance, studies have demonstrated that enlarged fat cells correlate far better with insulin resistance than any other measures of adiposity (Rangwala and Lazar 2004). As mentioned previously, the life cycle of adipocytes consists of the differentiation of fibroblasts/mesenchymal stem cells which then mature into adipocytes with lipid accumulation capacity. Accordingly, an improved understanding of the adipocyte differentiation process,

Table 33.1 Major findings, including the strengths and limitations, from the human, animal and laboratory studies regarding consumption of or exposure to isoflavones and obesity.

Study type	Major findings on effects of isoflavones	Strengths of findings	Weakness of findings	References
Feeding studies in humans	HDL and serum cholesterol ↑; LDL↓ Reduction of visceral and abdominal fat Reduction of total body weight and BMI	Positive effects on blood lipid levels found to be most consistent Addition of an exercise program appears to be important	Most studies small and of short duration Direct relationship between soy consumption and fat reduction not evident	Aubertin-Leheudre et al. 2007 Christie et al. 2010 Kohno et al. 2006 Liu et al. 2010 Maesta et al. 2007 Sites et al. 2007 Wong et al. 2009
Feeding studies using animals	Significant reductions in adipose weight, adipocyte size and overall body weight Positive effects on lipid metabolism, resulting in improved blood profiles: HDL↑; triglyceride levels ↓ Reduction of food intake Some positive effects observed on development of insulin resistance and metabolic syndrome	Choice of animal model e.g. Zucker rat, ovariectomized rodents and dietary induction of obesity Varied dietary approaches e.g. soy +/− added isoflavones, individual soy-derived components	Role of phenotype and any underlying genetic mutations on metabolism of soy and soy-derived components in specific model not yet clear for application of results to humans Comparison of *ad libitum* to controlled dietary intake effects in specific model	Cederroth et al. 2007 Davis et al. 2006 Kim et al. 2006 Naaz et al. 2003 Rachon et al. 2007 Rayalam et al. 2007 Yao et al. 2010
Tissue culture studies	Significant effects of isoflavones on the adipocyte life cycle in a variety of cell types	Enhanced positive effects observed with combinations of phytochemicals e.g. genistein + resveratrol	Role of phytochemical combinations and dose levels need further clarification with respect to effects on adipocytes	Rayalam et al. 2007 Relic et al. 2009 Yang et al. 2007 Yao et al. 2010
Tissue culture studies	Inhibition of differentiation Reductions in cell viability Decreases in lipid accumulation Increases in apoptosis			

specific gene regulation and the molecular basis of adipogenesis offers new opportunities to treat these common disorders resulting from aberrant fat deposition. Moreover, understanding the basic mechanisms of adipocyte growth, differentiation and function, in response to specific isoflavones, will provide clues and insights on the pathogenesis and treatment of obesity. This knowledge will provide us with valuable information that will identify molecular targets and the key signaling molecules with which soy isoflavones interact, and help us better understand the favorable effects of dietary soy isoflavones on obesity, insulin resistance and diabetes in both humans and animals, thereby contributing to knowledge for designing public health recommendations to prevent and treat diabetes-linked obesity.

The process of adipocyte differentiation has been extensively studied *in vitro* using both a mouse 3T3-L1 pre-adipocyte cell line, which is capable of differentiating into mature adipocytes with similar morphological and biochemical characteristics of adipocytes *in vivo* (Tontonoz et al. 1994). Treatment of confluent 3T3-L1 pre-adipocytes in culture with methylisobutyl xanthine, dexamethasone,and insulin (MDI) induces the differentiation of the committed pre-adipocytes into mature adipocytes within 6 to 7 days. Acquisition of the terminal adipocyte phenotype requires the expression of specific transcription factors including members of the CCAAT/enhancer-binding protein (C/EBP), peroxisome-proliferator-activated receptor γ (PPARγ) and sterol regulatory element-binding protein (SREBP) (Hallakou et al. 1997). These transcription factors in turn interact cooperatively and sequentially to trigger the subsequent induction of adipocyte-specific genes such as fatty acid synthase (FAS) and glucose transporter 4 (GLUT4), adipocyte fatty acid-binding protein (aP2) and stearoyl-CoA desaturase (SCD) (Hallakou et al. 1997). Additive effects due to combinations of naturally occurring phytochemicals, as observed with genistein, resveratrol and quercetin in human and 3T3-L1 adipocytes, suggest a potential for dietary intervention in dealing with the development and management of obesity (Park et al. 2008). Inhibition of lipid accumulation occurs in a dose-dependent manner at concentrations above 6.25 µM with dose-dependency also reported to vary based on age and gender (Dang 2009), and, as mentioned earlier, in a study by Relic et al. (2009) with genistein as the test component, with respect to lipid accumulation and cell viability, opposite effects were reported for low dose (3.25 µM) as compared with dose levels > 6.25 µM. Genistein treatment was shown to reverse the induced differentiation process at days 3–9 *via* suppression of phosphorylation of extracellular signal-regulated kinase 1 and 2 (ERK1/2) in mouse bone marrow-derived stem cell cultures (Liao et al. 2008).

33.3.2 A Facile Strategy for the Identification of Soy Isoflavones with Potential for Regulating Visceral Fat

While there is some evidence to support the anti-obesity effects of soy isoflavones, there remains a critical need to more completely define the direct target tissue(s) and the molecular pathways by which the specific soy constituents act in the control of visceral fat and adipogenesis. Herein, we propose a strategy based on the following considerations. Studies have shown that enlarged, hypertrophic adipocytes are associated with insulin resistance, T2DM and obesity. Separation of human adipocytes derived from adipose tissue biopsy specimens into small and large adipocyte cell populations and analysis of the expression profiles of the separated populations have revealed that NQO1 is significantly higher in large compared with small adipocytes, and the expression of NQO1 also correlates with adipocyte size, and with serum amyloid A (SAA) (Gaikwad et al. 2001; Jernas et al. 2006; Palming et al. 2007; Sjoholm et al. 2005). SAA has been implicated in inflammation and insulin resistance and NQO1 is known to be involved in oxidative stress, suggesting that these findings may provide novel insights into the connection between hypertrophic obesity and insulin resistance/T2DM. SAA, NQO1 and also the cell death-inducing effector A (CIDE-A) were predominantly expressed in human adipocytes as compared with a panel of 32 other human tissues and cell types. During diet-induced weight loss in obese subjects, adipose tissue expression of NQO1 was reduced and CIDE-A was elevated. NQO1 expression correlated to measures of adiposity, insulin and the markers of liver dysfunction aspartate aminotransferase (AST) and alanine transaminase (ALT) (Jernas et al. 2006; Palming et al. 2007; Sjoholm et al. 2005). These findings indicate a role for NQO1 in the metabolic complications of human obesity. Moreover, NQO1 is known to be involved in regulation of oxidative stress. These results suggest that NQO1 may be a novel target protein for providing insights into the connection and/or interplay between insulin resistance, T2DM and hypertrophic obesity.

Based on the information above, and with the goal of discovering novel single or grouped anti-obesity chemicals for global intervention initiatives in the future, we have designed a cost-effective approach, named differential displacement affinity chromatography (DAC) for identifying novel anti-obesity chemicals from existing chemical sets and from isoflavones. DAC is modified from a widely used biochemical technique known as affinity chromatography which, in its fundamental operation, covalently attaches a ligand to a solid matrix forming an immobilized high specificity and binding ligand–matrix platform suitable for capturing rare proteins and nucleic acids from complex mixtures and biological fluids. The improvement we have made in DAC is the aspect of differential displacement of a known obesity target protein called NQO1 bound to its inhibitor, dicumarol or diminutol (both high affinity inhibitors of NQO1) (Wignall et al. 2004), coupled to agarose as the basis of a novel screen for identifying chemicals containing anti-obesity potential (Figure 33.2). Competition and displacement of quantifiable NQO1 by known chemicals and unknown hits from soy extracts and single and grouped isoflavones could provide an unbiased and robust strategy for discovering small molecules with anti-obesity potential. Marginally active and highly potent NQO1-displacement chemicals may be combined to generate anti-obesity cocktails less likely to produce drug resistance and tolerance.

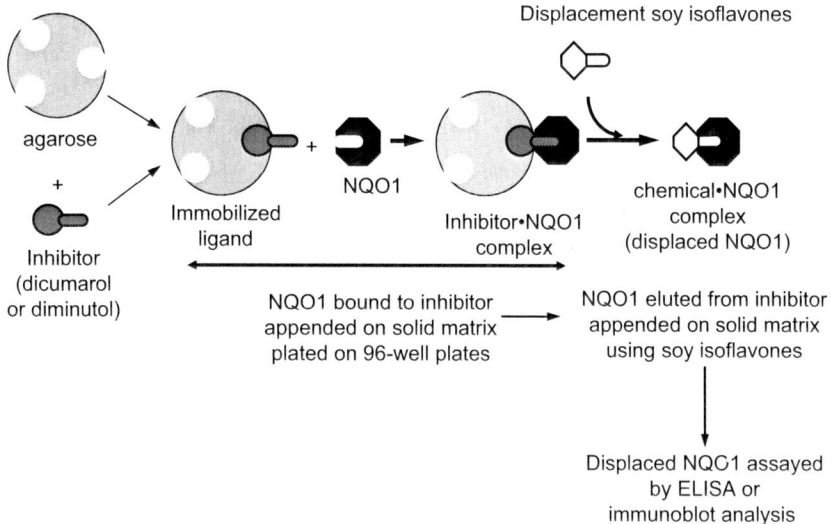

Figure 33.2 NQO1 displacement affinity chromatography as a strategy for screening anti-obesity chemicals from soy isoflavones.

Our proposed approach of the quantitative release of inhibitor-complexed NQO1 by increasing doses of chemicals including soy isoflavones is novel in the discovery of new anti-obesity chemical entities. First, the displacement affinity approach offers simplicity; a single step release may reveal low-abundance, high-potency anti-obesity chemicals. Second, an identified chemical highly efficacious in displacing NQO1 from bound inhibitor may be more likely to have a functional link to the role NQO1 plays in adipocyte hypertrophy and obesity. Third, the displacement strategy may enable a connection to be established between the candidate anti-obesity chemicals, the displaced NQO1 and the biology of obesity. In summary, the selective and differential displacement of NQO1 from NQO1 bound to its inhibitors, dicumarol or diminutol, affinity matrices can provide a facile approach to screen, discover and characterize novel single or grouped soy isoflavones with the potential for inhibiting or modulating inhibitors of adipogenesis and, by extrapolation, obesity.

33.3.3 Discovery of Proteins Targeted by Soy Isoflavones with a Regulatory Role in Adipogenesis

Once candidate isoflavones that are capable of displacing NQO1 have been identified and characterized, they can be immobilized on solid supports as biospecific platforms to discover proteins that could play an important role in the process of adipocyte differentiation and adipogenesis. In this strategy, the isoflavone-immobilized support will be used to fractionate extracts prepared from different stage differentiated 3T3 cells, as illustrated in Figure 33.3.

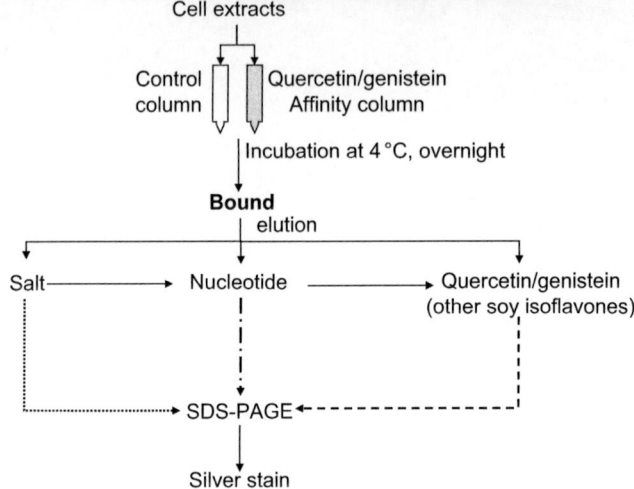

Figure 33.3 Soy isoflavone immobilized affinity chromatography as a strategy for discovering candidate anti-obesity target proteins using extracts from 3T3-L1 cells at different stages of adipocyte differentiation.

33.3.4 Candidate Molecular Targets and Mechanisms of Anti-obesity Isoflavones: Interplay Between NQO1 and p53 in Regulation of Adipogenesis

Emerging evidence has suggested consumption or enrichment of foods with high levels of isoflavones presents a promising approach for prevention and management of obesity. Isoflavones can function as phytoestrogens, compounds structurally similar to endogenous estrogens. As such, isoflavones elicit weak estrogenic effects *via* competitive binding to intranuclear estrogen receptors (ERs). A number of additional binding targets have been identified for isoflavones, *e.g.*, cyclin-dependent kinases, cyclins, mitogen-activated protein kinase (MAPK), HDL and LDL (prevent oxidation). Isoflavones have been reported to reduce LDL cholesterol, affect nitric oxide synthase and transcription factors, regulate stem cell differentiation in bone marrow, and to play a role in cell cycle control, induction of apoptosis and serum lipid contents.

The approach presented above highlights the focus on NQO1 as a molecular bait for the discovery of novel anti-obesity chemicals. Notably, NQO1 has also been shown to be a multi-tasking protein, with bioactivities that far exceed its assigned role in phase 2 detoxification reactions (Ross 2004). For example, NQO1 has been reported to play a role in determining the stability of tumor

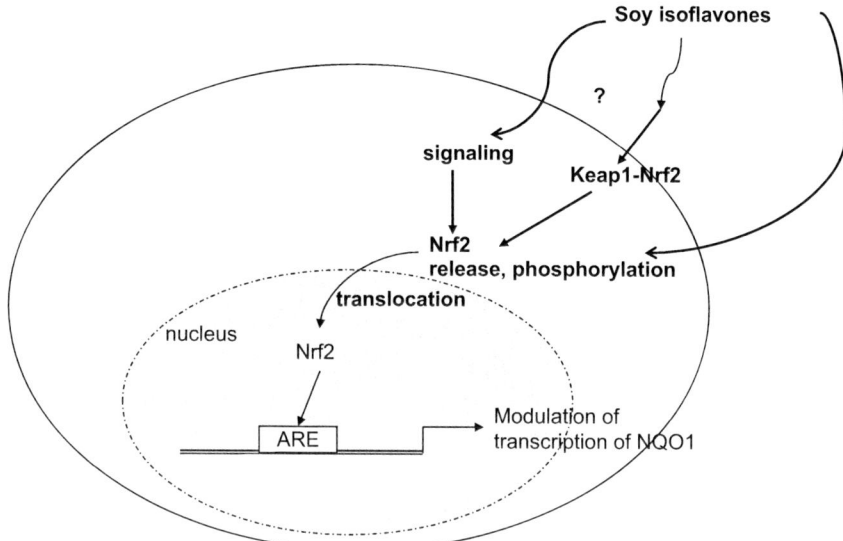

Figure 33.4 Proposed genomic activities of soy isoflavones may contribute to its anti-obesity activities. Soy isoflavones are hypothesized to target the ARE-dependent transcription regulation of NQO1 *via* modulation of translocation of Nrf2 from the cytosol-resident Nrf2–Keap1 complex, resulting in transcription commitment and expression of ARE-responsive genes including NQO1.

suppressor protein p53 (Asher *et al.* 2001), which has recently been suggested to counteract the effects of lipotoxicity, as well as control the induction of apoptosis and proliferation in adipocytes (Bazuine *et al.* 2009). Additionally, the expression of NQO1 is copiously induced by a multitude of chemicals and dietary agents, including soy isoflavones, by antioxidant response element (ARE)-dependent transcription mechanisms (Figure 33.4). On the basis of these considerations, we propose a mechanism whereby isoflavones, through transcription regulation of NQO1, control complex formation between NQO1 and p53, thereby affecting the stability of p53 and in turn the proliferation and genetic stability of adipocytes (Figure 33.5).

33.3.5 Role of PPARα, PPARγ, C/EBPα and Other Candidate Molecular Sensors/Effectors of Adipogenesis

PPARα and PPARγ (nuclear hormone receptors) have been known to play a role in fat metabolism for quite some time. PPARγ is predominantly expressed in adipose tissue (Tontonoz *et al.* 1994). Compared with PPARα and PPARδ, PPARγ shows preferential binding to two regulatory sequences derived from a fat-specific gene, suggesting that the adipogenic potential of PPARγ may be due to its ability to bind to fat-specific regulatory sequences. PPARγ and

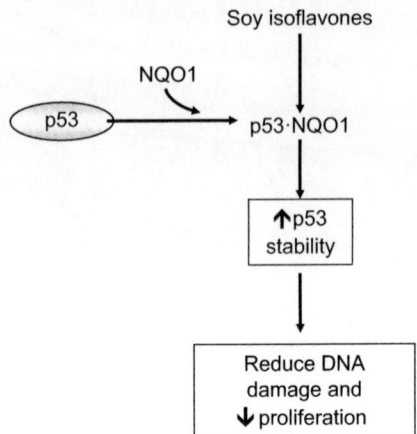

Figure 33.5 Proposed NQO1-mediated control of p53 and adipocyte proliferation by isoflavones. p53 is hypothesized to be a primary molecular determinant for control of adipocyte proliferation and genetic stability. The level of p53 is subject to control by formation of a complex with a multitude of proteins including NQO1. Soy isoflavones, in modulating the expression of NQO1 by a mechanism presented in Figure 33.4, affects NQO1–p53 complex formation and hence the stability and turnover of p53 and, in turn, p53-dependent inhibition of adipocyte genetic stability and proliferation.

C/EBPα are critical transcription factors in adipogenesis, but the precise roles of these proteins remain unknown because they positively regulate each other's expression (Rosen *et al.* 2002). The soy isoflavones genistein and daidzein at micromolar concentrations concurrently activate different amounts of ERs and PPARs, and the balance of the divergent action of ERs and PPARs may determine isoflavone-induced osteogenesis and adipogenesis. In *in vitro* studies utilizing a murine macrophage cell line, RAW264.7 cells, it was observed that treatment with a mixture of unconjugated soy isoflavones and the isoflavones genistein and daidzein positively influenced both PPARα- and PPARγ-directed gene expression (Mezei *et al.* 2003). Thus, it appears that soy isoflavones do have an *in vivo* activity on PPAR.

In addition to proposed mechanisms involving the activation of nuclear receptors, ERs and PPARs, and inhibition of various enzyme activities (Dang 2009), genistein also inhibits adipocyte differentiation *via* activation of adenosine monophosphate (AMP)-dependent protein kinase (Hwang *et al.* 2005).

A number of studies in humans and animals suggest that soy and soy isoflavones have beneficial effects on T2DM. The supplementation of soy together with isoflavones has been shown to reduce the fasting insulin level, insulin resistance and total cholesterol in T2DM subjects (Teixeira *et al.* 2004). Furthermore, the increased insulin sensitivity due to the soy consumption was also observed in ovariectomized cynomolgus monkeys fed soy protein with isoflavones (Wagner *et al.* 2006). In a recent well-controlled intervention study,

Jayagopal et al. (2002) evaluated the effects of soy isoflavones in 32 women with T2DM. Compared with placebo, isoflavone supplementation favorably altered insulin resistance, glycemic control and serum lipoproteins in postmenopausal women with T2DM, thereby improving their cardiovascular risk profiles.

33.4 Conclusions

As mentioned above, obesity is a medical condition that has reached epidemic proportions. In the United States, data from CDC the Center for Disease Control and Prevention (CDC) indicate that adult obesity has increased by 60% over the last two decades while obesity in children/adolescents has tripled in the past thirty years. Alarmingly, obesity-related disease development has become the second most prevalent cause of mortality in the United States, exceeded only by tobacco-related deaths. Consumption of unhealthy diets and sedentary lifestyles are considered to be the principal culprits for this fast-growing chronic disease. In general, there is lack of information on identity and mechanisms of molecular and cellular targets contributing to the establishment of an obese state. Equally in need of research is the identification of natural, anti-obesity dietary agents and the development of diet-based obesity preventive strategies.

Summary Points

- Obesity has become a leading metabolic disease posing a significant health problem globally including the United States.
- Obesity-associated metabolic disorders encompass a range of syndromes including insulin resistance, glucose intolerance and dyslipidemia.
- Naturally occurring dietary isoflavones may have anti-obesity functions.
- Our long term goal is to discover novel target proteins that might contribute to or participate in anti-obesity effects of isoflavones.
- A corollary objective is to better understand the molecular mechanisms by which soy isoflavones exert anti-obesity effects in order to find a safe and easily compliant intervention approach for the prevention and management of obesity and its associated metabolic syndromes.

Key Facts

- Obesity is an individual as well as clinical condition that is recognized as a serious health problem.
- Obesity is a major risk factor for the metabolic syndrome, a collective number of diseases including hypertension, high cholesterol, type 2 diabetes and predisposition for coronary heart disease.
- Existing medications are inadequate for controlling obesity; effective anti-obesity measures require major lifestyle changes comprising low fat diet, caloric restriction and regular exercise.

- Dietary isoflavones may offer promise in the armamentarium of yet-to-be-discovered naturally occurring diet-based anti-obesity compounds.
- Identifying molecular targets with which soy isoflavones interact will reveal how soy affects various signaling steps having a direct or indirect role in the development and progression of obesity.

Definitions of Words and Terms

Adipogenesis: The process of differentiating from a basic fibroblast cell type to the specific cell type making up adipose (fat) tissue.

Ad libitum: In an animal study where food is always available to be consumed as often or as much as is desired by the test animals.

Aspartate aminotransferase/alanine aminotransferase (AST/ALT): These are markers of liver dysfunction.

CCAAT: A specific nucleotide sequence present in the upstream region of eukaryotic genes involved with transcription factor CCAAT/enhancer-binding protein (C/EBP) interactions.

Displacement affinity chromatography (DAC): This is a modification of a biochemical technique known as affinity chromatography in which a ligand is covalently attached to a solid matrix to capture rare proteins and nucleic acids from complex mixtures and biological fluids.

Hyperplasia/hypertrophy: These are descriptive terms for increase in the number of cells and increase in the size of cells in a part of the body.

Insulin resistance: This is a decrease in the sensitivity to insulin which results in an inability of tissues such as liver, muscles and adipose to respond normally.

In vivo/in vitro: Studies which utilize living animals such as rodents as opposed to out of the body studies using such as test tubes or Petri dishes.

Isoflavones: These are naturally occurring chemicals found in soy and soy-based products such as genistein.

Metabolic syndrome: This is a term describing a group of symptoms such as elevated insulin levels, increased blood pressure, abnormal cholesterol readings and increased waist fat deposition. When these occur together, the risks of heart attack, stroke and diabetes are enhanced.

Obesity: Obesity is described as having a body mass index (BMI; determined by one's weight in kg/height in m^2) > 30.

Phytochemicals: These are natural plant components, usually with some protective role in the plant, which have been found through experimentation to have potential as disease-preventive agents in humans.

List of Abbreviations

ARE	antioxidant response element
BMI	body mass index
C/EBP	CCAAT/enhancer binding protein
CIDE-A	cell death-inducing effector A

DAC	differential displacement affinity chromatography
ER	estrogen receptor
HDL	high-density lipoprotein
LDL	low-density lipoprotein
NQO1	NAD(P)H:quinone oxidoreductase 1
PPAR	peroxisome-proliferator-activated receptor
SAA	serum amyloid A
T2DM	type 2 diabetes mellitus

References

Asher, G., Lotem, J., Cohen, B., Sachs, L., and Shaul, Y., 2001. Regulation of p53 stability and p53-dependent apoptosis by NADH quinine oxidoreductase 1. *Proceedings of the National Academy of Sciences USA*. 98(3): 1188–1193.

Aubertin-Leheudre, M., Lord, C., Khalil, A., and Dionne, I., 2007. Six months of isoflavone supplement increases fat-free mass in obese-sarcopenic postmenopausal women: a randomized double-blind controlled trial. *European Journal of Clinical Nutrition*. 61(12): 1442–1444.

Bazuine, M., Stenkula, K., Cam, M., Arroyo, M., and Cushman, S., 2009. Guardian of corpulence: a hypothesis on p53 signaling in the fat cell. *Clinical Lipidology*. 4: 231.

Cederroth, C., Vinciguerra, M., Kuhne, F., Madani, R., Doerge, D.R., Visser, M., Foti, T.J., Rohner-Jeanrenaud, F., Vassalli, J.D., and Nef, S., 2007. A phytoestrogen-rich diet increases energy expenditure and decreases adiposity in mice. *Environmental Health Perspective*. 115(10): 1467–1473.

Charles, C., Yuskavage, J., Carlson, O., John, M., Tagalicud, A.S., Maggio, M., Muller, D.C., Egan, J., and Basaria, S., 2009. Effects of high-dose isoflavones on metabolic and inflammatory markers in healthy postmenopausal women. *Menopause*. 16(2): 395–400.

Christie, D.R., Grant, J., Darnell, B.E., Chapman, V.R., Gastaldelli, A., and Sites, C.K., 2010. Metabolic effects of soy supplementation in postmenopausal Caucasian and African American women: a randomized, placebo-controlled trial. *American Journal of Obstetrics and Gynecology*. 203(2): 153e1–e9.

Coward, L., Smith, M., Kirk, M., and Barnes, S., 1998. Chemical modification of isoflavones in soyfoods during cooking and processing. *American Journal of Clinical Nutrition*. 68(Suppl. 6): 1486S–1491S.

Dang, Z.C., 2009. Dose-dependent effects of soy phyto-oestrogen genistein on adipocytes: mechanisms of action. *Obstetrics Review*. 10(3): 342–349.

Davis, D.A., Sarkar, S.H., Hussain, M., Li, Y., and Sarkar, F.H., 2006. Increased therapeutic potential of an experimental anti-mitotic inhibitor SB715992 by genistein in PC-3 human prostate cancer cell line. *BMC Cancer*. 6: 22.

Flegal, K.M., Carroll, M.D., Ogden, C.L., and Johnson, C.L., 2002. Prevalence and trends in obesity among US adults, 1999–2000. *Journal of the American Medical Association*. 288(14): 1723–1727.

Gaikwad, A., Long, II, D.J., Stringer, J.L., and Jaiswal, A.K., 2001. *In vivo* role of NAD(P)H:quinone oxidoreductase 1 (NQO1) in the regulation of intracellular redox state and accumulation of abdominal adipose tissue. *Journal of Biological Chemistry*. 276(25): 22559–22564.

Hallakou, S., Doare, L., Foufelle, F., Kergoat, M., Guerre-Millo, M., Berthault, M.F., Dugail, I., Morin, J., Auwerx, J., and Ferre, P., 1997. Pioglitazone induces *in vivo* adipocyte differentiation in the obese Zucker *fa/fa* rat. *Diabetes*. 46(9): 1393–1399.

Hutchins, A.M., Slavin, J.L., and Lampe, J.W., 1995. Urinary isoflavonoid phytoestrogen and lignan excretion after consumption of fermented and unfermented soy products. *Journal of the American Dietetic Association*. 95(5): 545–551.

Hwang, J.T., Park, I.J., Shin, J.I., Lee, Y.K., Lee, S.K., Baik, H.W., Ha, J., and Park, O.J., 2005. Genistein, EGCG, and capsaicin inhibit adipocyte differentiation process via activating AMP-activated protein kinase. *Biochemical and Biophysical Research Communications*. 338(2): 694–699.

Jayagopal, V., Albertazzi, P., Kilpatrick, E.S., Howarth, E.M., Jennings, P.E., Hepburn, D.A., and Atkin, S.L., 2002. Beneficial effects of soy phytoestrogen intake in postmenopausal women with type 2 diabetes. *Diabetes Care*. 25(10): 1709–1714.

Jernas, M., Palming, J., Sjoholm, K., Jennische, E., Svensson, P.A., Gabrielsson, B.G., Levin, M., Sjogren, A., Rudemo, M., Lystig, T.C., Carlsson, B., Carlsson, L.M., and Lonn, M., 2006. Separation of human adipocytes by size: hypertrophic fat cells display distinct gene expression. *Federation of American Societies for Experimental Biology Journal*. 20(9): 1540–1542.

Kim, H.K., Nelson-Dooley, C., Della-Fera, M.A., Yang, J.Y., Zhang, W., Duan, J., Hartzell, D.L., Hamrick, M.W., and Baile, C.A., 2006. Genistein decreases food intake, body weight, and fat pad weight and causes adipose tissue apoptosis in ovariectomized female mice. *Journal of Nutrition*. 136(2): 409–414.

Kohno, M., Hirotsuka, M., Kito, M., and Matsuzawa, Y., 2006. Decreases in serum triacylglycerol and visceral fat mediated by dietary soybean beta-conglycinin. *Journal of Atherosclerosis and Thrombosis*. 13(5): 247–255.

Liao, F.H., Shieh, M.J., Yang, S.C., Lin, S.H., and Chien, Y.W., 2007. Effectiveness of a soy-based compared with a traditional low-calorie diet on weight loss and lipid levels in overweight adults. *Nutrition*. 23(7–8): 551–556.

Liao, Q.C., Li, Y.L., Qin, Y.F., Quarles, L.D., Xu, K.K., Li, R., Zhou, H.H., and Xiao, Z.S., 2008. Inhibition of adipocyte differentiation by phytoestrogen genistein through a potential downregulation of extracellular signal-regulated kinases 1/2 activity. *Journal of Cellular Biochemistry*. 104(5): 1853–1864.

Liu, Z.M., Chen, Y.M., Ho, S.C., Ho, Y.P., and Woo, J., 2010. Effects of soy protein and isoflavones on glycemic control and insulin sensitivity: a 6-mo double-blind, randomized, placebo-controlled trial in postmenopausal Chinese women with prediabetes or untreated early diabetes. *American Journal of Clinical Nutrition.* 91(5): 1394–1401.

Maesta, N., Nahas, E.A., Nahas-Neto, J., Orsatti, F.L., Fernandes, C.E., Traiman, P., and Burini, R.C., 2007. Effects of soy protein and resistance exercise on body composition and blood lipids in postmenopausal women. *Maturitas.* 56(4): 350–358.

Mezei, O., Banz, W.J., Steger, R.W., Peluso, M.R., Winters, T.A., and Shay, N., 2003. Soy isoflavones exert antidiabetic and hypolipidemic effects through the PPAR pathways in obese Zucker rats and murine RAW 264.7 cells. *Journal of Nutrition.* 133(5): 1238–1243.

Naaz, A., Yellayi, S., Zakroczymski, M.A., Bunick, D., Doerge, D., Labahn, D.B., Helferich, W.G., and Cooke, P.S., 2003. The soy isoflavone genistein decreases adipose deposition in mice. *Endocrinology.* 144: 3315–20.

Orgaard, A., and Jensen, L., 2008. The effects of soy isoflavones on obesity. *Experimental Biology and Medicine (Maywood).* 233(9): 1066–1080.

Palming, J., Sjoholm, K., Jernas, M., Lystig, T.C., Gummesson, A., Romeo, S., Lonn, L., Lonn, M., Carlsson, B., and Carlsson, L.M., 2007. The expression of NAD(P)H:quinone oxidoreductase 1 is high in human adipose tissue, reduced by weight loss, and correlates with adiposity, insulin sensitivity, and markers of liver dysfunction. *Journal of Clinical Endocrinology and Metabolism.* 92(6): 2346–2352.

Park, H.J., Yang, J.Y., Ambati, S., Della-Fera, M.A., Hausman, D.B., Rayalam, S., and Baile, C.A., 2008. Combined effects of genistein, quercetin, and resveratrol in human and 3T3-L1 adipocytes. *Journal of Medicinal Food.* 11(4): 773–783.

Rachon, D., Vortherms, T., Seidlova-Wuttke, D., and Wuttke, W., 2007. Effects of dietary equol on body weight gain, intra-abdominal fat accumulation, plasma lipids, and glucose tolerance in ovariectomized Sprague–Dawley rats. *Menopause.* 14(5): 925–932.

Rangwala, S.M., and Lazar, M.A., 2004. Peroxisome proliferator-activated receptor γ in diabetes and metabolism. *Trends in Pharmacological Science.* 25(6): 331–336.

Rayalam, S., Della-Fera, M.A., Ambati, S., Boyan, B., and Baile, C.A., 2007. Enhanced effects of guggulsterone plus 1,25(OH)2D3 on 3T3-L1 adipocytes. *Biochemical and Biophysical Research Communications.* 364(3): 450–456.

Relic, B., Zeddou, M., Desoroux, A., Beguin, Y., de Seny, D., and Malaise, M.G., 2009. Genistein induces adipogenesis but inhibits leptin induction in human synovial fibroblasts. *Laboratory Investigation.* 89(7): 811–822.

Rosen, E.D., Hsu, C.H., Wang, X., Sakai, S., Freeman, M.W., Gonzalez, F.J., and Spiegelman, B.M., 2002. C/EBPα induces adipogenesis through PPARγ: a unified pathway. *Genes and Development.* 16(1): 22–26.

Ross, D., 2004. Quinone reductases multitasking in the metabolic world. *Drug Metabolism and Metabolic Reviews.* 36(3–4): 639–654.

Setchell, K.D., Brown, N.M., Desai, P., Zimmer-Nechemias, L., Wolfe, B.E., Brashear, W.T., Kirschner, A.S., Cassidy, A., and Heubi, J.E., 2001. Bioavailability of pure isoflavones in healthy humans and analysis of commercial soy isoflavone supplements. *Journal of Nutrition.* 131(Suppl. 4): 1362S–1375S.

Sites, C.K., Cooper, B.C., Toth, M.J., Gastaldelli, A., Arabshahi, A., and Barnes, S., 2007. Effect of a daily supplement of soy protein on body composition and insulin secretion in postmenopausal women. *Fertility and Sterilility.* 88(6): 1609–1617.

Sjoholm, K., Palming, J., Olofsson, L.E., Gummesson, A.P., Svensson, A., Lystig, T.C., Jennische, E., Brandberg, J., Torgerson, J.S., Carlsson, B., and Carlsson, L.M., 2005. A microarray search for genes predominantly expressed in human omental adipocytes: adipose tissue as a major production site of serum amyloid A. *Journal of Clinical Endocrinology and Metabolism.* 90(4): 2233–2239.

Skibola, C.F., and Smith, M.T., 2000. Potential health impacts of excessive flavonoid intake. *Free Radical Biology and Medicine.* 29(3-4): 375–383.

Teixeira, S.R., Tappenden, K.A., Carson, L., Jones, R., Prabhudesai, M., Marshall, W.P., and Erdman, Jr., J.W., 2004. Isolated soy protein consumption reduces urinary albumin excretion and improves the serum lipid profile in men with type 2 diabetes mellitus and nephropathy. *Journal of Nutrition.* 134(8): 1874–1880.

Tontonoz, P., Hu, E., Graves, R.A., Budavari, A.I., and Spiegelman, B.M., 1994. mPPARγ2: tissue-specific regulator of an adipocyte enhancer. *Genes and Development.* 8(10): 1224–1234.

Wagner, J.E., Kavanagh, K., Ward, G.M., Auerbach, B.J., Harwood, Jr., H.J., and Kaplan, J.R., 2006. Old world nonhuman primate models of type 2 diabetes mellitus. *Institute for Laboratory Animal Research Journal.* 47(3): 259–271.

Walls, H.L., Stevenson, C.E., Mannan, H.R., Abdullah, A., Reid, C.M., McNeil, J.J., and Peeters, A., 2010. Comparing trends in BMI and waist circumference. *Obesity (Silver Spring).* 19(1): 216–219.

Wang, Y., and Beydoun, M.A., 2007. The obesity epidemic in the United States – gender, age, socioeconomic, racial/ethnic, and geographic characteristics: a systematic review and meta-regression analysis. *Epidemiological Reviews.* 29: 6–28.

Wignall, S.M., Gray, N.S., Chang, Y.T., Juarez, L., Jacob, R., Burlingame, A., Schultz, P.G., and Heald, R., 2004. Identification of a novel protein regulating microtubule stability through a chemical approach. *Chemistry and Biology.* 11(1): 135–146.

Wong, B., Kyle, R., Croft, K., Quinn, C., Jessup, W., and Yeap, B., 2009. P732 hyperlipidemic serum alters macrophage lipid content, fatty acid composition and gene expression. *Atherosclerosis Supplements.* 10(2): e918.

World Health Organization, 2004. Facts related to chronic diseases – Fact sheet Obesity and Overweight. http//:www.who.int/hpr/gs.fs.obesity.shtml. Accessed 31 December 2010.

Xiao, R., Su, Y., Simmen, R.C., and Simmen, F.A., 2008. Dietary soy protein inhibits DNA damage and cell survival of colon epithelial cells through attenuated expression of fatty acid synthase. *American Journal of Physiology Gastrointestinal Liver Physiology*. 294(4): G868–G876.

Yang, J.Y., Della-Fera, M.A., Rayalam, S., Ambati, S., and Baile, C.A., 2007. Enhanced pro-apoptotic and anti-adipogenic effects of genistein plus guggulsterone in 3T3-L1 adipocytes. *Biofactors*. 30(3): 159–169.

Yao, Y., Li, X.B., Zhao, W., Zeng, Y.Y., Shen, H., Xiang, H., and Xiao, H., 2010. Anti-obesity effect of an isoflavone fatty acid ester on obese mice induced by high fat diet and its potential mechanism. *Lipids in Health and Disease*. 9: 49.

Zhan, S., and Ho, S.C., 2005. Meta-analysis of the effects of soy protein containing isoflavones on the lipid profile. *American Journal of Clinical Nutrition*. 81(2): 397–408.

CHAPTER 34
Soy Isoflavones and Testicular Function

BENSON T. AKINGBEMI

Department of Anatomy, Physiology and Pharmacology, Auburn University, Auburn, AL, USA
Email: akingbt@auburn.edu

34.1 Introduction

Dietary estrogens are diphenolic non-steroidal estrogen-like substances found in plants (*i.e.*, phytoestrogens) and present at high concentrations in leguminous plants. Soy beans contain predominantly the isoflavones genistin and daidzin, which are structurally similar to the female estrogen 17β-estradiol (E2) (Figure 34.1) and have the capacity to bind estrogen receptors (ESRs), acting as phytoestrogens or xenoestrogens. Phytoestrogens are classified as endocrine disruptors (EDs) due to their ability to interfere with the function of the endocrine axis. Due to their hormonal activity, isoflavones have been extensively investigated in the last couple of years due to their health beneficial effects. For example, the protective properties of isoflavones against cancers (*e.g.*, breast and prostate cancers) have been attributed to prevention of activation of ESRs by endogenous E2 due to competitive binding, inhibition of tyrosinase kinase activity, induction of cancer cell differentiation and/or direct antioxidant effects. In 1999, the Food and Drug Administration (United States) allowed food manufacturers to claim that eating foods high in soy protein can help lower a person's risk of heart disease (Food and Drug Administration 1999). Moreover,

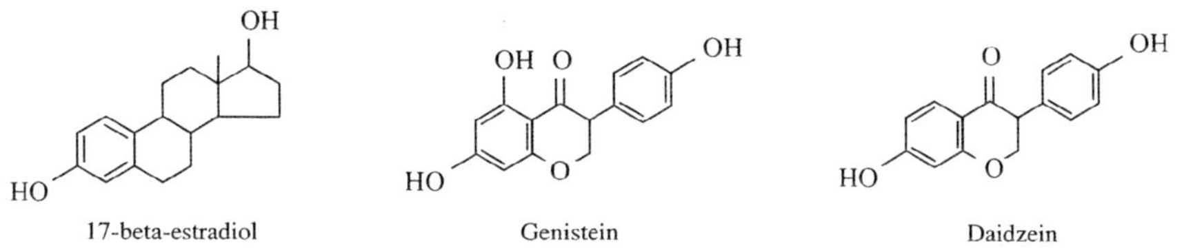

Figure 34.1 The soy isoflavones genistin and daidzin and their biologically active metabolites (genistein and daidzein) have similar structure to endogenous estrogen.

soy milk is the preferred milk product for individuals suffering from lactose intolerance.

Recent observations from laboratory studies indicating that prolonged exposures to estrogenic compounds caused developmental anomalies have brought attention to the biological actions of soy isoflavones in male reproductive tract tissues. Whereas reduced semen quality and fertility in laboratory species have been associated with several synthetic chemicals [*e.g.*, the plasticizer bisphenol A, phthalates, ethylene oxide, glycol ethers, tobacco smoke, pesticides, vinclozolin, polychlorinated biphenyls (PCBs) and dioxin], there is little information on reproductive toxicity due to consumption of soy-based food products. This Chapter provides an overview of the increasing body of evidence collected mostly from laboratory studies showing that soy isoflavones have the capacity to regulate testicular cells and affect male reproductive function.

34.2 Dietary Sources of Isoflavones

All parts of the population are exposed to isoflavones in the diet. However, the predominant isoflavones in soy beans are present in inactive forms (genistin and daidzin), which are hydrolyzed in the gastrointestinal tract to biologically active metabolites classified as aglycones (genistein and daidzein). Approximately 35% of bottle-fed babies in the United States receive significant amounts of their protein intake from soy-based formulas and are able to process genistein and daidzein as efficiently as adults consuming soy products (Irvine 1998). In this regard, exposure of growing individuals to high doses of isoflavones in the diet are concerning due to perceived sensitivity and vulnerability of developing tissues to the action of hormonally active chemicals. The adult population is exposed to significant amounts of soy isoflavones in the diet. While adults consuming soy-free diets have blood plasma isoflavone levels in the low nanomolar range, these levels are elevated to micromolar concentrations with increasing dietary soy portions intake (King and Bursill 1998).

34.3 Hormonal Regulation of Reproductive Tract Development

The testis is the predominant source of sex steroid hormones (mostly androgens) that support reproductive tract development and maintain the male phenotype and fertility. The testis consists of two main compartments: the interstitium contains hormone-secreting Leydig cells, and the seminiferous tubules containing germ cells at various stages of development in close association with supporting Sertoli cells (Figure 34.2). Leydig cells are the predominant source of male sex hormones, primarily testosterone (T), which is converted, in part, into E2 by aromatase enzyme action or reduced by 5α-reductase enzyme (5αR) activity to form the highly potent androgen dihydrotestosterone (DHT). On the other hand, the ability of the testis to produce sperm is directly related to the number of Sertoli cells. Whereas T stimulates Leydig cell development (autocrine regulation), Sertoli cell function and germ cell development is androgen-dependent (paracrine regulation). The action of DHT is critically required for virilization of the external genitalia, including development of the penis and scrotum, the prostate gland, and, in concert with the insulin-like growth factor 3, facilitates the descent of the testicles into the scrotum just prior to or shortly after birth.

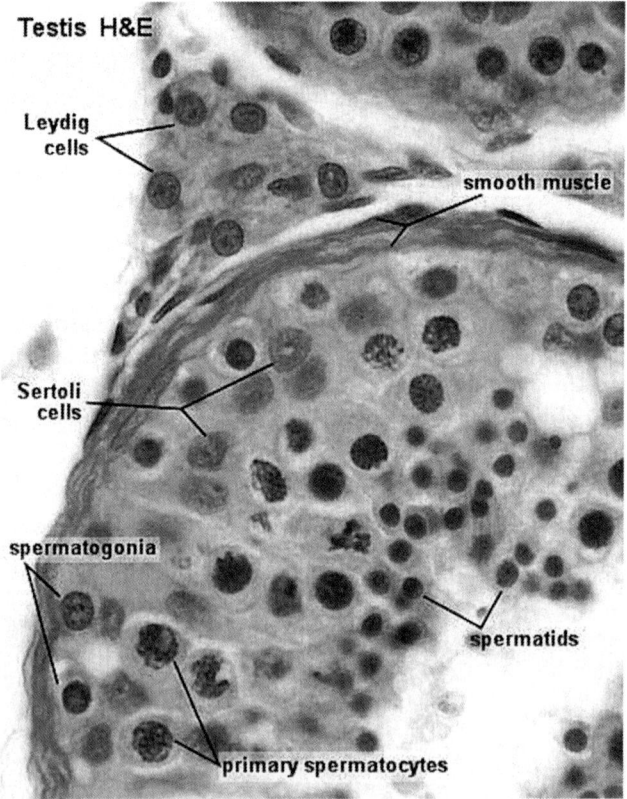

Figure 34.2 The testis consists predominantly of seminiferous tubules separated by small amounts of intertubular tissue or interstitium consisting of loose connective tissue, Leydig cells and blood capillaries.

Steroid hormones (T and E2) bind to their cognate receptors [androgen receptors (ARs) and ESRs], which are expressed in multiple cell types in the testis. The ESR occurs in most tissues as two subtypes: ESR1 and ESR2. However, in the rat, ESR1 is predominantly expressed in Leydig cells, whereas ESR2 is more abundant in Sertoli cells and germ cells in seminiferous tubules (Akingbemi 2005). However, the bulk of experimental evidence suggests that ESR2 is the predominant ESR subtype in the human testis (Saunders *et al.* 2001). More recently, a third ESR subtype (GPR30) was cloned and is thought to be involved in the mediation of estrogenic activity in several tissues (Thomas and Dong 2006). Although sex steroids (androgen and estrogen) are able to act directly in testicular cells, testicular function is primarily regulated by the hypothalamus and pituitary. Because ARs and ESRs are, in turn, expressed in several hypothalamic nuclei and pituitary cells (pituicytes), all levels of the hypothalamus–pituitary–gonadal (HPG) axis are subject to regulation by steroid hormones.

Evidence that estrogen and ESR-mediated activity are required for the physiological function of the reproductive tract was provided initially by observations indicating that fertility was impaired in male mice lacking ESR1 or the aromatase gene (Eddy *et al.* 1996). On the other hand, overexpression of aromatase enzyme increases E2 biosynthesis and the incidence of Leydig cell tumors (Fowler *et al.* 2000). These observations are similar to the finding that sexual impotence was present in males occupationally exposed to synthetic chemicals (Mattison *et al.* 1990). Because differentiation of reproductive tract tissues is regulated by steroid hormones, disruption of androgen production (T and DHT) and/or excessive E2 stimulation will affect reproductive tract function, including development of the external genitalia, descent of the testis into the scrotum and germ cell development and sperm production. The spectrum of testicular anomalies, which have been linked to environmental chemical exposures, are now collectively designated as testicular dysgenesis syndrome (TDS) (Bay *et al.* 2006).

34.4 Regulation of Gonadal Function

The gonadotropic hormones, *i.e.*, follicle stimulating hormone (FSH) and luteinizing hormone (LH), are the primary factors regulating testicular function and are secreted by the anterior portion or glandular part of the pituitary gland (adenohypophysis) (Figure 34.3). FSH receptors are expressed exclusively in Sertoli cells, whereas LH binding sites are located solely in Leydig cells. Somatic Sertoli cells provide physical and nutritional support to developing germ cells. The germ cells go through a series of developmental stages to form mature spermatozoa, and are arranged from the basement membrane toward the lumen of seminiferous tubules in a definite series: spermatogonia, spermatocytes, round and elongated spermatids, and spermatozoa. Although FSH is the primary regulator of spermatogenesis, T is required for spermatid development and maturation. Therefore, FSH acts synergistically with T to

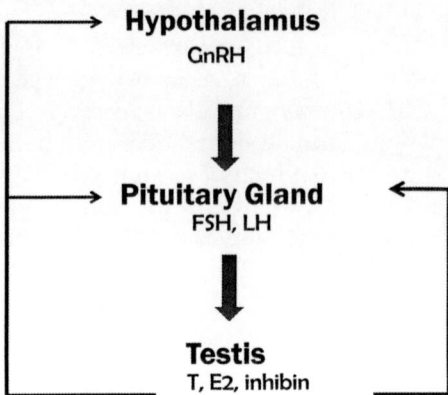

Figure 34.3 The hypothalamic–pituitary–gonadal (HPG) axis is integrated by gonadotropin releasing hormone (GnRH), FSH and LH, and gonadal products (T, E2 and inhibin).

increase the overall efficiency of the process of spermatogenesis. In the absence of T- or AR-mediated stimulation, germ cell development and the process of sperm release from Sertoli cells are impaired.

Adequate T production during embryonic development is critical for differentiation of the male reproductive tract and reaches peak levels by gestational day 18 in the rat and after the 8th week of gestation in humans. In general, adult Leydig cells develop from mesenchymal cells that commit to the Leydig cell lineage early in the neonatal period. In the rat testis, precursor Leydig cells are recognizable by postnatal day 14 and differentiate through an intermediate stage (immature Leydig cells), which transforms into highly steroidogenic adult Leydig cells by 56 days of age. The activity of several testicular enzymes [3β-hydroxysteroid dehydrogenase (HSD3B), 17α-hydroxylase/C17–20 lyase (CYP17A1) and 17β-hydroxysteroid dehydrogenase (HSD17B3)] increase gradually from day 20 postpartum and reach high levels by day 60. Testicular 5-alpha reductase activity, on the other hand, increases between days 20 and 40 and begins to decline between days 40 and 60. After secretion, T diffuses into the interstitial space to enter capillaries for distribution to peripheral tissues, but the greater portion gains access into seminiferous tubules to support the process of spermatogenesis.

34.5 Isoflavone Action in Testicular Cells

Multiple cell types in the testis express ESRs, and it is not surprising that genistein and daidzein, which are able to bind and activate ESRs, act directly in testicular cells. Overall, the bulk of experimental evidence infers that the action of isoflavones in testicular cells is mostly due to their estrogenic properties.Exposures to compounds with estrogenic activity have been shown to directly affect proliferation of Leydig cells (Abney and Myers 1991) and Sertoli cells (Atanassova et al. 1999) or act through ESR-mediated activity in the hypothalamus and pituitary gland to regulate gonadotropin secretion (Pinilla et al. 1992). For example, studies in a transgenic mouse model expressing the luciferase reporter linked to ESR transcriptional activity showed that feeding of genistein at 0, 5, 50, 500 and 5000 µg (kg of diet)$^{-1}$ induced ESR-mediated activity in a dose-dependent manner and was abrogated by co-treatment with the pure anti-estrogen ICI 182,780 (Montani et al. 2009).

Given extensive paracrine relationships among testicular cells, which integrate Sertoli cell–germ cell interactions with Leydig cell function, the action of isoflavones in any cell type invariably affects other cells. Paracrine regulators in the testis include T, Mullerian inhibiting substance (MIS), transferin and serum hormone binding globulin (SHBG). For example, secretion of MIS by Sertoli cells is subject to estrogen regulation. The MIS ligand acts through its receptors to prevent development of embryonic tissues that would otherwise form the female reproductive tract and regulates germ cell development and androgen secretion by Leydig cells in the postnatal period (Behringer et al. 1994). Moreover, there is evidence that estrogenic activity due to isoflavones is mediated in part by other transcription factors, e.g., growth factor receptors, which are known to cross-talk with ESRs (Cotroneo et al. 2005).

34.5.1 Effects on Steroidogenesis

ESR-mediated inhibition of androgen secretion appears to be conserved among mammalian species as indicated by recent observations that genistein caused direct inhibition of *Hsd17b3* activity both in the human and rat testis. ESR1-mediated activity was also specifically linked to inhibition of StAR function and *Hsd17b3* gene expression (Hu et al. 2010). Thus, there seems to be no ambiguity that soy isoflavones exert inhibitory effects on steroidogenic enzyme capacity in Leydig cells (Sherrill et al. 2010). For example, testicular T production was found to be decreased after male rats were administered moderate doses of daidzein (Huang et al. 2008). Similarly, perinatal exposure of prepubertal mice at 21 and 40 days of age to genistein [5 or 300 mg (kg feed)$^{-1}$] resulted in urogenital abnormalities, including decreased anogenital distance (AGD) and testis size, delayed preputial separation and decreased T concentration in adulthood (Wisniewski et al. 2003). The AGD is a known marker of reduced peripheral androgen action. Also, male rats receiving daily subcutaneous administration of 5000 µg of genistein from gestation days 16 to 20 exhibited decreased AGD (Levy et al. 1995). Perinatal exposures to 1000 mg of isoflavones (kg of diet)$^{-1}$ was found to increase serum T levels, but exposures to genistein at 12.5, 25, 50 and 100 mg (kg of body weight)$^{-1}$ in the neonatal period did not affect serum T levels in adult rats (Nagao et al. 2001). There is little or no information on the effects of phytoestrogens on testis function in farm animals. This lack of interest may be related to ruminal microbial action, which is thought to interfere with the bioavailability and pharmacokinetics of isoflavones in the blood. However, administration of phytoestrogen tablets mimicking varying levels of isoflavones in soy beans (4 mg of genistein, 3.5 of daidzein, 24.5 mg of biochanin A and 8 mg of formononetin) to 3-month-old goat kids increased serum T levels (Gunnarsson et al. 2009).

A critical evaluation of the literature implies discordant findings on serum T levels from several studies. These disparities are possibly related to the effects of isoflavones on cellular proliferation. In this regard, intense mitotic activity in Leydig cells occurs predominantly in the prepubertal and pubertal periods of development. There seems to be a general agreement that low doses of genistein ($\leq 10\,\mu M$) induce cell proliferation, whereas higher doses inhibit cell division and growth in estrogen-sensitive tissues. Since serum T levels represent the sum total of androgen secretion and is affected by the number of Leydig cells, the apparent discrepancy between decreased androgen secretion by Leydig cells and unchanged, and sometimes elevated, serum T levels may be related to changes in the population of Leydig cells. Indeed, results from our laboratory have demonstrated that unaltered or increased serum T levels occurring simultaneously with decreased steroidogenic enzyme activity indicates is associated with a greater number of Leydig cells contributing to the androgen pool.Consistent with these observations, the population of Leydig cells was increased in the testis of non-human primates (marmoset monkeys) raised on soy-based formula (Tan et al. 2006). Thus, it appears that deficits in androgen secretion by Leydig cells are alleviated by increased Leydig cell numbers and possibly alleviate androgen deficits and/or augment serum T levels (Figure 34.4). Therefore, interpretation of data describing measurements of serum T levels after exposures to estrogenic agents need to take into account the dose and period of exposure and effects on Leydig cell numbers.

34.5.2 Effects on Sperm Production

Perhaps the most demonstrated effect of isoflavones is their capacity to interfere with germ cell development and sperm production. For example, the

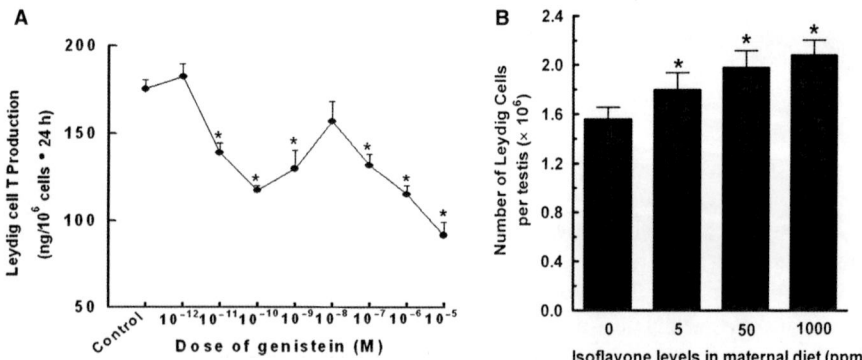

Figure 34.4 Treatment of immature Leydig cells with genistein decreased androgen secretion (A). Exposure of male rats to soy isoflavones in maternal diet in the perinatal period increased Leydig cell numbers in adulthood (B). [Akingbemi et al. (2007) Endocrinology, 148: 4475–4488; Sherrill et al. (2010) Biol. Reprod., 83: 488–501].

number of sperm in the testis was reduced in adult male rats maintained on a high phytoestrogen diet for a period of 24 days, and this effect was linked to a decrease in the numbers of round and elongated spermatids (Assinder et al. 2007). Similarly, adult male mice fed a soy-rich diet (150 ppm of daidzein and 190 ppm of genistein equivalents) exhibited normal male behavior and were fertile, but the number of sperm in the epididymis was decreased (Cederroth et al. 2010) (Figure 34.5). In contrast, gestational and lactational exposures to low doses of genistein (0, 0.1, 0.5, 2.5 and 10 mg kg^{-1} day^{-1}) did not affect sperm count and motility but sperm fertilizing ability was increased at the highest dose in sexually mature mice (Fielden et al. 2003).

Results from other studies indicated that administration of genistein (0, 5, 25, 100, 250, 625 or 1250 ppm) to pregnant dams starting on gestational day 7 through pregnancy and lactation and until day 50 postpartum decreased spermatogenesis in the 1250 ppm diet group (Delclos et al. 2001). Similarly, feeding of a diet containing 225 mg g^{-1} genistein, 180 mg g^{-1} daidzin and 60 mg g^{-1} glycitein caused a decrease in testicular and epididymal sperm numbers in adult male rats (Assinder et al. 2007). Also, sperm production and serum T levels were both decreased after life-time (chronic) exposures of male rats to a low dose genistein (1 mg kg^{-1} day^{-1}) (Eustache et al. 2009). Altogether, these observations imply that both low and high levels of dietary phytoestrogens affect germ cell development and sperm production.

34.5.3 Effects on Developing *versus* Mature Stages of Development

Several reports reinforce the view that agents with estrogenic activity cause greater effects in growing individuals than in the adult. For example, feeding of genistein to weanling ICR mice (2.5 and 5.0 mg kg^{-1} day^{-1}) for 5 weeks caused

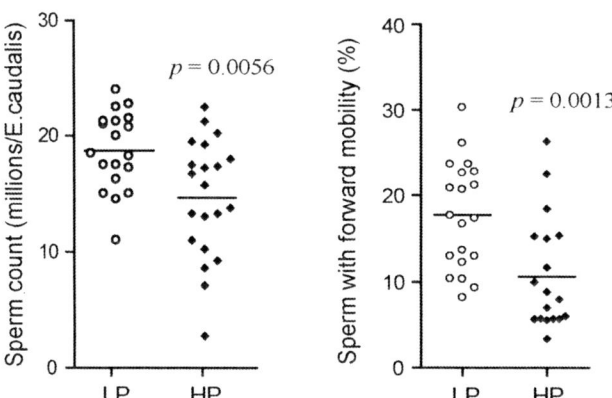

Figure 34.5 Feeding on a soy-based diet affects sperm count and motility in mice. HP, high phytoestrogen-fed mice; LP, low phytoestrogen-fed mice.
[Cederroth et al. (2010) *Mol. Cell. Endocrinol.* 321: 152–160]. (Used with permission.)

Leydig cell hyperplasia and decreased the number of sperm in the epididymis (Lee et al. 2004a), but the same exposure paradigm exerted no effects in adult animals (Lee et al. 2004b). Similarly, prenatal or prenatal plus adult exposures to low and high phytoestrogen doses caused a 25% reduction in epididymal sperm count and reduced FSH levels in 6-month-old CD-1 mice (Cederroth et al. 2010). In contrast, feeding of pregnant dams with 20, 200 and 1000 ppm of genistein in the perinatal period was found not to affect reproductive tract parameters measured in adult male rats (Masutomi et al. 2003), just as there were no adverse effects on reproductive tract parameters after a life-time feeding of soy-based diets to adult male rabbits (Cardoso and Báo 2009). Also, adult male rats fed a soy isoflavone mixture containing 45% genistein, 23% daidzein and 4% glycitein at 200 or 2000 mg (kg of diet)$^{-1}$ for 12 months showed no gross toxicity and with no effects on body, testis and epididymal weights, and sperm production and morphology (Faqi et al. 2004). Nevertheless, the preponderance of these conflicting reports seems to indicate that the adult compared with the developing reproductive tract is relatively insensitive to isoflavone toxicity. The discrepant observations from different laboratories are probably due to differences in experimental design as related to animal species and strains, dose and duration of exposure to isoflavones, and stage of reproductive tract development.

34.6 Mechanisms of Isoflavone Action

The action of isoflavones in testicular cells has been attributed to several mechanisms, including interference with the LH-stimulated signaling pathway or other signaling molecules regulating androgen biosynthesis, germ cell development and metamorphosis. For example, decreased androgen biosynthesis was linked to interference with cholesterol availability because perinatal exposures of male rats to isoflavones decreased StAR phosphorylation. The decrease in the rate of StAR phosphorylation was related to the capacity of genistein to uncouple LH from its receptors in Leydig cells. Also, genistein at 50 µM was found to decrease cAMP-stimulated progesterone secretion and inhibited StAR promoter activity in transiently transfected MA-10 tumor Leydig cells. In other studies, neonatal transient exposure of mice to genistein at 10, 100 or 1000 µg per mouse for up to 5 days of age decreased testicular ESR1 and AR mRNA expression levels in adulthood, indicating that neonatal exposure to genistein impacts molecular pathways mediating steroid hormone regulation of testicular function for extended periods of time (Shibayama et al. 2001).

Interference with growth factor stimulation of gonocyte development has been identified as a potential mechanism of isoflavone action. Gonocytes are precursors to spermatogonia in the testis, express platelet-derived growth factor receptor (PDGFR) and proliferate in response to PDGF (Li et al. 1997). There is evidence that isoflavones regulate early germ cell development through the PDGFR because expression of PDGFR-α and -β proteins were up-regulated in the rat testis after neonatal treatment with genistein (0.1–10 mg kg^{-1} day^{-1}) (Thuillier *et al.* 2003). Other reports have suggested that isoflavones induced apoptosis or programmed cell death in spermatocytes and round spermatids, which decreases total germ cell numbers.Similarly, life-time feeding of CD-1 mice with a diet containing 150 ppm of daidzein and 190 ppm of genistein equivalents caused a 30% reduction in the expression of the spermatid specific marker Gapd-s; this protein plays a critical role in spermatozoa glycolysis and is required for sperm motility and fertility (Cederroth *et al.* 2010). Moreover, isoflavones may interfere with sperm fertilizing ability because genistein stimulated or suppressed the acrosome reaction, which is required for sperm capacitation prior to egg fertilization, in a dose- and time-dependent manner (Kumi-Diaka and Townsend 2003).

A major concern with consumption of soy-based food items is the potential for altered serum E2 levels (Figure 34.6) and in the activity of the aromatase enzyme, which converts T into E2 in several tissues, including the testis. For example, exposure of male rats to genistein at 250 and 1000 mg (kg of diet)$^{-1}$ for 15 days was found to suppress testicular aromatase activity in the prepubertal period (Fritz *et al.* 2003). Suppression of E2 biosynthesis early in development potentially exerts adverse effects on spermatogenesis as was seen in transgenic mice deficient in the aromatase gene (ARKO mice) (Robertson *et al.* 1999). The impairment of spermatogenesis seen in ARKO mice maintained on soy-free diets occurred in the absence of decreased gonadotropin secretion, implying that the effects resulted from direct isoflavone action in the

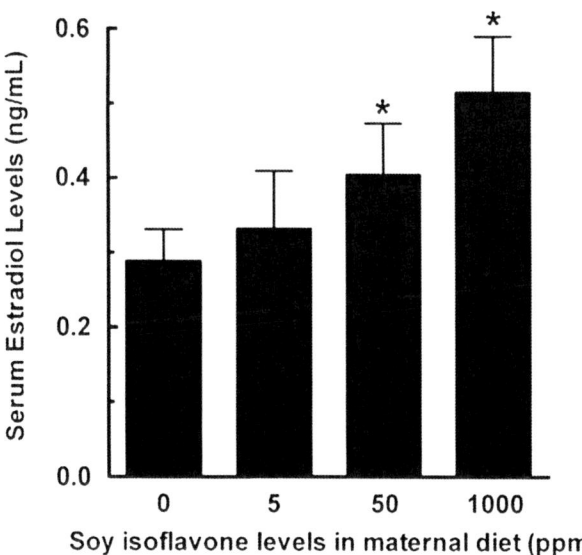

Figure 34.6 Exposure of male rats to soy isoflavones in maternal diet in the perinatal period increased serum estradiol (E2) concentrations in adulthood. [Sherrill *et al.* (2010) *Biol. Reprod.* 83: 488–501].

testis and are independent of the pituitary–gonadal axis. However, observations showing that isoflavones perturb serum E2 levels were collected mostly from studies in laboratory species and there are few reports from human studies. Although one report indicated that serum E2 concentrations were unaffected by consumption of soy products in men (Ostatníková *et al.* 2007), serum E2 levels were increased 4-fold in a male subject presenting with gynaecomastia who had been ingesting large amounts of soy milk in the previous 6 months (Martinez and Lewi 2008).

34.7 Isoflavone Action in the Human Testis

The effects on male fertility have not been described but there is little doubt that isoflavones can regulate biological action in human tissues, including the testis. Although a recent report did not find soy-related differences in reproductive organ size in infants at 4 months of age, testicular development appeared slower in infants raised on milk or soy formula compared with their peers that were raised on breast milk (Gilchrist *et al.* 2010). Studies in older individuals have provided somewhat mixed results. For example, healthy volunteers (18–35 years of age) ingesting a supplement containing 40 mg of isoflavones daily over a 2 month period and achieving blood concentrations of genistein and daidzein at 1 and 0.5 µM, respectively, exhibited no changes in endocrine measurements, testicular volume and semen parameters (Mitchell *et al.* 2001). Similarly, feeding of 27 peripubertal male rhesus monkeys with soy isolate (20% of feed

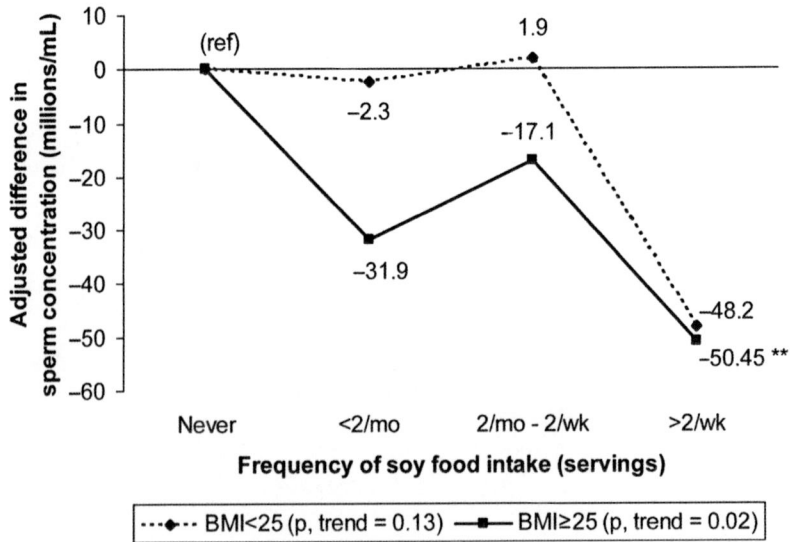

Figure 34.7 There is an indirect relationship between consumption of soy-based diets and sperm count in men. mo, month; wk, week.
[Chavarro *et al.* (2008) *Hum. Reprod.* 23: 2584–2590]. (Used with permission.)

by weight) for 6 months caused no adverse effects on the reproductive tract as reflected in reproductive organ weights and serum T levels (Anthony *et al.* 1996). On the other hand, a recent report concluded that consumption of soy foods was associated with decreased sperm counts in men (Chavarro *et al.* 2008) (Figure 34.7). Of interest, consumption of 40 g of soy protein daily for 3 months increased the serum levels of insulin-like growth factor-I in young and old men (Khalil 2002). Growth factors are known to affect growth and differentiation of several tissues, including the reproductive tract. Obviously, additional studies are warranted to characterize the effects of various exposure paradigms on endocrine function and sperm production capacity in the human testis.

34.8 Conclusion

The potential effects of xenoestrogens and EDs on reproductive health have received a great deal of public and scientific attention because these agents exert their effects largely through steroid hormone receptors. Because steroid hormone receptors are expressed at high levels in reproductive tract tissues, there is public concern with potential effects of environmental chemicals, including dietary estrogens, on reproductive health. Moreover, EDs may impair reproductive activity by indirect action through disruption of diverse metabolic and signaling pathways affecting energy homeostasis, the cardiovascular and immune systems and the brain. Unfortunately, data describing hormone-like activity over a wide spectrum of exposure paradigms for several phytoestrogens and EDs are not available. Therefore, there is need to continue to focus on the biological effects of long- and short-term exposures to low and high doses as well as the effects of simultaneous exposures to multiple chemicals. In addition, analysis of developmental changes in gene expression affecting organ function after exposures to hormonally active agents will facilitate the process of risk assessment of the population and identification of biomarkers. Given the complex paracrine or cell–cell interactions required to optimize endocrine function and sperm production in the testis, future studies investigating regulation of paracrine interactions will provide useful information describing ED effects on overall testis function than their effects in individual cell types. Despite the homology in reproductive physiology among mammalian species, there has been concern about extrapolating data collected in laboratory species and wildlife observations to humans. However, there are difficulties associated with performing human studies related to availability of research materials. Therefore, there is a need to develop experimental strategies to address this research challenge.

The use of soy-based foods and supplements is increasing in the population due to perceived health beneficial effects. However, government regulators are under increasing public pressure to limit the use of soy-based products and supplements in young individuals who are especially sensitive to the hormonal activity of sex steroids. Even if endocrine perturbations are reversed over time, tissue changes due to agents exhibiting estrogenic activity may persist for much longer periods. In order to assist regulatory agencies formulate appropriate policies to safeguard public health, the overall effects of soy-based products and supplements on the endocrine axis warrant continued investigation for the near future.

Summary Points

- Several chemicals present in the environment interfere with endocrine function.
- Soy isoflavones bind to estrogen receptors (ESRs) to exert estrogenic activity in testicular cells.
- Soy isoflavones act in testicular Leydig cells to regulate androgen secretion.
- Exposures to soy isoflavones may impair germ cell development and sperm production.
- The effects of isoflavone action in the human testis have not been determined.

Key Facts

- Infants are exposed to soy isoflavones mostly in soy-based infant formulas.
- The population feeds on soy protein present in baked food items, including breakfast cereals, pasta, beverages, toppings, meat, poultry and fish products.
- Soy supplements are marketed for their putative beneficial effects on health, such as improved cardiovascular health and treatment of menopausal symptoms.

Definitions of Words and Terms

Autocrine regulation: The regulation of a tissue by the action of its own product, *e.g.*, regulation of Leydig cells by testosterone.

Dietary phytoestrogens: Estrogenic compounds of plant origin.

Endocrine axis: Organ systems affected by the action of hormones.

Endocrine disruptors (EDs): Substances with the capacity to interfere with, mimic or antagonize endogenous hormonal activity and homeostasis.

Interstitial tissue: The smaller proportion of the testis (approx. 5%) containing loose connective tissue and Leydig cells.

Paracrine regulation: The regulation of a tissue by action of products from another tissue, *e.g.*, regulation of Sertoli cells by testosterone from Leydig cells.

Seminiferous tubules: The greater proportion of the testis (approx. 95%) containing Sertoli and germ cells.
Testicular dysgenesis syndrome (TDS): Multiple developmental anomalies due to action of environmental factors.
Virilization of external genitalia: The development and hormonal imprinting of the external genitalia.
Xenoestrogens: Chemicals in the environment (food, air or water) which cause estrogenic activity when ingested.

List of Abbreviations

AGD	anogenital distance
AR	androgen receptor
DHT	dihydrotestosterone
E2	17β-estradiol
ED	endocrine disruptor
ESR	estrogen receptors
FSH	follicle stimulating hormone
HSD17B3	17β-hydroxysteroid dehydrogenase
LH	luteinizing hormone
MIS	Mullerian inhibiting substance
PDGFR	platelet-derived growth factor receptor
T	testosterone
5αR	5α-reductase

References

Abney, T.O., and Myers, R.B., 1991. 17β-Estradiol inhibition of Leydig cell regeneration in the ethane dimethylsulfonate-treated mature rat. *Journal of Andrology*. 12: 295–304.

Akingbemi, B.T., 2005. Estrogen regulation of testicular function. *Reproduction Biology and Endocrinology*. 3: 51.

Akingbemi, B.T., Braden, T.D., Kemppainen, B.W., Hancock, K.D., Sherrill, J.D., Cook, S.J., He, X., and Supko, J.G., 2007. Exposure to phytoestrogens in the perinatal period affects androgen secretion by testicular Leydig cells in the adult rat. *Endocrinology*. 148: 4475–4488.

Anthony, M.S., Clarkson, T.B., Hughes, Jr, C.L., Morgan, T.M., and Burke, G.L., 1996. Soybean isoflavones improve cardiovascular risk factors without affecting the reproductive system of peripubertal rhesus monkeys. *Journal of Nutrition*. 126: 43–50.

Assinder, S., Davis, R., Fenwick, M., and Glover, A., 2007. Adult-only exposure of male rats to a diet of high phytoestrogen content increases apoptosis of meiotic and post-meiotic germ cells. *Reproduction*. 133: 11–19.

Atanassova, N., McKinnell, C., Walker, M., Turner, K.J., Fisher, J.S., Morley, M., Millar, M.R., Groome, N.P., and Sharpe, R.M., 1999.

Permanent effects of neonatal estrogen exposure in rats on reproductive hormone levels, Sertoli cell number, and the efficiency of spermatogenesis in adulthood. *Endocrinology.* 140: 5364–5373.

Bay, K., Asklund, C., Skakkebaek, N.E., and Andersson, A.M., 2006. Testicular dysgenesis syndrome: possible role of endocrine disrupters. *Best Practice and Research. Clinical Endocrinology and Metabolism.* 20: 77–90.

Behringer, R.R., Finegold, M.J., and Cate, R.L., 1994. Mullerian-inhibiting substance function during mammalian sexual development. *Cell.* 79: 415–425.

Cardoso, J.R., and Báo, S.N., 2009. Morphology of reproductive organs, semen quality and sexual behaviour of the male rabbit exposed to a soy-containing diet and soy-derived isoflavones during gestation and lactation. *Reproduction in Domestic Animals.* 44: 937–942.

Cederroth, C.R., Zimmermann, C., Beny, J.L., Schaad, O., Combepine, C., Descombes, P., Doerge, D.R., Pralong, F.P., Vassalli, J.D., and Nef, S., 2010. Potential detrimental effects of a phytoestrogen-rich diet on male fertility in mice. *Molecular and Cellular Endocrinology.* 321: 152–160.

Chavarro, J.E., Toth, T.L., Sadio, S.M., and Hauser, R., 2008. Soy food and isoflavone intake in relation to semen quality parameters among men from an infertility clinic. *Human Reproduction.* 23: 2584–2590.

Cotroneo, M.S., Fritz, W.A., and Lamartiniere, C.A., 2005. Dynamic profiling of estrogen receptor and epidermal growth factor signaling in the uteri of genistein- and estrogen-treated rats. *Food and Chemical Toxicology.* 43: 637–645.

Delclos, K.B., Bucci, T.J., Lomax, L.G., Latendresse, J.R., Warbritton, A., Weis, C.C., and Newbold, R.R., 2001. Effects of dietary genistein exposure during development on male and female CD (Sprague–Dawley) rats. *Reproductive Toxicology.* 15: 647–663.

Eddy, E.M., Washburn, T.F., Bunch, D.O., Goulding, E.H., Gladen, B.C., Lubahn, D.B., and Korach, K.S., 1996. Targeted disruption of the estrogen receptor gene in male mice causes alteration of spermatogenesis and infertility. *Endocrinology.* 137: 4796–4805.

Eustache, F., Mondon, F., Canivenc-Lavier, M.C., Lesaffre, C., Fulla, Y., Berges, R., Cravedi, J.P., Vaiman, D., and Auger, J., 2009. Chronic dietary exposure to a low-dose mixture of genistein and vinclozolin modifies the reproductive axis, testis transcriptome, and fertility. *Environmental Health Perspectives.* 117: 1272–1279.

Faqi, A.S., Johnson, W.D., Morrissey, R.L., and McCormick, D.L., 2004. Reproductive toxicity assessment of chronic dietary exposure to soy isoflavones in male rats. *Reproductive Toxicology.* 18: 605–611.

Fielden, M.R., Samy, S.M., Chou, K.C., and Zacharewski, T.R., 2003. Effect of human dietary exposure levels of genistein during gestation and lactation on long-term reproductive development and sperm quality in mice. *Food and Chemical Toxicology.* 41: 447–454.

Food and Drug Administration, 1999. Food labeling: health claims: soy protein and coronary heart disease. 21 CFR Part 101: (Docket No. 98-0683).

Fowler, K.A., Gill, K., Kirma, N., Dillehay, D.L., and Tekmal, R.R., 2000. Overexpression of aromatase leads to development of testicular leydig cell tumors: an *in vivo* model for hormone-mediated testicular cancer. *American Journal of Pathology*. 156: 347–353.

Fritz, W.A., Cotroneo, M., Wang, J., Eltoum, I.E., and Lamartiniere, C.A., 2003. Dietary diethylstilbestrol but not genistein adversely affects rat testicular development. *Journal of Nutrition*. 133: 2287–2293.

Gilchrist, J.M., Moore, M.B., Andres, A., Estroff, J.A., and Badger, T.M., 2010. Ultrasonographic patterns of reproductive organs in infants fed soy formula: comparisons to infants fed breast milk and milk formula. *Journal of Pediatrics*. 156: 215–220.

Gunnarsson, D., Selstam, G., Ridderstråle, Y., Holm, L., Ekstedt, E., and Madej, A., 2009. Effects of dietary phytoestrogens on plasma testosterone and triiodothyronine (T3) levels in male goat kids. *Acta Veterinaria Scandinavica*. 51: 51.

Hu, G.-X., Zhao, B.-H., Chu, Y.-H., Zhou, H.-Y., Akingbemi, B.T., Zheng, Z.-Q., and Ge, R., 2010. Effects of genistein and equol on human and rat testicular 3β-hydroxysteroid dehydrogenase and 17β-hydroxysteroid dehydrogenase 3 activities. *Asian Journal of Andrology*. 12: 519–526.

Huang, Y., Pan, L., Xia, X., Feng, Y., Jiang, C., and Cui, Y., 2008. Long-term effects of phytoestrogen daidzein on penile cavernosal structures in adult rats. *Urology*. 72: 220–224.

Irvine, C.H., 1998. Phytoestrogens in soy-based infant foods: concentrations, daily intake and possible biological effects. *Proceedings of the Society for Experimental Biology and Medicine*. 217: 247–253.

Khalil, D., 2002. Soy protein supplementation increases serum insulin-like growth factor-1 in young and old men but does not affect markers of bone metabolism. *Journal of Nutrition*. 32: 2605–2608.

King, R.A., and Bursill, D.B., 1998. Plasma and urinary kinetics of the isoflavones daidzein and genistein after a single soy meal in humans. *American Journal of Clinical Nutrition*. 67: 867–872.

Kumi-Diaka, J., and Townsend, J., 2003. Toxic potential of dietary genistein isoflavone and beta-lapachone on capacitation and acrosome reaction of epididymal spermatozoa. *Journal of Medicinal Food*. 6: 201–208.

Lee, B.J., Jung, E.Y., Yun, Y.W., Kang, J.K., Baek, I.J., Yon, J.M., Lee, Y.B., Sohn, H.S., Lee, J.Y., Kim, K.S., and Nam, S.Y., 2004a. Effects of exposure to genistein during pubertal development on the reproductive system of male mice. *Journal of Reproduction and Development*. 50: 399–409.

Lee, B.J., Kang, J.K., Jung, E.Y., Yun, Y.W., Baek, I.J., Yon, J.M., Lee, Y.B., Sohn, H.S., Lee, J.Y., Kim, K.S., and Nam, S.Y., 2004b. Exposure to genistein does not adversely affect the reproductive system in adult male mice adapted to a soy-based commercial diet. *Journal of Veterinary Science*. 5: 227–234.

Levy, J.R., Faber, K.A., Ayyash, L., and Hughes, Jr, C.L., 1995. The effect of prenatal exposure to the phytoestrogen genistein on sexual differentiation in rats. *Proceedings of the Society for Experimental Biology and Medicine*. 208: 60–66.

Li, H., Papadopoulos, V., Vidic, B., Dym, M., and Culty, M., 1997. Regulation of rat testis gonocyte proliferation by platelet-derived growth factor and estradiol: identification of signaling mechanisms involved. *Endocrinology*. 138: 1289–1298.

Martinez, J., and Lewi, J., 2008. An unusual case of gynecomastia associated with soy product consumption. *Endocrine Practice*. 14: 415–418.

Masutomi, N., Shibutani, M., Takagi, H., Uneyama, C., Takahashi, N., and Hirose, M., 2003. Impact of dietary exposure to methoxychlor, genistein, or diisononyl phthalate during the perinatal period on the development of the rat endocrine/reproductive systems in later life. *Toxicology*. 192: 149–170.

Mattison, D.R., Plowchalk, D.R., Meadows, M.J., al-Juburi, A.Z., Gandy, J., and Malek, A., 1990. Reproductive toxicity: male and female reproductive systems as targets for chemical injury. *Medical Clinics of North America*. 74: 391–411.

Mitchell, J.H., Cawood, E., Kinniburgh, D., Provan, A., Collins, A.R., and Irvine, D.S., 2001. Effect of a phytoestrogen food supplement on reproductive health in normal males. *Clinical Science*. 100: 613–618.

Montani, C., Penza, M., Jeremic, M., Rando, G., Ciana, P., Maggi, A., La Sala, G., De Felici, M., and Di Lorenzo, D., 2009. Estrogen receptor-mediated transcriptional activity of genistein in the mouse testis. *Annals of the New York Academy of Sciences*. 1163: 475–477.

Nagao, T., Yoshimura, S., Saito, Y., Nakagomi, M., Usumi, K., and Ono, H., 2001. Reproductive effects in male and female rats of neonatal exposure to genistein. *Reproductive Toxicology*. 15: 399–411.

Ostatníková, D., Celec, P., Hodosy, J., Hampl, R., Putz, Z., and Kúdela, M., 2007. Short-term soybean intake and its effect on steroid sex hormones and cognitive abilities. *Fertility and Sterility*. 88: 1632–1636.

Pinilla, L., Garmelo, P., Gaytan, F., and Aguilar, E., 1992. Hypothalamic-pituitary function in neonatally oestrogen-treated male rats. *Journal of Endocrinology*. 134: 279–286.

Robertson, K.M., O'Donnell, L., Jones, M.E., Meachem, S.J., Boon, W.C., Fisher, C.R., Graves, K.H., McLachlan, R.I., and Simpson, E.R., 1999. Impairment of spermatogenesis in mice lacking a functional aromatase (cyp 19) gene. *Proceedings of the National Academy of Science USA*. 96: 7986–7991.

Saunders, P.T., Sharpe, R.M., Williams, K., Macpherson, S., Urquart, H., Irvine, D.S., and Millar, M.R., 2001. Differential expression of oestrogen receptor alpha and beta proteins in the testes and male reproductive system of human and non-human primates. *Molecular Human Reproduction*. 7: 227–236.

Shibayama, T., Fukata, H., Sakurai, K., Adachi, T., Komiyama, M., Iguchi, T., and Mori, C., 2001. Neonatal exposure to genistein reduces expression of estrogen receptor alpha and androgen receptor in testes of adult mice. *Endocrinology Journal*. 48: 655–663.

Sherrill, J.D., Sparks, M., Dennis, J., Mansour, M., Kemppainen, B.W., Bartol, F.F., Morrison, E.E., and Akingbemi, B.T., 2010. Developmental exposures of male rats to soy isoflavones impact Leydig cell differentiation. *Biology of Reproduction*. 83: 488–501.

Tan, K.A., Walker, M., Morris, K., Greig, I., Mason, J.I., and Sharpe, R.M., 2006. Infant feeding with soy formula milk: effects on puberty progression, reproductive function and testicular cell numbers in marmoset monkeys in adulthood. *Human Reproduction*. 21: 896–904.

Thomas, P., and Dong, J., 2006. Binding and activation of the seven-transmembrane estrogen receptor GPR30 by environmental estrogens: a potential novel mechanism of endocrine disruption. *Journal of Steroid Biochemistry and Molecular Biology*. 102: 175–179.

Thuillier, R., Wang, Y., and Culty, M., 2003. Prenatal exposure to estrogenic compounds alters the expression pattern of platelet-derived growth factor receptors alpha and beta in neonatal rat testis: identification of gonocytes as targets of estrogen exposure. *Biology of Reproduction*. 68: 867–880.

Wisniewski, A.B., Klein, S.L., Lakshmanan, Y., and Gearhart, J.P., 2003. Exposure to genistein during gestation and lactation demasculinizes the reproductive system in rats. *Journal of Urology*. 169: 1582–1586.

CHAPTER 35
Equol and Cell Proliferation

ZHONG LI,* CAIYUN ZHONG AND CHUNYAN HU

Department of Nutrition and Food Hygiene, School of Public Health, Nanjing Medical University, 140 Hanzhong Rd, Nanjing 210029, People's Republic of China
*Email: uiuclz@126.com

35.1 Introduction

Phytoestrogens are plant-derived hundreds of molecules that structurally and functionally mimic the effects of estrogen; among them soy-derived isoflavones are the most abundant ones. The major soy isoflavonoids are genistein and daidzein, while equol is produced from daidzein by enteric bacterial metabolism. Unlike genistein and daidzein, equol is a chiral molecule and exists as two distinct enantiomeric forms, S-equol and R-equol. In humans, metabolism of daidzein to equol results in the production of S-equol only.

The chemical structure of equol was first identified by Marrian and Haslewood (1932). Equol can be produced in majority of animals, including rodents. However, only approximately 30–50% of individuals who consume soy foods produce equol. As a person's ability to produce equol is stable over time, individuals can be classified as either equol "producers" or "non-producers". Maximal responses to isoflavones' intake are observed in equol producers, who are at lower risk of breast cancer than equol non-producers.

Evidence has shown that equol has structural similarity to mammalian estrogen, binding and transactivating estrogen receptors (ERs), inducing proliferation in estrogen-dependent endometrial and breast tumor cells in culture, and eliciting estrogenic effects in rodent models. These studies have raised concern that high levels of equol intake may promote estrogen-dependent tumors in women. Here, we will overview the *in vitro* data in the context of evaluating equol as a possible compound promoting or inhibiting cancer growth. We will review, at the molecular level, the receptor binding properties of equol and its effects on cancer cell models. Other *in vitro* effects of equol will also be discussed, such as on nervous system, reproductive system, immune system, cardiovascular diseases (CVDs) and drug metabolism; these effects are considered as beneficial health effects.

35.2 Equol and Hormone-positive Cell Proliferation

35.2.1 Equol Binds to ERs with a Greater Affinity than Daidzein

The phenolic ring structures of isoflavones enable these compounds to bind ERs and exert endocrine-disrupting effects. It is well established that, due to their ability to mimic the structural confirmation of estradiol, genistein and other isoflavones have a higher affinity for ERβ than for ERα. The binding affinity of equol for human ERα and ERβ was found to be similar to that of genistein, but equol-induced transcription is stronger than its precursor daidzein, especially with ERβ. At non-cytotoxic concentrations, equol exhibited estrogenic effects with no significant antiestrogenic effects observed in the cells (Lehmann *et al.* 2005). However, daidzein showed poor affinity and transcriptional activity in these *in vitro* systems (Morito *et al.* 2001).

S-equol is the only equol isomer synthesized by human intestinal bacteria after soy consumption (Figure 35.1) (Setchell *et al.* 2005). S- and R-equol show very different behavior in terms of their binding affinities with ERα and ERβ. S-equol has a high affinity for binding to ERβ, whereas R-equol binds weakly to ERβ and with a preference for ERα. All equol isomers have higher affinity for both ERs than that of the biosynthetic precursor daidzein (Muthyala *et al.* 2004).

Equol also acts as ligand for other nuclear receptors than ERs such as estrogen-related receptor γ (ERRγ). ERRγ is an orphan nuclear receptor lacking identified natural ligands. Equol acts as an ERRγ agonist and stimulates the transcriptional activity of ERRγ (Hirvonen *et al.* 2011).

35.2.2 Estrogenic Activity and Breast Cancer

Matsumura *et al.* (2005) made a comparison between the estrogenic activities of eight phytoestrogens (genistein, daidzein, equol, miroestrol, deoxymiroestrol, 8-prenylnaringenin, coumestrol and resveratrol) in a variety of assays all based on the same MCF-7 cell system (Figure 35.2). Apart from resveratrol, seven of the eight phytoestrogens elicited similar maximal responses to that produced by 17β-estradiol (E2) (Matsumura *et al.* 2005).

Equol exerted estrogenic responses in cells derived from breast and ovary. Equol induced cell proliferation in the estrogen-sensitive breast cancer cell line

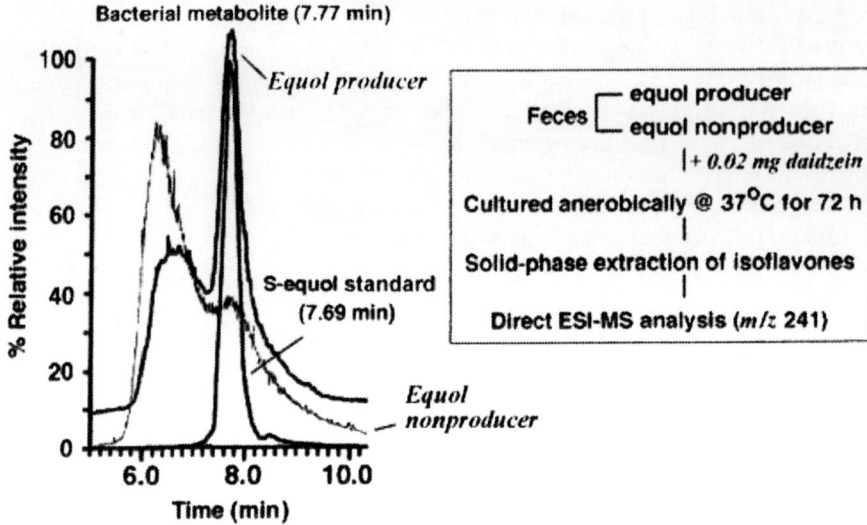

Figure 35.1 Definitive evidence for the enantiomer-specific synthesis of S-equol by cultured human fecal flora. Mass chromatograms obtained by using chiral-phase high-performance liquid chromatography (HPLC)–mass spectrometric analysis of a pure standard of S-equol (bottom trace) were compared with extracts from *in vitro* bacterial metabolism of daidzein by cultured fecal flora from a known equol producer (top trace) and an equol non-producer (middle trace). ESI-MS, electrospray ionization–mass spectrometry.
Data are from Setchell *et al.* (2005), with permission from the Publishers.

MCF-7 and the human ovarian cancer cell line BG-1, but not in the receptor-negative human breast cancer cell line MDA-MB-231. Equol exerted cell proliferation inhibiting effects at the highest concentration (10 µM) in MDA-MB-231 (Schmitt *et al.* 2001). Equol induced cell apoptosis (Figure 35.3) *via* caspase-9 and cytochrome *c*, independent of caspase-8, in human breast cancer MDA-MB-453 cells (Choi *et al.* 2009).

We also evaluated the effects of genistein and equol on cell proliferation and ERα transactivation *in vitro*. In MCF-7 human breast cancer cells, low concentrations of genistein and equol enhanced proliferation and induced MCF-7 cells to enter S-phase. Inhibition of extracellular-signal- regulated kinase 1/2 (ERK1/2) phosphorylation by U0126 led to complete suppression of genistein- and equol-induced estrogen response element reporter activity and suppression of the estrogen-responsive gene *pS2*. These results suggest that ERK1/2 activation is necessary for the transactivation of ERα by genistein and equol in estrogen-positive cancer cells (Liu *et al.* 2010).

Jiang *et al.* (2008) reported that equol induced the proteinase inhibitor 9 (PI-9) *via* ERα and inhibited NK92-induced apoptosis of MCF-7 breast cancer cells, although this ability was weaker than genistein. Thus equol is probably sufficient to reduce the ability of cytolytic lymphocytes to kill target breast

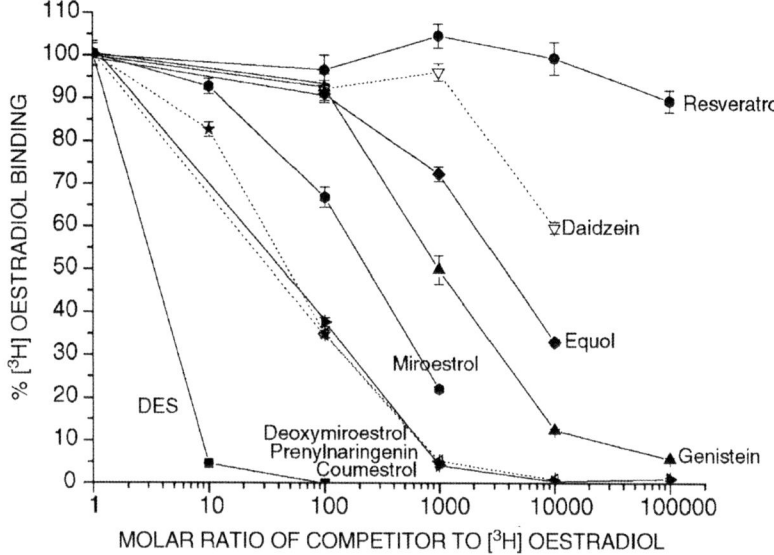

Figure 35.2 Competitive binding of eight phytoestrogens to ERα from MCF7 human breast cancer cells. In single point competitive binding assays, [2,4,6,7-^{3}H]estradiol was incubated with cytosol plus the stated molar excess of unlabelled diethylstilboestrol (DES), deoxymiroestrol, 8-prenylnaringenin, coumestrol, miroestrol, genistein, equol, daidzein or resveratrol. The results are means ± S.E.M. of triplicate assays.
Data are from Matsumura et al. (2005), with permission from the Publishers.

cancer cells. Because equol from dietary soy enhances the growth of pre-existing breast cancers, induction of PI-9 by equol may play a role in tumor progression in breast cancer patients (Jiang et al. 2008).

Based on current data, equol appears estrogenic in estrogen-sensitive reporter systems or cells in culture, and in ovariectomized rodents. However, in adult female ovariectomized monkeys, high doses of equol did not result in estrogenic effects in the breast or uterus. The lack of estrogenic proliferative effects in the primate with higher dose of equol is in contrast with certain data from *in vitro* and rodent models (Wood et al. 2006). Ju et al. (2006) also found (±)-equol stimulated MCF-7 cell growth *in vitro* but not *in vivo*.

35.2.3 Equol and Prostate Cancer

In recent decades, accumulating studies suggest that isoflavone-rich diets are associated with a lower risk of prostate cancer. Limited human studies indicate that the protective effect in equol producers is more evident than that in equol non-producers and the ethnic difference in daidzein metabolism may partly explain the variations of prostate cancer incidence between ethnic groups (Ozasa et al. 2004). Equol has potent anti-proliferative effects on benign and malignant prostatic epithelial cells at concentrations that can be obtained

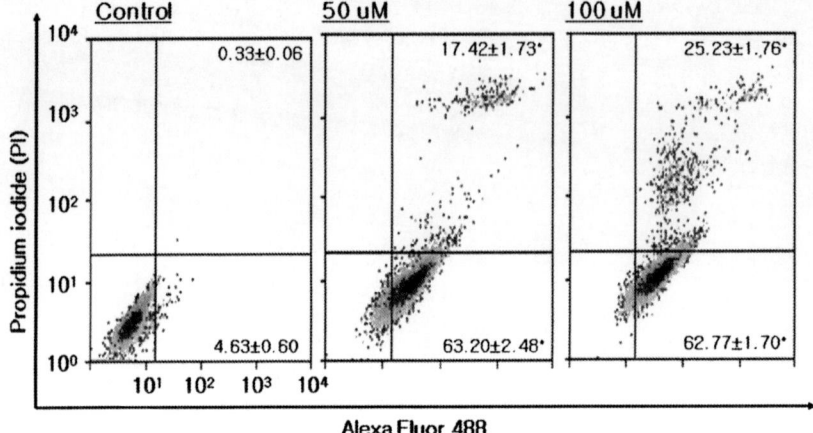

Figure 35.3 Equol-induced apoptosis in MDA-MB-453 cells. Cells were exposed to equol at high concentrations (50 and 100 μM) and incubated for 72 h. The results are means ± S.D. ($n=4$). Early apoptotic cells, right bottom; late apoptotic cells, right top; live cells, left bottom. *$P<0.05$, significantly different from the control group.
Data are from Choi *et al.* (2009), with permission from the Publishers.

Table 35.1 Summary of PrEC cell cycle distribution after 4 days of treatment with ethanol vehicle or 10^{-5} M genistein, equol or daidzein. Benign human prostatic epithelial cells (PrEC) were treated with genistein, equol, daidzein or ethanol. Cells were harvested and analyzed on a Coulter XL flow cytometer. Statistics were performed on 10000 events per sample. Data are from Hedlund *et al.* (2003), with permission from the Publishers.

	Sub G0/G1 (%)	*G0/G1 (%)*	*S phase (%)*	*G2/M (%)*
Ethanol control	0.0	46.01	34.29	19.69
Genistein	0.06	53.32	5.72	40.95
Equol	0.1	57.17	22.78	20.05
Daidzein	1.56	51.51	27.65	20.84

naturally through dietary soy consumption. Equol and daidzein caused an accumulation of cells in G0/G1, whereas genistein arrested cells in G2/M (Table 35.1). Isoflavonoids demonstrated differential effects on established prostate cancer cell lines 22Rv1, LNCaP, LAPC-4, PC-3 and DU 145; PC-3 cells showed the greatest resistance to isoflavonoids (Hedlund *et al.* 2003).

The effects of equol, daidzein and genistein on gene expression in the human androgen-responsive prostate cancer cell line LNCaP were examined by microarray technology. These compounds differentially modulated genes in multiple cellular pathways (Figure 35.4). However, they also exerted similar effects on the regulation of genes belonging to several other important cellular

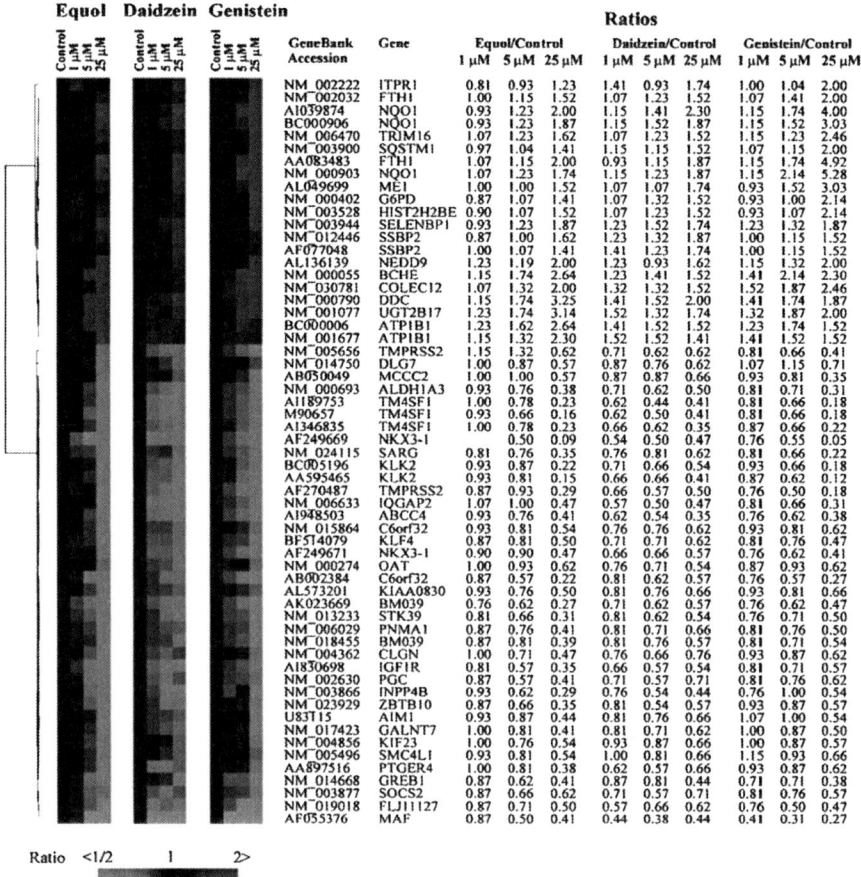

Figure 35.4 Genes similarly affected by equol, daidzein and genistein. LNCaP cells were treated with 0, 1, 5 or 25 µM equol, daidzein or genistein for 48 h. Total RNA was isolated and microarray analyses were performed. Mean fold changes relative to control are listed next to each gene.
Data are from Takahashi et al. (2006), with permission from the Publishers.

pathways (Takahashi et al. 2006). Mitchell et al. (2000) determined the effects of phytoestrogens on growth and DNA integrity in human prostate tumor cell lines. Cell growth tended to be inhibited at lower equol concentrations and/or to a greater extent in PC-3 than in LNCaP cells (Mitchell et al. 2000).

Equol administration appears to have potential beneficial effects for prostate health and other 5α-dihydrotestosterone (5α-DHT)-mediated disorders via its anti-androgen activity. The anti-androgen activity of equol is unique, as equol does not bind the androgen receptor (AR) but specifically binds 5α-DHT with high affinity and thereby prevents 5α-DHT from binding ARs (Lund et al. 2011).

35.2.4 Equol and the Reproductive System

The corpus luteum (CL) is a reproductive gland that plays a crucial endocrine role in the regulation of the estrous cycle, fertility and pregnancy in cattle. It has been shown that phytoestrogens may disrupt numerous reproductive functions at several levels of regulation and *via* different intracellular mechanisms. Using a cell-culture system of steroidogenic cells of the bovine CL, Woclawek-Potocka *et al.* (2006) investigated the effects of equol on prostaglandin $F_{2\alpha}$ ($PGF_{2\alpha}$), progesterone and testosterone synthesis in steroidogenic CL cells. Equol could stimulate $PGF_{2\alpha}$ in steroidogenic cells of the bovine CL *via* an ER-dependent genomic pathway (Woclawek-Potocka *et al.* 2006).

Selvaraj *et al.* (2004) further evaluated the effects of equol on reproductive and non-reproductive endpoints in animal model. Dietary equol did not significantly increase uterine weight in ovariectomized female C57BL/6 mice. Increasing dietary and injected equol doses caused a dose-dependent increase in vaginal epithelial thickness. Equol seems to act as a weak estrogen with modest effects on endpoints regulated by ERα, but its concentration in humans may not be sufficient to induce estrogenic effects (Selvaraj *et al.* 2004).

35.2.5 Equol and Neuroprotection

High intake of soy-derived phytoestrogens has been linked to low prevalence rate of Alzheimer's disease (AD) in Asia compared with the Western countries. Although this association lacks confirmation from randomized and controlled human studies, *in vivo* and *in vitro* studies have demonstrated powerful neuroprotective effects of estrogens against a variety of insults. These neuroprotective effects appear to utilize multiple mechanisms depending on the nature of injury and cell types.

Like estrogen, equol can be neuroprotective against cell apoptosis *in vitro*. This effect is mediated by an inhibition of caspase activity and depends on ER. These results further support a role for soy isoflavones as neuroprotectants and possible alternatives to estrogen (Schreihofer and Redmond 2009). Equol was effective to induce neuroprotective responses but at a much lower magnitude than those induced by E2. When combined with selected phytoestrogens that bind preferentially to ERβ over ERα, equol could enhance neural responses without affecting the reproductive system. It is suggested that ERβ plays a crucial role in mediating estrogenic activities to sustain neural defense against neurodegeneration and promote neural synaptic plasticity, learning and memory function (Zhao *et al.* 2009).

35.3 Equol and Other Cancer Cell Proliferation

35.3.1 Equol and Pancreatic Cancer

Pancreatic cancer is often detected in late stages of the carcinogenesis process and the cells are often highly resistant to various chemotherapeutic drugs, due to increased expression of the multidrug resistance gene (*mdr-1*). Because of their antiestrogenic and antiproliferative effects on a number of cancers, phytoestrogens have been proposed as potential chemopreventive agents. Synthetic antiestrogens, such as tamoxifen, have been used as chemotherapeutic agents for the treatment of pancreatic cancer. Thus, phytoestrogens with antiestrogenic effects may be potential agents for the prevention and treatment of pancreatic cancer.

Lyn-Cook *et al.* (1999) examined the chemoprotective effects of equol on human pancreatic adenocarcinoma cells *in vitro*. Two human adenocarcinoma cell lines, HPAF-11 from a male and Su 86.86 from a female, were used. Equol displayed marked differences in growth inhibition between cell lines. Equol inhibited the growth of the female pancreatic tumor cells; however, equol stimulated the growth of pancreatic tumor cells from the male. Equol decreased K-ras expression and had no effect on *mdr-1* expression in the female tumor cell line. Data from this study suggest that certain phytoestrogens may exert protective effects but cell line-dependent differences in response to these agents may exist (Lyn-Cook *et al.* 1999).

35.3.2 Equol and Colon Cancer

E2 regulates the transcription and expression of the vitamin D receptor (VDR) in rat colonocytes and duodenocytes and in HT29 human colon cancer cells by binding to ERβ and up-regulating signal transduction through ERK1/2 and the activator protein 1 (AP-1) site in the VDR promoter. The question arises as to whether phytoestrogens can induce similar activation. Gilad *et al.* (2006) investigated the effects of genistein, glycitein and equol on signaling pathways in HT29 cells, and determined whether these interactions affect VDR transcription and translation. Similar to E2, equol induced higher concentrations of intracellular free calcium in HT29 cells, an event that could provide evidence of the mechanism(s) triggered by E2 and equol as they initiate the signaling cascades, which result in the activation of ERK signaling pathways and modulation of Sp-1 sites of the VDR gene, and enhance the expression of VDR (Gilad *et al.* 2006).

35.4 Equol and CVDs

CVDs such as coronary heart disease, hypertension, atherosclerosis and diabetes are associated with increased generation of reactive oxygen species (ROS) and compromised endogenous antioxidant defenses. Increased generation of ROS in CVDs impairs endothelial function and reduces nitric oxide (NO) bioavailability. Soy isoflavones have been proposed to exert a number of beneficial effects on cardiovascular function. The molecular mechanisms by which soy isoflavones (genistein, daidzein and equol) afford protection against oxidative stress in CVDs remain to be investigated in large-scale clinical trials.

Soy isoflavones increased gene expression of endothelial nitric oxide synthase (eNOS) and antioxidant defense enzymes, resulting in improved endothelial function and lower blood pressure *in vivo*. Equol stimulated eNOS activity at basal cytosolic Ca^{2+} levels *via* ERK1/2 and phosphoinositide 3-kinase/Akt-dependent pathways (Joy *et al.* 2006). Equol inhibited the induction of apoptosis in response to oxidized low-density lipoprotein (OX-LDL) exposure in human umbilical vein endothelial cells. These results suggested that equol might contribute to a reduced level of OX-LDL-stimulated apoptosis linked to the reduced generation of intracellular ROS (Kamiyama *et al.* 2009).

Thromboxane A2 (TxA2), the main cyclo-oxygenase metabolite of arachidonic acid in platelets, acts through a membrane surface receptor to aggregate platelets and contract vascular smooth muscle. Muñoz *et al.* (2009) confirmed the antiplatelet activity of flavonoids, which selectively inhibited the effect of TxA2 at the receptor level. Equol was shown to be the most potent isoflavone that competes for TxA2 receptor binding (Muñoz *et al.* 2009).

35.5 Equol and Immune Functions

Estrogens are known to modulate immune functions in humans. T-cells express ERα and ERβ, and natural killer cells express ERβ. The expression patterns of ER in T-cells indicate these immune cells as targets of estrogens. A number of animal studies have determined the effects of isoflavones on the immune system. However, few studies have examined the immunomodulatory effect of equol. Gredel *et al.* (2008) investigated the effects of dietary isoflavones, together with their major metabolites, on various functions of human leukocytes *in vitro*. They found that genistein, daidzein and equol are the most potent inhibitors of leukocyte functions. Ten μM of genistein decreased proliferation, lytic activity of natural killer cells, and cytokine secretions. The latter proved to be the most sensitive marker of immune functions. Lignans and their metabolites had minor effects on the immune system. The antiestrogens tamoxifen and fulvestrant did not block the inhibition of cytokine secretion by genistein and equol (Gredel *et al.* 2008).

35.6 Equol and Genotoxicity

The genotoxicity of estrogens has been confirmed, and isoflavones may also have similar geneotoxic potential. Genistein carries only one hydroxy group more than daidzein, and its genotoxic and mutagenic activity are well known. However, daidzein has not been found to induce chromosomal damage *in vitro*, even when present in high concentrations. Schmitt *et al.* (2003) used an *in vitro* micronucleus assay in L5178Y mouse lymphoma cells to investigate the genotoxic potential of daidzein and its four metabolites known to be formed in humans. Daidzein did not induce the formation of micronuclei up to 100 μM. In contrast, equol caused an increase in micronucleus frequency. These results imply that ingestion of soy products may lead to the formation of potentially genotoxic metabolites *in vivo*. However, the genotoxic activity of equol has been found in routine genotoxicity testing, where metabolic activation is restricted to the addition of rat liver homogenate. Based on the evidence that E2 is clastogenic at micromolar concentrations and induces gene mutations at low concentrations, a mutagenic activity cannot be excluded for these phytoestrogens (Schmitt *et al.* 2003). Thus, further study for their genetic toxicology is needed.

35.7 Equol and Cell Invasion

Cancer cell invasion and metastasis are a major cause of mortality for cancer patients, and cancer cell invasion plays a crucial role in the metastatic process. Many dietary bioactive components have shown promising anticancer activities with little or no toxicity to normal cells. Investigations of how dietary bioactive components regulate adhesion, invasion and motility of cancer cells could play a significant role in the development of new agents for the prevention and treatment of cancer with low toxicity. So far, a very limited number of studies have investigated the inhibition of cancer cell invasion by equol. Magee *et al.* (2004) reported that equol induced a significant, dose-dependent inhibition of breast cancer MDA-MB-231 cell invasion through matrigel. Equol inhibited invasion by approximately 30% when present at a concentration of 10 µM. The effects of equol on invasion were not significantly different from that of daidzein. Inhibition of invasion induced by equol occurred without affecting cell viability, highlighting its possible chemoprotective effects (Magee *et al.* 2004).

35.8 Equol and Cancer Chemoprevention

Carcinogenesis is characterized as a multistage process that includes initiation, promotion and progression stages. Cancer prevention strategies that involve intervention at tumor promotion stage seem to be more practical than those intervening at the tumor initiation stage. About 40 plant-based foods which possess cancer preventive abilities have been identified by The National Cancer Institute (Bode and Dong 2005). Dietary phenolic phytochemicals originating from those foods have been suggested to have chemopreventive effects in carcinogenesis, particularly in the promotion stage. Accumulating evidence has showed that equol is a potent chemopreventive agent against carcinogenesis. Equol, but not daidzein, is a potent inhibitor of mitogen-activated protein kinase (MAPK) kinase (MEK) activity and subsequently inhibited AP-1 transactivation and cell transformation (Figure 35.5). Equol was more effective and less toxic in inhibiting 12-*O*-tetradecanoylphorbol-13-acetate-induced cell transformation of JB6 P+ cells. This inhibition was associated with suppression of MEK1 kinase activity (Kang *et al.* 2007). Equol also increased quinone reductase (QR) activity, protein and mRNA levels at physiological and supraphysiological concentrations in Hepa-1c1c7 cells. Equol modulates QR *via* both ERβ and nuclear factor erythroid 2-related factor 2 (Nrf2) binding to

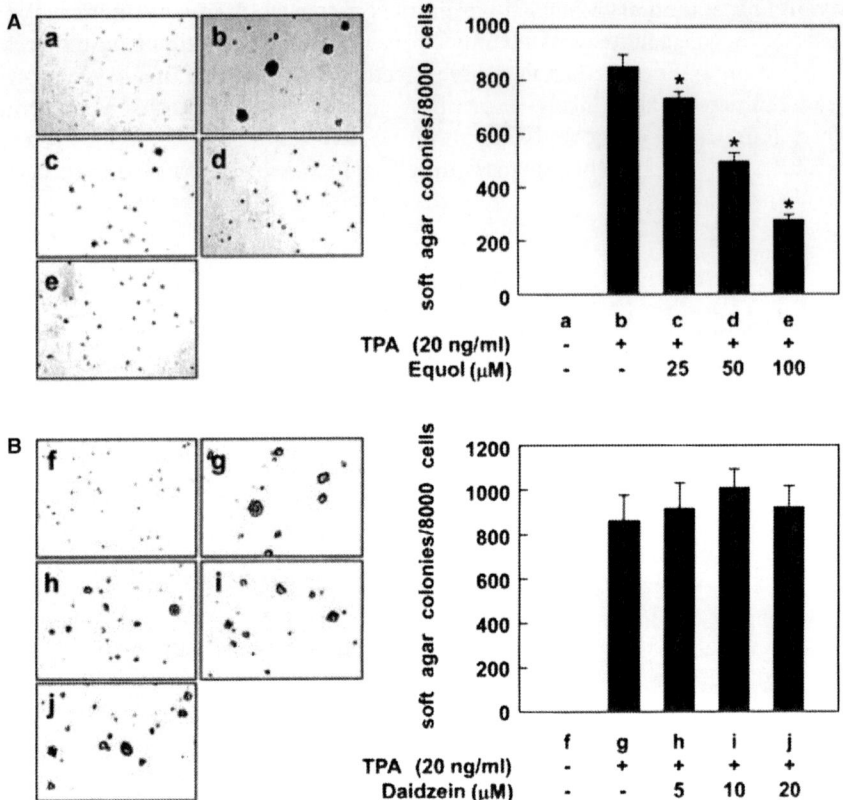

Figure 35.5 Comparison of the inhibitory effects of equol or daidzein on 12-O-tetradecanoylphorbol-13-acetate (TPA) induced neoplastic transformation in JB6 P+ cells. (**A**) Effects of equol on TPA-induced cell transformation. (**B**) Effects of daidzein on TPA-induced cell transformation. The results are means ± S.D. of the number of colonies as determined from three independent experiments. *$P<0.05$ between the group treated with TPA and daidzein or equol and the group treated with TPA alone. Reproduced with Permission from Kang et al. (2007).

the antioxidant response element, and genistein acts predominantly via Nrf2, with less involvement of ERβ (Froyen and Steinberg 2011).

35.9 Equol and Drug Metabolism

Pregnane X receptor (PXR) is a key regulator responsible for the metabolism of various xenobiotics. Human CYP3A4 and the murine homolog Cyp3A11 are the major CYPs regulated by PXR. Regulation of PXR by bioactive dietary factors is of considerable importance. Through interaction with and activation of PXR, dietary factors could modulate the pharmacokinetics and disposition of drugs. Li et al. (2009) investigated the ability of soy isoflavones to activate full-length mouse and human PXR and the subsequent effect on the expression of murine

Cyp3A11 and human CYP3A4 in primary hepatocytes, as well as the expression of hepatic Cyp3A11 expression in mice consuming a soy protein containing diet. They found that genistein and daidzein activated full-length, wild-type mouse PXR, but not a mutant form. In contrast, equol was a more potent activator of human PXR than genistein or daidzein in CV-1 cells and HepG2 cells. Equol induced recruitment of the co-activator steroid receptor co-activator 1 to PXR. Equol also increased CYP3A4 mRNA and protein expression in human hepatocytes. Genistein and daidzein induced the expression of CYP3A11 mRNA, whereas equol had no effect. Cyp3A11 mRNA was also induced in mice fed a soy protein-containing diet. These results suggest that there is a species-specific difference in the activation of PXR by isoflavones and equol (Li *et al.* 2009).

35.10 General Conclusions

Clinical and epidemiological studies have showed that equol intake is significantly correlated with a lower incidence of hormone-related cancer and CVDs. Equol is a more potent inducer of estrogenic activity than its precursor daidzein. The weak estrogenic activity of equol has been debated in terms of increased risk of breast and prostate cancer. S-equol is a unique non-steroidal estrogen that binds preferentially to ERβ, which also antagonizes the *in vivo* action of 5α-DHT. Future studies that clearly define the potential benefits of pure S-equol and R-equol will enable a better understanding of the extent to which there are advantages to producing equol from soy foods. The association of equol excretion and lowered certain cancer risk may largely reflect the tendency of equol producers to have more favorable hormonal profiles, not merely reflecting increased isoflavone intake. Future studies are needed that are designed to identify the specific bacterial species and strains capable of converting daidzein into equol or increasing equol production, and to address *a priori* the effect of the equol-producer phenotype on disease risk.

Summary Points

- This Chapter focuses on the effects of equol on cell proliferation.
- Equol, a metabolite of daidzein after soy food intake, exhibits more potent biological activity than its precursor.
- As the only isomer of equol produced in human intestine, S-equol is produced in approximately one-third of individuals consuming daidzein, and has selective affinity for the ERβ.
- Equol regulates cell proliferation in hormone-positive cells by binding ERs and other steroid receptors.
- Equol also possesses *in vitro* effects on nervous system, reproductive system, immune system, CVDs and drug metabolism.
- The equol-producer phenotype may be associated with reduced risk of certain diseases including breast and prostate cancers.
- However, its weak estrogenic activity has been considered as a potential risk.

Key Facts

Key Features of Hormone Replacement Therapy (HRT)

- As a female hormone, the function of estrogen is to maintain female sex characteristics and monthly menstrual cycles.
- Levels of estrogen usually drop in postmenopausal women.
- Low estrogen level is responsible for most menopausal symptoms, including hot flashes, infertility, low bone density, urogenital symptoms and other symptoms.
- Long-term low estrogen levels may cause osteoporosis in postmenopausal women.
- HRT could raise the levels of essential hormones, and help a woman control the symptoms of menopause and reduce the risk of developing osteoporosis.
- An estrogen/progesterone combination is often given to most postmenopausal women who take HRT.
- Women should start HRT as soon as menopausal symptoms begin.
- HRT treatment does slightly increase the risk of certain cancers and stroke.
- Consumption of phytoestrogens, such as soya beans, ginseng, block cohosh, red clover, and kava, also help with menopausal symptoms.

Definitions of Words and Terms

Agonist: A chemical that can bind with a receptor of a cell to trigger a physiological response typical of a naturally occurring substance.

Antagonist: A chemical that can bind with a receptor to block or inverse the action of an agonist on the receptor.

Athymic mouse: This is a laboratory mouse which is missing the thymus *via* genetic mutation technology. The athymic mouse has a deficiency in T-lymphocytes. Due to the absent immune response to tumor grafts, athymic mouse is valuable to cancer research.

Cell proliferation: This refers to growth of cell populations, where one cell grows and divides to produce two cells. This process is controlled by cell cycle, which is regulated by cyclins, cyclin-dependent kinases and their inhibitors.

17β-Estradiol: This is a naturally female sex hormone which is a critical to maintain female reproductive and sexual functions. It also affects other organs or systems, including the bones and cardiovascular system.

Genotoxicity: This is a deleterious action on a cell's genetic material which affects its integrity by genotoxic substances. Genotoxic substances can bind with DNA, resulting in the formation of DNA adduct and preventing accurate DNA replication. This process is believed to cause genetic mutation.

ICI 182,780: This is also known as fulvestrant and is a drug for the treatment of ER-positive breast cancer in postmenopausal women. It is a full antagonist of ER with no agonist effects.

Isoflavones: A class of natural compounds in plants. Many isoflavones act as antioxidants and phytoestrogens. Consumption of isoflavones, soy-derived isoflavones in particular, can protect against age-related diseases, certain cancers and postmenpausal symptoms.

Nuclear receptor: These are a class of proteins found within cells that are responsible for sensing steroid and thyroid hormones and certain other molecules. Following the complex formation of the nuclear receptor binding with its ligand, the receptor can translocate from cytoplasm to cell nucleus, directly bind to DNA and regulate the expression of specific genes, thereby controlling the development, homeostasis and metabolism of the organism.

Phytoestrogens: These are also called "dietary estrogens" and are a diverse group of naturally occurring non-steroidal plant compounds that function as the primary female sex hormone. Because of the structural similarity with estradiol, phytoestrogens exert estrogenic or antiestrogenic properties.

List of Abbreviations

AD	Alzheimer's disease
AP-1	activator protein 1
AR	androgen receptor
CL	corpus luteum
CVD	cardiovascular disease
5α-DHT	5α-dihydrotestosterone
E2	17β-estradiol
eNOS	endothelial nitric oxide synthase
ER	estrogen receptor
ERK	extracellular-signal-regulated kinase
ERRγ	estrogen-related receptor γ
mdr-1	multidrug resistance gene 1
MEK	mitogen-activated protein kinase (MAPK) kinase
NO	nitric oxide
Nrf2	nuclear factor erythroid 2-related factor 2
OX-LDL	oxidized low-density lipoprotein
PI-9	proteinase inhibitor 9
$PGF_{2\alpha}$	prostaglandin $F_{2\alpha}$
PXR	pregnane X receptor
QR	quinone reductase
ROS	reactive oxygen species
TxA2	thromboxane A2
VDR	vitamin D receptor

References

Bode, A.M., and Dong, Z., 2005. Signal transduction pathways in cancer development and as targets for cancer prevention. *Progress in Nucleic Acid Research and Molecular Biology*. 79: 237–297.

Choi, E.J., Ahn, W.S., and Bae, S.M., 2009. Equol induces apoptosis through cytochrome c-mediated caspases cascade in human breast cancer MDA-MB-453 cells. *Chemico-Biological Interactions*. 177: 7–11.

Froyen, E.B., and Steinberg, F.M., 2011. Soy isoflavones increase quinone reductase in hepa-1c1c7 cells via estrogen receptor β and nuclear factor erythroid 2-related factor 2 binding to the antioxidant response element. *Journal of Nutritional Biochemistry*. 22: 843–848.

Gilad, L.A., Tirosh, O., and Schwartz, B., 2006. Phytoestrogens regulate transcription and translation of vitamin D receptor in colon cancer cells. *Journal of Endocrinology*. 191: 387–398.

Gredel, S., Grad, C., Rechkemmer, G., and Watzl, B., 2008. Phytoestrogens and phytoestrogen metabolites differentially modulate immune parameters in human leukocytes. *Food and Chemical Toxicology*. 46: 3691–3696.

Hedlund, T.E., Johannes, W.U., and Miller, G.J., 2003. Soy isoflavonoid equol modulates the growth of benign and malignant prostatic epithelial cells *in vitro*. *Prostate*. 54: 68–78.

Hirvonen, J., Rajalin, A.M., Wohlfahrt, G., Adlercreutz, H., Wähälä, K., and Aarnisalo, P., 2011. Transcriptional activity of estrogen-related receptor γ (ERRγ) is stimulated by the phytoestrogen equol. *Journal of Steroid Biochemistry and Molecular Biology*. 123: 46–57.

Jiang, X., Patterson, N.M., Ling, Y., Xie, J., Helferich, W.G., and Shapiro, D.J., 2008. Low concentrations of the soy phytoestrogen genistein induce proteinase inhibitor 9 and block killing of breast cancer cells by immune cells. *Endocrinology*. 149: 5366–5373.

Joy, S., Siow, R.C., Rowlands, D.J., Becker, M., Wyatt, A.W., Aaronson, P.I., Coen, C.W., Kallo, I., Jacob, R., and Mann, G.E., 2006. The isoflavone equol mediates rapid vascular relaxation: Ca^{2+}-independent activation of endothelial nitric-oxide synthase/Hsp90 involving ERK1/2 and Akt phosphorylation in human endothelial cells. *Journal of Biological Chemistry*. 281: 27335–27345.

Ju, Y.H., Fultz, J., Allred, K.F., Doerge, D.R., and Helferich, W.G., 2006. Effects of dietary daidzein and its metabolite, equol, at physiological concentrations on the growth of estrogen-dependent human breast cancer (MCF-7) tumors implanted in ovariectomized athymic mice. *Carcinogenesis*. 27: 856–863.

Kamiyama, M., Kishimoto, Y., Tani, M., Utsunomiya, K., and Kondo, K., 2009. Effects of equol on oxidized low-density lipoprotein-induced apoptosis in endothelial cells. *Journal of Atherosclerosis and Thrombosis*. 16: 239–249.

Kang, N.J., Lee, K.W., Rogozin, E.A., Cho, Y.Y., Heo, Y.S., Bode, A.M., Lee, H.J., and Dong, Z., 2007. Equol, a metabolite of the soybean isoflavone daidzein, inhibits neoplastic cell transformation by targeting the MEK/ERK/p90RSK/activator protein-1 pathway. *Journal of Biological Chemistry*. 282: 32856–32866.

Lehmann, L., Esch, H.L., Wagner, J., Rohnstock, L., and Metzler, M., 2005. Estrogenic and genotoxic potential of equol and two hydroxylated metabolites of daidzein in cultured human Ishikawa cells. *Toxicology Letters*. 158: 72–86.

Li, Y., Ross-Viola, J.S., Shay, N.F., Moore, D.D., and Ricketts, M.L., 2009. Human CYP3A4 and murine Cyp3A11 are regulated by equol and genistein via the pregnane X receptor in aspecies-specific manner. *Journal of Nutrition*. 139: 898–904.

Liu, H., Du, J., Hu, C., Qi, H., Wang, X., Wang, S., Liu, Q., and Li, Z., 2010. Delayed activation of extracellular-signal-regulated kinase 1/2 is involved in genistein- and equol-induced cell proliferation and estrogen-receptor-alpha-mediated transcription in MCF-7 breast cancer cells. *Journal of Nutritional Biochemistry*. 21: 390–396.

Lund, T.D., Blake, C., Bu, L., Hamaker, A.N., and Lephart, E.D., 2011. Equol an isoflavonoid: potential for improved prostate health, *in vitro* and *in vivo* evidence. *Reproductive Biology and Endocrinology*. 9: 4.

Lyn-Cook, B.D., Stottman, H.L., Yan, Y., Blann, E., Kadlubar, F.F., and Hammons, G.J., 1999. The effects of phytoestrogens on human pancreatic tumor cells *in vitro*. *Cancer Letters*. 142: 111–119.

Magee, P.J., McGlynn, H., and Rowland, I.R., 2004. Differential effects of isoflavones and lignans on invasiveness of MDA-MB-231 breast cancer cells *in vitro*. *Cancer Letters*. 208: 35–41.

Marrian, G.F., and Haslewood, G.A., 1932. Equol, a new inactive phenol isolated from the ketohydroxyoestrin fraction of mares' urine. *Biochemical Journal*. 26: 1227–1232.

Matsumura, A., Ghosh, A., Pope, G.S., and Darbre, P.D., 2005. Comparative study of oestrogenic properties of eight phytoestrogens in MCF7 human breast cancer cells. *Journal of Steroid Biochemistry and Molecular Biology*. 94: 431–443.

Mitchell, J.H., Duthie, S.J., and Collins, A.R., 2000. Effects of phytoestrogens on growth and DNA integrity in human prostate tumor cell lines: PC-3 and LNCaP. *Nutrition and Cancer – An International Journal*. 38: 223–228.

Morito, K., Hirose, T., Kinjo, J., Hirakawa, T., Okawa, M., Nohara, T., Ogawa, S., Inoue, S., Muramatsu, M., and Masamune, Y., 2001. Interaction of phytoestrogens with estrogen receptors α and β. *Biological and Pharmaceutical Bulletin*. 24: 351–356.

Muñoz, Y., Garrido, A., and Valladares, L., 2009. Equol is more active than soy isoflavone itself to compete for binding to thromboxane A(2) receptor in human platelets. *Thrombosis Research*. 123: 740–744.

Muthyala, R.S., Ju, Y.H., Sheng, S., Williams, L.D., Doerge, D.R., Katzenellenbogen, B.S., Helferich, W.G., and Katzenellenbogen, J.A., 2004. Equol, a natural estrogenic metabolite from soy isoflavones: convenient preparation and resolution of R- and S-equols and their differing binding and biological activity through estrogen receptors alpha and beta. *Bioorganic and Medicinal Chemistry*. 12: 1559–1567.

Ozasa, K., Nakao, M., Watanabe, Y., Hayashi, K., Miki, T., Mikami, K., Mori, M., Sakauchi, F., Washio, M., Ito, Y., Suzuki, K., Wakai, K., and Tamakoshi, A., 2004. Serum phytoestrogens and prostate cancer risk in a nested case-control study among Japanese men. *Cancer Science*. 95: 65–71.

Schmitt, E., Dekant, W., and Stopper, H., 2001. Assaying the estrogenicity of phytoestrogens in cells of different estrogen sensitive tissues. *Toxicology In Vitro*. 15: 433–439.

Schmitt, E., Metzler, M., Jonas, R., Dekant, W., and Stopper, H., 2003. Genotoxic activity of four metabolites of the soy isoflavone daidzein. *Mutation Research*. 542: 43–48.

Schreihofer, D.A., and Redmond, L., 2009. Soy phytoestrogens are neuroprotective against stroke-like injury *in vitro*. *Neuroscience*. 158: 602–609.

Selvaraj, V., Zakroczymski, M.A., Naaz, A., Mukai, M., Ju, Y.H., Doerge, D.R., Katzenellenbogen, J.A., Helferich, W.G., and Cooke, P.S., 2004. Estrogenicity of the isoflavone metabolite equol on reproductive and non-reproductive organs in mice. *Biology of Reproduction*. 71: 966–972.

Setchell, K.D., Clerici, C., Lephart, E.D., Cole, S.J., Heenan, C., Castellani, D., Wolfe, B.E., Nechemias-Zimmer, L., Brown, N.M., Lund, T.D., Handa, R.J., and Heubi, J.E., 2005. S-equol, a potent ligand for estrogen receptor β, is the exclusive enantiomeric form of the soy isoflavone metabolite produced by human intestinal bacterial flora. *American Journal of Clinical Nutrition*. 81: 1072–1079.

Takahashi, Y., Lavigne, J.A., Hursting, S.D., Chandramouli, G.V., Perkins, S.N., Kim, Y.S., and Wang, T.T., 2006. Molecular signatures of soy-derived phytochemicals in androgen- responsive prostate cancer cells: a comparison study using DNA microarray. *Molecular Carcinogenesis*. 45: 943–956.

Woclawek-Potocka, I., Bober, A., Korzekwa, A., Okuda, K., and Skarzynski, D.J., 2006. Equol and para-ethyl-phenol stimulate prostaglandin $F_{2\alpha}$ secretion in bovine corpus luteum: intracellular mechanisms of action. *Prostaglandins and other Lipid Mediators*. 79: 287–297.

Wood, C.E., Appt, S.E., Clarksonm, T.B., Franke, A.A., Lees, C.J., Doerge, D.R., and Cline, J.M., 2006. Effects of high-dose soy isoflavones and equol on reproductive tissues in female cynomolgus monkeys. *Biology of Reproduction*. 75: 477–486.

Zhao, L., Mao, Z., and Brinton, R.D., 2009. A select combination of clinically relevant phytoestrogens enhances estrogen receptor β-binding selectivity and neuroprotective activities *in vitro* and *in vivo*. *Endocrinology*. 150: 770–783.

CHAPTER 36

Bone, Genistein, Daidzein and Equol

MARINA KOMRAKOVA,* EWA KLARA STUERMER,
KLAUS MICHAEL STUERMER AND
STEPHAN SEHMISCH

Department of Trauma Surgery and Reconstructive Surgery, University Medical Center Goettingen, Robert-Koch St. 40, Goettingen, Germany
*Email: komrakova@yahoo.com

36.1 Introduction

Isoflavones belong to a family of natural plant metabolites known as flavonoids and occur mostly in legumes (Grynkiewicz *et al.* 2005). Isoflavones have natural roles in plant defence and root nodulation. They are non-steroidal estrogens that are structurally similar to estrogen (Yuan *et al.* 2007) (Figure 36.1) and show estrogenic or anti-estrogenic effects, depending on the concentration of endogenous estrogen (Mueller *et al.* 2004).

Estrogen is an important regulator of skeletal growth, development and bone mass maintenance (Turner *et al.* 1994). The decline in the estrogen level in postmenopausal women is partially responsible for the development of osteoporosis, a disease characterised by enhanced bone resorption, accelerated bone loss and increased risk of fractures as well as impaired healing of fractured bone (Yingjie *et al.* 2007). Hormone replacement therapy has been used for the prevention and treatment of osteoporosis, although its use has declined recently because of many negative side effects. Treatment with isoflavones may be an

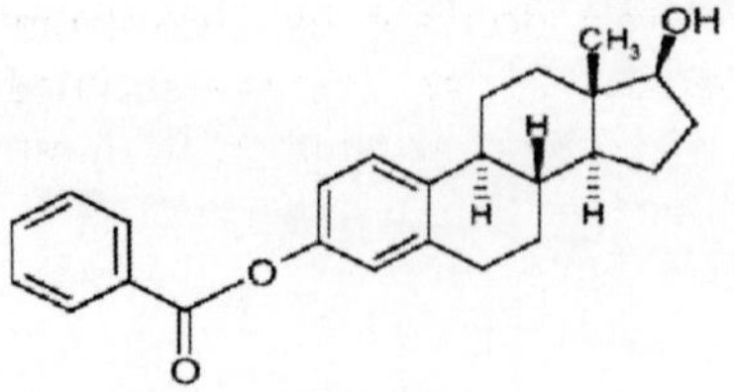

17β-Estradiol 3-benzoate

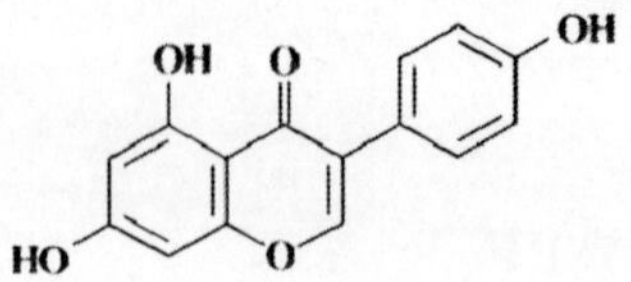

Genistein

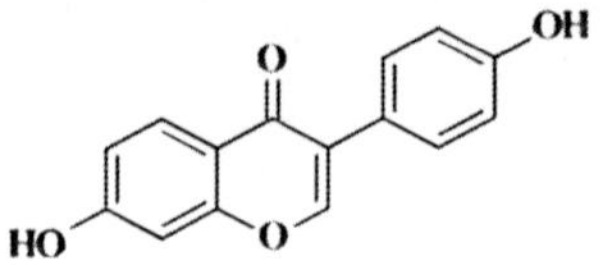

Daidzein

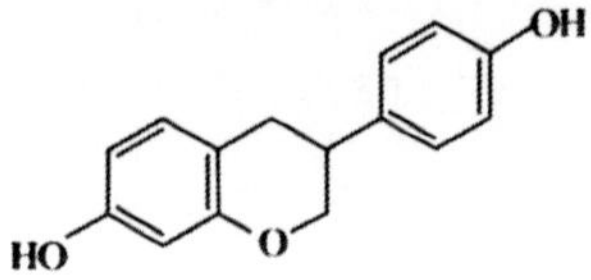

Equol

Figure 36.1 Chemical structure of 17β-estradiol 3-benzoate ($C_{25}H_{28}O_3$), genistein ($C_{15}H_{14}O_5$), daidzein ($C_{15}H_{10}O_4$) and equol ($C_{15}H_{14}O_3$).

alternative therapy for osteoporosis and osteoporotic fractures. There are many studies examining the effects of isoflavone-containing supplements and foods on health (Cornwell *et al.* 2004). A lower rate of osteoporosis has been documented in people who consume high amounts of isoflavones. Soy isoflavones have been shown to be somewhat effective in maintaining bone mineral density (BMD) in postmenopausal women (Cornwell *et al.* 2004). On the other hand, some authors reported no effects of isoflavones on human health (Cornwell *et al.* 2004). Soybeans are the major source of nutritional isoflavones in humans, and the biological effects of a soy diet are dependent on many factors, including the diversity of isoflavones in soy, dose, duration of use, individual metabolism and endogenous estrogen level (Cassidy 2003). Furthermore, soy-derived isoflavones may have the potential to effectively function as endocrine disruptors (Mueller *et al.* 2004). Nevertheless, extracts of isoflavone-containing plants are consumed without prescription as dietary supplements.

This Chapter focuses on the effect of purified genistein, daidzein and equol used as an osteoporosis therapy or for the treatment of osteoporotic bone fracture.

36.2 Bone and Estrogen

Bone tissue is a dense connective tissue that is continuously resorbed and rebuilt. In young adults, bone destruction and formation are balanced, and bone mass is maintained in a steady state. After the age of 40 years, bone resorption begins to exceed bone formation, causing bone loss, and this occurs more rapidly in women than in men. This bone loss is believed to be related to a rapid decrease in estrogen secretion during the postmenopausal period (Rodan and Martin 2000). During the menopause transition, serum 17β-estradiol levels decrease by 90% and serum estrone levels decrease by 70% from premenopausal levels (Clarke and Khosla 2010). Estrogen loss is associated with elevated bone resorption caused by a rise in osteoclast number. Both estrogen receptors (ERs) α and β are expressed in bone tissue (Ke *et al.* 2002). Sims *et al.* (2002) demonstrated that ERα regulated bone remodelling in male mice, whereas, in females, both ERα and ERβ affected this process and could, at least under basal knockout conditions, compensate for each other.

Estrogen has pleiotropic effects on bone tissue. Estrogen suppresses the production of receptor activator of nuclear factor κB ligand (RANKL), which is the final key molecule required for osteoclast development. Estrogen deficiency leads to an increase in RANKL that leads to increased osteoclast recruitment and activation and decreased osteoclast apoptosis. Estrogen also suppresses the production of bone-resorbing cytokines such as interleukin-1 (IL-1), IL-6, IL-11, tumour necrosis factor (TNF)-α, macrophage-colony stimulating factor (M-CSF) and prostaglandins. Estrogen increases the production of transforming growth factor (TGF)-β by osteoblast precursor cells, which induces osteoclast apoptosis. The lack of the estrogen leads to a rapid bone loss during the first 10 years after the onset of menopause (Clarke and Khosla 2010).

Bone loss can be effectively prevented by estrogen replacement therapy. Long-term estrogen treatment reduces the incidence of osteoporosis-related fractures by approximately 50% (Turner *et al.* 1994). Although estrogen replacement therapy is attractive because it corrects estrogen deficiency and prevents bone loss in postmenopausal women, it has many undesirable side effects such as endometrial, ovarian and breast cancer and venous thrombosis.

36.3 Bone and Soy Isoflavones

Recently, much attention has been focused on isoflavones as a potential alternative to estrogen (Glazier and Bowman 2001). Genistein and daidzein are principal isoflavones found in soybeans as glycosides genistin and daidzin, which are less estrogenic than their metabolites. The absorption of these glycosides requires their conversion into aglycones by intestinal β-glycosidase produced by bacteria that colonize the intestine. After the consumption of isoflavones, there is an early increase in genistein and daidzein concentrations in the plasma within 1–2 h, followed by a plateau and then a second peak at 4–8 h. Thereafter, the concentration decreases at 12–24 h and is not detectable at 48 h (Zubik and Meydani 2003). Genistein and daidzein are either absorbed by the intestine or further metabolized to several compounds by the intestinal microflora. The biological activities of isoflavones and their bacterial metabolites are different: genistein, daidzein and equol, a metabolic product of daidzein have relatively strong affinities for the ER, with preferences for ERβ, whereas other metabolites have weak affinities or are hormonally inert (Mueller *et al.* 2004). *In vitro* studies showed that soy isoflavones have a stimulatory effect on protein synthesis and on alkaline phosphatase release by osteoblast cells. In osteoclast cell culture, isoflavones suppress osteoclast activity (Setchell and Lydeking-Olsen 2003). Soy isoflavones may play an important role in bone remodelling by inhibiting bone turnover and might be used to prevent bone loss.

A number of observational studies were performed in women living in countries where the people have high isoflavone intake due to the consumption of soy protein-containing foods. These studies showed that postmenopausal women consuming high amounts of soy, and hence isoflavones, have the highest femoral and lumbar spine BMD (Setchell and Lydeking-Olsen 2003). A significant relationship between isoflavone consumption and serum/urinary markers of osteoblast and osteoclast activities was revealed that is indirectly consistent with reduced bone turnover (Setchell and Lydeking-Olsen 2003).

Soy dietary studies showed either the prevention of bone loss or an improvement or no change in BMD. These studies varied in design, had different durations and involved a variety of soy foods with different levels of isoflavones tested (Setchell and Lydeking-Olsen 2003). Therefore, it is difficult to reach a definitive conclusion on the effect of soy food-containing isoflavones on the prevention of bone loss in women. Furthermore, the extent of isoflavone metabolism is highly variable among individuals and is influenced by the activity of intestinal bacteria (Yuan *et al.* 2007). In individuals who consume soy products, only 35% had measurable quantities of equol (Kelly *et al.* 1995). In contrast, rodents were reported to biotransform soy isoflavones more efficiently, as all rats and mice fed different soy diets showed measurable amounts of equol in the blood (Brown and Setchell 2001). Soy protein in the diet has been reported to maintain BMD in ovariectomized (Ovx) rat at the level of estrogen-treated rats (Arjmandi 2001). In other animal studies, the treatment with pure isoflavones improved BMD (Dai *et al.* 2008; Picherit *et al.* 2000).

36.4 Bone and Purified Genistein, Daidzein and Equol

Genistein is one of the main soy isoflavones (Figure 36.1). In cell culture, genistein can stimulate the production of osteoprotegerin, which prevents bone resorption by modulating of osteoclast activity *via* osteoblasts. Genistein suppresses osteoclast activity by inducing apoptosis, inhibiting cytokines and tyrosine kinases, changes intracellular Ca^{2+} and depolarises membranes. Daidzein is the other main soy isoflavone (Figure 36.1). *In vitro*, daidzein has an effect that is similar, albeit weaker, to that exerted by genistein (Setchell and Lydeking-Olsen 2003). Equol is an active metabolite of daidzein produced by the intestinal microflora (Figure 36.1). Equol has a 100-fold higher affinity for ERs compared with daidzein and is more than twice as estrogenic as genistein (Setchell *et al.* 2002). The clinical importance of equol has been increasing recently. Tousen *et al.* (2011) investigated the effect of a 1 year trial with equol supplementation (2, 6 or 10 mg day^{-1}) in non-equol-producing postmenopausal Japanese women. The equol concentration in serum and urine increased, whereas the bone resorption marker urinary deoxypyridinoline decreased in a dose-dependent manner. Treatment with equol (10 mg day^{-1}) contributed to bone health without changing serum sex and thyroid hormone levels.

A 1 year treatment with genistein (54 mg day^{-1}) increased BMD at the lumbar spine and femoral neck and had no adverse effect on the breast and uterus in early postemopausal women in Italy (Morabito et al. 2002). An enhancement of BMD in the lumbar spine and femoral neck was reported after genistein treatment for 2 years (54 mg day^{-1}) in osteopenic postmenopausal women in Italy (Marini et al. 2007). Genistein had no effect on routine biochemical, liver function and haematological tests. However, genistein recipients experienced gastrointestinal side effects more often than placebo recipients.

The Ovx mice or rat model is often used to evaluate the influence of isoflavones on bone properties that have been diminished by ovarian hormone deficiency (Turner et al. 2001).

The effect on bone in 4-month-old Ovx rats fed a genistein-supplemented diet {6 or 60 mg [kg body weight (BW)]$^{-1}$} or 17β-estradiol-3-benzoate [0.7 mg (kg of BW)$^{-1}$] was studied over a 3 month period (Sehmisch et al. 2008). The comparisons were made with resveratrol and 8-prenylnaringenin, both of which are phytoestrogens. The BMD and biomechanical properties of the tibia were enhanced after genistein treatment in a dose-dependent manner, although to a lesser extent than those seen in estrogen- or 8-prenylnaringenin-treated animals. The high dose of genistein exerted a strong bone response, increased uterine weight and prevented BW increase in Ovx rats.

Dai et al. (2008) found that a 15 week subcutaneous (s.c.) treatment with genistein or 17β-estradiol [5 mg or 10 μg (kg of BW)$^{-1}$ day^{-1}] preserved the biomechanical quality of the vertebral body, whereas the microstructure and BMD were not affected in 7-month-old Ovx rats.

Fujioka et al. (2004) studied the effect of daily s.c. administration of equol (0.1 or 0.5 mg per mice) or 17β-estradiol (0.03 μg per mice) on bone in 2-month-old Ovx mice. Both doses of equol inhibited bone loss in a dose-dependent manner. The higher dose of equol maintained the BMD and histological parameters of the femur at the level seen in mice treated with estradiol. The plasma equol level was 432 nmol L^{-1} at the lower dose and 1551 nmol L^{-1} at the higher dose of equol. There was no effect on body and uterus weight, plasma low-density lipoprotein (LDL) cholesterol and triglyceride levels.

Sehmisch et al. (2010a) examined the effects of genistein and equol when used as osteoporosis prophylaxis in 2-month-old mice. After ovariectomy, mice were fed with genistein, equol or 17β-estradiol-3-benzoate (3 mg, 12 mg or 12.9 μg mice^{-1} day^{-1}, respectively) over 3 months. Genistein and estrogen enhanced the biomechanical properties and total BMD of the tibia, and equol had less of an effect, improving the cancellous BMD and cortical bone area. Neither genistein nor equol had an effect on body and uterus weight.

Picherit et al. (2000) compared the effect of daidzein or genistein [each 10 μg (g of BW)$^{-1}$ day^{-1}] on the lumbar spine and femur in 11-month-old Ovx rats treated over 3 months. Consumption of daidzein or 17α-ethinylestradiol [30 μg (g of BW)$^{-1}$ day^{-1}] was more efficient than genistein in preventing bone loss, and neither isoflavone exhibited any uterotrophic activity. The plasma concentrations of genistein, daidzein and equol were 560, 290 and 281 nmol L^{-1}, respectively. The enhanced bone response observed after daidzein treatment was explained by the presence of equol rather than daidzein or both equol and daidzein compounds.

Genistein is known to bind to both ER isoforms but preferentially to ERβ (Mueller *et al.* 2004). Hertrampf *et al.* (2007) tested properties of genistein as a weak ERα or potent ERβ agonist. Ovx 3-month-old rats were treated daily with s.c. genistein (10 mg/kg), 17β-estradiol (4 μg/kg), ERα-specific agonist 16α-lactone-estradiol (LE_2) (10 μg/kg) or the ERβ-specific agonist 8β-vinyl-estradiol (VE_2) (100 μg/kg) for 3 weeks. Animals either had free access to running wheels or no opportunity for voluntary wheel running. A decrease in trabecular BMD was detected in Ovx and 8β-VE_2-treated rats, whereas genistein, estrogen and 16α-LE_2 maintained BMD at a higher level. Genistein showed no bone-protective activity in the absence of a running wheel. Thus, the effect of genistein on bone was mediated *via* ERα, and physical activity had a strong impact on its bone-protective potency. Genistein treatment did not change body weight and food consumption but increased uterine weight.

In these animal studies, genistein, daidzein and equol were used for osteoporosis prophylaxis, and little is known of the potential of isoflavones administrated as osteoporosis therapy and/or for the treatment of osteoporotic fracture.

To determine the effect of isoflavones used for amelioration of osteoporotic changes in bone, 3-month-old female Sprague–Dawley rats were Ovx, developing severe osteopenia 2 months later (Figure 36.2). Thereafter, bilateral transverse osteotomy of the tibia metaphysis was performed, and rats were treated with equol, genistein (Kolios *et al.* 2009; Sehmisch *et al.* 2010b; Tezval *et al.* 2010) or daidzein (Komrakova *et al.* 2009; 2011). Equol or 17-estradiol-3-benzoate [20 or 0.3 mg (kg of BW)$^{-1}$] were reported to improve the biomechanical properties of the tibia, whereas genistein [40 (kg of BW)$^{-1}$] had only a minor effect after 35 day of treatment (Table 36.1). A reduction in trabecular and callus densities was observed after genistein treatment in the tibia metaphysis (Figure 36.3). (Kolios *et al.* 2009). Similar to the bone healing, equol and estrogen improved the biomechanical and histomorphometric properties of the proximal femur in these rats, whereas genistein had no effect (Tezval *et al.* 2010) (Tables 36.1 and 36.2; Figure 36.3). In the lumbar spine, estrogen treatment had a positive effect on both cortical and trabecular bone, and equol and genistein enhanced the biomechanical parameters; equol was more effective at increasing the number of trabecular nodes (Sehmisch *et al.* 2010b) (Tables 36.1 and 36.2; Figure 36.3). The differences were revealed in intact and injured bone in response to daidzein treatment. Osteopenic rats were fed either daidzein or 17β-estradiol-3-benzoate [50 or 0.4 mg (kg of BW)$^{-1}$]-supplemented diets over 35 or 70 days (Komrakova *et al.* 2009; 2011). Estrogen improved cancellous and cortical bone parameters in the lumbar spine and tibia, however, it was not favorable for osteotomy healing. Estrogen treatment has been reported to have diverse effects on osteoporotic fracture healing. Daidzein improved bone parameters to a lesser extend but facilitated

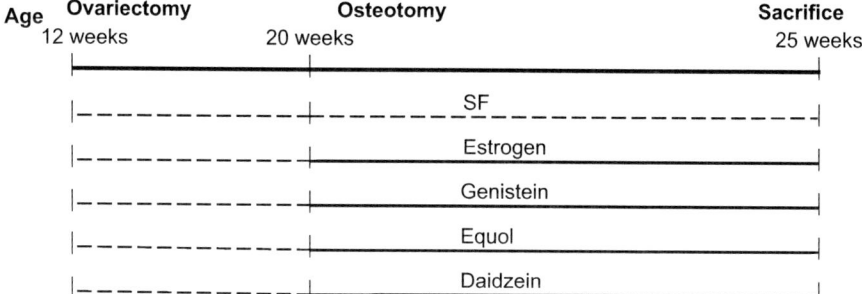

Figure 36.2 Schematic flowchart of the treatments. Twelve-week old female rats were Ovx; after 8 weeks both tibiae were osteotomized and rats were divided into groups ($n=15$) fed either with soy-free diet (SF), or diets supplimented with estradiol-17β-benzoate, genistein or equol for up to 35 days. Dashed line: Ovx + SF; continuous line: food supplementations.

Table 36.1 Serum analyses and biomechanical parameters of tibia, vertebral body and femur. Serum analyses and biomechanical parameters of tibia, the 4th lumbar vertebral body and proximal femur in osteopenic rats fed over 35 days either with soy free (SF) diet, SF with 17β-Estradiol 3-benzoate, SF with genistein or equol.

	SF		Estrogen		Genistein		Equol	
Parameters	Mean	STD	Mean	STD	Mean	STD	Mean	STD
Serum osteocalcin (ng/mL)	17.8	5.6	5.4*	1.8	19.6	6.2	15.2	1.8
Serum β-CrossLaps (ng/mL)	33.8	8.4	31.6	9.6	38.1	10.8	40.2	8.9
Tibia								
Stiffness (N/mm)	121.4	47.1	147.9	39.4	144.5	61.5	167.8*	59.9
Yield load (N)	76.4	18.8	93.8	29.7	84.8	20.0	102.6§	15.8
Vertebral body								
Stiffness (N/mm)	117.4	16.3	175.0*	54.2	178.3*	54.5	155.9*	42.5
Yield load (N)	170.3	23.1	153.9	8.6	159.0	27.8	170.6	27.4
Maximum load (N)	198.3	32.6	205.1	36.8	203.0	29.8	234.1*	13.1
Femur								
Stiffness (N/mm)	233.5	38.0	257.8*	27.4	222.7	35.4	264.7*	15.9
Yield load (N)	109.1	29.5	113.1	10.0	108.8	22.7	112.0	20.6
Maximum load (N)	162.8	34.7	188.4*#	8.0	165.7	21.7	177.3*#	35.5

*means differ significantly vs. SF.
#means differ significantly vs. genistein.
§mean differ from other means within row (P<0.05, Tukey-test).
STD, standard deviation.

osteotomy healing. The effect of daidzein on bone metabolism is often reported to be due at least in part to equol. In this study, serum daidzein concentration averaged 54 ng mL^{-1}, whereas equol was not detected (Komrakova et al. 2011). Serum osteocalcin and β-CrossLaps levels did not change after genistein and

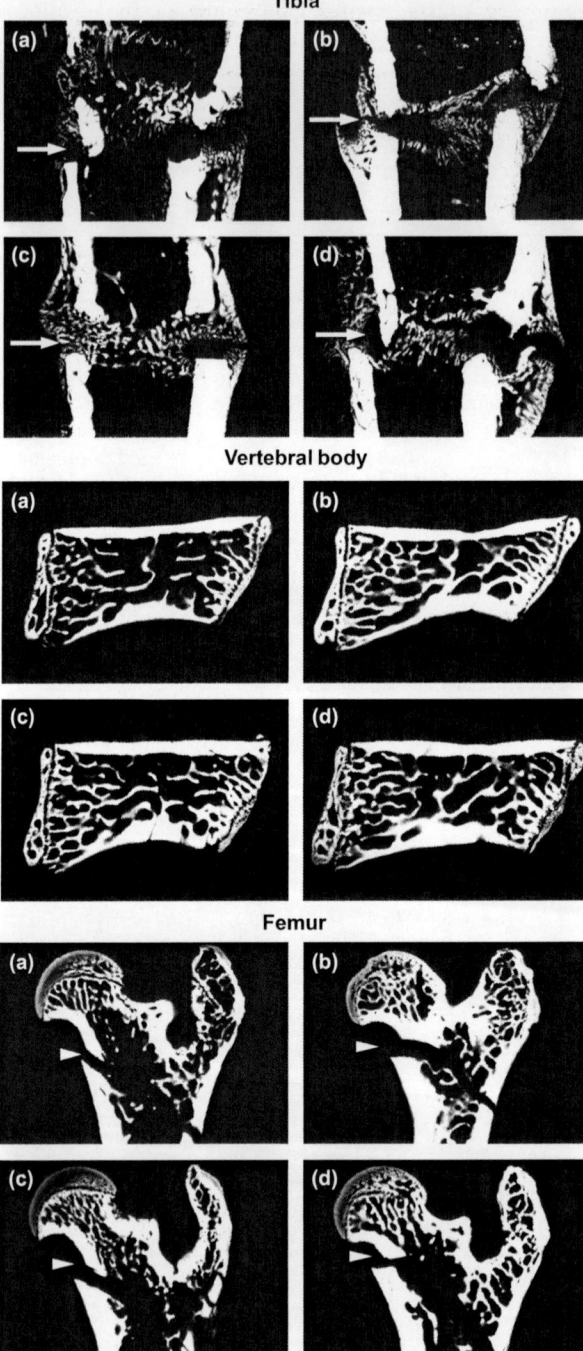

Table 36.2 Summary of the results. Summary of the results of the bone parameters of the vertebral body, femur and osteotomized tibia obtained in rats treated after osteotomy with estrogen, genistein or equol over 35 days and with daidzein over 35 or 70 days. ↔ No effect vs. soy free control (SF); ↑significant increase vs. SF; ↓significant decrease vs. SF.

Parameters	Estrogen	Genistein	Daidzein	Equol
Osteoporotic bone				
Stiffness	↑	↑	↑	↑
Yield load	↑	↔	↑	↑
Trabecular bone	↑	↔	↔	↑
Cortical bone	↑	↔	↔	↔
Osteoporotic fracture healing				
Stiffness	↔	↔	↔	↔
Yield load	↑	↔	↔	↑
Trabecular bone	↑	↓	↔	↔
Cortical bone	↔	↔	↔	↔
Callus	↔	↓	↑	↑

equol treatments (Table 36.1), and daidzein had no effect on serum alkaline phosphatase and osteocalcin levels (Komrakova et al. 2011). Neither genistein, equol nor daidzein exhibited estrogenic activity on the uterus, whereas estrogen increased uterine weight in Ovx rats (Kolios et al. 2009; Komrakova et al. 2009).

36.5 Conclusion

Interest in the extraction of soy isoflavones for further use as dietary supplements has been increasing recently. However, it is still contentious whether isoflavone supplementation has a beneficial effect on human health. Further research is required to clearly define the pharmacological and systemic effects of dietary isoflavones using standardized purified isoflavones.

Delayed bone healing is a well-known problem in osteoporotic patients. Studies have shown that daidzein and equol could be used to improve bone healing, whereas genistein impaired healing processes. Equol improved intact

Figure 36.3 Microradiographs of tibia, vertebral body and femur. Microradiographs of the tibia metaphysis, the 1st lumbar vertebral body and proximal femur made in osteopenic rats fed over 35 days (**a**) soy-free (SF) diet; (**b**) SF diet with 17β-estradiol-3-benzoate; (**c**) SF diet with equol; and (**d**) SF diet with genistein. Arrows, osteotomy lines; arrowheads, trochanteric fractures produced during biomechanical test. In SF- and genistein-treated rats (**a, d**), the callus of tibia is extended and less compact than in estrogen- and equol- treated rats (**b, c**). The trabecular network of vertebral body and femur is denser in estrogen- and equol-treated rats (**b, c**) than in SF- and genistein-treated rats (**a, d**).

bone, genistein had no effect and daidzein did not ameliorate spine properties in osteopenic rats.

Most of the studies have been conducted on rat models despite the differences in bone structure between rats and humans, and caution has to be taken in considering the results of animal studies for human health. However, animal studies may provide potential perspective for the treatment of osteoporotic patients with or without bone fractures. Further studies must consider possible responses to different isoflavones by injured and intact bones as well as different anatomical sites.

Summary Points

- This Chapter focuses on the soy isoflavones genistein and daidzein and the metabolite of daidzein equol.
- Reportedly, dietary soy isoflavones have osteoprotective effects in post-menopausal women.
- The effects of isoflavones depend on many factors, and it is difficult to determine their role in the prevention and treatment of osteoporosis.
- When purified isoflavones are administered in osteopenic rats, genistein impairs tibia healing but has no effect on the spine or femur, equol is favourable for the intact spine and femur and the injured tibia, and daidzein is more effective for tibia healing than for the spine.

Key Facts

- Isoflavones are similar in chemical structure and properties to estrogen and can be considered as a potential alternative to estrogen.
- Estrogen replacement therapy was used for the prevention of osteoporosis, however, it is not currently used because of negative side effects.
- Osteoporosis is a disease characterised by enhanced bone resorption, accelerated bone loss and an increased risk of fractures.
- The osteoprotective effect of isoflavones depends on dose, duration of use, individual metabolism and endogenous estrogen levels.
- Various responses to isoflavones by injured and intact bones as well as by different anatomical sites have been reported.

Definition of Words and Terms

Cancellous bone or trabecular bone: One of the two types of bone tissue; it has a porous spongy structure consisting of the trabecular network and has higher metabolic activity and is more affected by osteoporosis than cortical bone.

Cortical bone: One of the two types of bone tissue that forms the outer shell of the bone; it is denser, stronger and stiffer and less affected by osteoporosis than cancellous bone.

Daidzein: An isoflavone that is obtained following glycolysis of daidzin, which is found in soy, by intestinal microflora.
Equol: A metabolite of daidzein produced by the intestinal microflora.
Estrogen: Female sex hormone produced by the ovaries, which was previously used in estrogen replacement therapy for postmenopausal women.
Femur: The bone of the hip.
Genistein: An isoflavone that is obtained following glycolysis of genistin, which is found in soy, by intestinal microflora.
Isoflavones: Plant metabolites found mostly in legumes.
Lumbar spine: A lower part of the vertebral column that mostly consists of the trabecular bone; osteoporotic fractures of the spine occurs earlier than they do elsewhere.
Osteoblast: A bone cell responsible for bone resorption.
Osteoclast: A bone cell responsible for bone formation.
Postmenopausal osteoporosis: A disease characterised by progressive bone loss and increased bone fragility that is causes by a decline in estrogen production at menopause.
Tibia: The larger bone of the shin.
Transversal osteotomy: An operation during which bone is cut transversally.

List of Abbreviations

BMD	bone mineral density
BW	body weight
ER	estrogen receptor
IL	interleukin
Ovx	ovariectomized
RANKL	receptor activator of nuclear factor κB ligand
s.c.	subcutaneous
SF	soy free

References

Arjmandi, B.H., 2001. The role of phytoestrogens in the prevention and treatment of osteoporosis in ovarian hormone deficiency. *Journal of the American College of Nutrition*. 20: 398S–402S.

Brown, N.M., and Setchell, K.D.R., 2001. Animal model impacted by phytoestrogens in commercial chow: implications for pathways influences by hormones. *Laboratory Investigation*. 81: 735–747.

Cassidy, A., 2003. Potential risks and benefits of phytoestrogen-rich diets. *International Journal for Vitamin and Nutrition Research*. 73: 120–126.

Clarke, B.L., and Khosla, S., 2010. Physiology of bone loss. *Radiologic Clinics of North America*. 48: 483–495.

Cornwell, T., Cohick, W., and Raskin, I., 2004. Dietary phytoestrogens and health. *Phytochemistry*. 65: 995–1016.

Dai, R., Ma, Y., Sheng, Z., Jin, Y., Zhang, Y., Fang, L., Fan, H., and Liao, E., 2008. Effects of genistein on vertebral trabecular bone microstructure, bone mineral density, microcracks, osteocyte density, and bone strength in ovariectomized rats. *Journal of Bone and Mineral Metabolism*. 26: 342–349.

Fujioka, M., Uehara, M., Wu, J., Adlercreutz, H., Suzuki, K., Kanazawa, K., Takeda, K., Yamada, K., and Ishimi, Y., 2004. Equol, a metabolite of daidzein, inhibits bone loss in ovariectomized mice. *Journal of Nutrition*. 134: 2623–2627.

Glazier, M.G., and Bowman, M.A., 2001. A review of the evidence for use of phytoestrogens as a replacement for traditional estrogen replacement therapy. *Archives of Internal Medicine*. 161: 1161–1172.

Grynkiewicz, G., Ksycinska, H., Ramza, J., and Zagrodzka, J., 2005. Chromatographic quantification of isoflavones (why and how). *Acta Chromatographica*. 15: 31–65.

Hertrampf, T., Gruca, M.J., Seibel, J., Laudenbach, U., Fritzemeier, K.H., and Diel, P., 2007. The bone protective effect of the phytoestrogen genistein is mediated via ERα-dependent mechanisms and strongly enhanced by physical activity. *Bone*. 40: 1529–1535.

Ke, H.Z., Brown, T.A., Qi, H., Crawford, D.T., Simmons, H.A., Peterson, D.N., Allen, M.R., McNeish, J.D., and Thompson, D.D., 2002. The role of estrogen receptor-β in the early age-related bone gain and later age-related bone loss in female mice. *Journal of Musculoskeletal and Neuronal Interactions*. 2: 479–488.

Kelly, G.E., Joannou, G.E., Reeder, A.Y., Waring, M., and Nelson, C., 1995. The variable metabolic response to dietary isoflavones in humans. *Proceedings of the Society for Experimental Biology and Medicine*. 208: 40–43.

Kolios, L., Sehmisch, S., Daub, F., Rack, T., Tezval, M., Stuermer, K.M., and Stuermer, E.K., 2009. Equol but not genistein improves early metaphyseal fracture healing in osteoporotic rats. *Planta Medica*. 75: 459–465.

Komrakova, M., Werner, C., Wicke, M., Nguyen, B.T., Tezval, M., Semisch, S., Stuermer, K.M., and Stuermer, E.K., 2009. Effect of daidzein, 4-methylbenzylidene camphor or estrogen on gastrocnemius muscle of osteoporotic rats undergoing tibia healing period. *Journal of Endocrinology*. 201: 253–262.

Komrakova, M., Sehmisch, S., Mohammad, T., Frauendorf, H., Schmelz, U., Grueger, T., Wessling, T., Klein, C., Birth, M., Stuermer, K.M., and Stuermer, E.K., 2011. Impact of 4-methylbenzylidene camphor (4-MBC), daidzein and estrogen on intact and osteotomized bone in osteopenic rats. *Journal of Endocrinology*. 211: 1–12.

Marini, H., Minatoli, L., Polito, F., Bitto, A., Altavilla, D., Atteritano, M., Gaudio, A., Mazzaferro, S., Fristina, A., Fristina, N., Lebrano, C., Bonaiuto, M., D'Anna, R., Cannata, M.L., Corrado, F., Adamo, E.B., Wilson, S., and Squadrito, F., 2007. Effect of the phytoestrogen genistein on bone metabolism in osteopenic postmenopausal women. A randomized trial. *Annals of Internal Medicine*. 146: 839–847.

Morabito, N., Crisafulli, A., Vergara, C., Gaudio, A., Lasco, A., Frisina, N., D'Anna, R., Corrado, F., Pizzoleo, M.A., Cincotta, M., Altavilla, D.,

Ientile, R., and Squadrito, F., 2002. Effects of genistein and hormone-replacement therapy on bone loss in early postmenopausal women: a randomized double-blind placebo-controlled study. *Journal of Bone and Mineral Research*. 17: 1904–1912.

Mueller, S.O., Simon, S., Chae, K., Metzler, M., and Korach, K.S., 2004. Phytoestrogens and their human metabolites show distinct agonistic and antagonistic properties on estrogen receptor α (ERα) and ERβ in human cells. *Toxicological Science*. 80: 14–25.

Picherit, C., Coxam, V., Bennetau-Pelissero, C., Kati-Coulibaly, S., Davicco, M.-J., Lebecque, P., and Barlet, J.-P., 2000. Daidzein is more efficient than genistein in preventing ovariectomy-induced bone loss in rats. *Journal of Nutrition*. 130: 1675–1681.

Rodan, G.A., and Martin, T.J., 2000. Therapeutic approaches to bone diseases. *Science*. 289: 1508–1514.

Setchell, K.D.R., and Lydeking-Olsen, E., 2003. Dietary phytoestrogens and their effect on bone: evidence from *in vitro* and *in vivo*, human observational and dietary intervention studies. *The American Journal of Clinical Nutrition*. 78: 593S–609S.

Setchell, K.D.R., Brown, N.B., and Lydeking-Olsen, E., 2002. The clinical significance of the metabolite equol: a clue to the effectiveness of soy and its isoflavones. *Journal of Nutrition*. 132: 3577–3584.

Sehmisch, S., Hammer, F., Christoffel, J., Seidlová-Wuttke, D., Tezval, M., Wuttke, W., Stuermer, K.M., and Stuermer, E.K., 2008. Comparison of the phytohormones genistein and resveratrol and 8-prenylnaringenin as agents for preventing osteoporosis. *Planta Medica*. 74: 794–801.

Sehmisch, S., Uffenorde, J., Maehlmeyer, S., Tezval, M., Jarry, H., Stuermer, K.M., and Stuermer, E.K., 2010a. Evaluation of bone quality and quantity in osteoporotic mice – The effects of genistein and equol. *Phytomedicine*. 17: 424–430.

Sehmisch, S., Erren, M., Kolios, L., Tezval, M., Seidlová-Wuttke, D., Wuttke, W., Stuermer, K.M., and Stuermer, E.K., 2010b. Effects of isoflavones equol and genistein on bone quality in a rat osteopenia model. *Phytotherapy Research*. 24: S168–S174.

Sims, N.A., Dupont, S., Krust, A., Clement-Lacroix, P., Minet, D., Resche-Rigon, M., Gaillard-Kelly, M., and Baron, R., 2002. Detection of estrogen receptors reveals a regulatory role for estrogen receptor-β in bone remodeling in female but not in males. *Bone*. 30: 18–25.

Tezval, M., Sehmisch, S., Seidlová-Wuttke, D., Rack, T., Kolios, L., Wuttke, W., Stuermer, K.M., and Stuermer, E.K., 2010. Changes in the histomorphometric and biomechanical properties of the proximal femur of ovariectomized rats after treatment with the phytoestrogens genistein and equol. *Planta Medica*. 76: 235–240.

Tousen, Y., Ezaki, J., Fujii, Y., Ueno, T., Nishimuta, M. and Ishimi, Y., 2011. Natural S-equol decreases bone resorption in postmenopausal, non-equol-producing Japanese women: a pilot randomized, placebo-controlled trial. *Menopause* 18: 563–574.

Turner, R.T., Riggs, B.L., and Spelsberg, T.C., 1994. Skeletal effects of estrogen. *Endocrine Reviews.* 15: 275–300.

Turner, R.T., Maran, A., Lotinun, S., Hefferan, T., Evans, G.L., Zhang, M., and Sibonga, J.D., 2001. Animal models for osteoporosis. *Reviews in Endocrine and Metabolic Disorders.* 2: 117–127.

Yingjie, H., Ge, Z., Yisheng, W., Ling, Q., Hung, W.Y., Kwoksui, L., and Fuxing, P., 2007. Changes of microstructure and mineralized tissue in the middle and late phase of osteoporotic fracture healing in rats. *Bone.* 41: 631–638.

Yuan, J.-P., Wang, J.-H., and Liu, X., 2007. Metabolism of dietary soy isoflavones to equol by human intestinal microflora – implication for health. *Molecular Nutrition and Food Research.* 51: 765–781.

Zubik, L., and Meydani, M., 2003. Bioavailability of soybean isoflavones from aglycone and glucoside forms in American women. *The American Journal of Clinical Nutrition.* 77: 1459–1465.

CHAPTER 37

Isoflavones and Inflammation in Adipose Tissue and Implications for Health

MARIA TERESA BLAY,* MONTSERRAT PINENT AND ANNA ARDÉVOL

Departament de Bioquímica i Biotecnologia, Grup de Recerca en Nutrigenòmica, Universitat Rovira i Virgili, Tarragona, Spain
*Email: mteresa.blay@urv.cat

37.1 An Introduction to Inflammation in Adipose Tissue

Inflammation is considered a protective reaction of the body to irritation, injury, or infection, and it is involved in pathologies such as obesity, which entails "metaflammation" or metabolically triggered inflammation.

Obesity is characterized by an increase in adipose tissue mass (mainly by cell hypertrophy and hyperplasia) compared with total mass. In humans and animal models, obesity is characterized by a state of chronic sub-clinical inflammation at systemic and local level. Inflammation in obesity can also be localized in several tissues, of which the most important is white adipose tissue (WAT). WAT is a dynamic organ that changes mass throughout its life in response to the animal's metabolic needs. Adipose tissue depots are diverse and their metabolic functions vary. Subcutaneous or visceral tissues exert different functions in the body, with the visceral one being the most important in obesity metaflammation.

Food and Nutritional Components in Focus No. 5
Isoflavones: Chemistry, Analysis, Function and Effects
Edited by Victor R Preedy
© The Royal Society of Chemistry 2013
Published by the Royal Society of Chemistry, www.rsc.org

One of the main characteristics of adipose tissue depots in obesity is macrophage infiltration. Macrophages originate in the monocyte-circulating population in blood that enters the adipose tissue, where it differentiates into macrophages.

Macrophages are considered the major players in the body's response to immunogenical challenges, producing excess amounts of reactive oxygen species (ROS), high levels of nitric oxide (NO), and proinflammatory cytokines that can aggravate and propagate the local inflammation and deregulate the normal function of target cells.

Adipose tissue in the obese is thus a source of systemic inflammatory factors (Table 37.1). It accounts for one-third of interleukin-6 (IL-6) plasma concentration and produces IL-8 and macrophage chemoattractant protein-1 (MCP-1) that also reinforce the systemic proinflammatory state. However, macrophages by themselves are also a proinflammatory molecule source. In fact, previous studies performed with 3T3-L1 preadipocyte cells and RAW 264.7 murine macrophage cells have shown the importance of macrophages in the enhancement of the inflammatory state of adipocytes. It has been reported that macrophage-secreted factors, such as tumor necrosis factor-α (TNF-α) and IL-1β, among many others, have inflammatory effects on adipocytes producing a variety of adipokines, chemokines, and adhesion molecules at protein and expression levels, and nuclear factor κB (NF-κB) activation that can contribute significantly to the systemic inflammation and insulin resistance associated with obesity (Permana *et al.* 2006) (Figures 37.1 and 37.2).

Obesity-associated macrophage infiltration into adipose tissue is responsible for both local and systemic inflammation as seen in humans and animal models. Recent studies have pointed to the phenotypic change of macrophages in lean and obese adipose tissue; M1 or classically activated (proinflammatory) macrophages and M2 or alternatively activated (anti-inflammatory) macrophages. Adipocytes in lean adipose tissue produce humoral factors that induce M2 activation of macrophages that release anti-inflammatory mediators. Meanwhile, hypertrophied adipocytes secrete proinflammatory saturated fatty acids, cytokines, and chemokines to induce M1 polarization of macrophages,

Table 37.1 Inflammatory mediators in adipose tissue.

Name	Origin	Role
IL-6	Macrophages, adipocytes	Proinflammatory
Adiponectin	Adipocytes	Anti-inflammatory
TNF-α	Macrophages, adipocytes	Proinflammatory
Leptin	Adipocytes	Proinflammatory
FFA	Adipocytes	
Resistin	Adipocytes	Proinflammatory
MCP-1	Macrophage	Proinflammatory
CRP	Macrophages	Proinflammatory
IL-1β	Macrophages	Proinflammatory
MIP-1a	Macrophages	Proinflammatory
Pentraxin-3 (PTX-3)	Macrophages, adipocytes	Inflammatory marker

Isoflavones and Inflammation in Adipose Tissue and Implications for Health

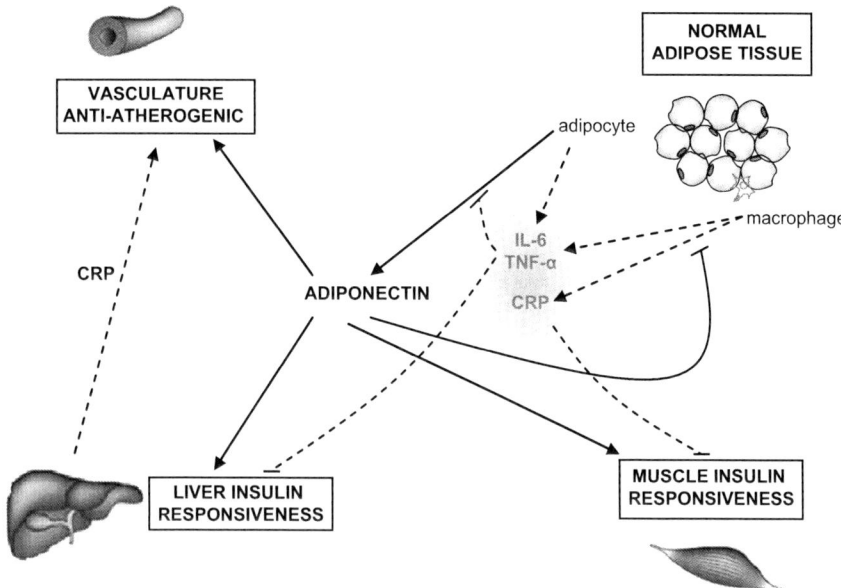

Figure 37.1 Inflammation in adipose tissue in normal weight individuals. In a lean state the adiponectin production by adipose tissue is normal and maintains muscle, liver, and vasculature functionality. The production of proinflammatory mediators is low and is inhibited by adiponectin.

which in turn produce proinflammatory cytokines and chemokines, thereby accelerating adipose tissue inflammation.

There are three main hypotheses that could explain the origin of inflammation or "metaflammation": (a) the hypoxia model (P. Trayhurn), (b) the endoplasmic reticulum stress model (G. Hotamisligil) and (c) the adipocyte apoptosis model (C. Keuper). They are all discussed below.

The basis for the link between increased adiposity and inflammation is unclear. It has been proposed by Trayhurn's group that hypoxia may occur in areas within adipose tissue in obesity as a result of adipocyte hypertrophy compromising effective O_2 supply from the vasculature, thereby instigating an inflammatory response through recruitment of the transcription factor hypoxic inducible factor-1. Studies in animal models (mutant mice, diet-induced obesity) and cell culture systems (mouse and human adipocytes) have provided strong support for a role for hypoxia in modulating the production of several inflammation-related adipokines, including increased IL-6, leptin and macrophage migration inhibitory factor (MIF) production, together with reduced adiponectin synthesis. Hypoxia also induces inflammatory responses in macrophages (Wood et al. 2009).

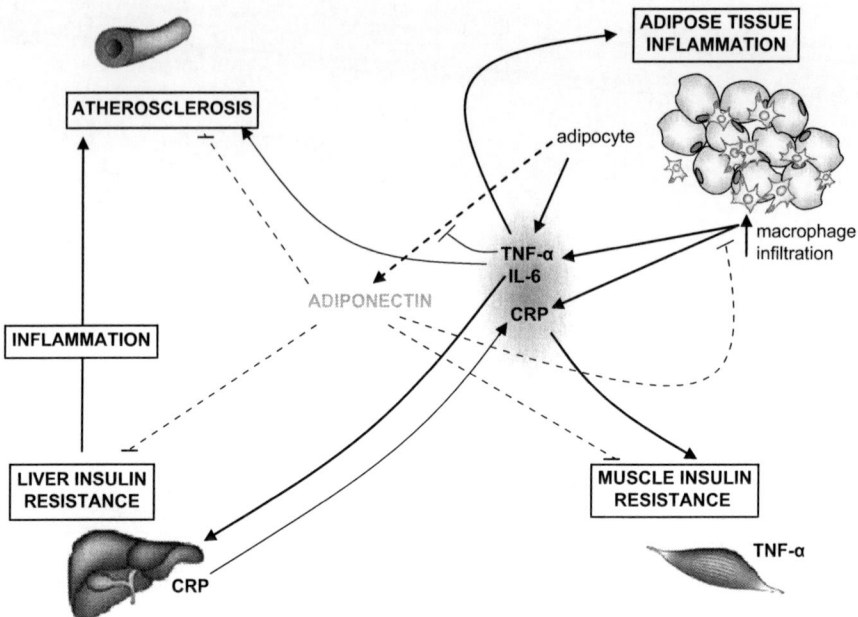

Figure 37.2 Inflammation in adipose tissue in obese individuals. Inflammation in obesity is initiated locally in adipose tissue by an increase of infiltrated macrophages and an increased production of proinflammatory molecules. Those molecules travel by blood from adipose tissue to target organs such as liver, muscles, and vasculature. Those target organs contribute actively to the maintenance and aggravation of systemic inflammatory state by producing more cytokines and acute phase reactants, such as the CRP production by the liver in response to adipose tissue-derived IL-6. On the other hand, the production of the anti-inflammatory adiponectin is decreased by the action of TNF-α.

According to the second model, there is a proinflammatory loop/crosstalk between resident macrophages and adipocytes in the adipose tissue *via* free fatty acids (FFA).

FFA originating in lipid lypolisis, which acts on the Toll-like receptor 4 (TLR-4) of macrophages, thus stimulating macrophage secretion of proinflammatory cytokines. This is mainly in the form of TNF-α, which in turn stimulates lipolysis in adipocytes. This is accompanied by endoplasmic reticulum stress induced by TNF-α in adipocytes (Hotamisligil 2006) (Figure 37.3).

Recent findings suggest fat cell apoptosis may initiate macrophage recruitment. Keuper *et al.* (2011) investigated the effects of an inflammatory microenvironment on fat cells using human THP-1 macrophages and SGBS human adipocytes. Macrophage-secreted factors induced insulin resistance and apoptosis of adipocytes. They describe a novel interaction of macrophages and fat cells, *i.e.* the induction of apoptosis, suggesting a feed-forward cycle in which

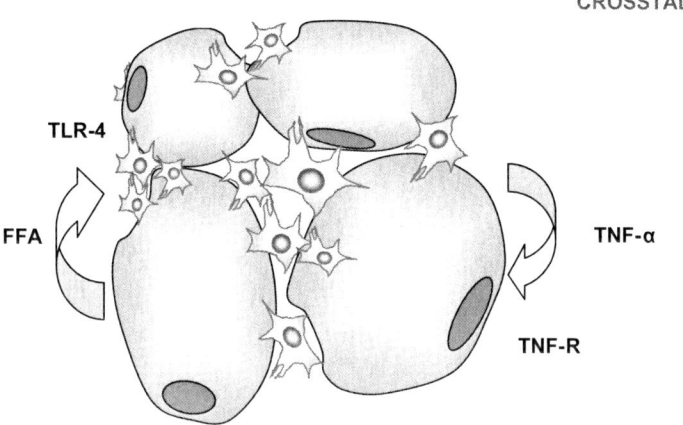

Figure 37.3 Crosstalk between macrophages and adipocytes. Molecular mechanism underlying adipose tissue inflammation. During the course of obesity, adipose tissue secretes several chemotactic factors to induce macrophage infiltration into adipose tissue. Circulating monocytes migrate and infiltrate into adipose tissue through adhesion process to endothelial cells. Macrophages enhance the inflammatory changes through the crosstalk with parenchymal adipocytes. The macrophage-derived TNF-α induces the release of saturated FFA from adipocytes *via* lipolysis once recognized by the TNF-α receptor (TNF-R), which, in turn, induces inflammatory changes in macrophages *via* TLR4. Such a paracrine loop between adipocytes and macrophages constitutes a vicious cycle, thereby further accelerating adipose tissue inflammation.

macrophages drive the inflammatory process forward by inducing insulin resistance and concomitant apoptosis of adipocytes (Keuper *et al.* 2011).

Some pathways in the cell modulate inflammation with signaling cascades involving several proteins in the cytoplasm that are phosphorylated and translocated to the nucleus to induce the expression of inflammation genes. One of the most important pathways in inflammation is the NF-κB signalling pathway. c-Jun N-terminal kinase (JNK)– activation protein-1 (AP-1) is also implicated in inflammation. When an extracellular stimulus arrives at the cell membrane, TLRs activate complex signaling cascades involving three major mitogen-activated protein kinase (MAPK) signaling pathways. These MAPKs activate the transcriptional activity of the AP-1 proteins such as Fos and Jun by phosphorylation, and they can then be translocated to the nucleus and induce the expression of inflammation target genes.

37.2 An Overview of Isoflavones as Anti-inflammatory Agents

Isoflavones, which belong to the flavan-3-ol group of phenolic compounds, are a variety of natural compounds in the plant kingdom which includes several compounds such as genistein (Gen), daidzein (Daid) and its metabolite equol (Eq), formononetin and puerarin. However, they are also present in some extracts such as chungkukjang extract, or red clover, and in isoflavone-rich foods such as soy protein, soy milk, and other soy-derived foods such as soy bean cake, all of which are commonplace in Asian cuisine. There are also synthetic isoflavones such as ipriflavone.

Isoflavones are phytoestrogens, which means that they can act as weak ER agonists (compared with animal estrogens) or as antagonists because of their similarity in structure to 17β-estradiol and their affinity for ERβ, thus blocking the ER or exerting mild estrogenicity or competing with the animal estrogens for the ER effect.

The perceived beneficial properties of isoflavones for health go beyond hormone-dependent cancers and osteoporosis, and include immunity and inflammation (Dixon 2004). The mechanism of anti-inflammatory action of isoflavones is now slightly clearer.

37.2.1 Isoflavones can Modulate Proinflamatory Factors in Macrophages and Adipocytes and Down-regulate Gene Expression of Proinflammatory Mediators

Isoflavones possess chemical properties such as free radical scavenging and antioxidant activity, and, in recent years, isoflavones have been demonstrated to have anti-inflammatory effects *in vivo* as well as *in vitro*. This anti-inflammatory activity of isoflavones uses several mechanisms. One of these is the decreased activation of inflammation-related transcription factors (NF-κB) and thus a down-modulation of a series of proinflammatory genes at TNF-α, IL-2, IL-6, IL-8 and IL-1β expression level in several mouse and human cell lines.

In a study by Blay *et al.* (2010), the expression pattern of proinflammatory genes was examined in order to determine the effect of inflammation on the RAW 264.7 cell model and to ascertain how isoflavones and their active functional metabolites alleviate the inflammatory burst in macrophages. The extent of gene modulation at a transcriptome level due to the presence of isoflavones was also evaluated. The results showed that Gen (20 µM) and Eq (10 µM) significantly inhibited the overproduction of NO and PGE_2 induced by lipopolysaccharide (LPS) plus interferon-γ (INF-γ) when a pre-activation treatment was performed, or when isoflavones were administered during activation with endotoxin (20 h). However, Daid did not exert similar effects. Moreover, both isoflavone treatments regulated the gene transcription of cytokines and inflammatory markers, among others (Table 37.2) (Blay *et al.* 2010).

In a similar study, but with a different cell line of macrophages, the analysis of proinflammatory cytokines in cell supernatants revealed that Gen displayed a dichotomous pattern. The *in vitro* production of TNF-α, macrophage inflammatory protein 1a (MIP-1a), and IL-6 on IC-21 macrophages stimulated with LPS revealed that at a concentration of 100 µM Gen, the proinflammatory

Table 37.2 Isoflavone effect on gene expression in endotoxin-inflammed RAW 264.7 mouse macrophages. RAW 264.7 macrophages were incubated for 24 h with 20 μM equol or 10 μM genistein and after treated with 1 μg mL^{-1} LPS or co-treated with isoflavones and LPS. Then cells were collected for RNA isolation and TaqMan low-density arrays (TLDAs) were performed. Results are calculated taking inflammed cells as control.

Gene symbol	Pretreatment with isoflavone (20 h)	
Cytokines and inflammatory factors	Equol	Genistein
Ptgs2	↓	↓
Nos2	↓	=
Crp	=	↓
Il1b	=	↓
Socs3	=	↓
Ppard	=	↓

Gene symbol	Co-treatment with isoflavone (20 h)	
Cytokines and inflammatory factors	Equol	Genistein
Ptgs1	↑	↑
Nos2	↓	↓
Crp	↓	=
Il1b	↓	↓
Socs3	↓	↓

cytokine levels decreased, whereas at a concentration of 10 μM, the production of TNF-α, MIP-1a, and IL-6 increased (Verdrengh et al. 2003).

In a relevant paper, Pinent et al. (2011) explored how the soy isoflavones Gen and Daid and their metabolite Eq affected inflammatory molecules in 3T3-L1 adipocytes, given that adipose tissue is a key organ in the development of obesity and insulin resistance in which isoflavones might have a beneficial role. Since obesity and insulin resistance have been associated with low grade chronic inflammation, they used 3T3-L1 adipocytes inflamed with TNF-α treatment as a model. The results showed that chronic exposure to isoflavones (24 h) prevented the secretion of the inflammatory factors PGE$_2$ and IL-6, with Eq being the most effective. The molecular basis of the isoflavone effect was the down-regulation of the cytokine gene expression and inflammatory factors (Table 37.3) (Pinent et al. 2011).

37.2.2 Isoflavones can Modulate Proinflamatory Factors in Adipose Tissue by a Decrease in Adipose Tissue Mass and/or Number

Natural products have potential for inducing apoptosis, inhibiting adipogenesis, and stimulating lipolysis in adipocytes. The decrease in adipose tissue mass is associated with a decrease in the proinflammatory state. Various dietary

Table 37.3 Isoflavone effect on gene expression in TNF-α-inflammed 3T3-L1 adipocytes. Mature 3T3-L1 adipocytes were incubated for 24 h with 20 µM Eq, 10 µM Daid or 10 µM Gen. Then cells were collected for RNA isolation and TLDAs were performed. Results are calculated taking non-treated cells as control.

Gene symbol	Chronic treatment with isoflavone (24 h)		
Cytokines and inflammatory factors	Daid	Eq	Gen
Mmp9	↓	↓	↓
Vcam-1	↓	↓	↓
Ccl2	↓	↓	↓
Adamts1	↓	↓	↓
Il6	↓	↓	=
Hsd11b1	↓	=	↓
Ptgs1	=	=	↓
Ptgs2	=	=	↑
Nos2	↓	=	=
NF-κB signalling cascade			
Ikbke	=	=	↓

bioactives such as Gen target different stages of the adipocyte life cycle. Different stages of adipocyte development include preadipocytes, maturing preadipocytes and mature adipocytes. Dietary bioactives such as Gen affect adipocytes during specific stages of development, resulting in either inhibition of adipogenesis or induction of apoptosis.

37.2.2.1 Modulation of Adipogenesis by Isoflavones

Negative regulators of adipogenesis have important health-related implications for anti-adiposity and anti-inflammation in obesity. Significant breakthroughs in the search for inhibitors of adipogenesis have been made in the last three decades. Gen is an inhibitor of the adipogenesis process, and is thus of great interest in obesity therapy (Harp 2004).

37.2.2.2 Modulation of Lipolysis by Isoflavones

Isoflavones can induce lipolysis in adipose tissue. Gen has a direct influence on adipocyte metabolism. Gen-induced impairment of the anti-lipolytic action of insulin has been found in adipocytes 3T3-L1. This ability of Gen may contribute to the decreased triglyceride accumulation in adipose tissue (Szkudelska *et al.* 2008). Gen also induces lipolysis in adipocytes. In adipocytes isolated from both ovariectomized (OVX) female and healthy male rats, incubation with Gen (20–300 µM) was shown to increase epinephrine-induced lipolysis, and, in fully differentiated 3T3-L1 adipocytes, Gen (100 µM) elevated basal and epinephrine-induced lipolysis. In 3T3-L1 preadipocytes, Gen (50–100 µM) inhibited differentiation, triglyceride accumulation, and peroxisome-proliferator-activated receptor-γ (PPAR-γ) expression. The lipolysis induction mechanism has been studied in depth and Gen has been shown toinhibit phosphodiesterase 3 (PDE3) and increase cAMP to induce lipolysis in isolated rat adipocytes in a mechanism dependent on protein kinase A. Indeed, Gen and Daid fit very well in the catalytic site of human PDE3B, PDE4B, and PDE4D (Peluso 2006).

37.2.2.3 The Proapoptotic Effect of Isoflavones

The proapoptotic effect of Gen has been demonstrated on adipose tissue *in vitro* and *in vivo*. *In vitro*, 3T3-L1 mature adipocytes were treated with 0, 1, 10, 100, and 400 μM Gen for 48 h and then assayed for apoptosis. Apoptosis occurred due to 400 μM Gen and *in vivo* in rats with 1500 mg kg^{-1} Gen in their diet (Kim et al. 2006).

37.3 What We Know from Animal Models and Isoflavone Effects on Adipose Tissue Inflammation and Health

Isoflavones have received prevalent usage due to their 'health benefits' of decreasing hormone-dependent cancers and postmenopausal symptoms. However, little is known about their effects on inflammation related to obesity or metaflammation.

37.3.1 The Effects of Isoflavones in Animal Models of Obesity: Effect of Isoflavones in Rats with High-fat Diet (HFD)-induced Obesity/Insulin Resistance and the Analysis of Inflammatory Factors

The link between obesity and insulin resistance largely accounts for the pathogenesis of metabolic syndrome and diabetes mellitus, in which adipokine expression plays a key role. Chen et al. (2006) performed an interesting experiment to explore the effects of soy isoflavone on low-grade inflammation in rats with HFD-induced insulin resistance and explore the mechanisms of soy isoflavones in improving insulin sensitivity. The rats with HFD-induced insulin resistance were randomly divided into one control group and three isoflavone groups gavaged with soy isoflavones at doses of 50, 150, and 450 mg kg^{-1} respectively. Fasting blood glucose, fasting insulin, and the inflammatory markers IL-6, TNF-α, C-reactive protein (CRP), resistin and adiponectin in the serum were measured 1 month after the treatment. In the 150 and 450 mg kg^{-1} soy isoflavone groups, fasting body weights, visceral adipose tissue deposition, fasting insulin, resistin, TNF-α in serum, and insulin resistance index (IR index) were lower in comparison with the control group, and in the 450 mg kg^{-1} soy isoflavone group, serum IL-6 levels were lower, and adiponectin had increased. No differences were found in the CRP protein levels between the three isoflavone groups. Soy isoflavone may ameliorate insulin sensitivity by decreasing visceral adipose deposition and adjusting low-grade inflammatory molecules derived from WAT (Chen *et al.* 2006).

The effects of soy isoflavone on insulin sensitivity and specifically on the adipocytokine profile in HFD-induced insulin resistant rats were studied later by Zhang *et al.* (2008). Male Sprague–Dawley rats (I = 80) were randomly assigned into a basal diet-fed group and an HFD-fed group. The HFD-induced insulin resistance rats were assigned into a control group and three soy isoflavones-treated groups with different dosages (50, 150, and 450 mg kg^{-1}). Fasting blood glucose, insulin, and adipocytokines in serum and mRNA expressions of adipocytokines in perirenal WAT were measured 30 days later. The administration of 450 mg kg^{-1} day^{-1} of soy isoflavones reduced the body weights and depositions of visceral adipose tissue as well as improved insulin resistance in HFD-induced insulin resistance rats. The mechanisms were associated with soy isoflavone, regulating the expression of adipocytokines, including increased adiponectin expression, a decrease in leptin, resistin, and TNF-α expression in WAT. The serum levels of adiponectin, leptin, resistin, and TNF-α were also dose-dependently modulated by soy intake in a corrective manner (Zhang *et al.* 2008).

The isoflavone puerarin has been studied due to its comprehensive biological actions. In one study, male Sprague–Dawley rats were fed on a normal control diet or HFD for 6 weeks, followed by an administration of puerarin (100 and 200 mg kg^{-1}) for up to 8 weeks. The effect of puerarin on HFD-induced insulin resistance and adipokine expression in rats was investigated. Serum levels of leptin and resistin were markedly increased by high fat and retarded by puerarin treatment. mRNA expression of leptin and resistin in epididymal WAT was modified by HFD and improved by puerarin in the same pattern as found in serum (Zhang *et al.* 2010).

Obesity has been strongly associated with fatty liver, visceral adiposity, adipose tissue inflammation, and a variety of adipocytokines. It has been reported that Gen inhibits fatty liver by enhancing fatty acid catabolism in the liver. In order to determine whether this anti-steatotic effect of Gen is linked to visceral adipocyte metabolism, C57BL/6J mice were therefore fed on a normal fat diet, a HFD or a HFD supplemented with Gen (1, 2, and 4 g kg^{-1} diet) for 12 weeks. The mice fed the HFD gained body weight, exhibited increased visceral fat mass and high levels of serum and liver lipids, and developed fatty liver disease, in contrast with the mice fed on the normal fat diet and Gen supplementation (2 and 4 g kg^{-1} diet), in which these alternations were normalised. In the linear regression analysis, visceral fat and the inflammatory adipocitokine TNFα were strongly correlated with fatty liver. Gen supplementation augmented adiponectin and reduced TNFα. Taken as a whole, these findings show that Gen may prevent fatty liver by regulating visceral adipocyte metabolism and inflammatory adipocytokines (Kim *et al.* 2010).

37.4 What We Know from Human Studies and the Effects of Isoflavone on Adipose Tissue Inflammation

37.4.1 Studies in Obese Postmenopausal Women

Postmenopausal women have been taken as a model of weight gain, because systemic loss of estrogen at menopause is associated with increased adiposity. Postmenopausal women can become obese, and their levels of systemic inflammation and oxidative stress consequently increase pathologically. The effect of isoflavones on this phenomenon has been studied.

The effect on abdominal fat and inflammatory markers of soy in obese postmenopausal women was studied. The women were Caucasian and African-American ($n = 39$) in a double-blind controlled trial. They were randomized to soy supplementation or to a casein placebo without isoflavones. Body composition and body fat distribution were measured at baseline and after 3 months, and serum levels of CRP, IL-6, TNF-α, leptin, and adiponectin were measured. The results showed that soy supplementation reduced total and subcutaneous abdominal fat and IL-6, but no difference was noted for CRP, TNF-α, leptin, or adiponectin in obese postmenopausal women (Christie et al. 2010). The simplest mechanism to modify CRP levels is probably weight reduction, as individuals increase from normal weight to overweight to obese, and the proportion of individuals with elevated CRP dramatically increases. These data are not surprising, because fat cells, or adipocytes, are a major source of IL-6 production and hence will lead to increased levels of CRP. Weight loss is probably the fundamental mechanism by which we can reduce the impact of the inflammatory process in the medium to long term. A study regarding the effect of isoflavones on inflammatory markers in postmenopausal women [body mass index (BMI) 18–40] evaluated the effects of soy isoflavones, both in soymilk and in supplement form, on markers of immunity and oxidative stress. Isoflavone treatment in postmenopausal women did not significantly influence concentrations of IFN-γ, IL-2, TNF-α, or CRP in plasma or of 8-isoprostane in urine (Ryan-Borchers et al. 2006).

37.4.2 Studies with Genistein (Gen) in Human Cell Inflammation

Gen has been tested with positive results as a putative anti-inflammatory molecule in human fat biopsies of obese subjects. First, in a study with human fat biopsies from obese patients that were ncubated with Gen, it was found that Gen inhibited TNF-α and that plasminogen activator inhibitor-1 (PAI-1) (an acute phase protein involved in inflammation in obesity) protein release was inhibited by Gen in a dose-dependent way. This inhibition reached 80% of the maximum inhibitory effect at 100 µg/ml. Gen down-modulates inflammatory factors in human adipose tissue cells (Cigolini et al. 1999).

Secondly, Gen inhibits the enzyme soluble phospholipase A_2 (sPLA$_2$) and PGE$_2$ secretion. PLA$_2$ catalyzes the hydrolysis of arachidonic acid from phospholipids, which is the precursor of prostaglandins and other eicosanoids. Prostaglandins are potent mediators of the inflammatory response identified in inflammatory sites. Gen inhibits the sPLA$_2$ of inflammatory exudates (human synovial fluid and human pleural fluid) in a concentration-dependent manner (Dharmappa et al. 2010).

Some studies have been performed with peripheral blood mononuclear cells (PBMCs) from healthy individuals post-stimulated with endotoxin (LPS) or proinflammatory cytokines (TNF-α) to ascertain the effect of Gen on inflammation with interesting results. Gen diminished secretion of TNF-α and IL-8 and down-regulated TNF-α and IL-8 expression but not PGE$_2$ production (Richard et al. 2005).

Summary Points

- Obesity is an inflammatory disease.
- Obesity involves "metaflammation" or a metabolically triggered inflammation.
- Adipose tissue in the obese is a source of systemic and local inflammatory factors.
- Isoflavones are natural food compounds mainly present in soy and soy derivates that resemble estrogen in structure and function.
- Isoflavones can bind estrogen receptors and act as mild estrogens.
- Isoflavones can act on estrogen receptors in an independent manner.
- Isoflavones can act as anti-inflammatories in adipose tissue by decreasing the production of inflammatory mediators in adipose tissue.
- Genistein is the most extensively studied isoflavone with beneficial effects for health.
- Some isoflavones have mild uterotropic effects and must be evaluated for safety when used in normal women.

Key Facts

Key Facts for Adipose Inflammatory Molecules

- Adipose tissue cells secrete proinflammatory factors in an autocrine, paracrine and endocrine way.
- The systemically relevant proinflammatory factors are IL-6, TNF-α, MCP-1, and CRP.
- The systemically relevant anti-inflammatory factors are adiponectin and insulin.
- Adipocytes secrete proinflammatory saturated fatty acids, cytokines, and chemokines to induce M1 polarization of infiltrating macrophages.

Key Facts for Isoflavone Effects in Adipose Tissue

- Equol administration attenuates weight gain and shows favorable metabolic effects. However, because of its mild uterotrophic activity, its use in women warrants further safety studies.
- Soy isoflavones decrease visceral adipose deposition and adjust low-grade inflammatory molecules derived from WAT in HFD-induced insulin resistant rats.
- Soy isoflavones regulate the gene expression of adipokines, including adiponectin, leptin, resistin, and TNF-α in HFD-induced insulin resistant rats.
- The chronic administration of $450\,\mathrm{mg\,kg^{-1}\,day^{-1}}$ of soy isoflavones decreased the body weights and depositions of visceral adipose tissue, as well as improving insulin resistance in HFD-induced insulin resistant rats.
- Soy supplementation reduces total and subcutaneous abdominal fat and IL-6 but not CRP, TNF-α, leptin, or adiponectin levels in obese postmenopausal women.

Definitions of Words and Terms

Acute-phase proteins: These are a type of proteins in which plasma concentrations increase (positive acute-phase proteins) or decrease (negative acute-phase proteins) in response to inflammation, and are mainly derived from the liver.

Adipocytes: These are fat cells that compose adipose tissue, and specialize in storing energy as fat.

Adipokines: These are cell-to-cell signalling molecules secreted by adipocytes.

Cytokines: These are small cell-signaling protein molecules that are secreted by numerous cells used in intercellular communication.

Eicosanoid: This is the collective term for oxygenated derivatives of three different 20-carbon essential fatty acids that act as inflammatory mediators.

Fatty acid: A carboxylic acid with an unbranched aliphatic chain, which is either saturated or unsaturated.

High-fat diet (HFD) rats: These are obese rats due to intake of a diet rich in fat content.

Inflammatory markers: These are molecules, normally plasmatic, that promote inflammation.

Macrophages: These are white blood cells produced by the differentiation of monocytes in tissues. Macrophages function in both innate immunity as well as specific defense mechanisms in vertebrates.

Subcutaneous adipose tissues: These are the fat deposits that are found just below the skin.

Transcription factor: A protein that binds to specific DNA sequences, thereby controlling the transcription of genetic information from DNA to mRNA.

Visceral adipose tissue, visceral or abdominal fat: This is located inside the abdominal cavity. Visceral fat is composed of several adipose depots including mesenteric, epididymal, and perirenal depots.

White adipose tissue (WAT): This is a tissue in mammals that accumulates energy in form of fat and also is a thermic insulator and an endocrine organ of the body.

List of Abbreviations

AP-1	activation protein-1
CRP	C-reactive protein
Daid	daidzein
Eq	equol
ER	estrogen receptor
FFA	free fatty acids
Gen	genistein
HFD	high-fat diet
IL	interleukin
IR index	insulin resistance index
LPS	lipopolysaccharide
MAPK	mitogen-activated protein kinase
MCP-1	macrophage chemoattractant protein-1
MIP	macrophage inflammatory protein
NF-κB	nuclear factor κB
NO	nitric oxide
PDE	phosphodiesterase
PGE_2	prostaglandin E_2
$(s)PLA_2$	(soluble) phospholipase A_2
PPAR	peroxisome-proliferator-activated receptor
TLDA	TaqMan low-density array
TLR	Toll-like receptor
TNF-α	tumor necrosis factor-α
WAT	white adipose tissue

References

Blay, M., Espinel, A.E., Delgado, M.A., Baiges, I., Bladé, C., Arola, L., and Salvadó, J., 2010. Isoflavone effect on gene expression profile and biomarkers of inflammation. *Journal of Pharmacologycal and Biomedical Analysis.* 51(2): 382–390.

Chen, S.W., Zhang, L.S., Zhang, H.M., and Feng, X.F., 2006. Effects of soy isoflavone on levels of low-grade inflammatory peptides in rats with insulin resistance. *Nan Fang Yi Ke Da Xue Xue Bao.* 26(10): 1484–1486.

Christie, D.R., Grant, J., Darnell, B.E., Chapman, V.R., Gastaldelli, A., and Sites, C.K., 2010. Metabolic effects of soy supplementation in postmenopausal

Caucasian and African American women: a randomized, placebo-controlled trial. *American Journal of Obstetrics and Gynecology*. 203(2): 153.

Cigolini, M., Tonoli, M., Borgato, L., Frigotto, L., Manzato, F., Zeminian, S., Cardinale, C., Camin, M., Chiaramonte, E., De Sandre, G., and Lunardi, C., 1999. Expression of plasminogen activator inhibitor-1 in human adipose tissue: a role for TNF-α? *Atherosclerosis*. 143(1): 81–90.

Dharmappa, K.K., Mohamed, R., Shivaprasad, H.V., and Vishwanath, B.S., 2010. Genistein, a potent inhibitor of secretory phospholipase A_2: a new insight in down regulation of inflammation. *Inflammopharmacology*. 18(1): 25–31.

Dixon, R.A., 2004. Phytoestrogens. *Annual Review in Plant Biology*. 55: 225–261.

Harp, J.B., 2004. New insights into inhibitors of adipogenesis. *Current Opinion in Lipidology*. 15(3): 303–307.

Hotamisligil, G., 2006. Inflammation and metabolic disorders. *Nature*. 444: 860–867.

Keuper, M., Blüher, M., Schön, M.R., Möller, P., Dzyakanchuk, A., Amrein, K., Debatin, K., Wabitsch, M., and Fischer-Posovszky, P., 2011. An inflammatory micro-environment promotes human adipocyte apoptosis. *Molecular and Celullar Endocrinology*. 339(1–2): 105–113.

Kim, H.K., Nelson-Dooley, C., Della-Fera, M.A., Yang, J.Y., Zhang, W., Duan, J., Hartzell, D.L., Hamrick, M.W., and Baile, C.A., 2006. Genistein decreases food intake, body weight, and fat pad weight and causes adipose tissue apoptosis in ovariectomized female mice. *Journal of Nutrition*. 136(2): 409–414.

Kim, M.H., Kang, K.S., and Lee, Y.S., 2010. The inhibitory effect of genistein on hepatic steatosis is linked to visceral adipocyte metabolism in mice with diet-induced non-alcoholic fatty liver disease. *British Journal of Nutrition*. 104(9): 1333–1342.

Peluso, M.R., 2006. Flavonoids attenuate cardiovascular disease, inhibit phosphodiesterase, and modulate lipid homeostasis in adipose tissue and liver. *Experimental Biology and Medicine*. 31(8): 1287–1299.

Permana, P.A., Menge, C., and Reaven, P.D., 2006. Macrophage-secreted factors induce adipocyte inflammation and insulin resistance. *Biochemistry and Biophysics Research Communications*. 341: 507–514.

Pinent, M., Espinel, A.E., Delgado, M.A., Baiges, I., Bladé, C., and Arola, L., 2011. Isoflavones reduce inflammation in 3T3-L1 adipocytes. *Food Chemistry*. 125: 513–520.

Richard, N., Porath, D., Radspieler, A., and Schwager, J., 2005. Effects of resveratrol, piceatannol, tri-acetoxystilbene, and genistein on the inflammatory response of human peripheral blood leukocytes. *Molecular Nutrition and Food Research*. 49(5): 431–442.

Ryan-Borchers, T.A., Park, J.S., Chew, B.P., McGuire, M.K., Fournier, L.R., and Beerman, K.A., 2006. Soy isoflavones modulate immune function in healthy postmenopausal women. *American Journal of Clinical Nutrition*. 83(5): 1118–1125.

Szkudelska, K., Nogowski, L., and Szkudelski, T., 2008. Genistein, a plant-derived isoflavone, counteracts the antilipolytic action of insulin in isolated rat adipocytes. *Journal of Steroid Biochemistry and Molecular Biology*. 109(1–2): 108–114.

Verdrengh, M., Jonsson, I., Holmdahl, R., and Tarkowski, A., 2003. Genistein as an anti-inflammatory agent. *Inflammation Research*. 52(8): 341–346.

Wood, I.S., de Heredia, F.P., Wang, B., and Trayhurn, P., 2009. Cellular hypoxia and adipose tissue dysfunction in obesity. *Proceedings of the Nutrition Society*. 68(4): 370–377.

Zhang, H.M., Chen, S.W., Zhang, L.S., and Feng, X.F., 2008. The effects of soy isoflavone on insulin sensitivity and adipocytokines in insulin resistant rats administered with high-fat diet. *Natural Product Research*. 22(18): 1637–1649.

Zhang, W., Liu, C.Q., Wang, P.W., Sun, S.Y., Su, W.J., Zhang, H.J., Li, X.J., and Yang, S.Y., 2010. Puerarin improves insulin resistance and modulates adipokine expression in rats fed a high-fat diet. *European Journal of Pharmacology*. 649(1–3): 398–402.

CHAPTER 38

Isoflavones for Menopausal Vasomotor Syndrome

RAFAEL BOLAÑOS DIAZ, MD, MSc*[a] AND
JUAN CARLOS ZAVALA GONZALES, MD, MSc[b]

[a] Calle 9, n° 241, Dpt. 101-A, Urb. Monterrico Norte, San Borja, Lima 41, Lima, Peru; [b] Av. Angamos-Este, Cdra. 26, F 47, San Borja (Lima 41), Lima, Perú
*Email: rbolanosd@correo.unmsm.edu.pe; rbolanosd@yahoo.es

38.1 Pharmacology Aspects

Isoflavones, coumestans, lignans and some resorcinol derivatives are considered to be phytoestrogens, molecules that perform important functions in certain vegetables and share properties similar to human estrogens. Because of this structural and functional similarity, they have been studied for many years to establish their clinical usefulness to manage menopausal syndrome. Isoflavones and coumestans (coumestrol, metoxycoumestrol) are flavonoid compounds, whereas lignans (enterodiol, enterolactone) and resorcinol derivatives are non-flavonoids (Figure 38.1). The latter abound in grapes and in grape products, such as wine. Greater emphasis has been placed on the study of this type of phytoestrogen because soy (*Glycine max*) isoflavones are distributed in many sources of human consumption and there are certain epidemiological associations described in populations with a high consumption of these foods.

There are various types of isoflavones but those most studied in connection with human health are genistein, daidzein and glycitein, which are the names given to the non-glycosidic forms to distinguish them from the corresponding

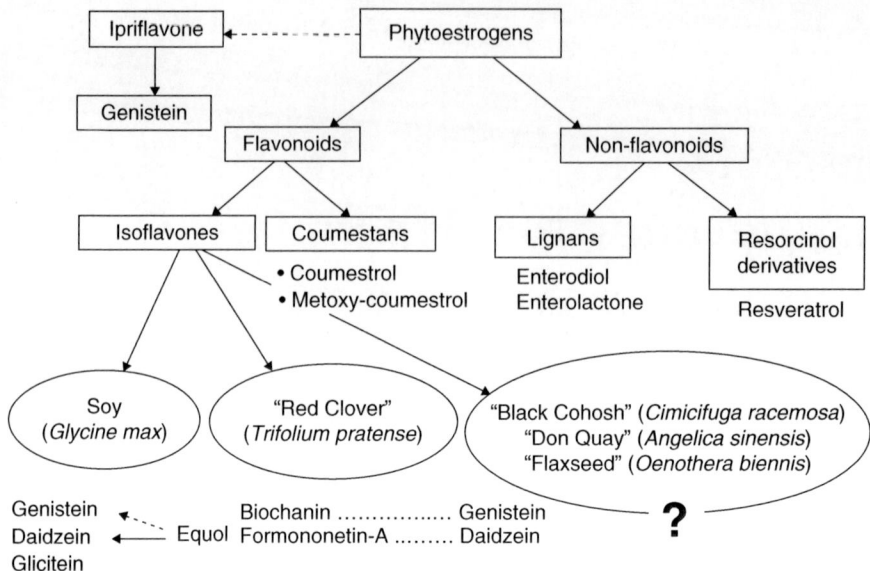

Figure 38.1 Chemical types of phytoestrogens.

glycosidic forms (linked to glucose): genistin, daidzin and glycitin. Other less active forms, such as formononetin and biochanin A, are partially metabolized by intestinal flora to obtain genistein and daidzein in a highly variable proportion that depends on individual intestinal enzymatic capacity. A synthetic form of isoflavone (ipriflavone) has been used in several clinical trials, as it is estimated that its intestinal metabolism produces genistein in approximately 10% of the total load contributed by ipriflavone. These compounds generally have a chemical structure based on two phenolic rings joined by a heterocyclic ring (forming the basic structure of isoflavones) and with various hydroxyl substitutions (-OH) at different points of the heterocyclic chain, generating a specific chemical name for each isoflavone (Figure 38.2). The presence of these hydroxyl radicals provides these compounds with a significant antioxidant capacity (Button and Patel 2004; Palacios 2002; Yildiz 2006).

Trifolium pratense (red clover), a vegetable species used to feed equines but not apt for human consumption, is a rich source of formononetin and biochanin A. Nevertheless, the clinical efficacy of this source on menopausal vasomotor syndrome has been evaluated for a long time, although the results were non-conclusive due to the high heterogeneity of the studies. Other sources, such as *Cimicífuga racemosa* (black cohosh), *Angelica sinensis* (dong quai) or *Oenothera bienis* (evening primrose), have been described, but their benefits on the climacteric syndrome are more based on experience than on formal scientific evidence. However, the phytoestrogen composition in these sources has not been well determined (Button and Patel 2004; Low Dog 2005).

Studies on phytoestrogens have accumulated enough evidence to support their alternative use in the management of climacteric vasomotor symptoms,

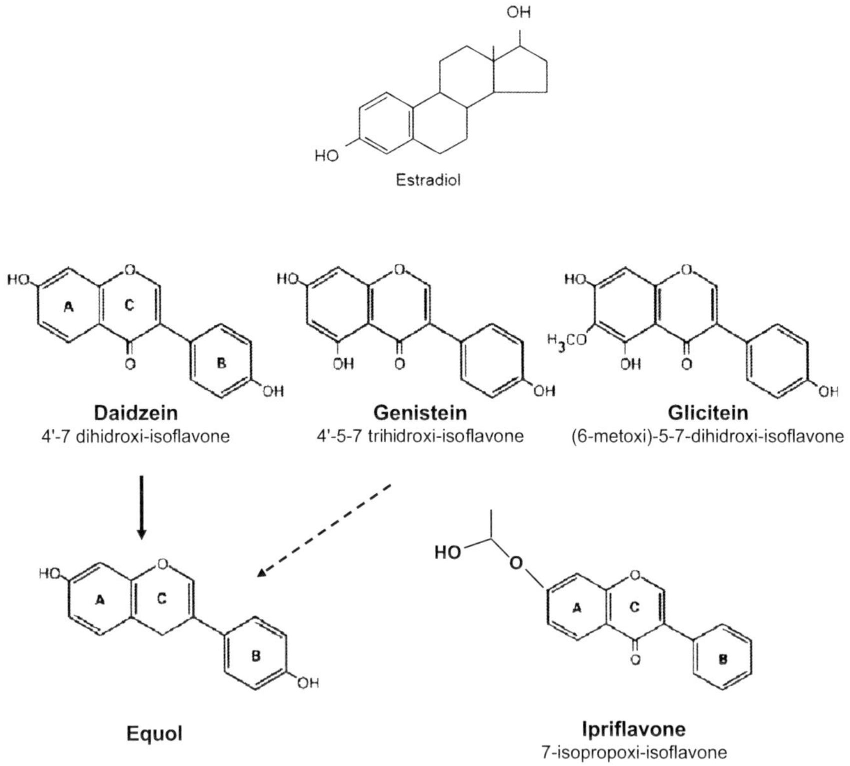

Figure 38.2 Chemical structures of isoflavones.

and such benefit seems to be supported by its isoflavone content. However, the main sources (soy, *T. pratense*, *C. racemosa* and *A. sinensis*) differ in isoflavone composition, and such difference may explain the heterogeneity in the results of the studies when analyzed as a whole (Button and Patel 2004; Low Dog 2005; Mahady 2003). Such heterogeneity can be minimized with subgroup analysis. For this reason, it is important that studies with phytoestrogens determine the exact composition of the intervention under study.

Genistein, daidzein and glycitein are the predominant isoflavones in soy (*G. max*), in contrast with the composition of red clover (*T. pratense*) based on formononetin and biochanin, its main flavonoids. Other sources, such as black cohosh (*C. racemosa*) and dong quai (*A. sinensis*), are popularly used but there is much uncertainty to their exact isoflavone composition (Mahady, 2003). For this reason, the results of studies that evaluate the efficacy of all these sources should not be extrapolated and it would more consistent to conduct separate analyses. Moreover, considering that the proportion of isoflavones in soy may vary in accordance with the preparation administered, we prefer to classify this source into three types, "dietary isoflavone intake", "soy extract" and "soy isoflavone concentrate" (genistein or daidzein), to try to reduce the heterogeneity observed in all the studies.

The study of equol has been a focus of attention. This active intestinal metabolite of daidzein and genistein has shown a significant biological activity. The clinical response has been related to the intestinal enzymatic capacity for generating equol. It is presumed that an important proportion of menopausal women could not produce sufficient equol, which represents a disadvantage to show a significant clinical result. This hypothesis is still under research and ongoing studies would enable us to draw more definitive conclusions in this regard. Other metabolites, such as O-desmethyl-angolensin (O-DMA) (Atkinson *et al.* 2005; Jou 2008; Yuan *et al.* 2007) have been described, but they do not perform an important biological function. Although it has been proposed that intestinal enzymatic capacity to form equol in the metabolism of phytoestrogens would be a plausible cause to predict the clinical response, some clinical studies have not been able to prove this hypothesis. New trials on the use of an equol concentrate are awaited to define the actual impact which the metabolism of isoflavones would have on the predicted clinical efficacy.

Some pharmacokinetic aspects in the metabolism of isoflavones must be highlighted. Although the intestinal absorption rate is still uncertain, preliminary studies indicate that it would be approximately $0.5\,\mathrm{mg\,kg^{-1}\,dose^{-1}}$, which would be consistent with most of the doses used in clinical trials. That is, loads above $0.5\,\mathrm{mg\,kg^{-1}\,dose^{-1}}$ would progressively saturate intestinal absorption receptors up to generating the fecal loss of the excess dose. However, there is some population variability about the calculation of this absorption rate, and today it is believed that the absorption range would be broader than believed up to now (Setchell 2001).

Isoflavones have affinity for the estrogen receptors (ER) α and β, but, for ERα, they behave like competitive antagonists, showing a blocking effect with regard to ovarian estrogens. Nevertheless, they have an intense activity in ERβ and their biological effects are shown through them. The distribution of ERs varies; thus, for example, ERα predominate in mammary and endometrial tissue, while ERβ predominate in bone, nervous and endothelial tissues. However, this distribution of receptors varies among individuals. This would (at least partially) explain the variability of their efficacy when evaluating the clinical effect of these compounds. Vaginal tissue does not have a clear predominance of any type of these receptors, making uncertain the use of local phytoestrogens to manage the vaginal dryness, dyspareunia or external dysuria of the climacteric syndrome (Palacios 2002; Yildiz 2006).

38.2 Clinical Evidence

The number of studies published on the efficacy of isoflavones to manage climacteric hot flushes has increased significantly since the 1990s (Usui 2006). Synthesis evidence currently reveals a clear tendency in favor of the effect of isoflavones on menopausal vasomotor syndrome, despite the existence of studies that have not proven a significant difference from placebo. Various hypotheses have been proposed to explain this difference in results, such as the capacity for metabolizing equol, the type of isoflavones or the critical quantity of some of them; however, new publications with better standardized sources of these compounds are awaited to homogenize the combined results of the studies. For instance, the meta-analysis by Williamson-Hughes *et al.* (2006) established a critical threshold of genistein to predict the efficacy of these phytoestrogens by observing that the trials which had found a significant clinical response associated to the use of phytoestrogens contained an intake of genistein above 15 mg/day in comparison with those studies in which the intake of isoflavones was smaller (Figure 38.3). However, this hypothesis was not strengthened by the observations of the Study of Women's Health Across the Nation (SWAN), which evaluated the relationship between vasomotor symptoms and race during menopausal transition in 3198 women, and where no relationship was found between the consumption of genistein and vasomotor symptoms (Gold *et al.* 2006).

Initially, the few trials that used *T. pratense* as sn isoflavone source did not yield convincing results about the efficacy of isoflavones (Knight *et al.* 1999; Tice *et al.* 2003); however, it is clear that the sources of these compounds may greatly differ in their composition. Thus, for instance a standardized extract of *T. pratense* (Promensil®) has a very different isoflavone composition from a standardized soy extract (the most currently used source in clinical trials). More than 80% of the composition of the first one is based on formononetin and

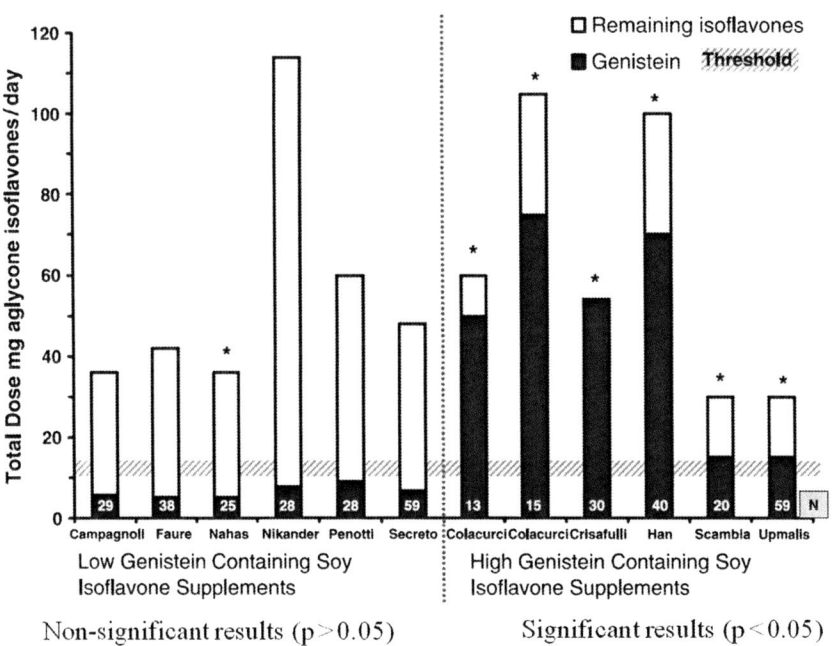

Figure 38.3 Meta-analysis by Williamson-Hughes *et al.* (2006).

biochanin A, whereas more than 90% of the composition of the latter is based on genistein and daidzein. These differences in composition generate part of the heterogeneity in published meta-analyses.

The importance and need to provide an appropriate treatment for menopausal vasomotor syndrome has been well proven, not only for the deterioration it causes to the quality of life in women but also for its probable association with other clinical conditions such as osteoporosis (Lee and Kanis 1994). The cumulative evidence proves the usefulness of isoflavones and their role in the management of menopausal vasomotor syndrome. A brief review of the main synthesis studies published in this regard is shown below.

The review of the Cochrane collaboration, in charge of Lethaby et al. (2007), grouped the studies conducted with isoflavones according to the isoflavone sources used. Thus, four groups of studies were analyzed: soy extract, dietary soy, T. pratense (red clover), and "other phytoestrogens" (C. racemosa, A. sinensis, etc.). A relevant aspect of this meta-analysis was the exclusion of studies conducted with women with breast cancer background, under the hypothesis that such background would alter hormonal physiology in some way. The studies that compose the analysis of each one of these groups are shown in Table 38.1, and those studies excluded from the analysis because of the participants' breast cancer background have been identified (Nahas et al. 2004; Nikander et al. 2005; Quella et al. 2000; Secreto et al. 2004; Van Patten et al. 2002).

Due to the presence of an important heterogeneity in the studies with "soy extract" (Bicca et al. 2004; Campagnolli et al. 2005; Duffy et al. 2003; Faure et al. 2002; Han et al. 2002; Kaari et al. 2006; Khaodhiar et al. 2008; Penotti et al. 2003; Upmalis et al. 2000), "dietary supplement" (Albertazzi et al. 1998; Balk et al. 2002; Brzezinski et al. 1997; Burke et al. 2003; Colacurci et al. 2004; Dalais et al. 1998; Knight et al. 2001; Kotsopoulos et al. 2000; Lewis et al. 2006; Murkies et al. 1995; St Germain et al. 2002) and "other phytoestrogens" (Crisafulli et al. 2004; Dalais et al. 1998; Dodin et al. 2005; Heyerick et al. 2006; Lewis et al. 2006; Woo et al. 2003), the Cochrane review only carries out the meta-analysis for the group of studies developed with T. pratense (Baber et al. 1999; Jeri 2002; Knight et al. 1999; Tice et al. 2003; Van der Weijer and Barentsen 2002), which shows a discrete tendency in favor of this isoflavone source versus placebo in the efficacy to treat the menopausal vasomotor syndrome (Figure 38.4). This result coincides with those of other published meta-analyses (Howes et al. 2006; Nelson et al. 2006; Thompson et al. 2007). Nevertheless, in this group of studies the heterogeneity reached 59.9% ($I^2 = 0.599$), and in the subgroup analysis according to the isoflavone dose (40 mg day^{-1} vs. 80 mg day^{-1}) it showed a reduction in the statistical tendency and an increase in heterogeneity (68.3% in the 40 mg day^{-1} subgroup). The trial carried out by Atkinson et al. (2004) was not included in this analysis; although nearly half of the women in each group of this study suffered from hot flushes, this was not a requirement to participate in the trial. The authors also excluded the trial by Hidalgo et al. (2005) from this analysis because they did not present sufficient statistical data (did not report P-values or measures of variation).

Table 38.1 Meta-analysis by Lethaby et al. (2007). Grouping of studies according to isoflavone source.

Trifolium pratense	Soy extract	Dietary soy	Other phytoestrogens	Excluded studies[c]
Baber et al. (1999)	Bicca et al. (2004)	Albertazzi et al. (1998)	Crisafulli et al. (2004)[b]	Nikander et al. (2005)
Jeri (2002)	Campagnolli et al. (2005)	Balk et al. (2002)	Dodin et al. (2005)	Van Patten et al. (2002)
Knight et al. (1999)	Duffy et al. (2003)	Brzezinski et al. (1997)	Heyerick et al. (2006)	Nahas et al. (2004)
Tice et al. (2003)	Faure et al. (2002)	Burke et al. (2003)	Woo et al. (2003)	Secreto (2004)
Van der Weijer and Barentsen (2002)	Han et al. (2002)	Dalais et al. (1998)	Dalais et al. (1998)	Quella et al. (2000)
Atkinson et al. (2004)[a]	Khaodhiar (2008)	Knight et al. (2001)	Lewis et al. (2006)	
Hidalgo et al. (2005)[a]	Penotti et al. (2003)	Kotsopoulos et al. (2000)		
	Upmalis et al. (2000)	Lewis et al. (2006)		
	Kaari et al. (2006)[a]	St Germain et al. (2001)		
		Colacurci et al. (2004)[a]		
		Murkies et al. (1995)[a]		

[a]Not considered in the analysis because of some exclusion criteria established by the authors.
[b]Genistein concentrate was used as the isoflavone source.
[c]Women had breast cancer antecedent.

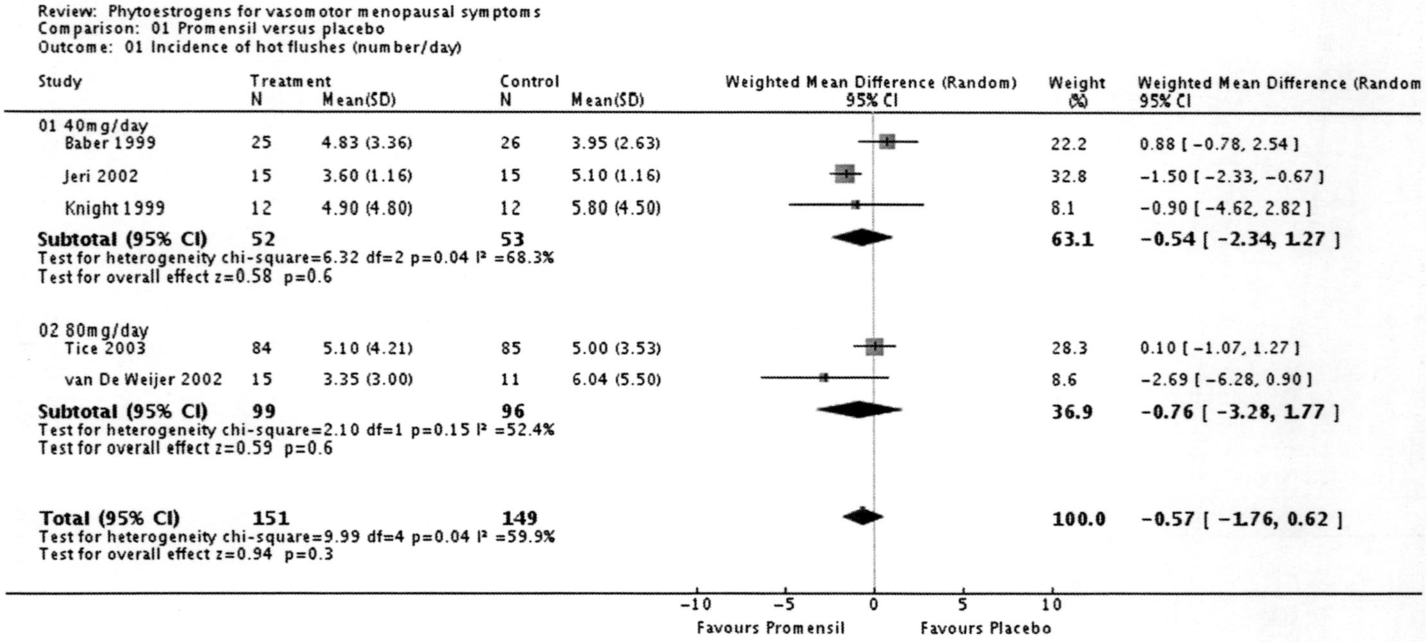

Review: Phytoestrogens for vasomotor menopausal symptoms
Comparison: 01 Promensil versus placebo
Outcome: 01 Incidence of hot flushes (number/day)

Study	Treatment N	Mean(SD)	Control N	Mean(SD)	Weighted Mean Difference (Random) 95% CI	Weight (%)	Weighted Mean Difference (Random) 95% CI
01 40mg/day							
Baber 1999	25	4.83 (3.36)	26	3.95 (2.63)		22.2	0.88 [-0.78, 2.54]
Jeri 2002	15	3.60 (1.16)	15	5.10 (1.16)		32.8	-1.50 [-2.33, -0.67]
Knight 1999	12	4.90 (4.80)	12	5.80 (4.50)		8.1	-0.90 [-4.62, 2.82]
Subtotal (95% CI)	**52**		**53**			**63.1**	**-0.54 [-2.34, 1.27]**
Test for heterogeneity chi-square=6.32 df=2 p=0.04 I²=68.3%							
Test for overall effect z=0.58 p=0.6							
02 80mg/day							
Tice 2003	84	5.10 (4.21)	85	5.00 (3.53)		28.3	0.10 [-1.07, 1.27]
van De Weijer 2002	15	3.35 (3.00)	11	6.04 (5.50)		8.6	-2.69 [-6.28, 0.90]
Subtotal (95% CI)	**99**		**96**			**36.9**	**-0.76 [-3.28, 1.77]**
Test for heterogeneity chi-square=2.10 df=1 p=0.15 I²=52.4%							
Test for overall effect z=0.59 p=0.6							
Total (95% CI)	**151**		**149**			**100.0**	**-0.57 [-1.76, 0.62]**
Test for heterogeneity chi-square=9.99 df=4 p=0.04 I²=59.9%							
Test for overall effect z=0.94 p=0.3							

-10 -5 0 5 10
Favours Promensil   Favours Placebo

Figure 38.4 Meta-analysis of *T. pretense* by Lethaby *et al.* (2007).

The prior meta-analysis by Nelson et al. (2006), which also grouped studies by isoflavone source and administered dose, did not find statistically relevant results in the group of studies that used *T. pratense*.

In the Cochrane review, five out of the eight studies that compared soy extract with placebo reported significant differences with respect to the frequency and severity of hot flushes (Bicca et al. 2004; Faure et al. 2002; Han et al. 2002; Khaodhiar et al. 2008; Upmalis et al. 2000); three studies found a reduction in the frequency of hot flushes (Bicca et al. 2004; Faure et al. 2002; Khaodhiar et al. 2008); and two studies observed a reduction in the severity of hot flushes (Han et al. 2002; Upmalis et al. 2000). The trial by Kaari et al. (2006) was not analyzed within this group of studies because it had a comparative design between soy extract and hormonal therapy and did not report significant differences between soy and estrogens in the decrease of hot flushes (at 6 months, $P = 0.74$, t test). The trial by Crisafulli et al. (2004) was not analyzed within this group as well, although the source used was a genistein concentrate; however, this decision from the authors contrasts with the matter of including the study by Khaodhiar et al. (2008) within the group that used soy extract, despite it used a daidzein concentrate (Table 38.1).

Crisafulli et al. (2004) had a parallel 3-arm design and contrasted the efficacy of genistein and hormonal therapy *versus* placebo. Although both alternatives surpassed the placebo, the effect of hormonal therapy was significantly better than that of genistein. These results contrast with Kaari et al. (2006), who did not find significant differences between soy and estrogen therapy. By combining direct and indirect evidence, it was recently proven in a meta-analysis that the effect of hormonal therapy is significantly distinguishable from that of soy when the purpose of the study corresponds to menopausal vasomotor syndrome (Bolaños et al. 2011). However, it is possible that the differences between both interventions shorten, depending on the effect studied; thus, for instance, this same difference was not found in a recent indirect meta-analysis when the effect evaluated was the risk of osteoporotic fracture (Bolaños et al. 2010-a).

Nelson et al. (2006) evaluated the studies conducted with soy under the perspective of the time of duration of the intervention. The authors of this meta-analysis made a separate evaluation of studies with durations of 4–6 weeks, 12–16 weeks and 24 weeks, but did not observe special changes in the combined result favorable to soy in these subgroups. Without making a special subgroup analysis, the meta-analysis by Thompson et al. (2007) found a marginal benefit in favor of a short-term use of *T. pratense*.

Other meta-analyses have been subsequently conducted with the same purpose of evaluating the efficacy of isoflavones *versus* placebo; however, the perspective of each one of these meta-analyses is varied, for the purpose of optimizing the analysis and not just updating the studies. Thus, the meta-analysis by Howes et al. (2006) only includes published studies and does not include those studies in which women with a breast cancer background have participated. The meta-analysis by Bolaños et al. (2010-b) only focused on published studies with soy or its by-products, regrouping them by the type of by-product used: soy extract, dietary isoflavone intake or soy isoflavone concentrate. As well as the Cochrane study, this meta-analysis did not include studies on women with a breast cancer background.

It is important to observe various studies included in each one of these meta-analyses to explain the change in the trend of the results. For example, unlike the Cochrane study, besides including studies on women with breast cancer background (Nahas *et al*. 2004; Nikander *et al*. 2005; Quella *et al*. 2000; Secreto *et al*. 2004; Van Patten *et al*. 2002), Howes *et al*. (2006) carries out subgroup analysis according to the source used: *T. pratense* and soy by-products; and it considers Atkinson *et al*. (2004), which was not included in the Cochrane meta-analysis, whereas it does not include Baber *et al*. (1999), which was included in the Cochrane meta-analysis. However, despite these modifications, the results obtained the same trend in both meta-analyses (Figure 38.5). As it was to be expected, Howes *et al*. (2006) includes a larger number of studies and their combined results show an increase in the tendency observed in the review of the Cochrane study for the group of studies with soy by-products. Another result to be highlighted in the analysis by Howes *et al*. (2006) is the following relationship that exists between the frequency of hot flushes and the clinical result, finding in the corresponding regression a higher clinical response in women with ≥ 4 hot flushes day^{-1} (Figure 38.6).

On the other hand, the meta-analysis by Bolaños *et al*. (2010-b) not only includes two studies in addition (D'Anna *et al*. 2007; Cheng *et al*. 2007) to the previously described studies, but it also has some peculiarities that distinguish it from previous analyses. The authors only focus on studies conducted with soy and its by-products and categorized these studies according to the isoflavone

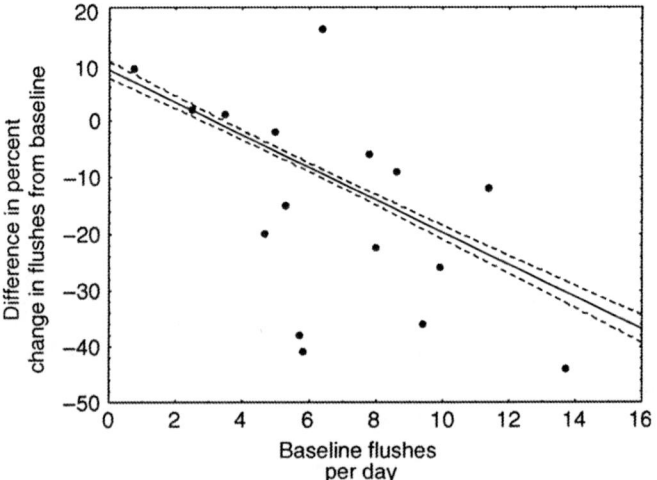

* The relationship was significant on multiple regression analysis. ($\beta = -0.48$, $P < 0.0001$).

Figure 38.5 Meta-analysis by Howes *et al*. (2006). Weighted regression analysis plot for the number of baseline flushes as a predictor of the percentage fall from baseline of flushes.

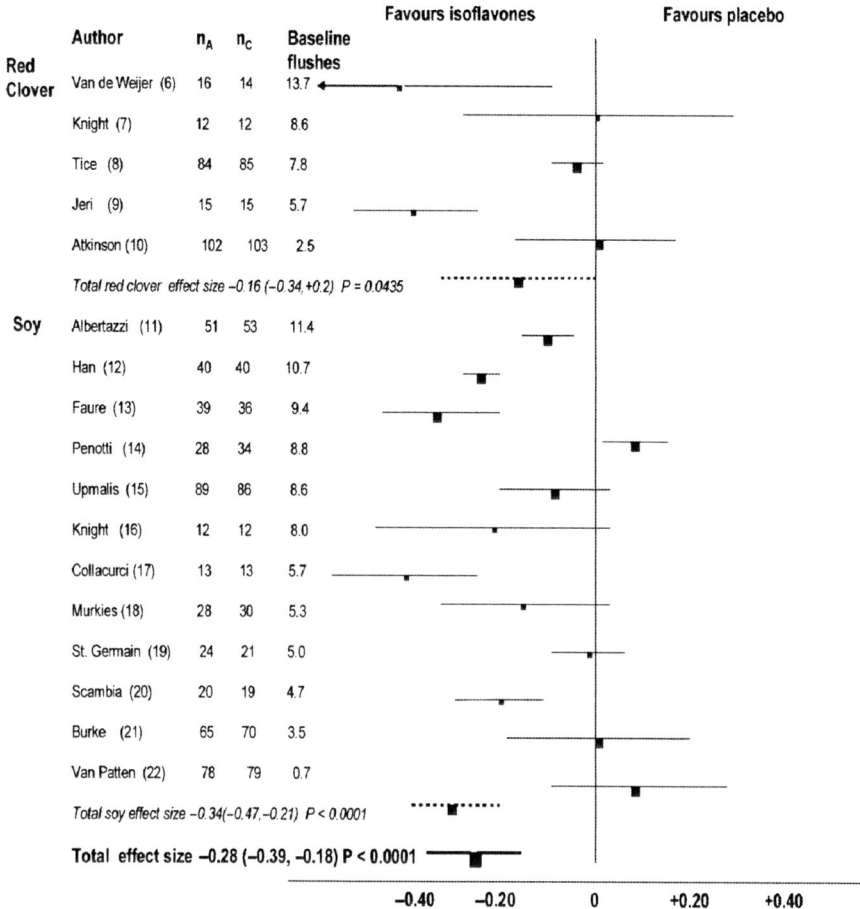

Figure 38.6 Meta-analysis by Howes *et al.* (2006).

source: soy extract, dietary isoflavone intake or soy isoflavone concentrate (Figure 38.7). The results from Crisafulli *et al.* (2004) are also included in the analysis (as this study used genistein concentrate), together with the trials by Khaodhiar *et al.* (2008) and D'Anna *et al.* (2007), which used daidzein and genistein concentrates, respectively. Similar to Howes *et al.* (2006), the authors of this work also considered the studies of Colacurci *et al.* (2004) (subgroup with soy extract) and Murkies *et al.* (1995) (subgroup with dietary isoflavone) to be acceptable for the analysis. Both trials were excluded from the Cochrane analysis.

The results of this meta-analysis are consistent with those of previous meta-analyses. The trend of the combined results favors soy and its by-products in the three subgroups analyzed; nevertheless, heterogeneity is important in "soy extract" and "dietary isoflavone intake" subgroups (42% and 56%, respectively). However, the isoflavone concentrate subgroup, made up by three

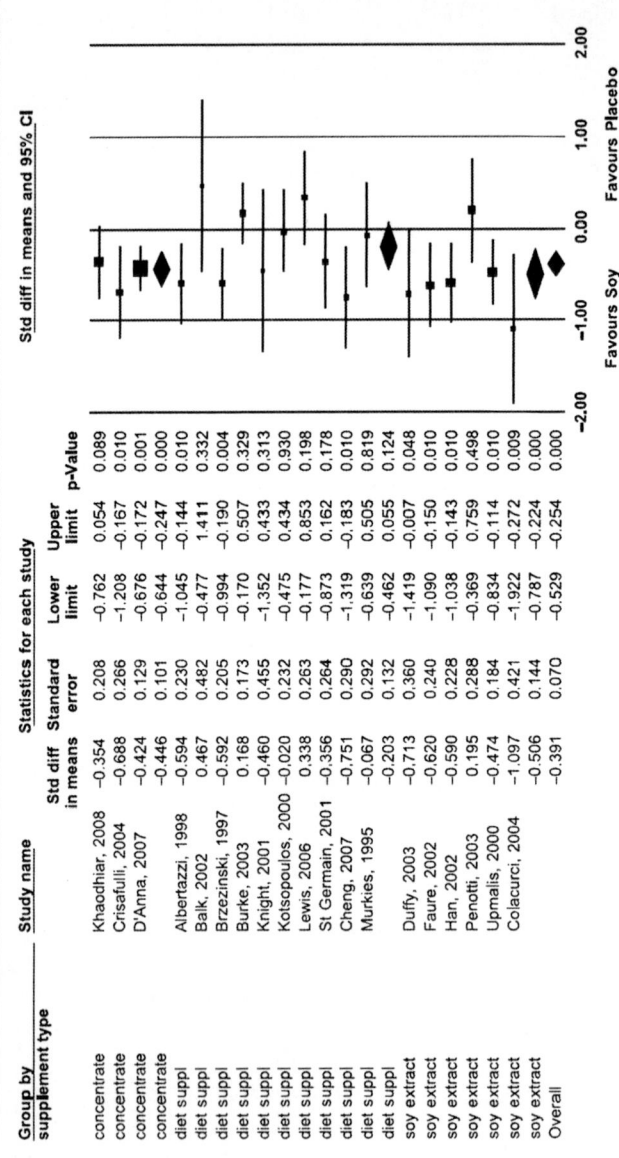

Figure 38.7 Meta-analysis by Bolaños *et al.* (2010-b).

clinical trials (Crisafulli *et al.* 2004; D'Anna *et al.* 2007; Khaodhiar *et al.* 2008), did not show a higher heterogeneity (0%). The global analysis of all the trials reached 53% heterogeneity. The same as found by Howes *et al.* (2006), in this study regression found a significant direct relationship between number of hot flushes and the response to the treatment with isoflavones.

Since the meta-analysis by Bolaños *et al.* (2010-b), new randomized and placebo-controlled trials have been published which have not been evaluated in a new synthesis review (Borges and Salazar 2009; Ferrari 2009). In the near future, this would provide higher statistical forcefulness to the results described in this review. It is important that the new studies for this same purpose standardize the main heterogeneity factors observed, such as isoflavone source, duration of treatment and measurement scales for results. On the other hand and taking into account that the updating of a meta-analysis may significantly increase the type I error (even more than the known publication bias), Bolaños *et al.* (2010-b) verified that including three additional studies to the analysis (not included in previous meta-analyses) would lead to an increase in said error. The *P*-value of the meta-analysis and the previously defined alpha value (0.05) made the updating possible without increasing the type I error (Borm and Donders 2009).

Although some meta-analyses exclude unpublished studies (unlike the Cochrane study), it is important to observe that, in general, such studies do not show the minimum methodological quality to include them in the final analysis. Because of this reason, many experts consider it more worthwhile investing in updating reviews than an exhaustive search for unpublished data. The authors also included studies with small samples and low statistical power, which increases the probability of obtaining non-significant results and reducing the probability of being published; however, this situation does not detract from the results, as the meta-analyses that include low power studies (50% on average) closely agree with the "actual" difference in the effect size of the intervention under study, even in presence of a publication bias. Previous studies have proven that the results of small trial meta-analyses are very similar to those of large trial meta-analyses on the same intervention. According to Van Driel *et al.* (2009), the key question should not focus on the existence of a publication bias but on the true impact of this bias.

It is important to observe some limitations in the synthesis studies published until now, as they limit the interpretation of results. First, in accordance with the current regulatory frame, a definite dose for dietary soy supplements has not been established. This hinders the comparison of studies that use said supplements without an appropriate standardization in their isoflavone composition. Besides, the role of intact soy protein on the effect size is still uncertain, and it has not yet been defined whether the presence of soy protein constitutes a heterogeneity factor across the studies. Second, despite the evidence that suggests that an intestinal metabolic trait could have a significant influence on the bioavailability and activity of isoflavones (a hypothesis that involves the formation of *equol*), most of the studies do not evaluate the participants' intestinal metabolic pattern, which could also be a source of heterogeneity in the results. Third, the studies analyzed often use different scores to measure the effect size (frequency and/or severity of hot flushes), which could generate another additional heterogeneity factor.

38.3 Conclusions

Studie on isoflavones (a type of phytoestrogen) have accumulated enough evidence to support their alternative use in the management of climacteric vasomotor symptoms, and such a benefit seems to be supported by its isoflavone content. However, the main sources of these compounds (soy, *T. pretense* and *C. racemosa*) differ in isoflavone content, and such differences may explain the heterogeneity in the results of the studies when analyzed as a whole.

Genistein, daidzein and glycitein are the predominant isoflavones in soy (*G. max*), in contrast with the composition of red clover (*T. pratense*) based on formononetin and biochanin A, its main flavonoids. Black cohosh (*C. racemosa*) is popularly used but evidence about its exact isoflavone content is uncertain.

Considering that the proportion of isoflavones in soy may vary in accordance with the preparation administered, this source could be classified into three types: "dietary isoflavone intake", "soy extract" and "soy isoflavone concentrate" (genistein or daidzein), to try to reduce the heterogeneity observed in all the studies.

Summary Points

- Isoflavones show a significantly higher efficacy than placebo on climacteric vasomotor syndrome.
- There are apparent differences in efficacy associated with the evaluated isoflavone source, as different sources vary in their isoflavone composition.
- The evidence does not support an effect equivalent to hormone therapy.
- A better response is obtained with isoflavones when the intensity of the vasomotor syndrome is mild or moderate, and when the number (frequency) of hot flushes is high (≥ 4 hot flushes day^{-1}).
- The use of isoflavones is pertinent when the patient does not wish to or cannot receive hormone therapy.
- Isoflavones show a safety level comparable with placebo.

Key Facts

- There has been some uncertainty about the true effect size of isoflavones. For a long time they were considered to hold a placebo effect.
- The evidence has increased in the last 20 years and several meta-analyses have demonstrated a significant effect size.
- Heterogeneity is still the key-point for the robustness of the conclusions, so it is necessary to standarize some criteria as type of supplement, patients follow-up and clinical scores to measure menopausal symptoms.

Definition of Words and Terms

Dietary isoflavone intake: Total amount of isoflavones that an individual intakes in their diet every day.

Soy extract: A genistein, daidzein and glicitein mixture that is isolated from soy plant.

Soy isoflavone concentrate: Total amount of genistein or daidzein administered as a formula.

List of Abbreviation

ER   estrogen receptor

References

Albertazzi, P., Pansini, F., Bonaccorsi, G., Zanotti, L., Forini, E., and Aloysio, D., 1998. The effect of dietary soy supplementation on hot flashes. *Obstetrics and Gynecology*. 91: 6–11.

Atkinson, C., Warren, R., Sala, E., Dowsett, M., Dunning, A., Healey, C., Runswick, S., Day, N., and Bingham, S., 2004. Red clover-derived isoflavones and mammographic breast density: A double-blind, randomized, placebo-controlled trial. *Breast Cancer Research*. 6: R170–R179.

Atkinson, C., Frankenfeld, C., and Lampe, J., 2005. Gut bacterial metabolism of the soy isoflavone daidzein: exploring the relevance to human health. *Experimental and Biological Medicine*. 230: 155–170.

Baber, R., Templeman, C., Morton, T., Kelly, G., and West, L., 1999. Randomized placebo-controlled trial of an isoflavone supplement and menopausal symptoms in women. *Climacteric*. 2: 85–92.

Balk, J.L., Whiteside, D.A., Naus, G., Deferrari, E., and Roberts, J.M., 2002. A pilot study of the effects of phytoestrogen supplementation on the postmenopausal endometrium. *Journal of the Society for Gynecological Investigation*. 9: 238–242.

Bicca, M.L., Horta, B.L. and Lethaby, A.E., 2004. Double-blind randomized clinical trial to assess the effectiveness of soy isoflavones in the relief of climacteric symptoms. Unpublished data.

Bolaños, R., and Francia, J., 2010-a. Isoflavones versus hormone therapy for reduction of vertebral fracture risk: indirect comparison. *Menopause*. 17: 1201–1205.

Bolaños, R., Del Castillo, A., and Francia, J., 2010-b. Soy isoflavones versus placebo in the treatment of climacteric vasomotor symptoms: Systematic review and meta-analysis. *Menopause*. 17: 660–666.

Bolaños, R., Zavala, J.C., Mezones, E., and Francia, J., 2011. Soy extracts versus hormone therapy for reduction of menopausal hot flashes: Indirect comparison. *Menopause*. 18: 825–829.

Borges, A.-M., and Salazar, V., 2009. Efectos de las isoflavonas de soya en el control de los síntomas perimenopáusicos. *Medicina Interna (Venezuela)*. 25: 111–127.

Borm, G., and Donders, R., 2009. Updating meta-analysis leads to larger type I errors than publication bias. *Journal of Clinical Epidemiology*. 62: 825–830.

Brzezinski, A., Adlercreutz, H., Shaoul, R., Rosler, A., Shmueli, A., Tanos, V., and Schenker, J.G., 1997. Short-term effects of phytoestrogen-rich diet on postmenopausal women. *Menopause*. 4: 89–94.

Burke, G.L., Legault, C., Anthony, M., Bland, D.R., Morgan, T.M., and Naughton, M.J., 2003. Soy protein and isoflavone effects on vasomotor in peri- and postmenopausal women: the Soy Estrogen Alternative Study. *Menopause*. 10: 147–153.

Button, B., and Patel, N., 2004. Phytoestrogens for osteoporosis. *Clinical Reviews in Bone and Mineral Metabolism*. 2: 341–356.

Campagnolli, C., Abba, C., Ambroggio, S., Peris, C., Perona, M., and Sanseverino, P., 2005. Polyinsaturated fatty acids (PUFAs) might reduce hot flushes: an indication from two controlled trials on soy isoflavones alone and with a PUFA supplement. *Maturitas*. 51: 127–134.

Cheng, G., Wilczek, B., Warner, M., Gustafsson, J., and Landgren, B.M., 2007. Isoflavone treatment for acute menopausal symptoms. *Menopause*. 14: 1–6.

Colacurci, N., Zarcone, R., Borrelli, A., Franciscis, P., Fortunato, N., Cirillo, M., and Fornaro, F., 2004. Effects of soy on menopausal neurovegetative symptoms. *Minerva Ginecologica*. 56: 407–412.

Crisafulli, A., Marini, H., Bitto, A., Altavilla, D., Squadrito, G., and Romeo, A., 2004. Effects of genistein on hot flushes in early postmenopausal women: a randomized, double–blind EPT-and placebo-controlled study. *Menopause*. 11: 400–404.

D'Anna, R., Cannata, M.L., Atteritano, M., Cancellieri, F., Corrado, F., Baviera, G., Onofrio, T., Antico, F., Gaudio, A., Frisina, N., Bitto, A., Polito, F., Minutoli, L., Altavilla, D., Marini, H., and Squadrito, F., 2007. Effects of the phytoestrogen genistein on hot flushes, endometrium, and vaginal epithelium in postmenopausal women: a 1-year randomized, double-blind, placebo-controlled study. *Menopause*. 14: 648–655.

Dalais, F.S., Rice, G.E., Wahlqvist, M.L., Grehan, M., Murkies, A.L., Medley, G., Ayton, R., and Strauss, BJG., 1998. Effects of dietary phytoestrogens in postmenopausal women. *Climacteric*. 1: 124–129.

Dodin, S., Lemay, A., Jacques, H., Legaré, F., Forest, J.-C., and Mâsse, B., 2005. The effects of flaxseed dietary supplement on lipid profile, bone mineral density, and symptoms in menopausal women: A randomized, double-blind, wheat germ placebo-controlled clinical trial. *Journal of Clinical Endocrinology and Metabolism*. 90: 1390–1397.

Duffy, R., Wiseman, H., and File, S.E., 2003. Improved cognitive function in postmenopausal women after 12 weeks of consumption of a soya extract containing isoflavones. *Pharmacology, Biochemistry and Behavior*. 75: 721–729.

Faure, E.D., Chantre, P., and Mares, P., 2002. Effects of a standardized soy extract on hot flushes: a multicentre, double-blind, randomized placebo-controlled study. *Menopause*. 9: 329–334.

Ferrari, A., 2009. Soy extract phytoestrogens with high dose of isoflavones for menopausal symptoms. *Journal of Obstetrics and Gynaecology Research*. 35: 1083–1090.

Gold, E.B., Colvin, A., Avis, N., Bromberger, J., Greendate, G.A., and Powell, L., 2006. Longitudinal analysis of the association between vasomotor symptoms and race/ethnicity across the menopausal transition: Study of Womens' health Across the Nation (SWAN). *American Journal of Public Health*. 96: 1226–1235.

Han, K.K., Soares, J.M., Haidar, M.A., Lima, G.R., and Bacarat, E.C., 2002. Benefits of soy isoflavone therapeutic regimen on menopausal symptoms. *Obstetrics and Gynecology*. 99: 389–394.

Heyerick, A., Vervarcke, S., Depypere, H., Bracke, M., and De Keukeleire, D., 2006. A first prospective, randomized, double-blind, placebo-controlled study on the use of a standardized hop extract to alleviate menopausal discomforts. *Maturitas*. 54: 164–175.

Hidalgo, L., Chedraui, P., Morocho, N., Ross, S., and San Miguel, G., 2005. The effect of red clover isoflavones on menopausal symptoms, lipids and vaginal cytology in menopausal women: A randomized, double-blind, placebo-controlled study.. *Gynecological Endocrinology*. 21: 257–264.

Howes, L., Howes, J., and Knight, D., 2006. Isoflavone therapy for menopausal flushes: A systematic review and meta-analysis. *Maturitas*. 55: 203–211.

Jeri, A., 2002. The use of an isoflavone supplement to relieve hot flushes. *Female Patient*. 27: 35–37.

Jou, H.-J., Wuc, S.-C., Chang, F.-W., Ling, P.-Y., Chu, K.S., and Wu, W.-H., 2008. Effect of intestinal production of equol on menopausal symptoms in women treated with soy isoflavones. *International Journal of Gynecology and Obstetrics*. 102: 44–49.

Kaari, C., Abi Haidar, M., Soares, J.M., Gaspar, M., Gerk de Azevedo, L., Kemp, C., Stavale, J.N., and Baracat, E., 2006. Randomized clinical trial comparing conjugated equine estrogens and isoflavones in postmenopausal women: A pilot study. *Maturitas*. 53: 49–58.

Khaodhiar, L., Ricciotti, H., Li, L., Pan, W., Schickel, Ch., Zhou, J., and Blackburn, G., 2008. Daidzein-rich isoflavone aglycones are potentially effective in reducing hot flashes in menopausal women. *Menopause*. 15: 125–132.

Knight, Krebs, E., Ensrud, K., MacDonald, R. and Wilt, T., 1999. Phytoestrogens for treatment of menopausal symptoms: a systematic review. Obstetrics and Gynecology. 104: 824–836.

Knight, D.C., Howes, J.B., Eden, J.A., and Howes, L.G., 2001. Effects on menopausal symptoms and acceptability of isoflavone-containing soy powder dietary supplementation. *Climacteric*. 4: 13–18.

Kotsopoulos, D., Dalais, F.S., Liang, Y.-L., McGranth, B.P., and Teede, H.J., 2000. The effects of soy protein containing phytoestrogens on menopausal symptoms in postmenopausal women. *Climacteric*. 3: 161–167.

Lee, S.J., and Kanis, J.A., 1994. An association between osteoporosis and premenstrual symptoms and postmenopausal symptoms. *Bone and Mineral*. 24: 127–134.

Lethaby, A.E., Brown, J., Marjoribanks, J., Kronenberg, F., Roberts, H. and Eden, J., 2007. Phytoestrogens for vasomotor menopausal symptoms (Cochrane Review). In: *The Cochrane Library*. Issue 4 (last update, July 2007).

Lewis, J.E., Nickell, L.A., Thompson, L.U., Szalai, J.P., Kiss, A., and Hilditch, J.R., 2006. A randomized controlled trial of the effect of dietary soy and flaxseed muffins on quality of life and hot flashes during menopause. *Menopause*. 13: 631–642.

Low Dog, T., 2005. Menopause: A review of botanical dietary supplements. *American Journal of Medicine*. 118: 98s–108s.

Mahady, G., 2003. Is black cohosh estrogenic?. *Nutrition Reviews*. 61: 183–186.

Murkies, A.L., Lombard, C., Strauss, B.J.G., Wilcox, G., Burger, H.G., and Morton, M.S., 1995. Dietary flour supplementation decrease postmenopausal hot flushes: effect of soy and wheat. *Maturitas*. 21: 189–195.

Nahas, E.P., Neto, J.N., Luca, L., Traiman, P., Pontes, A., and Dalben, I., 2004. Benefits of soy germ isoflavones in postmenopausal women with contraindication for conventional hormone replacement therapy.. *Maturitas*. 48: 372–380.

Nelson, H., Vesco, K., Haney, E., Fu, R., Nedrow, A., Miller, J., Nicolaidis, C., Walker, M., and Humphrey, L., 2006. Nonhormonal therapies for menopausal hot flashes: Systematic review and meta-analysis. *JAMA*. 295: 2057–2071.

Nikander, E., Rutanen, E.-M., Niemin, P., Wahlstrom, T., Ylikorkala, O., and Tiitinen, A., 2005. Lack of effect of isoflavonoids on the vagina and endometrium in postmenopausal women. *Fertility and Sterility*. 83: 137–142.

Palacios, S., 2002. *Phytoestrogens*. Ediciones Harcourt, Madrid, Spain.

Penotti, M., Fabio, E., Modena, A.B., Rinaldi, M., Omodei, U., and Vigano, P., 2003. Effect of soy–derived isoflavones on hot flushes, endometrial thickness and the pulsatility index of the uterine and cerebral arteries. *Fertility and Sterility*. 79: 1112–1117.

Quella, S.K., Loprinzi, C.L., Barton, D.L., Knost, J.A., Sloan, J.A., LaVasseur, B.I., Swan, D., Krupp, K.R., Miller, K.D., and Novotny, P.J., 2000. Evaluation of soy phytoestrogens for the treatment of hot flashes in breast cancer survivors: a North Central Cancer Treatment trial. *Journal of Clinical Oncology*. 18: 1068–1074.

Secreto, G., Chiechi, L.M., Amadori, A., Miceli, R., Venturelli, E., Valerio, T., and Marubini, E., 2004. Soy isoflavones and melatonin for the relief of climacteric symptoms: a multicenter, double blind, randomized study. *Maturitas*. 47: 11–20.

Setchell, K., 2001. Soy isoflavones – Benefits and risks from nature's selective estrogen receptor modulators (SERMs). *Journal of the American College of Nutritionists*. 20: 354s–362s.

St Germain, A., Peterson, C.T., Robinson, J.G., and Alekei, D.L., 2002. Isoflavone-rich or isoflavone-poor soy protein does not reduce menopausal symptoms during 24 weeks of treatment. *Menopause*. 8: 17–26.

Thompson, J., Pittler, M., and Ernst, E., 2007. *Trifolium pratense* isoflavones in the treatment of menopausal hot flushes: A systematic review and meta-analysis. *Phytomedicine*. 14: 153–159.

Tice, J., Ettinger, B., Ensrud, K., Wallace, R., Blackwell, T., and Cummings, S., 2003. Phytoestrogen supplements for the treatment of hot flashes: The Isoflavone Clover Extract (ICE) study. *JAMA*. 290: 207–214.

Upmalis, D.H., Lobo, R., Bradley, L., Warren, M., Cone, F.L., and Lamia, C.A., 2000. Vasomotor symptom relief by soy isoflavone extract tablets in postmenopausal women: multicenter double–blind randomized placebo-controlled study. *Menopause: Journal of the North American Menopause Society*. 7: 236–242.

Usui, T., 2006. Pharmaceutical prospects of phytoestrogens. *Endocrine Journal*. 53: 7–20.

Van der Weijer, P., and Barentsen, R., 2002. Isoflavones from red clover (Promensil®) significantly reduce menopausal hot flush symptoms compared with placebo. *Maturitas*. 42: 187–193.

Van Driel, M., De Sutter, De., Maeseneer, J., and Christiaens, T., 2009. Searching for unpublished trials in Cochrane reviews may not be worth the effort. *Journal of Clinical Epidemiology*. 62: 838–844.

Van Patten, C.L., Olivotto, I.A., and Chambers, G.K., 2002. Effects of soy phytoestrogens on hot flashes in postmenopausal women with breast cancer: a randomized controlled clinical trial. *Journal of Clinical Oncology*. 20: 1449–1455.

Williamson-Hughes, P., Flickinger, B., Messina, M., and Empie, M., 2006. Isoflavone supplements containing predominantly genistein reduce hot flash symptoms: a critical review of published studies. *Menopause*. 13: 831–839.

Woo, J., Lau, E., Ho, S., Cheng, F., Chan, C., Chan, A., Haines, C., Chan, T., Li, M., and Sham, A., 2003. Comparison of *Pueraria lobata* with hormone replacement therapy in treating the adverse health consequences of menopause. *Menopause*. 10: 352–361.

Yildiz, F., 2006. *Phytoestrogens in functional foods.* CRC Press, Taylor & Francis Group, Boca Raton, FL, USA.

Yuan, J.-P., Wang, J.-H., and Liu, X., 2007. Metabolism of dietary soy isoflavones to equol by human intestinal microflora – Implications for health. *Molecular and Nutrition Food Research*. 51: 765–781.

Subject Index

Illustrations and figures are in **bold**. Tables are in *italics*.

2HIS *see* 2-hydroxyisoflavanone synthase
4CL *see* 4-coumaroyl-CoA ligase

AB quartet 100, 111
absorption, dietary *see also* metabolism
  aglycones 4, 57, 366, 402, **440**, 544
  biochanin A (BCA) 439
  daidzein 439, 599
  formononetin (FMN) 439
  genistein 115–16, 439, 599
  genistin 116–17, 154
  glucosidases 341, 366, 439
  glucosides 4–5
  β-glucuronides 4
  glycitein 466–7
  glycosides 57, 152, 196–7, **440**, 482, 544, 599
  hydrolysis 439
  pharmacokinetics 630
  soy milk 117
ABX system 100, 111
accuracy of analysis
  definition 239, 327
  LC-ESI-MS/MS 226
  LC-MS/MS 228, 325
6″-*O*-acetyl-β-glycosides 50, 244
acetylcholine 302, 455
acetyldaidzin
  soaking 36
  structure **30**
  temperature 36

acetylgenistin
  structure **30**, **51**
  temperature 36
acetylglucosides *see also* acetyldaidzin; acetylgenistin; acetylglycitin
  plants 18
  soy products 21
  stability 36
  temperature 36
acetylglycitin
  soaking 36
  structure **30**
acid hydrolases 381
acid hydrolysis
  beverages 257–8
  definition 260, 377
  degradation issues 257, 259
  processing 36, **37**
  solid phase extraction (SPE) 257
acrosome 571
activator protein 1 (AP-1) 468, 474, 587, 589, 615
acute-phase proteins 623
acylation 288–9, 291
ad libitum definition 556
adenocarcinoma 523, 526, 587
adenosine monophosphate (AMP)-dependent protein kinase 554
S-adenosylhomocysteine (SAH) **162**
S-adenosylmethionine (SAM) **162**
adenylate cyclase 533, **535**, 537

adipocyte fatty acid-binding protein (aP2) 549
adipocytes
  apoptosis 546, 547, 553, 614–15, 619
  definition 623
  differentiation **542**, 546, 549
  genistein 618
  hypoxia 613
  macrophages 612, **615**
  peroxisome proliferator activated receptors (PPARs) 549, 553–4
  proliferation **542**
  size 546, 547, 550
  tumour necrosis factor-α (TNFα) 614
adipogenesis 556, 618
adipokines 620, 623
adiponectin 613, **614**, 619–20, 622, 623
adipose tissue *see also* obesity
  abdominal 545, 621
  diabetes risk 547
  equol 490
  genistein 544, 546–7, 549, 618–19
  isoflavones 623
  macrophages 612, **613**, 614
  quercetin 549
  resveratrol 547, 549
  subcutaneous 623
  visceral 545, 550, 620, 623, 624
  white (WAT) 546, 611, 620, 624
Adlercreutz, Dr Herman 494–5
adolescence 133
affinity chromatography 550–1, **552**, 556
AFP (alpha-fetoprotein) 486
age related differences 354, 357, 454–5, 458
ageing 452
aglycones *see also* daidzein; genistein; glycitein
  absorption, dietary 4, 57, 366, 402, **440**, 544
  acid hydrolysis 36
  definition 25
  dietary sources 356
  fermentation 54
  hydrolysis 226
  intestines 440
  metabolites 366
  soaking 38
  soy digestion 4
  stomach 152
  storage of soybeans 35
  structure **19**, *21*, **30**
  temperature 36
agonism
  definition 155, 592
  estrogen receptors (ER) 165, 471
  genistein 132, 151–2, 154, 426, 511
  glycitein 471
Akt–glycogen synthase kinase-3 (GSK-3) 133–4
alanine transaminase (ALT) 550
alcoholic beverages *371–2*, 376
alkaline hydrolysis
  beverages 255
  definition 260
  processing 36, **37**
alkaline phosphatase 599
allergies
  daidzein 305–6
  daidzin 305
  puerarin 305
almonds 402, 403, 404, 411
alpha-fetoprotein (AFP) 486
ALT (alanine transaminase) 550
Alzheimer's disease
  animal models 454, 455, 456
  glycitein 472, 473
  puerarin 304
  sex hormones 452
  soy 457, 586
AMPA (α-amino-3-hydroxy-5-methyl- 4-isoxazole propionic acid) 506, 508
β-amyloid 304, 454, 455, 472
amyloid fibrils 309
analysis 29–31, 198–204, 265, *319*, 367 *see also specific methods*
anaphylactic reactions 305

androgen receptors (AR) 484, 565, 585
Angelica sinensis (dong quai) 628–9
angina pectoris 309
angiogenesis
    daidzein 301
    definition 91, 309
    genistein 444
    isoflavones 139
animal foods *see* forage foods
animal models 592
    Alzheimer's disease 454–5, 456, 457
    coumestrol 519–21, 524–5
    development 485, *486*, 487–90, 493, 494
    menopause 456–7
    mucopolysaccharidoses (MPS) 392, **393**
    obesity 545–6, *548*, 619–20, 623
    spatial learning 453–4, 456–7
    steroid hormones 567–8
anogenital distance (AGD) 567
antagonism
    definition 592
    genistein 132, 154, 426
anthocyanidins **84**
anthocyanin synthase (ANS) **84**
anthocyanins **84**
anthroquinones 15
anti-bacterial effects 442
anti-glutathione S-transferase (GST) antibody 86–9
anti-inflammatories
    daidzein 302
    daidzin 305
    genistein 134–5
    peroxisome proliferator activated receptors (PPARs) 468
    puerarin 302, 305
antibodies 273
antimicrobial activity 149
antioxidant response element (ARE)-dependent transcription **553**
antioxidants
    daidzein 442
    definition 91, 377, 447
    equol 442, 483

    genistein 134, 426, 442, 531
    glycitein 472, 474
    heme oxygenase 1 (HO-1) 472
    8-hydroxydaidzein 442
    singlet oxygen 424
anxiety 501
AP-1 *see* activator protein 1
AP2 (clathrin adaptor protein 2) 503
APCI *see* atmospheric pressure chemical ionization
apigenin 290, 390
apigenin 7-*O*-glucoside **286**
Apios americana Medik *see* groundnuts
apoptosis
    adipocytes 546, 547, 553, 614–15, 619
    β-amyloid 454
    coumestrol 520, 524
    daidzein 301, 305
    definition 309, 447, 526
    equol 582, **584**, 588
    17β-estradiol 524
    genistein 133, 445, 547, 600, 619
    osteoclasts 599
    puerarin 298, 304, 454
    spermatocytes 571
    tumour suppressor protein p53 553
apple trees 15
arachidonic acid 622
Arachis hypogaea L., Leguminosae *see* peanuts
Argentina 41
arginase 1 (Arg1) 305
aromatase 275, 489, 518–19, 564, 565, 571
3-arylcoumarins 98–9
arylsulfatase 256
Asian diets *see also specific countries*
    cancer incidence 218
    cholesterol reduction 408
    compared to Western 168, 359, 360, 366, 400, 424
    soy intake levels 4, 7, 8, 245, 445, 490

aspartate aminotransferase/alanine aminotransferase (AST/ALT) 556
aspartate aminotransferase (AST) 550
AST (aspartate aminotransferase) 550
asthma 305
astrocytes 383–4, 472
ataxia telangiectasia and Rad3-related kinase (ATR) 137, **138**, 139
Ateleia herbert-smithii 96–8, **99**
atherosclerosis 6, 170, 303
atmospheric pressure chemical ionization (APCI) 214, 228
ATR *see* ataxia telangiectasia and Rad3-related kinase
Australia 350, 541
autocrine regulation 574
auxin transport 22

bacteria, intestinal *see* microflora, intestinal
barbiturates 504
basic hydrolysis *see* alkaline hydrolysis
BBB *see* blood–brain barrier
BCA *see* biochanin A
beer *371*, 376
Belamcanda chinensis 101–2
benzodiazepines (BZDs) 503, 504
benzopyran 79
beta cells 532–6, 537
beverages
  analysis methods *247–54*
  Brazil 351
  chromatography 258
  composition *32*, *34*, *368–72*
  consumption levels 245
  extraction 245–6, 255–8, 259
  hydrolysis 255–8
bias definition 448
bile acid 303
binding affinity
  biochanin A *88*
  coumestrol **583**
  daidzein *88*, **583**
  deoxymiroestrol **583**

diethylstilboestrol (DES) **583**
equol 6, 473, 580, 581, **583**, 600
estradiol *88*
estradiol (E2) 3, 5, 10, *88*, 438, 529
estriol 10
estrogen receptors (ER) 3, 5, 6, 10, 14, 85, 86–9, 170, 426
estrone 10
genistein 6, 10, 85, *88*, 163–4, 164–5, 500, 500–1, 529–30, **583**
genistin *88*
glycitein 471, 473
hydroxyl group effect 86
IC$_{50}$ values *88*
isoflavones 3, 5
miroestrol **583**
naringenin *88*
orobol *88*
preferences 163–4
8-prenylnaringenin **583**
prunetin *88*
prunetin (PRN) *88*
raloxifene 10
resveratrol **583**
5,7,3′,4′-tetrahydroxyisoflavone (THIF) *88*
bioavailability
  aglycones 152–3, 402
  definition 41–2, 57–8, 126, 214, 377
  factors 167
  glucosides 402
  microflora, intestinal 366
  soy products 401
  structure 197–8
biochanin A (BCA)
  absorption, dietary 439
  binding affinity *88*
  biosynthesis 159, **162**
  glucosides **290**
  metabolism 150, 220, 366, 628
  nomenclature 244
  red clover 21
  sources 4, 220
  spectra 225
  structure *21*, **219**
  toxicity 389

bioconversion 50, 53–4, 56–7, 58
    see also fermentation
biomarkers 525, 526
biosynthesis
    biochanin A (BCA) 159, **162**
    chorismate 159, **160**, 171
    p-coumaric acid 159, **161**
    p-coumaroyl-S-coenzyme A 159, **161**
    daidzein **20**, 150, 159, **161**, 387, 471, 628
    daidzein 7-O-(6″-malonyl) glucoside **20**
    O-desmethylangolensin (O-DMA) 5, 166, 439
    dihydrogenistein 141, 439
    equol 4, 5, 8, 428, 439, 580
    estradiol (E2) 275
    formononetin (FMN) 159, **162**
    genistein **20**, 150, 159, **161**, 387, 628
    glucosides 50
    glycitein 465, 474
    6′-hydroxy-*O*-demethylangolensin (DMA) 5
    6-hydroxy-*O*-demethylangolensin (DMA) 141
    isoliquiritigenin 159, **161**
    liquiritigenin 159, **161**
    naringenin 159, **161**, 387
    naringenin chalcone 159, **161**
    phenylalanine 159, **161**
    phenylpyruvate 159, **161**
    plants 18–19, **20**, 50, 158–9, **161**, 171
    puerarin 439
    5,7,3′,4′-tetrahydroxyisoflavone (THIF) 135–7, 141
bisphenol A (BPA)
    glucuronidation 123
    hydroxylation 124
    industrial uses 125
    placenta transfer 122, 124
    structure 123
    sulfation 123–4
black cohosh 628–9, 640
β-blockers 223–4

blood–brain barrier (BBB) 382, 387–8, 392, 393, 394
blood levels
    compared to estrogens 3
    conjugated metabolites 153, 544
    equol 484–5, 600
    ethnic variations 445
    genistein 511
    insulin-like growth factors 573
BMI *see* body mass index
body fat *see* adipose tissue
body mass index (BMI) 167, 541, 543, 545, 556
bone *see also* osteoporosis
    biochanin A 220
    cell proliferation 7
    coumestrol 520
    daidzein 600, 601–2, 605–6
    equol 600–1, 602, *603*, **604**, 605–6
    estradiol (E2) 601
    estrogen 597, 598–9, 602, *603*
    estrogen receptors (ER) 439, 599
    formation 318, 467, 598
    genistein 600, 601, 602, *603*, **604**, 605–6
    glycitein 467
    postmenopausal 7, 327
    resorption 316–17, 598–9, 600
    soy 599–600
    thyroid hormones 427
    types 606
bone mineral density (BMD) 7, 426, 598, 600, 601
BPA *see* bisphenol A
brain *see also* memory
    daidzein 455, **510**
    daidzin 455
    development 452, 453, 454
    estradiol 456, 459
    estrogen receptors (ER) 6–7
    genistein 453–4, **510**
    glutathione (GSH) 455
    hippocampus 453, 454, 456, 457, 458, 460, 471–2, **510**
    NMDA (*N*-methyl-D-aspartate) 506

oxidative stress 384
prefrontal cortex 456, 457, 458, 459, 460
puerarin 454
reproduction 6
soy protein isolate 456
Brazil
  coffee 377
  consumption levels 351–4
  soybean production 41
bread
  flavor 409, 414–15
  β-glucosidases 401–2
  HPLC analysis 404–5
  isoflavone content 409
  organoleptic properties 406–7, 413–16
  physiochemical properties 405–6, 410–13
  processing 418
  soy 401, 403–4
breast cancer
  blood equol levels 485
  cell proliferation 168–9
  coumestrol 524
  daidzein 301, 487–8
  daidzin 301
  equol 489, 580, 581–3
  ethnic variations 167–8, 170, 218
  genistein 132, 133, 135, 426, 488
  preventative mechanisms 7
  research history 4
  selective estrogen receptor modulators (SERMs) 90
  soy 7–8, 9, 359
  tamoxifen 90
  5,7,3′,4′-tetrahydroxyisoflavone (THIF) 137–9
  vasomotor symptom trials 632
Butea monosperma 318–20, 326, 327

C/EBP *see* CCAAT/enhancer-binding protein
c-Jun N-terminal kinase (JNK) 468, 615

C-reactive protein (CRP)
  adipose tissue 621
  atherosclerosis 303
  equol *617*
  genistein *617*
  inflammation 302, *612*, **613**, **614**
  insulin resistance 619–20
cabbage 483
cajanin (CJN) **318**
calcitonin 423
calcium sulfate 40
calibration range definition 239
cAMP *see* cyclic adenosine monophosphate
Canada 354
cancer *see also* breast cancer; prostate cancer
  anti-resorptive agents 317
  biochanin A 220
  cell proliferation 168–9
  chemoprevention 309
  colon 298, 301, 587
  coumestrol 523, 524, 525
  daidzein **300**, 301, 487–8
  daidzin 301
  diet 439, 523
  equol 489, 581–5, 587, 589–90
  estrogen receptors (ER) 562
  ethnic variations 167–8, 170
  extracellular-signal-regulated kinase (ERK) **469**, 582, 587
  gastric 439
  genistein 132, 133–4, **300**, 426, 488
  glycitein 467–9, 474
  inflammation 467–8
  inhibition 132, 359, 438
  kaempferol **299**
  luteolin **299**
  metastasis 301, 309–10, 468, 589
  mortality rates compared 356
  3-OH-flavone **299**
  ovarian 582–3
  pancreatic 586–7
  proteinase inhibitor 9 (PI-9) 582–3
  puerarin 298, **300**

cancer (*continued*)
  quercetin **299**
  recommended daily intake 31, 400
  research history 4
  selective estrogen receptor modulators (SERMs) 90
  soy food consumption 28, 85, 133, 220
  5,7,3′,4′-tetrahydroxyisoflavone (THIF) 137–9
capillary electrophoresis (CE) 271, 281
CAR (constitutive androstane receptor) 520–1
carbamylcholine **535**
carbohydrates 519–20
carbonyls 69
carcinogenicity 522, 526, 589–90
cardiovascular system
  daidzein 6, 303
  equol 6, 587–8
  estrogen receptor β (ERβ) 438
  extracellular-signal-regulated kinase (ERK) 588
  glycitein 470–1
  lipoproteins 170
  nitric oxide (NO) 170
  recommended daily intake 400
  thyroid hormones 427
case-control studies 445
caspases 301, 304, 520, 582, 586
catalase 134, 442
catechins
  definition 377
  structure **84**
catechol-*O*-methyltransferase (COMT) 166
CBG *see* cytosolic β-glucosidases
CCAAT/enhancer-binding protein (C/EBP) 549, 554, 556
CDK *see* cyclin-dependent kinases
CE *see* capillary electrophoresis
cell death-inducing effector A (CIDE-A) 550
cell membranes 504, 537

cell proliferation *see also* cancer
  daidzein 444
  definition 592
  equol 581–91
  genistein 5, 85, 133, 426, 444, 445, 582
  phytoestrogens 168–9, 172
  5,7,3′,4′-tetrahydroxyisoflavone (THIF) 137, 139
cell walls
  evaporation effect 184
  glycosaminoglycans (GAGs) 384
cellulase 256
Central American diets 4
central nervous system (CNS)
  GABA (γ-aminobutyric acid) 501–5, 509–11
  genistein 501
  glycitein 472
  mucopolysaccharidoses (MPS) 382, 383, 387, 394
  NMDA (*N*-methyl-D-aspartate) 505–11
  oxidative stress 472
cereals 350
chalcone isomerase (CHI) 20, **84**, 159, **161**, 387
chalcone reductase (CHR) 19, 20, **84**, 159, **161**, 387
chalcone synthase (CHS) 83–4, **84**, 159, **161**
chalcones
  daidzein synthesis 67
  structure **16**, **20**, **84**, **158**
chamomile tea 376, 377
chemical shift definition 111
chemiluminescence 269
chemokines 612–13
chemopreventive agent 126
chewiness *408*, 411, 418
CHI *see* chalcone isomerase
chickpeas 49, 79
children
  maximum daily intake 355
  obesity 542
  soy intake 8, 9

China
  consumption levels 354–5
  soybean production 41
Chinese medicine 294, 529 *see also* traditional medicine
Chk1 pathway **138**, 139
CHM *see* chorismate mutase
cholecystokinin **535**
cholesterol 303, 306, 309, 570
cholesterol reduction
  coumestrol 519–20
  glycitein 470
  recommended daily intake 31
  soy 408, 545, 546, 547, 554
chondroitin sulfate (CS) *383*, **385**
chorismate
  biosynthesis 159, **160**, 171
  isoflavonoids biosynthesis **161**, 171
  structure **160**
chorismate mutase (CHM) 159, **161**
chorismate synthase **160**
CHR *see* chalcone reductase
chromatograms
  groundnuts **336**
  soy milk **55**
chromatographic fingerprinting 317, **324**, 328
chromatography
  beverages 258
  definition 377
  method 275
  nuclear magnetic resonance (NMR) spectroscopy linking 101–2
  types 275
chrysoeriol 4′-glucosyl-7-xylosylglucosides 289
chungkukjang 354, 356, 616
CID *see* collision-induced dissociation
CIDE-A (cell death-inducing effector A) 550
cimetidine 135
Cimicifuga racemosa (black cohosh) 628–9, 640

cinnamate-4-hydroxylase (C4H) 83, **84**, 159, **161**
cinnamic acid **84**
cinnamoyl CoAS **84**
circulator 191
cladrin (CLN) **318**
Claisen-Schmidt condensation 64, 86
classification 3, 14, 29
clastogenicity 169, 519, 522, 526, 589
clathrin adaptor protein 2 (AP2) 503
climacteric syndrome 165–6, 169, 172, 628–9, 630 *see also* postmenopausal symptoms
clinical studies
  climacteric syndrome 630–1
  cognitive function 457–9
  equol producers 630
  functional foods **400**, 401, 407, 414, 416–17
  mucopolysaccharidoses (MPS) 392
  synthetic forms 628
  thyroid function 429, **430**, 431
CLN (cladrin) **318**
coagulation 40
Cochrane reviews 632, 635, 636, 639
coefficient of variation (CV)
  GC-MS 203
  HPLC-UV 202, 210
  isotope dilution methods 204
  LC-MS 202
  LC-MS/MS 203, 323
coffee
  analysis *248–9, 251–2, 254*
  brewing methods 375, 376, 377
  composition *369–70, 374–5*
  consumption levels 377
  extraction 256, 257
  production volumes 377
  types 375, 377
cognition
  age related differences 7
  estrogen receptor β (ERβ) 6–7

cognitive function
  endocrine disruptors (ED) 459
  estradiol 456
  genistein 392, 457
  glutathione (GSH) 455
  glycitein 471–2, 473
  postmenopausal women 452
  soy protein isolate 456
collagen 438
collision energy 283, *323*, 327
collision-induced dissociation (CID) 282–3, 285, 287, 288, 291
colon cancer 298, 301, 587
columns 270, 275–6
COMT *see* catechol-*O*-methyltransferase
conformations 102, 111, 151
β-conglycinin 545
conjugated metabolites
  blood levels 544
  definition 214
  excretion 544
  mass-to-charge ratios (m/z) *207*
  placenta 121–2
  tissue distribution 197
conjugation definition 448
constitutive androstane receptor (CAR) 520–1
consumer acceptance testing 406–7, 414–15, 417, 418, 419
consumption levels *see also* recommended daily intake
  adolescence 133
  age related differences 354, 357, 359
  Asian diets 445
  Australia 350
  Brazil 351–4
  Canada 354
  China 354–5
  education level 354, 359
  Hong Kong 355
  increase 366
  Indonesia 357–8
  Italy 355
  Japan 356
  Korea 356–7
  Singapore 357, 358
  Taiwan 358
  UK 350–1
  USA 358–9, 481, 544
convulsant drugs 503
cord blood 490, 491, 496
coronary artery disease 309
corpus luteum 586
correlation spectroscopy (COSY) 100, 111
costs
  extraction 267
  sample preparation 259
COSY *see* correlation spectroscopy
cotyledons 21, 31, 35, 42
coumarate-4-hydroxylase **84**
p-coumaric acid
  biosynthesis 159, **161**
  plants 290
  structure **84**, **161**
4-coumaroyl-CoA ligase (4CL) 83, **84**, 159, **161**
p-coumaroyl-S-coenzyme A
  biosynthesis 159, **161**
  structure **161**
coumestans **158**, 627
coumestrol
  animal models 519–20, 524–5
  binding affinity **583**
  dietary sources 525
  genotoxicity 519, 520–2, 523, 525
  health effects 519–20, 522, 524–5
  structure **16**, **99**, 100, 518
cow's milk **272**, 367, *369*, 373, 374, 377
COX-2 *see* cyclo-oxygenase-2
CPR *see* cytochrome P450 reductase
CRP (C-reactive protein)
  adipose tissue 621
  atherosclerosis 303
  equol *617*
  genistein *617*
  inflammation 302, *612*, **613**, **614**
  insulin resistance 619–20
crumb 405, *408*, 411, 412, **414**, 416

crust 405, *408*, 412, **416**
CS *see* chondroitin sulfate
CV *see* coefficient of variation
cyclic adenosine monophosphate (cAMP) 305, 532, **534**, **535**, 537, 570
cyclin B1 137, **138**, 445
cyclin-dependent kinases (CDK) 301
cyclo-oxygenase-2 (COX-2) 135
Cys-loop receptor superfamily 512
cytochrome P450
  biochanin A 220
  genistein 139, 140
cytochrome P450 enzymes
  daidzein 159, 440
  genistein 141, 159, 440
  genistein metabolism 119, 135
  isoflavonoids biosynthesis 163
cytochrome P450 reductase (CPR) 85–6, **87**
cytokines 612–13, 614, 616–17, 623
cytosolic β-glucosidases (CBG) 116–17

δ values 103, 107, *108*, 111
DAC *see* displacement affinity chromatography
DAD (diode array detectors) 184, 222
DAHP synthase **160**
daidzein
  absorption, dietary 439, 599
  antimicrobial activity 149
  binding affinity *88*, **583**
  biosynthesis **20**, 150, 159, **161**, 387, 471, 628
  dietary sources 4, 62, 350, 375
  dimethyldaidzein synthesis 74, **76**
  extraction 61–3
  fermentation 37, 54
  glycosaminoglycans (GAGs) 390
  health effects **300**, 302–3
  isoformononetin synthesis 73–4
  metabolism 5, 439
  metabolites 206
  nomenclature 149, 244, 500

peroxisome proliferator activated receptors (PPARs) 554
  soybean cultivar differences 35
  spectra **227**
  stability 36
  structure **16**, **20**, *21*, **30**, **53**, 79, 96
  synthesis 63–7
  toxicity 389
daidzein 7-*O*-(6″-malonyl) glucoside **20**
daidzein 7-*O*-glucoside-6″-*O*-malonate 62
daidzein 8-C-glucoside *see* puerarin
daidzin
  dietary sources 4, 296
  health effects 303
  stability 36
  structure **30**, **50**, **53**, **295**
de-phosphorylation definition 512
decarboxylation 21, **401**
declustering potential *323*, 328
deglycosylation
  definition 126
  intestines 116–17, 544
dendrites 454, 456, 460, 509
6′-deoxychalcones 19
5-deoxyisoflavones 19
deoxymiroestrol **583**
derivatization definition 260
dermatan sulfate (DS) *383*, **385**
descriptive analysis 407, 415–16, 419
O-desmethylangolensin (O-DMA)
  biosynthesis 5, 166, 439
  estrogenic activity 366
  excretion 402
development *see also* infants; maternal diet; placenta; reproductive system effects
  daidzein 487–8
  equol 488–9
  estrogenic activity 563
  genistein 488, 530
  humans compared to animal models 485–6
  perinatal 487–93
  placenta transfer 124–5

DHT *see* dihydrotestosterone
diabetes
  coumestrol 520
  daidzein 306–7
  development 537
  epidemiology 547
  genistein 530–7
  obesity 530, 541, 619
  puerarin 306
  retinopathy 135
  soy 549, 554
  symptoms 309
  type 2 (T2D) 530, 538, 547
dielectric material 180, 191
dietary sources
  beverages 245, 367, 376
  biochanin 4
  compared to supplements 9
  coumestrol 525
  daidzein 4, 62, 350, 375
  daidzin 4, 296
  equol 483–4
  fermented compared to non-fermented 359
  formononetin 4, 62, 373, 374, 375
  genistein 4, 149, 350, 374, 375, 426, 529
  genistin 4, 426
  glycitein 465
  glycitin 4
  isoflavones 3–4, 41, 481, 492, 543–4
  lignans 3, 500, 627
  puerarin 296
  soy products 148, 366, 563
  vegetables 245, 367
diethylstilboestrol (DES) **583**
diets
  Asian *see* Asian diets
  Central American 4
  early life exposure 6, 8, 9
  effect on epidemiological studies 4, 218–19
  maternal 453–4, 459, 489, 491, 569, 570 *see also* development

obesity 543, 544–5, 546
  soy 600
  Western *see* Western diets
differential scanning calorimetry (DSC) 406, 419
digestion **17**, 126, 439 *see also* absorption, dietary; intestines
dihydrodaidzein 5, 228, 439
dihydroflavanols **84**
dihydroflavonol reductase (DFR) **84**
dihydrogenistein
  biosynthesis 119–20, 121, 141, 439
  metabolism 5, 439
  structure **120**, **136**
dihydronaringenin chalcone *see* phloretin
dihydrotestosterone (DHT) 484–5, 490, 494, 564, 585
5,7-dihydroxy-3-(4-hydroxyphenyl)-4H-1-benzopyran-4-one *see* genistein
trans-4',7-dihydroxyisoflavan-4-ol 102
7,12-dimethylbenz(a)anthracene (DMBA) 496
dimethyldaidzein **62**, 74–8, **76**, 79
diode array detectors (DAD) 184, 222, 271
directional coupler 191
displacement affinity chromatography (DAC) 550–1, **552**, 556
distortionless enhancement by polarization transfer (DEPT) 111
DMA *see* 6'-hydroxy-O-demethylangolensin
DNA damage 169, 519, 522, **542**
doenjang 354, 356, 445, 446
dong quai 628–9
double quantum filtered COSY (DQFCOSY) 111, 340
drug discovery 221, 238
DSC *see* differential scanning calorimetry
dyer's broom (*Genista tinctoria*) 150

E2 *see* estradiol
ECM *see* extracellular matrix

ED *see* endocrine disruptors
education level 354, 359
EGFR *see* epidermal growth factor receptor
eicosanoids 622, 623
electrophile definition 79
electroporation 180, 191
electrospray ionization (ESI)
  advantages 224
  definition 328
  LC-MS/MS 203, 204, 224, 228
  mass spectrometry (MS) 272, 282, 283
  method 214
  tandem mass spectrometry (MS/MS) 321–2
  time-of-flight mass spectrometry (TOF) 103
elicitors definition 24
ELISA *see* enzyme-linked immunosorbent assay
EMS *see* enhanced mass scan
enantiomers definition 496
endocrine axis 574
endocrine disruptors (ED) 433, 459, 460, 562, 573, 574 *see also* xenoestrogens
endocytosis definition 512
endometrial tissue 9
endothelial cells 139–40, 587
endothelial NO synthase (eNOS) 303, 588
enhanced mass scan (EMS) 322, 328
enhanced product ion (EPI) scan 322, 328
enhanced resolution 322, 328
enolisation 107, **109**
enterodiol **17**
enterohepatic circulation 153, 167, 172, 440
enterolactone **17**
enzyme deficiency 381, 382, 394
enzyme hydrolysis
  beverages 255–6
  definition 260, 377
  food processing **37**, 39–40

plasma analysis 226
selective 198–203
selective estrogen receptor modulators (SERMs) 204
solid phase extraction (SPE) 256
urine analysis 223–4
enzyme-linked immunosorbent assay (ELISA) 273, 275
enzyme processing 40
enzyme replacement therapy (ERT) 382, 392, 395
enzymes
  definition 172
  inhibition 444
EPI *see* enhanced product ion (EPI) scan
epidemiology
  assays 228
  cancer 426, 445–6
  cholesterol reduction 547, *548*
  definition 448
  dietary variations 4, 426
  ethnic variations 4, 9, 10
  insulin sensitivity 547, *548*
  soybeans 41
epidermal growth factor receptor (EGFR) 391
epilepsy 501
epithelial cells 116
EPSP synthase **160**
equol
  binding affinity 6, 473, 580, 581, **583**, 600
  biosynthesis 4, 5, 8, 428, 439, 580
  dietary sources 483–4
  5α-dihydrotestosterone 484–5
  estrogenic activity 366
  excretion 5, 402
  genotoxicity 588–9
  health effects 5, 8, 9, 304, 483, 492, 496, 581–91
  isomers 484, 494, 580, 581
  lethal dose test (LD_{50}) 489
  nomenclature 483
  nuclear magnetic resonance (NMR) spectroscopy 106–7, *108*

equol (*continued*)
  research history 495
  spectra 95–6, **97**
  structure **95**, **482**, 484, 494
  urine levels 359
equol hypothesis 482, 495, 496
equol producers
  bone mineral density (BMD) 600
  breast cancer 580
  clinical studies 630
  definition 166, 483, 496
  ethnic variations 5, 7, 583
  hormones 591
  prostate cancer 583
  vasomotor symptoms 639–40
ERE *see* estrogen response element
ERK *see* extracellular-signal-regulated kinase
ERRγ (estrogen-related receptor γ) 581
ERs *see* estrogen receptors
ERT *see* enzyme replacement therapy
Escherichia coli 117
ESI *see* electrospray ionization
espresso *370*, *375*, *377*
ESRs *see* estrogen receptors
esterification 284, 289–90
estradiol (E2)
  binding affinity 3, 5, 10, *88*, 438, 529
  biosynthesis 275
  bone 601
  brain 456, 459
  clastogenicity 589
  dendrites 454, 509
  genistein competition 150
  glycosaminoglycans (GAGs) 391
  menopause 598–9
  soy milk 572
  structure **16**, **158**, 163, **482**, **530**, **563**
  testis 571–2
estriol 10
estrogen
  bone 597, 598–9
  GABA (γ-aminobutyric acid) 509–10

  isoflavone similarity 4, 132
  NMDA (*N*-methyl-D-aspartate) 509–10
estrogen receptor modification therapy 85, 90
estrogen receptors (ER)
  agonism 165
  antagonism 165
  binding affinity 3, 5, 6, 10, 14, 85, 86–9, 170, 426
  bone 439, 599
  cardiovascular system 438
  coactivators 165
  definition 91, 172–3
  drug research 150
  equol 580, 581
  genes 163
  genistein 150–2, 163–4, 165
  glycitein 470, 471, 473
  hippocampus 472
  human tissue distribution 5, 10
  hydroxyl group effect 86
  IC_{50} values *88*
  structure 163
  testis 565, 566
  thyroid function 426–7
  tissue distribution 630
estrogen-related receptor γ (ERRγ) 581
estrogen response element (ERE)
  equol 166, 582
  genistein 150, 165
estrogen synthetase *see* aromatase
estrogenic activity 265, 366, 424
estrogens
  definition 24
  environmental 157
  estrone 10, 599
  ethnic variations *see also* diets
    epidemiology 4, 9, 10, 218, 399, 445
    equol biosynthesis 5, 7, 583
p-ethyl-phenol 439
4-ethylphenol **120**, 121, 366
Eubacterium ramulus 121
evaporation 184

Subject Index

evening primrose 628
excitatory neurotransmitters 512–13
excretion
  calcium 520
  conjugated metabolites 544
  O-desmethylangolensin
    (O-DMA) 402
  equol 5, 402, 483
  genistein metabolites 119, 153
  glucuronides 167, 440, 466
  sulfate metabolites 167, 440
exercise 545
extracellular matrix (ECM),
  glycosaminoglycans (GAGs) 384,
  386
extracellular-signal-regulated kinase
  (ERK)
  cancer 468, **469**, 582, 587
  cardiovascular system 588
  daidzein **469**
  equol **469**, 582, 587, 588
  genistein 135, **469**, 549
  glycitein 468, **469**
  inflammation 135, 443
extraction *see also* acid hydrolysis;
  alkaline hydrolysis; enzyme
  hydrolysis; microwave-assisted
  extraction; solid phase extraction;
  Soxhlet extraction; supercritical
  fluid extraction
  beverages 245–6, 255–8, 259
  boiling water 51
  daidzein 61–3
  degradation issues 266
  direct 245–6, 255, 268
  food analysis 266
  freeze-drying 268
  groundnuts 334, 335
  hexane 51
  infant formula 267
  liquid-liquid 224, 228, 266
  method comparison 29–30, 266–7,
    274
  plant material 18
  plasma 224
  precipitation 267–8

rate constants *189*
solvents 267
sonication 246, 255, 257
soy products 246
temperature 186, 246, 267
time effects 187, 267
yields 186–9, 255, 256
extrusion 51
eyes 305

Fabaceae fm. 158
Fallopia japonica (Japanese
  knotweed) 15, **18**, 480
fatty acid synthase (FAS) 549
fatty acids 614, **615**, 620, 623
fatty liver 620
fermentation
  bread 402
  conversion of isoflavones 37, 50,
    53–4, 57
  definition 42
  reaction times 54–6
  soy bread 413, 418
fermented bean curd *see* tofu
fermented foods *see also* miso; natto;
  shoyu (soy sauce); tempeh
  aglycones 41, 50, 152–3
  bioavailability 544
  cancer 446, 447
fertility 265
ferulic acid 290
fetal development *see* development
FID *see* flame ionisation detectors
fingerprint analysis 317, **324**, 328
firmness *408*, 415, **416**
flame ionisation detectors (FID) 271
flavanones
  sources 15
  structure **16**, **20**, **84**, **158**
3-OH-flavone **299**
flavone synthases **84**
flavonoids
  classification 14, 294–6, 424
  definition 24–5
  health effects 296
  structure **16**, **158**

flavonol 3b-hydroxylase (FHT) **84**
flavonols **84**
flax seed 15, **17**
fluorescence assays 89
FMN *see* formononetin
follicle-stimulating hormone 132
follicle stimulating hormone (FSH) 565–6, 570
follitropin 275
food analysis
    capillary electrophoresis 271
    extraction 266
    gas chromatography (GC) 271
    immunoassays 273
    isoflavone levels 31, *32–3*, *34*
    liquid chromatography (LC) 268–9, 273–4
    mass spectrometry (MS) 272–3
Food and Drug Administration (United States) 562
food processing 401–2
forage foods 24 *see also* red clover
force 419
formononetin 7-*O*-glucoside-6″-*O*-malonate
    structure **62**
    vacuole storage 62
formononetin (FMN)
    absorption, dietary 439
    biosynthesis 159, **162**
    dietary sources 373, 374, 375
    equol biosynthesis 4
    metabolism 150, 366, 628
    nomenclature 244
    red clover 21
    sources 4, 62, 220
    spectra **227**, **320**
    structure *21*, **62**, **219**, **318**
    synthesis 67–73
    T-cells 79
    toxicity 389
    trade names 69
Föster resonance energy transfer (FRET) 89, 91
Fourier transform (FT) 282
fractalkine 302

fractures *see also* osteoporosis
    meta-analysis 635
    postmenopausal women 467, 599
    prevention 7
fragmentation 282, 285–7, 290, 291, *319*
freezable water (FW) 406, 413
freeze-drying 268
FRET *see* Föster resonance energy transfer
Friedel–Crafts acylation 63–4, 71, 79
fruit juice *369*, 373–4
FT *see* Fourier transform
fulvestrant 588, 592
functional foods 400, 407, 414, 416–17
fungi, mycorrhizal 22, 69

G-protein-coupled receptor **535**
GABA (γ-aminobutyric acid) 501–4, **505**, 509–11, 512, 513
$GABA_A$ receptor 501–3, **505**, 509–11
GAGs *see* glycosaminoglycans
gas chromatography (GC)
    beverages *251*, 258
    disadvantages 221, 271, 273
    food analysis 271
    mass spectrometry (MS) *201*, 203, 271
    sample preparation 271, 273
gastric cancer 439, 441–7
γ-GCS *see* γ-glutamyl cysteine synthetase
gender differences
    coumestrol 523
    spatial learning 452–3
gene expression 522
gene mutation 381
gene therapy 382
gene transcription 426, 444
genetically modified soybeans 355
Genista tenera 281
Genista tinctoria (dyer's broom) 150
genistein
    absorption, dietary 115–16, 439, 599
    binding affinity 6, 10, 85, *88*, 163–5, 500–1, 529–30, **583**

Subject Index

binding sites 163–5
biosynthesis **20**, 150, 159, **161**, 387, 628
blood–brain barrier (BBB) 393
cell growth 5, 85
cell membranes 504
dietary sources 4, 149, 350, 374, 375, 426, 529
17β-estradiol mimicry 123, 163–4
fermentation 37, 54
β-glucuronides 117–19
glycosaminoglycans (GAGs) 382, 390, 391–2
health effects 131, **300**, 337, 426, 501
hydrophilicity 14
metabolism 5, 119, 135, 152–3, 439
metabolites 103–5, 135, **136**, 206
natural killer cells 5
NMDA receptors 508–11
nomenclature *116*, 150, 244
peroxisome proliferator activated receptors (PPARs) 5, 154, 444, 554
protein tyrosine kinases (PTKs) 503–5
research history 150
spectra 95, **96**, **225**
stability 36
structure **16**, **20**, 115, 150–1, 529–30
substrate reduction therapy (SRT) 388–9
toxicity 389
tyrosine kinase inhibition 391
tyrosine protein kinase 5
genistein 6,8′-di-*C*-glucoside **288**
genistein-7-*O*-gentiobioside 335, **340**, 341–2, 343–4
genistein 7-*O*-glucoside **286**
genistein 7-*O*-malonylglucoside **20**, 288, **289**
genistein 7,4′-di-*O*-sulfate **105**
genistin
absorption, dietary 116–17, 154
binding affinity *88*
dietary sources 4, 53, 426
metabolism 152, 154, 336–7

nomenclature *116*
stability 36
structure **30**, **51**, **53**, **116**
genotoxicity definition 526, 592
genotypes definition 42
gentiobiose 344
gentiobioside 340
glabrene 103, **104**
glabridin **16**
glabrone **100**
glucagon 519
glucagon-like peptide-1 (GLP-1) 532, 537
glucokinase **536**
gluconeogenesis 306, 307
glucose **536**
glucose-stimulated insulin secretion (GSIS) 532–6
glucose transporters (GLUTs) 306, **536**, 549
glucosidases
absorption, dietary 341, 366, 439
bacteria, intestinal 4
bread 401–2
cytosolic (CBG) 116–17
function 42, 344, 419
genistin 337, 338
groundnuts 340–1, 342, 344
intestines 402, 544
reaction times *56*
sources 418
soy milk 54
soybeans 417–18
tofu 56–7
glucosides *see also* daidzin; genistin; glycitin
absorption, dietary 4–5
acid hydrolysis 36
biosynthesis 50
fermentation 54
plants 18
soy milk 373
spectra **284**, **286**, **288**, **290**
glucosinolates 424
glucosyltransferase 50
3-glucosyltransferase (3GT) **84**

β-glucuronidases 166, 198, 256
glucuronidation
  bisphenol A (BPA) 123
  definition 126
  enzymes 4–5
  excretion 440
  genistein 117–19, 121
  glycitein 466
  p-nonylphenol 123
β-glucuronides
  absorption, dietary 4
  chromatograms 209, **210**
  genistein 117–19
  structure **118**
glutamate 304, 507–8, 513
γ-glutamyl cysteine synthetase
  (γ-GCS) **138**, 140
glutathione (GSH)
  brain 455
  daidzein 140
  function 474
  genistein 134, 135–7, **138**, 139–40
  oxidative stress 473
glutathione peroxidase (GPx) 134,
  138, 140
glutathione reductase 474
glutathione S-transferase (GST)
  86–9, 134, 135–7, 139
gluten 361, 401
gluten intolerance 355
glycine 507
Glycine max see soybeans
glycitein
  absorption, dietary 466–7
  binding affinity 471, 473
  biosynthesis 4, 465, 474
  conjugates 465–6
  daidzein biosynthesis 471
  dietary sources 465
  health effects 465, 467–73
  industrial synthesis 474
  metabolism 471
  nomenclature 500
  research history 474
  stability 36
  structure *21*, **53**, **219**, **466**, **629**
  toxicity 389

glycitin
  dietary sources 4
  hydrolysis 439
  metabolism 440
  nomenclature 244
  structure **30**, **53**
glycoconjugates 284, 285, 287, 288–9, 291
glycogen 519
glycosaminoglycans (GAGs) 382–3,
  384–6, 388–90, 391–2, **393**, 394, 395
glycosides
  absorption, dietary 57, 152,
    196–7, **440**, 482, 544, 599
  definition 25
  differentiation 287
  food processing 244
  metabolism 482
  storage of soybeans 35
  structure **19**
glycosylation 89, 90, 141
7-*O*-glycosyltransferase 163
Glycyrrhiza glabra see licorice
goat's milk 374, 377
goitre 423, 425, 427–8, 433
gonadotropin 571
gonadotropin-releasing
  hormone (GnRH) 488, 521, **566**
gonocytes 570
GPx *see* glutathione peroxidase
grapes 15, **18**, 627
groundnuts
  analysis 334–6
  composition 333–4, 336–42
  description 344
  nutritional value 333
  processing 343
growth factors definition 154
Grubbs' catalyst 64, **65**
GSH *see* glutathione
guggulsterone 547

hardness *408*, *411*, 419
Hawaii 359
HDL *see* high-density lipoproteins
health effects *see also specific diseases*
  coumestrol 519–20, 522, 524–5
  daidzein **300**, 302–3

daidzin 303
equol 5, 8, 9, 304, 483, 492, 496, 581–91
estrogenic activity 264–5
flavonoids 296
genistein 131, **300**, 337, 426, 501
glycitein 465, 467–73
groundnuts 333
obesity 541–2
possible mechanisms 28, 218–20
puerarin 298, **300**, 302–5, 308–9
resveratrol 480, 547, 549
soy 85, 179, 263, 334, 349, 366, 399, 438, 554
xenoestrogens 122, 573
heart disease *see also* cardiovascular system
recommended daily intake 31
soy 562
hedonic scale 406, 418, 419
Helicobacter pylori 441–2, 447, 448
Helix 12 151, 165
Helix pomatia 256, 259
heme oxygenase 1 (HO-1) 472
heparan sulfate (HS) 383, **386**, 392
herbal extracts *see also* nutraceuticals
analysis 319–26
osteoporosis 317–18
red clover (*Trifolium pratense*) 220
traditional medicine 327
heteronuclear multiple bond correlation (HMBC) **96**, 98, 100, 106, 111, 340
heteronuclear multiple quantum correlation (HMQC) 111, 340
heteronuclear single quantum correlation (HSQC) **97**, 98, 111
hexane extraction 51
HID *see* 2-hydroxyisoflavanone dehydratase
high-density lipoproteins (HDL) 170, 173, 306, 545, 546
high performance liquid chromatography (HPLC)
advantages 238
automation 221, 238

benefits 30–1, 221
beverages *247–54*, 258, 259
definition 214
disadvantages 222
electrochemical detection (ECD) 202–3
groundnuts 334–5, 339–40
internal standards *247–50, 252–4*
mass spectrometry (MS) 58, 221, 224–8, *236–7*, 281
microwave-assisted extraction (MAE) 183–4
nuclear magnetic resonance (NMR) spectroscopy 101
plasma analysis *199–201*, 202–3, *229–31*
retention times *211*, **212**, 405
reverse phase 205, 209, 222, 223
simultaneous analysis 30–1
solvents 209, 222, 224
soy analysis 50, 223, *231–3*
soy bread 404–5
ultrafast 223, 226, *231, 233*
urine analysis *199–200*, 209–10
UV 205–12, 213, 222–4, *230*, 237
hippocampus 453, 454, 456, 457, 458, 460, 471–2, **510**
HMBC *see* heteronuclear multiple bond correlation
HMGR (3-hydroxy-3-methylglutaryl-CoA reductase) 303
HMQC *see* heteronuclear multiple quantum correlation
homology modeling 89
Hong Kong 355
hops 15
hormone replacement therapy (HRT) *see also* selective estrogen receptor modulators
alternative use of isoflavones 220, 238
bone 7, 597, 599, 606
function 592
vasomotor symptoms 635
hot flushes *see* vasomotor symptoms
HPLC *see* high performance liquid chromatography

HPRT (hypoxanthine guanine phosphoribosyl transferase) 522
HS *see* heparan sulfate
HSQC *see* heteronuclear single quantum correlation
Humulus lupulus L., Cannabaceae *see* hops
hyaluronates **385**
hyaluronic acid 386
hydrolysis **401**, 544
hydrophilicity 29, 58, 126
3-hydroxy-3-methylglutaryl-CoA reductase (HMGR) 303
7-hydroxy isoflavone **318**, 322
6′-hydroxy-*O*-demethylangolensin (DMA) 5, 119, **120**, 121, **136**, 141
6-hydroxybiochanin A 105, **106**
8-hydroxydaidzein 442
hydroxygenistein **51**
2-hydroxyisoflavanone dehydratase (HID) 19, **20**, 85–6, **87**
2-hydroxyisoflavanone synthase (2HIS) 19, **20**
hydroxylation
  daidzein 440
  definition 126
  genistein 119, **120**, 440
  glycitein 471
  xenoestrogens 124
4-hydroxyphenyl-2-propionic acid **120**, 121
hydroxysteroid dehydrogenase/ isomerase (3β-HSD) 132, 518–19
hypercholesterolemia 303
hyperglycemia 306–7, 531
hyperlipidemia 545
hyperplasia 556
hypertension
  daidzein 302–3
  glycitein 470–1, 473
  groundnuts 333
  puerarin 302
  soy 470–1
hypertrophy 556
hyphenated techniques definition 111
hypocotyls 31, 35, 42

hypoglycemia 532
hypothalamic–pituitary–gonadal (HPG) axis 565, **566**, 572
hypothalamus 488, 521, 565
hypothyroidism 423, 433
hypoxanthine guanine phosphoribosyl transferase (HPRT) 522
hypoxia 613

IBD *see* inflammatory bowel disease
ICAM-1 *see* intercellular adhesion molecule-1
IF7GT *see* UDP-glucose:isoflavone 7-*O*-glucosyltransferase
IF7MaT *see* isoflavone 7-*O*-malonyltransferase
IFMN *see* isoformononetin
IFNγ *see* interferon-γ
IFS *see* isoflavone synthase
IL-1β *see* interleukin-1β
IL-8 *see* interleukin-8
iminium cation 79
immune system 309, 588
immunoassays 268, 273, 274, 275
impotence 565
India 41
Indonesia 357–8
inducible nitric oxide synthase (iNOS) 304, 444
infant formula
  composition 351, 355, 359
  extraction 267
  goitre 428
  lactase insufficiency 360
  soy 264, 492, 530, 563, 572, 574
  testicular development 572
infants *see also* development
  intake levels 359
  iodine 428
  maximum daily intake 355
inflammation
  cancer 467–8
  daidzein 302
  daidzin 305
  equol 616, 617

extracellular-signal-regulated kinase (ERK) 135, 443
factors 612
function 309
genistein 134–5, 616, *617*, 621–2
glycitein 474
Helicobacter pylori 442–4
isoflavones 616–22
markers 616, 622, 623
neurodegenerative diseases 383–4
obesity 611–15, 616–22
puerarin 302, 305
serum amyloid A (SAA) 550
soy 619
inflammatory bowel disease (IBD) 135
information-dependent acquisition (IDA) 322, 326, 328
inhibitory neurotransmitters 512–13
iNOS (inducible nitric oxide synthase) 304, 444
inositol 1,4,5-triphosphate **535**
insulin 309, 519, 618
insulin-like growth factors 564, 573
insulin resistance
    definition 556
    macrophages 612, 614–15
    obesity 530, 619–20
    puerarin 620
    serum amyloid A (SAA) 550
    soy 546, 547, 554–5, 619–20, 623
insulin secretion 532–6
intercellular adhesion molecule-1 (ICAM-1) 134
interferon-γ (IFNγ) 134, 616
interleukin-1 (IL-1) 599
interleukin-1β (IL-1β) 134, 612, 616
interleukin-2 (IL-2) 523, 616
interleukin-6 (IL-6) 599, 612, 613, 616–17, 619–20, 621
interleukin-8 (IL-8) 134, 443, 444, 612, 616, 622
interleukin-11 (IL-11) 599
internal standards
    definition 239
    gas chromatography (GC) 258

high performance liquid chromatography (HPLC) *247–50, 252–4*
liquid chromatography (LC) 257–8
nuclear magnetic resonance (NMR) spectroscopy 110
requirements 257
tandem mass spectrometry (MS/MS) 322
ultrafast HPLC 226
interstitial tissue 574
intestines
    absorption 4–5, 197, 402
    bacteria 4
    biochanin A (BCA) 439
    daidzein 439
    deglycosylation 116–17, 544
    formononetin (FMN) 439
    genistein 115–16, 117–21, 125, **136**, 141, 439
    genistin 116–17
    glucuronidation 153, 166
    hydrolysis 439
    lactase-phlorizin hydrolase (LPH) 152
    lumen 119–21, 126
    methylation 166
    microflora *see* microflora, intestinal
    sulfation 153, 166
    uridine diphosphate glucosyltranferases (UGTs) 166
iodine
    daidzein synthesis **66**, 67
    deficiency 423–4
    infant formula 428
    intake requirements 432
    soy diets 425
    thyroid hormones 433
ion channels 501, 513, 533–4, **536**
ion-exchange chromatography 203
ion traps 282
ionization techniques 239, 272
ipriflavone 616, 628, **629**
irigenin **101**

irisflorentine 101
irisolidone 105, **106**
iristectorigenin 101
isoenzymes 440
trans-isoflavan-4-ol 102
isoflavans 23, **95**
isoflavone 7-*O*-malonyltransferase (IF7MaT) **20**
isoflavone dehydratase 387
isoflavone glucoside glucosidase 63
isoflavone malonylglucoside malonyleterase 63
isoflavone synthase (IFS)
  daidzein **20**, 159, **161**
  definition 19
  evolution 86
  function 24, 84–5, 89–90
  genistein **20**, 159, **161**, 387
  hydroxyl group effect 86
  mutagenesis 89
  reaction steps **87**
  semi-synthesis 85–6
  structure 89–90
isoflavonoid pathway reaction steps **84**
isoformononetin (IFMN)
  sources 79
  spectra **320**
  structure **62, 74, 318**
  synthesis 73–4, **75**
isoliquiritigenin
  biosynthesis 159, **161**
  glycitein biosynthesis 465, 474
  structure **20, 161**
isomers
  definition 496
  equol 484, 494, 580, 581
  genistein 4′,7-*O*-diglucosides 289
isotope dilution 203, 204
Italy 355

J definition 111
Japan
  breast cancer 218
  cancer mortality rates 356
  consumption levels 8, 356

Japanese knotweed *see Fallopia japonica*
JAR *see* just-about-right (JAR) scale
JNK (c-Jun N-terminal kinase) 468, 615
joint lubrication 384
just-about-right (JAR) scale 406, 414–15, 419

kaempferol **299**, 390
kainite 506
kakkalide 105, **106**
keratan sulfate (KS) *383*, 386, **386**
kinako 118, 210, **211, 212**
koji enzymes 40
Korea 356–7
KS *see* keratan sulfate
kudzu (*Puerariae radix*)
  anti-tumour activity 298–301
  Chinese medicine 294, 307–8
  composition 296–8
  optimum harvest 298
  roots **295**
  species 296

LAB *see* lactic acid bacteria
lactase insufficiency 360
lactase-phlorizin hydrolase (LPH) 4, 116–17, 152, 166, 544
lactate dehydrogenase (LDH) 304
lactic acid bacteria (LAB) 54–6, 58
lactose intolerance 366–7, 563
lactulose 56
lariciresinol **16**
lavendustin 504, 509
LBD *see* ligand-binding domain
LC-MS *see* liquid chromatography-mass spectrometry
LDH (lactate dehydrogenase) 304
LDL *see* low-density lipoproteins
learning 452, *460*, 586
lecithin *33*
legumes 22, 62, 85, 562
Leguminosae fm. 15, 19, 23, 24
leptin 613, 620, 623
lethal dose test (LD_{50}) 489, 496

leucocyanidin reductase (LAR) **84**
leuconathocyanidins **84**
leukocytes 135
Leydig cells 564, 565, 566–7, 568, 570, 574
LH (luteinizing hormone) 132, 488, 521, 565, 570
licorice 98–100, 354, 361, *371*, 376
ligand-binding domain (LBD) 151, 163
ligand-gated ion channels (LGICs) 501, 513
light levels effect on concentration 21
lignanes **158**
lignans
 definition 25, 627
 mammalian **17** *see also* enterodiol; enterolactone
 phytoestrogens 14
 sources 3, 500, 627
 structure **16**
limit of detection (LOD)
 definition 214, 239, 328
 ELISA 273
 HPLC-ECD 203
 HPLC-UV 202, 210, 213
 LC-MS 202
 LC-MS/MS *325*
 LC-UV 204
limit of quantification (LOQ)
 definition 214
 HPLC-UV 202, 210, 213, 223
 LC-ESI-MS/MS 226
 LC-MS 202
 LC-MS/MS 203, 213
linearity 325, 328
Linum usitatissimum L., Linaceae *see* flax seed
lipid metabolism
 atherosclerosis 6
 coumestrol 519–20
 genistein 531
lipochitin oligosaccharides 22
lipolysis 618–19
lipophilic definition 126

lipopolysaccharide (LPS) 134, 616
lipoproteins 170, 173
liquid chromatography (LC)
 atmospheric pressure chemical ionization (APCI) 228
 columns 270
 food analysis 268–9, 273–4
 mobile phases 270
 nuclear magnetic resonance (NMR) spectroscopy 101–2
 rapid resolution 223
 reverse phase 101
 UV 204
liquid chromatography-mass spectrometry (LC-MS)
 definition 344
 electrospray ionization (ESI) 224, 228
 isotope dilution 204
 plasma analysis 202, 226–8
 tandem mass spectrometry (MS/MS) 203, 204, 226–8, *229–30*, *233–5*, 319–25
liquiritigenin
 biosynthesis 159, **161**
 metabolism **162**
 structure **20**, **161**
liver
 absorption 5, 440
 fatty 620
 genistein 141
 glucuronidation 119, 153
 hydroxylation 119
 NAD(P)H:quinone oxidoreductase 1 (NQO1) 550
 sulfation 119, 153
 uridine diphosphate glucosyltranferases (UGTs) 166, 466
liverworts 15
LLOQ *see* lower limit of quantitation
LOD *see* limit of detection
long-term potentiation (LTP) 454, 508, 513
LOQ *see* limit of quantification

low-density lipoproteins (LDL)
  biochanin A 220
  cardiovascular system 170
  daidzein 303
  daidzin 303
  equol 588
  function 173
  recommended daily intake 447
  soy 545
lower limit of quantitation (LLOQ)
  definition 239, 328
  LC-MS/MS 228, *325*
  ultrafast HPLC 226
LPH *see* lactase-phlorizin hydrolase
LPS *see* lipopolysaccharide
LSDs *see* lysosomal storage diseases
lubrication, joints 384
Lupinus albus see white lupine
Lupinus angustifolious 281, 289
Lupinus hitoni **288**
Lupinus rotundiflorus **284**
Lupinus stipulatus **290**
luteinizing hormone (LH) 132, 488, 521, 565, 570
luteolin **299**
luteone *21*
lutropin 275
lyophilization 260
lysosomal storage diseases (LSDs) 381–2, 395
lysosomes 381

m/z *see* mass-to-charge ratios
maackiain 85
macrophage chemoattractant protein-1 (MCP-1) 612
macrophage-colony stimulating factor (M-CSF) 599
macrophage inflammatory protein 1a (MIP-1a) 616–17
macrophage migration inhibitory factor (MIF) 613
macrophages 612–15, 623
MAE *see* microwave-assisted extraction
magnetrons 181, 182, 191

Maillard browning 412
malonates 50, 264
$6''$-O-malonyl-β-glycosides 50, 244
malonyl-CoA 58, **84**
malonyl transferase 50
malonylation 288–9
malonyldaidzin
  stereoisomers 36–7
  structure **30**
malonylgenistin
  stereoisomers 36–7
  structure **30**, **51**
  temperature 36
malonylglucosides *see also*
  malonyldaidzin; malonylgenistin; malonylglycitin
  decarboxylation 21
  plants 18
  soybeans 21
  stability 36, 37
malonylglycitin
  soybean cultivar differences 33
  structure **30**
malonyltransferase 58
Malus sp., Rosaceae *see* apple trees
mannose 6-phosphate (M6P)
  receptors 387–8
mass spectrometry (MS)
  capillary electrophoresis 281
  electrospray ionization (ESI) 282, 283
  food analysis 270, 272–3
  fragmentation 282, 285–7, 290, 291, 320
  high performance liquid chromatography (HPLC) 58, 221, 224–8, *236–7*, 281
  ion sources 282
  ion traps 282
  ionization techniques 239, 272
  process 239, 272
  ultra-performance liquid chromatography (UPLC) 281
mass-to-charge ratios (m/z) *207*, 239, 281–2, 285, 292
mast cells 305

matairesinol 15, **16**
maternal diet 453–4, 459, 489, 491, 569, 570 *see also* development
matrix effects 266, 267, 367
matrix metalloproteinases (MMPs) 468
Maxwell's equations 181, 191
MCP-1 *see* monocyte chemoattractant protein-1
mediacarpin **318**
Medicago truncatula 290
medicarpin **16**, 85
memory *see also* brain
   β-amyloid 454, 455
   animal models 453–4, 456–7, *460*
   equol 586
   glycitein 471–2
   long-term potentiation (LTP) 454, 508, 513
   NMDA (N-methyl-D-aspartate) 506
   sex hormones 452
   spatial 452–4, 456
   working 457, 458, 460
menopause 8, 265, 429, 455–6, 457–9
menstrual cycle 132
meta-analysis
   breast cancer 7–8
   definition 448
   gastric cancer 446
   osteoporosis 635
   recommended daily intake 400
   vasomotor symptoms 631–9
metabolemics 292
metabolic diseases 381
metabolic profiling 280–1, 285
metabolic syndrome 546, 555, 556, 619
metabolism *see also* absorption, dietary
   biochanin A (BCA) 150, 220, 366, 628
   carbohydrates 519–20
   daidzein 5, 439
   dihydrodaidzein 5, 439
   dihydrogenistein 5, 439

formononetin (FMN) 150, 366, 628
genistein 5, 119, 135, 152–3, 439
genistin 152, 154, 336–7
glycitein 471
glycitin 440
glycosides 482
lipids 6, 519–20
liquiritigenin **162**
mammalian 166–7, 171–2
naringenin **162**
plants 159–63
metabolite profiling 292
metaflammation 611, 613, 622
metastasis 301, 309–10, 468, 589
method validation definition 377
methoxymethyl ether (MOM) 86, 91
N-methyl-D-aspartate (NMDA) 505–11, 512
microflora, intestinal
   bioavailability effect 167
   biochanin A (BCA) 366, 628
   deconjugation 440
   deglycosylation 117
   formononetin (FMN) 366, 628
   genistein 116, 117, 120–1
   glycitein 466–7
   hydrolysis 116, 166, 366
microglial activation 135, 383–4
microwave-assisted extraction (MAE)
   applicator 191
   cavities 181, 182
   design 180–3
   explanation 180, 190–1
   flow rate 182–3
   method 183–4
   rate constants *189*
   solvents 181, 190
   temperature 186, 190
   time effects 187, 190
   yields 184–9
microwave energy definition 191
middle cerebral artery occlusion (MCAO) 304
milk allergies 367

milk, cow's **272**, 367, *369*, 373, 374, 377
milk, goat's 374
milk, soy *see* soy milk
miroestrol **583**
miso *see also* fermented foods
  Brazil 354
  Canada 354
  composition *32*, *34*, 445
  consumption levels 446
  Japan 356
  Taiwan 358
  USA 358
mitogen-activated protein kinase (MAPK)
  daidzein 303
  equol 589
  genistein 133, 443, 444
  glycitein 468, 474
  Helicobacter pylori 443
  inflammation 615
MMPs (matrix metalloproteinases) 468
MOM *see* methoxymethyl ether
monocyte chemoattractant protein-1 (MCP-1) 134
monolithic columns 275–6
mood 472
motor function 304
MPO *see* myeloperoxidase
MPS *see* mucopolysaccharidoses
MRM *see* multiple reaction monitoring
MS/MS *see* tandem mass spectrometry
mucopolysaccharides *see* glycosaminoglycans (GAGs)
mucopolysaccharidoses (MPS)
  classification *383*, 394, 395
  clinical symptoms 383
  enzyme deficiency 381, *382*, 394
  pathophysiology 383–4, 394
  substrate reduction therapy (SRT) 387–93, 394
  transmission 382, 394

Mullerian inhibiting substance (MIS) 567
multidrug resistance 298, 310
multimode cavity 181, 191
multiple reaction monitoring (MRM) 204, 224, 226, 228, 319–20, 322, 328
mutagenesis 519, 522, 526
mycoestrogens 14, **16**
Mycoform® *see* formononetin
mycorrhizal fungi 22, 69
myelin 305
myeloperoxidase (MPO) 135
myocardial hypoxia 309

NAD *see* nicotinamide adenine dinucleotide
NAD(P)H:quinone oxidoreductase 1 (NQO1) 550–1, 552–3
naringenin
  binding affinity *88*
  biosynthesis 159, **161**, 387
  glycosaminoglycans (GAGs) 390
  metabolism **162**
  structure **20**, **161**
naringenin chalcone **20**, 159, **161**, 387
National Health and Nutrition Examination Board (NHANES) 541
natto *32*, 358, 445 *see also* fermented foods
natural killer cells
  equol 588
  genistein 5, 588
neuronal plasticity 452, 453, 459
neuroprotective effects 304–5, 472, 586
neurotransmitters 512–13
newborns *486*
NF-κB *see* nuclear factor-κB
NHANES (National Health and Nutrition Examination Board) 541
nicotinamide adenine dinucleotide (NAD) **160**
nicotinamide adenine dinucleotide phosphate (NADPH) oxidase 384

nitric oxide (NO) 170, 173, 444, 587, 612, 616
nitric oxide (NO) synthase 302
nitrogen fixation 22, **23**, 24, 25, 85, 159
NMDA (*N*-methyl-D-aspartate) 505–11, 512
NMDA receptors 506–8
NMR *see* nuclear magnetic resonance (NMR) spectroscopy
nodulation 22, **23**, 24, 25
NOESY *see* nuclear Overhauser enhancement spectroscopy
nomenclature
  biochanin A (BCA) 244
  daidzein 149, 244, 500
  equol 483
  formononetin 244
  genistein *116*, 150, 244
  genistin *116*
  glycitein 500
  glycitin 244
non-fermented foods *see also* soy milk; tofu
  bioavailability 544
  cancer 446, 447
  composition 41
  genistein 152
  β-glucosides 50
p-nonylphenol (NP)
  glucuronidation 123
  hydroxylation 124
  industrial uses 125
  placenta transfer 122, 124–5
  structure 123
  sulfation 123–4
NP *see p*-nonylphenol
NQO1 (NAD(P)H:quinone oxidoreductase 1) 550–1, 552–3
nuclear factor erythroid-derived 1 (Nrf1) **138**, 139, 140
nuclear factor erythroid-derived 2 (Nrf2) **138**, 139, 140, 472–3, 475, **553**, 590
nuclear factor-κB (NF-κB)
  coumestrol 523
  daidzein 301, 444, 447
  function 475
  genistein 134, 444, 447
  glycitein 468, 474
  Helicobacter pylori 443, 444
  inflammation 467–8, 612, 615, 616
  isoflavones 616
  macrophages 612
  puerarin 302, 305
nuclear hormone receptors 173
nuclear magnetic resonance (NMR) spectroscopy
  chromatography linking 101–2
  conformations 102
  definition 111
  development 94–5
  equol 106–7, *108*
  groundnuts 340
  metabolites 103–7, *208*
  solvents 101, 107, **109**
  structure elucidation 96–100, 102
  theory 110–11
nuclear Overhauser enhancement spectroscopy (NOESY) 98, 100, 103, 111–12
nuclear receptors 593
nucleotide-binding oligomerization domain 1 (NOD1) 443
nutraceuticals 15, 149, 154, 220, 424–5 *see also* herbal extracts; supplements

obesity *see also* adipose tissue
  animal models 545–6
  coumestrol 520
  definition 526, 541, 556
  diabetes risk 530, 547
  diets 543, 544–5, 546
  environmental factors 542
  exercise 545
  genistein 546–7, 620
  global trend 541
  guggulsterone 547
  health effects 541–2
  isoflavone research 550–6
  resveratrol 547
  treatment 543, 555

Oenothera bienis (evening primrose) 628
oestrogen *see* estrogen
omnivores intake levels 351, *353*
oncom *32*, 357, 361
one-carbon electrophiles 68–71
ontogenesis 25
optic nerves 305
orobol *see* 5,7,3',4'-tetrahydroxyisoflavone
osteoblasts 467, 473, 599–600, 607
osteoclasts 599, 600, 607
osteoporosis *see also* bone
    daidzein 602–3, 605–6
    definition 42, 316, 327
    equol 601, 602, 605–6
    estradiol (E2) 601
    estrogen 597, 607
    genistein 601, 602, 603, 605–6
    herbal extracts 326
    lumbar spine 607
    recommended daily intake 31
    red clover 317
    selective estrogen receptor modulators (SERMs) 90
    soy 317, 439
    vasomotor symptoms 632
osteoprotegerin 600
ovarian cancer 582–3
ovariectomy 460
ovaries 275
oxidation 50–1
oxidative rearrangement 67, **68**, 76, **77**
oxidative stress
    cardiovascular disease 587
    central nervous system (CNS) 472
    diabetes 531
    genistein 139, 442
    glutathione (GSH) 473
    glycitein 472–3
    lysosomal storage diseases (LSDs) 384
    NAD(P)H:quinone oxidoreductase 1 (NQO1) 550
    obesity **542**
    plants 483
    stomach 442, 447

p38 phosphorylation 135, 137, **138**
p53 pathway 137–9, **138**
Paecilomyces militaris 105
palatability 400, 401, 413–14, 415, 416–17
palladium catalysis 64, 67, 71, 74, 78
pancreas 466, 532–3
pancreatic cancer 586–7
Papilionideae fm. 15, 23
paracrine regulation 574
parathyroid hormone supplements 317
Parkinson's disease 304, 457
partial agonism
    definition 155
    genistein 151, 154
parvisoflavone **100**
pasteurization 374, 378
pattern profiling 317, 325, 328
PBMCs (peripheral blood mononuclear cells) 302, 303, 305
PCT *see* pharmacological chaperone therapy
PDA *see* photodiode array (PDA) detection
PDGFR (platelet-derived growth factor receptor) 570–1
peanut allergy 305
peanuts 15, **18**
peptidoglycan (PGN) 443
percolation 378
perinatal development 487–93
peripheral blood mononuclear cells (PBMCs) 302, 303, 305
peroxisome proliferator activated receptors (PPARs)
    adipocytes 549, 553–4, 618–19
    agonists 71
    daidzein 554
    definition 155
    function 468, 475
    genistein 5, 154, 444, 554
    glycitein 468, 474
peroxynitrite 303
Pezzuto, Dr John 495
pH effect on stability 36
pharmaceutical research 221, 238

pharmacokinetics
  β-blockers 224
  definition 214–15
  dose levels 630
  genistein 226–8
  herbal extracts 317
  plasma 226
pharmacological chaperone therapy (PCT) 382, 395
phenotype definition 42
phenylalanine 83, 159, **161**, 387, 465, 474
phenylalanine ammonia-lyase (PAL) 83, **84**, 159, **161**
phenylephrine 302
phenylpropanoic compounds 91
phenylpropanoid pathway 18–19, 83–4, **84**, 387, 465, 475
phenylpyruvate 159, **161**
phloretin 15, **16**
phosphatase and tensin homolog (PTEN) 133
phosphatidylinositol 3-kinase 133, **535**
3′-phosphoadenosine 5′-phosphosulfate (PAPS)-sulfotransferases 4–5
phosphodiesterase 3 (PDE3) 619
phosphoinositide 3-kinase/Akt-dependent pathways 588
phospholipases **535**, 622
phosphorylation 503, 508–9, 513, 549, 570, 615
photodiode array (PDA) detection 223, 269
phytoalexins 23, 25, 85, 91, 149
phytoanticipins 23, 25, 149
phytoestrogens
  administration route 152
  cancer research history 4
  classification 3, 14, **628**
  definition 126, 157
  sources **628**
  structure **16**, 150, **158**
PI-9 (proteinase inhibitor 9) 582–3
pisatin 85
Piscidia erythrina 103
pituitary gland 265, 423, 425, 428, 488, 521, 565

PKA *see* protein kinase A
placenta
  genistein 121–2, 124
  xenoestrogens 122, 124–5
planting season differences
  groundnuts 342
  kudzu (*Puerariae radix*) 298
  soybeans 35, 342
  white lupine 21–2
plants
  aglycones 283–4
  biosynthesis 18–19, **20**, 50, 158–9, **161**, 171
  cancer prevention 589
  environmental factors 280
  glycitein 465
  glycoconjugates 284
  glycosides 264, 284
  malonylglucosides 366
  maturity effects 21
  microbe interaction 22, 149, 158
  oxidative stress 483
  pathogen defence mechanisms 22–3, 158
  storage 62, 89, 465
  tissue distribution 19–22, 481
plasma
  aglycones 166, 198, 224–6, *229*
  biochanin A 226, *229–30*
  chromatograms **211**
  conjugated metabolites 153, 205, 213, 224, *229–30*
  coumestrol 226
  daidzein 202–3, 226, *230*, 491
  equol 489–90, 491
  formononetin 226, *230*
  genistein 117–18, 202–3, 226, *229–31*, 491, 511
  glucagon 519
  glucose 531
  glycitein 466
  insulin 519
  lipids 531
  lipoproteins 170
  maternal levels 490, 491
  retention times **212**
  zealarenone 226

plasma membrane 537
plasminogen activator inhibitor-1 (PAI-1) 621
platelet-derived growth factor receptor (PDGFR) 570–1
polar molecules 180, 181, 191, 281
Polygonaceae 15
Polygonum cuspidatum see *Fallopia japonica*
polyphenols definition 79
population studies 445–6 see also epidemiology
Portugal 377
post-harvest management 192
postmenopausal symptoms see also climacteric syndrome
  bone health 7, 467
  cognitive function 457–9
  definition 42
  estrogenic activity 439
  recommended daily intake 31
  self-medication 165–6, 171
postmenopausal women
  adiposity 621
  cognitive function 452
  genistein 429
  inflammation 621
  plasma glucose levels 531
  soy intake 133, 600
PPAR see peroxisome proliferator activated receptors
precipitation 267–8
precision
  definition 239, 328
  LC-ESI-MS/MS 226
  LC-MS/MS 228, 325
prefrontal cortex 456, 457, 458, 459, 460
pregnane X receptor (PXR) 520–1, 522, 590–1
8-prenylnaringenin **583**
6′-prenylpiscerythrone 103, **104**
prephenate 159, **161**
PRN see prunetin

processing
  composition effect 29, 36–40, 50–1, 343, 465
  conversion of isoflavones **37**, 366, 401–2
progesterone synthesis 132, 570, 586
prospective cohort studies 445
prostacyclin 438
prostaglandins 302, 586, 599, 616, 617, 622
prostate cancer
  daidzin 301
  equol 583–5
  genistein 132, 133–4
  glycitein 468–9
  preventative mechanisms 8
protein engineering 85, 89, 90, 91
protein kinase A (PKA) 503, 532, 534–6, 537–8, 619
protein kinase C (PKC) 503, **535**
protein kinase G (PKG) 503
protein tyrosine kinases (PTKs)
  GABA (γ-aminobutyric acid) 503–5
  genistein 444, 508–9, 511, 512, 529, 532–3, 538
  NMDA (*N*-methyl-D-aspartate) 508
protein tyrosine phosphates (PTPs) 503, 508
proteinase inhibitor 9 (PI-9) 582–3
proteoglycans 386
prunetin
  binding affinity *88*
  structure **318**
prunetin (PRN)
  binding affinity *88*
  structure **318**
  toxicity 389
prunetol see genistein
Prunus dulcis 418
pterocarpans 23
PTKs see protein tyrosine kinases
Pueraria lobata 105
Puerariae radix (kudzu)
  anti-tumour activity 298–301
  Chinese medicine 294, 307–8

Subject Index

composition 296–8
optimum harvest 298
roots **295**
species 296
puerarin
  biosynthesis 439
  dietary sources 296
  health effects 298, **300**, 302–5, 308–9, 454, 620
  structure **50**, **295**
  supplements 50
purification 344
PXR (pregnane X receptor) 520–1, 522, 590–1
pyran 79
pyruvate **536**

quercetin **299**, 549
quinone reductase 473, 475, 589–90
quinones 137, 159

radioimmunoassays 273
raloxifene 10, 151
RANKL (receptor activator of nuclear factor-κB (NF-κB) ligand) 599
RCM see ring-closing metathesis
reactive oxygen species (ROS)
  cardiovascular disease 587, 588
  diabetes 306, 531
  function 538
  Helicobacter pylori 442
  lysosomal storage diseases (LSDs) 384
  macrophages 612
receptor activator of nuclear factor-κB (NF-κB) ligand (RANKL) 599
recommended daily intake 31, 400, 447
  iodine 423–4
recommended limit (RL) 355
red clover (*Trifolium pratense*)
  analysis 230–1, 237, 289
  biochanin A (BCA) 21, 628
  composition 220, 616, 629, 640
  formononetin 21, 289

formononetin (FMN) 628
herbal extracts 361
infertility 149
isoformononetin 79
pasture grazing 149
supplements 4, 317, 354
tissue distribution 21
vasomotor symptoms 628–9, 631–2, **634**, 635–6
5α-reductase 564, 566
reference standards 205, 206, 321–2
reproductive system effects *see also* testis; uterus; vagina
  coumestrol 520, 525
  daidzein 487–8
  equol 485, 488, 489, 490, 586
  genistein 488, 530
  insulin-like growth factors 573
  phytoestrogens 265, 296
research history
  cancer 4
  genistein 150
  glycitein 474
resistin 619, 620, 623
resorcinol derivatives 627
resveratrol
  binding affinity **583**
  health effects 480, 547, 549
  research history 495
  sources 15, 480
  structure **16**, **18**
retention times 405
reverse phase 112, 205, 222, 260
Rhizobium spp. 22, **23**, 25, 85, 159, 171
ring-closing metathesis (RCM) 64, 79
RL see recommended limit
ROESY see rotating Overhauser effect spectroscopy
roots, nitrogen fixation 22
ROS see reactive oxygen species
rotating Overhauser effect spectroscopy (ROESY) 98, 99, 111–12

SAA (serum amyloid A) 550
Saccharomyces cerevisiae
  bread 402, 415
  semi-synthesis 85–6
SAH *see* S-adenosylhomocysteine
saliva 337, 366, 466
salt intake 441
SAM *see* S-adenosylmethionine
sample preparation *229–37 see also* extraction
  costs 259
  gas chromatography (GC) 271
  HPLC 334, 335, 404
  LC-MS 202
  LC-MS/MS 224
  U-HPLC 223–4
Sanfilippo disease 382 *see also* mucopolysaccharidoses (MPS)
SARs *see* structure-activity relationships
scanning electron microscopy (SEM) **186**
schizophrenia 501
scopolamine 455
seasonal planting differences
  groundnuts 342
  kudzu (*Puerariae radix*) 298
  soybeans 35, 342
  white lupine 21–2
secoisolariciresinol
  digestion **17**
  sources 15
  structure **16, 17**
secondary metabolites definition 91
selected ion monitoring (SIM) 203, 204, 224, 271
selected reaction monitoring (SRM) 203
selective estrogen receptor modulators (SERMs) *see also* raloxifene
  cancer 90
  definition 155
  genistein 151, 154, 426, 500–1
  osteoporosis 90
  phytoestrogens 5, 14

selectivity of analysis
  definition 240, 328
  HPLC 221, 222
  LC-UV 269–70
  mass spectrometry (MS) 272
  tandem mass spectrometry (MS/MS) 320–1
semi-synthesis 85–6, 91
seminiferous tubules 564, 565, 575
sensitivity of analysis
  electrospray ionization (ESI) 224
  HPLC 221, 222
  LC-MS/MS 325
  LC-UV 269–70
  mass spectrometry (MS) 272
  ultrafast HPLC 226
sensory analysis 417, 418
SERMs *see* selective estrogen receptor modulators
Sertoli cells 564, 565–6, 567, 574
serum amyloid A (SAA) 550
serum analysis 203, 204, *234–6*
sex hormone-binding globulin (SHBG) 132, 168, 486, 567
sex hormones 452, 491–2
sexual dimorphism 452, 460, 488
SFE *see* supercritical fluid extraction
SHBG *see* sex hormone-binding globulin
shikimate cascade 159, **160**
shoyu (soy sauce) 352, 356 *see also* fermented foods
signal transduction 389
SIM *see* selected ion monitoring
simultaneous analysis
  food products 268–9
  LC/APCI-MS 228
  nuclear magnetic resonance (NMR) spectroscopy 101–2
  ultrafast HPLC 223, 226
Singapore 357, 358
single mode cavity 181, 192
singlet oxygen 424
Smirnowia iranica 102
smoking 441

Subject Index

SNAP-25 (synaptosomal-associated protein of 25 kDa) 536, 538
soaking 37–8, **39**, 337, 413
SOD *see* superoxide dismutase
sodium-dependent glucose transporter 1 116–17
solid phase extraction (SPE)
  acid hydrolysis 257
  definition 215, 260
  enzyme hydrolysis 256
  food analysis 266
  LC/APCI-MS 228
  LC-MS/MS 203, 228
  nuclear magnetic resonance (NMR) spectroscopy 101–2
  plasma 207, **209**, 226
  simultaneous analysis 224
  soy products 246
solubility
  glucuronidation 466
  glycosylation 89
  groundnuts 341
  polarity effect 185–6
soluble phospholipase A_2 (sPLA_2) 622
solvent suppression 101
solvents
  direct extractiopn 245–6
  electrospray ionization (ESI) 224
  extraction 267
  HPLC 209, 222
  nuclear magnetic resonance (NMR) spectroscopy 107, **109**
  polarity 181
sonication 246, 255, 257, 260
Soxhlet extraction
  definition 79
  food analysis 274
  method 276
  soy analysis 223
  time 187
soy
  analysis 223, *231–3*, *237*
  cancer 7–8, 9
  cardiovascular system 6, 447
  composition 4, 148–9

food range 148, 149, 445–6, 481
genistein 149
health benefits 349
health effects 85, 179, 263, 334, 349, 366, 399, 438, 554
intake levels 7, 9, 400, 511
menopause 8
supplements 354
soy beverages 351
soy flour
  bread 402–4
  composition *34*, 51–3, 58
  processing 36
soy isolate
  composition *33*, *34*
  production 37
  reproductive system effects 572–3
soy lecithin *33*
soy milk *see also* infant formula
  absorption, dietary 117
  aglycones 373
  chromatograms **55**
  clinical studies 458
  composition *33*, *34*, 58, **351**, 357, 359, *368*, 465
  composition variability 373
  17β-estradiol 572
  extraction *247–52*
  fermentation 54–6
  glucosides 373
  Indonesia 357
  optimum ratios 40
  production 37, 38–40, 373
  stereoisomers 36–7
  Taiwan 358
  USA 358–9
soy protein isolate 359, 360, 456
soybeans
  cell walls **186**
  composition 29, 31–5, 41, 220, 334, 358, 481–2, 629
  cotyledons 21, 31, 35
  cultivar differences 31–3, 35
  fermentation 356, 358
  genetically modified 355

soybeans (*continued*)
  growth differences 33, 35, 131, 179–80, 220
  history of use 41, 49, 481
  hypocotyls 31, 35
  intake levels 132–3, 354, 356, 358, 425
  kinako 118, 210, **211**, **212**
  light levels effect on concentration 21
  metabolites 105, **107**
  processing 29, 152, 343, 349–50, 401–2
  production volumes 41, 179
  roots 21
  seasonal planting differences 21–2, 35
  soaking 37–8, **39**
  storage 35
  tissue distribution 21, 31
spatial learning 452–4, 456
SPE *see* solid phase extraction
specific loaf volume 405, 410–11, 415
specificity of analysis 221
spectra
  biochanin A (BCA) **225**
  daidzein **227**
  equol 95–6, **97**
  formononetin **227**, **320**
  genistein 95, **96**, **225**
  glucosides **286**, **288**, **290**
  isoformononetin **320**
  milk, cow's **272**
sperm 565–6, 568–9, 570, 571, 573
spinge 419
"spirodienone" model 159, **162**
springiness *408*, 411, 419
SREBP (sterol regulatory element-binding protein) 549
SRM *see* selected reaction monitoring
SRT *see* substrate reduction therapy
stability
  aglycones 259
  alkaline hydrolysis 255
  electrospray ionization (ESI) 224
  β-glycosides 259

  glycosylation 89
  pH 36
  processing 36, *38*, 350
  temperature 36
stearoyl-CoA desaturase (SCD) 549
Stechell, Dr Kenneth 495
STEPs (striatal-enriched tyrosine phosphatases) 508
stereoisomers
  definition 42
  malonyldaidzin 36–7
  malonylgenisitin 36–7
steroid hormones 491–2, 495, 565, 567–8, 573 *see also* testosterone
sterol regulatory element-binding protein (SREBP) 549
stilbene synthase (STS) **84**
stilbenes *see also* resveratrol
  definition 25
  sources 15
  structure **84**, **158**
stomach
  aglycones 152
  digestion 439
storage of soybeans 35
striatal-enriched tyrosine phosphatases (STEPs) 508
structure
  acetoxy groups 103
  acetyldaidzin 30
  acetylgenistin 30, **51**
  acetylglycitin 30
  aglycones **19**, *21*, 30
  anthocyanidins **84**
  anthocyanins **84**
  3-arylcoumarins **99**
  basic outline 29, 264
  bioavailability 197–8
  biochanin A (BCA) *21*, **219**
  bisphenol A (BPA) 123
  cajanin **318**
  catechins **84**
  chalcones 16, 20, **84**, **158**
  chondroitin sulfate (CS) **385**
  chorismate **160**
  cinnamic acid **84**

cinnamoyl CoAS **84**
cladrin **318**
compared to flavones 17
p-coumaric acid **84**, **161**
p-coumaroyl-S-coenzyme A **161**
coumestans **158**
coumestrol **16**, **99**, 100, 518
daidzein **16**, **20**, *21*, **30**, **53**, 79, 96
daidzein 7-*O*-glucoside-6″-*O*-malonate **62**
daidzin **30**, **50**, **53**, **295**
dermatan sulfate (DS) **385**
dihydroflavanols **84**
dihydrogenistein **120**, **136**
trans-4′,7-dihydroxyisoflavan-4-ol 102
dimethyldaidzein **62**, **76**
enterodiol 17
enterolactone 17
equol **95**, **482**, 484, 494
estradiol (E2) **16**, **158**, 163, **482**, **530**, **563**
estrogen receptors (ER) 163
4-ethylphenol **120**
flavanones **16**, **20**, **84**, **158**
flavone comparison 285
flavonoids **16**, **158**
flavonols **84**
formononetin 7-*O*-glucoside-6″-*O*-malonate **62**
formononetin (FMN) *21*, **62**, **219**, **318**
GABA$_A$ receptor 502
genistein **16**, **20**, *21*, 115, **150**–1, 529–30
genistein-7-*O*-gentiobioside **340**
genistein 7,4′-di-*O*-sulfate **105**
genistin **30**, **51**, **53**, 116
glabrene 103, **104**
glabridin **16**
glabrone **100**
β-glucuronides **118**
glycitein *21*, **53**, **219**, **466**, **629**
glycitin **30**, **53**
glycosaminoglycans (GAGs) 384
glycosides 19

glycosyl groups 103
heparan sulfate (HS) **386**
hyaluronates **385**
7-hydroxy isoflavone **318**
6′-hydroxy-*O*-demethylangolensin (DMA) **120**, **136**
6-hydroxybiochanin A **106**
hydroxygenistein **51**
hydroxyl groups 103
4-hydroxyphenyl-2-propionic acid **120**
ipriflavone **629**
irigenin **101**
irisflorentine **101**
irisolidone **106**
iristectorigenin A **101**
trans-isoflavan-4-ol 102
isoflavans **95**
isoflavone synthase (IFS) 89–90
isoformononetin **62**, **74**, **318**
isoformononetin (IFMN) **62**, **74**, **318**
isoliquiritigenin **20**, **161**
keratan sulfate (KS) **386**
lariciresinol **16**
leuconathocyanidins **84**
lignanes **158**
lignans **16**
liquiritigenin **20**, **161**
luteone *21*
malonyl-CoA **84**
malonyldaidzin **30**
malonylgenistin **30**, **51**
malonylglycitin **30**
matairesinol **16**
mediacarpin **318**
medicarpin **16**
methoxy groups 103, 106
modifications 17, **19**
mycoestrogens **16**
naringenin **20**, **161**
naringenin chalcone **20**, **161**
NMDA receptors 506–8
p-nonylphenol (NP) 123
parvisoflavone **100**
phenylalanine **161**
phenylpyruvate **161**

structure (*continued*)
  phloretin 16
  phytoestrogens 16, 150, **158**
  6′-prenylpiscerythrone 103, **104**
  prephenate **161**
  prunetin **318**
  prunetin (PRN) **318**
  puerarin 50, **295**
  resveratrol **16**, **18**
  secoisolariciresinol **16**, **17**
  stilbenes **84**, **158**
  sugars 17–18
  sulfate groups 103–5
  tectorigenin **101**, **106**
  5,6,7,3′-tetrahydroxy-4′-methoxyisoflavone **101**
  5,7,3′,4′-tetrahydroxyisoflavone (THIF) **136**
  variation 148–9, 283–4
  wighteone *21*
  xanthohumol 16
  zealarenone 16
  α-zealarenonol 16
structure-activity relationships (SARs) 157–8, 173
structure elucidation
  definition 112
  genistein-7-*O*-gentiobioside 340–1
  mass spectrometry (MS) 273
  nuclear magnetic resonance (NMR) spectroscopy 96–100, 102
substitution 103
substrate reduction therapy (SRT) 382, 388–92, 394, 395
sufu *33*, 358
sugars
  plant glycosides 264, 284
  polarity effect 281
  structure 17–18
sulfatases 198
sulfate groups 103–5
sulfation
  bisphenol A (BPA) 123–4
  definition 126
  excretion 440
  genistein 117, 119

intestines 4–5
NMR analysis 105
p-nonylphenol 123–4
sulfotransferases (SULTs)
  genistein 117, 119, 428
  intestines 166, 440
  placenta 122
  xenoestrogens 123–4
supercritical fluid definition 42
supercritical fluid extraction (SFE) 29–30, 181
superoxide 302
superoxide dismutase (SOD) 134, 442, 473, 475
supplemented food definition 276
supplements *see also* herbal extracts; nutraceuticals
  analysis 221, 228
  climacteric syndromes 169, 458–9
  clinical studies 458–9, 621
  combined with other drugs 425
  compared to dietary sources 9
  composition 50, 354, 391–2
  equol 600–1
  marketing 263–4
  red clover 4, 220
  soy 573–4, 621
  suppliers' recommended intake 424
Suzuki coupling **66**, **67**, **73**, 74, 79
symbiotic bacteria 25
synaptic plasticity 513, 586
synaptosomal-associated protein of 25 kDa (SNAP-25) 536, 538
syneresis 40
synthetic forms 85, 86, 616, 628 *see also specific isoflavone synthesis*
systems biology definition 292

T3 *see* 3,5,3′-tri-iodothyroine
T4 *see* thyroxine
Taiwan 358
tamoxifen 90, 169, 587, 588
tandem mass spectrometry (MS/MS)
  algae *233*
  herbal extracts 319–25, 326
  method 292

Subject Index

plants 286, 290, 291
serum analysis 203, 204, 226–8, *234–5*
urine analysis 203, *234–5*
validation 322–5
tea
  analysis *248–9, 251–2*
  composition *371*, 376, 377
  extraction 256
  thyroid function 425
tectorigenin **101**, 105, **106**
tempeh *see also* fermented foods
  composition *33, 34*, 357, 445
  digestion 544
  fermentation 37
  production 37
temperature
  conversion of isoflavones **37**, 152
  effect on groundnut composition 334, 337–9
  effect on soybean composition 35, **39**, 152, 401
  extraction 186, 246, 334, 337–9
  stability 36, 350
testicular dysgenesis syndrome (TDS) 575
testis
  anatomy 563–5, 574–5
  development 565–6, 567, 568, 569, 572–3
  function 563–5
  isoflavones 566–73
testosterone (T) 275, 564, 565–6, 567, 571, 586
3,5,3-5″-tetra-iodothyronine *see* thyroxine
5,6,7,3′-tetrahydroxy-4′-methoxyisoflavone **101**
5,7,3′,4′-tetrahydroxyisoflavone (THIF)
  binding affinity *88*
  biosynthesis 135–7, 141
  cancer 137–9
  endothelial cells 139–40
  structure **136**
texture profile analysis 405–6, 411, 419
thallium trinitrate (TTN) 67, **68**

theoretical maximum daily intake (TMDI) 355
thermogravimetric analysis 406, 419
thermoplastic extrusion 351, 361
THIF *see* 5,7,3′,4′-tetrahydroxyisoflavone
thromboembolism 317
thromboxane 588
thromboxane receptors 470–1, 473, 475
thyroglobulin 169, 173, 425, **427**, 428
thyroid
  daidzein 428–9
  equol 428
  function 169, 423, 429, 433
  genistein 426–31
thyroid peroxidase (TPO) 169, 425, 427–8, 433
thyroid replacement therapy 9
thyroid stimulating hormone (TSH) 169, 423, 425, 428, 433
thyroid stimulating releasing hormone (TRH) 433
thyrotropin 428
thyroxine (T4) 169, 423, 425, **427**, 428
time-of-flight mass spectrometry (TOF) 103, 282, 290
TLR (Toll-like receptors) 614, 615
TMDI *see* theoretical maximum daily intake
TNFα *see* tumour necrosis factor-α
toasting 51
tofu *see also* sufu
  Brazil 354
  China 354
  composition *33, 34*, 357, 360, 361, 445, 465
  fermented 357
  Hong Kong 355
  Indonesia 357
  optimum ratios 40
  processing 40, 56–7
  production 37
  Singapore 357
  soft 58
  USA 358

Toll-like receptors (TLR) 614, 615
topoisomerases 426, 442, 444, 501, 522
toxicity 389, 489, 490, 519, 520–1, 525
TPO see thyroid peroxidase
traditional medicine 327 see also Chinese medicine
transcription factors 623
transferin 567
transforming growth factor (TGF)-β 599
transversal osteotomy 602, 607
TRH see thyroid stimulating releasing hormone
3,5,3′-tri-iodothyroine (T3) 423, 425, **427**
tri-iodothyronine (T3) 169
tricarboxylic cycle **536**
Trifolium pratense see red clover
4′,5,7-trihydroxyisoflavone see genistein
trypsin inhibitor 54
TSH see thyroid stimulating hormone
tumour necrosis factor-α (TNFα)
  adipocytes 614, **615**
  estrogen 599
  fatty liver 620
  genistein 134, 135, 617, 620, 621, 622
  insulin resistance 619, 620
  isoflavones 614, 619, 620, 623
  macrophages 612, **613**, 614
  puerarin 304, 305
tumour suppressor protein p53 522, 552–3
tuning section 192
tyrosine ammonia-lyase (TAL) 83, **84**
tyrosine kinases
  function 433
  genistein 5, 133, 391, 426, 428, 501, 512

UDP-glucose:isoflavone 7-*O*-glucosyltransferase (IF7GT) **20**
UDP-glucuronyltransferases 4
UGTs see uridine diphosphate glucosyltranferases

UK
  consumption levels 350–1
  dietary sources *352*, *353*
ultra-performance liquid chromatography (UPLC) 270, 281
ultrasound 260
unfermented foods see non-fermented foods
uridine diphosphate glucosyltranferases (UGTs)
  function 90, 475
  genistein 117, 119, 141
  glycitein 466
  intestines 166
  liver 166
  placenta 122
  protein engineering 85, 89
urine analysis
  β-blockers 223, *234*
  daidzein 62, 203, 544
  enzymatic hydrolysis 198
  equol 359
  4-ethylphenol 121
  genistein 117, 119, 203, 544
  glucuronide conjugates 204, 209–10, *234*
  kakkalide metabolites 105
  simultaneous analysis 223
  sulfate conjugates 204, 209–10, *234*
USA
  consumption levels 358–9, 481, 544
  obesity 541
  soybean production 41
uterus 169
UV detection 205–12, 213, 222–4, *230*, *237*, 269–70, 285

vacuoles 62, 465
vagina 169, 630
validation procedures *229–37*
  beverage analysis *247–54*
  definition 240, 260
  herbal extracts 322–5

vascular endothelial growth factor (VEGF) 444, 447, 468, 469
vascular endothelium 6
vasodilation 170, 302, 470
vasomotor symptoms
  clinical studies 630–40
  isoflavones 8, 439
  phytoestrogens 628–9
  recommended daily intake 31
vegetables 245, 367, 424, 483
vegetarians 351, 352, *353*, 359, 366
VEGF (vascular endothelial growth factor) 444, 447, 468, 469
very low-density lipoproteins (VLDL) 170, 173
vine leaves **18**
virilization of external genitalia 575
vitamin D 7
vitamin D receptor 467, 587
Vitis vinifera L., Vitaceae *see* grapes
VLDL *see* very low-density lipoproteins
voltage-gated ion channels 513

Wacker-Cook tandem reactions 71, **73**
water content 419
water holding capacity (WHC) 405, 410, 412, 415
water load 192
waveguide 192

Western diets *see also specific countries*
  compared to Asian 168, 359, 360, 399, 409, 424
  soy food increase 366, 400
  soy intake levels 367
white lupine
  genistein 149
  planting season differences 21–2
  tissue distribution 21–2
wighteone *21*
wine *371*, 376, 627
Wittig reactions 64
women
  bone health 7
  thyroid diseases 429
  working memory 457, 458, 460

xanthohumol **16**
xenoestrogens *see also* coumestrol
  definition 122, 126, 157
  health effects 122, 573
  placenta 122, 124–5

yeast
  bread 402, 415
  semi-synthesis 85–6
  stabilising effect 54

zealarenone 14, **16**
α-zealarenonol **16**